T0310431

# MULTISCALE SIMULATIONS AND MECHANICS OF BIOLOGICAL MATERIALS

# MULTISCALE SIMULATIONS AND MECHANICS OF BIOLOGICAL MATERIALS

*Edited by*

**Shaofan Li**
*University of California at Berkeley, USA*

**Dong Qian**
*University of Texas at Dallas, USA*

A John Wiley & Sons, Ltd., Publication

This edition first published 2013


*Registered office*
John Wiley & Sons Ltd, The Atrium, Southern Gate, Chichester, West Sussex, PO19 8SQ, United Kingdom

For details of our global editorial offices, for customer services and for information about how to apply for permission to reuse the copyright material in this book please see our website at www.wiley.com.



*Library of Congress Cataloging-in-Publication Data*

Multiscale simulations and mechanics of biological materials / edited by Professor Shaofan Li, Dr Dong Qian.
pages cm
Includes bibliographical references and index.
ISBN 978-1-118-35079-9 (cloth)
1. Biomechanics. 2. Biomedical materials–Mechanical properties. 3. Multiscale modeling. I. Li, Shaofan, editor of compilation. II. Qian, Dong, editor of compilation.
QH513.M85 2013
612.7′6–dc23

2012040166

A catalogue record for this book is available from the British Library.

ISBN: 9781118350799

Typeset in 9/11pt Times by Aptara Inc., New Delhi, India
Printed and bound in Singapore by Markono Print Media Pte Ltd

# Contents

# About the Editors

**Dr. Shaofan Li** is currently a Professor of Applied and Computational Mechanics at the University of California–Berkeley. Dr. Li graduated from the Department of Mechanical Engineering at the East China University of Science and Technology (Shanghai, China) with a Bachelor Degree of Science in 1982; he also holds Master Degrees of Science from both the Huazhong University of Science and Technology (Wuhan, China) and the University of Florida (Gainesville, FL, USA) in Applied Mechanics and Aerospace Engineering in 1989 and 1993, respectively. In 1997 Dr. Li received a PhD degree in Mechanical Engineering from the Northwestern University (Evanston, IL, USA), and he was also a post-doctoral researcher at the Northwestern University during 1997–2000. In 2000 Dr. Li joined the faculty of the Department of Civil and Environmental Engineering at the University of California–Berkeley. Dr. Shaofan Li has also been a visiting Changjiang professor in the Huazhong University of Science and Technology, Wuhan, China (2007–2010). Dr. Shaofan Li is the recipient of the A. Richard Newton Research Breakthrough Award (2008) and an NSF Career Award (2003). Dr. Li has published more than 100 articles in peer-reviewed scientific journals, and he is the author and co-author of two research monographs/graduate textbooks. (li@ce.berkeley.edu)

**Dr. Dong Qian** is an associate professor in the Department of Mechanical Engineering at the University of Texas at Dallas. He obtained his BS degree in Bridge Engineering in 1994 from Tongji University in China. He came to the USA in 1996 and obtained an MS degree in Civil Engineering at the University of Missouri–Columbia in 1998. He continued his study at Northwestern University from 1998 and received his PhD in Mechanical Engineering in 2002. Shortly after his graduation, he was hired as an assistant professor at the University of Cincinnati. In 2008 he was promoted to the rank of associate professor with tenure and served as the Director of Graduate Studies from 2010. In the Fall of 2012 he joined the mechanical engineering department at the University of Texas at Dallas as an associate professor (tenured). His research interests include nonlinear finite-element and meshfree methods, fatigue and failure analysis and life prediction, surface engineering, residual stress analysis, and modeling and simulation of manufacturing processes (peening, forming, etc.) and nanostructured materials with a focus on mechanical properties and multiphysics coupling mechanisms. (dong.qian@utdallas.edu)

# List of Contributors

**Ashfaq Adnan,** Mechanical and Aerospace Engineering, University of Texas at Arlington, USA (aadnan@uta.edu)

**Facundo J. Bellomo,** INIQUI (CONICET), Faculty of Engineering, National University of Salta, Argentina (facundobellomo@yahoo.com.ar)

**Eduard Benet,** Department of Civil, Environmental and Architectural Engineering, University of Colorado, USA (ebenetcerda@gmail.com)

**Sagar Bhamare,** School of Dynamic Systems College of Engineering and Applied Science, University of Cincinnati, USA (bhamare.sagar@gmail.com)

**Jiun-Shyan Chen,** Department of Civil and Environmental Engineering, University of California, Los Angeles, USA (jschen@seas.ucla.edu)

**Sheng-Wei Chi,** Department of Civil and Environmental Engineering, University of Illinos at Chicago, USA (swchi@uic.edu)

**Jae-Hyun Chung,** University of Washington, USA (jae71@uw.edu)

**Suvranu De,** Department of Mechanical, Aerospace, and Nuclear Engineering, Rensselaer Polytechnic Institute, New York, USA (des@rpi.edu)

**Michel Devel,** FEMTO-ST Institute, Université de Franche-Comté, France (michel.devel@ens2m.fr)

**Khalil I. Elkhodary,** Department of Mechanical Engineering, Northwestern University, USA (k-elkhodary@northwestern.edu)

**Xavier Espinet,** Department of Civil, Environmental and Architectural Engineering, University of Colorado, USA (xavier.espinetalegre@colorado.edu)

**Leonora Felon,** X-spine Systems, Inc., USA (lfelon@x-spine.com)

**Sheikh F. Ferdous,** Mechanical and Aerospace Engineering, University of Texas at Arlington, USA (sheikh.ferdous@mavs.uta.edu)

**Jacob Fish,** Columbia University, USA (fishj@columbia.edu)

**Louis Foucard,** Department of Civil, Environmental and Architectural Engineering, University of Colorado, USA (louis.foucard@colorado.edu)

**Yao Fu,** Department of Mechanical Engineering and Materials Science, University of Pittsburgh, USA (yaf11@pitt.edu)

**Michael Steven Greene,** Theoretical & Applied Mechanics, Northwestern University, USA (greenes@u.northwestern.edu)

**Shaolie S. Hossain,** Institute for Computational Engineering and Sciences, The University of Texas at Austin, USA (sshossain@tmhs.org)

**Jia Hu,** Department of Mechanical Engineering and Mechanics, Lehigh University, USA; School of Mechanical Mechanics and Engineering, Southwest Jiaotong University, People's Republic of China (jih511@lehigh.edu)

**Daeyong Kim,** Korea Institute of Materials Science, South Korea (daeyong@kims.re.kr)

**Ji Hoon Kim,** Korea Institute of Materials Science, South Korea (kimjh@kims.re.kr)

**Jong-Hoon Kim,** University of Washington, USA (jhkim78@uw.edu)
**David Kirschman,** X-spine Systems, Inc., USA (dk@x-spine.com)
**Nikolay Kostov,** Mechanical Engineering, Rice University, USA (nmk1@tafsm.org)
**Hyun-Boo Lee,** University of Washington, USA (hyunboo@uw.edu)
**Myoung-Gyu Lee,** Pohang University of Science and Technology, South Korea (mglee@postech.ac.kr)
**Lisheng Liu,** Department of Civil and Environmental Engineering, The University of California at Berkeley, USA; Department of Engineering Structure and Mechanics, Wuhan University of Technology, People's Republic of China (liulish@mail.whut.edu.cn)
**Yaling Liu,** Department of Mechanical Engineering and Mechanics, Lehigh University, USA; Bioengineering Program, Lehigh University, USA (yal310@lehigh.edu)
**Seetha Ramaiah Mannava,** School of Dynamic Systems College of Engineering and Applied Science, University of Cincinnati, USA (mannavsr@ucmail.uc.edu)
**Virginia Monteiro,** International Center for Numerical Method in Engineering (CIMNE), Technical University of Catalonia, Spain (virginiamonteiro@yahoo.com)
**Liz G. Nallim,** INIQUI (CONICET), Faculty of Engineering, National University of Salta, Argentina (lgnallim@yahoo.com.ar)
**Devin O'Connor,** Department of Mechanical Engineering, Northwestern University, USA (devinoconnor2014@u.northwestern.edu)
**Sergio Oller,** International Center for Numerical Method in Engineering (CIMNE), Technical University of Catalonia, Spain (oller@cimne.upc.edu)
**Eugenio Oñate,** International Center for Numerical Method in Engineering (CIMNE), Technical University of Catalonia, Spain (onate@cimne.upc.edu)
**Harold S. Park,** Department of Mechanical Engineering, Boston University, USA (parkhs@bu.edu)
**Anthony Puntel,** Mechanical Engineering, Rice University, USA (anthony.puntel@tafsm.org)
**Farzad Sarker,** Mechanical and Aerospace Engineering, University of Texas at Arlington, USA (md.sarker@mavs.uta.edu)
**Kathleen Schjodt,** Mechanical Engineering, Rice University, USA (kms@tafsm.org)
**Daniel C. Simkins, Jr.,** University of South Florida, USA (dsimkins@eng.usf.edu)
**Kenji Takizawa,** Department of Modern Mechanical Engineering and Waseda Institute for Advanced Study, Waseda University, Japan (kenji.takizawa@tafsm.org) or (ktakiz@gmail.com)
**Shaoqiang Tang,** HEDPS, CAPT & Department of Mechanics, Peking University, People's Republic of China (maotang@pku.edu.cn)
**Tayfun E. Tezduyar,** Mechanical Engineering, Rice University, USA (tezduyar@gmail.com)
**Albert C. To,** Department of Mechanical Engineering and Materials Science, University of Pittsburgh, USA (albertto@pitt.edu)
**Vijay Vasudevan,** School of Dynamic Systems College of Engineering and Applied Science, University of Cincinnati, USA (vasudevk@ucmail.uc.edu)
**Franck J. Vernerey,** Department of Civil, Environmental and Architectural Engineering, University of Colorado, USA (franck.vernerey@colorado.edu)
**Gregory J. Wagner,** Sandia National Laboratories, USA (gjwagne@sandia.gov)
**Chu Wang,** Department of Mechanical, Aerospace, and Nuclear Engineering, Rensselaer Polytechnic Institute, New York, USA (wangc9@rpi.edu)
**Xiaodong Sheldon Wang,** College of Science and Mathematics, Midwestern State University, Texas, USA (sheldon.wang@mwsu.edu)
**Xingshi Wang,** Department of Mechanical, Aerospace, and Nuclear Engineering, Rensselaer Polytechnic Institute, Troy, New York, USA (wangxs165@gmail.com)
**Jie Yang,** Department of Mechanical Engineering and Mechanics, Lehigh University, USA (yangchenjie@home.swjtu.edu.cn)
**Judy P. Yang,** Department of Civil & Environmental Engineering, National Chiao Tung University, Taiwan (jpyang@nctu.edu.tw)

**Amir Reza Zamiri,** Department of Mechanical, Aerospace, and Nuclear Engineering, Rensselaer Polytechnic Institute, New York, USA (zamira@rpi.edu)
**Shahrokh Zeinali-Davarani,** Department of Mechanical Engineering, Boston University, USA (zeinalis@bu.edu)
**Yongjie Zhang,** Department of Mechanical Engineering, Carnegie Mellon University, USA (jessieaz@andrew.cmu.edu)
**Lucy Zhang,** Department of Mechanical, Aerospace, and Nuclear Engineering, Rensselaer Polytechnic Institute, New York, USA (zhanglucy@rpi.edu)
**Yanhang Zhang,** Department of Mechanical Engineering, Boston University, USA; Department of Biomedical Engineering, Boston University, USA (yanhang@bu.edu)
**Tarek Ismail Zohdi,** Department of Mechanical Engineering, University of California, Berkeley, USA (zohdi@me.berkeley.edu)
**Yihua Zhou,** Department of Mechanical Engineering and Mechanics, Lehigh University, USA (yiz311@lehigh.edu)

[illegible], Department of [illegible], [illegible]

[illegible], Department of [illegible], [illegible]

[illegible], Department of [illegible], [illegible]

[illegible], Department of [illegible], [illegible]

[illegible], Department of [illegible], [illegible]

[illegible], Department of [illegible], [illegible]

[illegible], Department of Mechanical Engineering, [illegible]

# Preface

This book is dedicated to Professor Wing Kam Liu (or Wing Liu for those who know him well) on the occasion of his 60th birthday.

In 1976, Professor Wing Kam Liu received a BS degree in Engineering Science from the University of Illinois at Chicago with honors. It was his time at UIC where Wing Liu met Ted Belytschko, then a young assistant professor, and took his graduate course on finite-element methods. After graduation from UIC, Wing Liu was admitted as a graduate assistant at the California Institute of Technology (Caltech) under the supervision of the young Thomas J.R. Hughes, who was beginning his academic career there. During his Caltech years, Wing Liu worked on a number of research topics, including finite-element shell elements, which is known today as the Hughes–Liu element.

Wing Liu received both his MS degree (1977) and PhD degree (1980) in Civil Engineering from Caltech, and he then came back to Chicago to become an assistant professor at Northwestern University, joining Ted Belytschko and kicking off a 30-year collaboration between them. In his 32-year academic career, Professor Liu has made numerous contributions to computational mechanics and micromechanics. Among his most noteworthy contributions are:

1. *Development of multiscale methods that bridge quantum to continuum mechanics.* Using these methods, he has developed software for the analysis and design of nanoparticles in materials, bio-sensing, and drug delivery.
2. *Development of new finite-element techniques.* These include introducing new shell elements, arbitrary Eulerian–Lagrangian methods, and explicit–implicit integration techniques that have significantly enhanced the accuracy and speed in software for crashworthiness and prototype simulations. Wing Liu was also the first to develop nonlinear probabilistic finite-element techniques that made nonlinear stochastic and reliability analyses possible.

3. *Development of meshfree formulations known as reproducing kernel particle methods.* These methods provide exceptional accuracy for the simulation of solids undergoing extremely large deformation and have been implemented in many commercial and laboratory software systems:
   (i) shell elements in DYNA3D, ABAQUS, LS-DYNA, ANSYS, and Argonne National Laboratory (ANL) software;
   (ii) explicit–implicit methods in US Ballistic Laboratory EPIC-2/EPIC-3 programs, and ANL software;
   (iii) Lagrangian–Eulerian methods adopted by ANL, Kawasaki, Mitsubishi, Ford Motors, and Grumman;
   (iv) various meshfree methods implemented by Sandia National Labs, Lawrence Livermore National Lab, General Motors, Ford Motors, Delphi, Ball Aerospace, and Caterpillar;
   (v) multiscale methods adopted by Goodyear for the design of tires and by Sandia in their TAHOE code for multiscale analysis.

Professor Wing Kam Liu is the recipient of numerous awards and honors that include: the 2012 Gauss–Newton Medal (IACM Congress Medal), the highest award given by IACM; the 2009 ASME Dedicated Service Award; the 2007 ASME Robert Henry Thurston Lecture Award; the 2007 USACM John von Neumann Medal, the highest honor given by USACM; the 2004 Japan Society of Mechanical Engineers (JSME) Computational Mechanics Award; the 2002 IACM Computational Mechanics Award; the 2001 USACM Computational Structural Mechanics Award; the 1995 ASME Gustus L. Larson Memorial Award; the 1985 ASME Pi Tau Sigma Gold Medal; the 1979 ASME Melville Medal (for best paper); the 1989 Thomas J. Jaeger Prize of the International Association for Structural Mechanics; and the 1983 Ralph R. Teetor Educational Award, American Society of Automotive Engineers. In 2001, he is listed by ISI as one of the most highly cited and influential researchers in engineering.

This large number of accolades highlights Wing Liu as a scholar and educator of extraordinary international reputation. This is also underlined by the fact that the present book comprises contributions from North American, Europe, and Asia, and from a very diverse group of people: colleagues, friends, collaborators, and former and current PhD students and post-docs. A wide range of topics is covered in this book: multiscale methods, atomistic simulations, micromechanics, and biomechanics/biophysics. These contributions represent either Wing Kam Liu's own research activities or topics he has taken an interest in over recent years. Moreover, the dedications of the contributing authors show that Wing Liu has represented more than just a scientist to a great number of people, to whom he also serves as friend, supporter, and source of inspiration. We are glad to have the opportunity of editing this book and would like to thank Wiley for its helpful collaboration, the authors for their contributions and making this book a success, and Wing Liu for his inspiring and initiating novel research in computational mechanics.

On behalf of the authors, we congratulate Wing Kam Liu to his 60th birthday and wish him happiness, health, success, and continued intellectual creativity for the years to come.

Shaofan Li and Dong Qian
*Houston, Texas*
*November 2012*

# Part I

# Multiscale Simulation Theory

Atomistic and multiscale simulation research is one of the current focuses of computational mechanics. In Part One we present a group of recent research studies in this active research area. Some of the chapters presented in this book contain research topics that are reported or released for the first time in the literature, and they touch almost every aspect of multiscale simulation research. In Chapter 1, Wagner presents an atomistic-based multiscale method to simulate heat transfer and energy conversion, which is a recent development of the bridging-scale method. In Chapter 2, Tang presents a detailed account on how to provide an accurate boundary treatment for concurrent multiscale simulation including the bridging-scale method. In Chapter 3, Liu and Li present for the first time a novel multiscale method called multiscale crystal-defect dynamics (MCDD), which is intended for simulation of dislocation motion, nanoscale plasticity, and small-scale fractures. In Chapter 4, Fu and To discuss their ingenious construction of a novel nonequilibrium molecular dynamics, and then Park and Devel, in Chapter 5, apply a coarse-grained multiscale method to study electromechanical coupling in surface-dominated nanostructures. In this part, Wagner, Tang, and Park were the main members of Wing Liu's research group in the early 2000s and have worked with Wing Liu in developing the bridging-scale method. The last chapter of this part is contributed by Dr. Fish, who presents a multiscale design theory and design procedure for general composite materials based on a multiscale asymptotic homogenization theory.

# 1

# Atomistic-to-Continuum Coupling Methods for Heat Transfer in Solids

Gregory J. Wagner
*Sandia National Laboratories, USA*

## 1.1 Introduction

New scientific and technical knowledge and advances in fabrication techniques have led to a revolution in the development of nanoscale devices and nanostructured materials. At the same time, improved computational resources and tools have allowed a continuously increased role for computational simulation in the engineering design process, for products at all scales. For many nano-mechanical or nano-electronic devices, models are sought that can accurately predict thermal and thermo-mechanical behavior under the range of conditions to which the devices will be subjected. However, at these small scales the limitations of continuum thermo-mechanical modeling techniques become apparent, as the effects of surfaces, grain boundaries, defects, and other deviations from a perfect continuum become important. Fourier's law, $\mathbf{q} = -\kappa \nabla T$ (where $\mathbf{q}$ is heat flux density, $\kappa$ is the thermal conductivity, and $\nabla T$ is the local temperature gradient) may not be applicable, nor may macroscale stress and strain laws; in fact, concepts like stress, strain, and even temperature may be difficult to define at the atomic scale.

Atomistic simulation techniques like molecular dynamics (MD) provide a way to simulate these small-scale behaviors, especially when combined with an accurate and efficient interatomic potential or force law that allows simulations of billions of atoms. However, even the very largest MD simulations may not be able to capture large enough length scales to simulate the interscale interactions important in real devices (since, typically, a nanoscale device must at some level be addressable from the macroscale in order to provide useful function). Classical MD has other shortcomings, as well, especially for real geometries at finite temperatures. For example, a number of approaches are available for holding an MD simulation at fixed, constant temperature [1, 2]; however, it is more difficult to regulate a spatially varying temperature, except through the use of discontinuous "blocks" of atoms held at different temperatures.

*Multiscale Simulations and Mechanics of Biological Materials*, First Edition. Edited by Shaofan Li and Dong Qian.

Limitations of MD have led to the development of atomistic-to-continuum coupling methods [3, 4], in which a continuum description (usually a finite-element discretization) of the material is used where valid, but a discrete atomistic representation of the material is used in regions where the continuum assumptions break down. Such a breakdown may occur near defects like cracks or dislocations, or in domains where the feature size is not much larger than the interatomic spacing. The atomistic and continuum descriptions are coupled together at an interface or overlap region, usually by combining the Hamiltonians of the two systems [5] or by ensuring that internal forces are properly balanced [6]. The resulting system couples the momentum equations (or, for statics, the force equilibria) of the two regions. Formation of a seamless coupling is nontrivial even for the static case (see Miller and Tadmoor [7] for an excellent review). For the dynamic case, an additional difficulty is the wave impedance mismatch at the MD–continuum boundary, leading to internal reflection of fine-scale waves back into the MD domain. Several approaches have been studied for removing these unwanted wave reflections, usually through some form of dissipation; typically, the goal in these situations is to completely remove the outgoing energy and minimize the reflected energy, optimally to zero [8–10].

Often in these coupled simulations, even for dynamics, the system is assumed to be initially quiescent, with no thermal vibrations about the mean atomic positions. This assumption is the basis of energy dissipation techniques that seek to completely remove internal wave reflections by zeroing the incoming energy. However, a more realistic environment for an actual nanoscale device, or nanostructured material, is finite nonzero temperature. At finite temperatures, atoms can be assumed to be vibrating about their mean positions with some thermal energy, and lattice waves can be recognized as energy-carrying phonons. Of course, additional thermal energy may be added to the system in a number of ways, and may propagate outward through the MD–continuum interface. However, in this case the goal is not to dissipate away all incoming energy; rather, the correct balance of incoming and outgoing phonons should be maintained, at least in some averaged sense.

Recently, methods have been developed that couple MD and continuum simulations, while allowing two-way coupling of thermal information. These methods are the subject of this chapter. Two-way coupling implies that heat in the simulation can be transferred both in the fine-to-coarse scale direction (MD to continuum), and in the coarse-to-fine direction. In the former case, this means that energy added to the MD domain as thermal vibrations may be transported to the neighboring continuum and lead to an increased internal energy. Phenomena that may lead to an increase in thermal vibrations in the MD region include friction, laser heating, fracture, and plastic failure. In the coarse-to-fine direction, it is required that internal energy in the continuum can be transported into the MD domain and lead to increased vibrational energy.

In addition to the spatial partitioning into MD and continuum regions used in these methods, another important type of MD-to-continuum coupling for thermal fields is required for the simulation of heat transfer in metals. For insulators, heat transfer is dominated by phonons, energy-carrying vibrations in the atomic lattice. For metals and some semiconductors, however, a large amount of the thermal energy is transported by electrons; in classical MD simulations, this contribution of the electron field to heat transfer is missing, and thermal conductivity cannot be accurately predicted for these materials. On the other hand, the contributions of the atomic lattice to thermal behavior cannot be ignored, especially since many of the same phenomena discussed above (like friction) add heat to the system initially through lattice vibrational degrees of freedom. To address this, several authors have developed coupled atomistic-to-continuum implementations of the so-called two-temperature model (TTM) in which the thermal energy is partitioned between lattice vibrations and electrons [11, 12]; lattice motion is simulated using MD, while the electron temperature is solved using a continuum heat equation. Jones *et al.* [13] built on this previous work, using energy-conserving techniques similar to those used in partitioned domain coupling methods.

The central idea in much of this work is the simultaneous definition of a solution field, in this case temperature, at multiple scales. This idea stems from the multiscale work by Wing Kam Liu and his research group at Northwestern University. In the context of atomistic-to-continuum coupling, these ideas are captured in the bridging-scale method developed in that group [9, 14, 15], with earlier roots in the multi-resolution and bridging-scale approaches developed for continuum simulations [16–18].

In the remainder of this chapter we will summarize these newly developed methods, including the development of the theory behind them, along with some demonstrations and applications.

## 1.2 The Coupled Temperature Field

An important step in developing a coupled method is to clearly define the relationship between the atomic motions and the macroscale temperature. From classical (i.e., non-quantum) statistical physics, for a system of atoms at equilibrium, the system temperature $T$ can be written in terms of the system kinetic energy [19]:

$$\frac{3}{2} n_{\mathrm{a}} k_{\mathrm{B}} T = \left\langle \sum_{\alpha} \frac{1}{2} m_{\alpha} |\mathbf{v}_{\alpha}|^2 \right\rangle, \tag{1.1}$$

where $k_{\mathrm{B}}$ is Boltzmann's constant, $n_{\mathrm{a}}$ is the number of atoms in the system, $m_{\alpha}$ and $\mathbf{v}_{\alpha}$ are the mass and velocity of atom $\alpha$, and the angle brackets represent an ensemble or time average. Here, for simplicity, we are assuming that the mean velocity of each atom is zero, but more generally the velocity used in the temperature definition can be some perturbation about the mean (although the precise definition of the "mean" must be determined). From Equation (1.1), we can identify an atomic temperature $T_{\alpha}$:

$$T_{\alpha} \equiv \frac{1}{3k_{\mathrm{B}}} m_{\alpha} |\mathbf{v}_{\alpha}|^2. \tag{1.2}$$

If it is assumed that the ensemble average in (1.1) is equivalent to an average over all atoms in an equilibrium system, then the system temperature at equilibrium is just an average of the atomic temperature $T_{\alpha}$ over the atoms.

In this work, we are primarily interested in nonequilibrium systems, where the temperature is not necessarily constant for the entire system, so that (1.1) does not apply directly. However, we can use the atomic temperature $T_{\alpha}$, together with appropriate spatial- and time-averaging operators, to define a spatially varying macroscale temperature field in terms of the atomic temperature.

### *1.2.1 Spatial Reduction*

First, we will assume that the spatially varying macroscale temperature can be represented by an interpolated field on a finite-element mesh (FEM), with an element size that is large compared with the interatomic spacing. This interpolation can be written as

$$T^{\mathrm{FE}}(\mathbf{X}, t) \equiv \sum_{I} N_I(\mathbf{X}) \theta_I(t). \tag{1.3}$$

In this expression, the sum is over the set of all nodes in the domain, $\theta_I$ is a temperature degree of freedom defined on node $I$, and $N_I(\mathbf{X})$ is the finite-element shape function associated with node $I$ evaluated at $\mathbf{X}$.

One way to relate the macroscale and atomic temperature fields is to use a least-squares projection to minimize the difference between the interpolated temperature $T^{\mathrm{FE}}(\mathbf{X}, t)$ and the atomic temperature at every atomic location $\mathbf{X}_\alpha$. This is attractive, but leads to a matrix equation that must be solved. To simplify this projection, we approximate the projection matrix with a row-sum lumping to get a diagonal matrix (see Wagner *et al.* [20] for details), leading to a simple relationship between the nodal temperatures $\theta_I$ and the atomic temperatures, through a scaled shape function $\hat{N}_I$:

$$\begin{aligned} \theta_I &= \sum_{\alpha \in \mathcal{A}} \hat{N}_{I\alpha} T_\alpha, \\ \hat{N}_{I\alpha} &\equiv \frac{N_{I\alpha} \Delta V_\alpha}{\sum_{\beta \in \mathcal{A}} N_{I\beta} \Delta V_\beta}, \end{aligned} \tag{1.4}$$

where $\Delta V_\alpha$ is the volume associated with atom $\alpha$ and $N_{I\alpha}$ is a shorthand notation for $N_I(\mathbf{X}_\alpha)$.

## *1.2.2 Time Averaging*

Equation (1.4) defines an efficient spatial reduction from the atoms to the finite-element nodes. However, even when the number of atoms is much larger than the number of nodes, experience shows that the resulting nodal temperatures fluctuate in time even for what should be statistically steady states. To help reduce these fluctuations, we can define a time filtering operation as

$$\langle f(t) \rangle_\tau \equiv \int_{-\infty}^{t} f(t') G(t - t')\, \mathrm{d}t', \tag{1.5}$$

where $G(t)$ is a kernel function of the form

$$G(t) = \frac{1}{\tau} \mathrm{e}^{-t/\tau} \tag{1.6}$$

and $\tau$ is the time scale of our filtering operation. This filtering operation commutes with time differentiation, and for this choice of the kernel function $G(t)$ the time derivative of a filtered function can be rewritten as a simple, first-order ordinary differential equation:

$$\frac{\mathrm{d}}{\mathrm{d}t} \langle f \rangle_\tau = \frac{f - \langle f \rangle_\tau}{\tau}. \tag{1.7}$$

The usefulness of this property should be apparent: by using (1.7) to update time-filtered values, the filtered value of any quantity $f$ can be computed without storing the time history of that quantity.

Ideally, the filtering timescale $\tau$ is much longer than the vibrational timescale of the atoms, but much shorter than the timescale associated with the expected macroscale temperature changes due to phonon heat transfer. If such a separation of scales is not possible, then a local macroscale temperature may be hard to define, and the only way to reduce fluctuations in the nodal temperatures may be to increase the number of atoms per node by coarsening the FEM.

By combining the spatial reduction (1.4) with the time filtering operator, we obtain a final expression relating the atomic motion to the finite-element temperature field:

$$\theta_I = \left\langle \sum_{\alpha \in \mathcal{A}} \hat{N}_{I\alpha} T_\alpha \right\rangle_\tau. \tag{1.8}$$

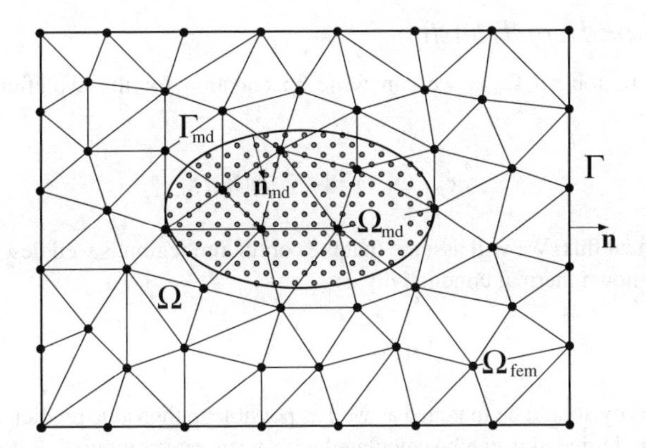


**Figure 1.1** Coupled domain geometry

## 1.3 Coupling the MD and Continuum Energy

### *1.3.1 The Coupled System*

The goal of our coupling strategy is to simulate domains decomposed as in Figure 1.1, in which a domain $\Omega$ is discretized with a FEM, while an internal portion of the domain $\Omega_{\mathrm{md}}$ is represented by a set of atoms $\mathcal{A}$. The remaining portion of the domain in which there are no atoms but only finite elements is denoted $\Omega_{\mathrm{fem}}$; the boundary between the two subdomains is given by $\Gamma_{\mathrm{md}}$, with normal vector $\mathbf{n}_{\mathrm{md}}$ oriented into the MD region. The purpose of this partitioning is to use classical MD, with atomic forces derived from an interatomic potential, in those regions where the heat flow and corresponding dynamics are too complex to be described by a simple continuum heat transfer law (like the Fourier law). Thus, $\Omega_{\mathrm{md}}$ will typically include defects, dislocation, grain boundaries, or any other nanoscale structures that may affect heat transfer.

To derive a coupled set of MD–FEM equations for this system, we assume that we can partition the total energy of the system into two parts that correspond to the two subdomains:

$$E^{\mathrm{tot}} = E^{\mathrm{md}} + E^{\mathrm{fem}}. \tag{1.9}$$

For simplicity, we will assume that strain energy and the mean velocity of the FEM region can be neglected, so that the energy of the finite-element region is just the thermal energy $\rho c_p T^h$ integrated over $\Omega_{\mathrm{fem}}$:

$$E^{\mathrm{fem}} = \int_{\Omega_{\mathrm{fem}}} \rho c_p T^h(\mathbf{X}, t)\,\mathrm{d}V, \tag{1.10}$$

where the density $\rho$ and the specific heat $c_p$ are intrinsic properties that can both be related to the properties of the atomic lattice [19]. The energy of the MD region is the sum of the potential and kinetic energies of the atoms, with time filtering applied:

$$E^{\mathrm{md}} = \left\langle U^{\mathrm{md}} + \frac{1}{2}\sum_{\alpha} m_\alpha |\mathbf{v}_\alpha|^2 \right\rangle_\tau, \tag{1.11}$$

where $U^{\mathrm{md}}$ is the potential energy of the atoms in the MD region.

### 1.3.2 Continuum Heat Transfer

Outside of the MD region, in $\Omega_{\text{fem}}$, we can write an equation for the evolution of the continuum temperature field:

$$\rho c_p \dot{T}(\mathbf{X}, t) = -\nabla \cdot \mathbf{q}(\mathbf{X}, t), \tag{1.12}$$

where $\mathbf{q}(\mathbf{X}, t)$ is the heat flux. We will assume that the continuum region is well described by the Fourier heat law with some known thermal conductivity $\kappa$:

$$\mathbf{q}(\mathbf{X}, t) = -\kappa \nabla T. \tag{1.13}$$

The thermal conductivity should be matched as well as possible to the value predicted by the MD system for a large, defect-free lattice; this can be calculated with a separate computation if needed.

The continuum heat equation applies only in $\Omega_{\text{fem}}$, but the FEM and the interpolated temperature field exist in the entire domain, including in $\Omega_{\text{md}}$. For nodes whose supports fully or partially overlap the MD region, the evolution of the nodal temperature includes a contribution from the atomic motion, so that heat information passes smoothly from the fine scale to the coarse.

### 1.3.3 Augmented MD

A common way of thermostatting an MD system is by adding a drag force to each atom that is proportional to the velocity of that atom [1, 2]. We can use a similar idea to include the effects of the continuum region temperature on the MD region in our coupled simulations. The equation of motion solved for each atom is

$$m_\alpha \dot{\mathbf{v}}_\alpha = \mathbf{f}_\alpha^{\text{md}} + \mathbf{f}_\alpha^{\lambda}, \tag{1.14}$$

where the classical MD force is computed from the potential energy

$$\mathbf{f}_\alpha^{\text{md}} = -\frac{\partial U^{\text{md}}}{\partial \mathbf{x}_\alpha} \tag{1.15}$$

and the drag force $\mathbf{f}_\alpha^{\lambda}$ is given by

$$\mathbf{f}_\alpha^{\lambda} = -\frac{m_\alpha}{2} \lambda_\alpha \mathbf{v}_\alpha. \tag{1.16}$$

The parameter $\lambda_\alpha$ may be different for every atom, and we assume that it is a field that can be interpolated from a set of nodal values $\lambda_I$ defined on the FEM, using the nodal shape functions:

$$\lambda_\alpha(t) = \sum_I N_{I\alpha} \lambda_I(t). \tag{1.17}$$

By enforcing conservation of total energy of the system (see Wagner *et al.* [20] for details), we can derive a matrix equation for these nodal coefficients $\lambda_I$:

$$\sum_J \left( \sum_{\alpha \in \mathcal{A}} N_{I\alpha} T_\alpha N_{J\alpha} \right) \lambda_J = \frac{2}{3k_{\text{B}}} \int_{\Gamma_{\text{md}}} N_I \mathbf{n}_{\text{md}} \cdot \kappa (\nabla T^h + \tau \nabla \dot{T}^h) \, \text{d}A \tag{1.18}$$

Templeton *et al.* [21] extended this approach to allow a prescribed heat flux to be applied to the boundary of an MD system through a modification of the equation for the $\lambda_I$ coefficients.

The right-hand-side integral in (1.18) is a surface integral on the interface between the MD and continuum regions. In effect, this expression relates the effect that the continuum has on the MD system (through $\lambda_I$ and the atomic drag forces) to the heat flux across the surface. This relation follows naturally from our conservation of total system energy, and completes the formulation. However, the numerical computation of this surface integral is not trivial, especially if the interface does not align with element boundaries. Wagner *et al.* [20] give a technique for approximating this integral using the divergence theorem together with a projection operation.

The final form of the evolution equation for the finite-element temperature field in the coupled system is [20]:

$$\begin{aligned}\sum_{J\in\mathcal{N}}\left(\int_\Omega N_I N_J \,\mathrm{d}V\right)\dot{\theta}_J &= \left\langle \frac{2}{3k_B}\sum_{\alpha\in\mathcal{A}} N_{I\alpha}\mathbf{v}_\alpha\cdot\left(\mathbf{f}^{\mathrm{md}}_\alpha + \frac{1}{2}\mathbf{f}^{\lambda}_\alpha\right)\Delta V_\alpha\right\rangle \\ &\quad + \sum_{J\in\mathcal{N}}\sum_{\alpha\in\mathcal{A}}\left(\nabla N_I\cdot\frac{\kappa}{\rho c_p}\nabla N_J\right)\bigg|_\alpha \theta_J\Delta V_\alpha \\ &\quad - \sum_{J\in\mathcal{N}}\left(\int_\Omega \nabla N_I\cdot\frac{\kappa}{\rho c_p}\nabla N_J\,\mathrm{d}V\right)\theta_J.\end{aligned} \tag{1.19}$$

## 1.4 Examples

### *1.4.1 One-Dimensional Heat Conduction*

A simple but useful demonstration of the coupled method is the transient, nonequilibrium heat flux through a quasi-one-dimensional domain [20]. Figure 1.2 shows the computational domain. The MD system represents solid argon, with $m_\alpha = 39.948$ amu and lattice constant $\ell = 5.406$ Å. Interatomic forces are computed from a Lennard-Jones potential with parameters $\varepsilon/k_\mathrm{B} = 119.8$ K and $\sigma = 3.405$ Å. A continuum thermal conductivity of $\kappa = 0.5$ W/(m K) was assumed in the finite-element region [22]. The computational domain comprises an MD region of size $(20\times 8\times 8)\ell$, centered in an overlapping finite-element domain of size $(48\times 8\times 8)\ell$. The entire domain was discretized using finite elements; Figure 1.2 shows a mesh size of $h = 4\ell$, giving 48 elements in the mesh.

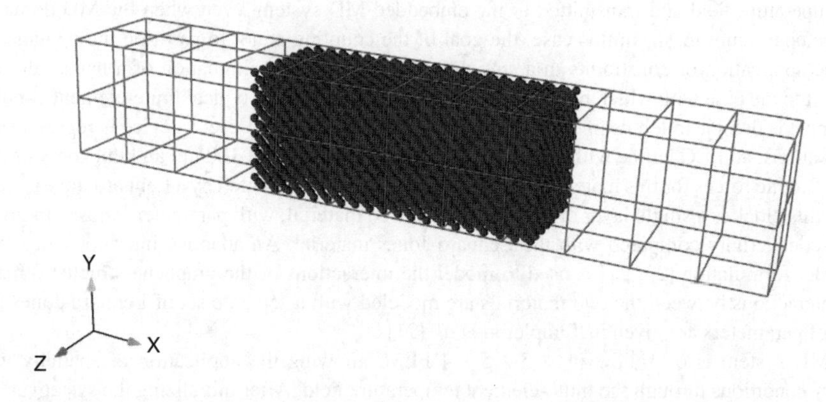


**Figure 1.2** One-dimensional heat conduction: mesh and atomic positions for $h = 4\ell$

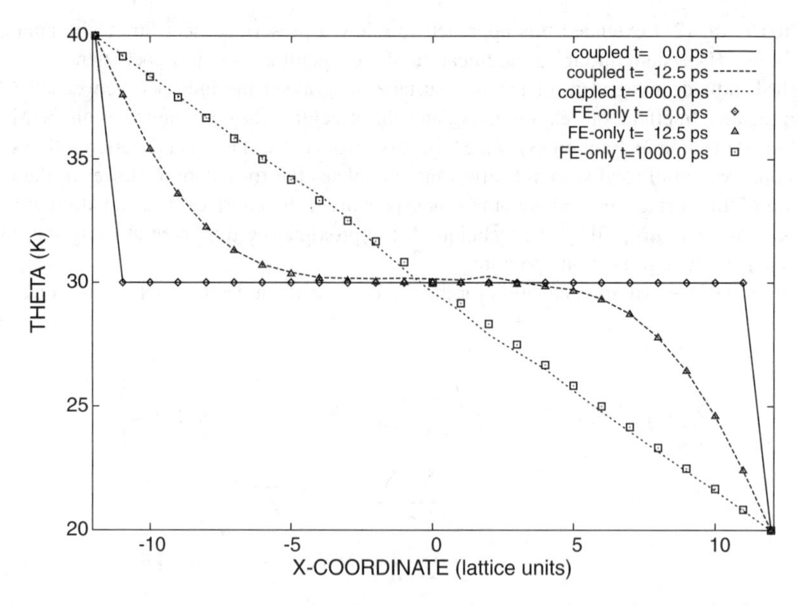


**Figure 1.3** Temperature profiles for $h = 2\ell$ and $\tau = 25$ ps at $t = 0.0$, 12.5, 1000.0 ps

Periodic boundary conditions are imposed on the lateral ($\pm y$ and $\pm z$) faces of the rectangular domain. The temperature of the entire system was initially brought to 30 K via rescaling and, immediately following the thermalization stage, the end temperatures were changed to 40 K and 20 K for the left ($-x$) and right ($+x$) ends, respectively. The longitudinal temperature traces, Figure 1.3, of this essentially one-dimensional problem show good agreement with the corresponding solution of the classical heat conduction equation. There is no discernible effect in this plot of the MD–finite-element interface on the temperature solution.

## *1.4.2 Thermal Response of a Composite System*

An extension of the coupled atomistic-to-continuum heat transfer method by Templeton *et al.* [21] allows the application of fixed temperature and heat flux boundary conditions to be specified on the continuum-scale temperature field and transmitted to the embedded MD system, even when the MD domain fills the entire continuum mesh. In this case, the goal of the coupling is not to partition the domain, but to allow the application of constraints in a straightforward way to the simulation of a nanoscale device, equivalent to the ease with which boundary conditions are applied in a typical finite-element simulation. An example is demonstrated using the domain pictured in Figure 1.4. Here, a device is represented by a block of atoms, an FCC lattice with approximate lateral dimensions of 100 nm and thickness of 25 nm. The interatomic forces for this material are assumed to be well-represented by a Lennard-Jones potential. Coating this block is a single layer of graphene-structured material, with parameters chosen to give high thermal conductivity compared with the Lennard-Jones material. An adaptive intermolecular reactive bond-order formulation [23, 24] is used to model the interactions of the graphene-structured material. Cross-interactions between the two materials are modeled with a separate set of Lennard-Jones forces. Details of parameters are given in Templeton *et al.* [21].

The MD system is overlaid with a $5 \times 5 \times 4$ FEM, allowing the application of spatially varying boundary conditions through the finite-element temperature field. After initializing the system at 300 K, a fixed boundary temperature is applied that varies linearly in the $x$ direction, and is also ramped linearly

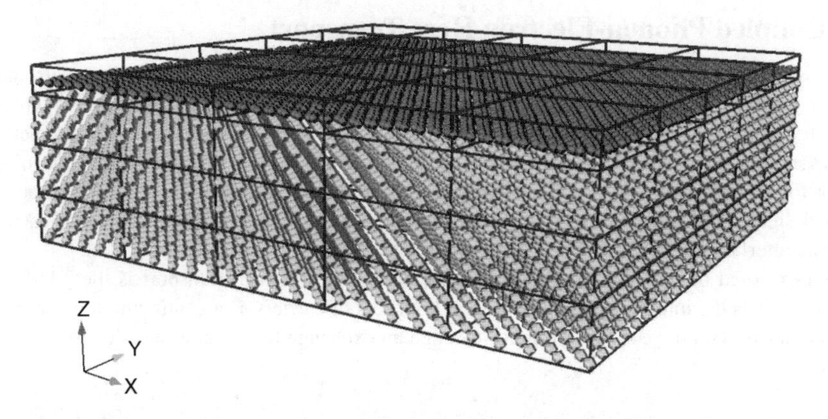


**Figure 1.4** Schematic of the composite system example problem. Lennard-Jones atoms are colored light gray and graphene-structured atoms are colored dark gray. The FEM is overlaid

in time over 0.1 ns; at the final time, the $+x$ surface has a fixed temperature of 400 K, while the $-x$ surface remains at 300 K. After the device reaches an equilibrium (but spatially varying) temperature, a spatially varying heat flux with a Gaussian profile is applied on the bottom ($-z$) face of the device, leading to further heat-up. This applied flux represents the intensity profile from laser heating of a nano-device. Figure 1.5 shows the continuum-scale temperature of top and bottom faces of the Lennard-Jones block, and of the graphene-structured layer, after 0.3 ns. An asymmetric temperature profile results in the slab from the linearly varying boundary condition. The graphene-structured layer is heated, but because of its high thermal conductivity the heat is well distributed through the layer. This example demonstrates how a coupled method like this might be used to simulate heat mitigation in a real nano-scale semiconductor device.

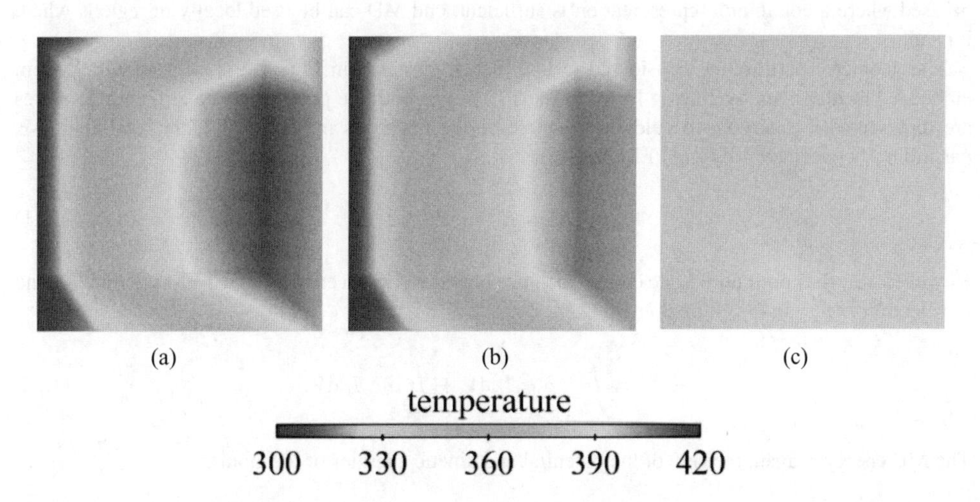


**Figure 1.5** Time-averaged temperature profiles of the Lennard-Jones (LJ) atoms as viewed from the bottom (a) and top (b), compared with the graphene-structured layer (c). Temperatures are in kelvin

## 1.5 Coupled Phonon-Electron Heat Transport

As discussed in Section 1.1, the role of electrons in the transport of heat is crucial in metals and semimetals. In classical MD simulations, this electron transport is missing. On the other hand, MD represents well the transport of heat by phonons, in both the ballistic and diffusive regimes [25–27]; a method that couples the phonon and electron modes of heat transport can allow more accurate simulations of thermal behavior for a number of applications, including laser processing of materials [28–31], thermoelectric material design [32–35], heat transport in conducting nanotubes and nanowires [32, 36], and heat transport at material interfaces [37].

A well-explored method for coupling phononic and electronic heat transport is the TTM [38–42]. This model treats the internal energy residing in both sets of carriers as a continuum temperature field; energy is transferred between the two fields through an exchange term that ensures conservation of the total energy:

$$\rho_e c_e \dot{T}_e = -\nabla \cdot \mathbf{q}_e - g + S_e, \tag{1.20a}$$

$$\rho_p c_p \dot{T}_p = -\nabla \cdot \mathbf{q}_p + g + S_p. \tag{1.20b}$$

The temperatures $T$, densities $\rho$, heat capacities $c$, heat fluxes $\mathbf{q}$, and source terms $S$ are subscripted with "e" or "p" to denote the electron or phonon quantities. The heat fluxes are typically modeled as functions of their respective temperatures and temperature gradients (e.g., $\mathbf{q}_p = -\kappa_p \nabla T_p$ in the Fourier model). The energy exchange $g$ is the rate of heat transferred from the electrons to the phonons, and is usually taken as a function of the temperature difference between the two sets of carriers, $g = g(T_e - T_p)$. In this model, both temperature fields exist everywhere in the solution domain.

Replacing the continuum representation of the phonon temperature (Equation (1.20b)) with an MD system requires a procedure for injecting energy into the phonons that is equal to the energy lost by the electrons, so that energy conservation is maintained. This energy exchange is still desired to be a function of the difference between the phonon and electron energy, where now the phonon energy must be computed through some spatial–temporal averaging on the MD system, as in earlier sections in this chapter. Jones *et al.* [13] developed a method that ensures this energy conservation, and at the same time allows a domain partitioning such that a finite-element representation of the phonon temperature can be used where a continuum representation is sufficient, and MD can be used locally in regions where it is not.

The geometry partitioning considered is like that represented in Figure 1.1, with an MD system embedded locally in an overlaying FEM. In the TTM case, both the phonon and electron temperatures are represented as interpolated fields everywhere on the FEM. As in Section 1.3, the total energy is partitioned between the MD and FEM parts:

$$E^{tot} = E^{md} + E^{fem}. \tag{1.21}$$

However, now the continuum-scale energy includes both the phonon energy in the FEM region $\Omega_{fem}$ and the electron energy in the entire domain $\Omega$:

$$E^{fem} = \int_{\Omega_{fem}} \rho_p c_p T_p \, dV + \int_{\Omega} \rho_e c_e T_e \, dV. \tag{1.22}$$

The MD energy is again the sum of the potential and kinetic energies of the atoms:

$$E^{md} = U^{md} + \frac{1}{2} \sum_{\alpha} m_\alpha |\mathbf{v}_\alpha|^2. \tag{1.23}$$

Note that Jones *et al.* [13] do not consider time filtering. As in previous sections (cf. Equation (1.14)), an augmented molecular dynamics force is used, where the added drag force is again

$$\mathbf{f}_\alpha^\lambda = -\frac{m_\alpha}{2}\mathbf{v}_\alpha \sum_I N_{I\alpha}\lambda_I. \tag{1.24}$$

An expression for the nodal coefficients $\lambda_I$ that define the drag force on the atoms can be derived through an energy conservation equation. For simplicity, let us assume that a heat flux $\bar{q}$ in the direction of the outward normal is defined everywhere on the outer boundary $\Gamma$ in Figure 1.1. The change in the total energy is determined by this boundary heat flux and the source terms $S_p$ and $S_e$:

$$\dot{E}^{tot} = \int_\Omega (S_p + S_e)\,dV - \int_\Gamma \bar{q}\,dA. \tag{1.25}$$

Taking the time derivatives of (1.22) and (1.23) and substituting into (1.25) gives

$$\begin{aligned}\int_\Omega (S_p + S_e)\,dV - \int_\Gamma \bar{q}\,dA = &-\frac{1}{2}\sum_\alpha m_\alpha |\mathbf{v}_\alpha|^2 \sum_J N_{J\alpha}\lambda_J \\ &+ \int_{\Omega_{fem}} (-\nabla\cdot\mathbf{q}_p + g + S_p)\,dV \\ &+ \int_\Omega (-\nabla\cdot\mathbf{q}_e - g + S_e)\,dV.\end{aligned} \tag{1.26}$$

In arriving at this expression we have assumed that Equation (1.20a) applies in $\Omega$ and applied in $\Omega_{fem}$ (i.e., outside the MD system); we have also used (1.14) and (1.24) to rewrite the rate of change of the MD energy. Using the divergence theorem on the heat flux integrals, noting that the boundary heat flux $\bar{q}$ on $\Gamma$ is composed of phonon and electron parts, and canceling some integrals gives a global energy conservation expression of the form

$$\frac{3k_B}{2}\sum_\alpha T_\alpha \sum_J N_{J\alpha}\lambda_J = -\int_{\Omega_{md}} (S_p + g)\,dV - \int_{\Gamma_{md}} \mathbf{n}_{md}\cdot\mathbf{q}_p\,dA. \tag{1.27}$$

The definition of the atomic temperature from (1.2) has been used to simplify this expression. The left-hand side of this expression is related to the energy removed from the MD system by the augmented force term, through the coefficients $\lambda_J$. The right-hand side represents the change in the phonon energy in $\Omega_{md}$ due to source and electron transfer terms, as well as heat flux across the boundary at the continuum interface. The minus sign on the right-hand side corresponds to the fact that positive values of the $\lambda_J$ represent a drag force that removes energy from the MD system.

In order to write an expression that can be solved for the $\lambda_J$ coefficients, we can take this global energy balance and localize it through the use of finite-element shape functions $N_I$:

$$\frac{3k_B}{2}\sum_J\sum_\alpha N_{I\alpha}T_\alpha N_{J\alpha}\lambda_J = -\int_{\Omega_{md}} N_I(S_p + g)\,dV - \int_{\Gamma_{md}} N_I\mathbf{n}_{md}\cdot\mathbf{q}_p\,dA. \tag{1.28}$$

Satisfaction of (1.28) implies satisfaction of (1.27), which can be verified by summing (1.28) over all nodes $I$ and noting that the finite-element shape functions are a partition of unity ($\sum_I N_I = 1$). However, (1.28) is not the only localized form that guarantees global energy conservation, and alternatives are possible that result in modified expressions.

Equation (1.28) defines a matrix equation for the nodal coefficients $\lambda_J$ in terms of known quantities that are either given or computable from the continuum electron and phonon temperature fields. Comparison with the similar Equation (1.18) derived for the case without electrons shows that the expressions are equivalent in the absence of source and electron transfer terms, if a Fourier law is used for the phonon heat transfer at the boundary and if time filtering is ignored ($\tau = 0$).

Jones *et al.* [13] also derived governing equations for the finite-element fields representing the electron and phonon temperatures. These fields are parameters by nodal coefficients $\theta_{eJ}$ and $\theta_{pJ}$, respectively, and satisfy the following weak forms of the energy equation:

$$\sum_J \left( \int_\Omega N_I \rho_e c_e N_J \, dV \right) (\dot{\theta}_e)_J = \int_\Omega [\nabla N_I \cdot \mathbf{q}_e + N_I (S_e - g)] \, dV - \int_\Gamma N_I \bar{q}_e \, dA, \tag{1.29a}$$

$$\begin{aligned} \sum_J \left( \int_\Omega N_I \rho_p c_p N_J \, dV \right) (\dot{\theta}_p)_J &= 2 \sum_\alpha N_{I\alpha} \mathbf{v}_\alpha \cdot (\mathbf{f}_\alpha^{md} + \mathbf{f}_\alpha^\lambda) \\ &+ \int_{\Omega_{fem}} [\nabla N_I \cdot \mathbf{q}_p + N_I (S + g)] \, dV - \int_\Gamma N_I \bar{q}_p \, dA. \end{aligned} \tag{1.29b}$$

In these equations, it is assumed that the prescribed heat flux $\bar{q}$ on the boundary $\Gamma$ can be decomposed into phonon ($\bar{q}_p$) and electron parts ($\bar{q}_e$). These two equations, together with (1.14) for each atom and (1.28) for the nodal $\lambda$ coefficients, completes the set of equations to be solved for the coupled MD–FEM, TTM system.

## 1.6 Examples: Phonon–Electron Coupling

### 1.6.1 Equilibration of Electron/Phonon Energies

A simple verification of the method for domain-partitioned MD–FEM with a TTM is obtained through simulation of a block of material in which the electrons have an initial temperature that is spatially varying and higher than the initial phonon temperature. Over time we expect that the two sets of carriers should equilibrate to a constant, common temperature. The test domain is shown in Figure 1.6. A square film of material 24 unit cells on each side and six unit cells deep is represented by a $12 \times 12 \times 1$ FEM; periodic boundary conditions are used in the out-of-plane direction. The central $12 \times 12$ unit cells of the domain are represented with atoms whose interatomic force is computed from a Lennard-Jones potential. In Figure 1.6, the dark atoms are "ghost" atoms used to constrain the inner MD system and provide a full complement of neighbors for all of the atoms. The exchange energy $g$ is assumed to be a linear function of the temperature difference between electrons and phonons: $g = g_{e\text{-}p}(T_e - T_p)$. Geometry, interatomic potential, and material parameters are taken from Jones *et al.* [13].

The initial temperature of the phonons is 20 K, and in the simulation an initial equilibration phase is used to thermostat the MD system at this temperature. The initial electron temperature is a radially varying function with a Gaussian profile, $T_e = 20[3\,e^{(r/25)^2} + 5]$ (in degrees kelvin), where $r = (x^2 + y^2)^{1/2}$ is the distance from the center of the domain in the $x$–$y$ plane. After initialization, the system is run with no other source terms and with adiabatic ($\bar{q}_e = \bar{q}_p = 0$) boundary conditions. In Figure 1.6, the FEM is colored by the initial electron temperature, while the atoms are colored by the instantaneous values of the atomic temperature $T_\alpha$ at time $t = 0$.

Figure 1.7a shows a time sequence of temperature profiles measured along the $x$-axis. A combination of diffusion and phonon-electron exchange drive the temperatures to a spatially constant common value. Spatial fluctuations in the phonon temperature are visible in the central region because of the stochastic nature of the MD system. In Figure 1.7b, the conservation of total energy is demonstrated; the energy of the electrons decreases while that of the phonons and MD system increase, such that the total remains constant in this adiabatic system.

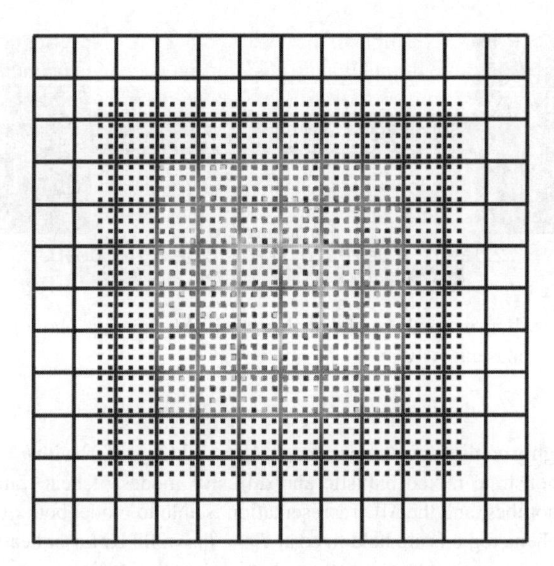

**Figure 1.6** Configuration of the MD lattice (colored by the atomic temperature $T_\alpha$) embedded in a finite-element grid (colored by the electron temperature). Note the five layers of ghosts (dark color) are fixed and, therefore, have zero kinetic temperature

## *1.6.2 Laser Heating of a Carbon Nanotube*

Another example application is the heating of a carbon nanotube [13]. A metallic (8, 8) armchair nanotube, with length 12.6 nm, is suspended with its ends embedded in solid graphite (Figure 1.8). The nanotube is modeled using a Tersoff potential for the interatomic forces [43], while the graphite is modeled as a continuum with the same thermal properties as the nanotube. For a nanotube, the phonon

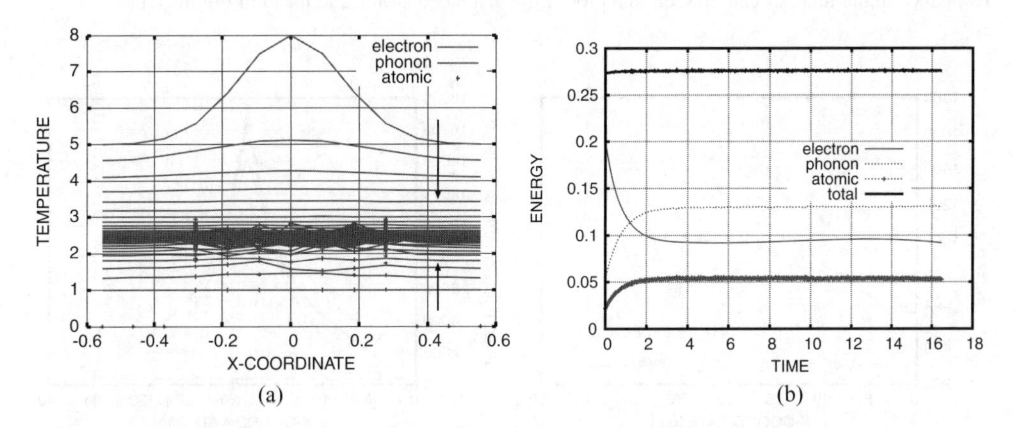


**Figure 1.7** (a) Sequence of temperature profiles through a cross-section aligned with the $x$-axis. The arrows indicate the progression of profiles with time. (b) Evolution of energy. The atomic thermal energy is calculated for the lattice, phonon energy is calculated in $\Omega_{\text{fem}}$, and the electron energy is integrated over the full domain $\Omega$

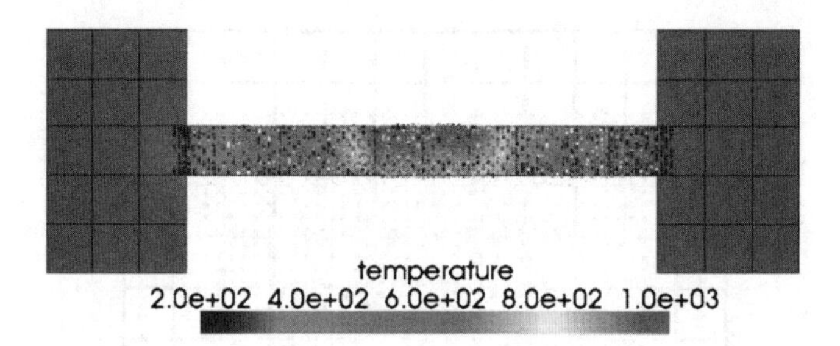


**Figure 1.8** Metallic carbon nanotube embedded in an FEM showing phonon temperature near the beginning of the heating phase, $t = 10$ ps

electron exchange is highly nonlinear in temperature [44], $g = h(T_e - T_p)^5$ with $h = 3.7 \times 10^4$ W/m$^3$K$^5$. Experiments [25] demonstrate mixed ballistic and diffusive modes of heat transport due to phonon transport in carbon nanotubes, and the MD representation is able to model both of these modes.

The ends of the graphene regions are held fixed at $T_e = T_p = 300$ K. Laser heating of the nanotube is modeled with a source term on the electron energy that is spatially varying:

$$S_e = (1.6 \times 10^{-12}\ \mathrm{W/m^3}) \exp\left(-\frac{x^2 + y^2}{0.01}\right), \tag{1.30}$$

where distances are measured in nanometers. This source term is applied for a time of 50 ps and the system is then allowed to relax.

Figure 1.9 shows sequences of temperature profiles over time for both the phonon and electron temperatures. The electrons, which are heated directly, show a very localized heating, while the indirectly heated phonons have a more diffuse temperature profile.

Figure 1.10 shows the average temperatures of the nanotube and the graphene reservoirs over time. The large oscillations in the nanotube temperature correspond to the excitation of a fundamental mode resonance in the tube, as can be seen in a plot of the displaced atoms and mesh in Figure 1.11.

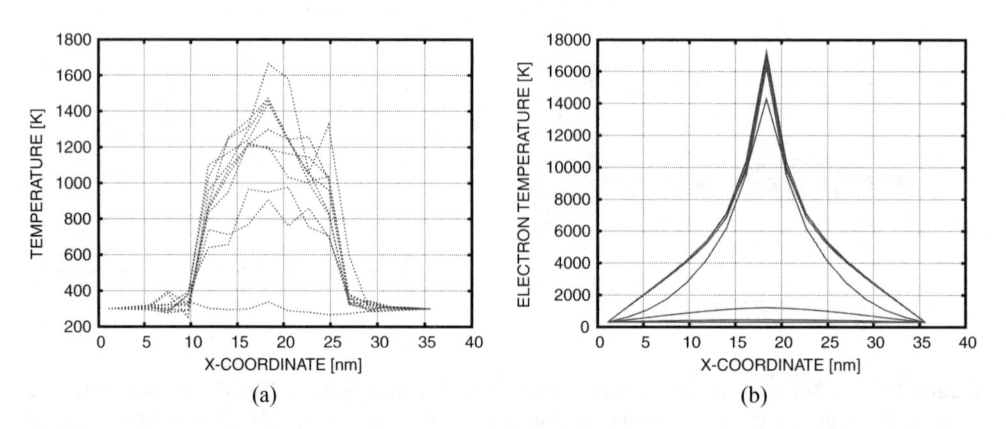


**Figure 1.9** Sequence of (a) temperature and (b) electron temperature profiles along the axis of the carbon nanotube. Note that the two plots show different temperature ranges

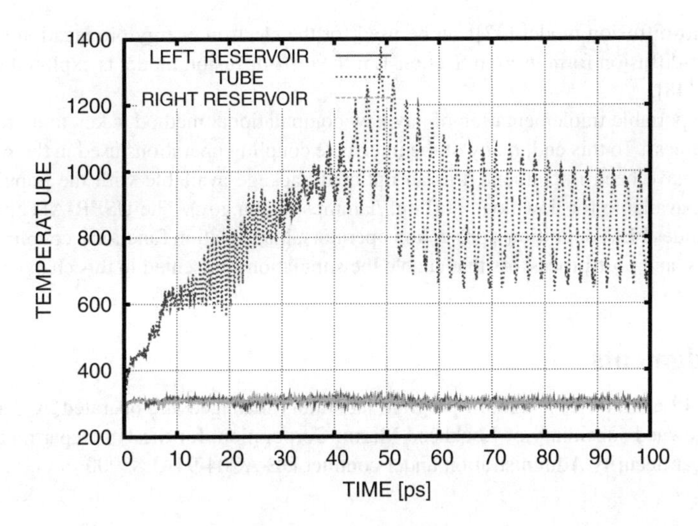


**Figure 1.10** Evolution of average temperatures of the two explicitly modeled reservoirs and the carbon nanotube

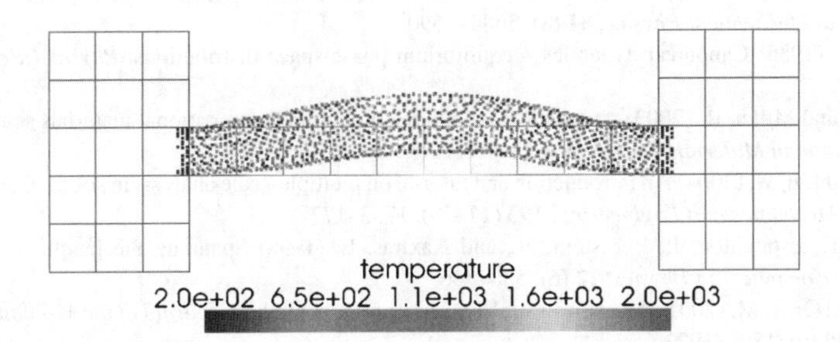


**Figure 1.11** Fundamental mode excited by focused irradiation. The atoms and the mesh are both colored by the phonon temperature

## 1.7 Discussion

The methods and demonstrations in this chapter represent steps toward the ability to simulate heat transfer in nanoscale systems, including non-continuum effects, with the same ease with which finite-element models can today be used to define and solve macroscale heat transfer problems. Through the coupled atomistic-to-continuum approach, we are able to apply spatially varying boundary and initial conditions on temperature, as well as spatially and temporally varying heat fluxes, to MD systems. At the same time, we can choose to use MD only in isolated regions of the domain, improving computational efficiency. Finally, we can use the coupling to a continuum field to capture effects, such as electronic heat transfer, that are otherwise missed by classical MD simulations.

It should be pointed out that many of the assumptions made in the continuum models presented in this chapter are easy to relax in the methods presented. For example, although we have assumed that a Fourier heat law describes the continuum region in Section 1.3.2, it is straightforward to replace this with the Maxwell–Cattaneo–Vernotte model typically employed to represent the finite speed of propagation of heat waves [45, 46]; the derivation of the coupled system proceeds in a similar way.

Likewise, a drift-diffusion model [47] can be used for the electron energy propagation rather than the simpler Fourier-diffusion form used in Section 1.5; the drift-diffusion model is explored in this context by Jones *et al.* [48].

An efficient, portable implementation of any new computational method is key to its future development and usefulness. To this end, many of the interscale coupling operations used in the work presented in this chapter have been implemented as an optional package available with the popular LAMMPS MD code [49], available for download at (http://lammps.sandia.gov). The USER-ATC package within LAMMPS includes FEM definitions, projection operators, interpolation functions, coupling thermostats, time integrators, and post-processors that enable the simulations presented in this chapter.

## Acknowledgments

Sandia National Laboratories is a multi-program laboratory managed and operated by Sandia Corporation, a wholly owned subsidiary of Lockheed Martin Corporation, for the US Department of Energy's National Nuclear Security Administration under contract DE-AC04-94AL85000.

## References

[1] Berendsen, H., Postma, J., Vangunsteren, W. *et al.* (1984) Molecular-dynamics with coupling to an external bath. *Journal of Chemical Physics*, **81** (8), 3684–3690.
[2] Hoover, W. (1985) Canonical dynamics – equilibrium phase-space distributions. *Physical Review A*, **31** (3), 1695–1697.
[3] Curtin, W. and Miller, R. (2003) Atomistic/continuum coupling in computational materials science. *Modelling and Simulation in Materials Science and Engineering*, **11** (3), R33–R68.
[4] Park, H. and Liu, W. (2004) An introduction and tutorial on multiple-scale analysis in solids. *Computer Methods in Applied Mechanics and Engineering*, **193** (17–20), 1733–1772.
[5] Abraham, F., Broughton, J., Bernstein, N., and Kaxiras, E. (1998) Spanning the length scales in dynamic simulation. *Computers in Physics*, **12** (6), 538–546.
[6] Knap, J. and Ortiz, M. (2001) An analysis of the quasicontinuum method. *Journal of the Mechanics and Physics of Solids*, **49** (9), 1899–1923.
[7] Miller, R. and Tadmor, E. (2009) A unified framework and performance benchmark of fourteen multiscale atomistic/continuum coupling method. *Modelling and Simulation in Materials Science and Engineering*, **17** (5), 053001.
[8] Cai, W., de Koning, M., Bulatov, V., and Yip, S. (2000) Minimizing boundary reflections in coupled-domain simulations. *Physical Review Letters*, **85** (15), 3213–3216.
[9] Wagner, G. and Liu, W. (2003) Coupling of atomistic and continuum simulations using a bridging scale decomposition. *Journal of Computational Physics*, **190** (1), 249–274.
[10] To, A. and Li, S. (2005) Perfectly matched multiscale simulations. *Phys Rev B*, **72** (3), 035414.
[11] Ivanov, D. and Zhigilei, L. (2003) Combined atomistic-continuum modeling of short-pulse laser melting and disintegration of metal films. *Physical Review B*, **68** (6), 064114.
[12] Xu, X., Cheng, C., and Chowdhury, I. (204) Molecular dynamics study of phase change mechanisms during femtosecond laser ablation. *ASME Journal of Heat Transfer*, **126** (5), 727–734.
[13] Jones, R., Templeton, J., Wagner, G. *et al.* (2010) Electron transport enhanced molecular dynamics for metals and semi-metals. *International Journal for Numerical Methods in Engineering*, **83** (8–9), 940–967.
[14] Qian, D., Wagner, G., and Liu, W. (2004) A multiscale projection method for the analysis of carbon nanotubes. *Computer Methods in Applied Mechanics and Engineering*, **193** (17-20), 1603–1632.
[15] Liu, W., Park, H., Qian, D. *et al.* (2006) Bridging scale methods for nanomechanics and materials. *Computer Methods in Applied Mechanics and Engineering*, **195** (13–16), 1407–1421.
[16] Liu, W., Uras, R., and Chen, Y. (1997) Enrichment of the finite element method with the reproducing kernel particle method. *Journal of Applied Mechanics, ASME*, **64** (4), 861–870.

[17] Uras, R., Chang, C., Chen, Y., and Liu, W. (1997) Multiresolution reproducing kernel particle methods in acoustics. *Journal of Computational Acoustics*, **5** (1), 71–94.

[18] Wagner, G. and Liu, W. (2001) Hierarchical enrichment for bridging scales and mesh-free boundary conditions. *International Journal for Numerical Methods in Engineering*, **50** (3), 507–524.

[19] Weiner, J. (2002) *Statistical Mechanics of Elasticity*, 2nd edition, Dover, Mineola, NY.

[20] Wagner, G., Jones, R., Templeton, J., and Parks, M. (2008) An atomistic-to-continuum coupling method for heat transfer in solids. *Computer Methods in Applied Mechanics and Engineering*, **197** (41–42), 3351–3365.

[21] Templeton, J., Jones, R., and Wagner, G. (2010) Application of a field-based method to spatially varying thermal transport problems in molecular dynamics. *Modelling and Simulation in Materials Science and Engineering*, **18** (8), 085007.

[22] Tretiakov, K. and Scandolo, S. (2004) Thermal conductivity of solid argon from molecular dynamics simulations. *Journal of Chemical Physics*, **120** (8), 3765–3769.

[23] Stuart, S., Tutein, A., and Harrison, J. (2000) A reactive potential for hydrocarbons with intermolecular interactions. *Journal of Chemical Physics*, **112** (14), 6472–6486.

[24] Brenner, D., Shenderova, O., Harrison, J. *et al.* (2002) A second-generation reactive empirical bond order (REBO) potential energy expression for hydrocarbons. *Journal of Physics: Condensed Matter*, **14** (4), 783–802.

[25] Wang, J. and Wang, J. (2006) Carbon nanotube thermal transport: ballistic to diffusive. *Applied Physics Letters*, **88** (11), 111909.

[26] Chen, R., Hochbaum, A., Murphy, P. *et al.* (2008) Thermal conductance of thin silicon nanowires. *Physical Review Letters*, **101** (10), 105501.

[27] Chang, C., Okawa, D., Garcia, H. *et al.* (2008) Breakdown of Fourier's law in nanotube thermal conductors. *Physical Review Letters*, **101** (7), 075903.

[28] Chrisey, D. and Hubler, G. (1994) *Pulsed Laser Deposition of Thin Films*, Wiley Interscience, New York, NY.

[29] Baerle, D. (2000) *Laser Processing and Chemistry*, Springer, Berlin.

[30] Sundaram, S. and Mazur, E. (2002) Inducing and probing non-thermal transitions in semiconductors using femtosecond laser pulses. *Nature Materials*, **1** (4), 217–224.

[31] Wang, Y., Xu, X., and Zheng, L. (2008) Molecular dynamics simulation of ultrafast laser ablation of fused silica film. *Applied Physics A: Materials Science & Processing*, **92** (4), 849–852.

[32] Kim, P., Shi, L., Majumdar, A., and McEuen, P. (2001) Thermal transport measurements of individual multiwalled nanotubes. *Physical Review Letters*, **87** (21), 215502.

[33] Yang, B., Liu, W., Liu, J. *et al.* (2002) Measurements of anisotropic thermoelectric properties in superlattices. *Applied Physics Letters*, **81** (19), 3588–3590.

[34] Li, D., Wu, Y., Fan, R. *et al.* (2003) Thermal conductivity of Si/SiGe superlattice nanowires. *Applied Physics Letters*, **83** (15), 3186–3188.

[35] Snyder, G. and Toberer, E. (2008) Complex thermoelectric materials. *Nature Materials*, **7** (2), 105–114.

[36] Ou, M., Yang, T., Harutyunyan, S. *et al.* (2008) Electrical and thermal transport in single nickel nanowire. *Applied Physics Letters*, **92** (6), 063101.

[37] Majumdar, A. and Reddy, P. (2004) Role of electron–phonon coupling in thermal conductance of metal–nonmetal interfaces. *Applied Physics Letters*, **84** (23), 4768–4770.

[38] Kaganov, M., Lifshits, I., and Tanatarov, L. (1956) Relaxation between electrons and the crystal lattice. *Zhurnal Eksperimental'noi i Teoreticheskoi Fiziki*, **31** (2(8)), 232–237.

[39] Anisimov, S., Kapeliov, B., and Perelman, T. (1998) Emission of electrons from the surface of metals induced by ultrashort laser pulses. *Zhurnal Eksperimental'noi i Teoreticheskoi Fiziki*, **66** (2), 776–781.

[40] Qiu, T. and Tien, C. (1993) Heat transfer mechanisms during short-pulse laser heating of metals. *Journal of Heat Transfer*, **115** (4), 835–841.

[41] Melikyan, A., Minassian, H., Guerra, III, A., and Wu, W. (1999) On the theory of relaxation of electrons excited by femtosecond laser pulses in thin metallic films. *Applied Physics B: Lasers and Optics*, **68** (3), 411–414.

[42] Jiang, L. and Tsai, H. (2005) Improved two-temperature model and its application in ultrashort laser heating of metal films. *Journal of Heat Transfer*, **127** (10), 1167–1173.

[43] Tersoff, J. (1988) Empirical interatomic potential for carbon, with applications to amorphous carbon. *Physical Review Letters*, **61** (25), 2879–2882.

[44] Hertel, T., Fasel, R., and Moos, G. (2002) Charge-carrier dynamics in single-wall carbon nanotube bundles: a time-domain study. *Applied Physics A: Materials Science & Processing*, **75** (4), 449–465.
[45] Joseph, D. and Preziosi, L. (1989) Heat waves. *Reviews of Modern Physics*, **61** (1), 41–73.
[46] Ho, J.R., Twu, C.J., and Hwang, C.C. (2001) Molecular-dynamics simulation of thermoelastic waves in a solid induced by a strong impulse energy. *Physical Review B*, **64** (1), 014302.
[47] Chen, G. (2005) *Nanoscale Energy Transport and Conversion*, Oxford University Press, Oxford.
[48] Jones, R., Templeton, J., Modine, N. *et al.* (2010) Molecular dynamics enhanced with electron transport, Technical Report SAND2010-6164, Sandia National Laboratories.
[49] Plimpton, S. (1995) Fast parallel algorithms for short-range molecular-dynamics. *Journal of Computational Physics*, **117** (1), 1–19.

# 2

# Accurate Boundary Treatments for Concurrent Multiscale Simulations

Shaoqiang Tang
*HEDPS, CAPT & Department of Mechanics, Peking University, People's Republic of China*

## 2.1 Introduction

Concurrent multiscale simulations play an important role in materials science and technology. In such a simulation, one first selects a subdomain where detailed dynamics are necessary to capture the correct physics. Away from this subdomain, a coarse-scale computation is performed to alleviate the heavy computing load.

To be specific, we confine ourselves to the coupling between a coarse scale and atomic dynamics. A concurrent multiscale algorithm contains several factors; namely, the atomic scheme and boundary treatment, the coarse-scale mesh generation and scheme, and the information exchange between the two scales. In this chapter, we focus on the boundary treatments for the atomic subdomain.

As is well known, accuracy and cost are two major indices to evaluate a multiscale algorithm. Efficiency may be defined as their ratio. The accuracy takes the lowest order among those of the above factors in an algorithm, whereas the cost takes the maximal order among those of the factors. For an accurate concurrent multiscale method, boundary treatments for the atomic subdomain are crucial and challenging.

In the following, we present several accurate boundary treatments. We start with the time history kernel treatment, which is exact for infinite lattices with equilibrium state outside of the atomic subdomain. Then, for a finite lattice, we explore the wave propagation and obtain the precise expression of the kernel functions. Afterwards, we describe several more effective approaches. In the first approach, we match a one-way wave operator factorized from the full dynamics and express the velocity at boundary atoms by linear combinations of the displacement at atoms near the boundary. The resulting boundary conditions are called velocity interfacial conditions. In the second approach, we include the velocity and displacement of atoms near the boundary to form a linear constraint, which in turn specifies the motion at the boundary atom. The constraint coefficients are determined by matching the dispersion relation. These types of boundary conditions are hence called matching boundary conditions (MBCs). Finally, a

*Multiscale Simulations and Mechanics of Biological Materials*, First Edition. Edited by Shaofan Li and Dong Qian.

more accurate way to get the constraint coefficients is by matching the kernel functions. We also describe a two-way boundary condition to include inputs toward the atomic subdomain. This may be used in loading processes as well as in finite temperature simulations.

Consider an infinite lattice in $\mathbb{R}^3$. The position of the $n$th atom at rest is $x_n$. Under suitable initial conditions, the displacement $u_n(t)$ is governed by the Newton equation:

$$m\ddot{u}_n = -\nabla_{u_n} U(u), \tag{2.1}$$

with the displacement vector $u$ for all atoms, the mass $m > 0$, and the interatomic potential $U$.

## 2.2 Time History Kernel Treatment

Time history kernel treatment was proposed first by Adelman and Doll [1], and adopted for multiscale computations by Cai *et al.* [2]. A displacement decomposition has been further incorporated in a bridging-scale method [3, 4] and a pseudo-spectral multiscale method [5]. There is a series of important works on the bridging-scale method for crystalline solids in one and multiple space dimensions; for example, see Refs [6–8]. For a comprehensive exposure of these works, please refer to Liu *et al.* [9]. Other numerical methods with similar strategy have also been developed [10].

There are two basic assumptions on the lattice away from the atomic subdomain for the time history kernel treatment. First, this part of lattice is at rest initially. Secondly, the lattice is infinite and takes a uniformly periodic structure and linear interaction among neighboring atoms. In real applications, there are two approximations made. First, to simplify the calculation of the kernel functions, a semi-infinite lattice is used to represent the finite atomic subdomain. Second, to alleviate the time-consuming convolution, the time history kernel function is truncated, making use of its decreasing feature. In this presentation, we focus on the these two approximations.

In the following, we first consider a harmonic chain in one space dimension to illustrate the basic ideas. Then we find the exact time history kernel functions for a square lattice in two space dimensions.

### 2.2.1 *Harmonic Chain*

We consider a harmonic lattice in one space dimension (see Figure 2.1). The position of the $n$th atom at rest is $x_n = nh_\mathrm{a}$ for $n \in \mathbb{N}$, with $h_\mathrm{a}$ the lattice constant. Under a potential $U = \sum_n k(u_{n+1} - u_n)^2/2$, the Newton equation reads

$$m\ddot{u}_n = k(u_{n-1} - 2u_n + u_{n+1}). \tag{2.2}$$

The atomic subdomain consists of atoms numbered from 1 through $n_\mathrm{b}$. Assuming that $u_n(0) = \dot{u}_n(0)$ for $n > n_\mathrm{b}$, we seek for an expression of $u_{n_\mathrm{b}+1}$ in terms of $u_{n_\mathrm{b}}$. This in turn provides the information to calculate the force exerted from the $(n_\mathrm{b} + 1)$th atom upon the $n_\mathrm{b}$th atom, and hence a boundary condition for the atomic subdomain.

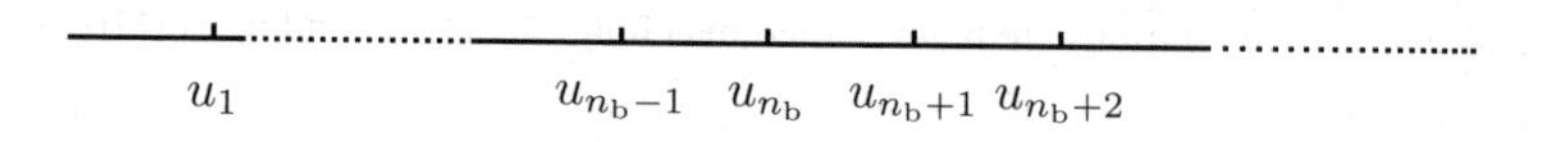


**Figure 2.1** Harmonic lattice in one space dimension

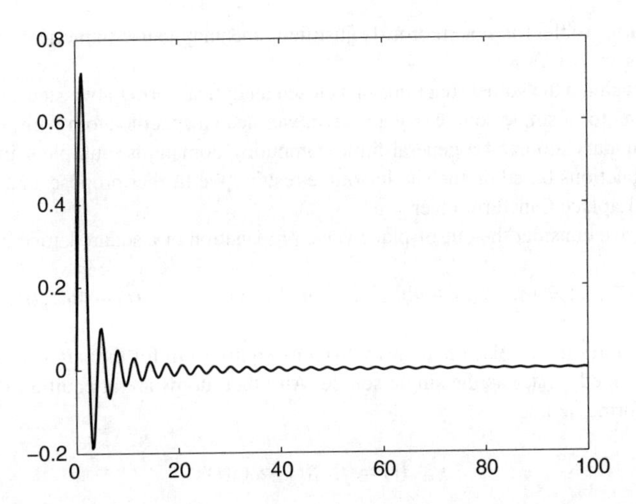


**Figure 2.2** Time history kernel function $\theta(t)$ for the harmonic chain

Rescaling time by $\sqrt{k/m}$, we obtain the governing equations for the semi-infinite chain $n > n_\mathrm{b}$ as follows:

$$\ddot{u} = Du + \begin{bmatrix} \tilde{u}_{n_\mathrm{b}} \\ 0 \\ \vdots \end{bmatrix}, \quad D = \begin{bmatrix} -2 & 1 & & \\ 1 & -2 & \ddots & \\ & \ddots & \ddots & \end{bmatrix}. \tag{2.3}$$

Taking the Laplace transform, we may convert the problem into a linear system. The Laplace inverse transform of the (1, 1) entry in $(s^2 I - D)^{-1}$ is $S = (s^2 + 2 - s\sqrt{s^2+4})/2$. So we obtain a simple and exact expression:

$$u_{n_\mathrm{b}+1} = \theta(t) * u_{n_\mathrm{b}}(t), \quad \theta(t) = \mathcal{L}^{-1}(S) = \frac{2J_2(2t)}{t}. \tag{2.4}$$

The kernel function $\theta(t)$ is displayed in Figure 2.2. When $t \to +\infty$, the Bessel function is approximately $J_n(t) \approx \sqrt{2/\pi t}\cos(t - n\pi/2 - \pi/4)$. Accordingly, we only keep a partial history for $\tau \in (t - T_\mathrm{c}, t)$ to alleviate the computing load. Here, $T_\mathrm{c}$ is the cut-off time, for which a certain choice may help improving the accuracy [11]. An upper bound for the numerical error is

$$\left| \int_0^{t-T_\mathrm{c}} \theta(t-\tau) g(\tau)\,\mathrm{d}\tau \right| \le \|g\|_{L^2([0,t-T_\mathrm{c}])} \|\theta\|_{L^2([T_\mathrm{c},t])} \le C\|g\|_{L^2([0,\infty))} T_\mathrm{c}^{-1}. \tag{2.5}$$

### 2.2.2 *Square Lattice*

In multiple space dimensions, the explicit expression for the time history kernel treatment does not exist in general. In the literature, the kernel functions are calculated numerically through a combination of the discrete Fourier transform and the Laplace transform [7, 8]. This approach gives exact kernel functions only for an infinite boundary; for example, a straight line for a two-dimensional square lattice. In fact, all real computations take finite boundaries. The direct application of the infinite boundary approximations to a finite boundary problem is not rigorously justified, and may cause numerical errors, particularly around the corners.

We suggest to evaluate the kernel functions in a closed analytical form in two steps. First, we calculate the kernel functions for a single source. By a source, we mean a specific atom with prescribed motion. Regarding the boundary atoms of a general finite computing domain as multiple sources, we compute the exact kernel functions based on the single-source results. We further propose an accurate numerical procedure for the Laplace transform inversion.

More precisely, we consider the out-of-plane wave propagation of a square lattice governed by

$$\ddot{u}_{m,n}(t) = u_{m-1,n}(t) + u_{m+1,n}(t) + u_{m,n+1}(t) + u_{m,n-1}(t) - 4u_{m,n}(t). \tag{2.6}$$

Here, $u_{m,n}(t)$ denotes the displacement away from its equilibrium for the $(m, n)$th atom. The $(0, 0)$th atom takes a prescribed $u_{0,0}(t)$ as the single source. All other atoms are at equilibrium initially. Using the Laplace transform, we find

$$u_{m,n}(t) = f_{m,n}(t) * u_{0,0}(t), \tag{2.7}$$

where the kernel function $f_{m,n}(t)$ is the inverse Laplace transform of

$$\bar{f}_{m,n}(s) = \frac{\int_0^\pi \int_0^\pi (\cos mx \cos ny)/\left(\frac{s^2}{2} + 2 - \cos x - \cos y\right) \mathrm{d}x\, \mathrm{d}y}{\int_0^\pi \int_0^\pi 1/\left(\frac{s^2}{2} + 2 - \cos x - \cos y\right) \mathrm{d}x\, \mathrm{d}y}. \tag{2.8}$$

Next, we consider a finite rectangular domain with $M \times N$ atoms, numbered by $(m, n)$ with $m = -1, \ldots, M-2; n = -1, \ldots, N-2$. The atoms in the first interior layer are indexed as $I(1), I(2), \ldots, I(L)$, where $L = 2M + 2N - 12$. The multiple index $I(l) = (I_1(l), I_2(l))$ identifies the position of the atom. See Figure 2.3.

To express the displacement at a boundary atom in terms of the first interior layer atoms, we first decompose the displacement $u_{I(l)}$ into two parts; namely, the independent source $u'_{I(l)}$ and the displacement due to the influence of other first interior layer atoms.

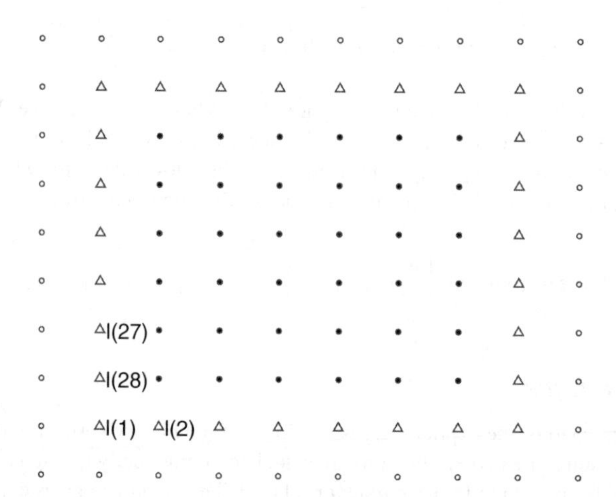


**Figure 2.3** Square lattice with $M = N = 10$: hollow circles for boundary and triangles for the first interior layer

Noticing that for each $l$, it holds that

$$\sum_{p=1}^{L} f_{I(l)-I(p)} * u'_{I(p)} = u_{I(l)}, \tag{2.9}$$

where $f_{0,0}(t) = \delta(t)$.

Taking the Laplace transform, we obtain an algebraic system:

$$\begin{bmatrix} \bar{f}_{I(1)-I(1)} & \cdots & \bar{f}_{I(1)-I(L)} \\ \vdots & \ddots & \vdots \\ \bar{f}_{I(L)-I(1)} & \cdots & \bar{f}_{I(L)-I(L)} \end{bmatrix} \begin{bmatrix} \bar{u}'_{I(1)} \\ \vdots \\ \bar{u}'_{I(L)} \end{bmatrix} = \begin{bmatrix} \bar{u}_{I(1)} \\ \vdots \\ \bar{u}_{I(L)} \end{bmatrix}. \tag{2.10}$$

At a boundary atom, we again use the single-source result:

$$\begin{aligned} \bar{u}_{m,n} &= \sum_{l=1}^{L} \bar{f}_{(m,n)-I(l)} \bar{u}'_{I(l)} \\ &= [\bar{f}_{(m,n)-I(1)}, \ldots, \bar{f}_{(m,n)-I(L)}] \begin{bmatrix} \bar{f}_{I(1)-I(1)} & \cdots & \bar{f}_{I(1)-I(L)} \\ \vdots & \ddots & \vdots \\ \bar{f}_{I(L)-I(1)} & \cdots & \bar{f}_{I(L)-I(L)} \end{bmatrix}^{-1} \begin{bmatrix} \bar{u}_{I(1)} \\ \vdots \\ \bar{u}_{I(L)} \end{bmatrix} \\ &\equiv [\bar{J}^{I(1)}_{m,n}, \ldots, \bar{J}^{I(L)}_{m,n}] \begin{bmatrix} \bar{u}_{I(1)} \\ \vdots \\ \bar{u}_{I(L)} \end{bmatrix}. \end{aligned} \tag{2.11}$$

The Laplace transform $\bar{J}^{I(l)}_{m,n}$ is analytic away from the line segment connecting $\pm 2\sqrt{2}\mathrm{i}$. The singular points are $z_0 = 0$, $z_{\pm 1} = \pm 2\mathrm{i}$, $z_{\pm 2} = \pm 2\sqrt{2}\mathrm{i}$. We take an integral path $\Gamma$ consisting of straight-line segments and semicircles around the singular points with radius $\varepsilon = 0.1$. We let $A_1 = -3.5\mathrm{i}$, $A_2 = -2.41\mathrm{i}$, $A_3 = -\mathrm{i}$, $A_4 = \mathrm{i}$, $A_5 = 2.41\mathrm{i}$, $A_6 = 3.5\mathrm{i}$. See Figure 2.4. For the path segments between every two subsequent points of $A_1$ to $A_6$ and around each singular point, there are series expansions available.

$$J^{I(l)}_{m,n}(t) = \frac{1}{2\pi\mathrm{i}} \int_{P-\mathrm{i}\infty}^{P+\mathrm{i}\infty} \mathrm{e}^{st} \bar{J}^{I(l)}_{m,n}(s)\, \mathrm{d}s. \tag{2.12}$$

The resulting kernel function $J^{0,0}_{-1,1}(t)$ for a $10 \times 10$ atomic subdomain is displayed in Figure 2.5. To make a comparison, the corresponding infinite boundary approximation is also shown. The difference is not small, which partly explains corner reflections when such an infinite boundary approximation is adopted.

We hence obtain the following exact boundary condition:

$$u_{m,n}(t) = \sum_{l=1}^{L} J^{I(l)}_{m,n}(t) * u_{I(l)}(t). \tag{2.13}$$

As a numerical test, we compute with initial data

$$u_{m,n}(t) = \varphi[\sqrt{(m-3.5)^2 + (n-3.5)^2}], \quad \dot{u}_{m,n}(t) = 0, \tag{2.14}$$

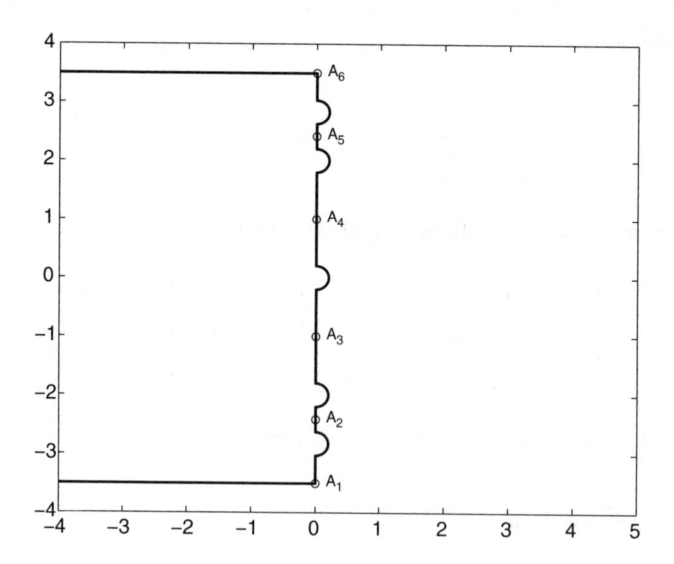


**Figure 2.4** Integral path in the complex plane to calculate $f_{m,n}(t)$

where

$$\varphi(x) = \begin{cases} e^{-1/(16-x^2)}, & |x| < 4, \\ 0, & |x| \geq 4. \end{cases} \tag{2.15}$$

The displacement $u_{2,3}(t)$ in Figure 2.6 clearly demonstrates the effectiveness of the proposed time history treatment in eliminating the corner reflection. In contrast, such a reflection occurs if an infinite boundary approximation is adopted.

For more details and discussions on the cubic lattice in three space dimensions, please refer to Pang and Tang [12].

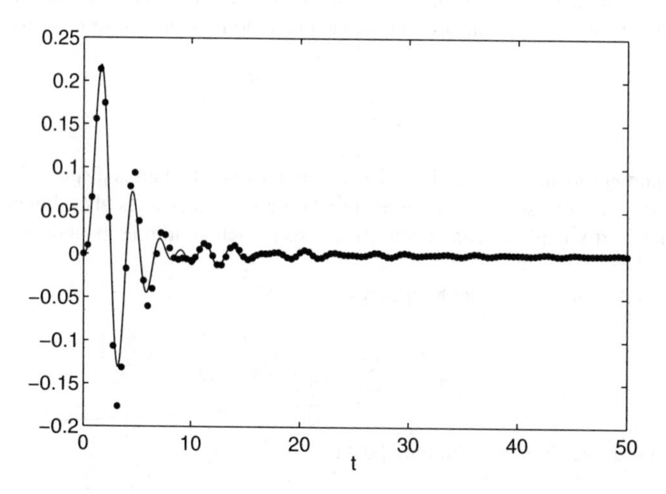


**Figure 2.5** Kernel functions in a 10 × 10 lattice: solid line for $J^{0,0}_{-1,1}(t)$ and dots for infinite boundary approximation

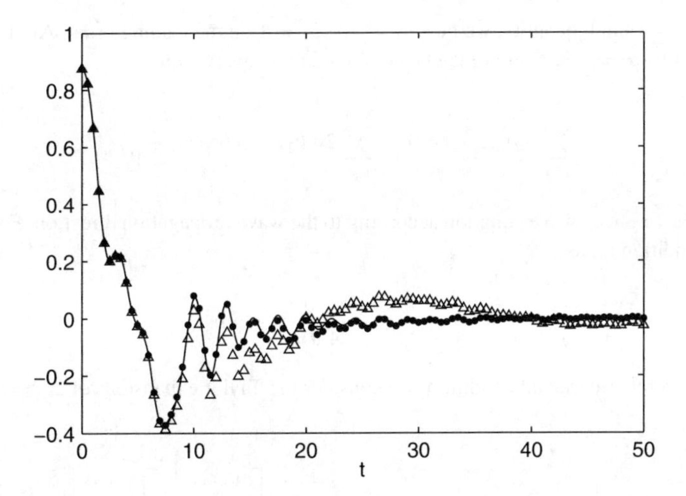


**Figure 2.6** Atomic displacement $u_{2,3}(t)$: solid line for exact, dots for finite boundary condition, and triangles for infinite boundary approximation

## 2.3 Velocity Interfacial Conditions: Matching the Differential Operator

The aforementioned time history kernel treatment, when no approximation is used, is an exact boundary condition. However, this treatment is accurate only for linear lattices. Because of the nonlocality in time, it does not admit a rigorous nonlinear generalization. For a crystalline solid with small deformation, one ignores the nonlinear effects around the interface in the coarse scale and adopts the interfacial condition corresponding to the lattice at equilibrium. This introduces error. When applied to nonlinear atomistic systems such as an anharmonic lattice, the reflection can be large [4, 5]. In addition, it is numerically demanding to perform the convolutions.

In contrast, for a boundary condition that is local in both space and time, we may regard that the fast fluctuations in the short length scale are carried by the long waves and that the long waves evolve in a slower time scale. Therefore, during each coarse-scale time step, the fine-scale waves around the interface are governed by a linearized system, with its parameters determined from the coarse grid solution. In this way, the condition applies directly to crystalline solids with relatively strong nonlinearity and large deformation. Here, we describe a class of velocity interfacial conditions where the linearization and velocity correction are designed with required accuracies.

The fundamental assumption in the velocity interfacial conditions, and also in the local boundary conditions in the next two subsections, is the one-way wave propagation out of the atomic subdomain.

The harmonic chain dynamics (2.2) takes a Taylor series form:

$$\ddot{z} = \sum_{m=1}^{\infty} \frac{2}{(2m)!} \frac{\partial^{2m} z}{\partial x^{2m}} h_{\rm a}^{2m}. \tag{2.16}$$

We factorize the differential operator as follows:

$$\left( \frac{\partial}{\partial t} + \sum_{i=1}^{\infty} a_i h_{\rm a}^i \frac{\partial^i}{\partial x^i} \right) \left( \frac{\partial}{\partial t} + \sum_{i=1}^{\infty} b_i h_{\rm a}^i \frac{\partial^i}{\partial x^i} \right) z = 0. \tag{2.17}$$

Regarding $h_a$ as a small quantity, we take $a_1 = -1, b_1 = 1$ on the leading order. After some manipulations, we find that $a_i = -b_i$, and $b_i$ may be computed recursively from

$$\sum_{i=1}^{n} b_i b_{2n+1-i} = 0, \quad \sum_{i=1}^{n-1} 2b_i b_{2n-i} + b_n^2 = \frac{2}{(2n)!}. \tag{2.18}$$

We pick up the one-way wave equation according to the wave propagation direction. For example, the fourth-order condition reads

$$\dot{z} = -h_a \frac{\partial}{\partial x} z - \frac{h_a^3}{24} \frac{\partial^3}{\partial x^3} z + o(h_a^4). \tag{2.19}$$

To design a discrete interfacial condition, we consider the Taylor expansions of $\Delta_i = u_{n_b - i} - u_{n_b}$ for $i = 1, 2, 3, 4$:

$$\begin{bmatrix} \Delta_1 \\ \Delta_2 \\ \Delta_3 \\ \Delta_4 \end{bmatrix} = \begin{bmatrix} -1 & 1 & -1 & 1 \\ -2 & 4 & -8 & 16 \\ -3 & 9 & -27 & 81 \\ -4 & 16 & -64 & 256 \end{bmatrix} \begin{bmatrix} h_a z_x \\ \frac{h_a^2}{2} z_{xx} \\ \frac{h_a^3}{6} z_{xxx} \\ \frac{h_a^4}{24} z_{xxxx} \end{bmatrix} + o(h_a^4). \tag{2.20}$$

It is solved by

$$\begin{bmatrix} h_a z_x \\ \frac{h_a^2}{2} z_{xx} \\ \frac{h_a^3}{6} z_{xxx} \\ \frac{h_a^4}{24} z_{xxxx} \end{bmatrix} = \begin{bmatrix} -4 & 3 & -4/3 & 1/4 \\ -13/3 & 17/4 & -7/3 & 11/24 \\ -3/2 & 2 & -7/6 & 1/4 \\ -1/6 & 1/4 & -1/6 & 1/24 \end{bmatrix} \begin{bmatrix} \Delta_1 \\ \Delta_2 \\ \Delta_3 \\ \Delta_4 \end{bmatrix} + o(h_a^4). \tag{2.21}$$

Together with (2.19), this leads to a fourth-order condition:

$$\dot{u}_{n_b} = -\frac{5}{16} u_{n_b-4} + \frac{13}{8} u_{n_b-3} - \frac{7}{2} u_{n_b-2} + \frac{35}{8} u_{n_b-1} - \frac{35}{16} u_{n_b}. \tag{2.22}$$

In the same way, we rectify velocities at the next two atoms as follows:

$$\dot{u}_{n_b-1} = \frac{1}{16} u_{n_b-4} - \frac{3}{8} u_{n_b-3} + \frac{5}{4} u_{n_b-2} - \frac{5}{8} u_{n_b-1} - \frac{5}{16} u_{n_b}, \tag{2.23}$$

$$\dot{u}_{n_b-2} = -\frac{1}{16} u_{n_b-4} + \frac{5}{8} u_{n_b-3} - \frac{5}{8} u_{n_b-1} + \frac{1}{16} u_{n_b}. \tag{2.24}$$

Velocities at atoms further inside the atomic subdomain are not corrected, owing to considerations on the wave propagation direction.

Nonlinearity exists in most applications. For a nonlinear system, an exact interfacial condition requires precise expression for the solution, which is not accessible in general. Because other existing interfacial treatments are less accurate even for the linear lattices, we discuss only the time history treatment here. If the nonlinearity is weak, one may adopt an interfacial condition corresponding to the linear lattice at equilibrium [2, 3, 5]. This introduces an error. When the nonlinearity is strong, the error can be large. Because a long time history is necessary for performing convolutions, one cannot update the time history kernel within the cut-off period. Moreover, the kernel function is hard to compute when the lattice is away from the uniform equilibrium.

In contrast, the proposed velocity interfacial conditions are local in both space and time. At any fixed time, an atom near the interface experiences the lattice approximately as a linear one with a uniform strain, which is determined by the coarse-scale displacement at these several grid points around the interface. This suggests a treatment using a linearized lattice:

$$\ddot{u}_n = c^2(t)(u_{n-1} - 2u_n + u_{n+1}), \tag{2.25}$$

where $c(t)$ is computed through the coarse grid displacements near the interface. There are various ways to compute $c(t)$. The velocity interfacial condition for the nonlinear lattice is taken as follows:

$$\begin{bmatrix} \dot{u}_{n_b-2} \\ \dot{u}_{n_b-1} \\ \dot{u}_{n_b} \end{bmatrix} = c(t) \begin{bmatrix} -1/16 & 5/8 & 0 & -5/8 & 1/16 \\ 1/16 & -3/8 & 5/4 & -5/8 & -5/16 \\ -5/16 & 13/8 & -7/2 & 35/8 & -35/16 \end{bmatrix} \begin{bmatrix} u_{n_b-4} \\ u_{n_b-3} \\ u_{n_b-2} \\ u_{n_b-1} \\ u_{n_b} \end{bmatrix}. \tag{2.26}$$

For an initial profile with relatively large deformation and, hence, strong nonlinearity, the numerical results by a finite-difference approach that incorporates the velocity boundary condition in Figure 2.7 agree well with the exact solution [13].

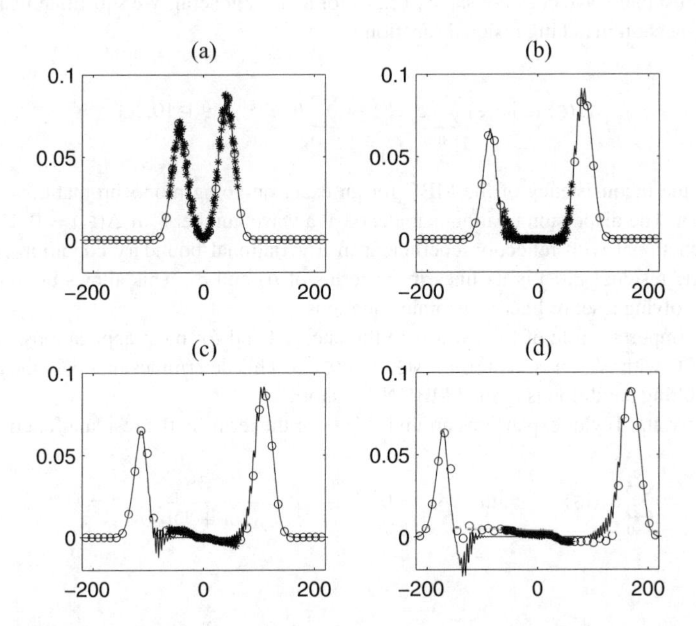


**Figure 2.7** Lattice with the Lennard-Jones potential by the finite-difference approach (solid lines for the exact solution, stars and circles for the multiscale solution): (a) $u(x, 5)$; (b) $u(x, 10)$; (c) $u(x, 13)$; (d) $u(x, 20)$

## 2.4 MBCs: Matching the Dispersion Relation

Noticing the linear form and locality in the velocity interfacial conditions, we may further consider a linear constraint of displacement and velocity at atoms near the boundary. A velocity interfacial condition falls into this category as a degenerate case with the velocity at non-boundary atoms and the displacement at the boundary atom dropped out. Fitting a set of suitable coefficients for the linear constraint, we may expect better efficiency in suppressing the spurious reflection at the boundary. The coefficients are selected via a dispersion residual function. After designing efficient boundary conditions in this manner, we adopt apparent wave propagation and operator multiplication to treat multiple dimensional lattices. We refer to these types of conditions as MBCs.

### *2.4.1 Harmonic Chain*

For the harmonic chain, we construct the following local boundary condition:

$$\sum_{j=0}^{N} c_j \dot{u}_{n_b-j} = \sum_{j=0}^{N} b_j u_{n_b-j}. \tag{2.27}$$

The coefficients $b_j$ and $c_j$ are to be determined.

As is well known, the dispersion relation

$$\omega(\xi) = 2\left|\sin\frac{\xi}{2}\right|, \quad \xi \in [-\pi, \pi] \tag{2.28}$$

relates the frequency $\omega$ to the wave number $\xi$ for the sinusoidal wave $u_j = e^{i(\omega t - j\xi)}$. Corresponding to the one-way wave equation, we only consider half of the first Brillouin zone $\xi \in [0, \pi]$ here.

The exact sinusoidal wave does not satisfy (2.28) for all $\xi$ in general. We substitute its form in (2.28) and define a dispersion matching residual function

$$\Delta(\xi) \equiv i\omega(\xi) \sum_{j=0}^{N} c_j e^{ij\xi} - \sum_{j=0}^{N} b_j e^{ij\xi}, \quad \xi \in [0, \pi]. \tag{2.29}$$

This measures the inconsistency of the MBC for an exact outgoing monochromatic wave solution of the infinite chain. The dispersion relation is matched at a wave number $\xi_s$ if $\Delta(\xi_s) = 0$. Compared with an objective functional with reflection coefficient in a variational boundary condition [10], the main advantage of the residual $\Delta(\xi)$ is its linearity in terms of $b_j$ and $c_j$. This allows us to determine the coefficients by solving a set of linear algebraic equations.

Noticing the important role of long waves in the energy band for most applications, we construct a Taylor-type MBC with $N$ atoms, requiring $\Delta(\xi) = o(\xi^{2N})$. This determines uniquely the coefficients in (2.39). The resulting condition is termed MBC$N$ for short.

More precisely, the Taylor expansions around $\xi = 0$ for the terms in the residual function (2.29) are

$$i\omega(\xi) = \sum_{n=0}^{\infty} a_n (i\xi)^n, \quad \text{with} \quad a_{2m} = 0, \quad a_{2m+1} = \frac{2^{-2m}}{(2m+1)!}, \quad m = 0, 1, 2, \ldots, \tag{2.30}$$

and

$$e^{ij\xi} = \sum_{n=0}^{\infty} h_{nj} (i\xi)^n, \quad \text{with} \quad h_{nj} = \frac{j^n}{n!}. \tag{2.31}$$

**Table 2.1** Coefficients for MBC$N$

| | $c_0$ | $c_1$ | $c_2$ | $c_3$ | $c_4$ | $c_5$ | $c_6$ | $b_0$ | $b_1$ | $b_2$ | $b_3$ | $b_4$ | $b_5$ | $b_6$ |
|---|---|---|---|---|---|---|---|---|---|---|---|---|---|---|
| MBC1 | 1 | 1 | | | | | | −2 | 2 | | | | | |
| MBC2 | 1 | 6 | 1 | | | | | −4 | 0 | 4 | | | | |
| MBC3 | 1 | 15 | 15 | 1 | | | | −6 | −14 | 14 | 6 | | | |
| MBC4 | 1 | 28 | 70 | 28 | 1 | | | −8 | −48 | 0 | 48 | 8 | | |
| MBC5 | 1 | 45 | 210 | 210 | 45 | 1 | | −10 | −110 | −132 | 132 | 110 | 10 | |
| MBC6 | 1 | 66 | 495 | 924 | 495 | 66 | 1 | −12 | −208 | −572 | 0 | 572 | 208 | 12 |

For example, MBC1 takes $N = 1$. We may approximate the dispersion relation up to $o(\xi^2)$. This gives

$$\begin{bmatrix} 1 & 0 & 0 & 0 \\ a_0 & a_0 & -1 & -1 \\ a_1 & a_1 + a_0 & 0 & -1 \\ a_2 & a_2 + a_1 + a_0/2 & 0 & -1/2 \end{bmatrix} \begin{bmatrix} c_0 \\ c_1 \\ b_0 \\ b_1 \end{bmatrix} = \begin{bmatrix} 1 \\ 0 \\ 0 \\ 0 \end{bmatrix}. \tag{2.32}$$

MBC1 then reads

$$\dot{u}_0 + \dot{u}_1 = -2u_0 + 2u_1. \tag{2.33}$$

The coefficients for MBC1 through MBC6 are listed in Table 2.1.

Alternatively, the dispersion relation may be approximated in the long wave limit up to $o(\xi^2)$. In addition, we enforce complete absorption at other $(N-1)$ selected wave numbers $\xi_2, \ldots, \xi_N$ with $N \geq 2$. Such a Taylor–Newton-type MBC is abbreviated as MBC$N(0, \xi_2, \ldots, \xi_N)$.

At a wave number $\xi_k \neq 0$, the real and imaginary parts of $\Delta(\xi_k) = 0$ give the following requirements:

$$\begin{cases} \omega(\xi_k) \sum_{j=0}^{N} c_j \sin j\xi_k + \sum_{j=0}^{N} b_j \cos j\xi_k = 0, \\ \omega(\xi_k) \sum_{j=0}^{N} c_j \cos j\xi_k - \sum_{j=0}^{N} b_j \sin j\xi_k = 0, \end{cases} \quad k = 2, \ldots, N. \tag{2.34}$$

Together with the matching condition at $\xi = 0$, we may determine the coefficients in the linear constraint. For example, the coefficients in MBC2(0, 1) may be computed from

$$\begin{bmatrix} 1 & 0 & 0 & 0 & 0 & 0 \\ a_0 & a_0 & a_0 & -1 & -1 & -1 \\ a_1 & a_1 + a_0 & a_1 + 2a_0 & 0 & -1 & -2 \\ a_2 & a_2 + a_1 + a_0/2 & a_2 + 2a_1 + 2a_0 & 0 & -1/2 & -2 \\ 0 & \omega(1)\sin(1) & \omega(1)\sin(2) & 1 & \cos(1) & \cos(2) \\ \omega(1) & \omega(1)\cos(1) & \omega(1)\cos(2) & 0 & -\sin(1) & -\sin(2) \end{bmatrix} \begin{bmatrix} c_0 \\ c_1 \\ c_2 \\ b_0 \\ b_1 \\ b_2 \end{bmatrix} = \begin{bmatrix} 1 \\ 0 \\ 0 \\ 0 \\ 0 \\ 0 \end{bmatrix}. \tag{2.35}$$

This yields MBC2(0, 1):

$$\dot{u}_0 + 5.5103\dot{u}_1 + \dot{u}_2 = -3.7552u_0 + 3.7552u_2. \tag{2.36}$$

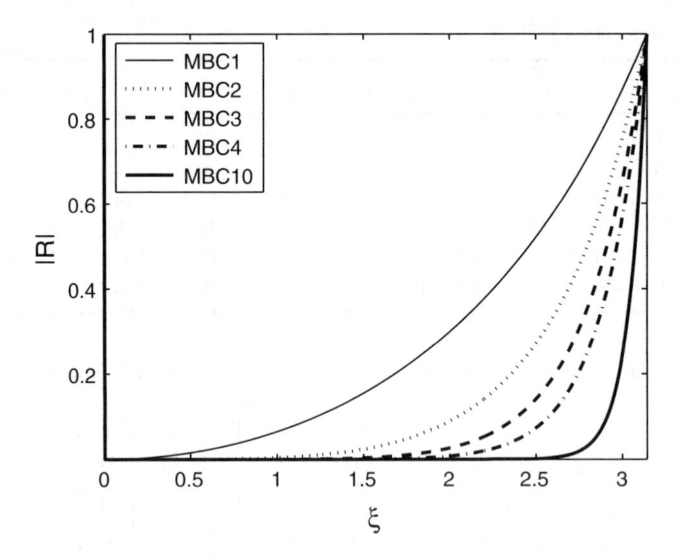


**Figure 2.8** Reflection coefficients of Taylor type MBCs

The effectiveness of MBCs may be observed from Figures 2.8 and 2.9. In most applications for the harmonic chain, MBC2 suffices to suppress effectively the reflections.

For a typical initial data, the performance of the boundary conditions may be judged by the energy evolution. As waves propagate away from the computing domain, ideally the energy keeps decreasing. As observed from Figure 2.10, MBC5 has a performance comparable to the time history kernel treatment before the cut-off time 50. After that, the spurious reflection increases rapidly in time history kernel treatment, whereas MBC5 maintains a decreasing energy. Noticing the difference in additional computing loads due to boundary treatments for these two methods, we clearly see the efficiency of the proposed MBCs.

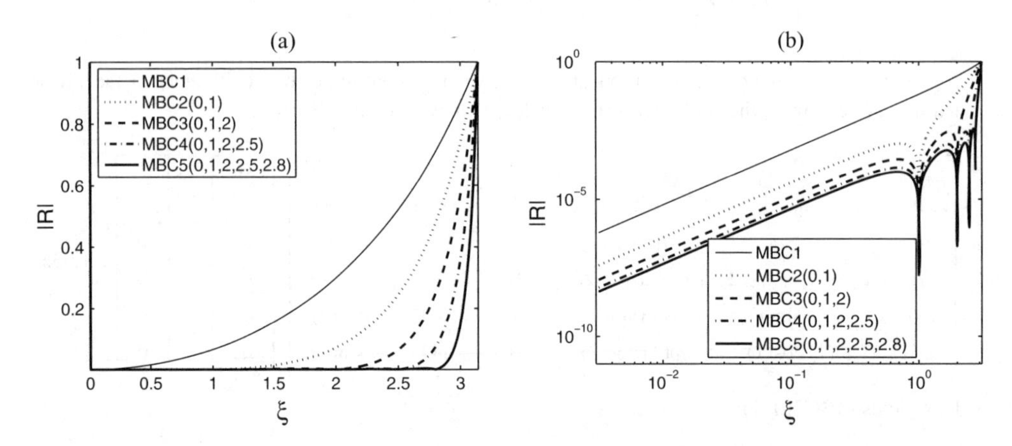


**Figure 2.9** Reflection coefficients of Taylor–Newton-type MBCs: (a) linear plots; (b) log–log plots

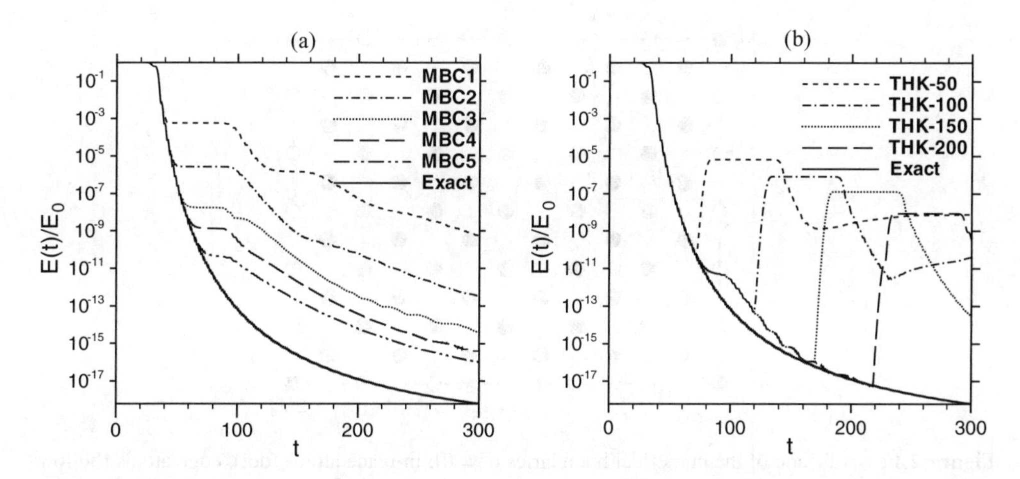


**Figure 2.10** Energy evolution in the segment with: (a) MBCs; (b) time history kernel treatment (with cut-off time $T_c$ in THK-$T_c$)

## 2.4.2 FCC Lattice

In the FCC lattice, an atom occupies the staggered position $(x, y, z) = (mh_a, nh_a, lh_a)$, where $h_a$ is the atomic spacing and $m + n + l$ is an even number. We denote such an atom by $(m, n, l)$ and its displacement as $u_{m,n,l}(t)$. See Figure 2.11.

For scalar waves in the FCC lattice with nearest neighboring linear interactions, the rescaled Newton equation reads

$$\begin{aligned}\ddot{u}_{m,n,l} &= u_{m+1,n,l+1} + u_{m-1,n,l-1} + u_{m-1,n,l+1} + u_{m+1,n,l-1} \\ &\quad + u_{m,n+1,l+1} + u_{m,n-1,l-1} + u_{m,n+1,l-1} + u_{m,n-1,l+1} \\ &\quad + u_{m+1,n+1,l} + u_{m-1,n-1,l} + u_{m+1,n-1,l} + u_{m-1,n+1,l} - 12u_{m,n,l}.\end{aligned} \tag{2.37}$$

For an atomic subdomain $[0, 2M_x h_a] \times [0, 2M_y h_a] \times [0, 2M_z h_a]$ there exist three kinds of artificial boundary atoms: the in-plane atoms, the edge atoms, and the corner atoms. See Figure 2.12. We illustrate the construction of MBCs for an in-plane atom, and adopt the infinite boundary approximation.

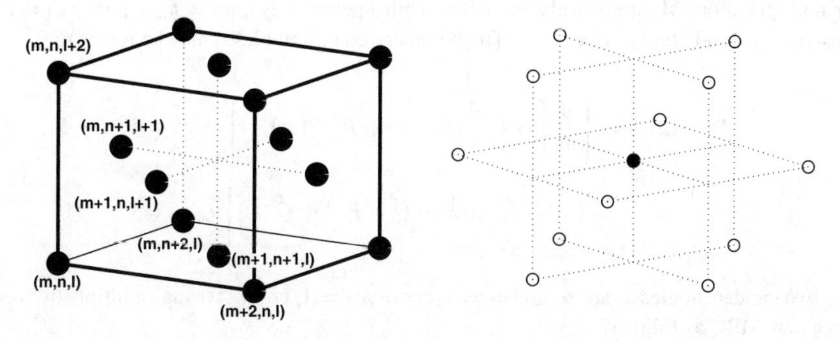


**Figure 2.11** Structure of the FCC lattice: the unit cell and numbering (left); the nearest neighbors represented by hollow circles of an atom represented by dot (right)

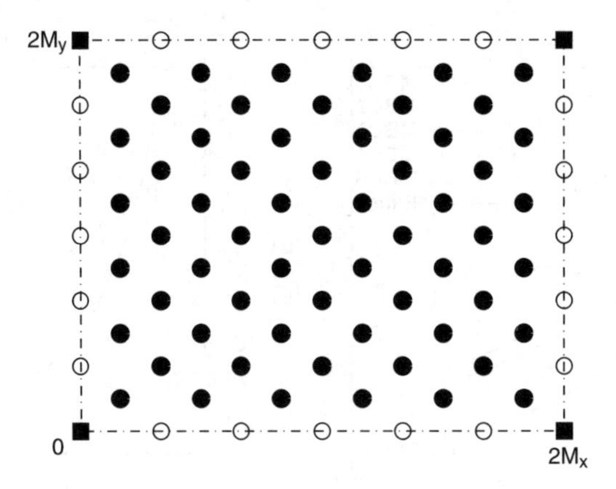


**Figure 2.12** A facade of the numerical boundaries ($l = 0$): in-plane atoms (dot), edge atoms (hollow circle) and corner atoms (square)

First, for normal incidence, namely with wave vector $(\xi_x, \xi_y, \xi_z) = (0, 0, 1)$, essentially the one-dimensional MBC applies. With considerations on the FCC structure, MBC1 reads

$$\begin{aligned}&\dot{u}_{m,n,0} + \frac{1}{4}(\dot{u}_{m,n+1,1} + \dot{u}_{m,n-1,1} + \dot{u}_{m+1,n,1} + \dot{u}_{m-1,n,1}) \\ &= -4u_{m,n,0} + (u_{m,n+1,1} + u_{m,n-1,1} + u_{m+1,n,1} + u_{m-1,n,1}).\end{aligned} \tag{2.38}$$

Next, for an incidence with a wave vector $(\xi_x, \xi_y, \xi_z)$ which forms a polar angle $\alpha$ with the normal direction (0, 0, 1), we utilize the notion of apparent wave propagation and construct the MBC as follows:

$$\begin{aligned}&\dot{u}_{m,n,0} + \frac{1}{4}(\dot{u}_{m,n+1,1} + \dot{u}_{m,n-1,1} + \dot{u}_{m+1,n,1} + \dot{u}_{m-1,n,1}) \\ &= a[-4u_{m,n,0} + (u_{m,n+1,1} + u_{m,n-1,1} + u_{m+1,n,1} + u_{m-1,n,1})],\end{aligned} \tag{2.39}$$

where $a = 1/\cos\alpha$ is the apparent wave propagation speed.

To further reduce reflections, we follow the experiences in [14] to design boundary conditions by operator multiplication. More precisely, we define shift operators $I_d u_{m,n,l} = u_{m,n,l}$, $P u_{m,n,l} = u_{m+1,n,l}$, $Q u_{m,n,l} = u_{m,n+1,l}$, and $R u_{m,n,l} = u_{m,n,l+1}$. The boundary condition (2.39) may be recast to

$$\begin{aligned}D(a)u_{m,n,0} = \Bigg\{&\frac{d}{dt}\left[I_d + \frac{R}{4}(P + Q + P^{-1} + Q^{-1})\right] \\ &+ a[4I_d - R(P + Q + P^{-1} + Q^{-1})]\Bigg\} u_{m,n,0} = 0.\end{aligned} \tag{2.40}$$

Taking two incident angles $\alpha_1$ and $\alpha_2$ and $a_1 = 1/\cos\alpha_1, a_2 = 1/\cos\alpha_2$, we may multiply two operators to obtain a new MBC as follows:

$$D(a_1)D(a_2)u_{m,n,0} = 0. \tag{2.41}$$

After straightforward calculations, we arrive at a two-angle MBC:

$$\begin{aligned}
\ddot{u}_{m,n,0} = & -\frac{1}{2}(\ddot{u}_{m,n+1,1} + \ddot{u}_{m,n-1,1} + \ddot{u}_{m+1,n,1} + \ddot{u}_{m-1,n,1}) \\
& -\frac{1}{16}(\ddot{u}_{m,n+2,2} + \ddot{u}_{m,n-2,2} + \ddot{u}_{m+2,n,2} + \ddot{u}_{m-2,n,2}) \\
& -\frac{1}{8}(\ddot{u}_{m+1,n+1,2} + \ddot{u}_{m-1,n+1,2} + \ddot{u}_{m+1,n-1,2} + \ddot{u}_{m-1,n-1,2} + 2\ddot{u}_{m,n,2}) \\
& +(a_1 + a_2)\left[-4\dot{u}_{m,n,0} + \frac{1}{4}(\dot{u}_{m,n+2,2} + \dot{u}_{m,n-2,2} + \dot{u}_{m+2,n,2} + \dot{u}_{m-2,n,2})\right. \\
& \left. + \frac{1}{2}(\dot{u}_{m+1,n+1,2} + \dot{u}_{m-1,n+1,2} + \dot{u}_{m+1,n-1,2} + \dot{u}_{m-1,n-1,2} + 2\dot{u}_{m,n,2})\right] \\
& -a_1 a_2[16u_{m,n,0} - 8(u_{m,n+1,1} + u_{m,n-1,1} + u_{m+1,n,1} + u_{m-1,n,1}) \\
& +(u_{m,n+2,2} + u_{m,n-2,2} + u_{m+2,n,2} + u_{m-2,n,2}) \\
& +2(u_{m+1,n+1,2} + u_{m-1,n+1,2} + u_{m+1,n-1,2} + u_{m-1,n-1,2} + 2u_{m,n,2})].
\end{aligned} \tag{2.42}$$

The two dominant angles of wave propagation are 0 and $\pi/4$, so we take $a_1 = 1$ and $a_2 = \sqrt{2}$.

The edge atoms and corner atoms are also treated in a similar fashion, with consideration of their different geometrical positions.

We perform atomic simulations with 64 000 unit cells ($-40 \leq m, n, l \leq 40$) and initial data

$$u_{m,n,l}(0) = \begin{cases} e^{-r^2/25}\{\cos[r(0.5 + \frac{\pi}{40})] + \cos[r(0.5 - \frac{\pi}{40})]\}, & |r| \leq 10, \\ 0, & |r| > 10, \end{cases} \tag{2.43}$$

where $r^2 = (\frac{m}{2})^2 + (\frac{n}{2})^2 + (\frac{l}{2})^2$. This is a sinusoidal wave enveloped by an exponential function.

At $t = 30$, the major waves propagate out of the computing domain, leaving a tail at the corner. This is a good time to monitor the performance of the boundary treatments. In Figure 2.13, we compare the exact solution and the MBC results. The two profiles are indiscernible.

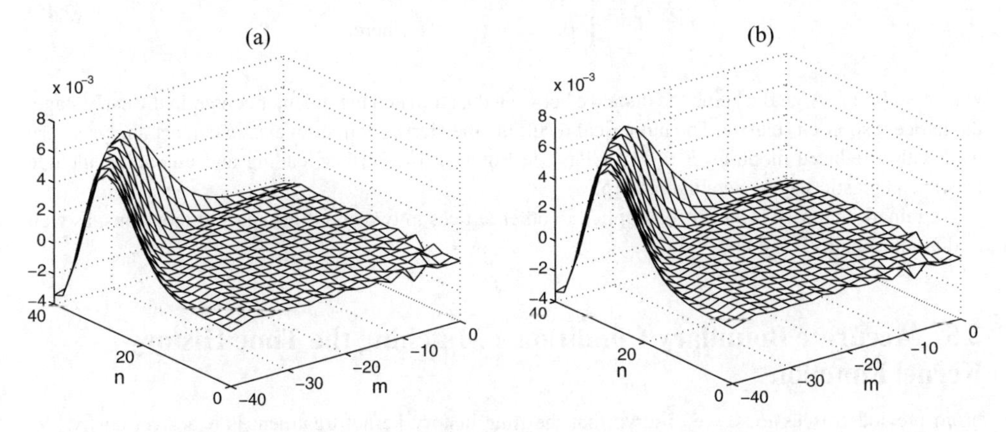


**Figure 2.13** Wave profile $u_{m,n,0}(30)$: (a) exact solution; (b) MBC

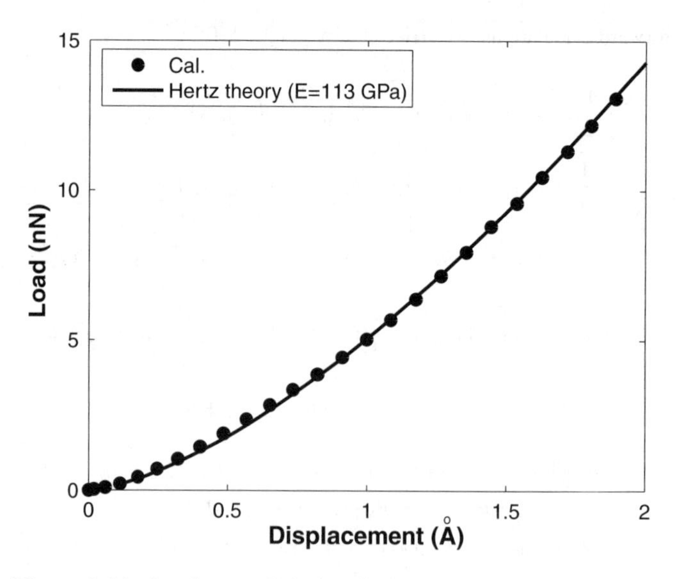


**Figure 2.14** Load versus indenter displacement for nanoindentation

This approach applies to vector waves as well. We verify the effectiveness by simulating nanoindentation in gold. We assume nearest-neighbor interaction with the pair-wise Morse potential for the atomic interactions:

$$\begin{aligned} V(r) &= D_e[e^{2\beta(\rho-r)} - 2e^{\beta(\rho-r)}], \\ D_e &= 0.560 \text{ eV}, \quad \beta = 1.637 \text{ Å}^{-1}, \quad \rho = 2.922 \text{ Å}. \end{aligned} \tag{2.44}$$

The indenter is modeled by a sphere with $R = 20$ Å. The interaction between the indenter and the gold atoms is described by a repulsive force

$$F(r) = \begin{cases} A(R-r)^2, & r < R, \\ 0, & \text{elsewhere}, \end{cases} \tag{2.45}$$

with $A = 10\,\text{eV}/\text{Å}^3$ and $r$ being the distance between the atom and the indenter center. In the early stage, the lattice response is elastic. The numerical result fits the Hertz continuum elastic contact analysis very well with a reduced modulus $E = 113$ GPa. See Figure 2.14. Further loading and unloading include elastic and plastic stages. See Figure 2.15.

We remark that MBCs have been applied to other lattices in two and three space dimensions, as well as diatomic chains [15–17].

## 2.5 Accurate Boundary Conditions: Matching the Time History Kernel Function

From previous discussions, we observe that the time history kernel treatment is exact yet costly. We thus aim at maintaining the accuracy while sharing the locality in both space and time as the velocity

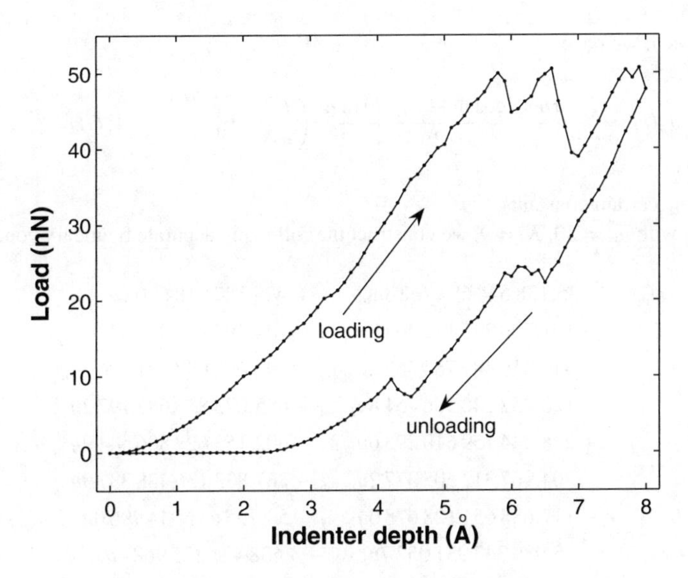


**Figure 2.15** Load versus indenter depth curve

interfacial conditions and matching boundary conditions. To this end, we consider a boundary condition for the harmonic chain in the form of a linear constraint

$$u_{n_b} = \sum_{p=1}^{K}(a_p u_{n_b-p} + b_p \dot{u}_{n_b-p}). \tag{2.46}$$

The coefficients are determined from a response function point of view. More precisely, we consider a delta function source at the zeroth atom. The responses for an infinite chain are then

$$u_j = \mathcal{L}^{-1}(S^j) = \frac{2jJ_{2j}(2t)}{t}. \tag{2.47}$$

Now we make expansion of the Bessel function $J_p(t)$ for three regions respectively as follows. First, for $|(t^2/n^2) - 1| < 256n^{-2/3}$, we have

$$J_p(t) = \sum_{l=0}^{\infty} \frac{H_l(p)}{l!}\left(\frac{t^2}{n^2} - 1\right)^l + O(p^{-4/3}), \tag{2.48}$$

with $H_l(p)$ a certain constant.

Second, for $t = n\sqrt{1+A}$ with $A \geq 0$, we have

$$J_p(t) = \sum_{l=0}^{\infty} \frac{H_{l,A}(p)}{l!}\left(\frac{t^2}{n^2} - 1 - A\right)^l + O(p^{-4/3}), \tag{2.49}$$

with $H_{l,A}(p)$ a certain constant.

Third, for $t \gg n$, we have

$$J_p(t) = \sum_{l=0}^{\infty} \frac{H_{l,1,p} \cos \Phi - H_{l,2,p} \sin \Phi}{l!} \left( \frac{t^2}{n^2} - 1 \right)^{-(l/2)-1/4} + O(p^{-3/2}). \tag{2.50}$$

with $H_{l,1,p}$, $H_{l,2,p}$ certain constants.

For instance, with $n_b = 50$, $K = 9$, we construct the following accurate boundary condition:

$$\begin{aligned} u_{n_b} = & -35.138\,644\,683\,762\,3u_{n_b-1} - 8.424\,587\,918\,690\,1\dot{u}_{n_b-1} \\ & - 191.700\,245\,143\,464\,9u_{n_b-2} - 104.492\,981\,842\,332\dot{u}_{n_b-2} \\ & - 341.410\,387\,784\,457\,2u_{n_b-3} - 396.696\,047\,171\,806\,3\dot{u}_{n_b-3} \\ & - 128.252\,246\,586\,451\,8u_{n_b-4} - 685.073\,853\,687\,797\,2\dot{u}_{n_b-4} \\ & + 258.344\,159\,610\,298\,6u_{n_b-5} - 602.153\,627\,806\,544\,9\dot{u}_{n_b-5} \\ & + 304.967\,312\,505\,377\,2u_{n_b-6} - 267.833\,136\,488\,337\,9\dot{u}_{n_b-6} \\ & + 117.688\,651\,128\,678\,6u_{n_b-7} - 53.735\,641\,044\,996\,0\dot{u}_{n_b-7} \\ & + 15.989\,347\,931\,052\,7u_{n_b-8} - 3.638\,421\,225\,968\,4\dot{u}_{n_b-8} \\ & + 0.511\,486\,125\,677\,3u_{n_b-9} - 0.034\,355\,802\,365\,7\dot{u}_{n_b-9}. \end{aligned} \tag{2.51}$$

The resulting approximation matches very well the exact kernel function (2.47) for the whole time span. Notice that we only use the current information at nine atoms near the boundary. See Figure 2.16.

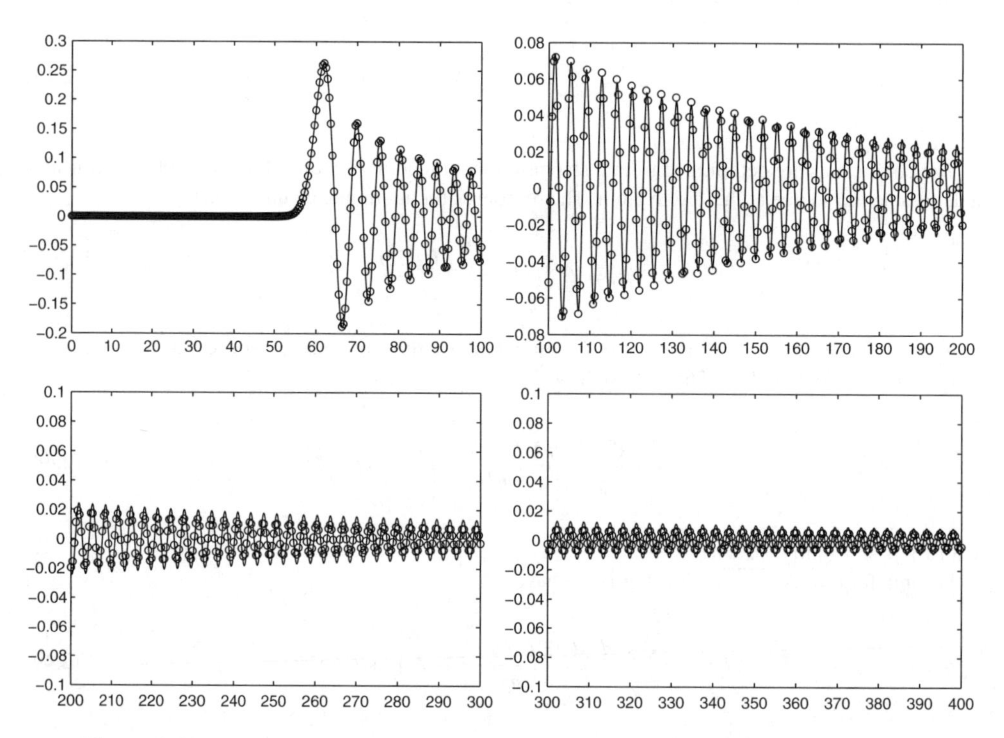


**Figure 2.16** Kernel function at the 50th atom (solid line) and its approximation (circles)

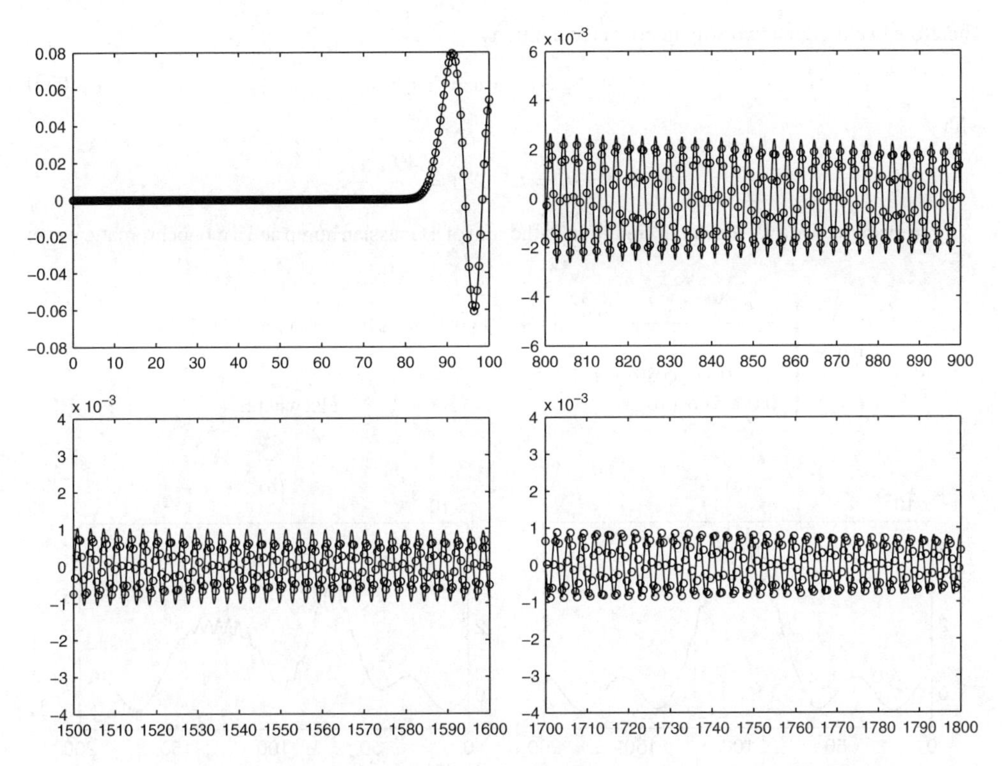


**Figure 2.17** Displacement at the 90th atom: solid line for the exact motion and circles for the numerical solution in [0, 100], [800, 900], [1500, 1600], and [1700, 1800]

Though we design the boundary condition with $n_b = 50$, we may apply that to the boundary of a chain with other lengths. For instance, using this boundary condition to a chain with 90 atoms and a source $u_1(t) = \sin 3t$, we display the resulting $u_{90}(t)$ in Figure 2.17. The numerical result and the exact motion are indiscernible, both in short and long runs. One nice property of this approach, besides taking a negligible additional numerical cost at the boundary, lies in removing the energy increment after cut-off time in the time history kernel approach.

## 2.6 Two-Way Boundary Conditions

In all aforementioned numerical interfacial treatments, fine fluctuations in the continuum region are not taken into account. No wave enters the atomic region unless it is resolved by the coarse grid. We notice that, based on the extended space–time finite element, efforts have been made to incorporate fine-scale oscillations into coarse-scale solutions [18]. In applications, however, fine-scale oscillations do exist in the continuum region and may propagate into the atomistic region. Thermal fluctuation is one such case with great importance. To our knowledge, there is as yet quite limited knowledge about how to treat accurately thermal fluctuations in a multiscale computation.

Taking the harmonic chain as an example, we first decompose the displacement into a left-going component $v_n(t)$ and a right-going component $w_n(t)$. This means $u_n = v_n + w_n$. Owing to the linearity, both components satisfy the same Newton's law. Hence, close to the left boundary atom $u_0$, it holds that

$$v_0(t) = v_1(t) * \theta_1(t), \quad w_1 = \theta_1 * w_0. \tag{2.52}$$

Therefore, we design a two-way interfacial condition

$$u_0 = \theta_1 * u_1 + w_0 - \theta_2 * w_0, \tag{2.53}$$

with

$$\theta_2 = \theta_1 * \theta_1 = \mathcal{L}^{-1}(S^2) = \frac{4J_4(2t)}{t}.$$

As an example, we take the initial condition as the sum of a Gaussian hump and a monochromatic wave:

$$u_n(0) = \begin{cases} h_a \dfrac{e^{-100(x_n-0.5)^2} - e^{-6.25}}{1 - e^{-6.25}}[1 + 0.1\cos(80\pi x_n)] & \text{for } |x_n - 0.5| \leq 0.25; \\ +0.1 h_a \cos(\omega x_n), & \\ 0.0005\cos(\omega x_n), & \text{elsewhere.} \end{cases} \tag{2.54}$$

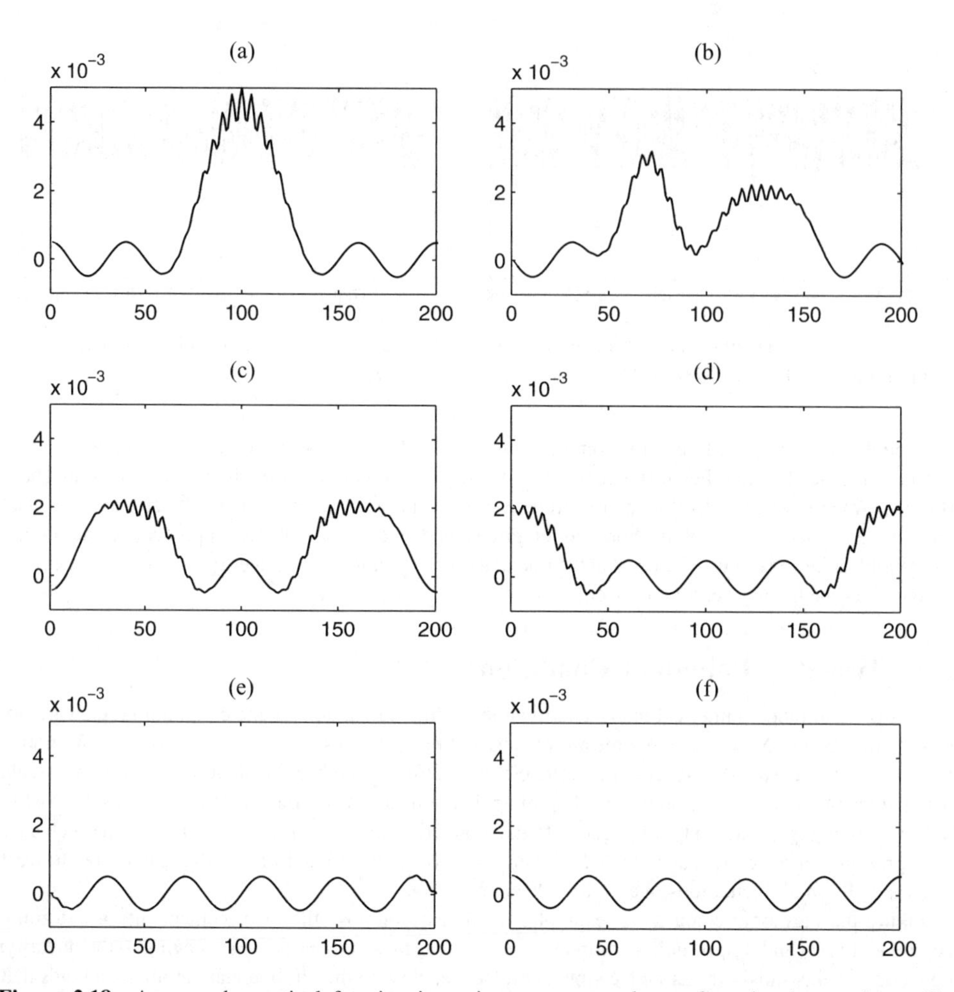


**Figure 2.18** A monochromatic left-going incoming wave overtakes a Gaussian hump: (a) $u_n(0)$; (b) $u_n(30)$; (c) $u_n(60)$; (d) $u_n(100)$; (e) $u_n(150)$; (f) $u_n(200)$

The initial velocity is taken as

$$\dot{u}_n(0) = 0.0005\lambda \sin(\omega x_n). \tag{2.55}$$

As shown in Figure 2.18, these waves simply superpose and pass across each other, without making any reflection at the boundaries. The Gaussian hump splits into a left-going part and a right-going part. Both parts propagate outward of the atomic region. Their influence disappears at $t$ between 150 and 200. Because the monochromatic wave has a definite propagation direction, the wave profiles are not symmetric. The monochromatic wave completely overtakes the Gaussian hump eventually.

## 2.7 Conclusions

In this chapter we have discussed several accurate boundary treatments for the atomic subdomain of a multiscale simulation. The time history kernel treatment is exact, yet demands much computing power. The evaluation of the kernel functions in multiple space dimensions is usually not trivial. Local boundary conditions such as the velocity interfacial conditions and the MBCs are more effective, and readily apply to general nonlinear lattices. The accurate boundary conditions designed by matching the kernel functions are most effective, yet rely heavily on mathematical understanding of the kernel functions, which is not available in general for most lattice structures. Moreover, the two-way boundary conditions provide a reasonable way to perform accurate simulations at finite temperature.

## Acknowledgments

I would like to thank Professor W.K. Liu and Professor T.Y. Hou for introducing me to the fascinating field of multiscale computations, and their support over the years. I would like to thank Professor M. Dreher, Dr. X. Wang, Dr. M. Fang, and Mr. G. Pang for collaborations. This research is supported partially by the NSFC under contract number 11272009 and the National Basic Research Program of China under contract number 2010CB731500.

## References

[1] Adelman, S. and Doll, J. (1974) Generalized Langevin equation appraoch for atom/solid-surface scattering: collinear atom/harmonic chain model. *Journal of Chemical Physics*, **61**, 4242–4245.

[2] Cai, W., de Koning, M., Bulatov, V., and Yip, S. (2000) Minimizing boundary reflections in coupled-domain simulations. *Physical Review Letters*, **85**, 3213–3216.

[3] Wagner, G. and Liu, W. (2003) Coupling of atomistic and continuum simulations using a bridging scale decomposition. *Journal of Computational Physics*, **190**, 249–274.

[4] Tang, S., Hou, T., and Liu, W. (2006) A mathematical framework of the bridging scale method. *International Journal for Numerical Methods in Engineering*, **65**, 1688–1713.

[5] Tang, S., Hou, T., and Liu, W. (2006) A pseudo-spectral multiscale method: interfacial conditions and coarse grid equations. *Journal of Computational Physics*, **213**, 57–85.

[6] Qian, D., Wagner, G., and Liu, W. (2004) A multiscale projection method for the analysis of carbon nanotubes. *Computer Methods in Applied Mechanics and Engineering*, **193**, 1603–1632.

[7] Karpov, E., Wagner, G., and Liu, W. (2005) A Green's function approach to deriving non-reflecting boundary conditions in molecular dynamics simulations. *International Journal for Numerical Methods in Engineering*, **62**, 1250–1262.

[8] Park, H., Karpov, E., Liu, W., and Klein, P. (2005) The bridging scale for two-dimensional atomistic/continuum coupling. *Philosophical Magazine*, **85**, 79–113.

[9] Liu, W., Karpov, E., and Park, H. (2005) *Nano Mechanics and Materials: Theory, Multiscale Methods and Applications*, John Wiley & Sons.

[10] Weinan, E. and Huang, Z. (2002) A dynamic atomistic–continuum method for simulation of crystalline materials. *Journal of Computational Physics*, **182**, 234–261.
[11] Dreher, M. and Tang, S. (2008) Time history interfacial conditions in multiscale computations of lattice oscillations. *Computational Mechanics*, **41**, 683–698.
[12] Pang, G. and Tang, S. (2011) Time history kernel functions for square lattice. *Computational Mechanics*, **48**, 699–711.
[13] Tang, S. (2008) A finite difference approach with velocity interfacial conditions for multiscale computations of crystalline solids. *Journal of Computational Physics*, **227**, 4038–406.
[14] Higdon, R. (1987) Absorbing boundary conditions for the wave equation. *Mathematics of Computation*, **49**, 65–90.
[15] Wang, X. and Tang, S. (2010) Matching boundary conditions for diatomic chains. *Computational Mechanics*, **46**, 813–826.
[16] Fang, M., Tang, S., Li, Z., and Wang, X. (2012) Artificial boundary conditions for atomic simulations of face-centered-cubic lattice. *Computational Mechanics*, doi: 10.1007/s00466-012-0696-8.
[17] Wang, X. and Tang, S. (2012) Matching boundary conditions for lattice dynamics. *International Journal for Numerical Methods in Engineering*, doi: 10.1002/nme.4426.
[18] Cirputkar, S. and Qian, D. (2008) Coupled atomistic/continuum simulation based on extended space–time finite element method. *Computer Modeling in Engineering & Sciences*, **24**, 185–202.

# 3

# A Multiscale Crystal Defect Dynamics and Its Applications

Lisheng Liu[a,b] and Shaofan Li[b]
[a]*Department of Civil and Environmental Engineering, The University of California, Berkeley, USA*
[b]*Department of Engineering Structure and Mechanics, Wuhan University of Technology, People's Republic of China*

## 3.1 Introduction

In order to design and manufacture reliable nanoscale or microscale electro-mechanical systems, there has been keen interest in developing multiscale methods to study or simulate defect dynamics (for instance, dislocation dynamics) between the nanoscale and sub-millimeter scale. In fact, nanoscale and sub-micrometer-scale plasticity has become a focal point in computational materials science. A critical issue is to find the precise atomistic origin of crystalline plasticity, which has long been speculated as the outcome of motions of aggregated dislocations and their evolution. A significant amount of research work focuses on using a first-principle-based approach to investigate. However, even with the state-of-the-art computing technology, the first-principle-based approach is still restricted by limited computing power. This difficulty may persist even if we had unlimited computational power, because without an adequate multiscale statistical methodology we may not be able to extrapolate useful information from a massive-sized simplistic microscale model to gain insights on many complex macroscale physical phenomena.

Towards a multiscale computational statistical paradigm, various multiscale models have been reported in the literature during last decade, and recent notable examples of such multiscale models include: multiscale field theory and modeling [1–3], concurrent electron-to-continuum method [4], the bridging-scale method [5, 6], molecular dynamics to continuum [7–9], and multiscale modeling for plasticity [10–13].

The multiscale continuum crystal defect dynamics model has two distinct novel features:

1. The multiscale crystal defect dynamics (MCDD) finite element mesh is built as a hierarchical multiscale element cluster that is an exact scale-up version of the lattice structure underneath. However, it consists of multi-levels (scale) of elements that has an increased order of interpolation ladder, and hence it can represent increased order of nonlinearity in finite deformation in the crystal lattice.

*Multiscale Simulations and Mechanics of Biological Materials*, First Edition. Edited by Shaofan Li and Dong Qian.

2. Correspondingly for lattice elements at different scales, atomistic-informed constitutive relations are built by using hierarchical Cauchy–Born rules to form a multiscale finite-element cluster or unit cell that will fill the whole crystal without any gaps. By doing so, we may scale out molecular dynamics into a continuum crystal dynamics that has inherent some basic fundamental properties of the atomistic lattice system.

The chapter is organized into six sections. In Section 3.2 we first construct the MCDD for each different element in different scales. In Section 3.3 we discuss how and why the method may work and in Section 3.4 we discuss the implementation of the finite-element formulation of MCDD. In Section 3.5 we present a few numerical examples by using MCDD to simulate dislocation motions and fracture at small scale. Finally, in Section 3.6 we close with a few remarks.

## 3.2 Multiscale Crystal Defect Dynamics

Consider a deformable FCC crystal lattice volume element in the (111) plane or an ideal hexagonal crystal in the basal plane (see Figure 3.1). The lattice structure in the FCC lattice structure in the (111) plane may be viewed as an assembly of many triangle lattice units. To coarse-grain such a lattice structure, we first scale up the FCC lattice to a set of super triangle elements with the same aspect ratio as in the atomistic scale and we then glue all these nanoscale or sub-micrometer-scale triangle elements with rectangular elements between any of the two triangle element pairs and a hexagonal element among six triangle elements at a shared vertex. By doing so, the whole lattice space will be filled up with a three-element assembly net without gaps and overlaps, as shown in Figure 3.2.

To simulate the crystal deformation and defect motion, we make the following kinematic assumptions:

*The triangle bulk element has only uniform deformation, and all the nonlinear deformations are confined in rectangular interphase elements and hexagonal vertex elements.*

We then set up the following atomistic informed multiscale constitutive relations in each element type via a hierarchical higher order Cauchy–Born rule:

1. In the triangle element, we employ the first-order Cauchy–Born rule to derive the mesoscale stress–strain relation; that is:

$$\mathbf{r} = \mathbf{F} \cdot \mathbf{R}, \quad \text{where } \mathbf{F} = \frac{\partial \mathbf{x}}{\partial \mathbf{X}}$$

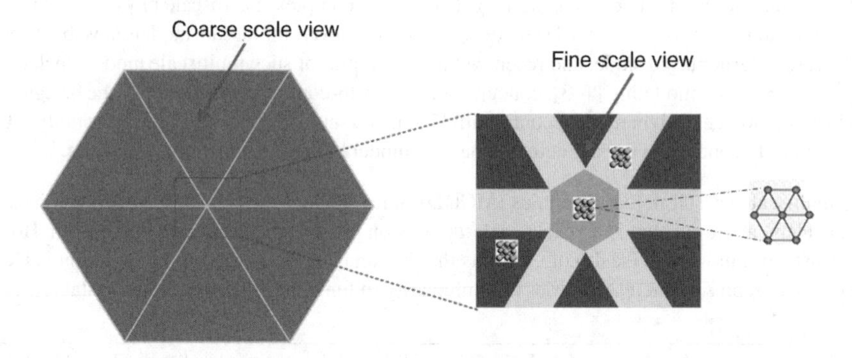


**Figure 3.1** Illustration of MCDD multiscale element configuration

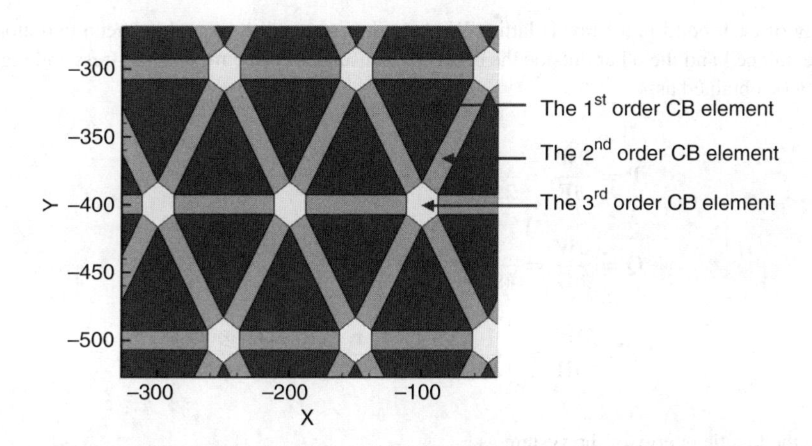


**Figure 3.2** MCDD finite-element mesh for FCC crystal on (111) plane

and where **R** is an original lattice vector, **F** is the deformation gradient in the triangle element, and **r** is the deformed lattice vector.

2. In the rectangular element, we employ the second-order Cauchy–Born rule to extrapolate the mesoscale strain-gradient-based constitutive relation; that is:

$$\mathbf{r} = \mathbf{F} \cdot \mathbf{R} + \frac{1}{2}\mathbf{G} : (\mathbf{R} \otimes \mathbf{R})$$

where

$$\mathbf{G} = \frac{\partial^2 \mathbf{x}}{\partial \mathbf{X} \otimes \partial \mathbf{X}} = \frac{\partial \mathbf{F}}{\partial \mathbf{X}}.$$

3. In the hexagonal vertex element, we employ the third-order Cauchy–Born rule to establish the second-order strain-gradient-based constitutive relation; that is:

$$\mathbf{r} = \mathbf{F} \cdot \mathbf{R} + \frac{1}{2!}\mathbf{G} : (\mathbf{R} \otimes \mathbf{R}) + \frac{1}{3!}\mathbf{H} \vdots (\mathbf{R} \otimes \mathbf{R} \otimes \mathbf{R}),$$

where

$$\mathbf{H} := \frac{\partial^3 \mathbf{x}}{\partial \mathbf{X} \otimes \partial \mathbf{X} \otimes \partial \mathbf{X}} = \frac{\partial^2 \mathbf{F}}{\partial \mathbf{X} \otimes \partial \mathbf{X}} = \frac{\partial \mathbf{G}}{\partial \mathbf{X}}.$$

The free-energy density inside the quasi-crystalline continuum is postulated as $W = W(\mathbf{F}, \mathbf{G}, \mathbf{H})$. For a Bravias lattice (FCC or ideal hexagonal), the free energy density in a Wigner–Seitz unit cell may be calculated as follows:

$$W_0 = \frac{1}{2\Omega_0} \sum_{i=1}^{n_{\text{bond}}} \phi(r_i) \tag{3.1}$$

where $\phi(r_i)$ is the atomistic potential, $r_i$ is the $i$th bound length, $N_{\text{bond}}$ is the number of total bonds in a unit cell, and $\Omega_0$ is the volume of the Wigner–Seitz cell. There is a factor 1/2 in Equation (3.1), because the energy of each bond in a Bravias lattice Wigner–Seitz unit cell is shared between two atoms, one inside the unit cell and the other outside the unit cell. Consequently, the mesoscale stress and high-order stress may be obtained as

$$\mathbf{P} = \frac{\partial W}{\partial \mathbf{F}} = \frac{1}{2\Omega_0} \sum_{i=1}^{n_{\text{bond}}} \phi'(r_i) \frac{\mathbf{r}_i \otimes \mathbf{R}_i}{r_i}, \tag{3.2}$$

$$\mathbf{Q} = \frac{\partial W}{\partial \mathbf{G}} = \frac{1}{4\Omega_0} \sum_{i=1}^{n_{\text{bond}}} \phi'(r_i) \frac{\mathbf{r}_i \otimes \mathbf{R}_i \otimes \mathbf{R}_i}{r_i}, \tag{3.3}$$

$$\mathbf{S} = \frac{\partial W}{\partial \mathbf{H}} = \frac{1}{12\Omega_0} \sum_{i=1}^{n_{\text{bond}}} \phi'(r_i) \frac{\mathbf{r}_i \otimes \mathbf{R}_i \otimes \mathbf{R}_i \otimes \mathbf{R}_i}{r_i}. \tag{3.4}$$

Define the kinetic energy of the system as

$$\mathcal{T} = \int_V \frac{1}{2} \rho \dot{\mathbf{u}} \cdot \dot{\mathbf{u}} \, \mathrm{d}V$$

where $\mathbf{u}(\mathbf{X})$ is the displacement field. We also denote the internal free-energy of crystal lattice continuum as

$$\mathcal{W}_{\text{int}} := \int_V W(\mathbf{F}, \mathbf{G}, \mathbf{H}) \, \mathrm{d}V.$$

Hamilton's principle may then be written in terms of displacement variation between the fixed time instance at $t_0$ and $t_1$:

$$\delta \int_{t_0}^{t_1} (\mathcal{T} - \mathcal{W}_{\text{int}}) \, \mathrm{d}t + \int_{t_0}^{t_1} \delta \mathcal{W}_{\text{ext}} \, \mathrm{d}t = 0, \tag{3.5}$$

where

$$\mathcal{T} = \frac{1}{2} \int_V \rho \dot{\mathbf{u}} \cdot \dot{\mathbf{u}} \, \mathrm{d}V$$

and $\mathcal{W}_{\text{ext}}$ is the external potential energy.

Successive integration by parts yields

$$\begin{aligned} \delta \mathcal{W}_{\text{int}} = & - \int_V \nabla_X \cdot [\mathbf{P} - \nabla_X \cdot (\mathbf{Q} - \nabla_X \cdot \mathbf{S})] \cdot \delta \mathbf{x} \, \mathrm{d}V \\ & + \int_{\partial V} \{[\mathbf{P} - \nabla_X \cdot (\mathbf{Q} - \nabla_X \cdot \mathbf{S})] \cdot \mathbf{N}\} \cdot \delta \mathbf{x} \, \mathrm{d}S \\ & + \int_{\partial V} [(\mathbf{Q} - \nabla_X \cdot \mathbf{S}) \cdot \mathbf{N}] : \delta \mathbf{F} \, \mathrm{d}S \\ & + \int_{\partial V} (\mathbf{S} \cdot \mathbf{N}) \vdots \delta \mathbf{G} \, \mathrm{d}S \end{aligned} \tag{3.6}$$

We assume that the crystalline solid under consideration is surrounded by a uniformly deformed environment, which implies that

$$\delta \mathbf{F} = 0, \quad \delta \mathbf{G} = 0, \quad \text{a.e.} \forall \mathbf{X} \in \partial V$$

and the last two terms of the internal work may be neglected.

We assume that the external virtual work of the system has the form

$$\delta \mathcal{W}_{\text{ext}} = \int_V \mathbf{b} \cdot \delta \mathbf{u} \, dV + \int_{\partial V_t} \bar{\mathbf{T}} \cdot \delta \mathbf{u} \, dS,$$

where $\mathbf{b}$ is the body force and $\bar{\mathbf{T}}$ is the traction vector on $\partial V_t$.

The dynamics equations of the continuous lattice system can be written as

$$\nabla_X \cdot [\mathbf{P} - \nabla_X \cdot (\mathbf{Q} - \nabla_X \cdot \mathbf{S})] + \mathbf{b} = \rho \ddot{\mathbf{u}}, \quad \forall \mathbf{X} \in V, \tag{3.7}$$

$$[\mathbf{P} - \nabla_X \cdot (\mathbf{Q} - \nabla_X \cdot \mathbf{S})] \cdot \mathbf{N} = \bar{\mathbf{T}}, \quad \forall \mathbf{X} \in \partial V_t. \tag{3.8}$$

Note that $\partial V = \partial V_t \bigcup \partial V_u$ and $\delta \mathbf{x} = 0, \forall \mathbf{X} \in \partial V_u$.

## 3.3 How and Why the MCDD Model Works

In MCDD, the bulk element, the triangle element in two dimensions, is the scale-up lattice cell unit (not unit cell) or block of the original fine-scale lattice structure. The higher order elements, the rectangular and hexagonal elements in the two-dimensional (2D) case, are either an interphase element that represents dislocation gliding zone or a vertex element that represents a vacancy or void zone.

Even though for all lattice elements at different scales they have the same atomistic potential, the kinematic constraints vary. That is, the bulk element has uniform deformation, the interphase element can accommodate the existence of strain gradients, and the vertex element allows even higher order strain gradients, namely the second-order strain gradient. Consequently, the mesoscale constitutive relation for elements at different scales will be different. In this work, we postulate a fundamental hypothesis of crystal defect mechanics:

*For a perfect crystal under external loads, the material will fail first at where the deformation state is inhomogeneous.*

Based on this assumption, we choose a subset of lattice planes a priori to form interphase elements and vertex elements so they form a network interphase zone that is susceptible to nonlinear finite deformation, and it kinematically allows defect nucleation, evolution, propagation, and growth.

## 3.4 Multiscale Finite Element Discretization

The multiscale continuum crystal dynamics is a continuum version of coarse-grain lattice molecular dynamics. It scales up the lattice structure in atomistic level to a nanoscale continuum dynamics so that the MCDD finite-element mesh structure has an exact resemblance to the discrete lattice structure.

In this chapter we mainly discuss the formulations and implementations of MCDD in 2D cases. The finite-element mesh displayed in Figure 3.2 is constructed for a crystal of FCC lattice in the (111) plane. From Figure 3.2, we can clearly see that we have three different types of elements: (1) triangle elements in the bulk, (2) rectangular elements as interphase (quadrilateral element), and (3) hexagonal elements as vertices.

For each type of element, a different type of finite-element method (FEM) interpolation function is used:

$$\mathbf{u}(\mathbf{X}, t) = \sum_{I=1}^{n_{\text{node}}} N_I(\mathbf{X})\mathbf{d}_I(t).$$

1. Since the deformation inside the bulk element is uniform, we employ the standard linear interpolation function for triangle elements:

$$N_I(X, Y) = \frac{D_{J,K}(X, Y)}{C_{I,J,K}}, \quad I \neq J \neq K, I, J, K = 1, 2, 3, \tag{3.9}$$

where

$$D_{J,K} = \det \begin{bmatrix} 1 & X & Y \\ 1 & X_J & Y_J \\ 1 & X_K & Y_K \end{bmatrix} \quad \text{and} \quad C_{I,J,K} = \det \begin{bmatrix} 1 & X_I & Y_I \\ 1 & X_J & Y_J \\ 1 & X_K & Y_K \end{bmatrix}.$$

2. Because the second-order Cauchy–Born rule is used in construction of the constitutive relation, the rectangular element has to capture the strain-gradient effects in the element interior. For this purpose, we adopt the Bonger–Fox–Schmit element, which is a 2D isoparametric Hermite-type element. However, when it matches with linear shape functions in the adjacent triangle elements, we only recover the $C^0$ continuity globally.

Specifically, we use the following interpolation functions in the local coordinates:

$$\mathbf{u}(X, Y, t) = \sum_{I=1}^{16} N_I(\xi, \eta)\tilde{\mathbf{u}}_I(t), \tag{3.10}$$

where

$$\tilde{\mathbf{u}}_I(t) = \mathbf{u}_I(t), \quad I = 1, 2, 3, 4,$$

$$\tilde{\mathbf{u}}_I(t) = \frac{\partial \mathbf{u}_{I-4}}{\partial \xi}(t), \quad I = 5, 6, 7, 8,$$

$$\tilde{\mathbf{u}}_i(t) = \frac{\partial \mathbf{u}_{I-8}}{\partial \eta}(t), \quad I = 9, 10, 11, 12,$$

$$\tilde{\mathbf{u}}_I(t) = \frac{\partial^2 \mathbf{u}_{I-12}}{\partial \xi \partial \eta}(t), \quad I = 13, 14, 15, 16.$$

**Remark** In the MCDD FEM solution, we only solve for displacement field $\mathbf{u}(\mathbf{X}, t)$. For the Bonger–Fox–Schmit interpolation, however, we need the nodal values of the first-order derivatives of displacements in order to calculate the displacement field inside the interphase element. To solve this problem, we first use the regular quadrilateral FEM interpolation function to calculate the nodal values of the first derivatives of the displacement field in computation and then use those nodal values as $\tilde{\mathbf{u}}_i$ and subsequently find $\mathbf{F}$ and $\mathbf{G}$ inside the interphase elements.

The global and local coordinate transformation between $(X, Y)$ and $(\xi, \eta)$ is as follows:

$$\begin{bmatrix} \xi \\ \eta \end{bmatrix} = \begin{bmatrix} \cos\theta & \sin\theta \\ -\sin\theta & \cos\theta \end{bmatrix} \begin{bmatrix} X - X_c \\ Y - Y_c \end{bmatrix}, \quad \text{where } \theta = 0, \pm\frac{\pi}{6},$$

and $X_c$, $Y_c$ are the center of the rectangular element.

The Bonger–Fox–Schmit interpolation functions are explicitly given as

$$\begin{aligned} N_1(\xi,\eta) &= \psi_1^0(\xi)\psi_1^0(\eta), \\ N_2(\xi,\eta) &= \psi_1^1(\xi)\psi_1^0(\eta), \\ N_3(\xi,\eta) &= \psi_1^0(\xi)\psi_1^1(\eta), \\ N_4(\xi,\eta) &= \psi_1^1(\xi)\psi_1^1(\eta), \end{aligned} \tag{3.11}$$

$$\begin{aligned} N_5(\xi,\eta) &= \psi_2^0(\xi)\psi_1^0(\eta), \\ N_6(\xi,\eta) &= \psi_2^1(\xi)\psi_1^0(\eta), \\ N_7(\xi,\eta) &= \psi_2^0(\xi)\psi_1^1(\eta), \\ N_8(\xi,\eta) &= \psi_2^1(\xi)\psi_1^1(\eta), \end{aligned} \tag{3.12}$$

$$\begin{aligned} N_9(\xi,\eta) &= \psi_2^0(\xi)\psi_2^0(\eta), \\ N_{10}(\xi,\eta) &= \psi_2^1(\xi)\psi_2^0(\eta), \\ N_{11}(\xi,\eta) &= \psi_2^0(\xi)\psi_2^1(\eta), \\ N_{12}(\xi,\eta) &= \psi_2^1(\xi)\psi_2^1(\eta), \end{aligned} \tag{3.13}$$

$$\begin{aligned} N_{13}(\xi,\eta) &= \psi_1^0(\xi)\psi_2^0(\eta), \\ N_{14}(\xi,\eta) &= \psi_1^1(\xi)\psi_2^0(\eta), \\ N_{15}(\xi,\eta) &= \psi_1^0(\xi)\psi_2^1(\eta), \\ N_{16}(\xi,\eta) &= \psi_1^1(\xi)\psi_2^1(\eta), \end{aligned} \tag{3.14}$$

and

$$\begin{aligned} \psi_1^0(\zeta) &= \tfrac{1}{4}(\zeta - 1)^2(\zeta + 2), \\ \psi_2^0(\zeta) &= \tfrac{1}{4}(\zeta + 1)^2(2 - \zeta), \\ \psi_1^1(\zeta) &= \tfrac{1}{4}(\zeta - 1)^2(\zeta + 1), \\ \psi_2^1(\zeta) &= \tfrac{1}{4}(\zeta + 1)^2(\zeta - 1). \end{aligned} \tag{3.15}$$

Subsequently, we can calculate deformation gradient $\mathbf{F}$ and the strain gradient $\mathbf{G}$:

$$\mathbf{F} = \mathbf{I} + \sum_{I=1}^{16} \tilde{\mathbf{u}}_I(t) \otimes \frac{\partial N_I}{\partial \mathbf{X}}, \tag{3.16}$$

$$\mathbf{G} = \frac{\partial^2 \mathbf{x}}{\partial \mathbf{X} \otimes \partial \mathbf{X}} = \sum_{I=1}^{16} \tilde{\mathbf{u}}_I(t) \frac{\partial^2 N_I}{\partial \mathbf{X} \otimes \partial \mathbf{X}}. \tag{3.17}$$

The third scale finite element is a planar hexagonal element, in which, we assume, the contribution of the second-order strain gradients to crystal defect dynamics is significant. This element provides an approximation model for micro-voids and vacancy aggregates.

We use a six-node honeycomb Wachspress element to model the crystal dynamics in this element. The interpolation field,

$$\mathbf{u}(X, Y, t) = \sum_{I=1}^{6} N_I(\xi, \eta)\mathbf{u}_I(t),$$

is facilitated by the following Wachspress shape function:

$$N_i(\boldsymbol{\zeta}) = c_i \frac{\lambda_{i+2}(\boldsymbol{\zeta})\lambda_{i+3}(\boldsymbol{\zeta})\lambda_{i+4}(\boldsymbol{\zeta})\lambda_{i+5}(\boldsymbol{\zeta})}{q(\boldsymbol{\zeta})}, \quad i = 1, 2, 3, 4, 5, 6,$$

and $\boldsymbol{\zeta} = (\xi, \eta)$.

$$c_i = \frac{q(\boldsymbol{\zeta}_i)}{\lambda_{i+2}(\boldsymbol{\zeta}_i)\lambda_{i+3}(\boldsymbol{\zeta}_i)\lambda_{i+4}(\boldsymbol{\zeta}_i)\lambda_{i+5}(\boldsymbol{\zeta}_i)},$$

where $\lambda_j(\boldsymbol{\zeta})$ is the line segment equation connecting vertex $j-1$ to vertex $j$ and

$$q(\boldsymbol{\zeta}) = \xi^2 + \eta^2 - R^2 = 0,$$

as shown in Figure 3.3. Note that, here, $R = 2na$, where $a$ is the lattice bond length and $n > 1$ is the scaling factor. For the dimension of the honeycomb shape element shown in Figure 3.3, we have the following exact values for $c_i$:

$$c_1 = \frac{\sqrt{3}}{3}, \quad c_2 = \frac{2}{3}, \quad c_3 = -\frac{\sqrt{3}}{3}, \quad c_4 = \frac{\sqrt{3}}{3}, \quad c_5 = \frac{2}{3}, \quad c_1 = -\frac{\sqrt{3}}{3}.$$

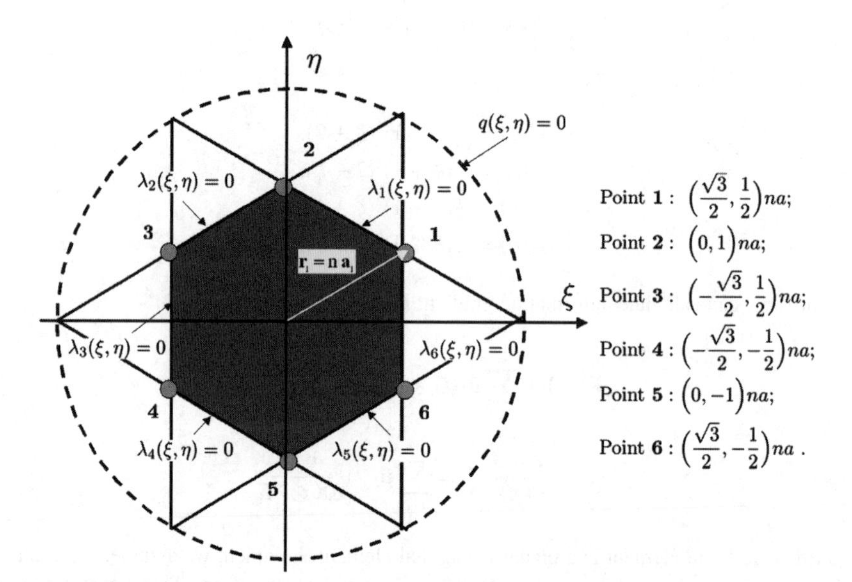


**Figure 3.3** Six-node honeycomb element

Based on the strong form of the crystal defect dynamics equations, we can derive the weak formulation of MCDD as

$$\begin{aligned}&\int_{V_0}\left(\mathbf{P}:\frac{\partial\delta\mathbf{u}}{\partial\mathbf{X}}+\mathbf{Q}\vdots\frac{\partial^2\delta\mathbf{u}}{\partial\mathbf{X}\otimes\partial\mathbf{X}}+\mathbf{S}::\frac{\partial^3\delta\mathbf{u}}{\partial\mathbf{X}\otimes\partial\mathbf{X}\otimes\partial\mathbf{X}}\right)dV\\&+\int_{V_0}\mathbf{b}\cdot\delta\mathbf{u}\,dV+\int_{\partial V_t}\bar{\mathbf{T}}\cdot\delta\mathbf{u}\,dS=\int_{V}\rho\ddot{\mathbf{u}}\cdot\delta\mathbf{u}\,dV.\end{aligned}\tag{3.18}$$

Considering the local FEM interpolation in an element

$$\mathbf{u}^h(\mathbf{X},t)=\sum_{I=1}^{n_{\mathrm{node}}}N_I(\boldsymbol{\zeta})\mathbf{d}_I(t),$$

we have

$$\begin{aligned}&\sum_{e=1}^{n_{\mathrm{elem}}}\sum_{I=1}^{n_{\mathrm{node}}}\int_{V_e}\left(\mathbf{P}\cdot\frac{\partial N_I}{\partial\mathbf{X}}+\mathbf{Q}:\frac{\partial^2 N_I}{\partial\mathbf{X}\otimes\partial\mathbf{X}}+\mathbf{S}\vdots\frac{\partial^3 N_I}{\partial\mathbf{X}\otimes\mathbf{X}\otimes\partial\mathbf{X}}\right)\cdot\delta\mathbf{d}_I\,dV\\&+\sum_{e=1}^{n_{\mathrm{elem}}}\sum_{I=1}^{n_{\mathrm{node}}}\int_{V_e}\mathbf{b}N_I(\mathbf{X})\cdot\delta\mathbf{d}_I\,dV+\sum_{e=1}^{n_{\mathrm{elem}}}\sum_{I=1}^{n_{\mathrm{node}}}\int_{\partial V_{et}}\bar{\mathbf{T}}N_I(\mathbf{X})\cdot\delta\mathbf{d}_I\,dS\\&+\sum_{e=1}^{n_{\mathrm{elem}}}\sum_{I,J=1}^{N_{\mathrm{node}}}\int_{V_e}\rho N_I(\mathbf{X})N_J(\mathbf{X})\ddot{\mathbf{d}}_J\cdot\delta\mathbf{d}_I\,dV=0.\end{aligned}\tag{3.19}$$

In a representative element $e$, we denote

$$\begin{aligned}\frac{\partial\mathbf{u}}{\partial\mathbf{X}}&=\sum_{I=1}^{n_{\mathrm{node}}}\frac{\partial N_I}{\partial\mathbf{X}}\mathbf{d}_I=\mathbf{B}_e(\mathbf{X})\mathbf{d}_e,\quad\mathbf{X}\in\Omega_e,\\\frac{\partial^2\mathbf{u}}{\partial\mathbf{X}\otimes\partial\mathbf{X}}&=\sum_{I=1}^{n_{\mathrm{node}}}\frac{\partial^2 N_I}{\partial\mathbf{X}\otimes\partial\mathbf{X}}\mathbf{d}_I=\mathbf{C}_e(\mathbf{X})\mathbf{d}_e,\quad\mathbf{X}\in\Omega_e,\\\frac{\partial^3\mathbf{u}}{\partial\mathbf{X}\otimes\partial\mathbf{X}\otimes\mathbf{X}}&=\sum_{I=1}^{n_{\mathrm{node}}}\frac{\partial^3 N_I}{\partial\mathbf{X}\otimes\partial\mathbf{X}\otimes\partial\mathbf{X}}\mathbf{d}_I=\mathbf{D}_e(\mathbf{X})\mathbf{d}_e,\quad\mathbf{X}\in\Omega_e,\end{aligned}$$

where $e=1,\ldots,n_{\mathrm{elem}}$,

$$\mathbf{d}_e(t)=\left[\mathbf{d}_1(t),\ldots,\mathbf{d}_{n_{\mathrm{node}}}(t)\right]^{\mathrm{T}}$$

is the element nodal displacement vector, and $\mathbf{B}_e$, $\mathbf{C}_e$, and $\mathbf{D}_e$ are the element interpolation gradient, the second gradient, and the third gradient tensors.

Following the standard finite-element discretization procedure (e.g., [14]), we have the following discrete dynamic equations of motion of crystal bundles:

$$\mathbf{M}\ddot{\mathbf{d}}+\mathbf{f}^{\mathrm{int}}(\mathbf{d})=\mathbf{f}^{\mathrm{ext}},\tag{3.20}$$

where

$$\mathbf{M} = \mathop{\mathrm{A}}_{e=1}^{n_{\mathrm{elem}}} \int_{B_0^e} \rho_0 \mathbf{N}_e{}^{\mathrm{T}} \mathbf{N}_e \, \mathrm{d}V, \tag{3.21}$$

$$\mathbf{f}^{\mathrm{int}} = \mathop{\mathrm{A}}_{e=1}^{n_{\mathrm{elem}}} \int_{B_0^e} (\mathbf{B}_e{}^{\mathrm{T}} \cdot \mathbf{P}_e + \mathbf{C}_e{}^{\mathrm{T}} \cdot \mathbf{Q}_e + \mathbf{D}_e{}^{\mathrm{T}} \cdot \mathbf{S}_e) \, \mathrm{d}V, \tag{3.22}$$

and

$$\mathbf{f}^{\mathrm{ext}} = \mathop{\mathrm{A}}_{e=1}^{n_{\mathrm{elem}}} \left( \int_{B_0^e} \mathbf{N}_e{}^{\mathrm{T}} \rho_0 \mathbf{b} \, \mathrm{d}V + \int_{\partial_t B_0^e} \mathbf{N}_e{}^{\mathrm{T}} \bar{\mathbf{T}}_e \, \mathrm{d}S, \right) \tag{3.23}$$

where

$$\mathbf{N}_e = \left[ N_1(\mathbf{X}), \ldots, \mathbf{N}_{n_{\mathrm{elem}}}(\mathbf{X}) \right]$$

is the element shape function matrix. Note that in lower order elements, the elemental shape function matrices, $\mathbf{C}_e$ and $\mathbf{D}_e$ are simply set to be zero.

## 3.5 Numerical Examples

We have employed the proposed multiscale crystal defect dynamics to simulate fracture at the scale. In the following example, we present a 2D simulation of crack propagation in a copper (Cu) single crystal. The elastic energy density in a unit cell can be found based on the embedded-atom method (EAM) [15]:

$$W = \frac{1}{\Omega} \left\{ \frac{1}{2} \sum_i \left[ \sum_{j \neq i} V(r_{ij}) + F(\bar{\rho}) \right] \right\}, \quad r_{ij} = |\mathbf{r}_i - \mathbf{r}_j|,$$

where $V$ is a pair potential and $F$ is the embedding energy.

For this particular example, a so-called EAM–Mishin potential for Cu [16] is employed. The pair potential is given by

$$V(r) = [E_1 M(r, r_0^{(1)}, \alpha_1) + E_2 M(r, r_0^{(2)}, \alpha) + \delta] \psi \left( \frac{r - r_{\mathrm{c}}}{h} \right) - \sum_{n=1}^{3} H(r_{\mathrm{s}}^{(n)} - r) S_n (r_{\mathrm{s}}^{(n)} - r)^4,$$

where

$$M(r, r_0, \alpha) = \exp[-2\alpha(r - r_0)] - 2\exp[-\alpha(r - r_0)]$$

is a Morse function, $H(x)$ is the Heaviside function, and the cutoff function is defined as

$$\psi(x) = \begin{cases} 0, & x \geq 0, \\ \dfrac{x^4}{(1 + x^4)}, & x < 0. \end{cases}$$

$F$ is the energy required to embed an atom into the electron density distribution, and the embedding function is given as

$$F(\bar{\rho}) = \begin{cases} F^0 + \dfrac{1}{2}F^{(2)}(\bar{\rho}-1)^2 + \displaystyle\sum_{n=1}^{4} q_n(\bar{\rho}-1)^{n+2}, & \bar{\rho} < 1, \\ \dfrac{F^0 + \frac{1}{2}F^{(2)}(\bar{\rho}-1)^2 + q_1(\bar{\rho}-1)^3 + Q_1(\bar{\rho}-1)^4}{1+Q_2(\bar{\rho}-1)^3}, & \bar{\rho} > 1. \end{cases}$$

The host electron density at the site $x_i$ is

$$\bar{\rho}_i = \sum_{j \neq i} \rho(r_{ij}),$$

where the electronic density function $\rho$ for each atom is given as

$$\rho(r) = \{a \exp[-\beta_1(r - r_0^{(3)})^2] + \exp[-\beta_2(r - r_0^{(4)})]\}\psi\left(\frac{r - r_c}{h}\right)$$

and for the Bravais lattice the host electron density $\bar{\rho}_i$ may be calculated in the group of coordinate shells such that

$$\bar{\rho}_i = \sum_m \sum_{j=1}^{N_m} \rho(r_{ij}).$$

All the parameters used in the computation are the same as listed in Mishin *et al.* [16] for single-crystal Cu.

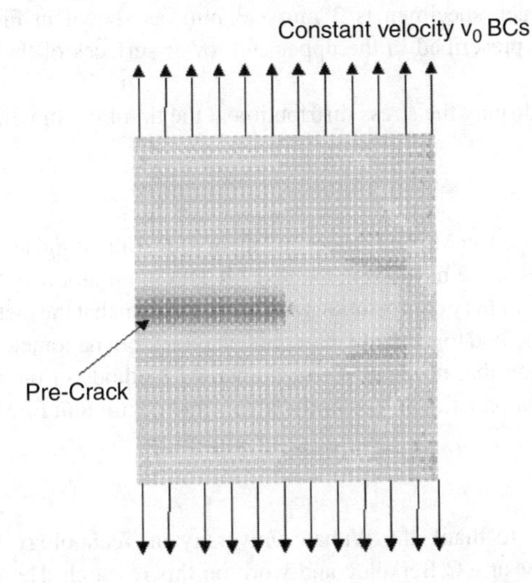


**Figure 3.4** Fracture test specimen, with permission from Surget and Barron (2005)

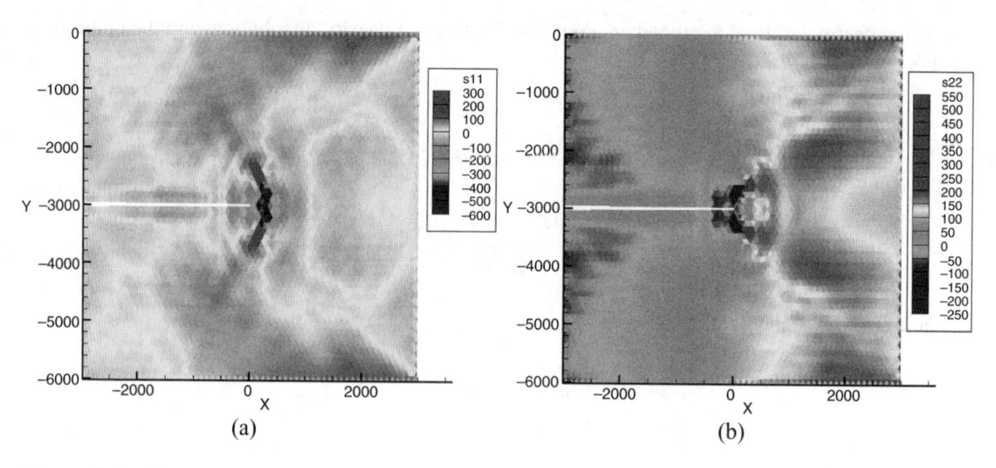


**Figure 3.5** The stress distribution in front of a moving crack: (a) $S_{11}$; (b) $S_{22}$, with permission from Leon *et al.* (2010)

Note that the function $\rho(r)$ is normalized so that $\bar{\rho} = 1$ in the equilibrium state; that is,

$$\bar{\rho}_0 = \sum_m N_m \rho_m(R_m) = 1, \tag{3.24}$$

where $m$ is the number of coordinate shells used in the computation. For the nearest-neighbor calculation, $m = 1$; and $R_m$ and $N_m$ are the radius and the number of atoms in the $m$th coordination shell. The pair potential $V$ also has to satisfy the mechanical equilibrium condition of the FCC crystal; that is,

$$\sum_m N_m R_m V'(R_m) = 0, \tag{3.25}$$

where Equations (3.24) and (3.25) can determine equilibrium bond length and the parameter $E_1$.

The dimension of the crack specimen is 3 mm × 3 mm, as shown in Figure 3.4. The displacement boundary condition is prescribed in the upper and lower surfaces of the specimen with velocity $v_0 = 50$ m/s.

In Figure 3.5a and b, we display the stress distribution at the tip of the moving crack.

## 3.6 Discussion

In this chapter we have proposed an MCDD model and theory for simulating defect motion and evolution in crystalline solids. Moreover, we have implemented MCDD in an atomistic-based multiscale finite-element framework. The preliminary computation results have shown that the method is robust, and it may provide a viable numerical method to simulate material behavior from nanometer scale to sub-millimeter scale, which is a spatial range that needs effective numerical methods in modeling and simulation in order to fill the gap between atomistic molecular dynamics and continuum FEMs.

## Acknowledgments

Dr. Lisheng Liu would like to thank the Wuhan University of Technology for providing a fellowship, which enables him to visit UC Berkeley and work on this research. He also thanks the financial support of "the Fundamental Research Funds for the Central Universities" and "A3 Foresight Program-51161140399."

## Appendix

Here, we list all the derivatives of the six-node honeycomb Wachspress FEM shape functions in explicit form. The first-order derivatives are

$$\begin{aligned}\frac{\partial N_1}{\partial \xi} &= f(\xi,\eta)[6\sqrt{3}\eta^5 + 3\eta^4(\sqrt{3}-4\xi) + (3\sqrt{3}+4\xi)(-3+\xi^2)^2 \\ &\quad -6\eta^3(5\sqrt{3}+8\xi+2\sqrt{3}\xi^2) + 2\eta^2(-9\sqrt{3}+6\xi+3\sqrt{3}\xi^2+4\xi^3) \\ &\quad +\eta(36\sqrt{3}+108\xi+30\sqrt{3}\xi^3-2\sqrt{3}\xi^4)],\end{aligned}$$

$$\begin{aligned}\frac{\partial N_1}{\partial \eta} &= f(\xi,\eta)[\eta^4(9+6\sqrt{3}\xi) - (-3+\xi^2)(9+12\sqrt{3}\xi+15\xi^2+2\sqrt{3}\xi^3) \\ &\quad +\eta^2(-72-42\sqrt{3}\xi+42\xi^2+20\sqrt{3}\xi^3) - 4\eta(9-15\xi^2+4\xi^4)],\end{aligned}$$

$$\frac{\partial N_2}{\partial \xi} = -4\xi f(\xi,\eta)[27\eta - 12\eta^3 - 6\eta^4 + 2(-3+\xi^2)^2 + \eta^2(6+4\xi^2)],$$

$$\frac{\partial N_2}{\partial \eta} = 2f(\xi,\eta)(-3+4\xi^2)[-3\eta^2 + 3(-3+\xi^2) + 4\eta(-3+\xi^2)],$$

$$\begin{aligned}\frac{\partial N_3}{\partial \xi} &= f(\xi,\eta)[-6\sqrt{3}\eta^5 - 3\eta^4(\sqrt{3}+4\xi) - (3\sqrt{3}-4\xi)(-3+\xi^2)^2 \\ &\quad +6\eta^3(5\sqrt{3}-8\xi+2\sqrt{3}\xi^2) + \eta^2(18\sqrt{3}+12\xi-6\sqrt{3}\xi^2+8\xi^3) \\ &\quad +2\eta(-18\sqrt{3}+54\xi-15\xi^2+\sqrt{3}\xi^4)],\end{aligned}$$

$$\begin{aligned}\frac{\partial N_3}{\partial \eta} &= f(\xi,\eta)[\eta^4(9-6\sqrt{3}\xi) + \eta^2(-72+42\sqrt{3}\xi+42\xi^2-20\sqrt{3}\xi^3) \\ &\quad +(-3+\xi^2)(-9+12\sqrt{3}\xi-15\xi^2+2\sqrt{3}\xi^3) - 4\eta(9-15\xi^2+4\xi^4)],\end{aligned}$$

$$\begin{aligned}\frac{\partial N_4}{\partial \xi} &= -f(\xi,\eta)[-6\sqrt{3}\eta^5 + 3\eta^4(\sqrt{3}+4\xi) + (3\sqrt{3}-4\xi)(-3+\xi^2)^2 \\ &\quad +6\eta^3(5\sqrt{3}-8\xi+2\sqrt{3}\xi^2) + 2\eta^2(-9\sqrt{3}-6\xi+3\sqrt{3}\xi^2-4\xi^3) \\ &\quad +2\eta(-18\sqrt{3}+54\xi-15\xi^2+\sqrt{3}\xi^4)],\end{aligned}$$

$$\begin{aligned}\frac{\partial N_4}{\partial \eta} &= f(\xi,\eta)[\eta^4(-9+6\sqrt{3}\xi) - (-3+\xi^2)(-9+12\sqrt{3}\xi-15\xi^2+2\sqrt{3}\xi^3) \\ &\quad +\eta^3(72-42\sqrt{2}\xi-42\xi^2+20\sqrt{3}\xi^3) - 4\eta(9-15\xi^2+4\xi^4)],\end{aligned}$$

$$\frac{\partial N_5}{\partial \xi} = -4f(\xi,\eta)[-27\eta + 12\eta^3 - 6\eta^4 + 2(-3+\xi^2)^2 + \eta^2(6+4\xi^2)],$$

$$\frac{\partial N_5}{\partial \eta} = 2f(\xi,\eta)(-3+4\xi^2)[9+3\eta^2-3\xi^2+4\eta(-3+\xi^2)],$$

$$\begin{aligned}\frac{\partial N_6}{\partial \xi} &= f(\xi,\eta)[-6\sqrt{3}\eta^5 + 3\eta^4(\sqrt{3}-4\xi) + (3\sqrt{3}+4\xi)(-3+\xi^2)^2 \\ &\quad +6\eta^3(5\sqrt{3}+8\xi+2\sqrt{3}\xi^2) + 2\eta^2(-9\sqrt{3}+6\xi+3\sqrt{3}\xi^2+4\xi^3) \\ &\quad +2\eta(-18\sqrt{3}-54\xi-15\sqrt{3}\xi^2+\sqrt{3}\xi^4)],\end{aligned}$$

and

$$\begin{aligned}\frac{\partial N_6}{\partial \eta} &= f(\xi,\eta)[-3\eta^4(3+2\sqrt{3}\xi) + \eta^2(72+42\sqrt{3}\xi-42\xi^2-20\sqrt{3}\xi^3) \\ &\quad +(-3+\xi^2)(9+12\sqrt{3}\xi+15\xi^2+2\sqrt{3}\xi^3) - 4\eta(9-15\xi^2+4\xi^4)],\end{aligned}$$

where

$$f(\xi, \eta) = \frac{1}{18}(-3 + \xi^2 + \eta^2)^{-2}.$$

The second-order derivatives of the MCDD FEM shape functions are

$$\frac{\partial N_1}{\partial \xi^2} = -2g(\xi, \eta)[3\eta^6 + 12\eta^5(1 + \sqrt{3}\xi) - 3\eta^2\xi^2(-15 + \xi^2) - (-3 + \xi^2)^3$$
$$-3\eta^4(4 + 5\xi^2) + 9\eta(9 + 9\sqrt{3}\xi + 9\xi^2 + \sqrt{3}\xi^2)$$
$$-\eta^3(63 + 63\sqrt{3}\xi + 36\xi^2 + 4\sqrt{4}\xi^3)],$$

$$\frac{\partial^2 N_1}{\partial \xi \partial \eta} = g(\xi, \eta)[3\sqrt{3}\eta^6 + 36\eta\xi(-3\xi^2) + 3\eta^4(-10\sqrt{3} + 8\xi + 7\sqrt{3}\xi^2)$$
$$+\eta^3(60\xi - 32\xi^3) - (-3 + \xi^2)(-18\sqrt{3} - 54\xi - 15\sqrt{3}\xi^2 + \sqrt{3}\xi^4)$$
$$-3\eta^2(-27\sqrt{3} - 18\xi + 12\sqrt{3}\xi^2 + 24\xi^3 + 5\sqrt{3}\xi^4)],$$

$$\frac{\partial^2 N_1}{\partial \eta^2} = 2g(\xi, \eta)(-3 + 4\xi^2)[\eta^3(3 + \sqrt{3}\xi) - 3\eta^2(-3 + \xi^2)$$
$$-3\eta(3 + \sqrt{3})(-3 + \xi^2) + (-3 + \xi^2)^2],$$

$$\frac{\partial^2 N_2}{\partial \eta^2} = 2g(\xi, \eta)[12\eta^5 + 6\eta^6 - 6\eta^2\xi^2(-15 + \xi^2) - 2(-3 + \xi^2)^3$$
$$+81\eta(1 + \xi^2) - 9\eta^3(7 + 4\xi^2) - 6\eta^4(4 + 5\xi^2)],$$

$$\frac{\partial^2 N_2}{\partial \xi \partial \eta} = -2g(\xi, \eta)(3 + 2\eta)\xi[6\eta^3 + \eta^2(21 - 16\xi^2) + 9(-3 + \xi^2) + 6\eta(-3 + \xi^2)],$$

$$\frac{\partial^2 N_2}{\partial \eta^2} = 2g(\xi, \eta)(-3 + 4\xi^2)[3\eta^3 - 9\eta(-3 + \xi^2) - 6\eta^2(-3 + \xi^2) + 2(-3 + \xi^2)^2],$$

$$\frac{\partial^2 N_3}{\partial \xi^2} = 2g(\xi, \eta)[3\eta^2 - 12\eta^5(-1 + \sqrt{3}\xi) - 3\eta^2\xi^2(-15 + \xi^2) - (-3 + \xi^2)^3$$
$$-3\eta^4(4 + 5\xi^2) - 9\eta(-9 + 9\sqrt{3}\xi - 9\xi^2 + \sqrt{3}\xi^3)$$
$$+\eta^3(-63 + 63\sqrt{3}\xi - 36\xi^2 + 4\sqrt{3}\xi^3)],$$

$$\frac{\partial^2 N_3}{\partial \xi \partial \eta} = g(\xi, \eta)[-3\sqrt{3}\eta^6 + 36\eta\xi(-3 + \xi^2) + \eta^4(30\sqrt{3} + 34\xi - 21\sqrt{3}\xi^2)$$
$$+\eta^3(60\xi - 32\xi^3) + (-3 + \xi^2)(-18\sqrt{3} + 54\xi - 15\sqrt{3}\xi^3 + \sqrt{3}\xi^4)$$
$$+3\eta^3(27\sqrt{3} + 18\xi^2 - 24\xi^3 + 5\sqrt{3}\xi^4)],$$

$$\frac{\partial^2 N_3}{\partial \eta^2} = -2g(\xi, \eta)(-3 + 4\xi^2)[\eta^3(3 - \sqrt{3}\xi) - 3\eta^2(-3 + \xi^2)$$
$$+3\eta(-3 + \sqrt{3}\xi)(-3 + \xi^2) + (-3 + \xi^2)^2],$$

$$\frac{\partial^2 N_4}{\partial \xi^2} = -2g(\xi, \eta)[3\eta^6 + 12\eta^5(-1 + \sqrt{3}\xi) - 3\eta^2\xi^2(-15 + \xi^2) - (-3 + \xi^2)^3$$
$$-3\eta^4(4 + 5\xi^2) + \eta^3(63 - 63\sqrt{3}\xi + 36\xi^2 - 4\sqrt{3}\xi^3)$$
$$+9\eta(-9 + 9\sqrt{3}\xi - 9\xi^2 + \sqrt{3}\xi^3)],$$

$$\frac{\partial^2 N_4}{\partial \xi \partial \eta} = g(\xi, \eta)[3\sqrt{3}\eta^6 + 36\eta\xi(-3 + \xi^2) + 3\eta^4(-10\sqrt{3} - \xi + 7\sqrt{3}\xi^2)$$
$$+\eta^3(60\xi - 32\xi^3) - (-3 + \xi^2)(-18\sqrt{3} + 54\xi - 15\sqrt{3}\xi^2 + \sqrt{3}\xi^4)$$
$$-3\eta^2(-27\sqrt{3} + 18\xi + 12\sqrt{3}\xi^2 - 24\xi^3 + 5\sqrt{3}\xi^4)],$$

$$\frac{\partial^2 N_4}{\partial \eta^2} = 2g(\xi,\eta)(-3+4\xi^2)[\eta^3(-3+\sqrt{3}\xi) - 3\eta^2(-3+\xi^2) \\ -3\eta(-3+\sqrt{3}\xi)(-3+\xi^2) + (-3+\xi^2)^2],$$

$$\frac{\partial^2 N_5}{\partial \xi^2} = 2g(\xi,\eta)[-12\eta^5 + 6\eta^6 - 6\eta^2\xi^2(-15+\xi^2) - 2(-3+\xi^2)^3 \\ -81\eta(1+\xi^2) + 9\eta^3(7+4\xi^2) - 6\eta^4(4+5\xi^2)],$$

$$\frac{\partial^2 N_5}{\partial \xi \partial \eta} = 2g(\xi,\eta)(-3+2\eta)\xi[6\eta^3 - 9(-3+\xi^2) + 6\eta(-3+\xi^2) + \eta^2(-21+16\xi^2)],$$

$$\frac{\partial^2 N_5}{\partial \eta^2} = 2g(\xi,\eta)(-3+4\xi^2)[-3\eta^3 + 9\eta(-3+\xi^2) - 6\eta^2(-3+\xi^2) + 2(-3+\xi^2)^2],$$

$$\frac{\partial^2 N_6}{\partial \xi^2} = -2g(\xi,\eta)[3\eta^6 - 12\eta^5(1+\sqrt{3}\xi) - 3\eta^2\xi^2(-15+\xi^2) \\ -(-3+\xi^2)^3 - 3\eta^4(4+5\xi^2) - 9\eta(9+9\sqrt{3}\xi + 9\xi^2 + \sqrt{3}\xi^3) \\ +\eta^3(63+63\sqrt{3}\xi + 36\xi^2 + 4\sqrt{3}\xi^3)],$$

$$\frac{\partial^2 N_6}{\partial \xi \partial \eta} = g(\xi,\eta)[-3\sqrt{3}\eta^6 + 36\eta\xi(-3=\xi^2) - 3\eta^4(-10\sqrt{3} + 8\xi + 7\sqrt{3}\xi^2) \\ +\eta^3(60\xi - 32\xi^3) + (-3+\xi^2)(-18\sqrt{3} - 54\xi - 15\sqrt{3}\xi^2 + \sqrt{3}\xi^4) \\ +3\eta^2(-27\sqrt{3} - 18\xi + 12\sqrt{3}\xi^2 + 24\xi^3 + 5\sqrt{3}\xi^4)],$$

and

$$\frac{\partial^2 N_6}{\partial \eta^2} = 2g(\xi,\eta)(-3+4\xi^2)[\eta^3(3+\sqrt{3}\xi) + 3\eta^2(-3+\xi^2) \\ -3\eta(3+\sqrt{3}\xi)(-3+\xi^2) - (-3+\xi^2)^2],$$

where

$$g(\xi,\eta) = \frac{1}{9}(-3+\eta^2+\xi^2)^{-3}.$$

# References

[1] Chen, Y. and Lee, J. (2005) Atomistic formulation of a multiscale field theory for nano/micro solids. *Philosophical Magazine*, **85**, 4095–4126.

[2] Lee, J., Wang, X., and Chen, Y. (2009) Multiscale material modeling and its application to a dynamic crack propagation problem. *Theoretical and Applied Fracture Mechanics*, **51**, 3–40.

[3] Chen, J., Wang, X., Wang, H., and Lee, J. (2010) Multiscale modeling of dynamic crack propagation. *Engineering Fracture Mechanics*, **77**, 736–743.

[4] Lu, G., Tadmor, E., and Kaxiras, E. (2006) From electrons to finite elements: a concurrent multiscale approach for metals. *Physical Review B*, **73**, 024108.

[5] Wagner, G. and Liu, W. (2003) Coupling of atomic and continuum simulations using a bridging scale decomposition. *Journal of Computational Physics*, **190**, 249–274.

[6] Park, H., Karpov, E., Liu, W., and Klein, P. (2005) The bridging scale for two-dimensional atomistic/continuum coupling. *Philosophical Magazine*, **85** (1), 79–113.

[7] Horstemeyer, M., Baskes, M., Prantil, V., and Philliber, J. (2003) A multiscale analysis of fixed-end simple shear using molecular dynamics, crystal plasticity, and a macroscopic internal state variable theory. *Modelling and Simulation in Materials Science and Engineering*, **11** (3), 265–286.

[8] Clayton, J.D. and Chung, P.W. (2006) An atomistic-to-continuum framework for nonlinear crystal mechanics based on asymptotic homogenization. *Journal of Mechanics and Physics of Solids*, **54**, 1604–1639.

[9] Li, S., Liu, X., Agrawal, A., and To, A. (2006) The perfectly matched multiscale simulations for discrete systems: extension to multiple dimensions. *Physical Review B*, **74**, 045418.
[10] Clayton, J., McDowell, D., and Bammann, D. (2004) A multiscale gradient theory for single crystalline elastoviscoplasticity. *International Journal of Engineering Science*, **42**, 427–457.
[11] McDowell, D.L. (2007) Simulation-assisted materials design for the concurrent design of materials and products. *JOM Journal of the Minerals, Metals and Materials Society*, **9**, 21–25.
[12] McDowell, D.L. and Olson, G.B. (2009) Concurrent design of hierarchical materials and structures. *Scientific Modeling and Simulations*, **68**, 207–240.
[13] McDowell, D.L. (2010) A perspective on trends in multiscale plasticity. *International Journal of Plasticity*, **26**, 1280–1309.
[14] Hughes, T. (1987) *The Finite Element Method: Linear Static and Dynamic Finite Element Analysis*, Prentice Hall.
[15] Daw, M. and Baskes, M. (1983) Semiempirical, quantum mechanical calculation of hydrogen embrittlement in metals. *Physical Review Letters*, **50**, 1285–1288.
[16] Mishin, Y., Mehl, M., Papaconstantopoulos, D. *et al.* (2001) Structural stability and lattice defects in copper: ab initio, tight-binding, and embedded-atom calculations. *Physics Review B*, **63**, 224106.

# 4

# Application of Many-Realization Molecular Dynamics Method to Understand the Physics of Nonequilibrium Processes in Solids

Yao Fu and Albert C. To
*Department of Mechanical Engineering and Materials Science, University of Pittsburgh, USA*

## 4.1 Chapter Overview and Background

Many systems of great interest are nonequilibrium thermomechanical processes, which involve mass, momentum, energy transport, and even chemical reactions. Understanding the underlying mechanisms of the phenomena occurring in thermomechanical nonequilibrium systems is crucial for material performance and new materials design. Shock impact on materials can trigger a variety of interesting nonequilibrium phenomena, such as plastic deformation [1–6], phase transformation [7–10], and chemical reactions [11–15], which depend on shock loading conditions and the state of material. Molecular dynamics (MD) simulation [16–18], which is based on solving the classical equations of motion for interacting atoms, is a powerful way to study shocked materials [13, 19–21]. Its capability of simulating real-time evolution naturally meets the requirement of capturing the rapidity and locality of shock waves. The first nonequilibrium molecular dynamics (NEMD) simulation of a shock-wave-related phenomenon was carried out by George Vineyard and his group at Brookhaven National Laboratory [22]. In that study, the problem of radiation damage in a crystal was investigated, where a high-energy neutron collided with a primary knock-on atom, which then violently ejected from its lattice site. This initiated a chain of collisions with its neighboring atoms, which created a miniature spherical shock wave. Since then, NEMD has been widely used to study shocked induced plastic flow and phase transformation. A noteworthy example was demonstrated by Holian and co-workers using NEMD simulations to study shock induced plasticity in a three-dimensional 10-million-atom face-centered cubic (fcc) crystal with cross-sectional dimensions of $100 \times 100$ unit cells [3]. They discovered that the system slips along all

*Multiscale Simulations and Mechanics of Biological Materials*, First Edition. Edited by Shaofan Li and Dong Qian.

of the available slip planes and the shock waves propagating through the system can instantly create a large number of stacking faults. For shock compression, the Hugoniot relation is widely used to provide structural and thermodynamic information of a material subjected to varying shock strengths, giving insight into the performance and behavior of a material under shock impact. The "uniaxial Hugoniostat molecular dynamics" proposed by Maillet *et al.* [23] provides a way to use MD simulation to generate the Hugoniot relation. By reformulating the equations of motion for shock problems, this method can efficiently enable single small-scale MD simulation to produce time-averaged properties that lie on the shock Hugoniot curve.

Shock-initiated chemical reactions in energetic materials belong to another class of widely investigated phenomena using NEMD. Many useful energetic materials (PETN, TATB, TNT, RDX, and HMX) are molecular crystals which are bonded together by covalent bonds within individual molecular and by noncovalent bonds between adjacent molecules [24–27]. Stretching the covalent bonds gives rise to excitations termed intramolecular vibrations or simply vibrations, while stretching the intermolecular bonds gives rise to collective delocalized excitations termed phonons. The transfer of substantial amounts of mechanical energy from the shock front to the internal vibrational states of the molecules [28–33] is termed "multiphonon up-pumping." In this process, the energy from the shock deposited in the phonon modes can be further transferred to the internal vibrational degrees of a molecule, heating them to a temperature at which a chemical bond can be broken [14]. The conversion of the energetic material to its final products is the mix of various fundamental chemical and physical steps. Elucidating the important fundamental steps would contribute to the identification and understanding of the whole process. MD simulations can provide atomic-level details at a short time scale that cannot be obtained through experiments or the continuum-type models. Recent developments in reactive force-field molecular dynamics (ReaxFF-MD) combined with advances in parallel computing have paved the way for accurately simulating the reaction pathway as well as the structure of shock fronts. It has been used to study the chemical reactions at shock fronts by mechanical stimuli prior to detonation in energetic materials, revealing a transition from well-ordered molecular dipoles to a disordered dipole distribution behind the shock front [15]. In addition, MD simulation can also be used to characterize and follow details of important time-dependent processes occurring within the localized region of microscopic defects that might result in the generation of hot spots, which is believed to be closely related to the chemical initiation of energetic materials and its sensitivity to impact [14, 34–37].

However, a computational approach based on MD simulations is still lacking that is able to determine accurately the thermal and mechanical states of individual atoms in a solid in which a nonequilibrium process occurs. This method is crucial to the understanding of the underlying physics of certain phenomena associated with a nonequilibrium process. For example, it may be applied to learn whether an atom in a shocked solid is in local equilibrium or nonequilibrium and what temperature the atom has at a certain time instant. Also, it may be utilized to further elucidate the fundamental relationship between the local thermal and mechanical states of individual atoms and the chemical kinetics in shocked energetic materials.

In this chapter we review a "many-realization" method that can determine accurately the thermal and mechanical states of individual atoms modeled by NEMD simulations [35]. The approach is proposed based on the "ensemble" concept in statistical mechanics. As an example, it will be applied to analyze a shock-compressed nickel (Ni) plate and reveal unprecedented details on the interaction between heat and mechanical deformation.

## 4.2 Many-Realization Method

The many-realization method is proposed based on the "ensemble" concept in statistical mechanics [38]. An ensemble is a collection of all the possible microscopic states that a system could be in, consistent with its observed macroscopic properties. A microstate of a system of atoms is a complete specification

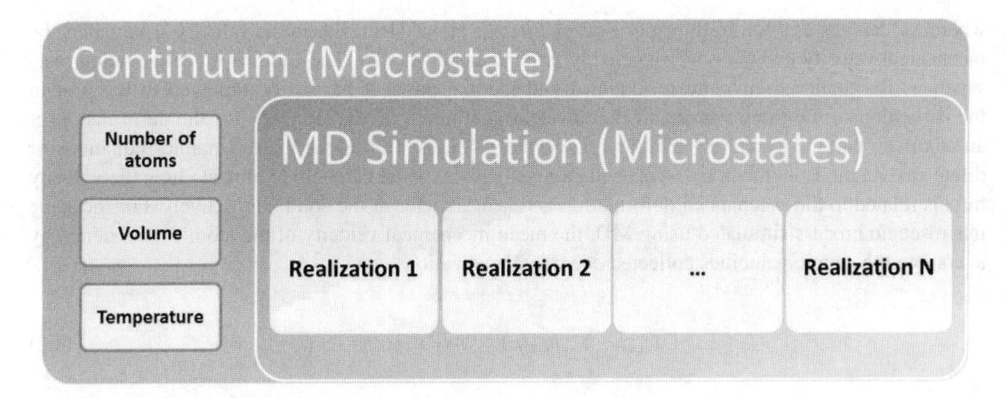


**Figure 4.1** Schematic illustration of the many-realization method

of all positions and momenta of the atoms. For example, a microstate of system consisting of $N$ particles can be specified by the $3N$ canonical coordinates and their conjugate momenta. In contrast, a macrostate is a statistical description of the microstates and some macroscopic quantities are temperature, density, and total energy. As long as the extensive quantities defined for the macrostate are satisfied, an ensemble could be defined as a collection of all the ways a set of atoms could be arranged in a system. Hence, in the many-realization method, the local macrostates can be obtained by first sampling or realizing the microstates from an ensemble using MD simulations many times and then applying one of the atomistic-to-continuum theorems such as the Virial theorem on the microstates [39–41] (Figure 4.1). In each MD realization, the microstates (i.e., atomic velocities) corresponding to the same initial macrostate in the ensemble are sampled and then the atoms in the system are allowed to evolve into new microstates. For nonequilibrium processes, it is not clear at all if the ensemble concept is still valid and hence that the many-realization method is applicable to analyze these processes. Typically, the velocity time histories of selected individual atoms are collected over time from the $N$ realizations.

In the study of nonequilibrium processes, the procedure to implement the many-realization method is as follows:

1. An NVT ensemble with constant number of atoms, volume, and temperature is adopted.
2. Each of the $N$ realizations of the same ensemble (macrostate) is initiated by assigning each atom a velocity sampled from the Maxwell–Boltzmann distribution at the prescribed temperature.
3. The number of atoms and the initial position of the atoms are set to be the same in each of the $N$ realizations.
4. With the assigned atomic velocity and position, each realization is left to evolve and equilibrate in parallel under control of the thermostat.
5. After equilibration, the thermostat is turned off and an external stress is applied to the material model to drive the system to nonequilibrium.
6. The atomic velocities and positions from selected atoms are collected from the $N$ realizations over time as the system evolves.

The collected atomic velocity can be assumed to be the sum of a mean velocity $\bar{v}$ and a random velocity $\tilde{v}$ at time $t$:

$$v_\alpha = \bar{v}(x_\alpha, t) + \tilde{v}(x_\alpha, t) \tag{4.1}$$

where $x_\alpha$ denotes the equilibrium position of the $\alpha$th atom. Here, the mean velocity $\bar{v}$ is called the mechanical velocity and the random velocity $\tilde{v}$ the thermal velocity. The additive model equation (4.1) separates the total velocity into mechanical and thermal parts. It has been employed in the past in the derivation of the thermomechanical equivalent continuum (TMEC) theory for the deformation of an atomistic particle system of arbitrary size scales [42], as well as the hydrodynamics equations of dilute gas via the Boltzmann transport equation [38]. This model can help to eluciate how the velocity field is related to the mechancial deformation and heat transfer in the solid, respectively. For the same macroscopic process simulated using MD, the mean mechanical velocity of the atom $\alpha$ is obtained by averaging the atomic velocities collected over the $N$ realizations:

$$\bar{v}_{\alpha,i} = \sum_{n=1}^{N} v_{\alpha,i}[n], \quad i = x, y, z, \tag{4.2}$$

where $i$ denotes the direction in the Cartesian coordinate system. The thermal velocity of the $n$th realization is then obtained through Equations (4.1) and (4.2) by subtracting the resulting mechanical velocity from the atomic velocity.

When an atom is in local equilibrium, the temperature of the atom is related to its thermal velocity through the Maxwell–Boltzmann distribution based on the definitions in statistical mechanics [43–45]. The definition of temperature in a nonequilibrium state is not well defined, since the velocity distribution of the atom in thermal nonequilibrium does not follow the Maxwell–Boltzmann distribution. The definition of quasi-temperature or effective temperature established by Dlott and co-workers [30, 46, 47] and Chen [45, 48] is adopted here to measure the temperature based on the Maxwell–Boltzmann distribution even when the atom is in nonequilibrium. The quasi-temperature of an atom is computed by

$$\tilde{T}_{\alpha,i} = \frac{1}{Nk_{\mathrm{B}}} \sum_{n=1}^{\mathrm{N}} m|\tilde{v}_{\alpha,i}[n]|^2, \quad i = x, y, z, \tag{4.3}$$

where $k_{\mathrm{B}}$ is the Boltzmann constant. Note that the quasi-temperature is a vector quatity rather than a scalar quantity, as the quasi-temperature can differ in different directions when the atom is in nonequilibrium. This will be just called temperature in the rest of this chapter.

It is worthwhile mentioning that local macroscopic quantities such as the thermal velocity and temperature in (4.3) in nonequilibrium MD simulations or ensemble averaged quantities are traditionally obtained by local spatial and/or temporal averaging [42, 49, 50], which has a tradeoff between resolution and accuracy in choosing the size of the averaging window. Since the averaging via the many-realization method is performed over many realizations for the same atom, the resolution is not lost as long as the number of realizations is sufficiently large. Hence, it is a much more accurate way of computing local macroscopic quantities, but the disadvantage is the higher computational cost in performing the many realizations. To compute the probability distribution of atomic velocity and thermal state (equilibrium versus nonequilibrium) of an atom, the many-realization method seems to be the feasible way.

## 4.3 Application of the Many-Realization Method to Shock Analysis

Here, we apply the "many-realization" approach to the analysis of shock wave propagation, which is a well-known nonequilibrium thermomechanical process. An Ni plate is created, with dimensions of $L_x = 16[100]$, $L_y = 25[011]$, and $L_z = 2.8[0\bar{1}1]$ in the $x$, $y$, and $z$ directions. Periodic boundary conditions are applied to the $x$- and $z$-directions, while free-surface conditions are employed in the $y$-direction (Figure 4.2). The interaction of the Ni atoms is described by the embedded atom method (EAM) interatomic potential [51–55]. Applying the many-realization method, 1000 realizations with the same

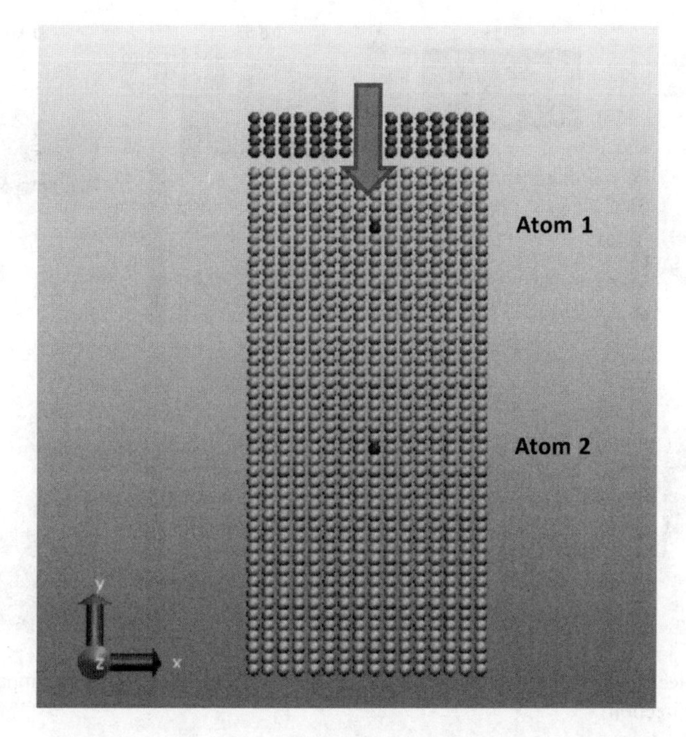


**Figure 4.2** An Ni crystal plate subjected to shock loading from the dropping Ni hammer on top

simulation setup described above are carried out. Each realization is initialized with atomic velocities randomly sampled at 300 K. After global thermal equilibrium is achieved for the system, the thermostat is turned off. A smaller Ni plate is assigned with an initial constant velocity in the −y-direction and subsequently induces an impact loading on top of the larger Ni plate, which results in a shock wave propagating in the plate. The initial velocity applied on the smaller Ni plate is set to be 1000, 1500, 2000, 2500 m/s in the $-y$-direction to create shock strengths of different levels. For computational efficiency, only the middle plane of atoms is selected to perform the statistical analysis.

Figure 4.3, Figure 4.4, Figure 4.5, and Figure 4.6 show the mechanical velocity and temperature at selected time instants as computed based on Equation (4.2) at shock impact velocities of 1000 or 2500 m/s. The plots corresponding to 1500 and 2000 m/s are not shown because of the similarity with the other two velocities. The shock impact introduced by the moving smaller Ni plate in the $-y$-direction creates a shock front that consists of a few bands instead of a single band. This can be shown from the mechanical velocity plot that the dark bands are moving downward as the main shock front. Compression of atoms occurs at the shock front as the shock wave steadily propagates, which results in tensile deformation behind the shock front. These atoms in tension cause aftershocks by interacting with the atoms behind them. The springback motions of atoms behind the shock front can create much weaker secondary shock waves. This can be seen from the lighter color and thinner bands moving upward identified as the secondary waves after $t = 0.6$ ps in the mechanical energy plot. Similar phenomena can also be observed for the other shock velocities. The bands that constitute the main shock front evolve into sharper and thinner bands as the shock impact velocity increases from 1000 to 1500 m/s, and this feature is maintained for higher shock impact velocities. The velocity of the shock front is increased in magnitude as the shock impact velocity increases and the shock wave propagates at higher velocity. It

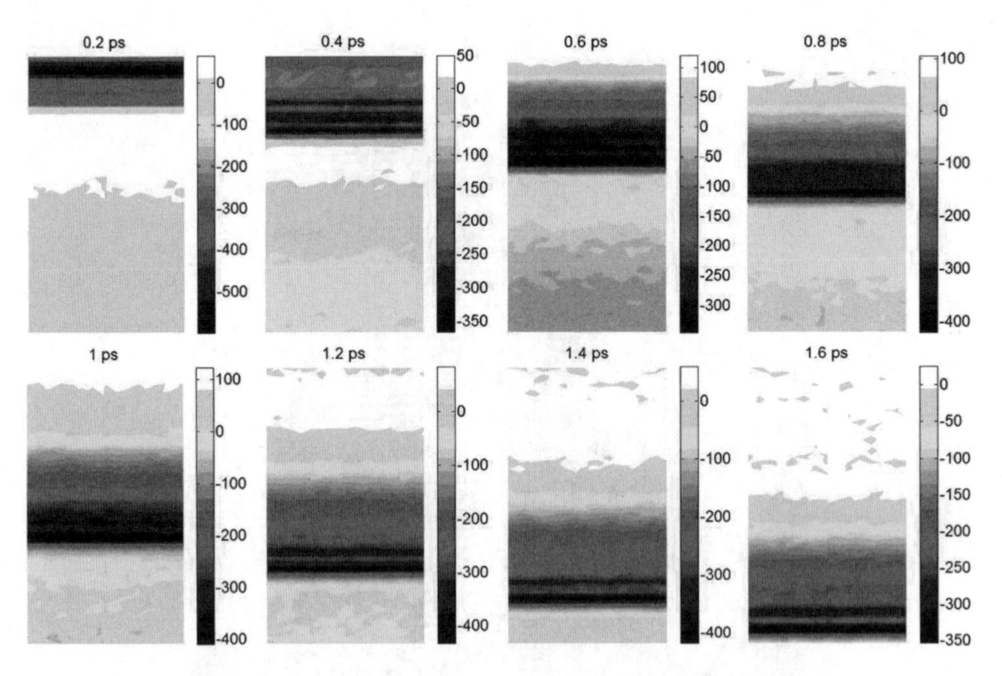


**Figure 4.3** Mechanical velocity (shock velocity in the material) plot for shock impact velocity of 1000 m/s ($-y$-direction)

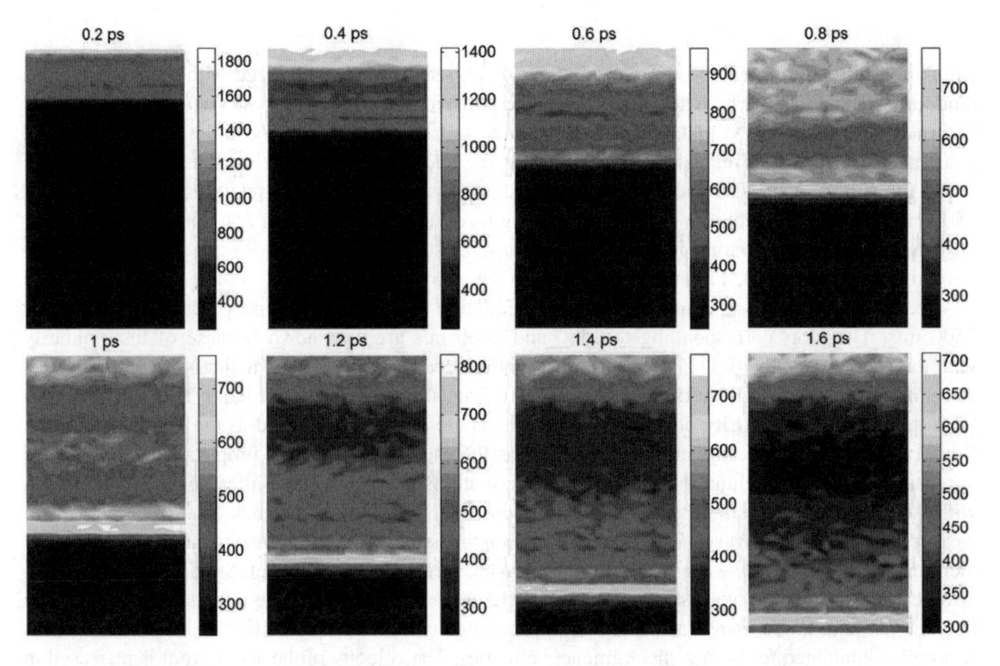


**Figure 4.4** Temperature plot for shock impact velocity of 1000 m/s ($-y$-direction)

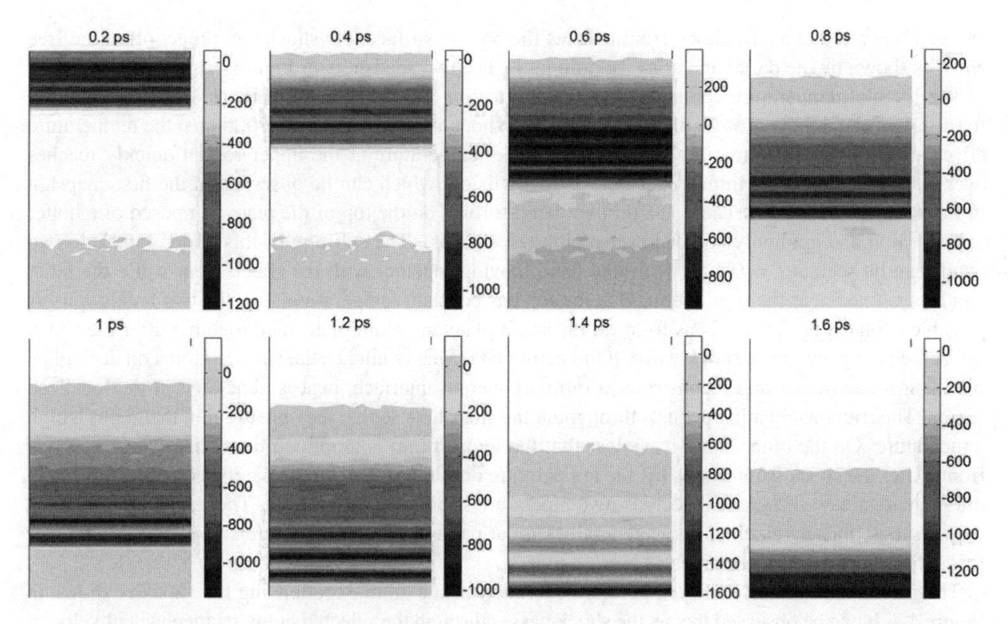


**Figure 4.5** Mechanical velocity plot for shock impact velocity of 2500 m/s ($-y$-direction)

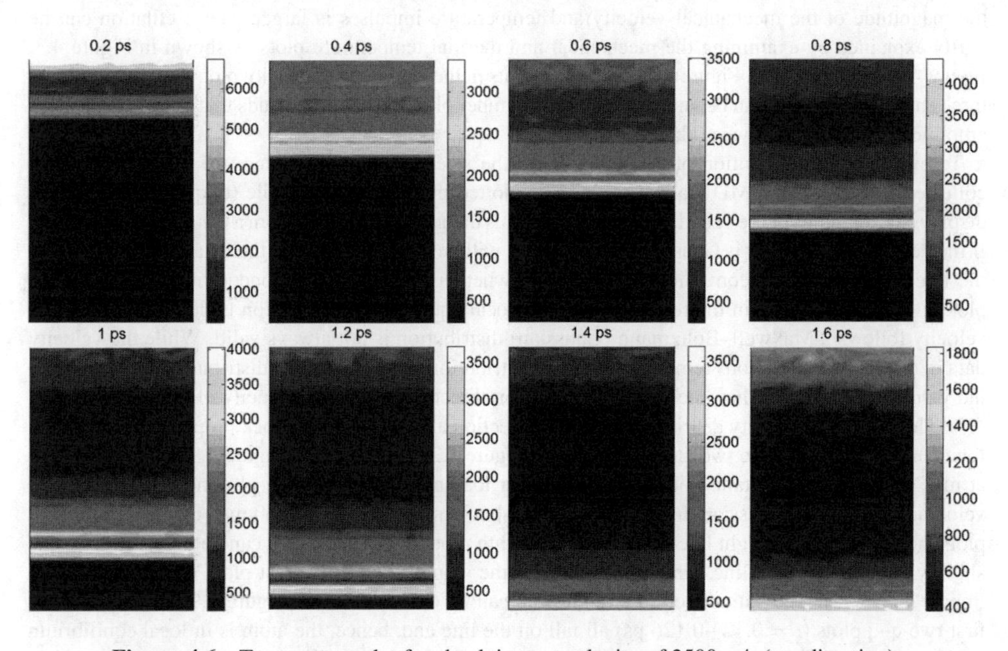


**Figure 4.6** Temperature plot for shock impact velocity of 2500 m/s ($-y$-direction)

can be observed that as the shock front reaches the bottom surface, the shock can reflect off at the free ends, as shown by the dark band at the bottom in the last two snapshots of Figure 4.5.

The thermal temperature computed based on Equation (4.3) is shown in Figure 4.4 and Figure 4.6 for different shock impact velocities. For the lowest shock impact velocity of 1000 m/s, the temperature arises after the upper plate hits the main plate. The temperature at the upper region quickly reaches over 1000 K from 300 K initially set before the collision, which can be observed in the first snapshot in Figure 4.4 ($t = 0.2$ ps). There is a thin hot band formed on the top of the plate composed of a hotter upper region and gradually cooler lower region ($t = 0.2$ ps). The splitting of this wider band into two bands can be seen at $t = 0.4$ ps, with one band moving together with the shock front while the other band stays fixed near the top. The band at the top has constant temperature after the two bands split. A possible explanation for this is as follows: the small plate and main plate stick together after the shock impact, which generates a shock wave in the main plate. Meanwhile, as the surface atoms on each plate rub against each other and reconstruct to form a coherent interface, heat is generated in the localized region. The frictional heating persists throughout the snapshots so that the upper region maintains a high temperature. On the other hand, it is clear that the temperature is amplified dramatically at the shock front. After the shock front passes by, the temperature does not reduce immediately to its original value, thus leaving areas with gradual decreasing temperature behind the shock front. The temperature plots at higher shock impact velocities show generally similar patterns. The temperature amplification is greater for higher shock impact velocities.

The mechanical velocity and temperature evolution of the atom experiencing the shock is shown in Figure 4.7. It can be observed that as the shock passes through the selected atom, its mechanical velocity and temperature drastically change. The mechanical velocity and temperature oscillate a few cycles before they gradually die down to their original values. As observed, the mechanical velocity typically oscillate four or five times and the temperature two or three times. As shock impact velocity increases, the magnitude of the mechanical velocity and temperature impulses is larger. The oscillation can be partly explained by examining the mechanical and thermal temperature plots as shown in Figure 4.3, Figure 4.4, Figure 4.5, and Figure 4.6: the shock fronts in the mechanical velocity plots usually consist of three sharper bands that can be visualized and the number of corresponding bands that show temperature amplification are around two at the shock front.

To examine the distribution of the velocity for the selected atom with time, the velocity statistics collected from the many MD realizations can be plotted in a quantile–quantile (q–q) plot. A q–q plot displays the quantiles of the velocity distribution versus theoretical quantiles from a Gaussian distribution. If the velocity distribution is Gaussian, then the plot will be linear; otherwise, the data will deviate from the line. Therefore, it is convenient to determine whether the atom is in nonequilibrium by the q–q plot. It will be discussed in the following that the local equilibrium assumption under which the atom velocity follows a Maxwell–Boltzmann (Gaussian) distribution is not always valid. While the velocity data in the $x$- and $z$-directions remain a Gaussian distribution, a non-Gaussian distribution is observed in the $y$-direction, which is the shock wave propagation direction at a shock velocity of 2500 m/s (Figure 4.8). Therefore, the velocity distribution in the $y$-direction along which the shock propagates is the main focus of this study. For the two atoms selected in Figure 4.2, the q–q plots (Figure 4.9 and Figure 4.10) at nine different time instants following the impact loading are obtained by plotting the quantiles of velocities against the Gaussian distribution. At the shock impact velocity 1000 m/s, the data in the q–q plots mostly fall on a straight line. In the third to eighth plots of Figure 4.9a, it can be observed that data slightly deviate from the line, which goes back to the straight line in the last plot. With higher shock impact velocity, the deviation from a straight line can be clearly seen. In Figure 4.9b, the data in the first two q–q plots ($t = 0.12-0.126$ ps) all fall on the line and, hence, the atom is in local equilibrium at those times. The data points start deviating from the straight line from the third plot. Some data points in the fourth plot ($t = 0.186$ ps) fall off the line considerably, and hence the atom is in a local nonequilibrium as the initial shock wave just passes through. After that, the atom gradually relaxes back to a local equilibrium state, as shown in the fifth to ninth plots ($t = 0.216-0.42$ ps). This deviation from

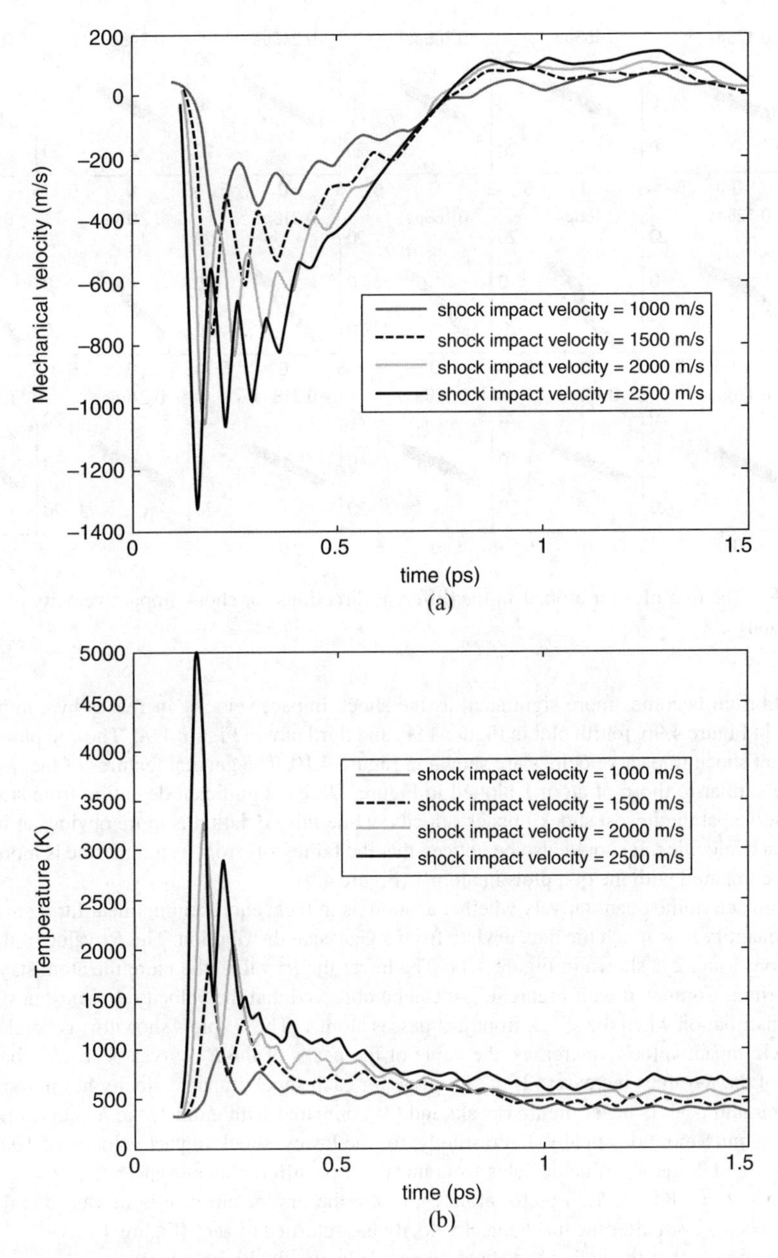


**Figure 4.7** Time evolution of (a) mechanical velocity and (b) temperature of atom 1 for different shock velocities

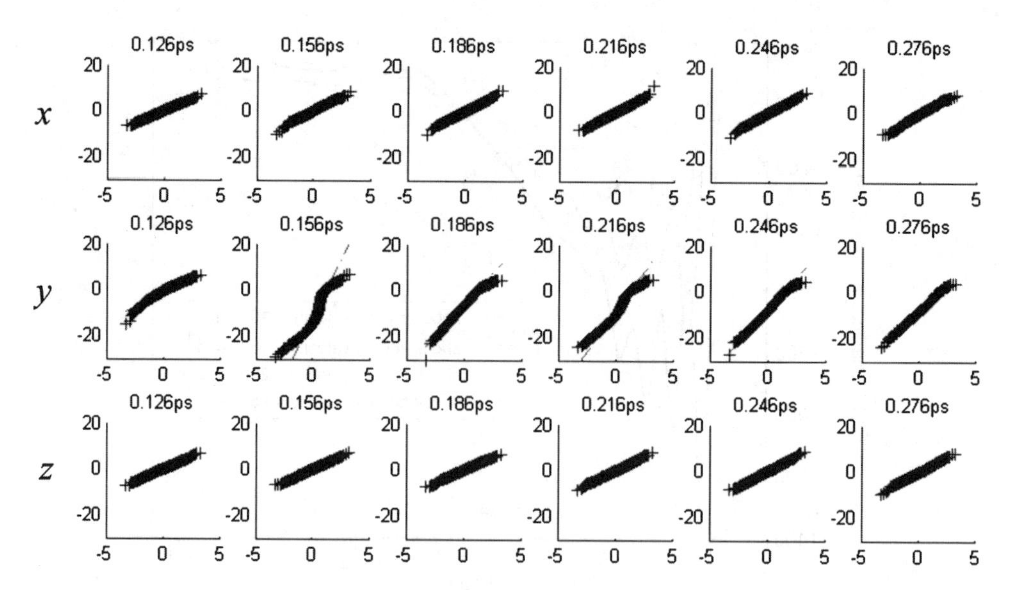


**Figure 4.8** The q–q plot for atom 1 in the different directions for shock impact velocity of 2500 m/s ($-y$-direction)

local equilibrium becomes more significant as the shock impact velocity increases by comparing the fourth plot in Figure 4.9b, fourth plot in Figure 4.9c, and third plot in Figure 4.9d. The q–q plots for atom 2 at different shock impact velocities are shown in Figure 4.10. The general features of the q–q plot for atom 2 are similar to those of atom 1 plotted in Figure 4.9. No significant deviation from straight line can be observed at the lowest shock impact velocity, while this deviation is more obvious at increasing shock impact velocities. It should also be noticed that the falling off from a straight line is more obvious for atom 2 compared with the q–q plots in atom 1 (Figure 4.9).

In order to determine quantitatively whether an atom is in local equilibrium, linear fitting to q–q plots is used to quantify how much the data deviate from a Gaussian distribution. The $R^2$ value of the fit with time for atom 1 and 2 is shown in Figure 4.11. The lower the $R^2$ value, the more the atom stays at local nonequilibrium. Compared with Figure 4.7, it can be observed that the velocity deviates mostly from a Gaussian distribution when the shock front just passes atom 1. The $R^2$ plots show this general trend: (1) as the shock impact velocity increases, the value of the peaks of the $R^2$ curves decreases, but the rate of change of the $R$-values increase; (2) the $R^2$ curve for each shock impact velocity has multiple peaks, but the minimum peak is never the first peak; and (3) compared with atom 1, the $R^2$ curves for atom 2 deviate from unity more drastically. Interestingly, for the lowest shock impact velocity of 1000 m/s for both atoms 1 and 2, the $R^2$ value deviates from unity at two different occurrences (i.e., $t = 0.2$ and 0.5 ps for atom 1; $t = 0.85$ and 1.25 ps for atom 2), where the first occurrence is shorter than the second one, which occurs even after the mechanical velocity has returned to zero (Figure 4.7).

Trend 1 suggests that the higher the shock impact velocity, the higher the degree of nonequilibrium an atom achieves and the higher the surge and relaxation rates are. The competition between these two quantities determines the time duration in which the shocked atom stays in nonequilibrium. To gain more insight, the nonequilibrium time duration for each shock impact velocity is calculated by adding up the time interval at which the $R^2$ value is below the threshold value of 0.99. The threshold value is set based on the observation of the minimum $R^2$ value at local equilibrium state. Table 4.1 lists the time durations calculated for atoms 1 and 2. Note that there is no clear trend for each atom with respect to the shock impact velocity. This is quite surprising, because as the shock impact velocity increases, the

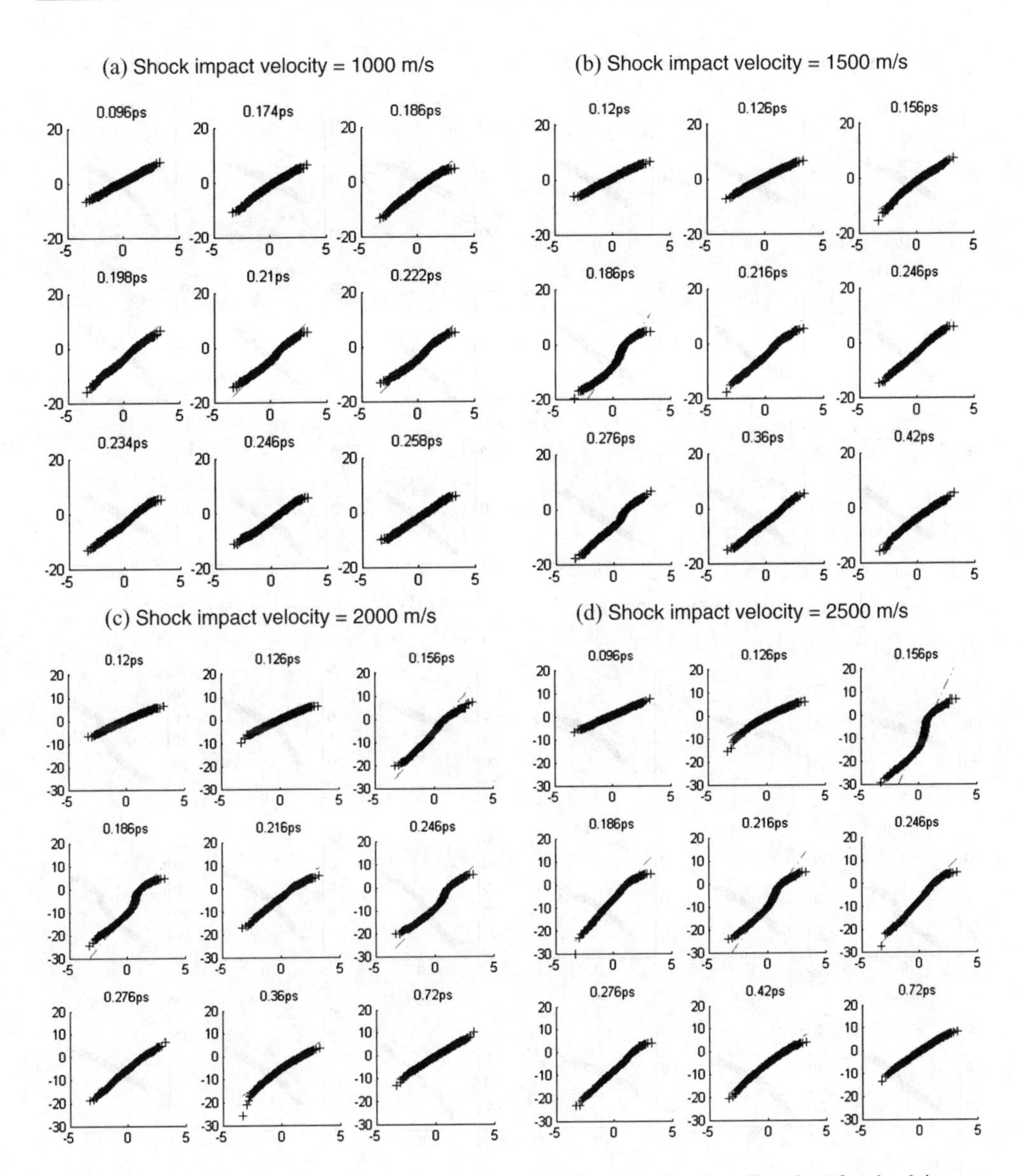


**Figure 4.9** The q–q plot for atom 1 in the shock propagation direction ($-y$-direction) for shock impact velocity of (a) 1000 m/s, (b) 1500 m/s, (c) 2000 m/s, and (d) 2500 m/s

atom affected should stay in nonequilibrium longer because of the stronger perturbation. Apparently, this effect is offset by the higher frequency oscillations of the shock wave as the rebounds assist the atom to relax back to equilibrium faster in the case of higher shock wave velocity. This hypothesis will become more evident when the frequency analyses of the mechanical and thermal velocities are presented.

Regarding Trend 2, it is somewhat surprising that the first peak of any one of the $R^2$ curves is not the lowest peak (see Figure 4.11), because the first peak of any shock wave is always the maximum peak, as shown in the mechanical velocity plot in Figure 4.7. This observation implies that the phonons perturb

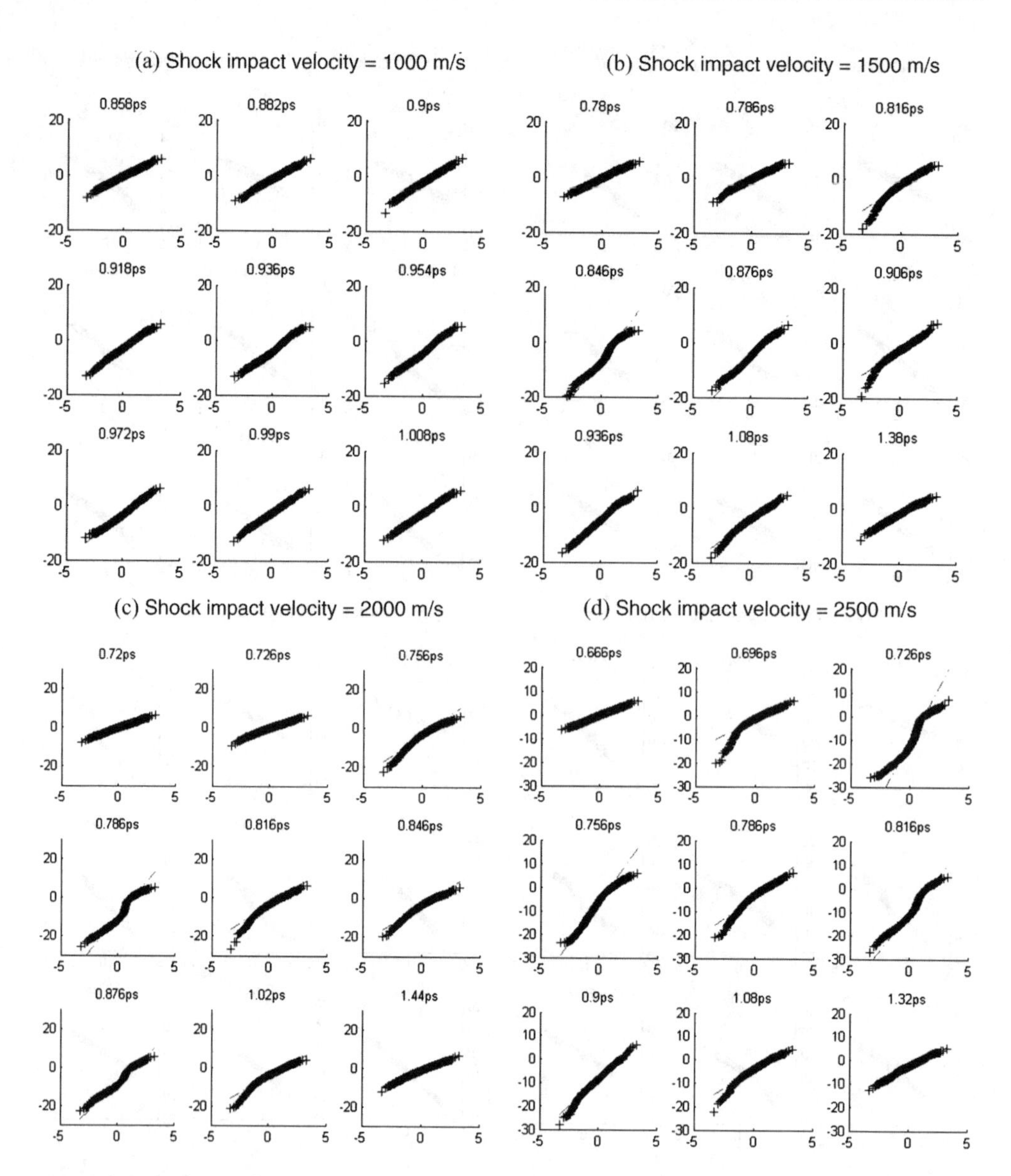


**Figure 4.10** The q–q plot for atom 2 in the shock propagation direction ($-y$-direction) for different shock impact velocity of (a) 1000 m/s, (b) 1500 m/s, (c) 2000 m/s, and (d) 2500 m/s, © Guan *et al.*, 2011

by the first peak of the shock have not completely relaxed back to equilibrium before the second shock peak arrives and perturbs additional phonons away from equilibrium. The degree of nonequilibrium eventually reaches a maximum value as the subsequent weaker shock peaks pass by and the respective magnitudes of perturbation and relaxation balance out.

The observation of Trend 3 that atom 2 spends a much longer time in a local nonequilibrium state compared with atom 1 is also confirmed by the quantitative analysis of the $R^2$ values in Table 4.1. It may

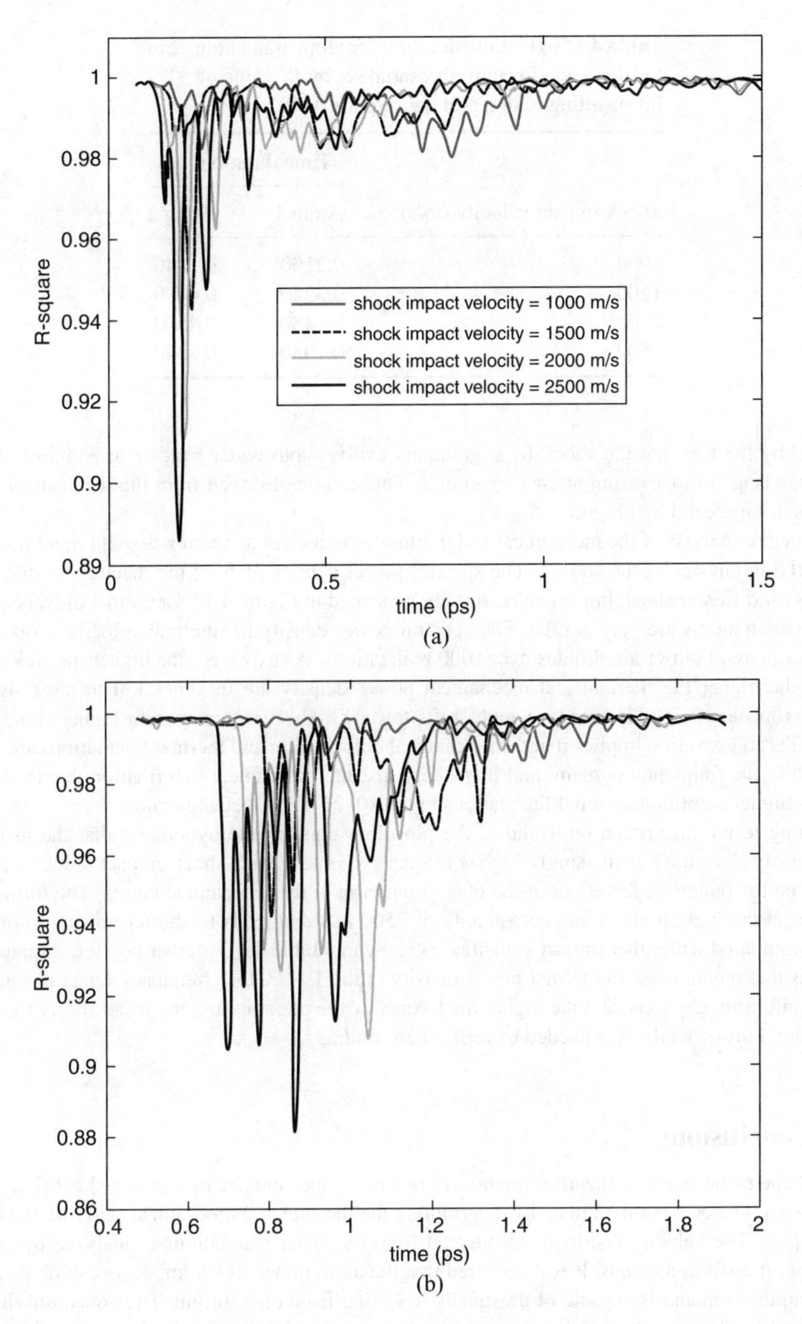


**Figure 4.11** Goodness of linear fitting to the q–q plot data for (a) atom 1 and (b) atom 2, © Guan *et al.*, 2011

**Table 4.1** The time duration for atom 1 and atom 2 in local nonequilibrium (threshold set by $R^2$ value of linear fitting to q–q plot data equals 0.99)

| | Time duration (ps) | |
|---|---|---|
| Shock impact velocity (m/s) | Atom 1 | Atom 2 |
| 1000 | 0.2100 | 0.1140 |
| 1500 | 0.3240 | 0.4080 |
| 2000 | 0.3480 | 0.4500 |
| 2500 | 0.1980 | 0.4440 |

be caused by the fact that the shock front gradually evolves into wider bands consisting of separated sharper bands as it moves from atom 1 to atom 2. This can be observed from the mechanical velocity snapshots in Figure 4.3 and Figure 4.5.

A frequency analysis of the mechanical and thermal velocities of an atom will yield more insight into the underlying physics of the shocks. The spectral power density plots of the mechanical and thermal velocities for different shock impact velocities are presented in Figure 4.12 for atom 1 only, because the trends for both atoms are very similar. The spectral power density for thermal velocity is obtained by the mean-squared Fourier amplitudes over 1000 realizations. As expected, the higher the shock impact velocity, the higher the thermal and mechanical power density are in general. The most significant increase is in the 10–23 THz range, which shifts toward higher frequency with higher shock impact velocity. This observation implies that the mechanical deformation and thermal fluctuations are strongly correlated in the frequency domain, and it is likely that the mechanical deformation due to the shock generates higher amplitude thermal fluctuations in the 10–23 THz frequency range.

Returning to the discussion on Trend 1, the plot does confirm the hypothesis that the mechanical power density does have increasingly higher frequency content with shock impact velocity and may have helped to counteract the effect of the overall increase in the mechanical power. The former effect dominates at the highest shock impact velocity of 2500 m/s and leads to shorter nonequilibrium time duration compared with other impact velocities as shown in Table 4.1. Another possible explanation of Trend 1 is that the increase in thermal power density in the 10–23 THz frequency range is contributed by nonequilibrium phonons, and the higher the frequency the phonons are, the faster they will return to equilibrium. Further analysis is needed to verify these findings.

## 4.4 Conclusions

In this chapter, the many-realization method is reviewed and applied to analyze shock impact with different shock velocities on Ni plate by determining the thermal and mechanical states of atoms under shock impact. The velocity distribution collected from the many realizations is analyzed by q–q plots and Fourier transform analysis. It is discovered that the atom under shock impact does not always stay in local equilibrium and is capable of drastically deviating from equilibrium. The nonequilibrium state is quantified by the $R^2$ value via linear fitting to the q–q plot. Increasing shock impact velocities can drive the atom to deviate further from the local equilibrium state. In this study, the typical time that the atom stays away from local equilibrium state is less than 1 ps. The duration that a shocked atom stays in nonequilibrium depends both on the shock impact velocity and the frequency content of the shock waveform, but the underlying mechanisms cannot be clearly identified. Future work will be devoted to investigate the velocity distribution in the nonequilibrium state.

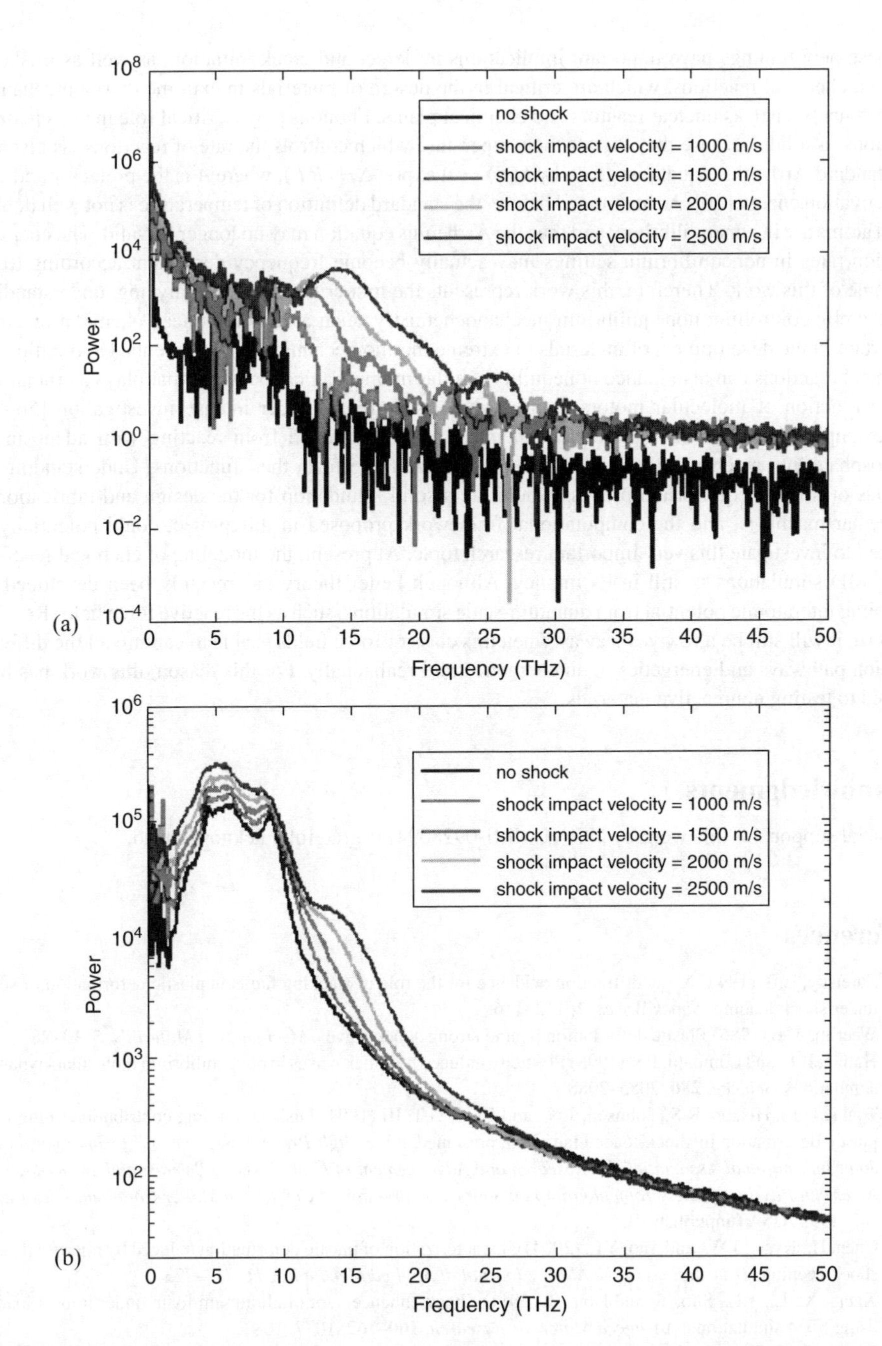


**Figure 4.12** Fast-Fourier transform of the (a) mechanical and (b) thermal velocities for atom 1 ($t$ = 0.066–12 ps at a time interval of 0.006 ps)

These new findings have important implications in defect and crack initiation, as well as in shock-induced chemical reactions, which are critical in the design of materials in extreme thermomechanical environments, such as nuclear reactors and chemical plants. Phonons play a critical role in any chemical reactions in solids because they dictate the temperature which controls the rate of reactions, as given by the standard Arrhenius equations given by $k(T) = A\exp(-\Delta H/RT)$, where $A$ is the prefactor and $\Delta H$ the activation energy [30]. As mentioned before, the standard definition of temperature is not well defined when atoms are in nonequilibrium, and thus the Arrhenius equation may no longer be valid. The chemical reaction rates in nonequilibrium settings may actually become frequency dependent according to the outcome of this work. Therefore, this work represents the first critical step in analyzing, understanding, and thereby controlling nonequilibrium mechanochemistry, such as shock-induced chemical reactions important in the development of materials in extreme thermomechanical environments. On the flip side, chemical reactions can also induce nonequilibrium thermomechanical behavior that plays a crucial role in the function of molecular motors in life, which is currently under intense investigation [56–58]. For example, molecular motors convert chemical energy obtained from reacting with adenosine-5-triphosphate into mechanical energy, which they utilize to perform their functions. Understanding the physics of this nonequilibrium process would lay a solid foundation for the design and fabrication of molecular machines, and the computational framework proposed in this project could potentially be utilized to investigate this very important research topic. At present, the modeling of chemical reactions using MD simulations is still in its infancy. Although better theory has recently been developed for modeling interatomic potential from quantum-scale simulations, such as the reactive force field (ReaxFF) [59, 60], it will still be a few years away when this class of force fields is able to capture all the different reaction pathways and energetics in an energetic solid realistically. For this reason, this work has been limited to testing nonreactive materials.

## Acknowledgments

Financial support for this work by NSF (CMMI-0928094) is gratefully acknowledged.

## References

[1] Zaretsky, E.B. (1992) X-ray diffraction evidence for the role of stacking faults in plastic deformation of solids under shock loading. *Shock Waves*, **2**, 113–116.

[2] Weertman, J. (1986) Plastic deformation behind strong shock waves. *Mechanics of Materials*, **5**, 13–28.

[3] Holian, B.L. and Lomdahl, P.S. (1998) Plasticity induced by shock waves in nonequilibrium molecular-dynamics simulations. *Science*, **280**, 2085–2088.

[4] Tonks, D.L., Hixson, R.S., Johnson, J.N., and Gray, G.T. III (1994) Dislocation-drag contribution to high-rate plastic deformation in shock-loaded tantalum, presented at the *High-Pressure Science and Technology – 1993. Joint International Association for Research and Advancement of High Pressure Science and Technology and American Physical Society Topical Group on Shock Compression of Condensed Matter Conference*, 28 June–2 July 1993, USA (unpublished).

[5] Chen, H., Kysar, J.W., and Yao, Y.L. (2004) Characterization of plastic deformation induced by microscale laser shock peening. *Transactions of the ASME, Journal of Applied Mechanics*, **71**, 713–723.

[6] Xiang, X., Li, Y.L., Suo, T., and Hou, B. (2011) Shock enhancement of aluminum foam under impact loading using FEM simulations. *Advanced Materials Research*, **160–162**, 1077–1082.

[7] He, H., Sekine, T., Kobayashi, T., and Kimoto, K. (2001) Phase transformation of germanium nitride ($Ge_3N_4$) under shock wave compression. *Journal of Applied Physics*, **90**, 4403–4406.

[8] Manaa, M.R. (2003) Shear-induced metallization of triamino-trinitrobenzene crystals. *Applied Physics Letters*, **83**, 1352–1354.

[9] Cerreta, E., Gray, G.T., Lawson, A.C. *et al.* (2006) The influence of oxygen content on the phase transformation and shock hardening of titanium. *Journal of Applied Physics*, **100**, 13530.

[10] Li, J., Zhou, X., Zhu, W. *et al.* (2007) A shock-induced phase transformation in a $LiTaO_3$ crystal. *Journal of Applied Physics*, **102**, 083503.
[11] Sharma, J., Armstrong, R.W., Elban, W.L. *et al.* (2001) Nanofractography of shocked RDX explosive crystals with atomic force microscopy. *Applied Physics Letters*, **78**, 457–459.
[12] Strachan, A., van Duin, A.C.T., Chakraborty, D. *et al.* (2003) Shock waves in high-energy materials: the initial chemical events in nitramine RDX. *Physical Review Letters*, **91**, 098301.
[13] Zybin, S.V., Xu, P., An, Q., and Goddard, W.A. (2011) ReaxFF reactive molecular dynamics: coupling mechanical impact to chemical initiation in energetic materials, presented at the *2010 DoD High Performance Computing Modernization Program Users Group Conference, HPCMP UGC 2010*, June 14, 2010–June 17, 2010, Schaumburg, IL, USA (unpublished).
[14] Tokmakoff, A., Fayer, M.D., and Dlott, D.D. (1993) Chemical reaction initiation and hot-spot formation in shocked energetic molecular materials. *Journal of Physical Chemistry*, **97**, 1901–1913.
[15] Nomura, K.-i., Kalia, R.K., Nakano, A. *et al.* (2007) Dynamic transition in the structure of an energetic crystal during chemical reactions at shock front prior to detonation. *Physical Review Letters*, **99**, 148303.
[16] Frenkel, D. and Smit, B. (2001) *Understanding Molecular Simulation: From Algorithms to Applications*, vol. 1 of *Computational Science Series*, vol. 1, Academic Press, New York, NY.
[17] Haile, J.M. (1992) *Molecular Dynamics Simulation Elementary Methods*, John Wiley & Sons, Inc., New York, NY.
[18] Rapaport, D.C. (2004) *The Art of Molecular Dynamics Simulation*, Cambridge University Press, London.
[19] Holian, B.L. (1988) Modeling shock-wave deformation via molecular dynamics. *Physical Review A*, **37**, 2562–2568.
[20] Brenner, D.W., Robertson, D.H., Elert, M.L., and White, C.T. (1993) Detonations at nanometer resolution using molecular dynamics. *Physical Review Letters*, **70**, 2174–2177.
[21] Bickham, S.R., Kress, J.D., and Collins, L.A. (2000) Molecular dynamics simulations of shocked benzene. *Journal of Chemical Physics*, **112**, 9695–9698.
[22] Gibson, J.B., Goland, A.N., Milgram, M., and Vineyard, G.H. (1960) Dynamics of radiation damage. *Physical Review*, **120**, 1229–1253.
[23] Maillet, J.B., Mareschal, M., Soulard, L. *et al.* (2000) Uniaxial Hugoniostat: a method for atomistic simulations of shocked materials. *Physical Review E*, **63**, 016121.
[24] Kitaigorodskii, A.I. (1973) *Molecular Crystals and Molecules*, Academic Press, New York, NY.
[25] Califano, S., Schettino, V., and Neto, N. (1981) *Lattice Dynamics of Molecular Crystals*, Springer-Verlag, Berlin.
[26] Pertsin, A.J. and Kitaigorodsky, A.I. (1987) *The Atom–Atom Potential Method: Applications to Organic Molecular Solids*, Springer-Verlag, Berlin.
[27] Kohler, J. and Meyer, R. (1993) *Explosives*, 4th edn, VCH, New York, NY.
[28] Coffey, C.S. and Toton, E.T. (1982) A microscopic theory of compressive wave-induced reactions in solid explosives. *Journal of Chemical Physics*, **76**, 949–954.
[29] Trevino, S.F. and Tsai, D.H. (1984) Molecular dynamical studies of the dissociation of a diatomic molecular crystal. II. Equilibrium kinetics. *Journal of Chemical Physics*, **81**, 248–256.
[30] Dlott, D.D. and Fayer, M.D. (1990) Shocked molecular solids: vibrational up pumping, defect hot spot formation, and the onset of chemistry. *Journal of Chemical Physics*, **92**, 3798–3812.
[31] Zerilli, F.J. and Toton, E.T. (1984) Shock-induced molecular excitation in solids. *Physical Review B*, **29**, 5891–5902.
[32] Walker, F.E. (1988) Physical kinetics. *Journal of Applied Physics*, **63**, 5548–5554.
[33] Wei, T.-G. and Wyatt, R.E. (1990) Semiclassical dynamics of shock wave propagation in molecular crystals: application to the morse lattice. *Journal of Physics: Condensed Matter*, **2**, 9787–9795.
[34] Menikoff, R. (2011) Hot spot formation from shock reflections. *Shock Waves*, **21**, 141–148.
[35] Fu, Y., Kirca, M., and To, A.C. (2011) On determining the thermal state of individual atoms in molecular dynamics simulations of nonequilibrium processes in solids. *Chemical Physics Letters*, **506**, 290–297.
[36] Tarver, C.M., Chidester, S.K., and Nichols, A.L. (1996) Critical conditions for impact- and shock-induced hot spots in solid explosives. *Journal of Physical Chemistry*, **100**, 5794–5799.
[37] Qi, A., Zybin, S.V., Goddard, W.A. III *et al.* (2011) Elucidation of the dynamics for hot-spot initiation at nonuniform interfaces of highly shocked materials. *Physical Review B (Condensed Matter and Materials Physics)*, **84**, 220101.

[38] Huang, K. (1987) *Statistical Mechanics*, John Wiley & Sons, Inc., New York, NY.
[39] Clausius, R.J.E. (1870) On a mechanical theorem applicable to heat. *Philosophical Magazine*, **40**, 122–127.
[40] Maxwell, J.C. (1870) On reciprocal figures, frames and diagrams of forces. *Transactions of the Royal Society of Edinburgh*, **XXVI**, 1–40.
[41] Maxwell, J.C. (1874) Van der Waals on the continuity of the gaseous and liquid states. *Nature*, **10** (259), 477–480.
[42] Zhou, M. (2005) Thermomechanical continuum representation of atomistic deformation at arbitrary size scales. *Proceedings of the Royal Society A: Mathematical, Physical and Engineering Science*, **461**, 3437–3472.
[43] Joshi, A.A. and Majumdar, A. (1993) Transient ballistic and diffusive phonon heat transport in thin films. *Journal of Applied Physics*, **74**, 31–39.
[44] Chen, G. (1998) Thermal conductivity and ballistic-phonon transport in the cross-plane direction of superlattices. *Physical Review B*, **57**, 14958–14973.
[45] Chen, G. (1996) Nonlocal and nonequilibrium heat conduction in the vicinity of nanoparticles. *Journal of Heat Transfer*, **118**, 539–546.
[46] Dlott, D.D. (1999) Ultrafast spectroscopy of shock waves in molecular materials. *Annual Review of Physical Chemistry*, **50**, 251–278.
[47] Wen, X., Tolbert, W.A., and Dlott, D.D. (1993) Ultrafast temperature jump in polymers: phonons and vibrations heat up at different rates. *Journal of Chemical Physics*, **99**, 4140–4151.
[48] Chen, G. (2000) Particularities of heat conduction in nanostructures. *Journal of Nanoparticle Research*, **2**, 199–204.
[49] Zimmerman, J.A., Webb, E.B. III, Hoyt, J.J. *et al.* (2004) Calculation of stress in atomistic simulation. *Modelling and Simulation in Materials Science and Engineering*, **12**, S319–S332.
[50] Zimmerman, J.A., Jones, R.E., and Templeton, J.A. (2010) A material frame approach for evaluating continuum variables in atomistic simulations. *Journal of Computational Physics*, **229**, 2364–2389.
[51] Foiles, S.M., Baskes, M.I., and Daw, M.S. (1986) Embedded-atom-method functions for the fcc metals Cu, Ag, Au, Ni, Pd, Pt, and their alloys. *Physical Review B*, **33**, 7983–7991.
[52] Daw, M.S., Foiles, S.M., and Baskes, M.I. (1993) The embedded-atom method: a review of theory and applications. *Materials Science Reports*, **9**, 251–310.
[53] Daw, M.S. and Baskes, M.I. (1983) Semiempirical, quantum mechanical calculation of hydrogen embrittlement in metals. *Physical Review Letters*, **50**, 1285–1288.
[54] Daw, M.S. and Baskes, M.I. (1984) Embedded-atom method: derivation and application to impurities, surfaces, and other defects in metals. *Physical Review B*, **29**, 6443–6453.
[55] Foiles, S.M. (1985) Calculation of the surface segregation of Ni–Cu Alloys with the use of the embedded-atom method. *Physical Review B*, **32**, 7685–7693.
[56] Bustamante, C., Chemla, Y.R., Forde, N.R., and Izhaky, D. (2004) Mechanical processes in biochemistry. *Annual Review of Biochemistry*, **73**, 705–748.
[57] Spudich, J.A. (2011) Molecular motors, beauty in complexity. *Science*, **331**, 1143–1144.
[58] Fletcher, S.P., Dumur, F., Pollard, M.M., and Feringa, B.L. (2005) A reversible, unidirectional molecular rotary motor driven by chemical energy. *Science*, **310**, 80–82.
[59] Van Duin, A.C.T., Dasgupta, S., Lorant, F., and Goddard, W.A. (2001) ReaxFF: a reactive force field for hydrocarbons. *Journal of Physical Chemistry A*, **105**, 9396–9409.
[60] Van Duin, A.C.T., Strachan, A., Stewman, S. *et al.* (2003) ReaxFFSiO reactive force field for silicon and silicon oxide systems. *Journal of Physical Chemistry A*, **107**, 3803–3811.

# 5

# Multiscale, Multiphysics Modeling of Electromechanical Coupling in Surface-Dominated Nanostructures

Harold S. Park[a] and Michel Devel[b]

[a]*Department of Mechanical Engineering, Boston University, USA*

[b]*FEMTO-ST Institute, Université de Franche-Comté, France*

## 5.1 Introduction

Developing a fundamental understanding of how nanomaterials deform and respond mechanically to externally applied electromagnetic fields will be critical to advancing various aspects of nanoscale science and engineering. For example, recent research has shown that carbon nanotubes (CNTs) exhibit giant electrostriction [1], in which the nanotubes show significant electromechanical energy conversion potential by undergoing extremely large deformations in response to an applied electric field. Furthermore, many nanoelectromechanical systems (NEMS), which are used for a wide range of applications, including force [2], displacement [3, 4], and mass [5, 6] sensing, are actuated by means of external electromagnetic fields [6, 7]. Finally, electric fields are also used by experimentalists to grow and align nanostructures [8], to drive nanostructures to resonance to measure their size-dependent elastic properties [9, 10], and also to deform nanostructures in order to examine the instability and failure mechanisms that occur with increasing deformation [1, 11–13].

The multiphysics computational modeling of electromechanical phenomena has focused in recent years on developing techniques to study microelectromechanical systems (MEMS). A common approach to this problem has been to utilize an electrostatic analysis to compute the electrostatic forces, while a mechanical analysis is required to compute the deformation of the structure in response to the applied electrostatic forces. The semi-Lagrangian approach is a common technique for solving the coupled electromechanical system of equations in which the mechanical analysis is performed using the finite-element method (FEM) in the undeformed configuration using a Lagrangian description [14–16], while the boundary-element method (BEM) is typically used to perform the electrostatic analysis on the deformed geometry. Self-consistency between the mechanical and electrostatic simulations is achieved using relaxation

*Multiscale Simulations and Mechanics of Biological Materials*, First Edition. Edited by Shaofan Li and Dong Qian.

techniques [17]. However, these approaches also have certain deficiencies; these include the fact that the geometry of the structure must be remeshed before an electrostatic analysis is performed [18]; the BEM interpolation functions must be recalculated during each relaxation iteration [19]; convergence rates for the coupled analysis using the relaxation scheme can falter when deformations of the mechanical domain are large [19]. These issues can be alleviated using the recently developed full Lagrangian approach of Aluru and co-workers and others [17, 19–21]; however, this approach still requires the implicit and coupled solution of both the mechanical and electrostatic equations, as well as the calculation of the electromechanical coupling matrices, leading to considerable computational expense, particularly for three-dimensional (3D) problems. We note that other approaches, including Lagrange multiplier-based staggered techniques for solving the coupled electromechanical system of equations [22, 23], also require the solution of both the mechanical and electrostatic governing equations in addition to electromechanical coupling terms.

Moreover, there are additional challenges that must be overcome to extend the previously discussed computational electromechanical techniques for MEMS down to the nanoscale; these challenges pertain to capturing the appropriate surface physics that occur in either the mechanical or electrostatic domains. For example, owing to nanoscale surface effects [24] that arise from the undercoordinated nature of atomic bonding at surfaces, the mechanical properties of nanostructures have been shown both experimentally [9, 25–28] and theoretically [29–33] to deviate from the expected bulk values, particularly when the characteristic size of the nanostructure decreases below about 50 nm [25].

In addition to the surface effects on the mechanical properties of nanostructures, atoms that lie at surfaces respond, and in particular polarize differently due to applied electric fields compared with atoms that lie within the material bulk. While this has not been studied extensively in the literature, recent atomistic calculations have found that surface atoms for various materials exhibit a significantly different polarizability in response to applied electric fields [34–37] compared with bulk atoms.

Among Professor Wing Kam Liu's many important contributions to the field of computational solid mechanics was his work on the bridging-scale method. In that pioneering work [38], a concurrent multiscale formulation that coupled atomistics and continua was presented, and the method was later applied to dynamic lattice fracture in two [39] and three dimensions [40]. A key foundation of that work was in developing a coarse-scale, continuum constitutive model based on the underlying interatomic potential through the application of the Cauchy–Born model [41, 42].

Recently, Park and co-workers have further developed the Cauchy–Born model to account for the fact that the surfaces of nanomaterials have a different bonding environment, and thus mechanical properties, than do atoms that lie within the material bulk. Because the classical Cauchy–Born hypothesis views the crystal as infinite and repetitive, it does not admit nanoscale surface effects. Because of this, Park and co-workers have recently developed the surface Cauchy–Born (SCB) model [43–45], which extends the bulk Cauchy–Born (BCB) model such that the system total energy includes contributions from not only the bulk, but also the surface atoms. In doing so, they were able to develop multiscale, finite-element-based computational models that enabled the 3D solution of nanomechanical boundary-value problems while accounting for nanoscale surface stress and surface elastic effects within a nonlinear, finite deformation framework. The SCB model was previously developed for both face-centered cubic (FCC) metals [43, 44] and non-centrosymmetric diamond cubic lattices such as silicon [45], and applied to nanomechanics and NEMS-related problems ranging from nanoscale resonant mass sensing [46], thermomechanical coupling [47], resonant frequencies, and elucidating the importance of nonlinear, finite deformation kinematics on the resonant frequencies of both FCC metal [33] and silicon nanowires [48, 49], strain sensing [50], bending of FCC metal [51] and silicon nanowires [52] and length/time-scale bridging for dynamic multiscale simulations [53].

In the present work, there are two major objectives with regard to further developing the SCB model to study multiphysics, electromechanical coupling in surface-dominated nanomaterials, as summarized in the work of Park *et al.* [54]. First, we wish to significantly reduce the computational expense of analyzing coupled electromechanical problems by solving only one governing equation, rather than equations for both the mechanical and electrostatic domains. Second, we wish to incorporate the appropriate surface-driven mechanical and electrostatic physical phenomena into our computational model. To accomplish the first task, we follow the recent developments of Wang and co-workers [55, 56], who, within a fully atomistic context, used an atomistic total energy that combined both mechanical and electrostatic energies. In their approach, the mechanical potential energy was obtained from the adaptive intermolecular reactive bond order (AIREBO) interatomic potential for carbon [57], while the electrostatic potential energy in response to an external electrostatic field was obtained from either the Gaussian dipole [58] (for semiconducting nanostructures) or Gaussian charge-dipole [59, 60] (for metallic nanostructures) approaches. In these papers [55, 56], the self-consistent calculation of the effective charges, dipoles, energies, and electrostatic forces on every atom at each iteration was made computationally tractable through analytic expressions that involved the inversion of a single matrix, thus eliminating the need for numerical derivations to compute the forces. Indeed, this allowed them to find equilibrium positions by minimizing the total atomistic potential energy fully self-consistently with a single total energy functional.

The second task is accomplished through proper extensions of the SCB model. The major result of this article is then a multiscale, finite deformation FEM-based methodology that enables us to study how surface-dominated nanostructures respond to externally applied electric fields. For the sake of clarity we restrict this work to one-dimensional (1D) systems. We therefore begin by developing the total atomistic potential energy for 1D systems, then arrive at a single governing electromechanical finite element equation that captures external electric field effects on nanostructures. We further derive from the Gaussian dipole model (GDM) of electrostatics, using Cauchy–Born kinematics, the bulk and surface electrostatic Piola–Kirchoff stresses that are required for the coupled electromechanical finite element governing equations. Our numerical examples in one dimension validate both the bulk electrostatic stress and also the surface electrostatic stress by comparison to fully coupled electromechanical atomistic simulations.

## 5.2 Atomistic Electromechanical Potential Energy

We briefly describe in this section the atomistic electromechanical potential energy that was previously developed by Wang and co-workers [55, 56] to study the effect of an external electrostatic field on the deformation of semiconducting and metallic CNTs. Specifically, they wrote the total energy of the nanostructure as the sum of the mechanical and electrostatic energies as

$$U^{\text{total}}(r_{ij}) = \sum_{i=1}^{N} U_i^{\text{elec}}(r_{ij}) + \sum_{i=1}^{N} U_i^{\text{mech}}(r_{ij}), \tag{5.1}$$

where $N$ is the total number of atoms in the system and $r_{ij}$ is the distance between two atoms $i$ and $j$. In their approach, the mechanical potential energy $U^{\text{mech}}$ was obtained using a standard interatomic potential (i.e., the AIREBO potential for carbon [57]), while the supplementary electrostatic potential energy $U^{\text{elec}}$ due to an externally applied electric field was obtained using either the Gaussian dipole-only model for semiconducting CNTs [55] or the Gaussian charge–dipole model for metallic CNTs [56], which we will discuss in further detail in Section 5.2.1. It is worth noting that there are no restrictions on the choice of the mechanical interatomic potential as long as it accurately represents the behavior of the material to be studied in the absence of an externally applied electric field. The minimum energy configuration (i.e., the mechanical deformation) of the nanotubes in the presence of the externally applied electric field was then obtained by direct minimization of the total energy (5.1).

Because the calculation of mechanical interatomic forces is routinely done in the literature, we will not overview that here, and instead refer the reader to classic texts on molecular simulation [61, 62]. However, the analytical calculation of electrostatic forces from the GDM has been much less publicized. We will therefore discuss it in the next chapter, emphasizing the aspects of computational difficulty and expense that motivate the present work.

### 5.2.1 Atomistic Electrostatic Potential Energy: Gaussian Dipole Method

#### Background

In this section we describe the theory underlying the calculation of the atomistic electrostatic potential energy arising in response to an external field via the GDM, and provide motivation for why it is critical to develop computationally efficient, multiscale techniques for calculating the electrostatic stress. The GDM can be viewed as an extension of the point dipole interaction (PDI) model, which was originally developed as a semi-phenomenological model to describe the interaction of matter at the atomistic scale with an external electric field [63–66]. This model simply states that the response of a dielectric to an externally applied field can be described to first order by the fact that the nuclei and electrons are attracted in opposite directions. Hence, the macroscopic polarization of the material is accounted for by the creation of elementary dipoles on every atom. The total field on a given atom is then computed self-consistently by stating that it is the sum of the external field plus the fields created by the other dipoles which are themselves created by the total fields at their positions. Furthermore, the (mesoscopic) discretization of the volume integrals occurring in electromagnetic waves scattering by finite objects leads to the same kind of equations, so that this method has been used in many branches of science with so many variants bearing different names that we can give only examples here: astrophysics [67, 68], where it is known as the discrete dipole approximation (DDA); local probe microscopies, where it is known as the generalized field susceptibility technique [69, 70] or electric field propagators (EFP) technique [71] or Green's dyadic function technique, as in other domains of electromagnetism [72], biophysics, and organic molecular simulations [73–77].

The fact that the PDI/DDA model has been successful across this range of length scales, from atoms to interstellar dust grains, indicates its robustness and physical correctness. It is, however, only a special case of a more general multipolar approach [69, 78, 79]. Indeed, for metallic materials or organic molecules with delocalized electrons coming from conjugated double bounds, a charge + dipole model is better suited [59, 60, 80, 81]. However, in all of these models, there are numerical difficulties coming from the self-energy of the atoms and the fact that, when atoms are too closely bound, the point charge or dipole approximation is not a good approximation for nearest neighbors, leading to so-called "polarization catastrophes." Several techniques have been developed to avoid these divergences [58–60, 81–86], which allowed one to consider metallic and semiconducting nanostructures, including CNTs [13, 60, 87–89] and small metallic [35, 37] and silicon nanoclusters [34]. Among them, the GDM simply states that the interaction of the atoms with an external field does not create point dipoles but Gaussian dipoles (i.e., dipoles created from the shift of Gaussian distributions of charges). This results in interaction tensors between dipoles which are the convolution of the classical gradients of Green's function for Poisson's equation by one [58] or two [59] Gauss normalized distributions. This is, in fact, very similar to the Gaussian orbitals used in quantum chemistry and to the spatial part of the Ewald summation method for dipoles [90], and requires only knowledge of the deformed atomic positions to calculate the induced dipole moments $p_i$ on each atom due to an external electric field $E_0$.

#### Computation of the Dipoles, Energies and Forces

In one dimension, according to the GDM, the electrostatic potential energy $U^{\text{elec}}$ corresponding to the response of the system to an external field can be written as, for a system of $N$ atoms that has zero net charge (i.e., a semiconductor),

$$U^{\text{elec}} = \left( -\frac{1}{2} \sum_{i=1}^{N} \sum_{j=1}^{N} p_i T^{(2)}(r_{ij}) p_j \right) - \sum_{i=1}^{N} p_i E_0(r_i), \tag{5.2}$$

where $E_0$ is the externally applied electric field (in units of V/Å), $p_i$ is the dipole moment for atom $i$ (homogeneous to a charge multiplied by a distance), $\epsilon_0$ is the permittivity of free space, and where $T^{(2)}$ is the Gaussian dipole–Gaussian dipole interaction tensor. The 1D vacuum dipole–dipole interaction tensor $T_0^{(2)}$ is obtained from Equation (A3) of Langlet *et al.* [58] or Equation (3) of Mayer [60] (with $a_{i,j} = \sqrt{R_i^2 + R_j^2} = \sqrt{2}R$ if all atoms are of the same chemical nature):

$$T_{i,j}^{(2)} = \frac{2}{(4\pi\epsilon_0)r_{ij}^3}\left[\operatorname{erf}\left(\frac{r_{ij}}{a_{i,j}}\right) - \frac{2}{\sqrt{\pi}}\frac{r_{ij}}{a_{i,j}}\left(1 + \frac{r_{ij}^2}{a_{i,j}^2}\right)\exp^{-(r_{ij}/a_{i,j})^2}\right], \tag{5.3}$$

where $T_{i,j}^{(2)} = T^{(2)}(r_{ij})$. Thanks to the convolution with the Gaussian(s), the quantities $T_{i,i}^{(2)}$ are well defined ($T_{i,i}^{(2)} = -1/[(3\pi\epsilon_0)\sqrt{\pi}a_{i,j}^3]$)), which is not the case for the classical point dipole–point dipole propagator $-\nabla\nabla'[1/(4\pi\epsilon_0|\mathbf{r} - \mathbf{r}'|)]$ which diverges when $|\mathbf{r} - \mathbf{r}'| \to 0$. Indeed, coupled to Mayer's crucial remark [81] that the diagonal terms $-(1/2)p_i T_{i,i}^{(2)} p_i$ should be included in the sum of Equation (5.2) and physically interpreted as the self-energy of the atom $(1/2)\alpha_i E_i^2 = (1/2)p_i^2/\alpha_i$, the isotropic polarizability is related to the width of the Gaussian distribution of charge on each atom by Equation (5.2) of Mayer *et al.* [59], and takes the form

$$\frac{\alpha_i}{4\pi\epsilon_0} = 3\frac{\sqrt{\pi}}{4}a_{i,i}^3 = 3\sqrt{\frac{\pi}{2}}R_i^3, \tag{5.4}$$

where $R_i$ is the width of the Gaussian distribution for atom $i$, with units of Å. If $\alpha_i/(4\pi\epsilon_0)$ is given in Å$^3$, as is commonly the case, one can forget the factor $(4\pi\epsilon_0)$ in $T^{(2)}$ and fall back to the CGS-Gauss expression for it.

The distribution of dipoles is then determined by enforcing that the actual dipole distribution $\{p_j^*\}$ should correspond to the minimum value of $U^{\text{elec}}$. Following Wang *et al.* [55], this can be written as

$$\forall i = 1, \ldots, N \quad \frac{\partial U^{\text{elec}}}{\partial p_i}(\{p_j^*\}) = 0. \tag{5.5}$$

Enforcing these conditions on Equation (5.2), the actual dipole moments on each atom $p_j^*$ can be found by solving the following dense linear system:

$$\forall i = 1, \ldots, N \quad \sum_{j=1}^{N} T_{i,j}^{(2)} p_j^* = -E_{0,i}, \tag{5.6}$$

where $E_{0,i} = E_0(r_i)$. One can then define an $N \times N$ matrix $\widehat{T^{(2)}}$ and the corresponding vectors $\widehat{p^*}$ and $\widehat{E_0}$ and restate the linear system to solve as

$$\widehat{T^{(2)}}\widehat{p^*} = -\widehat{E_0}. \tag{5.7}$$

By substituting (5.6) into (5.2), one gets

$$U^{\text{elec}} = -\frac{1}{2}\sum_{i=1}^{N} p_i^*(-E_{0,i}) - \sum_{i=1}^{N} p_i^* E_{0,i} = -\frac{1}{2}\sum_{i=1}^{N} p_i^* E_{0,i} = -\frac{1}{2}\widehat{p^*} \cdot \widehat{E_0}. \tag{5.8}$$

Using Equations (5.7) and (5.8), the electrostatic forces on each atom can be written as

$$\forall i = 1, \ldots, N \quad f_i^{\text{elec}} = -\frac{\partial U^{\text{elec}}}{\partial r_i} = \frac{\mathrm{d}E_0(r_i)}{\mathrm{d}r_i} p_i^* - \frac{1}{2}\widehat{E_0} \cdot \frac{\partial [\widehat{T^{(2)}}]^{-1}}{\partial r_i} \widehat{E_0}, \tag{5.9}$$

Then, using the fact that

$$\frac{\partial \{\widehat{T^{(2)}}[\widehat{T^{(2)}}]^{-1}\}}{\partial r_i} = 0 \Rightarrow \frac{\partial [\widehat{T^{(2)}}]^{-1}}{\partial r_i} = -[\widehat{T^{(2)}}]^{-1} \frac{\partial \widehat{T^{(2)}}}{\partial r_i} [\widehat{T^{(2)}}]^{-1}, \tag{5.10}$$

we get

$$\forall i = 1, \ldots, N \quad f_i^{\text{elec}} = \frac{\mathrm{d}E_0(r_i)}{\mathrm{d}r_i} p_i^* + \frac{1}{2}\widehat{p^*} \cdot \frac{\partial \widehat{T^{(2)}}}{\partial r_i} \widehat{p^*}. \tag{5.11}$$

If we now define

$$\begin{aligned} T^{(3)}(r_{ij}) &= T^{(3)}(r_i - r_j) \\ &= \frac{\partial T^{(2)}(r_i - r_j)}{\partial r_i} \\ &= -\frac{r_i - r_j}{(4\pi\epsilon_0) r_{ij}^5} \left[ 6 \operatorname{erf}\left(\frac{r_{ij}}{a_{i,j}}\right) - \frac{4}{\sqrt{\pi}} \left( 3\frac{r_{ij}}{a_{i,j}} + 2\frac{r_{ij}^3}{a_{i,j}^3} + 2\frac{r_{ij}^5}{a_{i,j}^5} \right) \exp\left( -\frac{r_{ij}^2}{a_{i,j}^2} \right) \right], \end{aligned} \tag{5.12}$$

one can show that

$$\forall i, j, k = 1, \ldots, N \qquad \frac{\partial [\widehat{T^{(2)}}]_{j,k}}{\partial r_i} = T_{i,k}^{(3)} \delta_{i,j} - T_{j,i}^{(3)} \delta_{i,k}. \tag{5.13}$$

Using this result and the antisymmetry of $T^{(3)}$ with respect to the interchange of the two positions, the forces can now be written as

$$\forall i = 1, \ldots, N \quad f_i^{\text{elec}} = p_i^* \left( \frac{\mathrm{d}E_0(r_i)}{\mathrm{d}r_i} + \sum_{j=1}^{N} T_{i,j}^{(3)} p_j^* \right). \tag{5.14}$$

We note that this result would simply be $f_i^{\text{elec}} = p_i^* \, \mathrm{d}E_i/\mathrm{d}r_i$ with $E_i = E_0(r_i) + \sum_{j=1}^{N} T_{i,j}^{(2)} p_j^*$ the total electric field at $r_i$, if the $p_j^*$ were not functions of $r_i$, which they are, due to (5.6).

From Equations (5.9) and (5.14), we can see that the computation of the $p_j^*$ and thus the inverse of the $N \times N$ system in (5.9) ($3N \times 3N$ in three dimensions) is the limiting step during the calculation of the electrostatic forces with this GDM model. As the inverse of this equation occurs *for each iteration* during the energy minimization process, practical use of it [55, 56] was limited to CNTs with the number of atoms of the order of 5000. This is still far from what would be needed for realistic nanomaterials with sizes exceeding 10 nm in which nanoscale surface effects still lead to unexpected mechanical properties compared with bulk materials [25].

### 5.2.2 *Finite Element Equilibrium Equations from Total Electromechanical Potential Energy*

To significantly alleviate the computational expense incurred in a fully atomistic calculation of the electrostatic forces using the GDM, we propose an alternative approach in the present work, whereby we will obtain an FEM solution to minimizing the total energy $U^{\text{total}}$ in (5.1). In essence, we will transform the problem of minimizing $U^{\text{total}}$ from one that must be done for each and every atom in the system to one in which the mechanical energy $U^{\text{mech}}$ and electrostatic energy $U^{\text{elec}}$ are evaluated only at FEM integration points. Furthermore, we will demonstrate below that (in one dimension) the calculation of the bulk electrostatic stress does not require the inverse of an $N \times N$ system, while the calculation of the surface electrostatic stress requires the inverse of an $N_{\text{surf}} \times N_{\text{surf}}$ system only, where $N_{\text{surf}}$ is on the order of 10 for sufficient accuracy compared with benchmark fully atomistic calculations.

We accomplish this by applying the standard FEM displacement approximation to (5.1); doing so leads to FEM equilibrium equations which naturally reflect the self-consistent competition between mechanical and electrostatic forces. The details regarding the transition from the mechanical potential energy in (5.1) while delineating the bulk and surface energies to the FEM equilibrium equations is given in detail by Park *et al.* [43]; we note that the electrostatic potential energy including surface effects in (5.1) can be treated similarly. The final electromechanical FEM equilibrium equations that we implement and solve numerically are

$$\begin{aligned}\frac{\partial U^{\text{total}}}{\partial \mathbf{u}_I} &= \int_{\Omega_0^{\text{bulk}}} \mathbf{B}^{\mathrm{T}}\mathbf{P}^{\text{mech}}\,\mathrm{d}\Omega + \int_{\Gamma_0} \mathbf{B}^{\mathrm{T}}\tilde{\mathbf{P}}^{\text{mech}}\,\mathrm{d}\Gamma - \int_{\Gamma_0} N_I\mathbf{T}\,\mathrm{d}\Gamma \\ &\quad + \int_{\Omega_0^{bulk}} \mathbf{B}^{\mathrm{T}}\mathbf{P}^{\text{elec}}\,\mathrm{d}\Omega + \int_{\Gamma_0} \mathbf{B}^{\mathrm{T}}\tilde{\mathbf{P}}^{\text{elec}}\,\mathrm{d}\Gamma,\end{aligned} \tag{5.15}$$

where $N_I$ are the FEM shape functions, $\mathbf{B}^{\mathrm{T}} = (\partial N_I/\partial \mathbf{X})^{\mathrm{T}}$, $\mathbf{T}$ are externally applied tractions, $\mathbf{P}^{\text{mech}}$ is the bulk mechanical first Piola–Kirchoff stress, $\mathbf{P}^{\text{elec}}$ is the bulk electrostatic first Piola–Kirchoff stress, $\tilde{\mathbf{P}}^{\text{mech}}$ is the surface mechanical first Piola–Kirchoff stress, and $\tilde{\mathbf{P}}^{\text{elec}}$ is the surface electrostatic first Piola–Kirchoff stress. Thus, the boundary-value problem in (5.15) can be stated as: given an applied electric field $\mathbf{E}_0$ that is considered to be homogeneous in a given finite element and the applied external mechanical forces, find the atomic bond lengths $r_{ij}$ that minimize the total electromechanical potential energy.

Because Park and co-workers have previously discussed how to obtain the bulk and surface mechanical stresses using the SCB model [43], we will not cover that in this chapter, and will focus only on what is new; that is, calculation of the bulk and surface electrostatic stresses. Furthermore, the method of evaluating both the bulk and surface integrals for the electrostatic stresses is identical to how the bulk and surface integrals are evaluated for the mechanical stresses; details are again given in Park *et al.* [43].

The FEM governing equations in (5.15) are novel and important because they demonstrate that: (1) A coupled electromechanical analysis of surface-dominated nanomaterials is possible by solving only one governing equation, which fully accounts for the external electric-field-induced polarization–strain coupling between mechanical and electrostatic forces, including surface effects, and which obtains the required mathematical relationships for both the mechanical and electrostatic fields directly from underlying atomistic principles. (2) By changing the problem formulation from one that is computationally intractable (i.e., fully atomistic and involving millions of degrees of freedom) to one that is tractable (i.e., based on well-established FEM techniques with a significantly reduced number of degrees of freedom) we will achieve significant computational savings, as the FEM element size is typically 100 to 1000 times larger than the atomic spacing [44]. (3) The coupled electromechanical FEM governing equations are cast naturally in a total Lagrangian formulation, which ensures that no remeshing of the domain is required. (4) Because the FEM governing equation in (5.15) eliminates the need for any atomistic equations of motion, we will avoid both the time-step and time-scale issues that arise when using atomistic simulation techniques, though this implies that the externally applied electric field is constrained to be homogeneous within each finite element. We note that because the GDM approach, and specifically the electrostatic potential energy is, similar to mechanical interatomic potentials, dependent only on the distance $r_{ij}$ between two atoms $i$ and $j$, there is no issue applying the GDM approach to dynamic, or finite temperature problems so long as an appropriate parameterization of the atomic polarizabilities is available.

We now discuss how, using multiscale Cauchy–Born-based techniques, the bulk and surface mechanical and electrostatic stresses that are needed to solve (5.15) can be calculated in an efficient and accurate manner.

## 5.3 Bulk Electrostatic Piola–Kirchoff Stress

### *5.3.1 Cauchy–Born Kinematics*

After having derived the expression for the electrostatic potential energy in (5.8), we utilize Cauchy–Born principles to first convert the electrostatic potential energy into an electrostatic energy density, which we can then differentiate to obtain the electrostatic Piola–Kirchoff stresses.

The Cauchy–Born model [41–43, 91] has been utilized in recent years as a multiscale, hyperelastic material model that originates from a given interatomic potential energy function. The multiscale link between atomistics and continua is achieved by normalizing the interatomic potential by a representative volume and enforcing that the bond lengths between atoms $r_{ij}$ are constrained to deform via the local value of the continuum deformation gradient $\mathbf{F}$ or stretch tensor $\mathbf{C}$. In doing so, continuum stress and stiffness, which are needed for nonlinear FEM simulations, can be directly obtained by taking one and two derivatives, respectively, of the strain energy with respect to the continuum deformation measure. Because of the fact that continuum stress and stiffness can be derived directly from an atomistic interatomic potential, the Cauchy–Born model is regarded as a hierarchical multiscale constitutive model. However, it should be noted that while the Cauchy–Born hypothesis does lead to a multiscale link between atomistics and continua, it does also place a key restriction upon the deformation of the underlying crystal. Specifically, it enforces a locally homogeneous deformation assumption upon the underlying crystal by defining the deformed bond length $\mathbf{r} = \mathbf{F}\mathbf{r}_0$ between two atoms to be a function of the deformation gradient or stretch tensor at the corresponding finite-element integration point.

To briefly overview Cauchy–Born kinematics, we note that in Green elastic theory, stress is derived by differentiating the material strain energy density function. In order to satisfy material frame indifference, the strain energy density must be expressed as a function of the right stretch tensor $\mathbf{C}$,

$$W(\mathbf{F}) = \Phi(\mathbf{C}), \tag{5.16}$$

where

$$\mathbf{C} = \mathbf{F}^{\mathrm{T}}\mathbf{F}, \tag{5.17}$$

and $\mathbf{F}$ is the deformation gradient. From the strain energy density, one can obtain the first ($\mathbf{P}$) and second ($\mathbf{S}$) Piola–Kirchoff stresses as

$$\mathbf{P} = \frac{\partial W(\mathbf{F})}{\partial \mathbf{F}^T} \quad \text{and} \quad \mathbf{S} = 2\frac{\partial \Phi(\mathbf{C})}{\partial \mathbf{C}}, \tag{5.18}$$

where the Piola–Kirchoff stresses are related by

$$\mathbf{P} = \mathbf{S}\mathbf{F}^{\mathrm{T}}. \tag{5.19}$$

In the previous Cauchy–Born literature [42, 43, 91], the strain energy density was assumed to be a mechanical energy density which was obtained from an underlying atomistic interatomic potential; for example, an embedded atom (EAM) potential for FCC metals [42, 44], Brenner potentials for carbon [91, 92], or Tersoff potentials for silicon [45, 93].

In contrast to previous research that has utilized the Cauchy–Born hypothesis for mechanical problems, we utilize it here to obtain the electrostatic Piola–Kirchoff stresses. To do so, we first consider the 1D bulk case, which can be interpreted as consideration of an infinite bulk chain of atoms in which all atoms have the same bonding environment, such that surface effects are not considered; we note that this is standard for Cauchy–Born approximations and that it is the reason why the SCB model of Park and co-workers [43–45] was developed to account for the fact that the standard Cauchy–Born mechanical model does not admit surface effects.

Therefore, within this 1D electrostatic bulk idealization, all atoms have the same dipole moment $p$, and thus we can derive the electrostatic energy density by normalizing (5.2) by the equilibrium atomic lattice spacing $r_0$, and by enforcing the Cauchy–Born hypothesis upon the deformed atomic lattice spacing $r = r_0\sqrt{C}$ to give

$$\Phi^{\text{elec}}(C) = \frac{1}{r_0}\left[-\frac{1}{2}\sum_{n=-\infty}^{\infty}(p_0 T^{(2)}(C)p_n) - p_0 E_0\right], \tag{5.20}$$

where for all $n$, $p_n = p_0 = p^*$ at electrostatic equilibrium for an infinite, periodic bulk system of identical atoms. Therefore, the electrostatic equilibrium condition, $(\mathrm{d}\Phi^{\text{elec}}/\mathrm{d}p)(p^*) = 0$, reduces to

$$-p^*\sum_{n=-\infty}^{\infty} T^{(2)}(nr_0\sqrt{C}) - E_0 = 0. \tag{5.21}$$

Therefore, the bulk dipoles $p^*$ take the value

$$p^*(C) = \frac{-E_0}{\sum_{n=-\infty}^{\infty} T^{(2)}(nr_0\sqrt{C})} \tag{5.22}$$

and the electrostatic energy density can then be written using (5.20) as

$$\Phi^{\text{elec}}(C) = \frac{E_0^2}{2r_0\sum_{n=-\infty}^{\infty} T^{(2)}(nr_0\sqrt{C})}. \tag{5.23}$$

From (5.23), we can obtain the bulk electrostatic second Piola–Kirchoff stress as

$$S^{\text{elec}}(C) = 2\frac{\mathrm{d}\Phi^{\text{elec}}}{\mathrm{d}C} = -\frac{E_0^2}{r_0}\frac{\mathrm{d}}{\mathrm{d}C}\frac{\sum_{n=-\infty}^{\infty} T^{(2)}(nr_0\sqrt{C})}{\left[\sum_{n=-\infty}^{\infty} T^{(2)}(nr_0\sqrt{C})\right]^2}, \tag{5.24}$$

which, using definition (5.12), can be rewrittten as

$$
\begin{aligned}
S^{\mathrm{elec}}(C) &= -\frac{E_0^2}{r_0}\frac{\sum\limits_{n=-\infty}^{\infty} T^{(3)}(nr_0\sqrt{C})\dfrac{nr_0}{2\sqrt{C}}}{\left[\sum\limits_{n=-\infty}^{\infty} T^{(2)}(nr(C))\right]^2} \\
&= -\frac{E_0^2 r_0}{2r}\frac{\sum\limits_{n=-\infty}^{\infty} T^{(3)}(nr(C))n}{\left[\sum\limits_{n=-\infty}^{\infty} T^{(2)}(nr(C))\right]^2} \\
&= -\frac{E_0^2 r_0}{r}\frac{\sum\limits_{n=1}^{\infty} T^{(3)}(nr(C))n}{\left[\sum\limits_{n=-\infty}^{\infty} T^{(2)}(nr(C))\right]^2}.
\end{aligned}
\tag{5.25}
$$

Finally, the bulk electrostatic first Piola–Kirchoff stress can be obtained by using the standard continuum mechanics relationship in (5.19), while noting that, in one dimension, $F = F^{\mathrm{T}} = \sqrt{C} = r/r_0$ and

$$
\begin{aligned}
P^{\mathrm{elec}} &= -F\frac{E_0^2 r_0}{2r}\frac{\sum\limits_{n=-\infty}^{\infty} T^{(3)}(nr(F))n}{\left[\sum\limits_{n=-\infty}^{\infty} T^{(2)}(nr(F))\right]^2} \\
&= -E_0^2\frac{\sum\limits_{n=1}^{\infty} nT^{(3)}(nr(F))}{\left[\sum\limits_{n=-\infty}^{\infty} T^{(2)}(nr(F))\right]^2}.
\end{aligned}
\tag{5.26}
$$

### *5.3.2 Comparison of Bulk Electrostatic Stress with Molecular Dynamics Electrostatic Force*

It is important here to compare the computational advantages gained in the expression for the bulk electrostatic stress in (5.25) in comparison with the electrostatic forces (5.14). First, we note that no matrix inverse is required to compute the bulk electrostatic stress in (5.25). Instead, only the calculation of discrete sums for the dipole interaction tensors $T^{(2)}$ and $T^{(3)}$ are required, where we have found that extending both summations to about $n = 20$ is sufficient to obtain a converged response. These sums of interaction tensors will be called periodized interaction tensors. We note that, for 3D problems, the inverse of an $3N_{\mathrm{per}} \times 3N_{\mathrm{per}}$ system will be needed at each FEM integration point, where $N_{\mathrm{per}}$ is the number of atoms in the 3D lattice unit cell (i.e., one for an FCC crystal), provided periodized 3D interaction tensors are used.

Second, we note that the computational expense in calculating the electrostatic stress in (5.25) is reduced dramatically compared with the computation of the forces in the full system of $N$ atoms, not only because the system size is significantly smaller, but also because the electrostatic stress is calculated only at FEM integration points, such that it is evaluated at far fewer points in the domain than the electrostatic force is, which is evaluated (at significantly greater computational expense) for every atom in the domain.

## 5.4 Surface Electrostatic Stress

The final aspect to the coupled electromechanical formulation is to calculate the surface electrostatic stress; we now present the derivation in one dimension. In nanomechanics, surface effects play a significant role in causing nanostructures to exhibit non-bulk mechanical properties compared with the corresponding bulk material [25]. The surface effects arise because atoms at or near the surfaces of the material have fewer bonding neighbors than do atoms in the bulk, which alters their elastic properties compared with the bulk atoms.

Similar effects occur in electrostatics; specifically, because atoms at the surface have fewer bonding neighbors, their effective dipolar polarizability tensor $\alpha_i$ is different, which leads to atoms at or near the surface polarizing differently and, thus, having a different dipole moment in response to an applied electric field compared with bulk atoms [34–37].

In the present work, we assume that, for each surface finite-element integration point, we have $n_{\text{surf}}$ surface atoms with indices $i \leq 0$, while atoms $i > 0$ are bulk atoms; that is, they all have the same dipole moment $p^*$, as computed using (5.22). We assume that the atoms initially have positions $x_i = ir_0$ before the homogeneous deformation due to the deformation gradient $F$ and that the external electrostatic field $E_0$ is also homogeneous in the small volume corresponding to the integration point. Then, the system of equations to solve for the surface dipoles $p_j^{\text{s}}$ is based upon a modification of the bulk dipole solution in (5.22) to give, for all $i \leq 0$,

$$\sum_{j=-n_{\text{surf}}+1}^{0} T^{(2)}[(i-j)r_0F]p_j^{\text{s}} = -E_0 - \sum_{n=1}^{\infty} T^{(2)}\left((i-n)r_0F\right)p^*. \tag{5.27}$$

The surface energy density (energy per unit surface area) is then written as

$$\begin{aligned}\tilde{\Phi}^{\text{elec}}(F) &= \left\{-\frac{1}{2}\sum_{i=-n_{\text{surf}}+1}^{0}\ \sum_{m=-n_{\text{surf}}+1}^{\infty} p_i^{s}T^{(2)}[(i-m)Fr_0]p_m\right\} \\ &\quad - \sum_{i=-n_{\text{surf}}+1}^{0} p_i^{\text{s}}E_0 \\ &= -\frac{1}{2}\sum_{i=-n_{\text{surf}}+1}^{0} p_i^{\text{s}}(F)E_0.\end{aligned} \tag{5.28}$$

Consequently, we define the surface electrostatic first Piola–Kirchhoff stress as

$$\tilde{P}^{\text{elec}}(F) = \frac{\mathrm{d}\tilde{\Phi}^{\text{elec}}}{\mathrm{d}F^{\text{T}}} = -\frac{E_0}{2}\sum_{i=-n_{\text{surf}}+1}^{0}\frac{\mathrm{d}p_i^{\text{s}}}{\mathrm{d}F^{\text{T}}}, \tag{5.29}$$

where for all $i \leq 0$, by solving (5.27),

$$p_i^{s} = \sum_{j=-n_{\text{surf}}+1}^{0}\left([\widehat{T^{(2)}}]_{i,j}^{-1}\left\{-E_0 - \sum_{n=1}^{\infty} T^{(2)}[(j-n)Fr_0]p^*\right\}\right), \tag{5.30}$$

and thus the surface electrostatic first Piola–Kirchoff stress is

$$\tilde{P}^{\text{elec}}(F) = -\frac{E_0}{2} \times \sum_{i=-n_{\text{surf}}+1}^{0} \sum_{j=-n_{\text{surf}}+1}^{0} \frac{\mathrm{d}}{\mathrm{d}F}\left( [\widehat{T^{(2)}}]^{-1}_{i,j} \left\{ -E_0 - \sum_{n=1}^{\infty} T^{(2)}[(j-n)Fr_0]p^* \right\} \right). \tag{5.31}$$

To evaluate the derivative in (5.31) we use an approach similar to what was done for the computation of the forces in Section 5.2.1. Using Equation (5.12), we first get

$$\frac{\mathrm{d}T^{(2)}_{i,j}}{\mathrm{d}F} = \frac{\mathrm{d}T^{(2)}[(i-j)Fr_0]}{\mathrm{d}F} = (i-j)r_0 T^{(3)}[(i-j)Fr_0] \tag{5.32}$$

Now, using this and the fact that

$$\widehat{T^{(2)}}[\widehat{T^{(2)}}]^{-1} = \mathbf{I} \Rightarrow \frac{\mathrm{d}[\widehat{T^{(2)}}]^{-1}}{\mathrm{d}F} = -[\widehat{T^{(2)}}]^{-1} \frac{\mathrm{d}\widehat{T^{(2)}}}{\mathrm{d}F} [\widehat{T^{(2)}}]^{-1}, \tag{5.33}$$

we get

$$\frac{\mathrm{d}[\widehat{T^{(2)}}]^{-1}}{\mathrm{d}F} = -r_0 [\widehat{T^{(2)}}]^{-1} [\widehat{(i-j)T^{(3)}_{i,j}}] [\widehat{T^{(2)}}]^{-1}. \tag{5.34}$$

Furthermore, since $p^* = -E_0 / \sum_{n=-\infty}^{\infty} T^{(2)}(nFr_0)$, we have

$$\begin{aligned} &\frac{\mathrm{d}\left\{ -E_0 - \sum_{n=1}^{\infty} T^{(2)}[(j-n)Fr_0]p^* \right\}}{\mathrm{d}F} = E_0 \frac{\mathrm{d}}{\mathrm{d}F} \left\{ \frac{\sum_{n=1}^{\infty} T^{(2)}[(j-n)Fr_0]}{\sum_{n=-\infty}^{\infty} T^{(2)}(nFr_0)} \right\} \\ &= E_0 r_0 \left\{ \frac{\sum_{n=1}^{\infty} (j-n) T^{(3)}[(j-n)Fr_0]}{\sum_{n=-\infty}^{\infty} T^{(2)}(nFr_0)} \right\} - \left\{ \frac{\sum_{n=1}^{\infty} T^{(2)}[(j-n)Fr_0] \sum_{n=-\infty}^{\infty} nT^{(3)}(nFr_0)}{\left[ \sum_{n=-\infty}^{\infty} T^{(2)}(nFr_0) \right]^2} \right\}. \end{aligned} \tag{5.35}$$

Putting (5.34) and (5.35) into (5.31) and using (5.27), the surface electrostatic stress can be written as

$$\begin{aligned} \tilde{P}^{\text{elec}}(F) = {} & \frac{E_0 r_0}{2} \sum_{i=-n_{\text{surf}}+1}^{0} \sum_{k=-n_{\text{surf}}+1}^{0} \sum_{l=-n_{\text{surf}}+1}^{0} [\widehat{T^{(2)}}]^{-1}_{i,k} (k-l) T^{(3)}[(k-l)Fr_0] p^s_l \\ & + \frac{E_0 r_0 p^*}{2} \sum_{i=-n_{\text{surf}}+1}^{0} \sum_{j=-n_{\text{surf}}+1}^{0} [\widehat{T^{(2)}}]^{-1}_{i,j} \sum_{n=1}^{\infty} (j-n) T^{(3)}[(j-n)r_0 F] \\ & + \frac{r_0 (p^*)^2}{2} \left[ \sum_{n=-\infty}^{\infty} nT^{(3)}(nFr_0) \right] \\ & \times \sum_{i=-n_{\text{surf}}+1}^{0} \sum_{j=-n_{\text{surf}}+1}^{0} \left\{ [\widehat{T^{(2)}}]^{-1}_{i,j} \sum_{n=1}^{\infty} T^{(2)}[(j-n)r_0 F] \right\}. \end{aligned} \tag{5.36}$$

There are several relevant points to be discussed before moving onto the numerical examples. First, it is seen in both (5.36) and (5.30) that the inverse of the dipole–dipole interaction tensor $T^{(2)}$ is required to calculate both the surface dipoles and, thus, the electrostatic stress, where the size of the tensor is related to the number of surface atoms $n_{surf}$ that are considered. We will show in the numerical examples that, in one dimension, a value of $n_{surf} = 10$ or 15 gives results with accuracy that is comparable to the fully atomistic calculations. However, in three dimensions, limiting the value of $n_{surf}$ will clearly be critical in keeping the computational expense to a minimum; we plan to investigate techniques pioneered in truncating Ewald sums to finite distances [94, 95] for this purpose in future research.

## 5.5 One-Dimensional Numerical Examples

### 5.5.1 *Verification of Bulk Electrostatic Stress*

We first validate the bulk electrostatic stress. To do so, we compare results obtained from applying an external electric field ranging between 0.025 and 0.5 V/Å to an infinite 1D chain of atoms, where the physical parameters for the GDM were taken to mimic the values for CNTs [55, 60] (i.e., the equilibrium lattice spacing $r_0 = 3$ Å, and the isotropic polarizability $\alpha = 5.5086$ Å$^3$), while the width of the distribution $R$ can be found from Equation (5) of Mayer [60] to be 1.135 76 Å.

For the mechanical response, we utilized a standard Lennard-Jones 6–12 potential, which takes the form

$$U_{\mathrm{LJ}}(r_{ij}) = 4\epsilon\left[\left(\frac{\sigma}{r_{ij}}\right)^{12} - \left(\frac{\sigma}{r_{ij}}\right)^{6}\right], \tag{5.37}$$

where the parameters for the Lennard-Jones potential were $\sigma = 2.6727$ Å, and $\epsilon = 0.4096$ eV; the choice of $\sigma$ ensured that the equilibrium lattice spacing for the mechanical Lennard-Jones potential was also 3 Å. Finally, in the mechanical problem, each atom was assumed to interact only with its nearest neighbors.

The minimum energy configuration of the fully atomistic system was systematically obtained by varying the deformation gradient $F$ between 0.96 and 1.04 while evaluating both the electrostatic potential energy in (5.8) and the mechanical potential energy in (5.37) for a single, representative bulk atom. The varying deformation gradient is used to deform the surrounding bulk atoms so as to evaluate changes in both the electrostatic and mechanical potential energies. The electrostatic potential energy was calculated for a single bulk atom that was surrounded by 100 nearest neighbors on each side to represent a bulk, infinite crystal.

Similarly, the minimum energy configuration of the BCB-based FEM model was obtained by evaluating the weak form in (5.15), while neglecting both the surface electrostatic and mechanical stresses for an arbitrary domain size with a single linear finite element. Note that, because all representative bulk atoms are identical, a single element is sufficient to capture the deformation due to the externally applied electric field.

The results are summarized in Table 5.1, while the mechanical, electrostatic, and total potential energies are shown for the $E_0 = 0.25$ V/Å fully atomistic case in Figure 5.1. As can be seen in Table 5.1,

**Table 5.1** Comparison of strain (in percent) for bulk, infinite 1D chain of atoms subject to an externally applied electric field of $E_0 = 0.25$ V/Å

| $E_0$ (V/Å) | MD strain (%) | BCB strain (%) | Error in strain (%) |
|---|---|---|---|
| 0.025 | −0.006 59 | −0.006 56 | 0.54 |
| 0.05 | −0.026 5 | −0.026 24 | 0.96 |
| 0.1 | −0.105 6 | −0.104 8 | 0.79 |
| 0.25 | −0.649 5 | −0.646 3 | 0.48 |
| 0.5 | −2.461 7 | −2.473 8 | 0.49 |

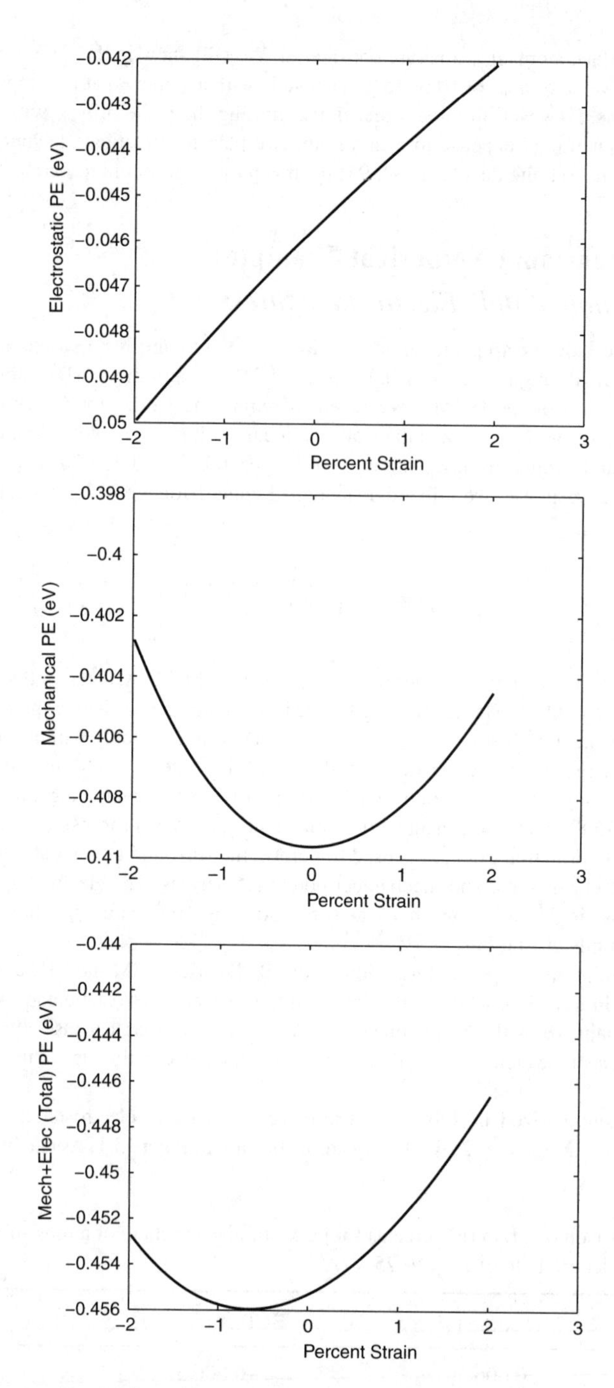


**Figure 5.1** Example of mechanical, electrostatic, and total potential energy for an infinite 1D chain of atoms subject to an externally applied electric field of $E_0 = 0.25$ V/Å

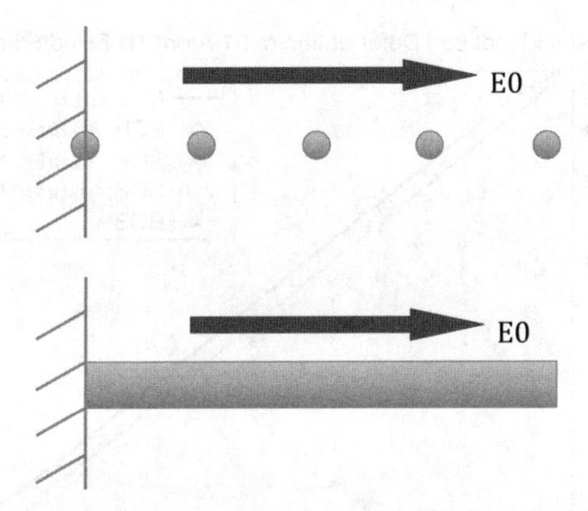


**Figure 5.2** Schematic of 1D chain of atoms subject to externally applied electric field $E_0$. *Top*: discrete, atomistic model; *bottom*: the equivalent continuum model

the amount of strain that is incurred by subjecting an infinite, bulk 1D chain of atoms increases with an increase in the electric field intensity. Furthermore, the BCB results compare extremely well to the benchmark MD simulations for the wide range of applied electric field values that we have considered that result in both very small ($E_0 = 0.025$ V/Å), and finite ($E_0 = 0.5$ V/Å) strains. It is also interesting to point out that while the error remains essentially constant with an increase in electric field strength, Table 5.1 shows that the strain is proportional to $E_0^2$, which is called the electrostriction effect. This is expected because the bulk first Piola–Kirchoff stress is also proportional to $E_0^2$, as shown in (5.25).

The good agreement between MD and BCB simulations enables us to move forward with confidence to including surface effects through the surface electrostatic stress, which we discuss next.

### 5.5.2 *Verification of Surface Electrostatic Stress*

We now discuss numerical validation of the proposed coupled electromechanical SCB model compared with benchmark fully atomistic calculations. To do so, we considered 1D atomic chains of various length with fixed/free boundary conditions, as illustrated in Figure 5.2. The externally applied electric field $E_0$ was applied in the positive $x$-direction parallel to the chain of atoms. We took two values of the external electric field, $E_0 = 0.1$ V/Å, and to test the nonlinear, finite deformation response, $E_0 = 0.5$ V/Å. A summary of all simulations and comparisons with fully atomistic calculations is shown in Table 5.2.

We first considered a 1D chain of 51 atoms subject to an applied electric field of 0.1 V/Å, which was equivalently modeled using the SCB approach using three linear finite elements. While the bulk and

**Table 5.2** Comparison of percentage error in free-end displacement for 51- and 101-atom 1D chains for various applied electric fields between SCB for various values of $n_{surf}$ and fully atomistic calculations. All displacements are in Ångströms; values in parentheses are the percentage error compared with the benchmark MD solution

| Atoms | $E_0$ (V/Å) | MD | SCB $n_{surf} = 5$ | SCB $n_{surf} = 10$ | SCB $n_{surf} = 15$ |
|---|---|---|---|---|---|
| 51 | 0.1 | −0.1439 | −0.1496 (3.9%) | −0.1458 (1.3%) | −0.1447 (0.5%) |
| 51 | 0.5 | −3.376 | −3.528 (4.5%) | −3.4281 (1.54%) | −3.397 (0.6%) |
| 101 | 0.1 | −0.303 | −0.308 (1.86%) | −0.3044 (0.6%) | −0.303 (0.23%) |

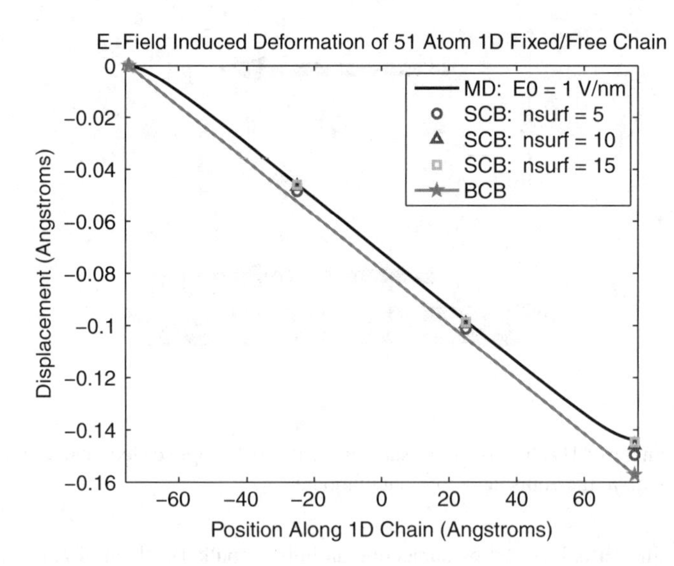


**Figure 5.3** Comparison between full MD and SCB results for a 51-atom 1D chain subject to externally applied electric field of $E_0 = 0.1$ V/Å. The BCB result is also shown to indicate how surface effects impact the deformation of a 1D chain to an electric field

surface electrostatic stresses were calculated using the formulation presented in this chapter, the bulk and surface mechanical stresses were calculated using the previously developed SCB model; details on the SCB implementation for a Lennard-Jones potential in one dimension was presented by Park *et al.* [43].

The numerical results are shown in Figure 5.3, where the MD results, the SCB results for various values of $n_{\text{surf}}$, and BCB results are shown. First, we notice that the displacement field of the 1D chain is negative, or compressive, though the electric field is applied in the positive $x$-direction. Physically, this occurs because neighboring dipoles attract each other when they are aligned in the direction of the applied electric field.

There are other interesting effects that can be observed in comparing the finite-chain (MD) and infinite-chain (BCB) results. Specifically, it can be seen that the slope of the compressive displacement field becomes smaller near the free surfaces, and thus the deformation of the finite chain is smaller than the infinite chain by more than 9%, which demonstrates the importance of accounting for the electromechanical surface effects. This occurs due to the reduced polarization at the free surfaces.

It is also worth emphasizing here that for this, and all comparisons between the MD, BCB, and SCB results, the key value that must be validated is that the SCB and MD results agree well, which is indeed demonstrated in Figure 5.3. Because the problem is 1D, and because the mechanical potential energy is modeled using the Lennard-Jones potential, which does not represent the behavior of any real material, the actual value of the surface effect is irrelevant; instead, the important point is that the MD and SCB results agree well, since they are both based upon the same mechanical and electrostatic potential energies.

For the SCB/MD comparison, we took the number of elements to be fixed at three, and instead varied between 5 and 15 the number of surface atoms ($n_{\text{surf}}$ in (5.27)–(5.36)) for which we computed dipoles. The results demonstrate convergence of the SCB solution to the full MD solution as the number of surface atoms $n_{\text{surf}}$ is increased, in accordance with expectation. Overall, while the SCB solution captures the surface effects, it is clear in Figure 5.3 that the accuracy of the solution, particularly near the free end of the chain, increases with increasing $n_{\text{surf}}$. The error in the free-end displacement decreases from 3.9% when $n_{\text{surf}} = 5$ to 1.3% when $n_{\text{surf}} = 10$ to 0.5% when $n_{\text{surf}} = 15$. We also note that quantifying the error

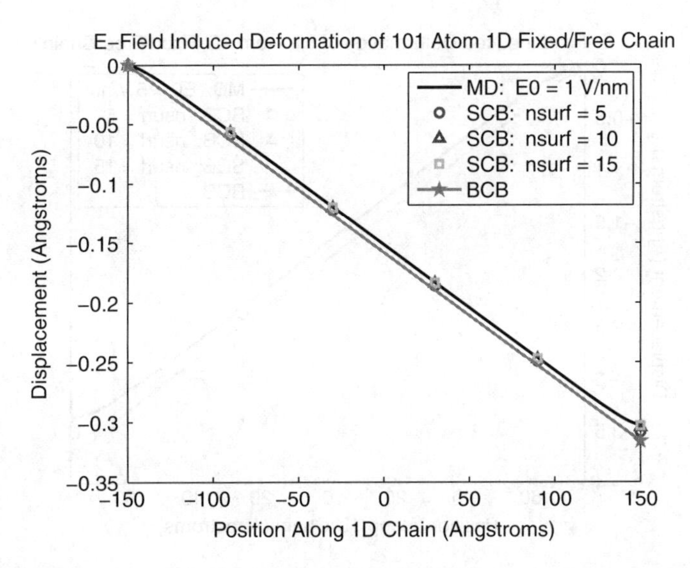


**Figure 5.4** Comparison between full MD and SCB results for a 101-atom 1D chain subject to externally applied electric field of $E_0 = 0.1$ V/Å. The BCB result is also shown to indicate how surface effects impact the deformation of a 1D chain to an electric field

introduced in truncating the surface dipoles is critical, because the inverse of an $n_{\text{surf}} \times n_{\text{surf}}$ matrix is required at the surface to obtain the surface stress; thus, it is important to have a small number of atoms $n_{\text{surf}}$ to keep computational efficiency compared with the benchmark MD solution, particularly as future extensions to three dimensions are made.

We also tested the accuracy of the SCB model for a larger system, that of a 101-atom 1D chain with the same fixed/free boundary conditions, where the results for the BCB, MD, and SCB simulations are given in Figure 5.4. There are noticeable differences between the MD, BCB, and SCB results compared with the 51-atom case in Figure 5.3. Most notably, the error between the BCB and MD solutions is only about 4.4% for the free-end displacement, which is about half that of the 51-atom case. This is anticipated because the surface effects become less significant for the 101-atom chain.

The error introduced using the SCB model for the 101-atom case was also significantly smaller than for the 51-atom case. Specifically, the error in the end-atom displacement was 1.86% when $n_{\text{surf}} = 5$, 0.6% when $n_{\text{surf}} = 10$, and 0.23% when $n_{\text{surf}} = 15$. The error is smaller than in the 51-atom case because the surface electrostatic effects are minimized due to the fact that more atoms are contained within the bulk of the material. We also quantified the surface effect by calculating the dipole moments using fully atomistic simulations on a very long 1D chain. We found that the values for the dipole moment did not approach 99% of the bulk dipole moment until about 14 atoms into the bulk from the free surface. Thus, for the 51-atom chain, more than half of the atoms can be considered to have a non-bulk dipole moment, thus leading to the much stronger surface effects and difference compared with the BCB value than was found for the 101-atom chain.

To further investigate both the larger surface effects and the finite deformation induced by the electric field, we reconsidered the 51-atom case with a larger externally applied electric field of $E_0 = 0.5$ V/Å, with the results shown in Figure 5.5. We see that the compressive displacement is significantly larger with the enhanced electric field; that is, the compressive displacement of the free end of the 1D chain is nearly 4 Å, for a strain of nearly 2.5%. Despite this, the SCB results show similar trends as for the previous 51-atom case subject to smaller electric field, in that, with an increase in $n_{\text{surf}}$, the SCB solution

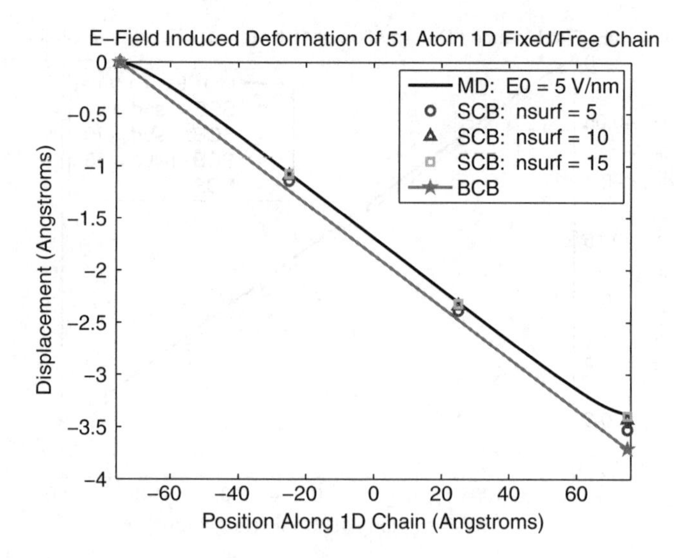


**Figure 5.5** Comparison between full MD and SCB results for a 51-atom 1D chain subject to externally applied electric field of $E_0 = 0.5$ V/Å. The BCB result is also shown to indicate how surface effects impact the deformation of a 1D chain to an electric field

converges to the fully atomistic result despite the highly nonlinear deformation. Specifically, the error between the SCB and MD solutions is about 4.5% when $n_{\text{surf}} = 5$, but decreases to 0.6% when $n_{\text{surf}} = 15$. Furthermore, the effect of elastic nonlinearity can be observed by comparing the free-end displacement for the 51-atom chain subject to the fields of 0.5 and 0.1 V/Å, as seen in Table 5.2. Specifically, the free-end displacement for the 0.5 V/Å case is only 23.5 times the 0.1 V/Å case, which suggests, as expected, that the ideal linear electrostriction relationship, which is proportional to $E_0^2$, breaks down at larger strains.

## 5.6 Conclusions and Future Research

We have presented in one dimension a novel multiscale, finite deformation finite-element approach to solving coupled electromechanical boundary-value problems in which surface-dominated nanostructures are subject to externally applied electric fields. The key step was to create a multiscale total electromechanical potential energy based upon previous, purely atomistic electromechanical total energies [55, 56]. In doing so, and in utilizing standard Cauchy–Born kinematics, we were able to derive a new coupled electromechanical variational form that has significant advantages compared with previous electromechanical coupling approaches. Specifically, the coupled electromechanical response is obtained solving only one governing equation using the FEM, which was made possible through derivation of new analytic formulas for the electrostatic forces, as well as both the bulk and surface electrostatic Piola–Kirchoff stresses. In doing so, previous issues including the solution of the equations of both the mechanical and electrostatic domains are avoided, as are issues with deriving the coupling matrix between mechanical and electrostatic domains, while issues related to remeshing either the electrostatic and mechanical domains are also eliminated. The method was validated and shown to be accurate in comparison with fully atomistic electromechanical calculations in one dimension.

While the approach was based upon finite-deformation Cauchy–Born kinematics, the utilization of the Cauchy–Born hypothesis indicates that defect nucleation cannot be captured without modifications to the method. Indeed, defect nucleation can be captured by adopting the quasi-continuum technique

[42], whereby the finite-element mesh is meshed down to the atomic scale at or near regions where defect nucleation may occur. However, the Cauchy–Born kinematics are valid for large, nonlinear elastic deformations, which are known not only to occur in nanowires [31], but also to have a significant effect on their mechanical behavior and properties [33].

The key challenge in extending this approach to three dimensions lies solely with the electrostatic domain, as SCB models for the mechanical domain have already been developed for both FCC metals [44] and semiconductors, such as silicon [45]. Specifically, the challenge is first to generalize the analytic formulas for the electrostatic forces and Piola–Kirchoff stresses to the 3D case and then to keep to a minimal size the number of surface atoms $n_{surf}$ that are needed to calculate surface dipoles, and thus the surface electrostatic stress, because the size of the matrix that is needed to be inverted to obtain the surface electrostatic stress is directly proportional to $n_{surf}$. Techniques for accomplishing this are currently under evaluation.

## Acknowledgments

HSP gratefully acknowledges the support of NSF grant CMMI-0856261 for this research, and valuable discussions with Patrick A. Klein. MD and HSP thank Zhao Wang for valuable discussions on this subject. This chapter is reproduced with permission from Harold S. Park, Michel Devel and Zhao Wang, A New Multiscale Formulation for the Electromechanical Behavior of Nanomaterials, Computer Methods in Applied Mechanics in Engineering, 200, 2447–2457, © 2011 Elsevier.

## References

[1] Guo, W. and Guo, Y. (2003) Giant axial electrostrictive deformation in carbon nanotubes. *Physical Review Letters*, **91** (11), 115501.
[2] Mamin, H.J. and Rugar, D. (2001) Sub-attonewton force detection at millikelvin temperatures. *Applied Physics Letters*, **79** (20), 3358–3360.
[3] LaHaye, M.D., Buu, O., Camarota, B., and Schwab, K.C. (2004) Approaching the quantum limit of a nanomechanical resonator. *Science*, **304**, 74–77.
[4] Knobel, R.G. and Cleland, A.N. (2003) Nanometre-scale displacement sensing using a single electron transistor. *Nature*, **424**, 291–293.
[5] Jensen, K., Kim, K., and Zettl, A. (2008) An atomic-resolution nanomechanical mass sensor. *Nature Nanotechnology*, **3**, 533–537.
[6] Ekinci, K.L. and Roukes, M.L. (2005) Nanoelectromechanical systems. *Review of Scientific Instruments*, **76**, 061101.
[7] Unterreithmeier, Q.P., Weig, E.M., and Kotthaus, J.P. (2009) Universal transduction scheme for nanomechanical systems based on dielectric forces. *Nature*, **458**, 1001–1004.
[8] Zhang, Y., Chang, A., Cao, J. *et al.* (2001) Electric-field-directed growth of aligned single-walled carbon nanotubes. *Applied Physics Letters*, **79** (19), 3155–3157.
[9] Chen, C.Q., Shi, Y., Zhang, Y.S. *et al.* (2006) Size dependence of the Young's modulus of ZnO nanowires. *Physical Review Letters*, **96**, 075505.
[10] Poncharal, P., Wang, Z.L., Ugarte, D., and de Heer, W.A. (1999) Electrostatic deflections and electromechanical resonances of carbon nanotubes. *Science*, **283**, 1513–1516.
[11] Wei, Y., Xie, C., Dean, K.A., and Coll, B.F. (2001) Stability of carbon nanotubes under electric field studied by scanning electron microscopy. *Applied Physics Letters*, **79** (27), 4527–4529.
[12] Guo, Y. and Guo, W. (2003) Mechanical and electrostatic properties of carbon nanotubes under tensile loading and electric field. *Journal of Physics D: Applied Physics*, **36**, 805–811.
[13] Wang, Z. and Philippe, L. (2009) Deformation of doubly clamped single-walled carbon nanotubes in an electrostatic field. *Physical Review Letters*, **102**, 215501.
[14] Aluru, N.R. and White, J. (1997) An efficient numerical technique for electromechanical simulation of complicated microelectromechanical structures. *Sensors and Actuators A*, **58**, 1–11.
[15] Senturia, S.D., Harris, R.M., Johnson, B.P. *et al.* (1992) A computer-aided design system for microelectromechanical systems (MEMCAD). *Journal of Microelectromechanical Systems*, **1** (1), 3–13.

[16] Shi, F., Ramesh, P., and Mukherjee, S. (1996) Dynamic analysis of micro-electro-mechanical systems. *International Journal for Numerical Methods in Engineering*, **39**, 4119–4139.
[17] De, S.K. and Aluru, N.R. (2004) Full-Lagrangian schemes for dynamic analysis of electrostatic MEMS. *Journal of Microelectromechanical Systems*, **13** (5), 737–758.
[18] Soma, A., Bona, F.D., Gugliotta, A., and Mola, E. (2001) Meshing approach in non-linear FEM analysis of microstructures under electrostatic loads. *Proceedings of the SPIE*, **4408**, 216–225.
[19] Li, G. and Aluru, N.R. (2002) A Lagrangian approach for electrostatic analysis of deformable conductors. *Journal of Microelectromechanical Systems*, **11** (3), 245–254.
[20] Telukunta, S. and Mukherjee, S. (2006) Fully Lagrangian modeling of MEMS with thin plates. *Journal of Microelectromechanical Systems*, **15** (4), 795–810.
[21] Sumant, P.S., Aluru, N.R., and Cangellaris, A.C. (2009) A methodology for fast finite element modeling of electrostatically actuated MEMS. *International Journal for Numerical Methods in Engineering*, **77**, 1789–1808.
[22] Kim, G.H. and Park, K.C. (2008) A continuum-based modeling of MEMS devices for estimating their resonant frequencies. *Computer Methods in Applied Mechanics and Engineering*, **198**, 234–244.
[23] Kim, G.H. (2006) Partitioned analysis of electromagnetic field problems by localized lagrange multipliers, Technical report, Center for Aerospace Structures, University of Colorado at Boulder.
[24] Cammarata, R.C. (1994) Surface and interface stress effects in thin films. *Progress in Surface Science*, **46** (1), 1–38.
[25] Park, H.S., Cai, W., Espinosa, H.D., and Huang, H. (2009) Mechanics of crystalline nanowires. *MRS Bulletin*, **34** (3), 178–183.
[26] Agrawal, R., Peng, B., Gdoutos, E., and Espinosa, H.D. (2008) Elasticity size effects in ZnO nanowires – a combined experimental–computational approach. *Nano Letters*, **8** (11), 3668–3674.
[27] Cuenot, S., Frétigny, C., Demoustier-Champagne, S., and Nysten, B. (2004) Surface tension effect on the mechanical properties of nanomaterials measured by atomic force microscopy. *Physical Review B*, **69**, 165410.
[28] Zhu, Y., Xu, F., Qin, Q. *et al.* (2009) Mechanical properties of vapor–liquid–solid synthesized silicon nanowires. *Nano Letters*, **9** (11), 3934–3939.
[29] Miller, R.E. and Shenoy, V.B. (2000) Size-dependent elastic properties of nanosized structural elements. *Nanotechnology*, **11**, 139–147.
[30] Zhou, L.G. and Huang, H. (2004) Are surfaces elastically softer or stiffer? *Applied Physics Letters*, **84** (11), 1940–1942.
[31] Liang, H., Upmanyu, M., and Huang, H. (2005) Size-dependent elasticity of nanowires: nonlinear effects. *Physical Review B*, **71**, 241403(R).
[32] Diao, J., Gall, K., and Dunn, M.L. (2004) Atomistic simulation of the structure and elastic properties of gold nanowires. *Journal of the Mechanics and Physics of Solids*, **52**, 1935–1962.
[33] Park, H.S. and Klein, P.A. (2008) Surface stress effects on the resonant properties of metal nanowires: the importance of finite deformation kinematics and the impact of the residual surface stress. *Journal of the Mechanics and Physics of Solids*, **56**, 3144–3166.
[34] Guillaume, M., Champagne, B., Begue, D., and Pouchan, C. (2009) Electrostatic interaction schemes for evaluating the polarizability of silicon clusters. *Journal of Chemical Physics*, **130**, 134715.
[35] Mayer, A., Gonzales, A.L., Aikens, C.M., and Schatz, G.C. (2009) A charge–dipole interaction model for the frequency-dependent polarizability of silver clusters. *Nanotechnology*, **20**, 195204.
[36] Jensen, L.L. and Jensen, L. (2008) Electrostatic interaction model for the calculation of the polarizability of large nobel metal nanostructures. *Journal of Physical Chemistry C*, **112**, 15697–15703.
[37] Jensen, L.L. and Jensen, L. (2009) Atomistic electrodynamics model for optical properties of silver nanoclusters. *Journal of Physical Chemistry C*, **113**, 15182–15190.
[38] Wagner, G.J. and Liu, W.K. (2003) Coupling of atomistic and continuum simulations using a bridging scale decomposition. *Journal of Computational Physics*, **190**, 249–274.
[39] Park, H.S., Karpov, E.G., Liu, W.K., and Klein, P.A. (2005) The bridging scale for two-dimensional atomistic/continuum coupling. *Philosophical Magazine*, **85** (1), 79–113.
[40] Park, H.S., Karpov, E.G., Klein, P.A., and Liu, W.K. (2005) Three-dimensional bridging scale analysis of dynamic fracture. *Journal of Computational Physics*, **207**, 588–609.
[41] Klein, P.A. (1999) A virtual internal bond approach to modeling crack nucleation and growth, PhD thesis, Stanford University.
[42] Tadmor, E., Ortiz, M., and Phillips, R. (1996) Quasicontinuum analysis of defects in solids. *Philosophical Magazine A*, **73**, 1529–1563.

[43] Park, H.S., Klein, P.A., and Wagner, G.J. (2006) A surface cauchy–born model for nanoscale materials. *International Journal for Numerical Methods in Engineering*, **68**, 1072–1095.

[44] Park, H.S. and Klein, P.A. (2007) Surface cauchy–born analysis of surface stress effects on metallic nanowires. *Physical Review B*, **75**, 085408.

[45] Park, H.S. and Klein, P.A. (2008) A surface cauchy–born model for silicon nanostructures. *Computer Methods in Applied Mechanics and Engineering*, **197**, 3249–3260.

[46] Yun, G. and Park, H.S. (2008) A finite element formulation for nanoscale resonant mass sensing using the surface cauchy–born model. *Computer Methods in Applied Mechanics and Engineering*, **197**, 3324–3336.

[47] Yun, G. and Park, H.S. (2008) A multiscale, finite deformation formulation for surface stress effects on the coupled thermomechanical behavior of nanomaterials. *Computer Methods in Applied Mechanics and Engineering*, **197**, 3337–3350.

[48] Park, H.S. (2008) Surface stress effects on the resonant properties of silicon nanowires. *Journal of Applied Physics*, **103**, 123504.

[49] Park, H.S. (2009) Quantifying the size-dependent effect of the residual surface stress on the resonant frequencies of silicon nanowires if finite deformation kinematics are considered. *Nanotechnology*, **20**, 115701.

[50] Park, H.S. (2008) Strain sensing through the resonant properties of deformed metal nanowires. *Journal of Applied Physics*, **104**, 013516.

[51] Yun, G. and Park, H.S. (2009) Surface stress effects on the bending properties of fcc metal nanowires. *Physical Review B*, **79**, 195421.

[52] Yun, G. and Park, H.S. (2012) Bridging the gap between experimental measurements and atomistic predictions of the elastic properties of silicon nanowires using multiscale modeling. *Finite Elements in Analysis and Design*, **49**, 3–12.

[53] Park, H.S. (2010) A multiscale finite element method for the dynamic analysis of surface-dominated nanomaterials. *International Journal for Numerical Methods in Engineering*, **83**, 1237–1254.

[54] Park, H.S., Devel, M., and Wang, Z. (2011) A new multiscale formulation for the electromechanical behavior of nanomaterials. *Computer Methods in Applied Mechanics and Engineering*, **200**, 2447–2457.

[55] Wang, Z., Devel, M., Langlet, R., and Dulmet, B. (2007) Electrostatic deflections of cantilevered semiconducting single-walled carbon nanotubes. *Physical Review B*, **75**, 205414.

[56] Wang, Z. and Devel, M. (2007) Electrostatic deflections of cantilevered metallic carbon nanotubes via charge–dipole model. *Physical Review B*, **76**, 195434.

[57] Stuart, S.J., Tutein, A.B., and Harrison, J.A. (2000) A reactive potential for hydrocarbons with intermolecular interactions. *Journal of Chemical Physics*, **112**, 6472–6486.

[58] Langlet, R., Arab, M., Picaud, F. *et al.* (2004) Influence of molecular adsorption on the dielectric properties of a single wall carbon nanotube: a model sensor. *Journal of Computational Physics*, **121**, 9655–9665.

[59] Mayer, A., Lambin, Ph., and Langlet, R. (2006) Charge–dipole model to compute the polarization of fullerenes. *Applied Physics Letters*, **89**, 063117.

[60] Mayer, A. (2007) Formulation in terms of normalized propagators of a charge–dipole model enabling the calculation of the polarization properties of fullerenes and carbon nanotubes. *Physical Review B*, **75**, 045407.

[61] Haile, J.M. (1992) *Molecular Dynamics Simulations*, John Wiley & Sons.

[62] Leach, A. (2001) *Molecular Modelling: Principles and Applications*, Pearson Education Limited.

[63] Silberstein, L. (1917) Dispersion and the size of molecules of hydrogen, oxygen, and nitrogen. *Philosophical Magazine*, **33** (193–198), 215–222.

[64] Silberstein, L. (1917) Molecular refractivity and atomic interaction. *Philosophical Magazine*, **33** (193–198), 92–128.

[65] Silberstein, L. (1917) Molecular refractivity and atomic interaction. II. *Philosophical Magazine*, **33** (193–198), 521–533.

[66] Applequist, J., Carl, J.R., and Fung, K.K. (1972) An atom dipole interaction model for molecular polarizability. Applicaction to polyatomic molecules and determination of atom polarizabilities. *Journal of the American Chemical Society*, **94** (9), 2952–2960.

[67] Purcell, E.M. and Pennypacker, C.R. (1973) Scattering and absorption of light by nonspherical dielectric grains. *Astrophysics Journal*, **1986**, 705–714.

[68] Draine, B.T. and Flatau, P.J. (1994) Discrete-dipole approximation for scattering calculations. *Journal of the Optical Society of America A*, **11** (4), 1491–1499.

[69] Girard, C. (1986) The electronic structure of an atom in the vicinity of a solid body: application to the physisorption on metal surface. *Journal of Computational Physics*, **85** (11), 6750–6757.

[70] Martin, O.J.F., Girard, C., and Dereux, A. (1995) Generalized field propagator pf electromagnetic scattering and light confinement. *Physical Review Letters*, **74** (4), 526–529.
[71] Devel, M., Girard, C., and Joachim, C. (1996) Computation of electrostatic fields in low-symmetry systems: application to STM configurations. *Physical Review B*, **53** (19), 13159–13168.
[72] Tai, C.T. (1994) *Dyadic Green Functions in Electromagnetic Theory*, Institute of Electrical & Electronics Engineering.
[73] Torrens, F., Ruiz-Lopez, M., Cativiela, C. *et al.* (1992) Conformational aspects of some asymmetric Diels–Alder reactions. A molecular mechanics + polarization study. *Tetrahedron*, **48**, 5209–5218.
[74] Torrens, F., Sanchez-Martin, J., and Nebot-Gill, I. (1999) Interacting induced dipoles polarization model for molecular polarizabilities. Reference molecules. *Journal of Molecular Structure: THEOCHEM*, **463**, 27–39.
[75] Torrens, F. (2000) Polarization force fields for peptides implemented in ECEPP2 and MM2. *Molecular Simulation*, **24**, 391–400.
[76] Kaminski, G.A., Friesner, R.A., and Zhou, R. (2003) A computationally inexpensive modification of the point dipole electrostatic polarization model for molecular simulations. *Journal of Computational Chemistry*, **24**, 267–276.
[77] Stillinger, F.H. (1979) Dynamics and ensemble averages for the polarization models of molecular interactions. *Journal of Chemical Physics*, **71** (4), 1647–1651.
[78] Girard, C. (1992) Multipolar propagators near a corrugated surface: implication for local probe microscopy. *Physical Review B*, **45** (4), 1800–1810.
[79] Stone, A.J. (1996) *The Theory of Intermolecular Forces*, Oxford University Press.
[80] Olson, M.L. and Sundberg, K.R. (1978) An atom monopole–dipole interaction model with charge transfer for the treatment of polarizabilities of pi-bonded molecules. *Journal of Chemical Physics*, **69** (12), 5400–5404.
[81] Mayer, A. (2005) A monopole–dipole model to compute the polarization of metallic carbon nanotubes. *Applied Physics Letters*, **86**, 153110.
[82] Birge, R.B. (1980) Calculation of molecular polarizabilities using an anisotropic point dipole interaction model which includes the effect of electron repulsion. *Journal of Chemical Physics*, **72** (10), 5312–5319.
[83] Thole, B.T. (1981) Molecular polarizabilities calculated with a modified dipole interaction. *Chemical Physics*, **59**, 341–350.
[84] Shanker, B. and Applequist, J. (1996) Atom monopole–dipole interaction model with limited delocalization length for polarization of polyenes. *Journal of Physical Chemistry*, **100**, 10834–10836.
[85] Jensen, L., Åstrand, P.-O., and Mikkelsen, K.V. (2001) An atomic capacitance–polarizability model for the calculation of molecular dipole moments and polarizabilities. *International Journal of Quantum Chemistry*, **84**, 513–522.
[86] Jensen, L., Åstrand, P.-O., Osted, A. *et al.* (2002) Polarizability of molecular clusters as calculated by a dipole interaction model. *Journal of Computational Physics*, **116** (10), 4001–4010.
[87] Wang, Z., Zu, X., Yang, L. *et al.* (2007) Atomistic simulation of the size, orientation, and temperature dependence of tensile behavior in GaN nanowires. *Physical Review B*, **76**, 045310.
[88] Mayer, A. (2005) Polarization of metallic carbon nanotubes from a model that includes both net charges and dipoles. *Physical Review B*, **71**, 235333.
[89] Langlet, R., Devel, M., and Lambin, P. (2006) Computation of the static polarizabilities of multi-wall carbon nanotubes and fullernes using a Gaussian regularized point dipole interaction model. *Carbon*, **44**, 2883–2895.
[90] De Leeuw, S.W., Perram, J.W., and Smith, E.R. (1980) Simulation of electrostatic systems in periodic boundary conditions. I. Lattice sums and dielectric constants. *Proceedings of The Royal Society of London. Series A, Mathematical and Physical Sciences*, **373**, 27–56.
[91] Arroyo, M. and Belytschko, T. (2002) An atomistic-based finite deformation membrane for single layer crystalline films. *Journal of the Mechanics and Physics of Solids*, **50**, 1941–1977.
[92] Zhang, P., Huang, Y., Geubelle, P.H. *et al.* (2002) The elastic modulus of single-wall carbon nanotubes: a continuum analysis incorporating interatomic potentials. *International Journal of Solids and Structures*, **39**, 3893–3906.
[93] Tang, Z., Zhao, H., Li, G., and Aluru, N.R. (2006) Finite-temperature quasicontinuum method for multiscale analysis of silicon nanostructures. *Physical Review B*, **74**, 064110.
[94] Wolf, D., Keblinski, P., Phillpot, S.R., and Eggebrecht, J. (1999) Exact method for the simulation of coulombic systems by spherically truncated, pairwise $r^{-1}$ summation. *Journal of Chemical Physics*, **110** (17), 8254–8282.
[95] Fennell, C.J. and Gezelter, J.D. (2006) Is the Ewald summation still necessary? Pairwise alternatives to the accepted standard for long-range electrostatics. *Journal of Chemical Physics*, **124**, 234104.

# 6

# Towards a General Purpose Design System for Composites

Jacob Fish
*Columbia University, USA*

## 6.1 Motivation

To meet our future energy needs in an environmentally friendly manner, polymer matrix composites (PMCs) that are safe, lightweight, and cost-effective will need to be developed. A recent DOT report [1] indentified weight reduction through lightweight materials as one of the best ways to achieve the reduction of energy consumption, lower emissions, and better safety. With 75% of vehicle gas (energy) consumption directly related to factors associated with vehicle weight, the potential benefits of weight reduction will enable smaller powerplant (engine, turbine, fuel cells, etc.) and energy storage (battery, flywheel, etc.) systems, with corresponding cost and/or performance benefits. In all cases, the safety and crashworthiness of lighter weight vehicles is a significant consideration, as well as other environmental, safety, and hazards issues associated with new materials and processing technologies.

The enormous gains offered by advanced high-temperature polymer matrix composites (HTPMCs), such as PMR-15 and Renegade Materials' FreeForm14 Polyimide (http://renegadematerials.com/), due to their light weight, high specific strength and stiffness, and property tailorability are well known [2]. If these materials could be utilized in turbine engines, exhaust wash structures, and high-speed aircraft skins, where structural components are exposed to harsh service conditions, the application window would be significantly larger. The HTPMC materials in these environments generally experience coupled thermal oxidation reaction, microstructural damage, and thermo-mechanical loading, leading to mechanical and other physical property degradation, and consequently limited life.

To develop nearly optimal PMC and HTPMC material systems, a truly predictive model is urgently needed. While, candidate material systems that could retain their mechanical properties in extreme environments exist, their insertion in mission critical applications is hindered by their *lack of predictability*. This lack of predictability has often resulted in overdesign, and thus limited their use since

*Multiscale Simulations and Mechanics of Biological Materials*, First Edition. Edited by Shaofan Li and Dong Qian.

the overdesigned component may not yield any design advantages. There are several reasons for this state of affairs:

1. existence of multiple spatial and temporal scales;
2. existence of multiple physical processes, such as coupled thermo–mechano–oxidation–fatigue processes;
3. lack of robustness due to loss of accuracy in the vicinity of high-gradient regions, large representative volume element (RVE) distortions, and lack of periodicity;
4. lack of integrated design tools; and
5. lack of material characterization and experimental data at the microscale.

The *first barrier* is primarily computational. The behavior and degradation mechanisms of PMC and HTPMC composites components are determined by phenomena operating on a range of time and spatial scales. Yet, the supporting modeling and simulation tools that attempt to deal with this tyranny of scales range from the rule of mixtures to various effective medium and self-consistent models [3–7]. To enable design of these composites components to their potential, in particular for use in extreme environments, such simplistic models must be replaced by multiscale–multiphysics (coupled thermo–chemo–mechanical–fatigue) approaches, which can reliably predict long-term durability and performance.

Emerging computational homogenization methods based on the mathematical homogenization theory [8–15] has had so far little or no impact on practitioners due to the enormous computational complexity involved. To illustrate the computational complexity involved, consider a macro-problem with $N_{\text{cells}}$ Gauss points, $n$ load increments at the macroscale, and $I_{\text{coarse}}$ and $I_{\text{fine}}$ average iterations at the macro- and micro-scales, respectively. The total number of linear solves of a micro-problem is thus $N_{\text{cells}} n I_{\text{coarse}} I_{\text{fine}}$ – a formidable computational complexity if the number of RVEs and degrees of freedom in the RVEs is substantial. This tyranny of scales can be effectively addressed by a combination of parallel methods and by introducing an intermediate mesomechanical model, as shown in Figure 6.1.

Development of mesomechanical or reduced-order models for heterogeneous continua has been an active research area in the past decade. Perhaps one of the oldest reduced-order approaches is based on the purely kinematical Taylor hypothesis, which assumes uniform deformation at the fine scale; it satisfies compatibility but fails to account for equilibrium across microconstituents' boundaries. Major progress in mesomechanical modeling (at the expense of computational cost) has been made by utilizing the Voronoi cell method [16], the spectral method [17], the network approximation method [18], the fast Fourier transforms [19, 20], and transformation field analysis (TFA) [21, 22]. Despite significant

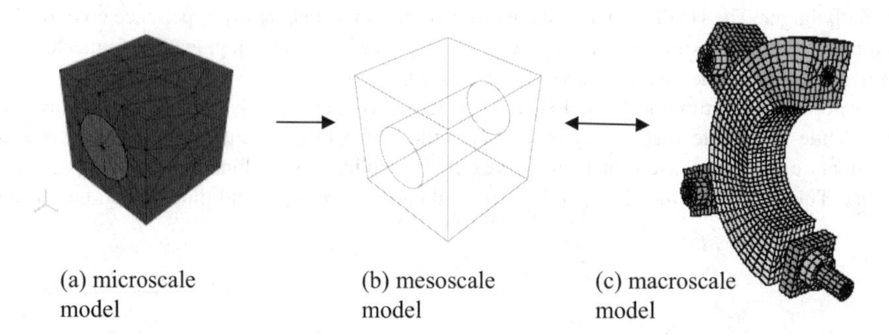


**Figure 6.1** Linking micromechanical and macromechanical problems through a mesomechanical (reduced-order) model

progress, the need for a flexible low-cost mesomechanical approach, which can be easily adapted to meet accuracy needs, still remains.

TFA [23–26] is based on a brilliant idea that allows precomputing certain information (localization operators, concentration tensors, transformation influence functions) in the preprocessing phase prior to nonlinear analysis, which consequently can be carried out with a small subset of unknowns. By this approach the effect of eigenstrains, representing inelastic strains, thermal strains, or phase transformation strains, is accounted for by solving a sequence of linear elasticity problems. The salient feature of TFA-based approaches is that RVE equilibrium equations, which have to be solved $N_{\text{cells}} n I_{\text{coarse}} I_{\text{fine}}$ times in direct homogenization approaches (see earlier discussion) are satisfied a priori, in the preprocessing stage.

Yet, despite the promise, TFA has found limited success in practice. There are several reasons for this:

1. it does not account for interface failure;
2. it is not hierarchical, in the sense that model improvement is not possible when higher accuracy is needed, such as in the regions of boundary layers;
3. it is limited to two scales; and
4. it lacks a rigorous mathematical framework.

The above deficiencies have recently been circumvented by reduced-order homogenization [27–31]. The reduced-order homogenization, which serves as the basis for the present work, has the following characteristics:

1. It accounts for interface failure in addition to failure of microconstituents; interface failure is modeled using so-called eigenseparations – a concept similar to eigenstrains used for modeling inelastic deformation of phases.
2. It is equipped with a hierarchical model improvement capability; that is, it incorporates a hierarchical sequence of computational homogenization models where the most inexpensive member of the sequence represents uniform deformation within each microphase (inclusion, matrix and interface), whereas the most comprehensive model of the hierarchical sequence coincides with a direct homogenization approach.
3. Like TFA, it constructs residual-free stress fields and, thus, de facto eliminates the need for costly solution of discrete equilibrium equations in the RVE. This removes a major computational bottleneck and thus permits accounting for arbitrary micromechanical details at a cost comparable to phenomenological modeling of nonlinear heterogeneous medium.

However, the basic reduced-order homogenization formulation does not account for size effect, high gradients, large RVE distortions, and nonperiodic solutions. *These limitations will be addressed in this chapter.*

The *second barrier*, lack of a predictive multiphysics model that couples thermo–mechano–oxidative–fatigue degradation processes at multiple scales, is by and large responsible for the fact that, to date, life prediction of critical composites components is based on experiments. There seems to be a disconnect between academic community and practicing world. While the academic community continues to pursue various idealized environmental degradation and Paris-like fatigue models, very few, if any, complex composites components were designed that way.

*In this chapter we will describe a space–time multiscale fatigue model.*

The *third barrier*, lack of accuracy in the vicinity of high-gradient regions, is concerned with the principal limitations of the homogenization approach: periodicity and uniformity of macroscale fields. Various hierarchical techniques [32, 33], higher order homogenization techniques [34], and nonperiodic homogenization methods [15, 35– 37] have been proposed to extend the range of validity of the computational homogenization approaches. While higher order theories [14, 38, 39] are equipped with

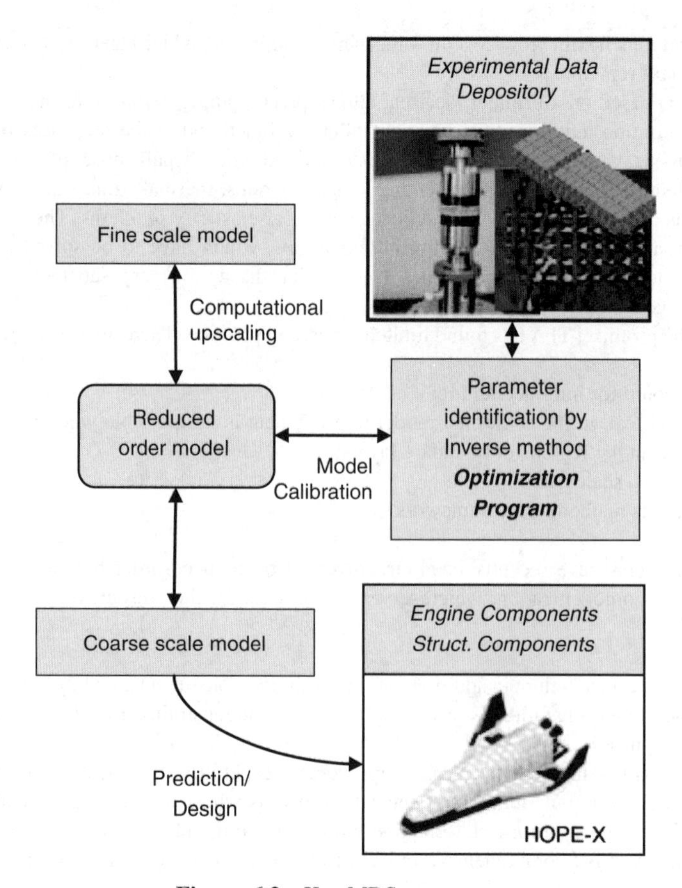


**Figure 6.2** Key MDS components

a capability of subjecting the RVE to "true" macroscopic deformation, they have not found their way into the commercial arena because they require consideration of coupled tractions and $C^1$ continuous formulation.

*In this chapter we will outline the reduced-order computational continua ($RC^2$) approach that possesses the accuracy of second-order homogenization methods but involves no coupled tractions and $C^1$ continuity.*

The *fourth barrier* is concerned with absence of integrated design tools. An experimentally validated multiscale design system (MDS) that would account for phenomena at multiple scales to predict the behavior of composites components has long been on the wish list of practitioners. Such a design system would be indispensable in systematic exploration of alternative designs at the material and structural scales. Development of such an MDS would advance the state of the art in the field far beyond an equivalent investment in its comprising components (materials, mechanics, testing, and computations).

The multiscale design process consists of several interlinked stages schematically illustrated in Figure 6.2. The technologies and modules involved in this process are:

1. *Mathematical upscaling:* derivation of coupled coarse-scale equations from fine-scale equations using mathematical homogenization theory [13, 14, 16, 36, 37].

2. *Computational upscaling:* reducing the complexity of solving the RVE problem to a manageable size that can be adapted based on available computational resources and error estimates in the quantities of interest [27–29].
3. *Model calibration:* solving an inverse problem for constitutive parameters (interfaces, fibers/tows, matrix) by minimizing the error between experimental data at a coupon level [40].

The *fifth barrier* is concerned with scale-specific measurements of in-situ properties, uncertainty quantification, and indirect calibration by inverse methods [41]. It is noteworthy to mention that while small random variations in the composite microstructure are unlikely to significantly affect linear properties, they are critical in predicting evolution of failure with any confidence.

*In this chapter we will study how the uncertainties caused by random fields affect component behavior.*

## 6.2 General Purpose Multiscale Formulation

### 6.2.1 The Basic Reduced-Order Model

The reduced homogenization approach [27– 29] hinges on construction of the residual-free displacement field in the RVE domain that removes the bottleneck of solving discrete equilibrium equations. The residual-free field is defined as

$$u_i^{(1)}(x, y) = H_i^{kl}(y)\varepsilon_{kl}^c(x) + \int_\Theta \tilde{h}_i^{kl}(y, \tilde{y})\mu_{kl}(x, \tilde{y})\, d\tilde{\Theta} + \int_S \breve{h}_i^{\breve{n}}(y, \breve{y})\delta_{\breve{n}}(x, \breve{y})\, d\breve{S}, ) \tag{6.1}$$

where $H_i^{kl}$, $\tilde{h}_i^{kl}$, and $\breve{h}_i^{\breve{n}}$ are so-called transformation influence functions for the coarse-scale strain, the fine-scale eigenstrain, and the fine-scale eigenseparation, respectively; $x$ and $y = x/\zeta$ are macroscopic and microscopic position vectors with $0 < \zeta \ll 1$. Model reduction is accomplished by discretizing the eigenstrains $\mu_{kl}$ and eigenseparations $\delta_{\breve{n}}$ in terms $C^{-1}(\Theta)$ and $C^0(S)$ continuous shape functions in the RVE interior $\Theta$ and at the interfaces $S$, respectively

$$\mu_{ij}(x, y) = \sum_{\alpha=1}^{\tilde{M}} \tilde{N}^{(\alpha)}(y)\mu_{ij}^{(\alpha)}(x) \qquad \delta_{\breve{n}}(x, y) = \sum_{\xi=1}^{\breve{M}} \breve{n}^{(\xi)}(y)\delta_{\breve{n}}^{(\xi)}(x), \tag{6.2}$$

where $\tilde{M}$ and $M$ are the number of phases and interfaces modes (or partitions), respectively, and $\mu_{ij}^{(\alpha)}$ and $\delta_{\breve{n}}^{(\xi)}$ are the corresponding coefficients. At the minimum, there is one mode (partition) for each inelastic phase or interface that is needed. The most accurate member of the hierarchy has the number of modes (partitions) equal to the number of finite elements in the RVE including the interface elements.

The resulting macroscopic constitutive model is given by

$$\sigma_{ij}^c(x) = L_{ijkl}^c\varepsilon_{kl}^c(x) + \sum_{\alpha=1}^{\tilde{M}} A_{ijkl}^{c(\alpha)}\, \mu_{kl}^{(\alpha)}(x) + \sum_{\xi=1}^{\breve{M}} B_{ij\breve{n}}^{c(\xi)}\, \delta_{\breve{n}}^{(\xi)}(x), \tag{6.3}$$

where $L_{ijkl}^c$, $A_{ijkl}^{c(\alpha)}$, and $B_{ij\breve{n}}^{c(\xi)}$ are precomputed elastic and inelastic influence functions; $\sigma_{ij}^c(x)$ and $\varepsilon_{kl}^c(x)$ are the macroscopic stress and strain components; and $\delta_{\breve{n}}^{(\xi)}(x)$ and $\mu_{kl}^{(\alpha)}(x)$are the eigenseparation in the interface partition $\zeta$ and the eigenstrain in the interior partition $\alpha$ computed by solving a nonlinear reduced-order RVE problem at each macroscopic Gauss point [27–30].

Computational savings compared with the direct homogenization range from factor of 30 in the impact of sandwich plates and crack growth in composite ships, to over two orders of magnitude in composite cars and compression molding of GE90 fan blades, to over three orders of magnitude in fatigue crack growth.

### 6.2.2 *Enhanced Reduced-Order Model*

The basic reduced-order model described in Section 6.2.1 suffers from the following limitations:

1. The basic model is based on the classical double-scale asymptotic expansion method, which assumes that macroscopic fields are constant over the RVE domain. In the case of large macroscale gradients, the aforementioned classical assumption is no longer valid.
2. The model assumes that the influence functions $H_i^{kl}$, $\tilde{h}_i^{kl}$, and $\breve{h}_i^{\breve{n}}$ (see Equation (6.1)) can be precomputed prior to nonlinear analysis. This is no longer true for large RVE distortions.
3. The model takes advantage of periodic boundary conditions.

To circumvent the above limitations we generalize the above formulation in the following respects.

The *first salient feature* of the proposed generalization is designated to allow for large RVE distortions, which gives rise to coupling between fine and coarse scales. Coupling between the two scales will be introduced by expanding the leading order displacement $u_i^0(\boldsymbol{X})$ in the Taylor series around the unit cell centroid $\boldsymbol{X} = \hat{\boldsymbol{X}}$ in the initial unit cell configuration.

To account for existence of high macroscale gradients we assume

$$u_i^0(\boldsymbol{X}) = u_i^0(\hat{\boldsymbol{X}}) + \left.\frac{\partial u_i^0}{\partial X_j}\right|_{\hat{X}} (X_j - \hat{X}_j) + \frac{1}{2}\left.\frac{\partial^2 u_i^0}{\partial X_j \partial X_k}\right|_{\hat{X}} (X_j - \hat{X}_j)(X_k - \hat{X}_k) + \cdots \tag{6.4}$$

$$\frac{\partial^{n+1} u_i^0}{\partial X_j \ldots \partial X_k} = \frac{O(\zeta^{-n\alpha})}{l'^n} \quad \text{for } n = 0, 1 \text{ and } 0 < \alpha \leq 1 \tag{6.5}$$

and/or the existence of large unit cells

$$\left.\frac{\partial u_i^0}{\partial X_j}\right|_{\hat{X}} (X_j - \hat{X}_j) = l' O(\zeta^\alpha); \quad \left.\frac{\partial^2 u_i^0}{\partial X_j \partial X_k}\right|_{\hat{X}} (X_j - \hat{X}_j)(X_k - \hat{X}_k) = l' O(\zeta^\alpha), \tag{6.6}$$

where $l'$ is a dimensional characteristic parameter and $\boldsymbol{Y} \equiv \boldsymbol{X}/\zeta^\alpha$. For the second-order theory considered hereafter we assume that higher order terms in Equation (6.4) remain of order $O(\zeta^{2\alpha})$ and higher.

It can be shown that combining the above with governing equilibrium equations stated on the fine scale yields the weak form of the coarse-scale equations:

$$\begin{aligned}\int_{\Omega_X} \frac{\partial w_i^0}{\partial X_k} \bar{P}_{ik}\, d\Omega + \zeta^\alpha \int_{\Omega_X} \frac{\partial^2 w_i^0}{\partial X_j \partial X_k} \bar{Q}_{ikj}\, d\Omega &= \int_{\partial\Omega_X^t} w_i^0 \bar{T}_i\, d\Gamma \\ &+ \int_{\Omega_X} w_i^0 \bar{B}_i\, d\Omega + \zeta^\alpha \int_{\partial\Omega_X^m} \frac{\partial w_i^0}{\partial X_k} \bar{T}_{ij}\, d\Gamma + \zeta^\alpha \int_{\Omega_X} \frac{\partial w_i^0}{\partial X_k} \bar{B}_{ij}\, d\Omega,\end{aligned} \tag{6.7}$$

where $\Omega_X$ is the macro-problem domain and $\partial\Omega_X$ its boundary; $P_{ik}$ and $F_{ik}$ are the first Piola–Kirchhoff stress and deformation gradient, respectively; $\bar{T}_i$ and $\bar{B}_i$ are coarse-scale tractions and body forces, respectively; $\boldsymbol{w}^0(\boldsymbol{X}) \in W^1_{\Omega_X}$ is the leading-order asymptotic expansion of the weight (or test) function, defined by

$$W^1_{\Omega_X} = \left\{ \boldsymbol{w}^0 \text{ defined in } \Omega_X,\ \boldsymbol{w}^0 \in C^1(\Omega_X),\ \boldsymbol{w}^0 = 0 \text{ on } \partial\Omega_X^{u\zeta} \right\}. \tag{6.8}$$

$\bar{P}_{ij}(\hat{\boldsymbol{X}})$and $\bar{Q}_{ikj}(\hat{\boldsymbol{X}})$ are the overall first Piola–Kirchoff stress and coupled stress, respectively, given by

$$\bar{P}_{ij}(\hat{\boldsymbol{X}}) = \frac{1}{|\Theta_Y|}\int_{\Theta_Y} P_{ij}(\boldsymbol{F}(\hat{\boldsymbol{X}}, \boldsymbol{Y}))\,\mathrm{d}\Theta; \quad \bar{Q}_{ikj}(\hat{\boldsymbol{X}}) = \frac{1}{|\Theta_Y|}\int_{\Theta_Y} P_{ik}(\boldsymbol{F}(\hat{\boldsymbol{X}}, \boldsymbol{Y}))Y_j\,\mathrm{d}\Theta. \tag{6.9}$$

The resulting strong form obtained by appropriate integration by parts of Equation (6.7) gives rise to a second-order continuum formulation, which is capable of capturing size effects, but suffers from anomalies associated with coupled traction boundary conditions and the need for $C^1$ continuous formulation.

Thus the *second salient feature* is to introduce the reduced order computational continua (RC[2]) description [42, 43] that is equivalent to Equation (6.7), but is free of higher order derivatives and higher order boundary conditions. This will be accomplished by introducing an additive decomposition of stress $P_{ik}(\hat{\boldsymbol{X}}, \boldsymbol{\chi})$ into the coarse-scale stress $P^{\mathrm{C}}_{ik}(\hat{\boldsymbol{X}}, \boldsymbol{\chi})$ and the fine-scale perturbation $P^*_{ik}(\hat{\boldsymbol{X}}, \boldsymbol{\chi})$ as

$$P^0_{ik}(\hat{\boldsymbol{X}}, \boldsymbol{\chi}) = P^{\mathrm{C}}_{ik}(\hat{\boldsymbol{X}}, \boldsymbol{\chi}) + P^*_{ik}(\hat{\boldsymbol{X}}, \boldsymbol{\chi}), \tag{6.10}$$

$$P^{\mathrm{C}}_{ik}(\hat{\boldsymbol{X}}, \boldsymbol{\chi}) = \bar{P}_{ik}(\hat{\boldsymbol{X}}) + Q_{ikj}(\hat{\boldsymbol{X}})\chi_j, \tag{6.11}$$

where $\boldsymbol{\chi} = \zeta^\alpha \boldsymbol{Y} = \boldsymbol{X} - \hat{\boldsymbol{X}}$ denotes the physical coordinates in the undeformed physical unit cell domain denoted by $\Theta_\chi$. The first term in Equation (6.11) describes the constant part of the coarse-scale stress, whereas the second term denotes its linear variation. The two constants $\bar{P}_{ik}(\hat{\boldsymbol{X}})$ and $Q_{ikj}(\hat{\boldsymbol{X}})$ are defined to satisfy the following conditions:

$$\int_{\Theta_\chi} P^*_{ik}(\hat{\boldsymbol{X}}, \boldsymbol{\chi})\,\mathrm{d}\Theta = 0 \quad \text{and} \quad \int_{\Theta_\chi} P^*_{ik}(\hat{\boldsymbol{X}}, \boldsymbol{\chi})\chi_j\,\mathrm{d}\Theta = 0. \tag{6.12}$$

The aforementioned definitions (6.10) and (6.12) can be shown to result in the coupled two-scale weak form, which states

$$\int_{\Omega_X}\left(\frac{1}{|\Theta_\chi|}\int_{\Theta_\chi}\frac{\partial w_i^{\mathrm{C}}}{\partial X_j}(\boldsymbol{X}, \boldsymbol{\chi})P^{\mathrm{C}}_{ik}(\boldsymbol{X}, \boldsymbol{\chi})\,\mathrm{d}\Theta\right)\mathrm{d}\Omega = \int_{\partial\Omega_X^t} w_i^{\mathrm{C}}T_i^{\zeta\Omega}\,\mathrm{d}\Gamma + \int_{\Omega_X} w_i^{\mathrm{C}}B_i^{\zeta}\,\mathrm{d}\Omega \quad \forall \boldsymbol{w}^{\mathrm{C}} \in W^{\mathrm{C}}_{\Omega_X}, \tag{6.13}$$

$$\int_{\Theta_Y}\frac{\partial w_i^1}{\partial Y_k}P_{ik}\,\mathrm{d}\Theta = 0 \quad \forall \boldsymbol{w}^1 \in W_{\Omega_X\times\Theta_Y},$$

where

$$\begin{aligned}
W_{\Theta_\chi\times\Omega_{\hat{X}}} &= \{\boldsymbol{u}^1(\boldsymbol{\chi}, \hat{\boldsymbol{X}}), w^1(\boldsymbol{\chi}, \hat{\boldsymbol{X}}) \text{ defined in } \Theta_\chi \times \Omega_{\hat{X}},\ C^0(\Theta_\chi),\ \text{weakly } \boldsymbol{\chi}\text{-periodic}\},\\
U_{\Omega_X} &= \{\boldsymbol{u}^{\mathrm{C}}(\boldsymbol{X}) \text{ defined in } \Omega_X,\ C^0(\Omega_X),\ \boldsymbol{u}^{\mathrm{C}} = \bar{\boldsymbol{u}} \text{ on } \partial\Omega_X^{u\zeta}\},\\
W^{\mathrm{C}}_{\Omega_X} &= \{\mathrm{w}^{\mathrm{C}}(\boldsymbol{X}) \text{ defined in } \Omega_X,\ C^0(\Omega_X),\ \boldsymbol{w}^{\mathrm{C}} = \boldsymbol{0} \text{ on } \partial\Omega_X^{u\zeta}\}.
\end{aligned} \tag{6.14}$$

It is noteworthy to mention that the RC[2] formulation hinges on the concept of *macro-stress function* as opposed to macro-stress being constant that averages micro-stresses. The left-hand side of (6.13) can be implemented through a specialized Gauss quadrature, where a quadrature point placed at the

RVE centroid is replaced by $2^{nsd}$ quadrature points required to integrate the product of bilinear macro-stress function $P^{C}_{ik}(\boldsymbol{X}, \boldsymbol{\chi})$ and the virtual deformation gradient $(\partial w^{C}_{i}/\partial X_j)(\boldsymbol{X}, \boldsymbol{\chi})$, where nsd denotes the number of space dimensions.

This leads us to the *third salient feature* of the enhanced model, which is consideration of weakly periodic boundary conditions. For problems involving nonzero gradients (that is, the macroscale deformation gradient $F^{C}_{jk}(\hat{\boldsymbol{X}}, \boldsymbol{Y})$ is a function of $\boldsymbol{Y}$), $u^{1}_{i}(\hat{\boldsymbol{X}}, \boldsymbol{Y})$ is no longer periodic even though some investigators [35, 44] assumed periodicity for this case as well. This lack of periodicity can be shown to exist due to the following. Consider the usual decomposition of the perturbation $u^{1}_{i} = H_{ijk}(\boldsymbol{Y})F^{C}_{jk}(\hat{\boldsymbol{X}}, \boldsymbol{Y})$, where $H_{ijk}(\boldsymbol{Y})$ is a periodic function. If $F^{C}_{jk}(\hat{\boldsymbol{X}}, \boldsymbol{Y}) = \bar{F}_{jk}(\hat{\boldsymbol{X}})$ is constant, then $u^{1}_{i}(\hat{\boldsymbol{X}}, \boldsymbol{Y})$ is periodic. Otherwise $u^{1}_{i}(\hat{\boldsymbol{X}}, \boldsymbol{Y})$ is no longer periodic.

The *fourth salient feature* is concerned with construction of the *residual free higher order* fields [44] for the RVE problem. The expansion (6.1) does not account for high gradients in the RVE domain or finite RVE rotations and assumes constant influence functions. To circumvent these limitations the residual free field (6.1) is generalized as follows:

$$\dot{u}^{\mathrm{cor}}_{i}(\boldsymbol{x}, \boldsymbol{y}) = H^{kl}_{i}(\boldsymbol{y})\dot{\bar{\varepsilon}}_{kl}(\boldsymbol{x}) + \boxed{h^{klm}_{i}(\mathbf{y})\dot{\bar{\varepsilon}}_{kl,m}(\boldsymbol{x})} + \int_{\Theta} \tilde{h}^{kl}_{i}(\boldsymbol{y}, \tilde{\boldsymbol{y}})\dot{\mu}_{kl}(\boldsymbol{x}, \tilde{\boldsymbol{y}})\,\mathrm{d}\tilde{\Theta} + \boxed{\int_{\Theta} \tilde{h}^{klm}_{i}(\boldsymbol{y},\tilde{\boldsymbol{y}})\dot{\mu}_{kl,m}(\boldsymbol{x},\tilde{\boldsymbol{y}})\mathbf{d}\tilde{\Theta}} + \int_{S} \breve{h}^{\hat{n}}_{i}(\boldsymbol{y}, \breve{\boldsymbol{y}})\dot{\delta}_{\hat{n}}(\boldsymbol{x}, \breve{\boldsymbol{y}})\,\mathrm{d}\breve{S} \tag{6.15}$$

It is noteworthy to mention the following attributes of the above generalization (6.15):

1. It is stated in the rate form, allowing the influence function to vary it time.
2. It includes the contribution of the residual free high-order gradients of the rate of deformation $\dot{\bar{\varepsilon}}_{kl,m}$ and eigenstrain $\dot{\mu}_{kl,m}$.
3. It is defined in the corotational frame placed at the RVE centroid, so that macroscopic rotations can be accounted for in the usual way [45, 46] (superscript "cor" on $\dot{u}^{\mathrm{cor}}_{i}$ denotes corotational displacements).

The computational efficiency of the higher order residual free formulation depends on how often the influence functions have to be recomputed.

## 6.3 Mechanistic Modeling of Fatigue via Multiple Temporal Scales

In the following we outline fatigue life prediction methodology for composites based on the adaptive cycle block technique for temporal upscaling [47, 48] and reduced-order homogenization for spatial upscaling [28–30] that does not rely on *S–N* and *ε–N* curves or Paris-law-like fatigue models [49]. The temporal scales exist due to slow degradation of material properties on the one hand and rapidly oscillatory loading on the other hand. For temporal upscaling, an efficient time integrator has been developed by discretizing the cyclic loading history into a series of load cycle blocks where each block consists of several load cycles. Since the increment of damage accumulation in a single load cycle is typically very small, the derivative of the damage parameter with respect to the number of load cycles can be approximated reasonably well. Consequently, damage evolution is modeled as a first-order initial-value problem with respect to number of load cycles:

$$\left.\frac{\mathrm{d}\dot{\omega}^{(m)}}{\mathrm{d}N}\right|_{K} \approx \int_{t}^{t+\tau_0} \dot{\omega}^{(m)}\,\mathrm{d}t = \left.\Delta\omega^{(m)}\right|_{K} = \left.\omega^{(m)}\right|_{K} - \left.\omega^{(m)}\right|_{K-1}, \tag{6.16}$$

where $N$ denotes the number of load cycles and $\omega^{(m)}|_K$ is phase damage (matrix of fiber) at the end of load cycle $K$. It is obtained by the incremental finite-element analysis for this cycle with the initial damage of $\omega^{(m)}|_{K-1}$ and the corresponding initial strain/stress conditions. The block size can be adaptively controlled using predictor–corrector time integrators and can be selected to keep the maximal phase damage increment sufficiently small.

Although the above adaptive block cycle scheme has been shown to be effective for simulation of evolution of state variables, the governing equations are not guaranteed to be satisfied since not all the cycles are simulated. In the following, we describe an alternative space–time multiscale fatigue life prediction methodology that satisfies the governing equations at each cycle. The space–time multiscale fatigue life prediction method takes advantage of the fact that accumulation of damage is a relatively slow temporal process compared with rapid fluctuations within each load cycle. The disparity between the two characteristic time scales suggests introduction of *multiple temporal scales* that decompose the original initial boundary-value problem into a *micro-chronological* (temporal RVE) problem, governed by fast time coordinate $\tau$, and a *macro-chronological* (homogenized) problem, governed by intrinsic time coordinate $t$. These two time scales are related through a scaling parameter:

$$\tau = \frac{t}{\zeta}; \quad 0 < \zeta \ll 1. \tag{6.17}$$

The resulting response fields $\phi^{\zeta}(\boldsymbol{X}, t)$ are assumed to depend on two spatial and two temporal scales:

$$\phi^{\zeta}(\boldsymbol{X}, t) = \phi(\boldsymbol{X}, \boldsymbol{Y}, t, \tau). \tag{6.18}$$

The space–time mathematical homogenization theory accounts for nonperiodicity in the time domain, which is a by-product of irreversible processes, such as damage accumulation caused by thermal (or mechanical) fatigue that violates the temporal periodicity condition. The response fields are assumed to be *almost periodic* in the time domain [50, 51] to account for irreversible processes of damage accumulation. The almost periodicity implies that, at the neighboring points in a spatial or temporal domain homologous by periodicity, the change in the response function is finite (in contrast to the case of local periodicity, in which the change in response functions is vanishing). A function $\phi_{\text{ap}}$ is said to be almost periodic if it belongs to the space of almost periodic functions $\Im$, defined as

$$\Im := \{\phi_{\text{ap}} | \phi_{\text{ap}}(\boldsymbol{X}, \boldsymbol{Y}, t, \tau + k\kappa) - \phi(\boldsymbol{X}, \boldsymbol{Y}, t, \tau) = O(\zeta)\}. \tag{6.19}$$

To construct the macro-chronological equations, define an almost periodic temporal homogenization (APTH) operator $\wp(\phi_{\text{ap}})$ as

$$\frac{\partial \wp(\phi_{\text{ap}})}{\partial t} = \frac{1}{|\kappa|} \int_{\kappa} \frac{\partial \phi_{\text{ap}}}{\partial t} \mathrm{d}\tau. \tag{6.20}$$

Figure 6.3 compares the fatigue life as a function of temperature and load amplitude as obtained with the space–time multiscale model and experimental results.

## 6.4 Coupling of Mechanical and Environmental Degradation Processes

### *6.4.1 Mathematical Model*

Chemical changes occurring in composites during oxidation result in mechanical response changes and in local loss of mass associated with outgassing of oxidation by-products. Severe surface oxidation degradation results in complex microstructure changes, formation of a nonuniform skin-core structure, and fiber–matrix disbonds coalescing into transverse surface cracks. These cracks not only reduce strength, but also create enhanced pathways for oxygen to penetrate deeper into the composite [52].

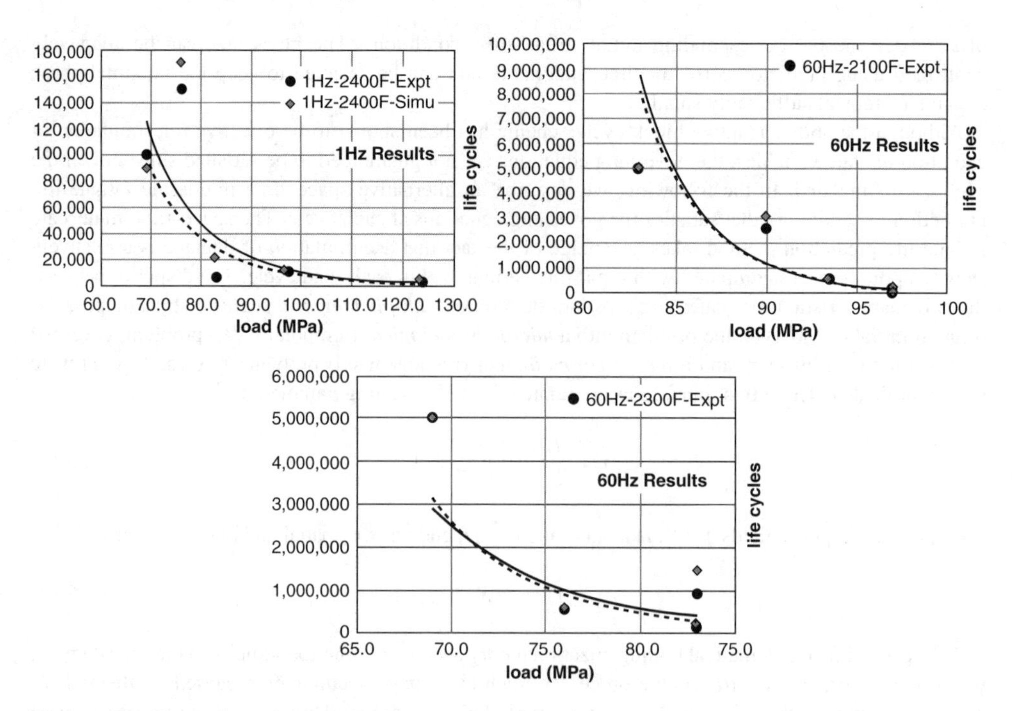


**Figure 6.3** Dependence of fatigue life on temperature and cycling load amplitude. Circles denote experimental results

Unlike most of the existing literature, where either a single physical process of environmental degradation at multiple scales is assumed [53, 54] or multiple physical processes on a single scale are considered [55], we will postulate the governing equations of the coupled mechanical, thermal oxidation, and reaction processes at the microscale of interest and will systematically derive the macroscale equations by means of mathematical and computational upscaling. The premise of the multiscale–multiphysics approach is that at the fine scale of interest the physical processes are better understood than at the coarse scale.

The governing equations to be considered at the fine scale are

$$\frac{\partial J_i}{\partial x_i} - R^\zeta(c^\zeta) = \dot{c}^\zeta \quad \text{and} \quad \frac{\partial \sigma_{ij}^\zeta}{\partial x_j} + b_i^\zeta = \rho^\zeta \ddot{u}_i^\zeta \quad \text{on} \quad \Omega^\zeta,$$

$$J_i^{\nabla\zeta} = D_{ij}^\zeta\left(c^\zeta, \bar{\eta}_{\mathrm{VBO}}^\zeta\right)\frac{\partial \dot{c}^\zeta}{\partial x_i} \quad \text{and} \quad \sigma_{ij}^{\nabla\zeta} = L_{ijkl}^\zeta\left(\dot{\varepsilon}_{kl}^\zeta - \dot{\mu}_{kl}^\zeta\left(\bar{\eta}_{\mathrm{VBO}}^\zeta, c^\zeta\right)\right) \quad \text{on} \quad \Omega^\zeta, \tag{6.21}$$

$$\dot{\varepsilon}_{ij}^\zeta = \dot{u}_{(i,j)}^\zeta \equiv \frac{1}{2}\left(\frac{\partial \dot{u}_i^\zeta}{\partial x_j} + \frac{\partial \dot{u}_j^\zeta}{\partial x_i}\right) \quad \text{on} \quad \Omega^\zeta,$$

where superscript $\zeta$ denotes existence of material heterogeneity; temperature-dependent diffusivities will be assumed to follow the Arrhenius law $D_{ij}^\zeta = D_{ij}^0 \exp(-E_\mathrm{a}/RT)$, where $D_{ij}^0$ is the pre-exponent and $E_\mathrm{a}$ is the activation energy; $c^\zeta$ is the oxygen concentration; $R^\zeta(c^\zeta)$ is the oxygen consumption rate; $J_i^{\nabla\zeta}$ and $\sigma_{ij}^{\nabla\zeta}$ are objective measures of oxygen flux and stress, respectively; $\bar{\eta}_{\mathrm{VBO}}^\zeta = \bar{\eta}_{\mathrm{VBO}}^\zeta(\sigma_{ij}^\zeta, \varepsilon_{ij}^\zeta, \mathbf{s}_{\mathrm{VBO}}^\zeta)$, the equivalent inelastic strain in the matrix phase, will be assumed to be governed by the viscoplasticity theory based on overstress (VBO model) [55, 56]; $\bar{\eta}^{\zeta}_{\text{VBO}}$ is a function of strain, stress, and additional state variables $\mathbf{s}^{\zeta}_{\text{VBO}}$; $L^{\zeta}_{ijkl}$ is the tensor of elastic moduli obeying conditions of symmetry and positivity. The boundary sorption on the exposed boundaries is given by Henry's equation. Experimental data for oxygen consumption rate as a function of oxygen concentration for PMR 15 has been reported by Abdeljaoued and co-workers [57, 58]. These experiments suggest a linear relation between normalized material weight loss $\phi = m/m_0$ and oxygen consumption rate $r^{\zeta}(c^{\zeta})$:

$$\frac{\mathrm{d}\phi^{\zeta}}{\mathrm{d}t} = -\alpha r^{\zeta}(c^{\zeta}), \tag{6.22}$$

where parameter $\alpha$ and $r^{\zeta}(c^{\zeta})$ can be calibrated experimentally (see [52] for $r^{\zeta}(c^{\zeta})$ expression). Material weight loss ranges from $\phi^{\text{ox}}$ (weight loss in oxidized region) to unity (weight loss in unoxidized region). The weight loss parameter $\phi^{\zeta}$ controls the availability of an active polymer site for reaction. The reaction is assumed to take place only when and where there are active polymer sites $\phi^{\zeta} > \phi^{\text{ox}}$ and availability of oxygen $c^{\zeta} > 0$, which defines the so-called active zone.

The total eigenstrain $\mu^{\zeta}_{ij}$, which is required in the eigendeformation homogenization, consists of a sum of mechanical $\mu^{\text{me}}_{ij}$, thermal $\mu^{\text{th}}_{ij}$, and oxidation reaction $\mu^{\text{ox}}_{ij}$ eigenstrains:

$$\mu^{\zeta}_{ij} = \mu^{\text{me}}_{ij} + \mu^{\text{th}}_{ij} + \mu^{\text{ox}}_{ij}. \tag{6.23}$$

The oxidation reaction eigenstrain is directly related to material weight loss as

$$\frac{\mathrm{d}\mu^{\text{ox}}_{ij}}{\mathrm{d}t} = \gamma_{ij}(T, m)\frac{\mathrm{d}m^{\zeta}}{\mathrm{d}t}, \tag{6.24}$$

where $\gamma_{ij}(T, m)$ are the coefficients of oxidation reaction contraction. For the VBO-based viscoplastic material, the mechanical eigenstrain is given by

$$\mu^{\text{me}}_{ij} = \eta^{\zeta}_{ij(\text{VBO})} = \sqrt{\frac{3}{2}}\bar{\eta}^{\zeta}_{\text{VBO}} n_{ij}, \tag{6.25}$$

where $n_{ij}$ is the normalized tensor of the overstress deviator and the thermal eigenstrain is defined in the usual manner as a product of thermal expansion and temperature.

## 6.4.2 *Mathematical Upscaling*

The primary objective of mathematical upscaling is to derive coupled mechano–diffusion–reaction equations at the macroscale from the fine-scale equations using mathematical homogenization theory. Here, we focus on upscaling of diffusion–reaction equations only. Following the mathematical upscaling approach for mechanical equations, the oxygen concentration is first expanded in the double-scale asymptotic expansion followed by introduction of coupling between the two scales that will give rise to the following expansion of oxygen concentration:

$$c^{\zeta}(\boldsymbol{X}) = c(\hat{\boldsymbol{X}}, \boldsymbol{Y}) = c^0_i(\hat{\boldsymbol{X}}) + \zeta c^1_i(\hat{\boldsymbol{X}}, \boldsymbol{Y}) + \zeta^2 c^2_i(\hat{\boldsymbol{X}}, \boldsymbol{Y}) + O(\zeta^3). \tag{6.26}$$

To account for size effect and the existence of macroscopic oxygen concentration gradients within the RVE domain, we will make a similar assumption to Equation (6.6):

$$\frac{\partial c^0}{\partial X_j} = O(1); \quad \frac{\partial^2 c^0}{\partial X_j \partial X_k} = O(\zeta^{-1}) \tag{6.27}$$

and the oxygen flux $\bar{q}_i(\hat{\mathbf{X}})$ at a material point $\hat{\mathbf{X}}$ placed at the RVE centroid is replaced by the corresponding *linear flux field*

$$q_i^{\mathrm{C}}(\hat{\mathbf{X}}, \mathbf{Y}) = \bar{q}_i(\hat{\mathbf{X}}) + k_{ij}(\hat{\mathbf{X}})Y_j, \tag{6.28}$$

where $k_{ij}$ will be defined to satisfy variational equivalence between the fine- and coarse-scale fields. Combining the above asymptotic expansions with the governing equations (6.21) will decompose the source fine-scale problem (6.21) into the coupled fine-scale problem in the RVE domain $\Theta$ and the macro problem in $\Omega$.

The bottleneck of solving the two-scale coupled problem is in the computational complexity of solving the discrete RVE problem at each time step and each integration (material) point. This issue is addressed by computational upscaling described below.

### 6.4.3 Computational Upscaling

The primary objective of computational upscaling is to reduce the computational complexity of solving the RVE problem to a manageable size that can be adapted based on available computational resources. Owing to the different nature of the mechanical and oxygen diffusion reaction equations, different computational upscaling strategies will be employed. Here, we extend the computational upscaling strategy based on construction of residual free mechanical fields to the diffusion–reaction problem. Assuming scale separation, the coupled mechano–diffusion problem can be decomposed into fine- and coarse-scale problems as

$$\mathrm{O}(\zeta^{-1}): \quad \begin{cases} J_{i,y_i}(\mathbf{x}, \mathbf{y}) = 0 \\ \sigma_{ij,y_j}(\mathbf{x}, \mathbf{y}) = 0 \end{cases} \quad \mathbf{y} \in \Theta, \tag{6.29}$$

$$\mathrm{O}(\zeta^{0}): \quad \begin{cases} \bar{J}_{i,x_i}(\mathbf{x}) - \bar{R} = \dot{\bar{C}} \\ \bar{\sigma}_{ij,x_j}(\mathbf{x}) + \bar{b}_i = \bar{\rho}\ddot{\bar{u}}_i \end{cases} \quad \mathbf{x} \in \Omega, \tag{6.30}$$

where $\zeta$ denotes the dependence on the fine-scale heterogeneities; $\mathbf{x}$ and $\mathbf{y} = \mathbf{x}/\zeta$ are the coarse- and fine-scale position vectors, respectively; $J_i$ is the fine-scale diffusion flux; $\bar{J}_i$ and $\bar{R}$ are the average flux and reaction rate at the coarse scale, respectively; and $\dot{\bar{C}}$ is the time derivative of the coarse-scale concentration. $\sigma_{ij}$ is the fine-scale stress; $\bar{\sigma}_{ij}$ and $\bar{b}_i$ are the average stress and body force at the coarse scale, respectively.

First, consider the diffusion problem. Fick's law is rewritten in such a way that the nonlinear diffusivity is expressed in terms of the additive decomposition of the concentration gradient field and the so-called eigen-concentration gradient $\eta_k$:

$$\begin{aligned} J_i &= -D_{ij}C_{,j} \\ &= -D_{ik}^{\mathrm{un}}(D_{kl}^{\mathrm{un}})^{-1}D_{lj}C_{,j} \\ &= -D_{ik}^{\mathrm{un}}[\delta_{kj} + (D_{kl}^{\mathrm{un}})^{-1}D_{lj} - \delta_{kj}]C_{,j} \\ &= -D_{ik}^{\mathrm{un}}(C_{,k} - \eta_k), \end{aligned} \tag{6.31}$$

with eigen-concentration gradient $\eta_k$ defined as

$$\eta_k \equiv [(D_{kl}^{\mathrm{un}})^{-1}D_{lj} - \delta_{kj}]C_{,j}, \tag{6.32}$$

where $D_{ik}^{\mathrm{un}}$ is the unoxidized (undamaged) diffusivity.

By introducing the concentration gradient and the eigen-concentration gradient influence functions and by utilizing a piecewise constant approximation of the eigen-concentration gradient $\eta_k^{(\alpha)}$ over $\Theta^{(\alpha)}$, the residual-free diffusion flux can be expressed in terms of the partitioned eigen-concentration gradient at the fine scale as follows:

$$J_i = -D_{ij}^{\text{un}}(\mathbf{y})\left[A_{jk}(\mathbf{x},\mathbf{y})\bar{C}_{,k}(\mathbf{x}) + \sum_{\alpha=1}^{n} S_{jk}^{(\alpha)}(\mathbf{x},\mathbf{y})\eta_k^{(\alpha)}(\mathbf{x})\right], \tag{6.33}$$

with

$$A_{ij}(\mathbf{x},\mathbf{y}) = \delta_{ij} + G_{ij}(\mathbf{x},\mathbf{y}), \tag{6.34}$$

$$S_{ij}^{(\alpha)}(\mathbf{x},\mathbf{y}) = P_{ij}^{(\alpha)}(\mathbf{x},\mathbf{y}) - \delta_{ij}^{(\alpha)}(\mathbf{x},\mathbf{y}), \tag{6.35}$$

$$\eta_i^{(\alpha)}(\mathbf{x}) = \frac{1}{\left|\Theta^{(\alpha)}\right|}\int_{\Theta^{(\alpha)}} \eta_i(\mathbf{x},\mathbf{y})\,\mathrm{d}\Theta, \tag{6.36}$$

in which the total volume of the unit cell $\Theta$ at the fine scale is partitioned into $n$ nonoverlapping sub-domains $\Theta^{(\alpha)}$. The coefficient tensors ($A_{ij}$, $S_{ij}^{(\alpha)}$) are governed by the following equations:

$$\{D_{ij}(\mathbf{x},\mathbf{y})[\delta_{jk}(\mathbf{x},\mathbf{y}) + G_{jk}(\mathbf{x},\mathbf{y})]\}_{,y_i} = 0 \quad \mathbf{y} \in \Theta, \tag{6.37}$$

$$\{D_{ij}(\mathbf{x},\mathbf{y})[P_{jk}^{(\alpha)}(\mathbf{x},\mathbf{y}) - \delta_{jk}^{(\alpha)}(\mathbf{x},\mathbf{y})]\}_{,y_i} = 0 \quad \mathbf{y} \in \Theta. \tag{6.38}$$

The resulting residual-free diffusion flux at the coarse scale is given as

$$\bar{J}_i = \bar{D}_{ik}^{\text{un}}(\mathbf{x}) \cdot \bar{C}_{,k}(\mathbf{x}) + \sum_{\alpha=1}^{n} \bar{E}_{ik}^{(\alpha)}(\mathbf{x}) \cdot \eta_k^{(\alpha)}(\mathbf{x}), \tag{6.39}$$

where

$$\bar{D}_{ik}^{\text{un}}(\mathbf{x}) = \frac{1}{|\Theta|}\int_{\Theta} D_{ij}^{\text{un}}(\mathbf{y})A_{jk}(\mathbf{x},\mathbf{y})\,\mathrm{d}\Theta, \tag{6.40}$$

$$\bar{E}_{ik}^{(\alpha)}(\mathbf{x}) = \frac{1}{|\Theta|}\int_{\Theta} D_{ij}^{\text{un}}(\mathbf{y})S_{jk}^{(\alpha)}(\mathbf{x},\mathbf{y})\,\mathrm{d}\Theta. \tag{6.41}$$

Next, consider the mechanical problem remains the same as in described in Section 6.2.

Figure 6.4 compares oxidation thickness as a function of aging time and temperature as predicted with the reduced-order multiscale–multiphysics model and experimental results.

## 6.5 Uncertainty Quantification of Nonlinear Model of Micro-Interfaces and Micro-Phases

One of the major hurdles in successful utilization of the proposed RC$^2$ approach is in uncertainty quantification of nonlinear behavior of micro-interfaces and micro-phases and their effect on the component response. Preliminary studies [59, 60] indicated that these are the dominant variables affecting the coarse-scale quantities of interest, such as critical stresses and strains, mass loss, depth of oxidized region, and critical inelastic strain. To quantify how the oxygen diffusion affects the critical overall stress (i.e., how the fine-scale parameters in the diffusion–reaction problem affect the coarse-scale fields in the

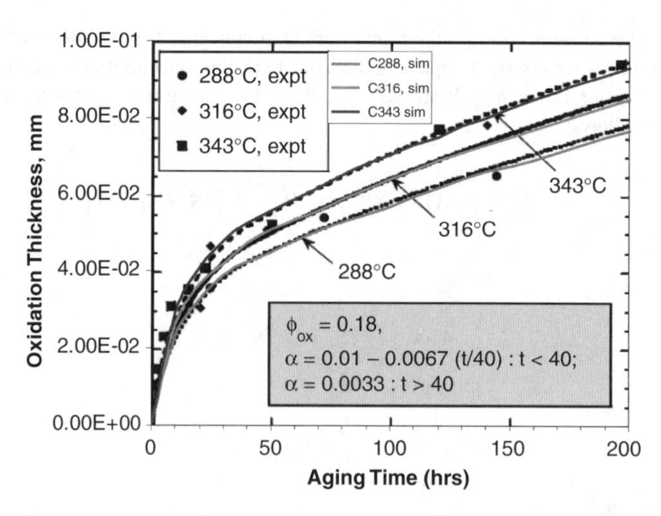


**Figure 6.4** Prediction of oxidation thickness in PMR-15 at 343°C, 316°C and 288°C

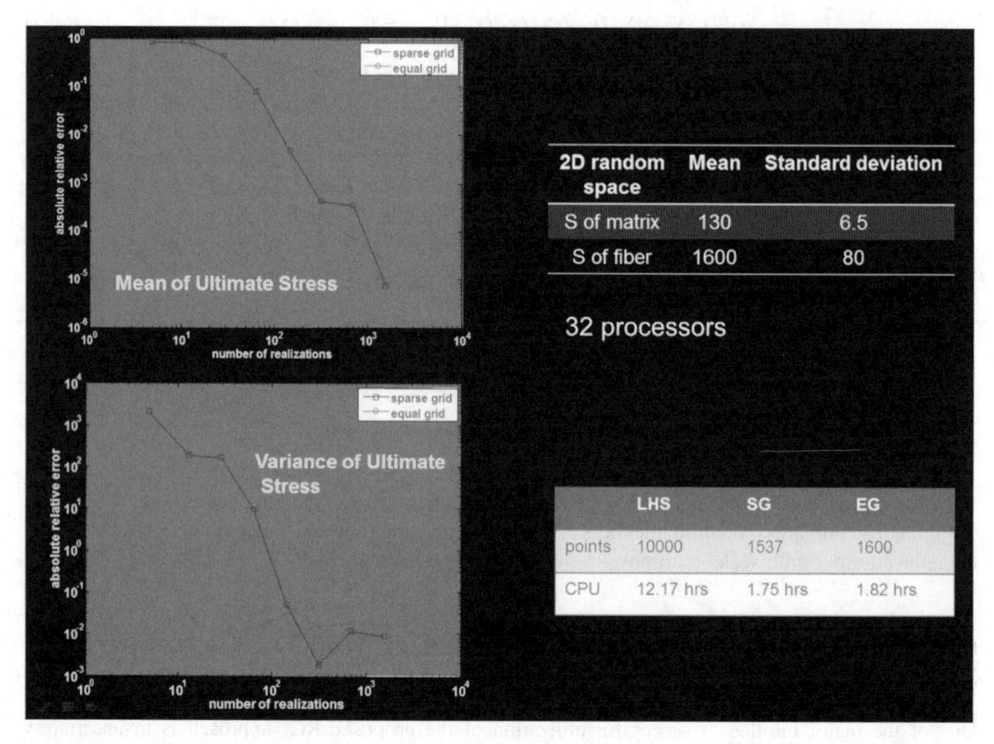


**Figure 6.5** Error in the mean and standard deviation of composite plate strength as a function of number of realizations for composite plate problem [59, 60]. Reproduced with permission from [59] © 2010 Begellhouse; [60] © 2011 John Wiley & Sons

mechanical problem and, vice versa, how the mechanical deformation, which may give rise to failure of interfaces, enhances oxygen diffusion along the interfaces) it is necessary to quantify the dominant input uncertainties and to propagate them.

This poses stochastic forward and inverse problems. Here, we only address the forward problem. The forward problem involves the study of how the coarse-scale quantities of interest respond to variations in the input data. We will represent parameters describing cohesive laws and nonlinear constitutive laws of micro-phases as random fields, denoted by $\eta(\boldsymbol{X}, \boldsymbol{\omega})$; that is, we will assume they vary randomly in the macro-domain with a given covariance structure. Although such random fields are properly described only by means of an infinite number of random variables, the Karhunen–Loeve expansion [61] can be employed to describe the random fields in terms of a small number of uncorrelated random variables. Here, the sparse grid collocation method [62–64] in combination with the multiscale approach is employed to solve the two-scale stochastic partial differential equation. This gives rise to a nonintrusive multiscale method.

Here we show preliminary results obtained using a combination of the sparse grid collocation method and reduced-order multiscale approach outline in Section 6.2.1. The mean and standard deviation of matrix and fiber strengths are given and the goal is to find the mean and standard deviation of composite plate strength (Figure 6.5).

## References

[1] Brecher, A. (2007) A safety roadmap for future plastics and composites intensive vehicles. DOT NHTSA Report DOT HS 810 863, DOT-VNTSC-NHTSA-07-02.
[2] Brinson, L. (2005) *Meeting the Emerging Demand for Durable Polymer Matrix Composites*, The National Academies Press, Washington, DC.
[3] Eshelby, J.D. (1957) The determination of the elastic field of an ellipsoidal inclusion, and related problems. *Proceedings of the Royal Society of London A*, **241**, 376–396.
[4] Hashin, Z. (1962) The elastic moduli of heterogeneous materials. *Journal of Applied Mechanics*, **29**, 143–150.
[5] Mori, T. and Tanaka, K. (1973) Average stress in matrix and average elastic energy of materials with misfitting inclusions. *Acta Metallurgica*, **21**, 571–574.
[6] Hill, R. (1965) Elastic properties of reinforced solids: some theoretical principles. *Journal of the Mechanics and Physics of Solids*, **13**, 357–372.
[7] Christensen, R.M. and Lo, K.H. (1979) Solutions for effective shear properties in the three phase sphere and cylinder models. *Journal of the Mechanics and Physics of Solids*, **27**, 315–330.
[8] Babuska, I. (1975) Homogenization and its applications: mathematical and computational problems, in *Numerical Solution of Partial Differential Equations – III, SYNSPADE* (ed. B. Hubbard), Academic Press, pp. 89–147.
[9] Benssousan, A., Lions, J.L., and Papanicolaou, G. (1978) *Asymptotic Analysis for Periodic Structures*, North-Holland, Amsterdam.
[10] Suquet, P.M. (1987) in *Homogenization Techniques for Composite Media* (eds E. Sanchez-Palencia and A. Zaoui), vol. 272 of *Lecture Notes in Physics*, Springer-Verlag.
[11] Sanchez-Palencia, E. (1980) *Non-Homogeneous Media and Vibration Theory*, vol. 127 of Lecture Notes in Physics, Springer-Verlag, Berlin.
[12] Guedes, J.M. and Kikuchi, N. (1990) Preprocessing and postprocessing for materials based on the homogenization method with adaptive finite element methods. *Computer Methods in Applied Mechanics and Engineering*, **83**, 143–198.
[13] Terada, K. and Kikuchi, N. (1995) Nonlinear homogenization method for practical applications, in *Computational Methods in Micromechanics* (eds S. Ghosh and M. Ostoja-Starzewski), vol. AMD-212/MD-62, ASME, New York, NY, pp. 1–16.
[14] Fish, J., Nayak, P., and Holmes, M.H. (1994) Microscale reduction error indicators and estimators for a periodic heterogeneous medium. *Computational Mechanics*, **14**, 1–16.
[15] Yuan, Z. and Fish, J. (2008) Towards realization of computational homogenization in practice. *Journal for Numerical Methods in Engineering*, **73** (3), 361–380.

[16] Ghosh, S. and Moorthy, S. (1995) Elastic–plastic analysis of heterogeneous microstructures using the Voronoi cell finite element method. *Computer Methods in Applied Mechanics and Engineering*, **121** (1–4), 373–409.

[17] Aboudi, J. (1982) A continuum theory for fiber-reinforced elastic–viscoplastic composites. *International Journal of Engineering Science*, **20** (55), 605–621.

[18] Berlyand, L.V. and Kolpakov, A.G. (2001) Network approximation in the limit of small inter-particle dispersed composite. *Archive for Rational Mechanics and Analysis*, **159**, 179–227.

[19] Moulinec, H. and Suquet, P. (1994) A fast numerical method for computing the linear and nonlinear properties of composites. *Comptes Rendus de l'Académie des Sciences. Série II*, **318**, 1417–1423.

[20] Moulinec, H. and Suquet, P. (1998) A numerical method for computing the overall response of nonlinear composites with complex microstructure. *Computer Methods in Applied Mechanics and Engineering*, **157**, 69–94.

[21] Dvorak, G.J. (1992) Transformation field analysis of inelastic composite materials. *Proceedings of the Royal Society of London A*, **437**, 311–327.

[22] Bahei-El-Din, Y.A., Rajendran, A.M., and Zikry, M.A. (2004) A micromechanical model for damage progression in woven composite systems. *International Journal of Solids and Structures*, **41**, 2307–2330.

[23] Dvorak, G.J. and Benveniste, Y. (1992) On transformation strains and uniform fields in multiphase elastic media. *Proceedings of the Royal Society of London A*, **437**, 291–310.

[24] Laws, N. (1973) On the thermostatics of composite materials. *Journal of the Mechanics and Physics of Solids*, **21**, 9–17.

[25] Willis, J. (1981) Variational and related methods for the overall properties of composite materials, in *Advances in Applied Mechanics* (ed. C.S. Yih), vol. 21, Academic Press, New York, NY, pp. 1–78.

[26] Dvorak, G.J. (1990) On uniform fields in heterogeneous media. *Proceedings of the Royal Society of London A*, **431**, 89110.

[27] Oskay, C. and Fish, J. (2008) Fatigue life prediction using 2-scale temporal asymptotic homogenization. *Computational Mechanics*, **42** (2), 181–195.

[28] Oskay, C. and Fish, J. (2007) Eigendeformation-based reduced order homogenization. *Computer Methods in Applied Mechanics and Engineering*, **196**, 1216–1243.

[29] Yuan, Z. and Fish, J. (2009) Multiple scale eigendeformation-based reduced order homogenization. *Computer Methods in Applied Mechanics and Engineering*, **198** (21–26), 2016–2038.

[30] Yuan, Z. and Fish, J. (2009) Hierarchical model reduction at multiple scales. *International Journal for Numerical Methods in Engineering*, **79**, 314–339.

[31] Fish, J. and Yuan, Z. (2009) *N*-scale model reduction theory, in *Multiscale Methods: Bridging the Scales in Science and Engineering* (ed. J. Fish), Oxford University Press, pp. 57–90.

[32] Zohdi, T.I., Oden, J.T., and Rodin, G.J. (1996) Hierarchical modeling of heterogeneous bodies. *Computer Methods in Applied Mechanics and Engineering*, **138**, 273–298.

[33] Ghosh, S., Lee, K., and Raghavan, P. (2001) A multi-level computational model for multi-scale damage analysis. *International Journal of Solids and Structures*, **38**, 2335–2385.

[34] Geers, M.G.D., Kouznetsova, V.G., and Brekelmans, W.A.M. (2003) Multi-scale second-order computational homogenization of microstructures towards continua. *International Journal for Multiscale Computational Engineering*, **1** (4), 371–386.

[35] E, W. and Engquist, B. (2003) The heterogeneous multiscale methods. *Communications in Mathematical Sciences*, **1** (1), 87–132.

[36] Fish, J., Shek, K., Pandheeradi, M., and Shephard, M. (1997) Computational plasticity for composite structures based on mathematical homogenization: theory and practice. *Computer Methods in Applied Mechanics and Engineering*, **148**, 53–73.

[37] Fish, J. and Fan, R. (2008) Mathematical homogenization of nonperiodic heterogeneous media subjected to large deformation transient loading. *International Journal for Numerical Methods in Engineering*, **76** (7), 1044–1064.

[38] Ostoja-Starzewski, M., Boccara, S., and Jasiuk, I. (1999) Couple-stress moduli and characteristic length of a two-phase composite. *Mechanics Research Communications*, **26** (4), 387–396.

[39] Van der Sluis, O., Vosbeek, P., Schreurs, P., and Meijer, H. (1999) Homogenization of heterogeneous polymers. *International Journal of Solids and Structures*, **36**, 3193–3214.

[40] Botkin, M., Johnos, N., Zywicz, E., and Simunovic, S. (1988) Crashworthiness simulation of composite automotive structures, in *Proceedings of the 13th Annual Engineering Society of Detroit Advanced Composites Technology*, Detroit, MI.

[41] Jardak, M. and Ghanem, R. (2004) Spectral stochastic homogenization of divergence-type PDEs. *Computer Methods in Applied Mechanics and Engineering*, **193** (6–8), 429–447.
[42] Fish, J. and Kuznetsov, S. (2010) Computational continua. *International Journal for Numerical Methods in Engineering*, **84**, 774–802.
[43] Fish, J., Filonova, V., and Yuan, Z. (2012) Reduced order computational continua. *Computer Methods in Applied Mechanics and Engineering*, **221–222**, 104–116.
[44] Kouznetsova, V., Brekelmans, W.A., and Baaijens, F.P.T. (2001) An approach to micro–macro modeling of heterogeneous materials. *Computational Mechanics*, **27**, 37–48.
[45] Belytschko, T., Liu, W., and Moran, B. (2003) *Nonlinear Finite Elements for Continua and Structures*, John Wiley & Sons, Ltd, Chichester.
[46] Simo, J. and Hughes, T. (1998) *Computational Inelasticity*, vol. 7 of Interdisciplinary Applied Mathematics, Springer-Verlag, New York, NY.
[47] Fish, J. and Yu, Q. (2002) Computational mechanics of fatigue and life predictions for composite materials and structures. *Computer Methods in Applied Mechanics and Engineering*, **191**, 4827–4849.
[48] Gal, E., Yuan, Z., Wu, W., and Fish, J. (2007) A multiscale design system for fatigue life prediction. *International Journal of Multiscale Computational Engineering*, **5** (6), 435–446.
[49] Paris, P.C. and Erdogan, F. (1963) A critical analysis of crack propagation laws. *Journal of Basic Engineering*, **85**, 528534.
[50] Oskay, C. and Fish, J. (2004) Fatigue life prediction using 2-scale temporal asymptotic homogenization. *International Journal for Numerical Methods in Engineering*, **61** (3), 329–359.
[51] Fish, J. and Oskay, C. (2005) Nonlocal multiscale fatigue model. *Mechanics of Advanced Materials and Structures*, **12** (6), 485–500.
[52] Schoeppner, G., Tandon, G., and Ripberger, E. (2007) Anisotropic oxidation and weight loss in PMR-15 composites. *Composites: Part A*, **38**, 890–904.
[53] Tandon, G., Pochiraju, K., and Schoeppner, G. (2006) Modeling of oxidative development in PMR-15 resin. *Polymer Degradation and Stability*, **91**, 1861–1869.
[54] Pochiraju, K.V. and Tandon, G. (2006) Modeling thermo-oxidative layer growth in high-temperature resins. *Journal of Engineering Materials and Technology*, **128**, 107–116.
[55] McClung, A.J.W. and Ruggles-Wrenn, M.B. (2009) Effects of prior aging at 288°C in air and in argon environments on creep. *Journal of Pressure Vessel Technology*, **131**, 031405.
[56] Gomaa, S., Sham, T.-L., and Krempl, E. (2004) Finite element formulation for finite deformation, isotropic viscoplasticity theory based on overstress (FVBO). *International Journal of Solids and Structures*, **41**, 3607–3624.
[57] Abdeljaoued, K., Bellanger, V., Desarmot, G. *et al.* (1998) Etude de l'oxydation de la matrice PMR-15 [Thermal oxidation of PMR-15 matrix], in *JNC 11-Journees Nationales sur les Composites*, Arachon (France), 18–20 novembre.
[58] Abdeljaoued, K. (1999) Etude de l'oxydation thermoque de la matrice dans les composites fibres de carbone/PMR-15, PhD Thesis, ENSAM, Paris.
[59] Wu, W. and Fish, J. (2010) Towards nonintrusive stochastic multiscale design system for composite materials. *International Journal of Multiscale Computational Engineering*, **8**, 549–559.
[60] Fish, J. and Wu, W. (2011) Nonintrusive stochastic multiscale solver. *International Journal for Numerical Methods in Engineering*, **88**, 862–879.
[61] Ghanem, R. and Spanos, P. (1991) *Stochastic Finite Elements: A Spectral Approach*, Springer-Verlag, New York, NY.
[62] Xiu, D. (2007) Efficient collocational approach for parametric uncertainty analysis. *Communications in Computational Physics*, **2**, 293–309.
[63] Nobile, F., Tempone, R., and Webster, C. (2008) A sparse grid collocation method for elliptic partial differential equations with random input data. *SIAM Journal of Numerical Analysis*, **45**, 2309–2345.
[64] Ma, X. and Zabaras, N. (2009) An adaptive hierarchical sparse grid collocation algorithm for the solution of stochastic differential equations. *Journal of Computational Physics*, **228**, 3084–3113.

# Part II

# Patient-Specific Fluid-Structure Interaction Modeling, Simulation and Diagnosis

Patient-specific modeling integrates computational modeling, experimental procedures, imagine clinical segmentation, and mesh generation with the finite element method (FEM) to solve clinical problems such as diagnosis and disease modeling by using various techniques in computational biomedicine and bioengineering. Since its inception, it has truly become one of the greatest triumphs of bio-engineering as well as computational fluid mechanics. In this part, Dr. Tezduyar, a world renowned expert in patient-specific modeling, and his co-workers present their latest developments in this exciting subject, and subsequently two young new stars, Hossain and Zhang, discuss how to link the patient-specific modeling to drug delivery. Chapter 9 is an interesting contribution by T. Zodhi, who presents a new type of patient-specific modeling by utilizing computational electromagnetic to detect and diagnose blood diseases. In a similar way, Chung and his co-workers discuss an electrohydrodynamic assembly of nanoparticles, which has broad biomedical implications from biosensors to drug delivery.

A key challenge of bio-fluid modeling and simulation, especially cardiovascular fluid modeling, is how to deal with fluid–structural interactions. Over the last two decades, various computational algorithms and methods have been developed to solve this problem; for example, the arbitrary-Lagrangian–Eulerian (ALE) method and the immersed FEM (IFEM). In the past decade, Wing Liu and his co-workers have developed the IFEM and extensively applied it to treat fluid–structure interactions in bio-fluid simulation. L. Zhang and X.S. Wang were the original collaborators of Wing Liu in IFEM research. Both of them either present their recent results or summarize their recent research on how to apply the IFEM to model fluid–structural interaction problems.

# 7

# Patient-Specific Computational Fluid Mechanics of Cerebral Arteries with Aneurysm and Stent

Kenji Takizawa[a], Kathleen Schjodt[b], Anthony Puntel[b], Nikolay Kostov[b], and Tayfun E. Tezduyar[b]

[a]*Department of Modern Mechanical Engineering and Waseda Institute for Advanced Study, Waseda University, Japan*

[b]*Mechanical Engineering, Rice University, USA*

## 7.1 Introduction

Cardiovascular fluid mechanics modeling now has a very significant place in computational mechanics; for example, see [1–12]. Many of these articles are on patient-specific modeling of arteries, especially those with aneurysm, much of it focusing on the fluid–structure interaction (FSI) between the blood flow and arterial walls. The FSI aspect of the cardiovascular fluid mechanics modeling has been receiving this level of attention because a large portion of the computational mechanics techniques developed in recent decades targeted FSI or flows with moving interfaces. The arbitrary Lagrangian–Eulerian (ALE) finite-element method [13] is still the most popular interface-tracking (moving-mesh) technique used in computation of this class of problems. However, the deforming-spatial-domain/stabilized space–time (DSD/SST) formulation [14–17], which is also an interface-tracking technique, now is a strong alternative to the ALE method. The DSD/SST formulation is based on the streamline-upwind/Petrov–Galerkin (SUPG) [18] and pressure-stabilizing/Petrov–Galerkin (PSPG) [14] stabilizations.

The DSD/SST formulation has been one of the most widely used techniques in patient-specific arterial FSI computations reported in the literature [1–12]. A number of special techniques for arterial fluid mechanics were developed in conjunction with the DSD/SST technique. These include techniques for calculating an estimated zero-pressure (EZP) arterial geometry [5, 7, 9, 12], a special mapping technique for specifying the velocity profile at an inflow boundary with noncircular shape [7], techniques for using variable arterial wall thickness [7, 9], mesh-generation techniques for building layers of refined

*Multiscale Simulations and Mechanics of Biological Materials*, First Edition. Edited by Shaofan Li and Dong Qian.

fluid mechanics mesh near the arterial walls [6, 7, 9], techniques [9] for the projection of fluid–structure interface stresses, calculation of the wall shear stress (WSS) and calculation of the oscillatory shear index (OSI), techniques [10] for extracting the arterial-lumen geometry from three-dimensional rotational angiography (3DRA) and generating a mesh for that geometry, and a new scaling technique [11, 12] for specifying a more realistic volumetric flow rate.

A new version of the DSD/SST method with turbulence modeling features was introduced by Takizawa and Tezduyar [17, 19]. This DSD/SST formulation is a space–time version [17, 19] of the residual-based variational multiscale (VMS) method [20]. It was named "DSD/SST–VMST" (i.e., the version with the VMS turbulence model) [17], which was also called "ST-VMS" (meaning "space–time VMS") in Ref. [19]. It has been successfully tested on wind-turbine rotor aerodynamics [21], simple FSI problems [19], and aerodynamics of flapping wings in [22]. Because of its space–time nature, as can be discerned from the derivation presented by Takizawa and Tezduyar [17, 19], the DSD/SST formulation provides a more comprehensive framework for the VMS method and, as pointed out by Takizawa and Tezduyar [17, 19], has a number desirable features, which are now better articulated and demonstrated. For example, it was shown [17, 19] that using linear or higher order basis functions for the temporal representation in a space–time computation gives us better solution accuracy.

While modeling the FSI between the blood flow and arterial walls is one of the most challenging problems in cardiovascular fluid mechanics, there are other complex problems that are comparably challenging. Patient-specific computation of unsteady blood flow in an artery with aneurysm and stent is one of them. In this chapter we present the special arterial fluid mechanics techniques we have developed for such computations. These techniques are used with the DSD/SST–VMST method, which serves as the core computational technique. The special techniques include using NURBS for the spatial representation of the surface over which the stent mesh is built, mesh-generation techniques for both the finite- and zero-thickness representations of the stent, techniques for generating refined layers of mesh near the arterial and stent surfaces, and models for representing double stent.

We compute the unsteady flow patterns in the aneurysm and investigate how those patterns are influenced by the presence of single and double stents. We also compare the flow patterns obtained with the finite- and zero-thickness representations of the stent. The arterial-lumen geometries were extracted from the 3DRA images that were provided to us while carrying out the arterial FSI research reported in [10–12]. In the computations, we use the lumen geometries obtained after the artery goes through the EZP process [11, 12] and is inflated to a pressure corresponding to the pressure at the start of our computation cycle (cardiac cycle), which is approximately 80 mmHg. We do that instead of directly using the geometries extracted from the 3DRA so that we have a consistent basis for comparison with future FSI computations with these lumen geometries.

In Section 7.2 we describe how, given one of those lumen geometries and the stent, we generate the mesh used in the fluid mechanics computations. The computational results are presented in Section 7.3 and the concluding remarks are given in Section 7.4.

## 7.2 Mesh Generation

Mesh generation of the cerebral artery with aneurysm and stent requires numerous steps that include taking the flat stent and lumen geometry and generating a fluid volume mesh representative of a stented artery with aneurysm. We begin by mapping the flat stent to the deformed stent, which fits across the aneurysm neck. The artery is split into two segments: parent and aneurysm. Layers of refined mesh are generated at the stent and arterial walls in both segments. After the remaining volume mesh in each segment is generated, the two segments are merged on the interior-boundary mesh containing the stent.

1. Prepare lumen geometry and flat-stent model as shown in Figure 7.1. We extract the arterial surface geometry from 3DRA images and generate a lumen geometry reflective of the inflated arterial-wall

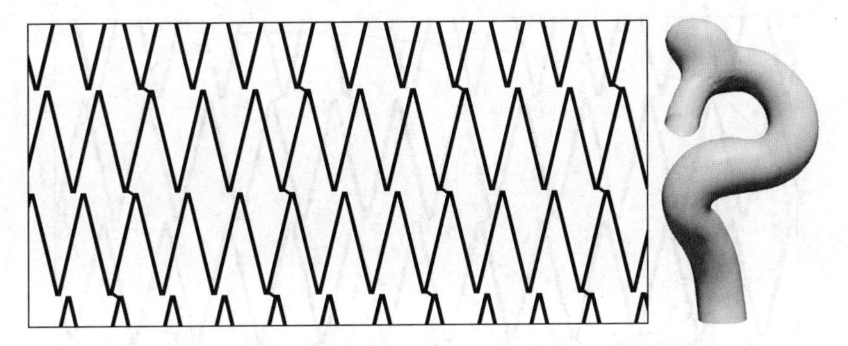

**Figure 7.1** Flat-stent geometry (left) and arterial lumen geometry (right)

structure through the process reported in [10]. The flat-stent model was generated using the geometry of a Cordis Precise Pro Rx nitinol self-expanding stent (PC0630RXC) with a wire diameter of about 0.1 mm.

2. Generate a NURBS surface slightly larger than the artery such that the surface intersects the lumen geometry as shown in Figure 7.2. We swept a NURBS surface following the curvature of the parent artery and extending slightly beyond the aneurysm neck. To simplify the mesh-generation process, we only model the portion of the stent crossing the neck of the aneurysm. The intersection of the NURBS surface and lumen geometry is the periphery of the interior boundary containing the stent.
3. Map the periphery of the interior boundary, described above, to the flat stent and mesh as seen in Figure 7.3. We generate a triangular surface mesh using ANSYS and the geometry defined by the flat stent and interior-boundary periphery. With the maximum element size specified in mesh generation, the width of the stent wire is meshed with three or four elements. This ensures sufficient refinement on the stent. The flat-stent mesh is then mapped from the flat NURBS surface to the deformed NURBS surface to form the interior-boundary mesh positioned across the neck of the aneurysm.
4. Use the periphery of the interior-boundary mesh as a predefined set of element edges, splitting the lumen geometry into parent and aneurysm segments as shown in Figure 7.4. This reduces complexity in mesh generation. We use ANSYS to generate the triangular surface meshes on the parent and aneurysm segments.

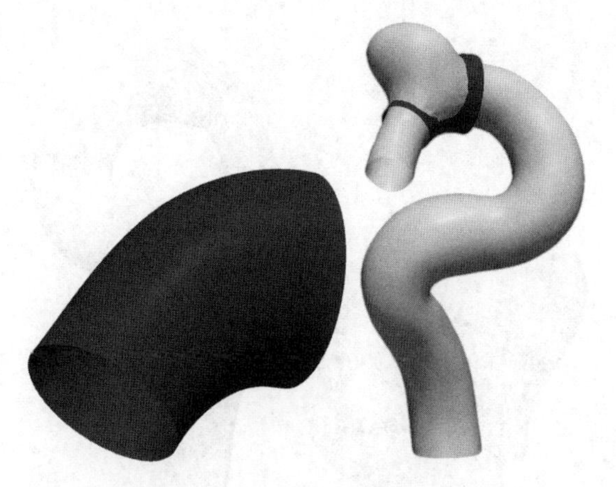

**Figure 7.2** Deformed stent (left) and split lumen geometry with the stent (right)

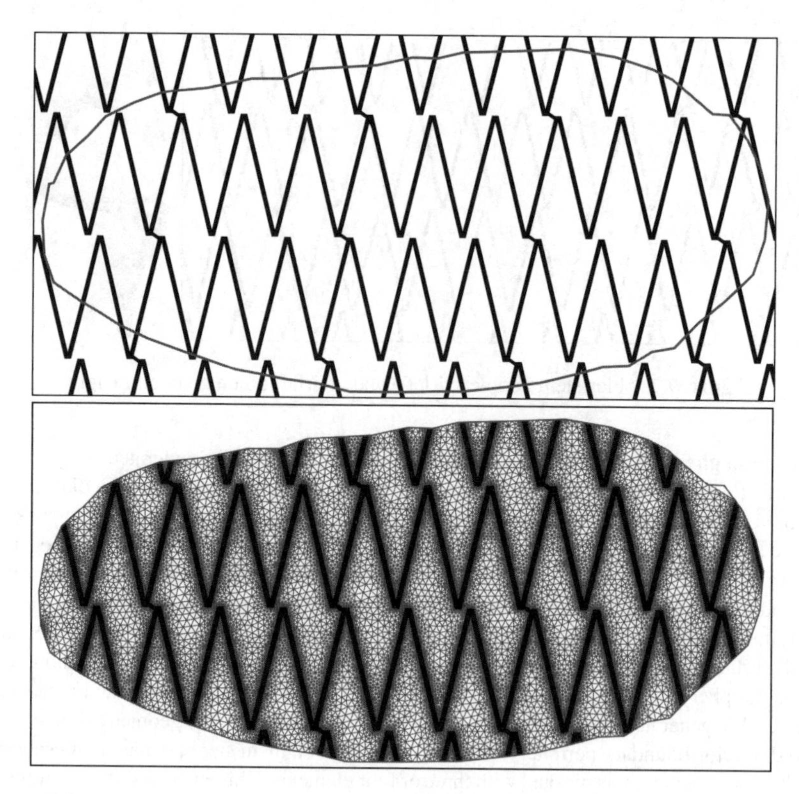

**Figure 7.3** Flat stent with the periphery of the interior-boundary (top) stent mesh (bottom)

5. Using the surface meshes for the parent and aneurysm segments, we generate layers of refined mesh on either side of the stent and near the arterial walls. We use the process reported by Takizawa *et al.* [10] to generate the layers in the parent segment. In generating the layers in the aneurysm segment, we first start by separating the surface mesh into different regions as shown in Figure 7.5. Due to the sharp angle of the geometry, no layers are explicitly generated in the red region. We specify a

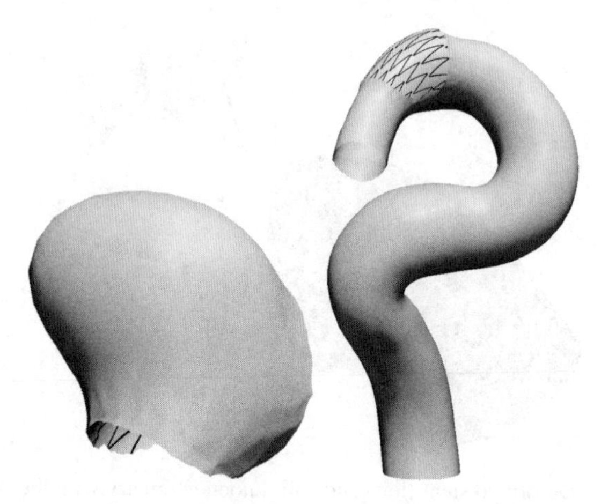

**Figure 7.4** Aneurysm (left) and parent (right) artery segments, separated by the stent

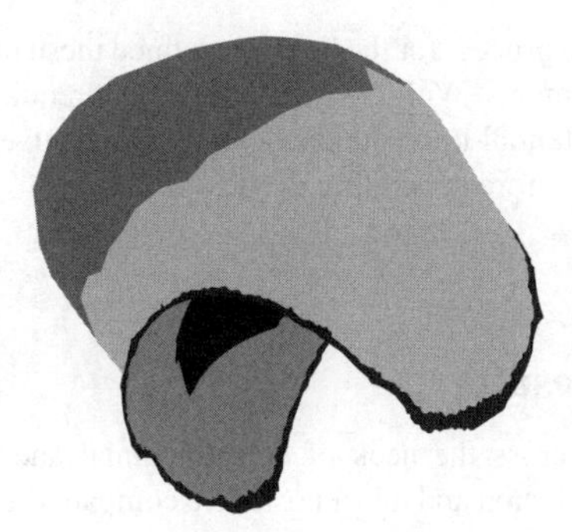

**Figure 7.5** Aneurysm artery segment showing regions of different thickness for the layers of refined mesh

uniform thickness for the layers of refined mesh in the blue regions. The thickness of the first layer is approximately equal to the first layer of refined mesh in the parent segment. There are a total of four layers, each increasing in thickness using a progression ratio of 1.75 (the same number of layers and progression ratio used by Takizawa *et al.* [10]). To prevent elements tangling, the Laplace equation is solved over the green region of the surface mesh to determine the thickness growth from essentially zero at the boundary with the red region to the desired layer thickness at the blue region boundary. We generate each of the four layers in the aneurysm segment separately and merge the layers (see Remark 2).

6. The rest of the fluid mechanics volume mesh is generated using ANSYS. The innermost surface of the layers of refined mesh is extracted from the volume mesh and used as the surface mesh for generating the volume mesh in both the parent and aneurysm segments. The inner volume mesh is then merged to the refined layers.
7. The parent and aneurysm fluid-volume mesh segments are merged on the interior-boundary mesh containing the stent. For the no-stent cases, all nodes are merged on that interior-boundary mesh. For the single- and double-stent cases, the nodes on the stent portion of the interior-boundary mesh are not merged and instead colocated.

**Remark 1** We generate the double stent by overlaying two single flat-stent geometries and translating one of the geometries in two directions. We map the intersection of the deformed NURBS surface and lumen geometry, which is again the periphery of the interior boundary, to the flat double-stent geometry and mesh the double stent as one mesh. The double-stent mesh is treated the same as the single-stent mesh in the remaining mesh generation steps. Figure 7.6 shows the full single and double stents.

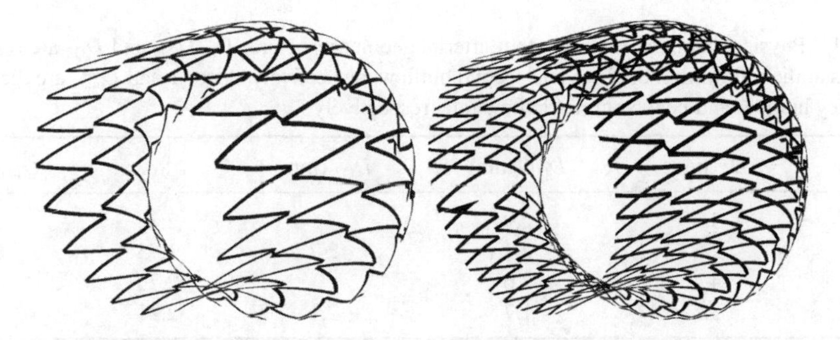

**Figure 7.6** Surface for single stent (left) and double stent (right)

**Remark 2** The mesh-generation process for the layers of refined mesh in the aneurysm segment presents challenges regarding tangled elements. With the mesh refinement required by the problem, building the layers into the artery has the potential to create elements with negative Jacobians. Each layer must be checked for the Jacobian values before generating the next layer.

## 7.3 Computational Results

Endovascular stent placement across the neck of an intracranial aneurysm can cause hemodynamic changes leading to aneurysm occlusion and thrombosis. We compare the flow field of arterial geometries before and after virtual "stenting" to assess the changes. Select aneurysms require treatment using two or more stents to sufficiently alter the flow field allowing for thrombosis. The test computations for each geometry include a before-stenting case and after-stenting cases for both single- and double-stent treatments to compare the effectiveness of stenting with multiple stents. Section 7.3.1 details the parameters for the four arterial geometries used in the computations. In Section 7.3.2 we compare hemodynamic values before and after stenting. In Section 7.3.3 we compare the results from modeling the stent with zero- and finite-thickness representations.

### *7.3.1 Computational Models*

As was done for the computations reported in Torii *et al.* [1], the blood is assumed to behave like a Newtonian fluid (see Section 2.1 in Tezduyar *et al.* [5]). The density and kinematic viscosity are set to 1000 kg/m$^3$ and $4.0 \times 10^{-6}$ m$^2$/s, respectivly. Other computational conditions, including the structural properties used in the EZP and inflation processes and the boundary conditions used in the flow computations, can be found in Tezduyar *et al.* [11].

Four patient-specific cerebral arteries with aneurysms are studied at three states: before stenting, after stenting with a single stent deployed, and after stenting with two stents deployed. The physical parameters for the four arterial models are listed in Table 7.1 and the lumen geometries are shown in Figure 7.7.

The fluid mechanics meshes for the single-stent case for each model are shown in Figures 7.8–7.11. The cross-section view shows the refined mesh at the aneurysm neck on either side of the boundary separating the aneurysm from the parent artery. The node and element numbers for the 12 fluid mechanics meshes are given in Table 7.2. We note that all computations presented in Section 7.3.2 are for zero-thickness representation of the stent. All computations are carried out using the DSD/SST–VMST technique (see Takizawa and Tezduyar [17, 19] for the terminology), with the stabilization parameters as given by Eqs. (7–11) in Tezduyar and Sathe [16] for $\tau_M = \tau_{SUPG}$ and Eq. (37) in Takizawa *et al.* [22] for $\nu_C$.

**Table 7.1** Physical parameters for the four arterial geometries. Here, $D_I$, $D_{O1}$, and $D_{O2}$ are the diameters at the inflow, first outflow, and second outflow, respectively. Also, $\alpha$ and $Q_{max}$ are the Womersley number and peak volumetric flow rate, respectively

| Model | $D_I$ (mm) | $D_{O1}$ (mm) | $D_{O2}$ (mm) | $\alpha$ | $Q_{max}$ (ml/s) |
|---|---|---|---|---|---|
| Model 1 | 3.7 | 2.9 | | 2.33 | 2.05 |
| Model 2 | 2.8 | 2.4 | 2.7 | 1.75 | 0.78 |
| Model 3 | 4.4 | 2.6 | | 2.73 | 3.40 |
| Model 4 | 3.5 | 1.7 | 2.1 | 2.21 | 1.63 |

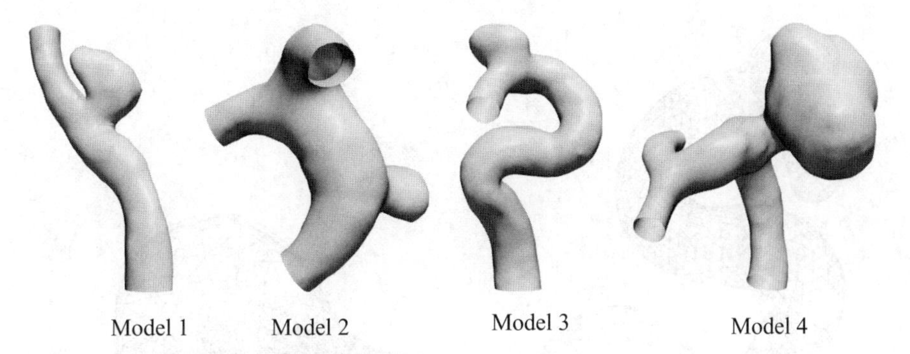


**Figure 7.7** Arterial lumen geometry obtained from voxel data for the four models studied

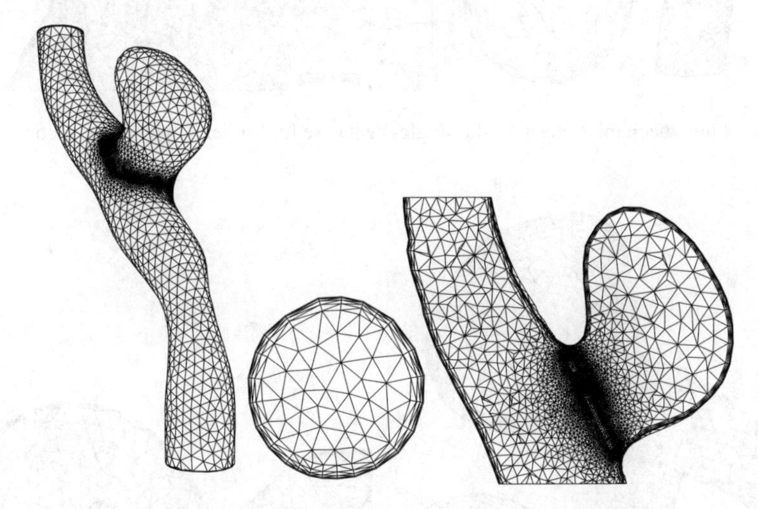

**Figure 7.8** Fluid mechanics mesh for the single-stent case for Model 1, with cross-section and inflow plane views

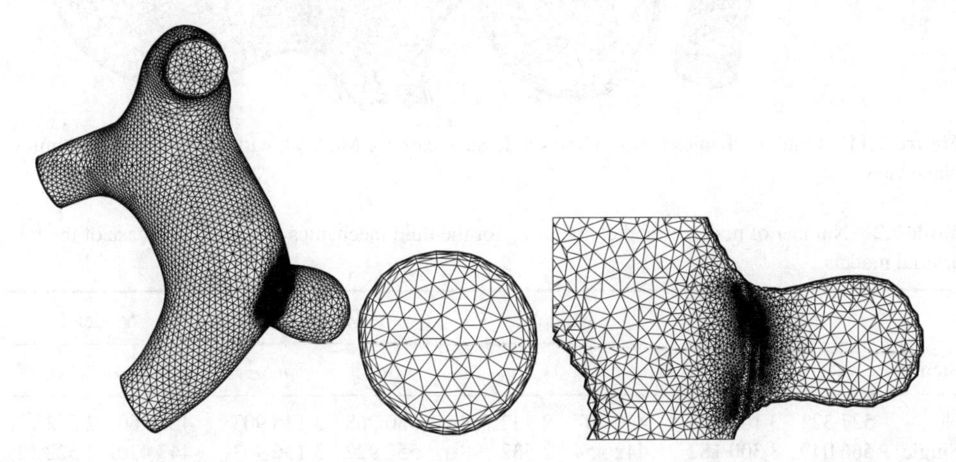

**Figure 7.9** Fluid mechanics mesh for the single-stent case for Model 2, with cross-section and inflow plane views

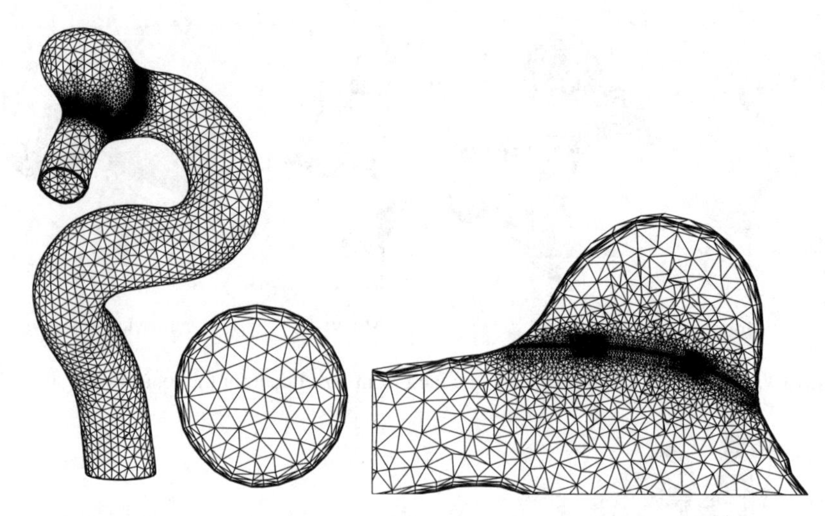

**Figure 7.10** Fluid mechanics mesh for the single-stent case for Model 3, with cross-section and inflow plane views

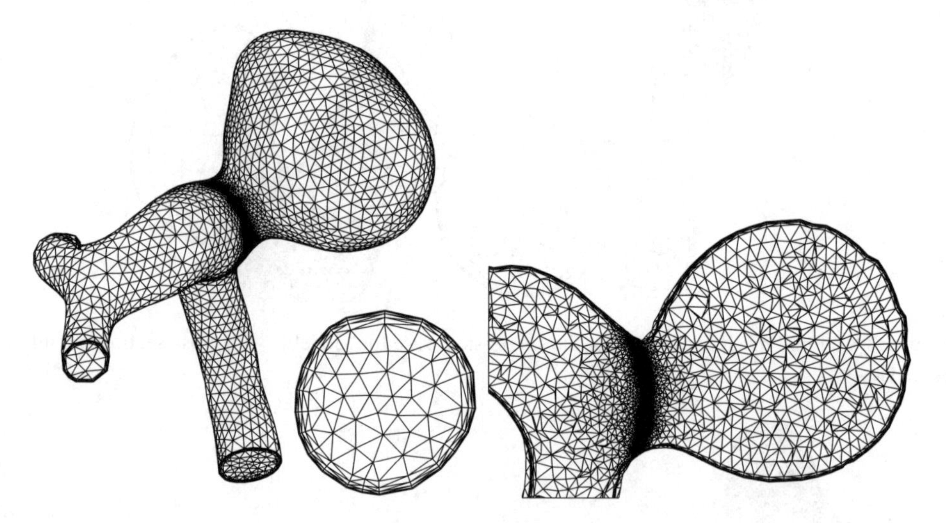

**Figure 7.11** Fluid mechanics mesh for the single-stent case for Model 4, with cross-section and inflow plane views

**Table 7.2** Number of nodes $n_n$ and elements $n_e$ for the fluid mechanics mesh for each case of the four arterial models

| | Model 1 | | Model 2 | | Model 3 | | Model 4 | |
|---|---|---|---|---|---|---|---|---|
| Stent | $n_n$ | $n_e$ | $n_n$ | $n_e$ | $n_n$ | $n_e$ | $n_n$ | $n_e$ |
| No | 527 323 | 3 168 305 | 430 497 | 2 532 798 | 530 268 | 3 136 903 | 428 260 | 2 522 129 |
| Single | 566 049 | 3 300 182 | 442 454 | 2 532 798 | 552 922 | 3 136 903 | 447 430 | 2 522 129 |
| Double | 662 431 | 3 736 603 | 503 819 | 2 823 729 | 930 403 | 5 261 467 | 867 500 | 4 916 931 |

**Table 7.3** GMRES iterations per nonlinear iteration for each case of the four models

| Stent | Model 1 | Model 2 | Model 3 | Model 4 |
|---|---|---|---|---|
| No | 1000 | 1000 | 1500 | 1000 |
| Single | 1500 | 1000 | 1500 | 1000 |
| Double | 1500 | 1200 | 2000 | 1700 |

The time step size is $3.333 \times 10^{-3}$ s. The number of nonlinear iterations per time step is 4 and the number of GMRES iterations per nonlinear iteration for each model is shown in Table 7.3. We check the mass balance as one of the indicators of numerical convergence. Sufficient mass balance is reached when the difference between the inflow and outflow rates essentially equals zero. Figures 7.12–7.23 show the mass balance for all cases.

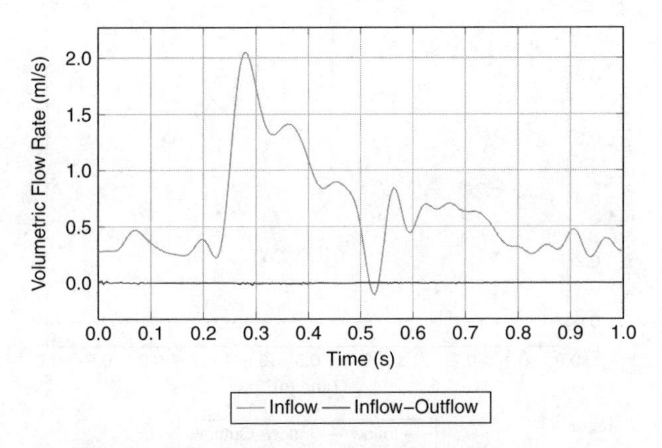


**Figure 7.12** Mass balance for Model 1 before stenting

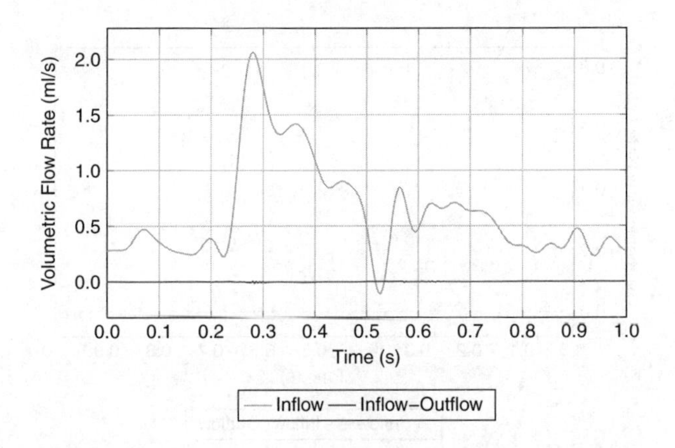


**Figure 7.13** Mass balance for Model 1 with single stent deployed

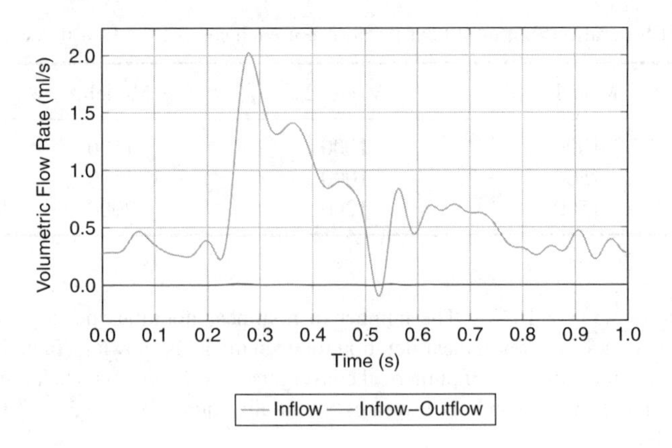


**Figure 7.14** Mass balance for Model 1 with double stent deployed

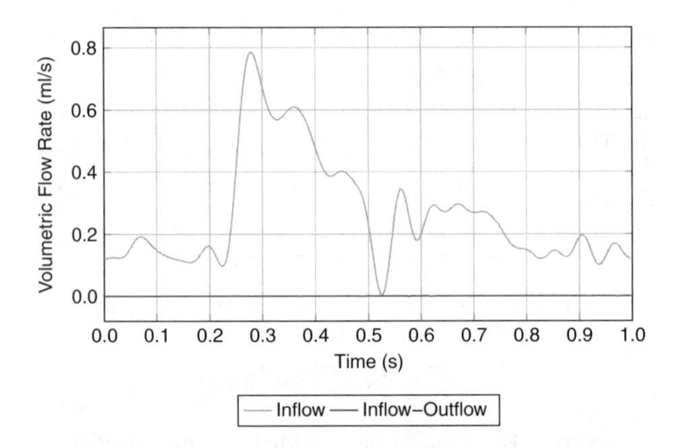


**Figure 7.15** Mass balance for Model 2 before stenting

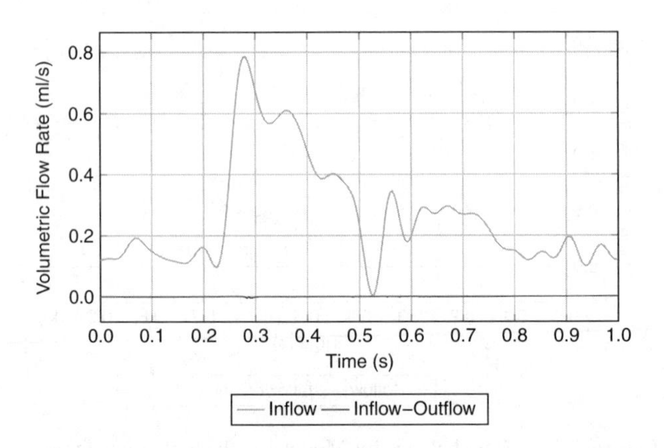


**Figure 7.16** Mass balance for Model 2 with single stent deployed

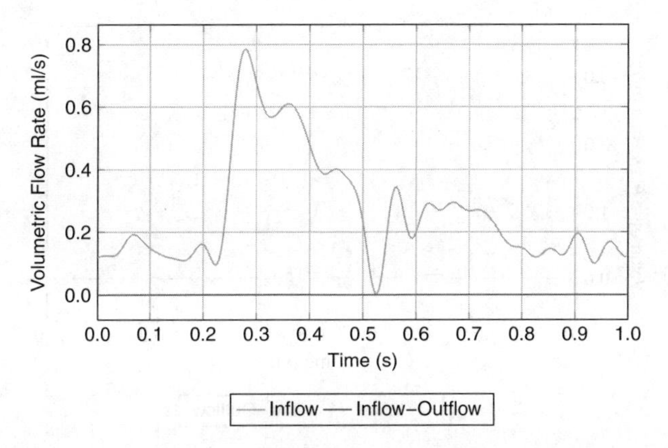


**Figure 7.17** Mass balance for Model 2 with double stent deployed

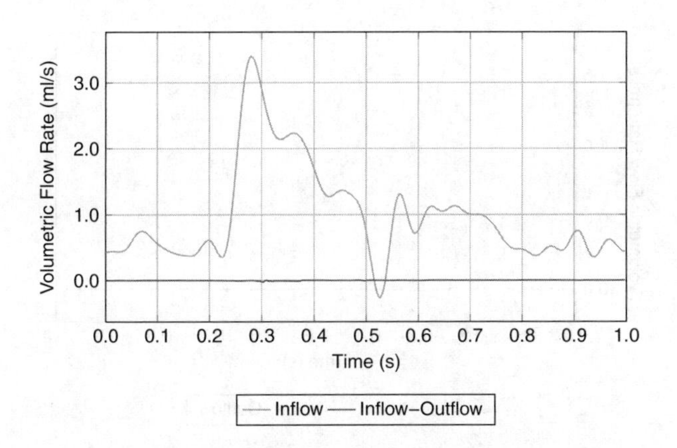


**Figure 7.18** Mass balance for Model 3 before stenting

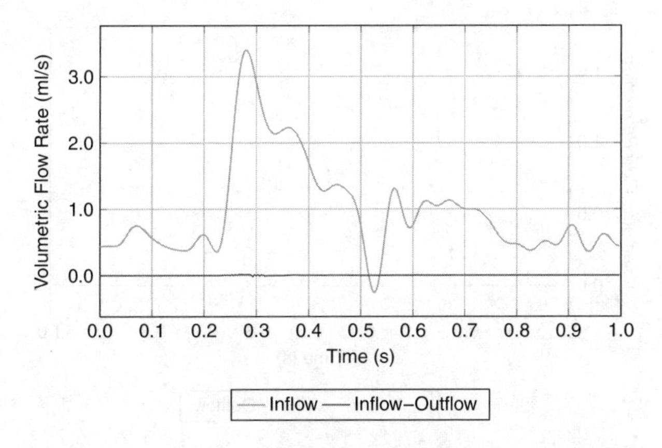


**Figure 7.19** Mass balance for Model 3 with single stent deployed

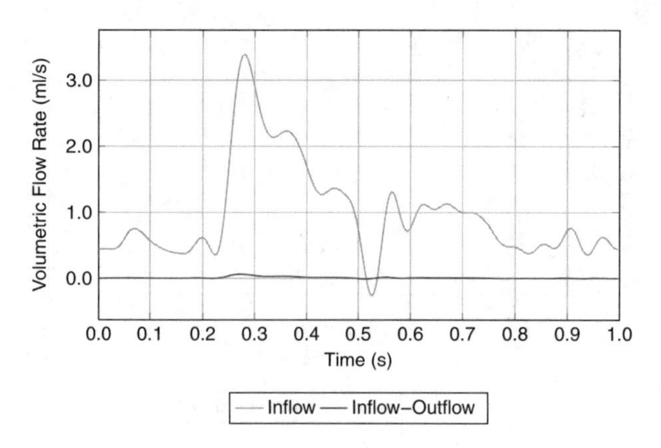


**Figure 7.20** Mass balance for Model 3 with double stent deployed

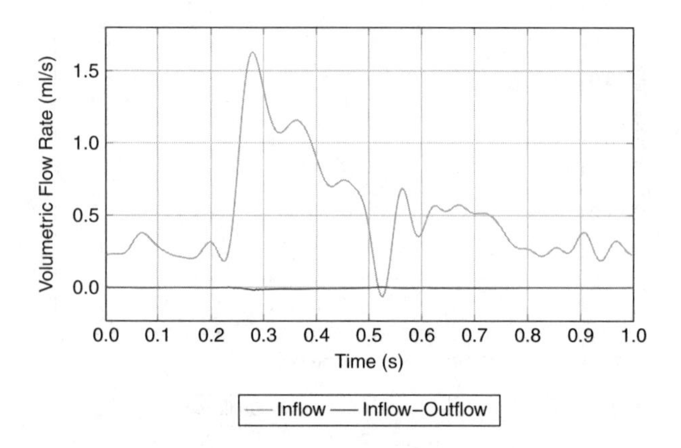


**Figure 7.21** Mass balance for Model 4 before stenting

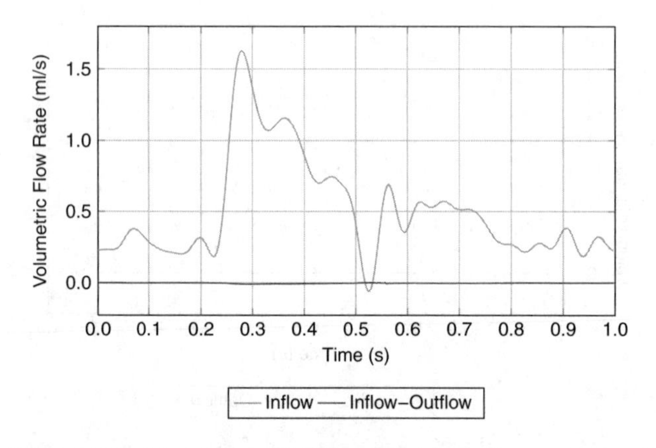


**Figure 7.22** Mass balance for Model 4 with single stent deployed

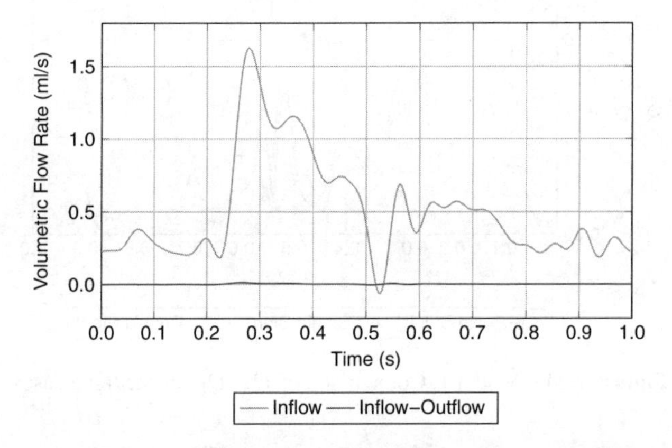


**Figure 7.23** Mass balance for Model 4 with double stent deployed

## *7.3.2 Comparative Study*

Inducing thrombosis in an aneurysm requires altering the hemodynamics at the aneurysm. Inserting a stent changes the pattern and amount of blood flow from the parent artery to the aneurysm, influencing stasis within the aneurysm. The stent-free area at the neck of the aneurysm is reduced to approximately 85% and 71% in the single- and double-stent cases, respectively, for all four models. In the following sections, we compare the fluid mechanics before and after stenting in each of the four models by analyzing the ratio of the aneurysm-inflow rate to the time-averaged parent-artery-inflow rate $Q_A/Q_P$, the spatially averaged kinetic energy and vorticity in the aneurysm, and OSI. The aneurysm-inflow rate is calculated by integrating the magnitude of the normal component of the velocity over the interior-boundary mesh containing the stent and dividing that by 2. The effectiveness of stenting using either the single or double stent depends on the degree to which the flow characteristics were altered and also the arterial geometry and size of the aneurysm. The higher OSI observed in stent cases for all models follows the belief that regions with increased OSI prompt thrombus formation [23, 24].

### Model 1

The aneurysm in Model 1 has a volume of 0.10 cm$^3$ and approximate neck area of 0.47 cm$^2$. The total area in the neck blocked by the stent in the single- and double-stent cases is 0.07 cm$^2$ and 0.13 cm$^2$, respectively. Figures 7.24–7.27 show the reduction in blood flow into and within the aneurysm caused by stenting. The parent artery has an average inflow rate of 0.62 ml/s. The peak blood flow into and within the aneurysm occurs approximately 0.02 s before peak inflow rate in the parent artery. The time-averaged $Q_A/Q_P$ decreases by 22% and 78% in the single- and double-stent cases, respectively. Similarly, the kinetic energy averaged in space and time decreases by 72% in the single-stent case and 92% in the double-stent case. The reduction in vorticity in the aneurysm caused by stenting is shown in Figures 7.28 and 7.29. The vorticity, averaged in space and time, is reduced by 47% and 72% in the single- and double-stent cases, respectively. Figure 7.30 shows the OSI for all three cases.

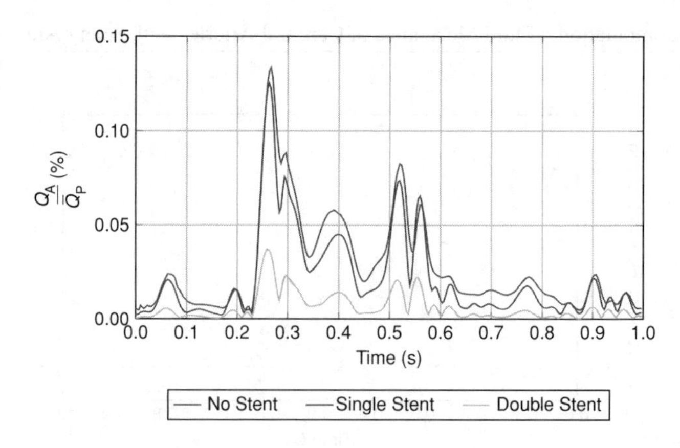


**Figure 7.24** Model 1. Comparison of $Q_A/Q_P$ for the three cases

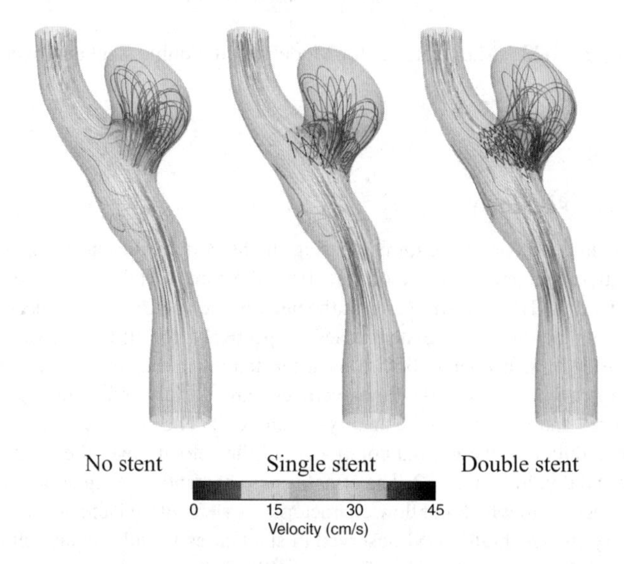


**Figure 7.25** Model 1. Streamlines showing changes in blood flow patterns and velocity induced by stenting at peak flow into the parent artery

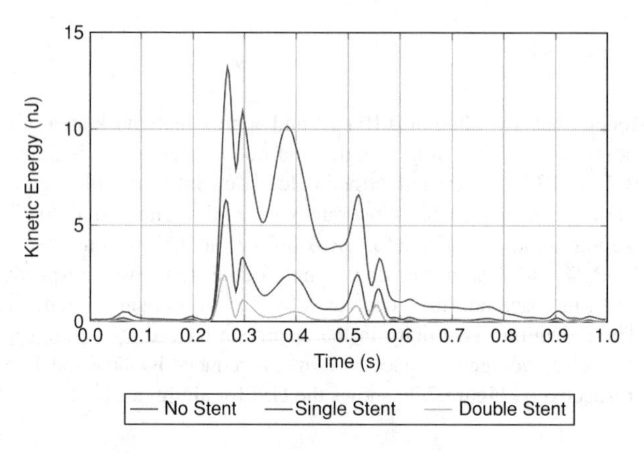


**Figure 7.26** Model 1. Comparison of spatially averaged kinetic energy in the aneurysm

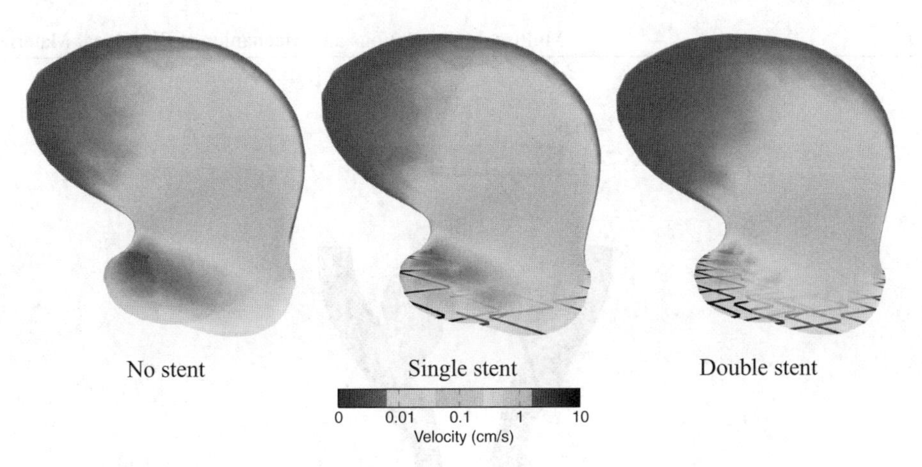


**Figure 7.27** Model 1. Volume rendering of aneurysm velocity magnitude at peak flow into the aneurysm

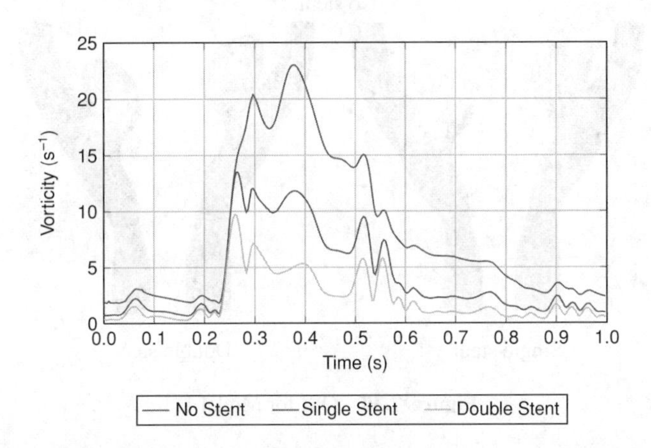


**Figure 7.28** Model 1. Comparison of spatially averaged vorticity magnitude in the aneurysm

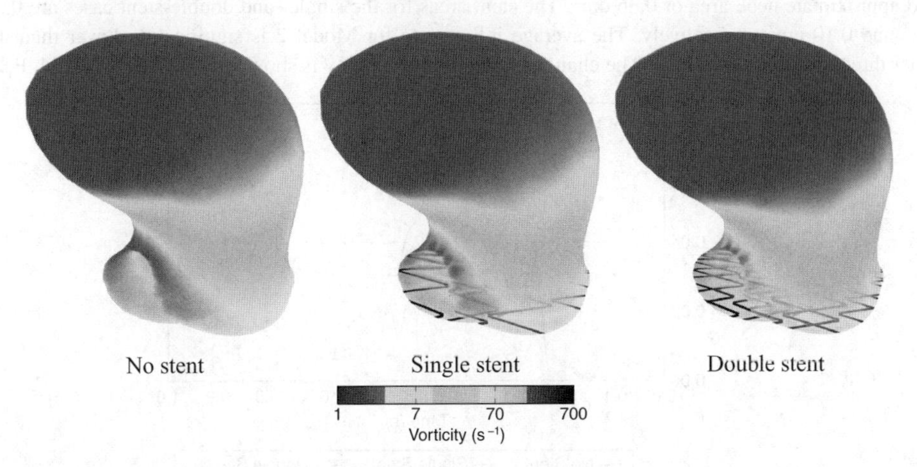


**Figure 7.29** Model 1. Volume rendering of aneurysm vorticity magnitude at peak flow into the aneurysm

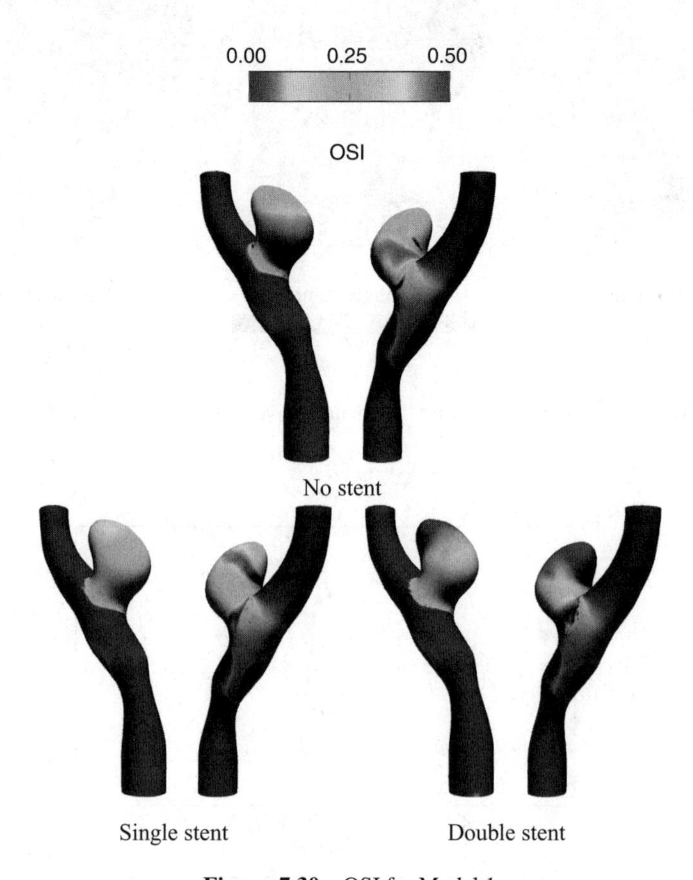


**Figure 7.30** OSI for Model 1

## Model 2

The aneurysm in Model 2 is the smallest of the four models, computed with a volume of 0.04 cm$^3$ and approximate neck area of 0.36 cm$^2$. The stent areas for the single- and double-stent cases are 0.05 cm$^2$ and 0.10 cm$^2$, respectively. The average inflow rate for Model 2 is significantly lower than the other three models at 0.26 ml/s. The change in blood flow velocity is shown in Figures 7.31–7.34. Peak

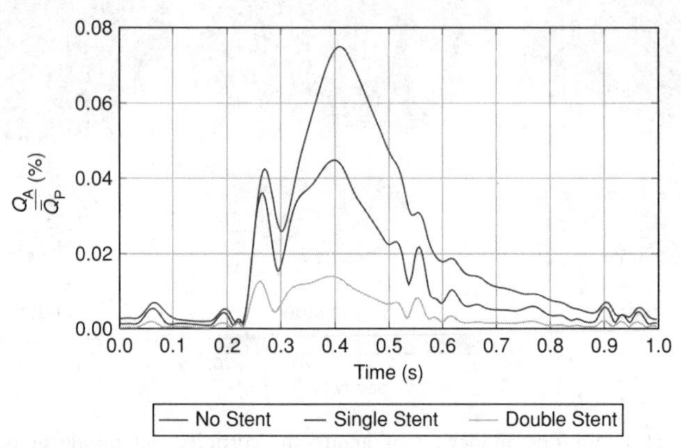


**Figure 7.31** Model 2. Comparison of $Q_A/Q_P$ for the three cases

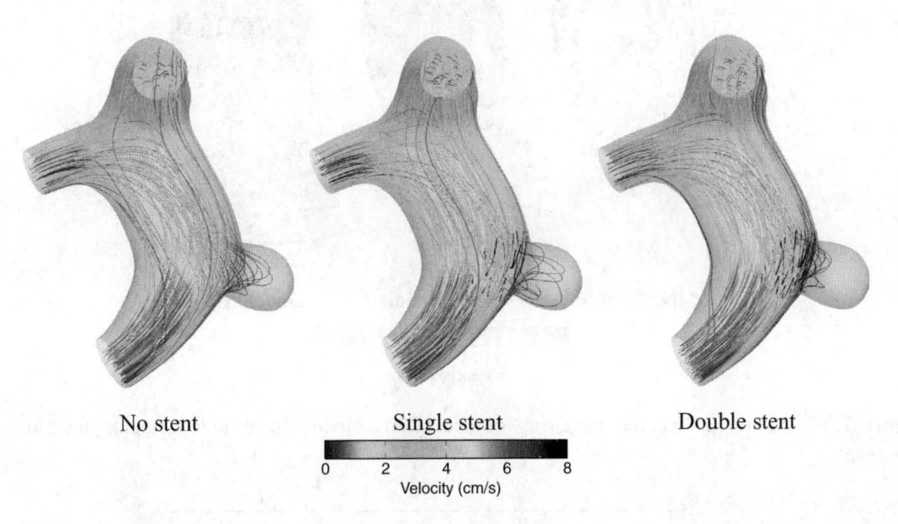


**Figure 7.32** Model 2. Streamlines showing changes in blood flow patterns and velocity induced by stenting at peak flow in the parent artery

blood flow into the aneurysm, along with peak kinetic energy, occurs approximately 0.13 s after peak inflow rate into the parent artery. In Figure 7.31, the time-averaged flow-rate ratio is reduced by 41% in the single-stent case and 81% in the double-stent case. As can be seen in Figure 7.33, the kinetic energy within the stent is significantly reduced with the single stent alone. The kinetic energy within the aneurysm averaged in space and time decreases by 83% and 95% in the single- and double-stent cases, respectively. The change in vorticity between the three cases is shown in Figures 7.35 and 7.36. The single-stent case has a 50% reduction in vorticity averaged in space and time while the double-stent case shows a 69% reduction. Figure 7.37 shows the increase in OSI from before stenting to stenting with two stents.

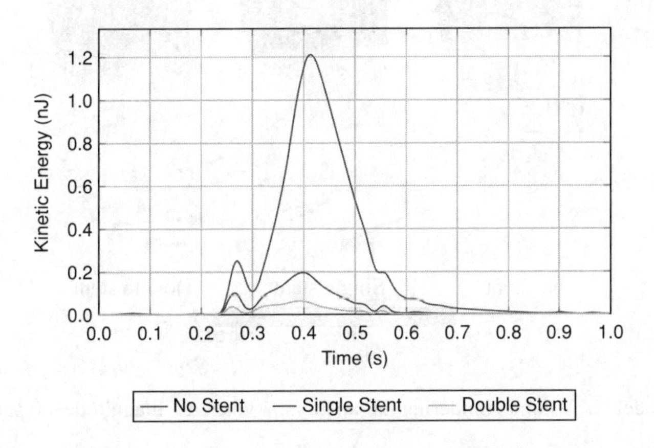


**Figure 7.33** Model 2. Comparison of spatially averaged kinetic energy in the aneurysm

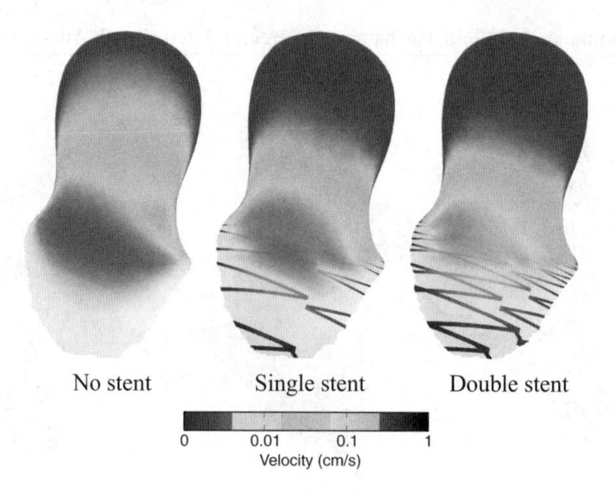


**Figure 7.34** Model 2. Volume rendering of aneurysm velocity magnitude at peak flow into the aneurysm

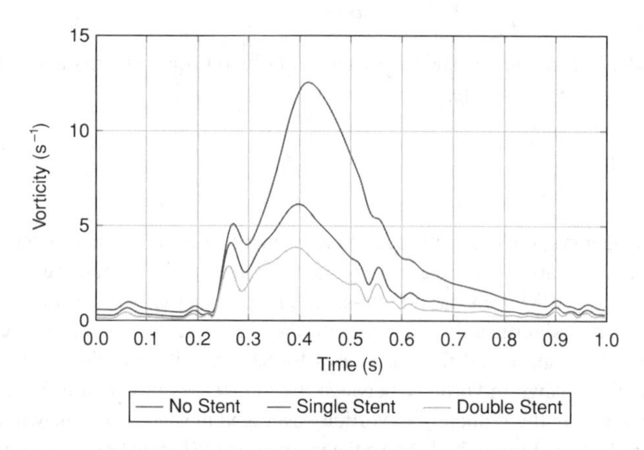


**Figure 7.35** Model 2. Comparison of spatially averaged vorticity magnitude in the aneurysm

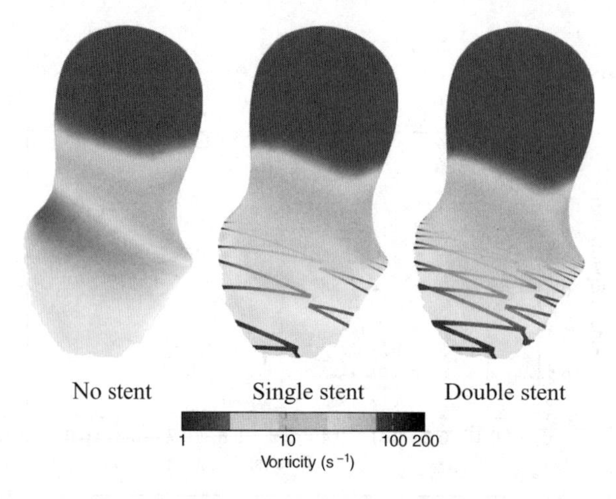


**Figure 7.36** Model 2. Volume rendering of aneurysm vorticity magnitude at peak flow into the aneurysm

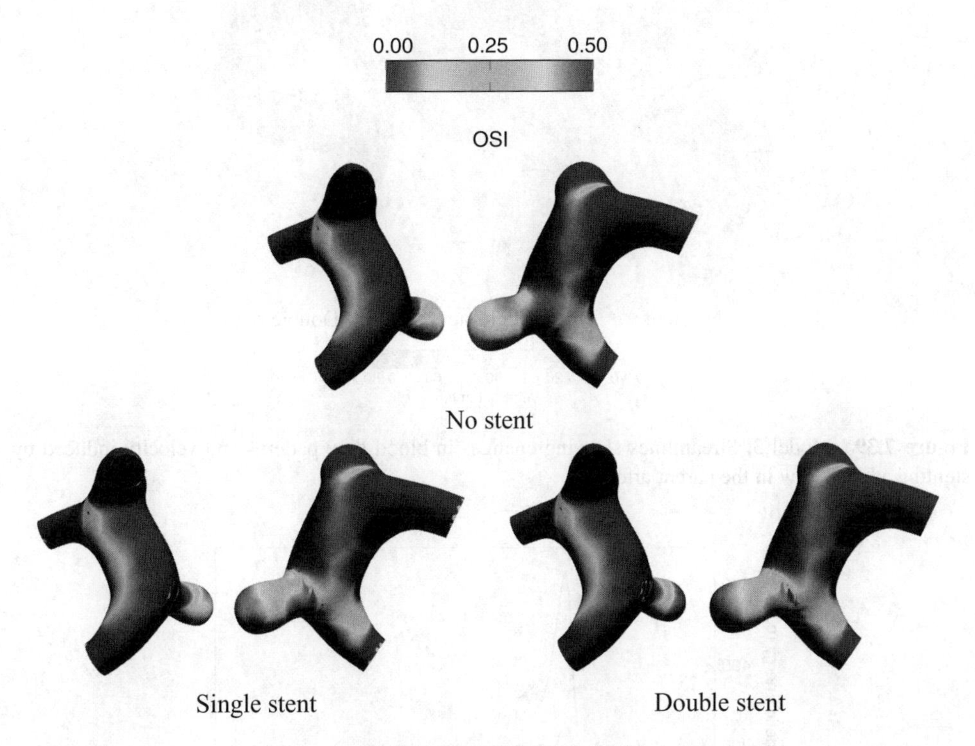


**Figure 7.37** OSI for Model 2

## Model 3

The geometry of Model 3 has a pronounced curvature just prior to and at the aneurysm location. The small size of the aneurysm, 0.05 $cm^3$ in volume, coupled with the significant curvature, results in greater blood flow into and within the aneurysm compared with the other models. The neck of the aneurysm spans an area of 0.60 $cm^2$, the largest neck area of the four models, contrary to the overall small size of the aneurysm. The single stent covers an area of 0.09 $cm^2$ and the double stent nearly doubles the area covered to 0.17 $cm^2$. Figures 7.38–7.41 show the changes induced by deploying a stent to treat the aneurysm.

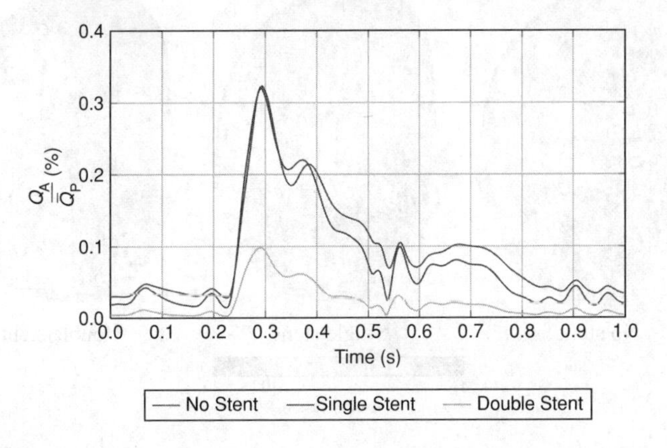


**Figure 7.38** Model 3. Comparison of $Q_A/Q_P$ for the three cases

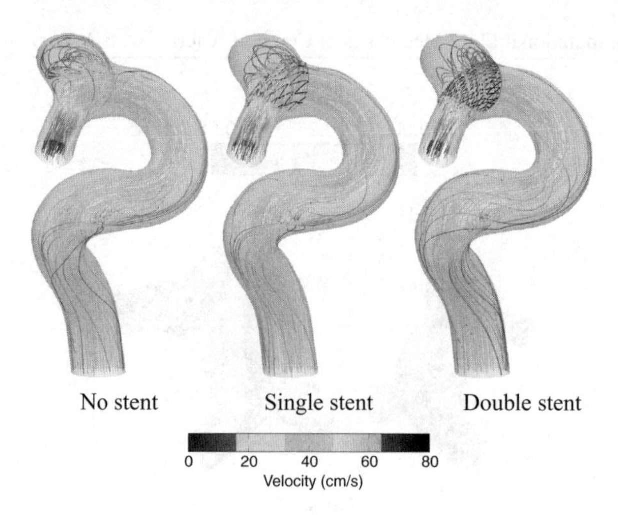


**Figure 7.39** Model 3. Streamlines showing changes in blood flow patterns and velocity induced by stenting at peak flow in the parent artery

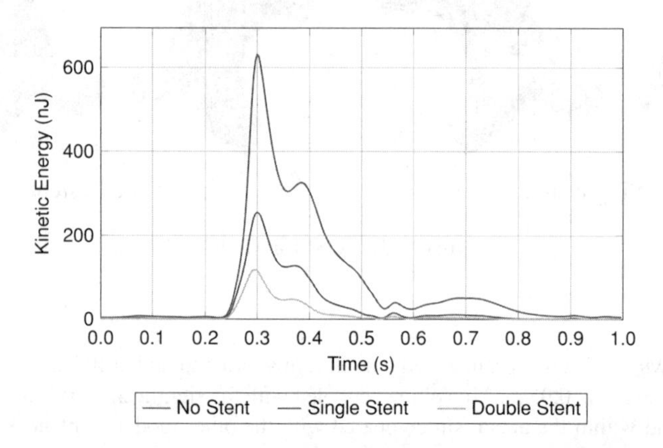


**Figure 7.40** Model 3. Comparison of spatially averaged kinetic energy in the aneurysm

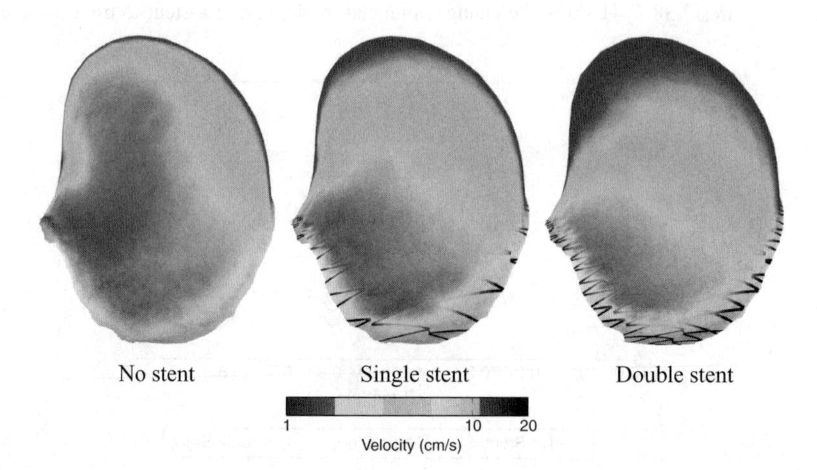


**Figure 7.41** Model 3. Volume rendering of aneurysm velocity magnitude at peak flow into the aneurysm

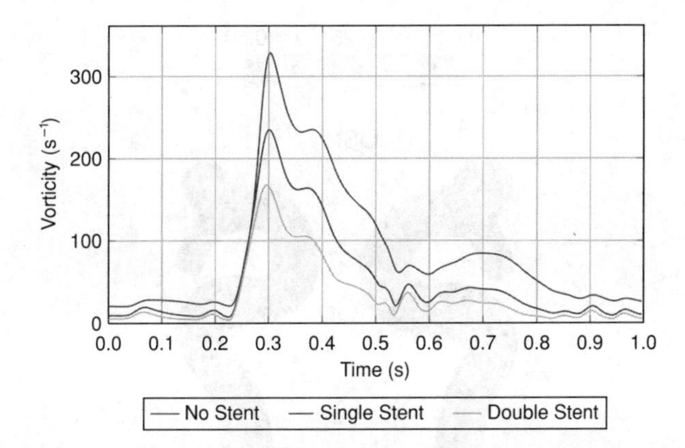


**Figure 7.42** Model 3. Comparison of spatially averaged vorticity magnitude in the aneurysm

The average inflow rate for the parent artery is slightly higher than the other models at 0.97 ml/s. The peak flow into the aneurysm occurs approximately 0.02 s after peak flow into the parent artery. On average, $Q_A/Q_P$ decreases by 16% and 78% in the single- and double-stent cases, respectively. The single-stent case has a significant drop in average kinetic energy at 66% and the double-stent case is reduced about 87%. The vorticity in the aneurysm follows a similar pattern to kinetic energy and the reduction is shown in Figures 7.42 and 7.43. The average vorticity decreases 41% and 62% in the single- and double-stent cases, respectively. Figure 7.44 shows the OSI for all three cases.

## Model 4

Model 4 has the largest aneurysm of the four arteries studied with a volume of 0.61 $cm^3$ and has the second largest neck with an area spanning 0.53 $cm^2$. The single and double stents cover areas of 0.08 $cm^2$ and 0.15 $cm^2$, respectively. For the no-stent case, we report the results from the computation of the second cardiac cycle, because the first cycle is not enough to obtain temporally periodic values for the quantities displayed. A gradual decrease in flow going into the aneurysm is observed from the no-stent case to the

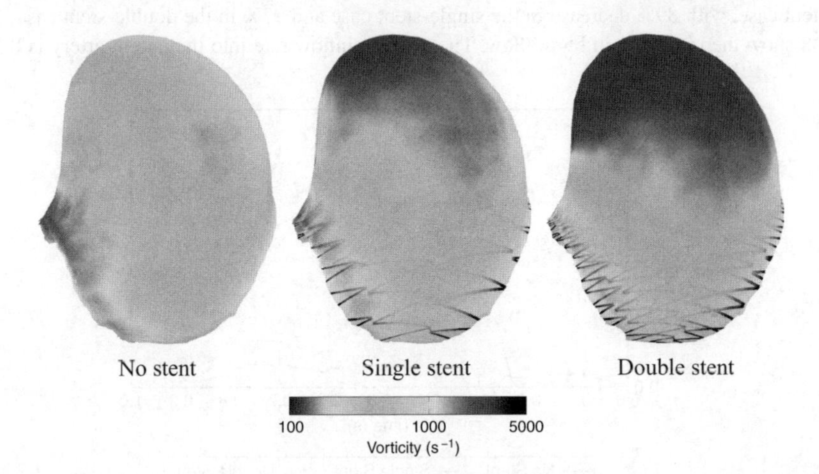


**Figure 7.43** Model 3. Volume rendering of aneurysm vorticity magnitude at peak flow into the aneurysm

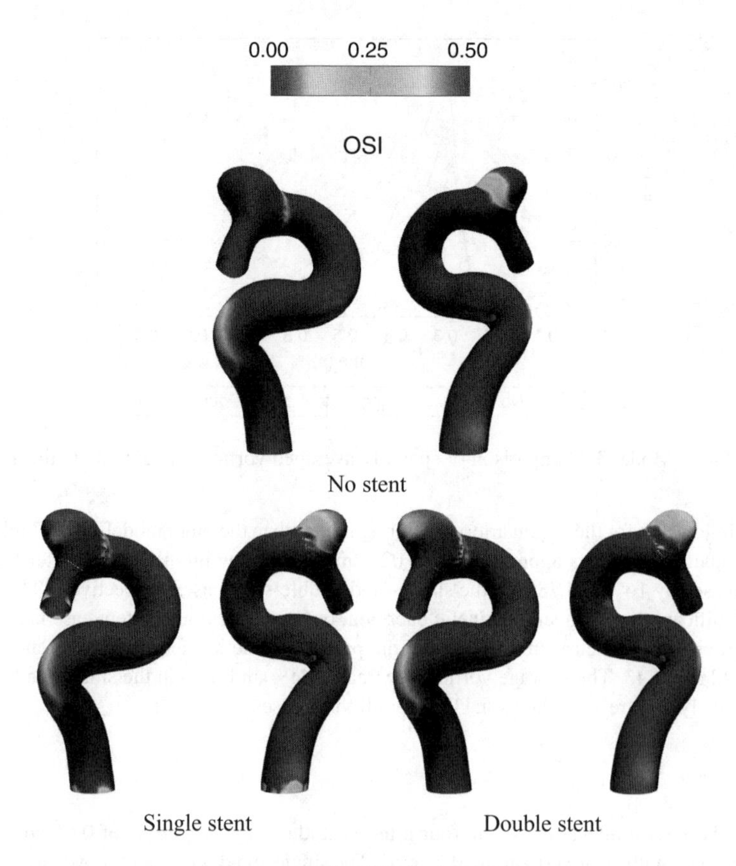


**Figure 7.44** OSI for Model 3

double-stent case. The single-stent case has a reduction of 37% in $Q_A/Q_P$ and the double-stent case has a 82% reduction. A dramatic reduction in kinetic energy occurs in both of the stent cases compared to the no-stent case, with 89% decrease in the single-stent case and 97% in the double-stent case. Figures 7.45–7.48 show the reduction in blood flow. The average inflow rate into the parent artery is 0.50 ml/s

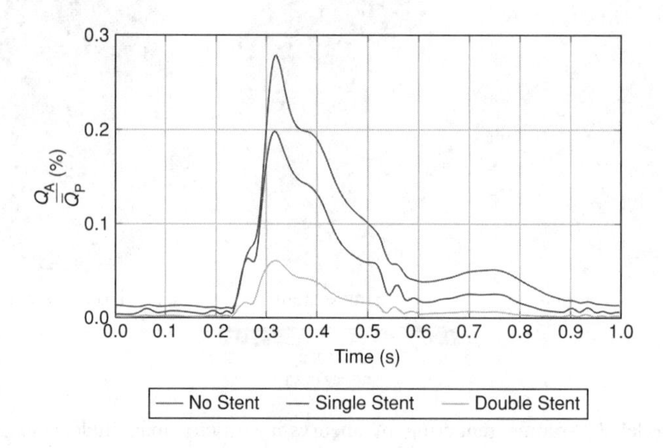


**Figure 7.45** Model 4. Comparison of $Q_A/Q_P$ for the three cases

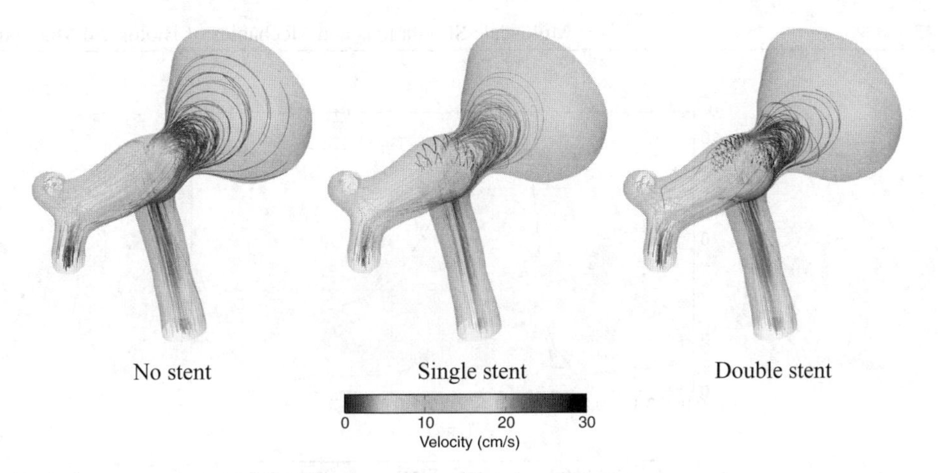


**Figure 7.46** Model 4. Streamlines showing changes in blood flow patterns and velocity induced by stenting at peak flow in the parent artery

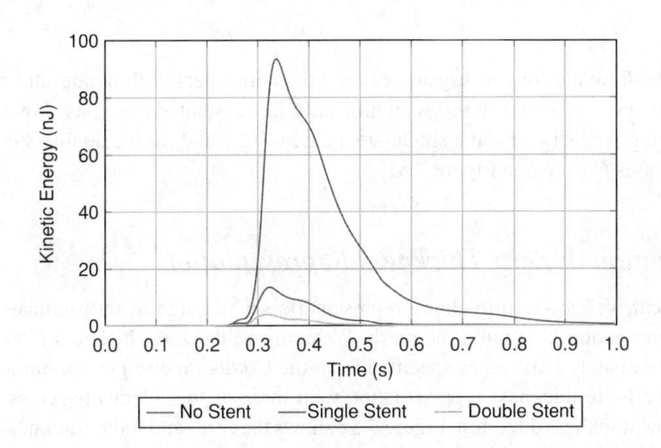


**Figure 7.47** Mode 4. Comparison of spatially averaged kinetic energy in the aneurysm

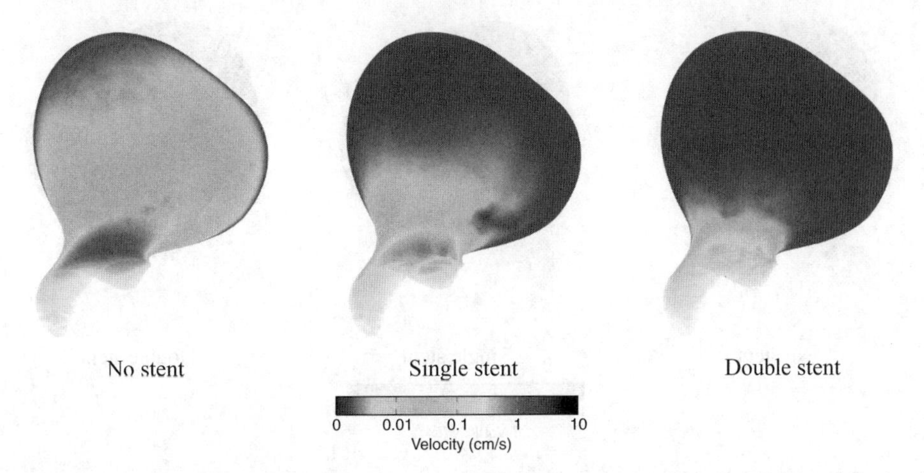


**Figure 7.48** Model 4. Volume rendering of aneurysm velocity magnitude at peak flow into the aneurysm

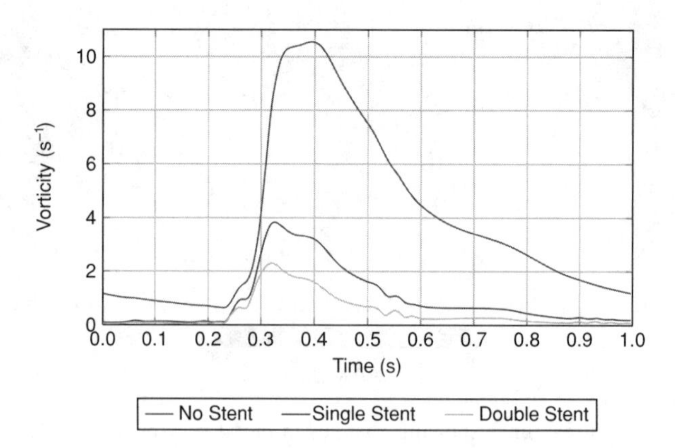


**Figure 7.49** Model 4. Comparison of spatially averaged vorticity magnitude in the aneurysm

and the peak blood flow into the aneurysm occurs 0.04 s after peak inflow rate into the parent artery. Similarly to kinetic energy, vorticity drops significantly in the stented cases, as shown in Figures 7.49 and 7.50. The average reduction for the single-stent case is 72% and for the double-stent case 86%. The OSI for all three cases is shown in Figure 7.51.

## *7.3.3 Evaluation of Zero-Thickness Representation*

In modeling the stent, we use a zero-thickness representation, which significantly reduces mesh generation complexity and time required to build the mesh. We compare the zero-thickness representation with a thickness of approximately 0.01 cm as specified from the Cordis Precise Pro Rx nitinol self-expanding stent. We mesh the finite-thickness representation with three or four elements across the width of the stent wire and in the thickness direction. Figure 7.52 shows the zero- and finite-thickness representations.

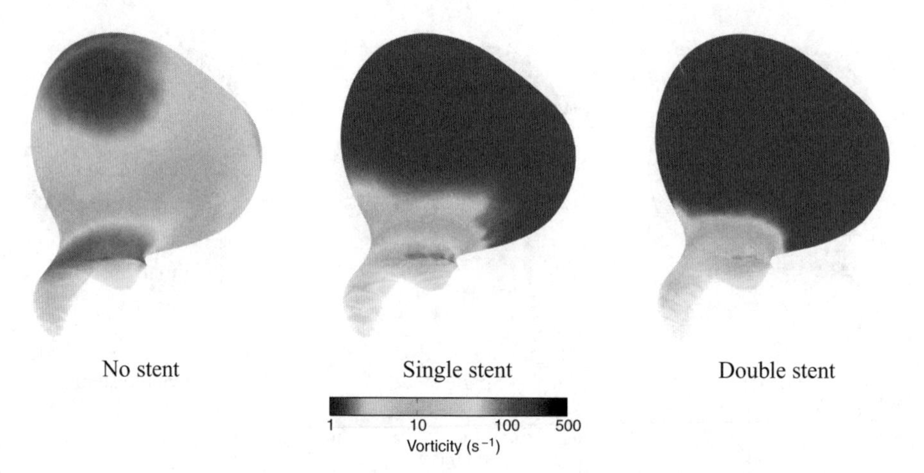


**Figure 7.50** Model 4. Volume rendering of aneurysm vorticity magnitude at peak flow into the aneurysm

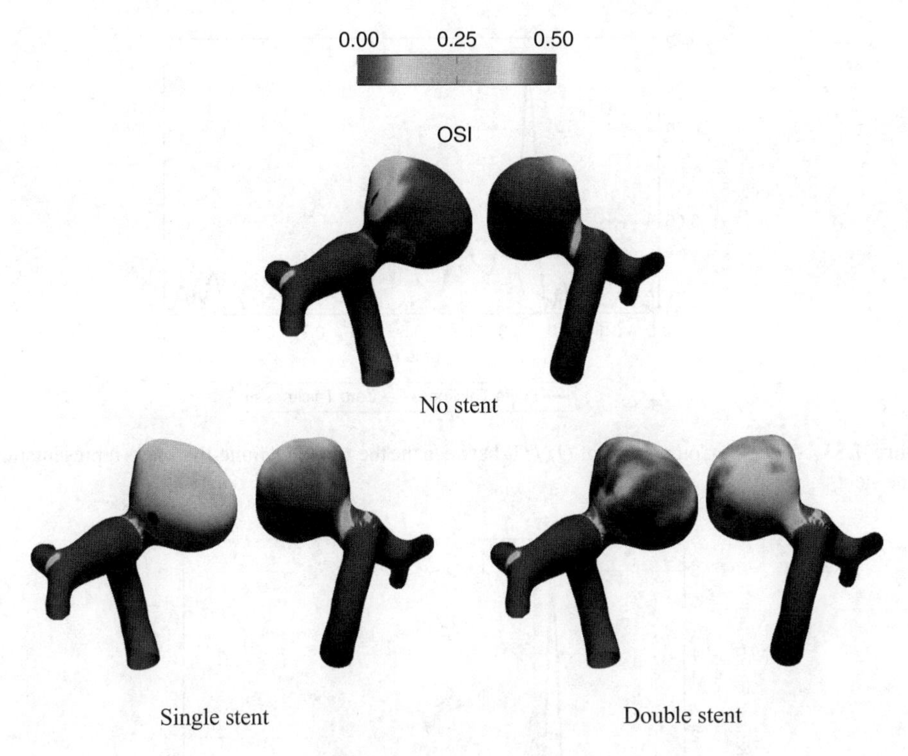


**Figure 7.51** OSI for Model 4

Model 1 geometry is used for the comparison of stent representations. Figures 7.53–7.55 and Table 7.4 show, for the zero- and finite-thickness representations, the blood flow in and out of the aneurysm, kinetic energy, and vorticity. Overall, the zero-thickness representation results in a slightly greater flow into and within the aneurysm and slightly more vorticity within the aneurysm than the finite-thickness representation does. On average, the peak values are 9% higher and the average values are 19% higher for the zero-thickness representation. Figure 7.56 shows the comparison of OSI for the zero- and

**Figure 7.52** Stent surface for zero-thickness (left) and finite-thickness (right) representations

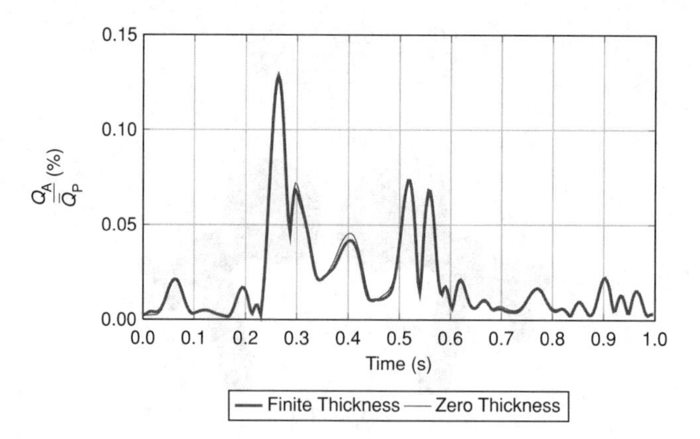


**Figure 7.53** Model 1. Comparison of $Q_A/Q_P$ between the the zero- and finite-thickness representations of the stent

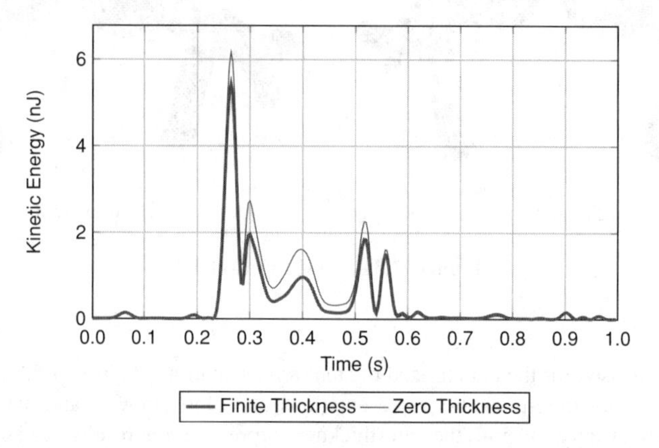


**Figure 7.54** Model 1. Comparison of spatially averaged kinetic energy in the aneurysm between the zero- and finite-thickness representations of the stent

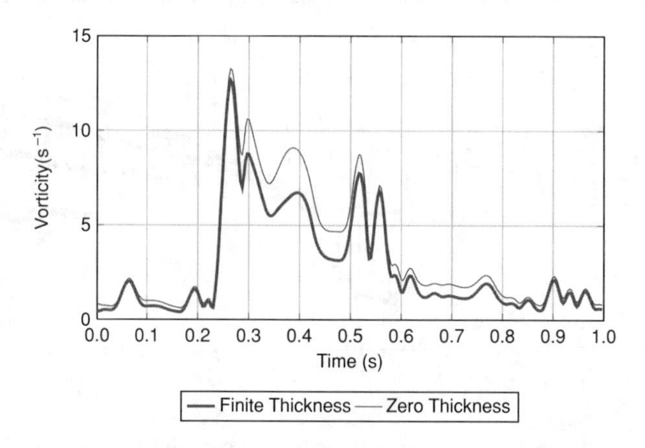


**Figure 7.55** Model 1. Comparison of spatially averaged vorticity magnitude in the aneurysm between the zero- and finite-thickness representations of the stent

**Table 7.4** Model 1. Average and peak values for the zero- and finite-thickness representations of the stent

| | Average | | Peak | |
|---|---|---|---|---|
| | Zero thickness | Finite thickness | Zero thickness | Finite thickness |
| $Q_A/Q_P$ (%) | 0.022 | 0.021 | 0.125 | 0.129 |
| Kinetic energy (nJ) | 0.53 | 0.39 | 6.16 | 5.43 |
| Vorticity ($s^{-1}$) | 4.76 | 3.84 | 17.96 | 17.12 |

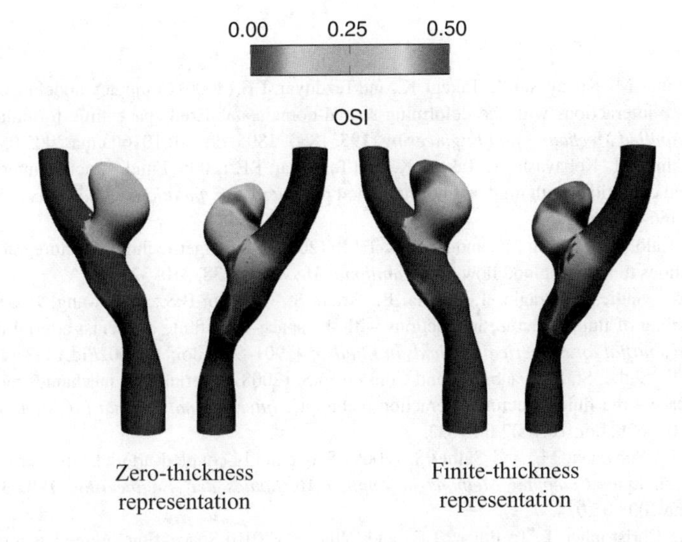


**Figure 7.56** Model 1. OSI for the zero- and finite-thickness representations of the stent

finite-thickness representations of the stent. The spatially averaged OSI in the aneurysm segment differs by 31.6%, with the zero-thickness representation at 0.27 and the finite-thickness representation at 0.38.

## 7.4 Concluding Remarks

We presented the special arterial fluid mechanics techniques we have developed for patient-specific computation of unsteady blood flow in an artery with aneurysm. The special techniques include using NURBS for the spatial representation of the surface over which the stent mesh is built, mesh-generation techniques for both the finite- and zero-thickness representations of the stent, techniques for generating refined layers of mesh near the arterial and stent surfaces, and models for representing double stent. These techniques are used in conjunction with the DSD/SST–VMST method, which is the core computational technique. We computed the unsteady flow patterns in the aneurysm and investigated how those patterns are influenced by the presence of single and double stents. We also compared the flow patterns obtained with the finite- and zero-thickness representations of the stent. The computations show that the stent substantially reduces the flow circulation in the aneurysm, the double stent does so even more, and the zero-thickness representation of the stent yields results that are close in peak values and comparable in average values to those obtained with the finite-thickness representation.

## Acknowledgments

This work was supported in part by a seed grant from the Gulf Coast Center for Computational Cancer Research funded by the John & Ann Doerr Fund for Computational Biomedicine. It was also supported in part by the Program to Disseminate Tenure Tracking System, Ministry of Education, Culture, Sports, Science and Technology, Japan. Supercomputing resources were provided in part by the Rice Computational Research Cluster funded by NSF Grant CNS-0821727. We thank Dr. Ryo Torii (University College London) for the inflow velocity waveform used in the computations. Dr. Peng R. Chen (University of Texas Medical School at Houston) provided the stent sample and the arterial images from 3DRA.

## References

[1] Torii, R., Oshima, M., Kobayashi, T., Takagi, K., and Tezduyar, T.E. (2006) Computer modeling of cardiovascular fluid–structure interactions with the deforming-spatial-domain/stabilized space–time formulation. *Computer Methods in Applied Mechanics and Engineering*, **195**, 1885–1895, doi: 10.1016/j.cma.2005.05.050.

[2] Torii, R., Oshima, M., Kobayashi, T., Takagi, K., and Tezduyar, T.E. (2006) Fluid–structure interaction modeling of aneurysmal conditions with high and normal blood pressures. *Computational Mechanics*, **38**, 482–490, doi: 10.1007/s00466-006-0065-6.

[3] Bazilevs, Y., Calo, V.M., Zhang, Y., and Hughes, T.J.R. (2006) Isogeometric fluid–structure interaction analysis with applications to arterial blood flow. *Computational Mechanics*, **38**, 310–322.

[4] Tezduyar, T.E., Sathe, S., Cragin, T., Nanna, B., Brian, S., Conklin, B.S., Pausewang, J., and Schwaab, M. (2007) Modeling of fluid–structure interactions with the space–time finite elements: arterial fluid mechanics. *International Journal for Numerical Methods in Fluids*, **54**, 901–922, doi: 10.1002/fld.1443.

[5] Tezduyar, T.E., Sathe, S., Schwaab, M., and Conklin, B.S. (2008) Arterial fluid mechanics modeling with the stabilized space–time fluid–structure interaction technique. *International Journal for Numerical Methods in Fluids*, **57**, 601–629, doi: 10.1002/fld.1633.

[6] Tezduyar, T.E., Schwaab, M., and Sathe, S. (2009) Sequentially-coupled arterial fluid–structure interaction (SCAFSI) technique. *Computer Methods in Applied Mechanics and Engineering*, **198**, 3524–3533, doi: 10.1016/j.cma.2008.05.024.

[7] Takizawa, K., Christopher, J., Tezduyar, T.E., and Sathe, S. (2010) Space–time finite element computation of arterial fluid–structure interactions with patient-specific data. *International Journal for Numerical Methods in Biomedical Engineering*, **26**, 101–116, doi: 10.1002/cnm.1241.

[8] Tezduyar, T.E., Takizawa, K., Moorman, C., Wright, S., and Christopher, J. (2010) Multiscale sequentially-coupled arterial FSI technique. *Computational Mechanics*, **46**, 17–29, doi: 10.1007/s00466-009-0423-2.

[9] Takizawa, K., Moorman, C., Wright, S., Christopher, J., and Tezduyar, T.E. (2010) Wall shear stress calculations in space–time finite element computation of arterial fluid–structure interactions. *Computational Mechanics*, **46**, 31–41, doi: 10.1007/s00466-009-0425-0.

[10] Takizawa, K., Moorman, C., Wright, S., Purdue, J., McPhail, T., Chen, P.R., Warren, J., and Tezduyar, T.E. (2011) Patient-specific arterial fluid–structure interaction modeling of cerebral aneurysms. *International Journal for Numerical Methods in Fluids*, **65**, 308–323, doi: 10.1002/fld.2360.

[11] Tezduyar, T.E., Takizawa, K., Brummer, T., and Chen, P.R. (2011) Space–time fluid–structure interaction modeling of patient-specific cerebral aneurysms. *International Journal for Numerical Methods in Biomedical Engineering*, **27**, 1665–1710, doi: 10.1002/cnm.1433.

[12] Takizawa, K., Brummer, T., Tezduyar, T.E., and Chen, P.R. (2012) A comparative study based on patient-specific fluid–structure interaction modeling of cerebral aneurysms. *Journal of Applied Mechanics*, **79**, 010908, doi: 10.1115/1.4005071.

[13] Hughes, T.J.R., Liu, W.K., and Zimmermann, T.K. (1981) Lagrangian–Eulerian finite element formulation for incompressible viscous flows. *Computer Methods in Applied Mechanics and Engineering*, **29**, 329–349.

[14] Tezduyar, T.E. (1992) Stabilized finite element formulations for incompressible flow computations. *Advances in Applied Mechanics*, **28**, 1–44, doi: 10.1016/S0065-2156(08)70153-4.

[15] Tezduyar, T.E. (2003) Computation of moving boundaries and interfaces and stabilization parameters. *International Journal for Numerical Methods in Fluids*, **43**, 555–575, doi: 10.1002/fld.505.

[16] Tezduyar, T.E. and Sathe, S. (2007) Modeling of fluid–structure interactions with the space–time finite elements: solution techniques. *International Journal for Numerical Methods in Fluids*, **54**, 855–900, doi: 10.1002/fld.1430.

[17] Takizawa, K. and Tezduyar, T.E. (2011) Multiscale space–time fluid–structure interaction techniques. *Computational Mechanics*, **48**, 247–267, doi: 10.1007/s00466-011-0571-z.

[18] Brooks, A.N. and Hughes, T.J.R. (1982) Streamline upwind/Petrov–Galerkin formulations for convection dominated flows with particular emphasis on the incompressible Navier–Stokes equations. *Computer Methods in Applied Mechanics and Engineering*, **32**, 199–259.

[19] Takizawa, K. and Tezduyar, T.E. (2012) Space–time fluid–structure interaction methods. *Mathematical Models and Methods in Applied Sciences*, **22** (Supp02), 1230001, doi: 10.1142/S0218202512300013.

[20] Hughes, T.J.R. (1995) Multiscale phenomena: Green's functions, the Dirichlet-to-Neumann formulation, subgrid scale models, bubbles, and the origins of stabilized methods. *Computer Methods in Applied Mechanics and Engineering*, **127**, 387–401.

[21] Takizawa, K., Henicke, B., Tezduyar, T.E., Hsu, M.-C., and Bazilevs, Y. (2011) Stabilized space–time computation of wind-turbine rotor aerodynamics. *Computational Mechanics*, **48**, 333–344, doi: 10.1007/s00466-011-0589-2.

[22] Takizawa, K., Henicke, B., Puntel, A., Spielman, T., and Tezduyar, T.E. (2012) Space–time computational techniques for the aerodynamics of flapping wings. *Journal of Applied Mechanics*, **79**, 010903, doi: 10.1115/1.4005073.

[23] Rhee, K., Han, M.H., Cha, S.H., and Khang, G. (2001) The changes of flow characteristics caused by a stent in fusiform aneurysm models, in *Engineering in Medicine and Biology Society, 2001. Proceedings of the 23rd Annual International Conference of the IEEE*, vol. 1, pp. 86–88, doi: 10.1109/IEMBS.2001.1018852, http://ieeexplore.ieee.org/stamp/stamp.jsp?tp=&arnumber=1018852&isnumber=21918.

[24] Jou, L.D. and Mawad, M.E. (2011) Hemodynamic effect of neuroform stent on intimal hyperplasia and thrombus formation in a carotid aneurysm. *Medical Engineering and Physics*, **33** (5), 573–580, doi: 10.1016/j.medengphy.2010.12.013, http://www.sciencedirect.com/science/article/pii/S1350453310003000.

# 8

# Application of Isogeometric Analysis to Simulate Local Nanoparticulate Drug Delivery in Patient-Specific Coronary Arteries

Shaolie S. Hossain[a] and Yongjie Zhang[b]
[a]*Institute for Computational Engineering and Sciences, The University of Texas at Austin, USA*
[b]*Department of Mechanical Engineering, Carnegie Mellon University, USA*

## 8.1 Introduction

Cardiovascular disease is the number one killer in the USA. Each year about 1.1 million Americans suffer heart attack, and almost half of them die from it, accounting for 1-in-5 deaths in the USA [1, 2]. Heart attacks are caused by blockages in the coronary arteries that supply oxygen-rich blood to the heart's muscular wall (the myocardium). These blockages occur due to an inflammatory disease called atherosclerosis, the gradual build up of plaque in the arterial wall that develops from low-density lipoprotein (LDL) cholesterol. Patients with blockage of more than 60–70% are generally considered at a higher risk for heart attacks, and candidates for clinical interventions such as coronary angioplasty and stent deployment [3]. But studies using angiographic images repeatedly show that the vast majority of these cardiac events occur in patients with less than 50% blockage [4], pointing to another mechanism. In fact, most acute coronary syndromes can be attributed to coronary thrombosis resulting from plaque rupture. Plaque disruption exposes its lipid content into the lumen, creating blood clots and blockage of the arterial lumen causing heart attacks (see Figure 8.1). These plaques that are at high risk for rupture or erosion are referred to as vulnerable plaques. Researchers have identified

*Multiscale Simulations and Mechanics of Biological Materials*, First Edition. Edited by Shaofan Li and Dong Qian.

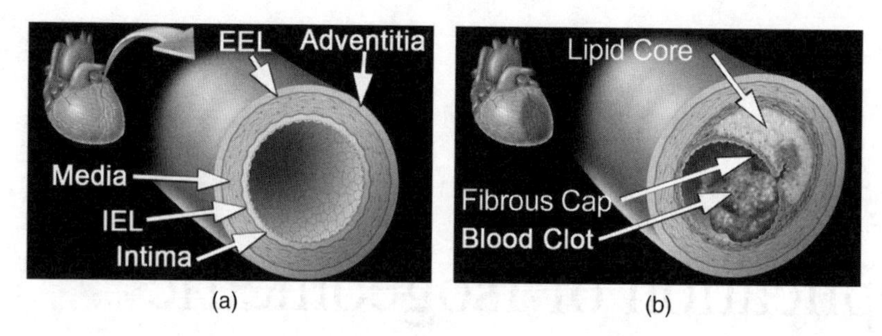


**Figure 8.1** Schematic of the disease process in an artery wall. (a) A cross-section of a normal coronary artery; (b) formation of a vulnerable plaque with a necrotic core encased by a fibrous cap, and subsequent plaque rupture leading to thrombus formation and blockage of the arterial lumen, causing heart attacks (images courtesy of Abbott Vascular Inc.)

specific structural characteristics that distinguish vulnerable plaques from stable plaques, including (i) a large lipid-rich core (>40% of plaque volume), (ii) a thin fibrous cap (<65 μm), and (iii) inflammation. Proposed clinical strategies for diffusing vulnerable plaques mainly involve (i) reducing the propensity to rupture by changing plaque composition or by improving endothelial function and (ii) reducing blood thrombogenicity. By decreasing lipid content and reducing inflammation in atherosclerotic plaque through local or systemic administration of drugs, or by controlling diet, one can achieve such plaque stabilization. These conclusions, supported by animal studies [5–7], provide the rationale for treatment of vulnerable plaques.

While systemic drugs such as statins (lipid-lowering agent) and aspirins (antithrombotic agent) are preventing a substantial portion of events in patients [8–11], most are still left to occur. It has been postulated that a treatment approach incorporating both local and systemic administration of a combination of anti-inflammatory, anti-thrombotic, and lipid-lowering drugs may provide a more effective pharmacokinetic profile and rapid treatment of vulnerable plaques, as well as long-term prevention of heart attacks.

In a typical local drug-delivery mechanism, a catheter is inserted in the groin. With the aid of live angiographic images, the physician guides the catheter through the artery into the heart until it reaches the target site of the diseased coronary artery. Drug delivery to the region of diffuse lesion in the target vessel can then be achieved by releasing a formulation containing a fluid carrier medium (e.g., polymeric nanoparticle (NP)), encapsulated drug, and a freely dissolved or suspended drug in the carrier medium. These NPs are usually 20–500 nm polymeric spherical particles that have the drug encapsulated within the polymer. Upon release of the NP formulation from the catheter into the bloodstream, mass transport into the arterial lumen occurs, and is influenced by the following factors: (a) formulation viscosity; (b) formulation miscibility properties with blood; (c) NP size, shape, and surface properties; (d) flow property in the targeted region of the vessel; and (e) delivery system geometry and proximity to the target lesion. Following their tissue uptake, the NPs transport through the wall mainly via diffusion and advection while releasing the encapsulated drug. The released drug then propagates through the wall and exerts a therapeutic effect to the targeted region (see Figure 8.2). Such a luminal drug delivery device concept can provide regional treatment of the atherosclerotic plaque in a timely manner by reducing plaque progression or inducing plaque regression, thereby adding value over systemic therapy.

The application of these medical devices in treating patients with high-risk lesions in a clinical setting can be challenging. Key design issues need to be resolved on a case-by-case basis prior to

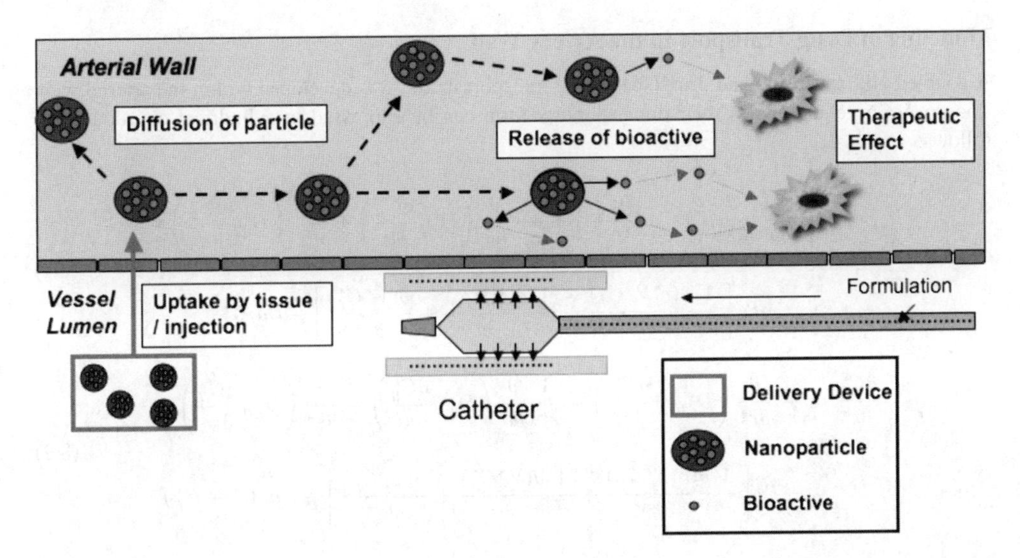


**Figure 8.2** Schematic of the lumen-side drug delivery concept (artwork created by Dariush Davalian, Abbott Vascular Inc.)

treatment administration. This requires performing specific tasks with some degree of reliability; for example, quantitative prediction of temporal and spatial drug distribution in the artery wall, evaluation of the effect of physiological forces on local drug distribution, and tuning of the drug release rate (RR) profile to the temporal sequence of the target biological process in the affected region. Numerical tools are particularly useful in this regard because they can provide crucial information that is otherwise unavailable from experiments, which are often deemed infeasible. Therefore, there is a pressing need for building technologies that will use mathematical models in conjunction with patient-specific geometries to determine optimum drug design parameters, such as delivery location, delivery mechanism, drug RR profile, vehicle size, vehicle surface properties, and drug viscosity for a desired drug tissue concentration in the arterial wall. This work aims to address this need by developing a computational tool-set to support the design and analysis of a catheter-based local drug delivery system, and thus contribute toward the prevention of heart attacks.

## 8.2 Materials and Methods

### *8.2.1 Mathematical Modeling*

In the catheter-based lumen-side delivery system, the radial release of NPs into the bloodstream takes place over a few cardiac cycles, while the drug transport in the artery wall can continue for days, if not weeks. Because of such a significant difference in time-scales, the artery wall can be decoupled from the lumen. This work mainly focuses on the coupled transport of drug and drug-encapsulated NPs in the coronary artery wall. Blood flow calculations were carried out independently, and the resulting lumen-side contribution was incorporated in the wall model through appropriate boundary conditions. In the following section, we present the developed modeling theory in four parts.

## Modeling of Drug Transport in the Artery Wall

A coupled NP (species $I$) and drug (species $II$) transport model was developed within the arterial wall. The governing equations in their dimensionless form can be expressed in cylindrical coordinates as follows:

$$\begin{aligned}\frac{\partial C_I^*}{\partial t^*} &= \frac{1}{r^*}\frac{\partial}{\partial r^*}\left(D_{r,I}^* r^* \frac{\partial C_I^*}{\partial r^*}\right) + \frac{1}{r^{*2}}\frac{\partial}{\partial \theta}\left(D_{\theta,I}^* \frac{\partial C_I^*}{\partial \theta}\right) + \frac{\partial}{\partial z^*}\left(D_{z,I}^* \frac{\partial C_I^*}{\partial z^*}\right) \\ &\quad - Pe\left[\frac{1}{r^*}\frac{\partial (r^* V_r^* C_I^*)}{\partial r^*} + \frac{1}{r^*}\frac{\partial (V_\theta^* C_I^*)}{\partial \theta} + \frac{\partial (V_z^* C_I^*)}{\partial z^*}\right] - Da_I C_I^*\end{aligned} \tag{8.1}$$

$$\begin{aligned}\frac{\partial C_{II}^*}{\partial t^*} &= \frac{1}{r^*}\frac{\partial}{\partial r^*}\left(D_{r,II}^* r^* \frac{\partial C_{II}^*}{\partial r^*}\right) + \frac{1}{r^{*2}}\frac{\partial}{\partial \theta}\left(D_{\theta,II}^* \frac{\partial C_{II}^*}{\partial \theta}\right) + \frac{\partial}{\partial z^*}\left(D_{z,II}^* \frac{\partial C_{II}^*}{\partial z^*}\right) \\ &\quad - Pe\left[\frac{1}{r^*}\frac{\partial (r^* V_r^* C_{II}^*)}{\partial r^*} + \frac{1}{r^*}\frac{\partial (V_\theta^* C_{II}^*)}{\partial \theta} + \frac{\partial (V_z^* C_{II}^*)}{\partial z^*}\right] - Da_{II} C_{II}^* + C_I^* f^*\end{aligned} \tag{8.2}$$

where the Péclet number is

$$Pe = \frac{V_{r,0}b}{D_{r,II}^0} \tag{8.3}$$

and the Damkohler numbers are

$$Da_I = \frac{\sigma_I b^2}{D_{r,II}^0} \quad \text{and} \quad Da_{II} = \frac{\sigma_{II} b^2}{D_{r,II}^0}. \tag{8.4}$$

In deriving the dimensionless equations, the following definitions of nondimensional space and time were adopted:

$$r^* = \frac{r}{b}, \quad t^* = \frac{tb^2}{D_{r,II}^0}. \tag{8.5}$$

Here, $b$ denotes the thickness of the artery wall. The concentration of the NPs $C_I$, was nondimensionalized by the initial concentration of NPs in the formulation $C_{I,0}$, while that for the drug $C_{II}$ was nondimensionalized by the initial concentration of free drug in the formulation $C_{II,0}$. The nondimensionalized variables used in the computation are listed in Table 8.1. Omitting subscripts $I$ and $II$ for simplicity, $D_r$, $D_\theta$, and $D_z$ are the diffusivities of the concerned species in radial, circumferential, and axial directions, respectively, such that $D_{i,j} = D_{i,j}^0 \varphi_i(r,t)$, where $i = r, \theta, z$ and $j = I, II$. Similarly, $V_r$, $V_\theta$, and $V_z$ are the corresponding advective velocities in the artery wall such that $V_i = V_{i,0}\varphi_i(r,t)$, where $i = r, \theta, z$. Here, $D_i^0$ and $V_{i,0}$ are the reference values of the diffusivity and the advective velocity, respectively, and $\varphi_i(r,t)$ is a space- and time-varying function accounting for the spatial and temporal variation of these quantities. A physical assumption has been made here; that is, the diffusivity tensor has principal directions in the $r$, $\theta$, and $z$ coordinate in accordance with the experimental data available in the literature, where it has been cited that diffusion in the planar (circumferential and axial) direction is faster than that in the radial direction for typical drugs of interest [12, 13].

**Table 8.1** Non-dimensional variables used

| Notation | Definition |
|---|---|
| $r^*$ | $r/b$ |
| $t^*$ | $tb^2/D^0_{r,II}$ |
| $C^*_I$ | $C_I/C_{I,0}$ |
| $C^*_{II}$ | $C_{II}/C_{II,0}$ |
| $D^*_{r,I}$ | $D_{r,I}/D^0_{r,II}$ |
| $D^*_{\theta,I}$ | $D_{\theta,I}/D^0_{r,II}$ |
| $D^*_{z,I}$ | $D_{z,I}/D^0_{r,II}$ |
| $D^*_{r,II}$ | $D_{r,II}/D^0_{r,II}$ |
| $D^*_{\theta,II}$ | $D_{\theta,II}/D^0_{r,II}$ |
| $D^*_{z,II}$ | $D_{z,II}/D^0_{r,II}$ |
| $V^*_r$ | $V_r/V_{r,0}$ |
| $V^*_\theta$ | $V_\theta/V_{r,0}$ |
| $V^*_z$ | $V_z/V_{r,0}$ |
| $C^*_I f^*$ | $\frac{C_I}{C_{I,0}} f \frac{C_{I,0}}{C_{II,0}} \frac{b^2}{D^0_{r,II}}$ |

## Modeling of Drug Release Rate from the NPs

Drug release from the NPs was assumed to occur by diffusion. A biphasic diffusion model [14], originally devised for predicting drug release from a drug eluting stent (DES) coating [15], was adopted and extended to polymeric NPs based on the assumption that transport through the dispersed drug phase within the NPs takes place via two modes – the fast mode and the slow mode. The fast mode is the release of drug from a highly percolated structure of drug phase within the polymer, and the slow mode is the release of drug from a non-percolated polymer-encapsulated phase of the drug. This biphasic drug RR model developed for the monodisperse case (same-sized particles) was further enhanced and extended to the polydisperse case (NPs having a range of particle sizes). The mathematical expression for the polydisperse case is

$$\frac{M}{M_0} = \sum_i \alpha_i f_1 \left(1 - \sum_n \frac{6}{n^2\pi^2} e^{-n^2\pi^2 t D_1/R_i^2}\right) + \sum_i \alpha_i (1 - f_1)\left(1 - \sum_n \frac{6}{n^2\pi^2} e^{-n^2\pi^2 t D_2/R_i^2}\right) \tag{8.6}$$

where $M/M_0$ is the percentage release of drug from all the NPs at time $t$ with $M_0$ denoting the total weight of drug encapsulated in all the NPs, $N_0$ (total number of NPs) at $t = 0$. Consequently, the nondimensionalized drug RR is given by

$$\begin{aligned} f^* &= \frac{dM/M_0}{dt^*} = f \frac{C_{I,0}}{C_{II,0}} \frac{b^2}{D^0_{r,II}} \\ &= \frac{6}{D^0_{r,II}/b^2}\left[\sum_i \alpha_i f_1 \frac{D_1}{R_i^2} \sum_n e^{-n^2\pi^2 t D_1/R_i^2} + \sum_i \alpha_i (1 - f_1)\frac{D_2}{R_i^2} \sum_n e^{-n^2\pi^2 t D_2/R_i^2}\right] \end{aligned} \tag{8.7}$$

There are three major design parameters: effective diffusivity of slow drug $D_1$, effective diffusivity of fast drug $D_2$, and fraction of drug in slow phase $f_1$, which by manipulating one can obtain a desired RR profile. Additionally, the drug RR can be tailored by mixing different sizes of NPs in the polydisperse case, which provides two additional design parameters: $R$, the size of the $i$th NP in the mixture, and $\alpha_i$, the fraction by weight of the $i$th NP in the formulation. This adds to the flexibility in designing an optimum RR for a desired drug distribution.

## Boundary Conditions

To determine appropriate boundary conditions for the lumen side, the following factors were considered: (1) the loss of particles and free drug at the wall due to their exposure to blood flow that tends to dislodge the NPs back into the bloodstream; (2) the propensity of the NP to adhere or stick to the wall, depending on the surface reaction rate (sticking coefficient) for the particular species; and (3) drug solubility in the tissue compared with the blood through the partition coefficient. The above physiological phenomena were modeled as follows.

*Lumen side (at location 1 in Figure* 8.3*):* the number of NPs that will stick to the artery wall tissue will depend on the competing influences of sticking and unsticking due to their special surface characteristics. The unsteady state mass balance of the NPs at the lumen-side boundary volume leads to

$$\frac{\partial C_I^{+*}}{\partial t^*} = K_I' C_I^{-*} - K_{II}'' C_{II}^{+*}, \tag{8.8}$$

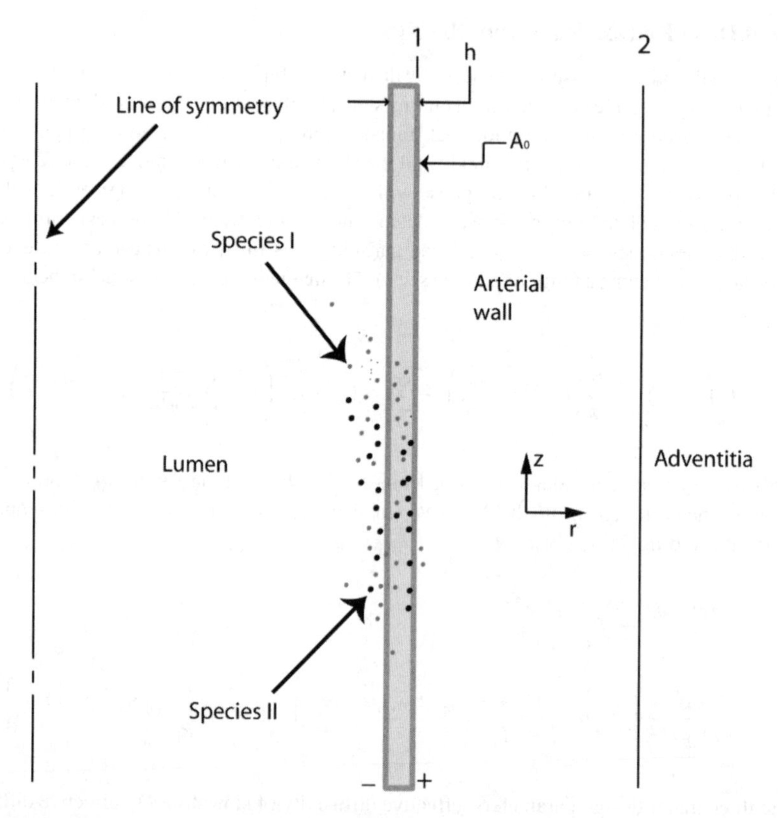


**Figure 8.3** Schematic diagram of the cross-section of an artery to explain the derivation of boundary conditions. Here, the leftmost discontinuous line is the symmetry line

where $K'$ (1/s) is the forward reaction rate, $K''$ (1/s) is the backward reaction rate, and $C_I^{-*}$ is the loss of NPs to blood flow expressed as

$$C_I^{-*} = \frac{C_I^-}{C_{I,0}^{-*}} = \exp\left[\left(-\frac{k_{\mathrm{m},I} b^2}{D_{r,II}^0}\frac{\mathrm{Factor}}{h}\right)t^*\right], \tag{8.9}$$

with $k_{\mathrm{m}}$ (cm/s) denoting the mass transfer coefficient. Here, Factor $= (A_0 + A_0')/A_0 \geq 1$, since it scales flux area from the sides $A_0'$ of the hollow cylinder, in addition to the circumferential area $A_0$ (see Hossain *et al.* [16] for a detailed derivation).

Solving for $C_I^{+*}$ gives

$$C_I^*\big|_1 = C_I^{+*} = \frac{K_I'}{K_I'' - K_I^+}[\exp(-K_I^+ t^*) - \exp(K_I'')],$$

where

$$K_I^+ = \frac{k_{\mathrm{m},I} b^2}{D_{r,II}^0}\frac{\mathrm{Factor}}{h} \tag{8.10}$$

such that $C_I^*\big|_1 = C_I^{+*}$ is the NP concentration at the lumen-side boundary. Similarly, the loss of drug to blood flow at the lumen-side boundary is modeled as

$$C_{II}^*\big|_1 = \bar{K}_{II} C_{II}^{+*} = \frac{\bar{K}_{II} C_{II}^-}{C_{II,0}^-} = \bar{K}_{II}\exp\left[\left(\frac{k_{\mathrm{m},I} b^2}{D_{r,II}^0}\frac{\mathrm{Factor}}{h}\right)t^*\right], \tag{8.11}$$

where $\bar{K}_{II}$ is the drug-tissue partition coefficient.

*Adventitia side (at location 2 in Figure 8.3):*

$$-D_{r,I}^* \left.\frac{\partial C_I^*}{\partial r^*}\right|_2 = \frac{k_{\mathrm{m},I} b}{D_{r,II}^0}(C_I^*|_2 - C_{I,\infty}^*), \tag{8.12}$$

$$-D_{r,II}^* \left.\frac{\partial C_{II}^*}{\partial r^*}\right|_2 = \frac{k_{\mathrm{m},II} b}{D_{r,II}^0}(C_{II}^*|_2 - C_{II,\infty}^*). \tag{8.13}$$

Here, $C_\infty$ (g mol/cm$^3$) is the concentration of species in the adventitia, with the subscripts $I$ and $II$ representing NP and drug, respectively.

## Modeling Solute Permeation across Tissue Barrier

The species transport across the internal elastic lamina (IEL) between the two homogeneous parts, the intima and the media, was simulated by implementing Kedem and Katchalsky's model for solute transport across a tissue barrier [17]:

$$J_{\mathrm{s}} = J_{\mathrm{v}}(1-\sigma_i)\frac{C_1 + C_2}{2} + P(C_1 - C_2), \tag{8.14}$$

where $J_{\mathrm{s}}$ (cm/s) is the solute flux per unit area, $J_{\mathrm{v}}$ (cm/s) is the filtration rate of solute, $\sigma_i$ is the reflection factor, $C_1$ is the solute concentration in solvent on the intima side (normalized), $C_2$ is the solute concentration in solvent on the media side (normalized), and $P$ (cm/s) is the diffusive permeability of the membrane to the solute. When the Péclet number *Pe* is less than 1.2, the advective contribution in Equation (8.14) can be neglected [17], such that

$$J_s = P(C_1 - C_2). \tag{8.15}$$

## 8.2.2 Parameter Selection

### Coronary Artery Wall Composition and Dimensions

Coronary artery walls are multilayered with mainly three homogeneous parts, namely the intima, the media and the adventitia, separated by the IEL and the external elastic lamina (EEL) (see Figure 8.1a). In this work, two homogeneous layers, the intima and the media, separated by the IEL, were considered [18]. The lumen- and adventitia-side contributions were incorporated through appropriate boundary conditions. When an artery becomes diseased, it undergoes significant changes morphologically that affect material, and hence transport, properties. Figure 8.1 illustrates these morphological changes that lead to the formation of vulnerable plaques characterized by two distinct regions within the intima: a large lipid-rich necrotic core (>40% of plaque volume) and a thin fibrous cap (<65 μm) that separates it from the lumen. Table 8.2 lists the relevant wall dimensions used in the simulations.

### Transport Properties

The solution of the coupled transport model relies on a number of physical parameters, each of which may vary over a wide range depending on physiological conditions. The parameters have been selected in accordance with experimental evidence or some rationalization based on previous experiences that are consistent with typical drug transport scales. A more detailed analysis of the major parameters used and the rationale behind their selection can be found in Hossain *et al.* [16].

#### *Advective Velocity*

A filtration velocity of $V_r = 1.78 \times 10^{-6}$ cm/s is prescribed according to Meyer *et al.* [19], while neglecting the lateral components of the convective velocity in the artery wall [16]. That is,

$$V_r = V_0 \quad \text{and} \quad V_\theta = V_z = 0. \tag{8.16}$$

#### *Metabolic Reaction Rate*

A typical value for the consumption rate for LDL is $10^{-4}$/s [20]. For the hydrophobic drug Everolimus, in-house data indicate a consumption rate of 0.006/h. For the NPs, this parameter is set to zero, as these NPs were not considered to be biodegradable in our calculations.

#### *Effective Diffusivity and Permeability*

Drug-encapsulated polymeric NPs are typically 20–500 nm hydrophobic spherical particles. The transport properties of these NPs are comparable to those of macromolecules like LDL (MW ≈ 2500 kDa), which are 25–50 nm in diameter and are likewise hydrophobic/lipophilic (with affinity to lipid) in nature.

**Table 8.2** Coronary artery wall dimensions

| | Healthy | Diseased |
|---|---|---|
| Intima (mm) | <0.3 (choose 0.15) | 0.65 |
| Media (mm) | 0.35 | 0.35 |

**Table 8.3** Transport properties of NPs and the drug in the artery wall

| Species | Properties | Intima | IEL | Media |
|---|---|---|---|---|
| NP | Effective diffusivity $D_{\mathrm{eff}}$ (cm²/s) | $4.8 \times 10^{-8}$ | | $7.5 \times 10^{-11}$ |
| | Permeability $P$ (cm/s) | | $1.59 \times 10^{-7}$ | |
| Drug | Effective diffusivity $D_{\mathrm{eff}}$ (cm²/s) | $9.6 \times 10^{-9}$ | | $1.5 \times 10^{-11}$ |
| | Permeability $P$ (cm/s) | | $3.14 \times 10^{-8}$ | |

**Table 8.4** Transport properties of NPs and the drug in the diseased parts of the artery wall

| | Effective diffusivity $D_{\mathrm{eff}}$ (cm²/s) | |
|---|---|---|
| Species | Fibrous cap | Lipid core |
| NP | $4.5 \times 10^{-9}$ | $1.13 \times 10^{-10}$ |
| Drug | $6.23 \times 10^{-8}$ | $1.45 \times 10^{-13}$ |

The parameters selected for normal and diseased artery cases are presented in Table 8.3 and Table 8.4, respectively.

## Boundary Conditions

Using a mass transfer coefficient[1] of $10^{-5}$ cm/s for the NPs and $2.44 \times 10^{-7}$ cm/s for the hydrophobic drug paclitaxel [12] in the lumen-side boundary condition equations (8.10) and (8.11), the resulting time-varying concentrations for the NPs and the free drug that may be present in the formulation are shown in Figure 8.4a and b, respectively. We see that the NPs penetrate the artery wall and reach a maximum concentration within a few minutes of administration, after which it depletes to zero in approximately 21 days. This is a consequence of competition between NP forward ("sticking") and backward reaction ("unsticking") rates along with their loss to blood flow. Based on experiences in the medical device industry, a forward reaction rate $K_I' = \frac{1}{10}$/min with a time constant of 10 min and backward reaction rate $K_{II}'' = \frac{1}{10\,080}$/min with a time constant of 7 days were chosen as an example case.

The formulation released into the bloodstream may contain free drug in addition to the drug encapsulated within the NPs. A formulation of 20 mg of free drug for 200 mg of encapsulated drug was used. Figure 8.4b depicts the exponential loss of the free drug to blood flow at the lumen-side boundary, where it takes about 6 h for the drug to be completely washed away. A zero-concentration sink boundary condition was employed at the adventitia side for both species [18].

## Drug Release Rate

For the application under consideration, the NPs are typically designed to release 30% of the entire encapsulated drug by the end of day 1, and a 70% release is achieved by the end of the first week. Manipulating the design parameters $D_1$, $D_2$, and $f_1$ in the biphasic drug release model presented in Equation (8.6), we obtain the drug RR profile shown in Figure 8.5, which was used in the simulations.

[1] These are assumed properties based on typical drug transport time-scales; for example, the LDL mass transfer coefficient through the endothelium is reported to be $2.2 \times 10^{-6}$ cm/s [21].

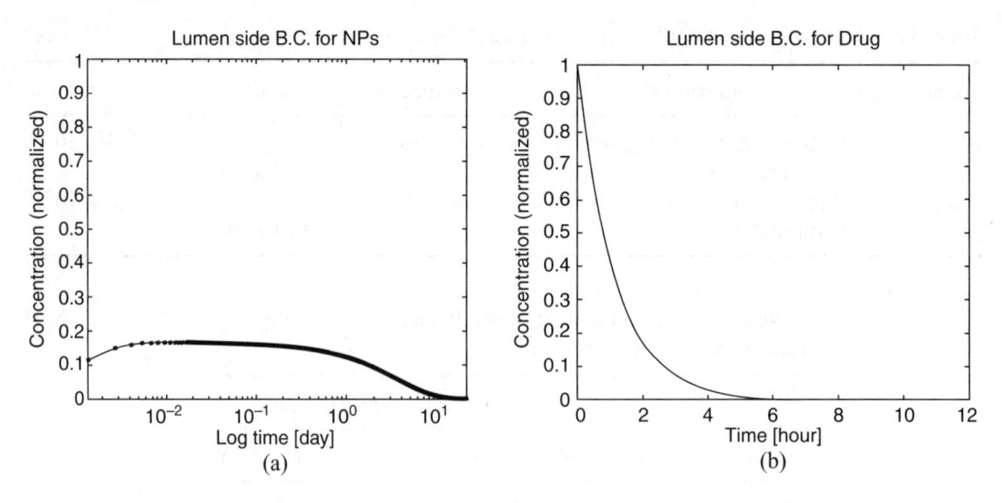


**Figure 8.4** Lumen-side boundary condition for (a) the NPs and (b) the free drug

## *8.2.3 Mesh Generation from Medical Imaging Data*

A chain of specific procedures, referred to as the vascular modeling pipeline [22], has been utilized to construct a hexahedral solid nonuniform rational B-spline (NURBS) model for a patient-specific left coronary artery (LCA) directly from CT imaging data of a healthy over-55 volunteer. Figure 8.6 illustrates the relevant steps in generating this three-dimensional (3D) geometric model. In scanned images, the intensity contrast sometimes is not clear enough, noise exists, and the luminal surface may be

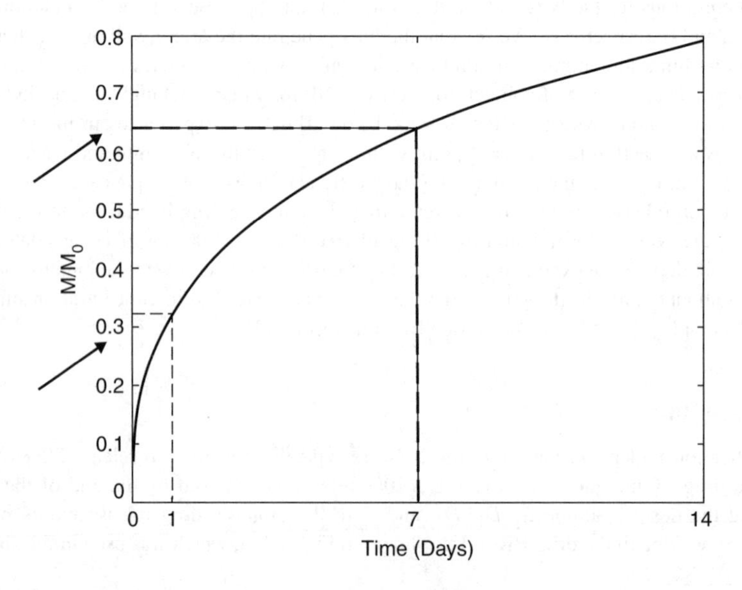


**Figure 8.5** Tailored biphasic (monodisperse case) drug RR profile used in the simulations approximately achieves 30% drug release by day 1 and 65% drug release by day 7

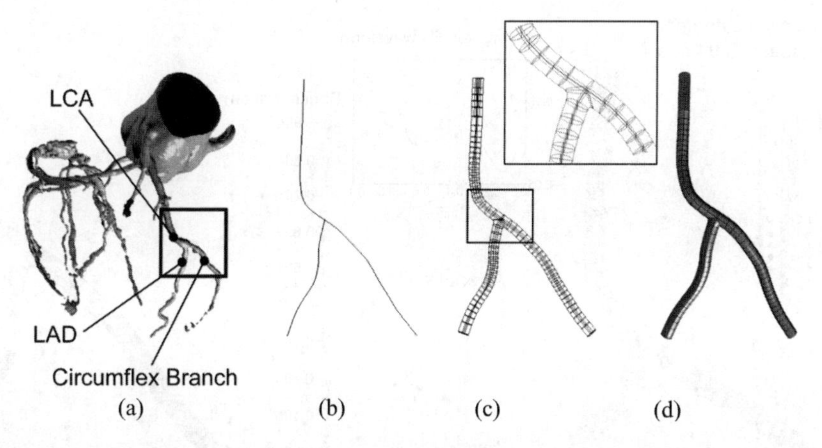


**Figure 8.6** Skeleton-based sweeping method and meshing: (a) isocontour of LCA; (b) path; (c) sweeping along the path–templated circle translated and rotated to each cross-section; (d) solid NURBS mesh

blurred. Therefore, we first utilized image-processing techniques, such as contrast enhancement, filtering, classification, and segmentation, to improve the quality of the input image. Then, luminal surfaces (Figure 8.6a) were extracted by isocontouring the processed data. A portion of the coronary arterial tree was picked and its vascular skeleton (Figure 8.6b) was extracted via Voronoi and Delaunay diagrams. This skeletonization scheme is suitable for noisy input and creates one-dimensional clean skeletons for vascular structures. Next, the skeleton-based sweeping method was used to construct hexahedral control meshes (Figure 8.6c) by sweeping a templated quadrilateral mesh of a circle along the arterial path. Here, templates were designed for the bifurcation configuration to decompose the geometry into mapped meshable patches. Each patch was then meshed using one-to-one sweeping techniques, and boundary vertices were projected to the luminal surface. Finally, hexahedral solid NURBS were constructed for the lumen (Figure 8.6d), which was modified to include an artery wall of uniform thickness (~0.6 mm) surrounding the lumen using isogeometric modeling techniques [23].

## 8.3 Results

### *8.3.1 Extraction of NP Wall Deposition Data*

The NP wall deposition data for the patient-specific coronary artery was obtained from complementary work carried out by Calo *et al.* [24]. As a first approximation, the arterial wall was assumed rigid for simplicity. The simulation setup is shown in Figure 8.7a. A catheter releases a solution containing drug-encapsulated NPs radially into the lumen, with/without free drug in the solution, over six cardiac cycles. A Navier–Stokes solver coupled to the scalar advection–diffusion equation was used to determine the time evolution of NP concentration at the lumen–wall interface. This NP wall deposition data was then normalized against the highest NP concentration to get the spatial distribution of NPs at the lumen-side boundary of the artery wall (see Figure 8.7b). Quite expectedly, the NPs accumulate near the bifurcation area. Flow through this recirculation zone is delayed as the residence time of circulating particles is prolonged. Such flow separation has also been reported to enhance platelet deposition [25]. Simulations were then run on a multilayered patient-specific coronary artery wall segment under both normal and diseased conditions. Finite-element-based isogemetric analysis was employed with quadratic NURBS used for spatial discretization. A residual-based multiscale method [26] was utilized to solve the coupled equations with the generalized-$\alpha$ method [27] adopted for time advancing. The highly advective nature

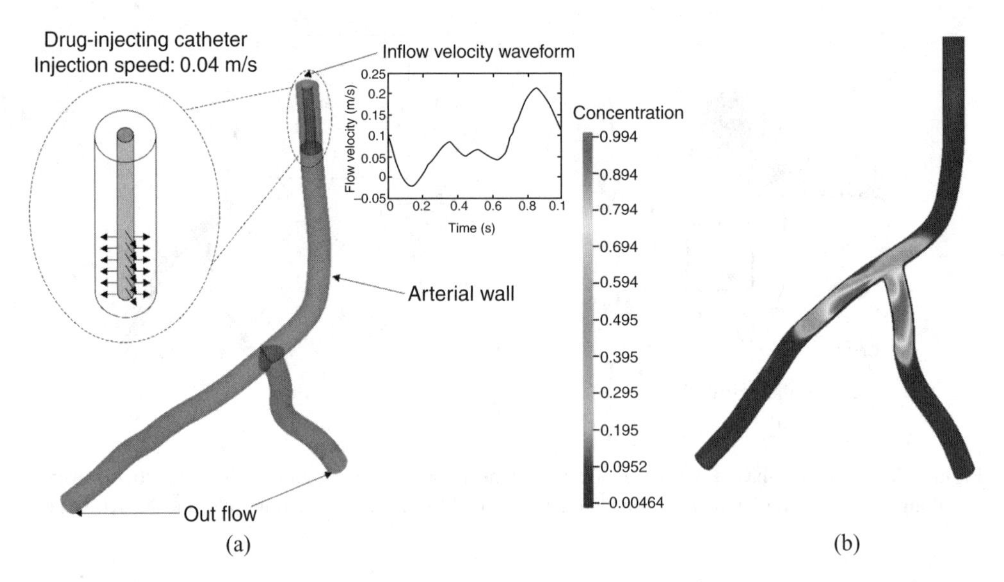


**Figure 8.7** (a) A schematic describing the problem setup for the simulation of catheter-based local drug-encapsulated NP delivery [24]. (b) The NP wall deposition data at $t = 0$ that acts as the lumen-side boundary condition for the artery wall transport model

of the blood flow necessitated implementation of stabilization schemes such as the YZβ discontinuity capturing method [28]. The methodology developed was applied to a multilayered patient-specific coronary artery wall segment under both normal and diseased conditions.

## 8.3.2 Drug Distribution in a Normal Artery Wall

Figure 8.8 provides a side-by-side comparison of the time evolution of NP and drug distribution in terms of concentration at one-third radial depth into the normal artery wall. Similar to the NP wall deposition data (see Figure 8.7b), the highest concentration of NPs occurs near the bifurcation, indicating that the wall deposition data has a considerable influence on the spatial distribution of NPs, and, consequently, drug tissue concentration because of the strongly coupled nature of transport. At this depth (i.e., roughly the outer edge of the intima), the NP concentration rapidly increases with time due to its fast diffusion ($Pe < 1$) through the intima before decreasing to about 1/10 of the peak concentration by day 7. The spatial distribution of drug is noticeably different from this NP distribution due to a much faster planar drug diffusion compared with its radial counterpart. However, the temporal behavior of the NPs and the drug are comparable. Much like the NPs, the highest drug intensity occurs around $t = 24$ h. This is because drug RR from the NPs was tailored such that approximately 30% of the drug is released by day 1. Drug availability from the traveling NPs is at its peak during this period. As a result, drug tissue concentration reaches a maximum as well. This highlights the importance of drug RR as a design parameter and its role in influencing local pharmacokinetics.

## 8.3.3 Drug Distribution in a Diseased Artery Wall with a Vulnerable Plaque

The diseased artery model contains a vulnerable plaque, characterized by a large lipid core and a thin fibrous cap, placed in the circumflex branch close to the bifurcation area. The plaque is approximately

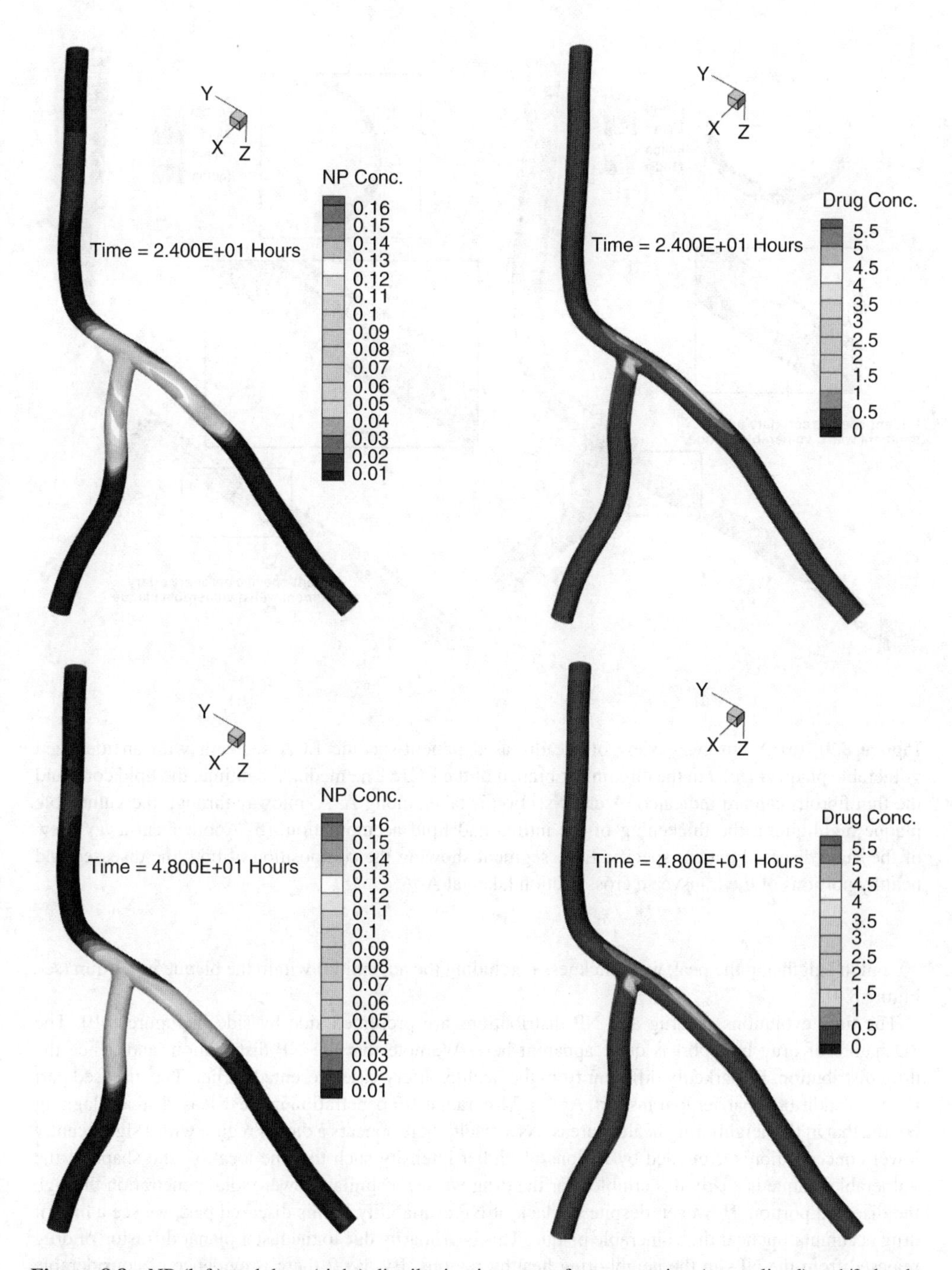


**Figure 8.8** NP (left) and drug (right) distribution in terms of concentration (normalized) at 1/3 depth from the lumen side of the healthy patient-specific coronary artery wall segment at times $t = 24$ h (top) and $t = 48$ h (bottom). N.B. Concentration scales used for NPs and drug differ here

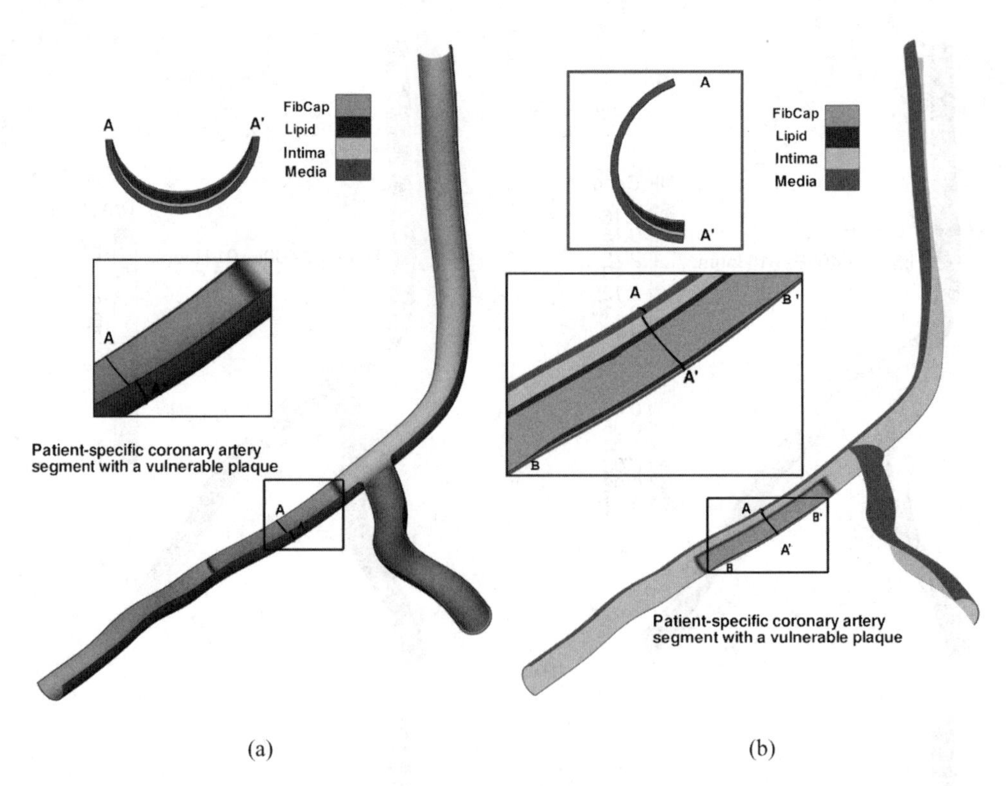


**Figure 8.9** (a) A cut-away view of the diseased patient-specific LCA segment with an idealized vulnerable plaque placed in the circumflex branch of the LCA. The media, the intima, the lipid core, and the thin fibrous cap are indicated. A cross-section is taken along A–A′ midway through the vulnerable plaque highlighting the thickening of the intima and lipid accumulation. (b) Another cut-away view of the same diseased patient-specific LCA segment showing the composition of both the diseased and healthy portions of the transverse cross-section taken at A–A′

40 mm in length and the peak wall thickness (excluding the adventitia) within the plaque is 1.1 mm (see Figure 8.9).

The time evolutions of drug and NP distributions are presented side by side in Figure 8.10. The 3D nature of drug transport is quite apparent here. We note that the NP distribution, and hence the drug distribution, is markedly different from the healthy artery case presented earlier. The diseased part poses an additional barrier to transport. At $t = 24$ h, radial NP penetration in the diseased area is lagging behind that in the neighboring healthy areas. As a result, there appears a darker region with a significantly lower concentration surrounded by regions of higher intensity such that the location and shape of the vulnerable plaque is easily discernible. For the drug we see a similar slower radial penetration through the diseased portion. However, despite the lack of NP availability in this diseased part, we see a hint of drug accumulating near the vulnerable plaque. This is primarily due to the faster planar diffusion of drug released from the NPs in the neighboring healthy regions. By day 2 there is evidence of considerable radial penetration of the NPs through the lipid core leading to a higher NP concentration at this location in the circumflex branch. Consequently, drug concentration also reaches a maximum before diminishing to about 1/20 of this highest intensity by the end of day 7.

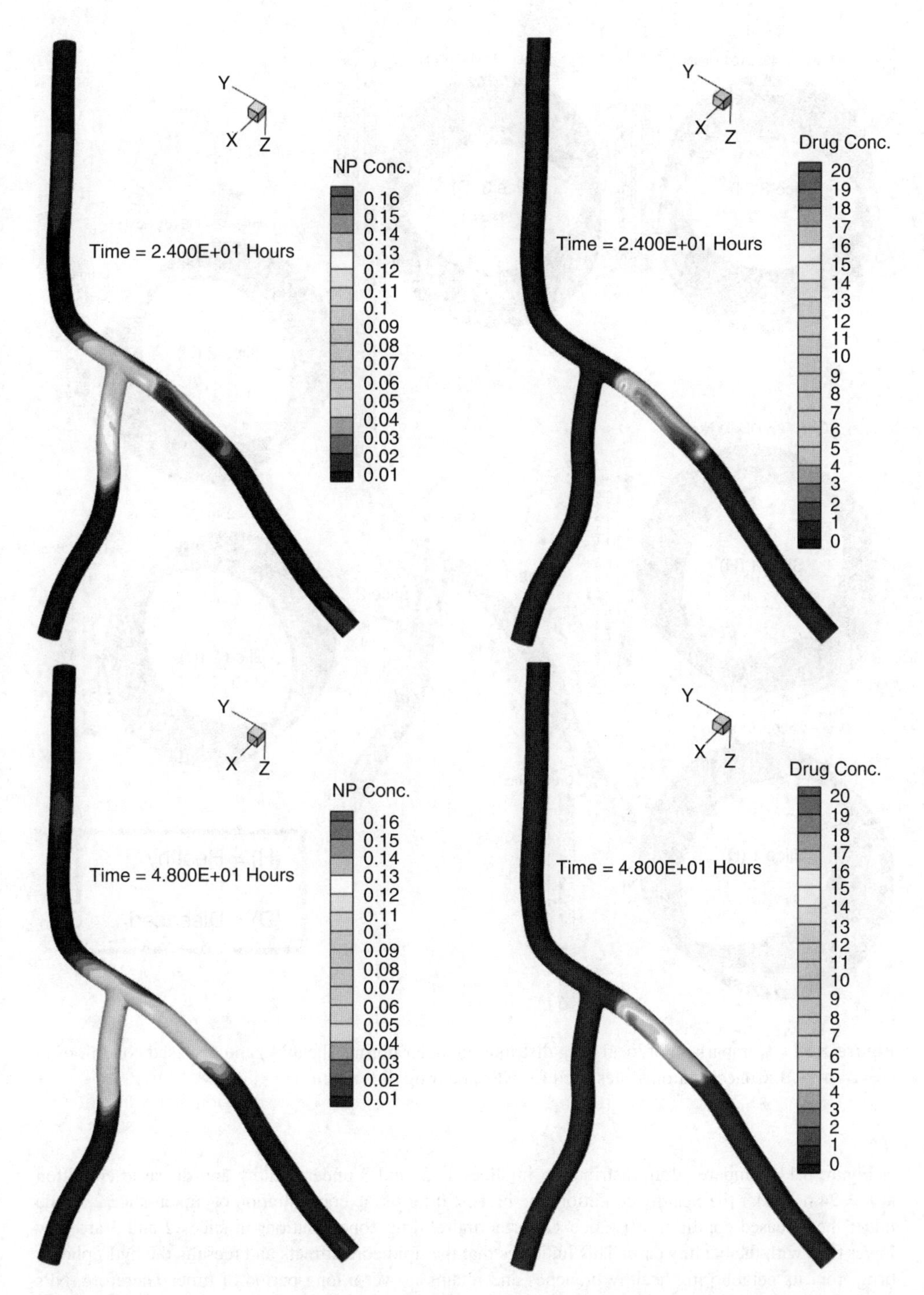


**Figure 8.10** NP (left) and drug (right) distribution in terms of concentration (normalized) at one-third depth from the lumen side of the diseased patient-specific coronary artery wall at times $t = 24$ h (top) and $t = 48$ h (bottom). N.B. Concentration scales used for NPs and drug differ here

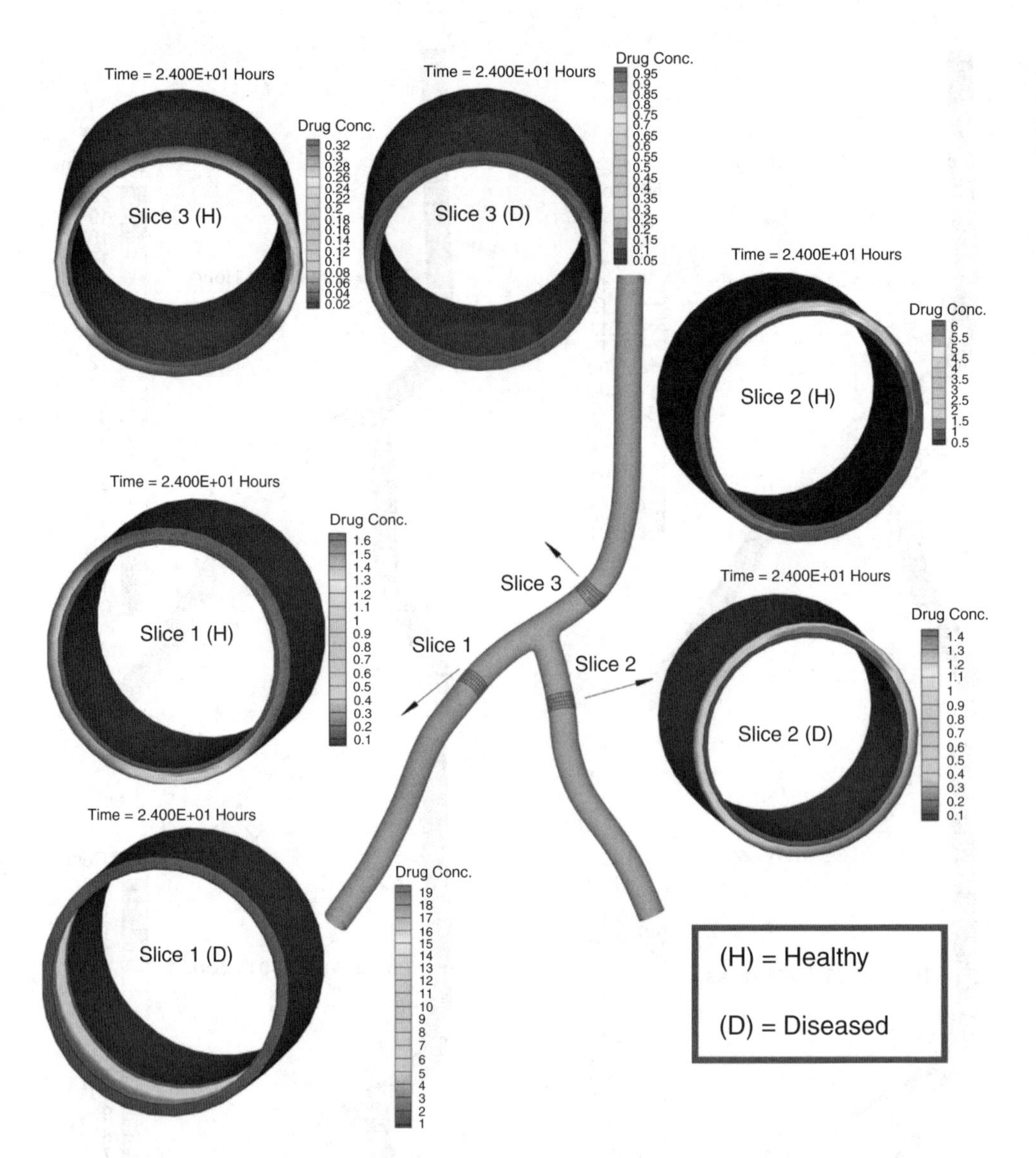


**Figure 8.11** Comparison of local drug distribution under normal (healthy) and diseased conditions at $t = 24$ h. N.B. Concentration scales used for NPs and drug differ here

Figure 8.11 compares drug distribution in slices 1, 2, and 3 under healthy and diseased condition at $t = 24$ h. Under the healthy condition, the highest drug tissue concentration occurs at slice 2, while under the diseased condition it is slice 1. Furthermore, drug concentrations in slices 2 and 3 are now lower than with the healthy case. This indicates that the lipid core attracts and recruits the hydrophobic drug from its neighboring healthy branches and retains it over a long period of time. Therefore, NPs that are off target may still contribute significantly to the overall drug tissue concentration in the target region.

## 8.4 Conclusions and Future Work

Design and evaluation of local drug delivery systems involve understanding the complex interplay between a series of physiological processes, along with the properties and the interactions of the compounds delivered. Concepts and technologies applicable to these configurations have been investigated herein in order to realistically simulate transport phenomena and commensurate physiological processes necessary for device design and optimization that is otherwise difficult to achieve in an experimental setting. The simulation results greatly help narrow down the design and parameter space of the drug delivery system, including formulation. The computational tool, therefore, reduces the number of in vivo and in vitro testing schemes. This translates into significant cost saving and timeline acceleration for companies developing breakthrough drug delivery technologies for unmet clinical need. The specific NP formulation addressed here has the potential to revolutionize therapies for refractory cardiovascular diseases such as diffuse lesions, multivessel disease, and vulnerable plaque. It can also be used for other drug delivery therapies outside the cardiovascular domain; for example, cancer, diabetic nephropathy, congestive heart failure, as well as oral biologics delivery.

In developing such a computational tool, the biggest challenge lies in the selection of parameters that are reliable and physiologically relevant, clinically applicable geometry, and appropriate boundary conditions. A number of assumptions have been made that are deemed reasonable; however, their effects should be further investigated. The goal was to demonstrate that the computational tool developed was capable of revealing relevant trends even though all the parameters may not be measurable in a precise way and, therefore, is suitable for use in helping optimize drug delivery system design parameters. The coupling of patient-specific anatomy into the drug transport phenomena enhances the computational tool's specificity and applicability in a clinical sense.

The methodology developed was applied to a few cases of physical interest in order to demonstrate its capabilities. The IEL acts as a significant transport barrier to NPs and drug, although the resistance appears to be more pronounced for the drug, as drug transport in the intima is essentially diffusive in nature. This is consistent with the findings of Penn *et al.* [29]. Also, the diseased artery provides more impedance to drug transport, as reported by Hwang and Edelman [30]. Owing to the highly coupled nature of transport between NPs and drug, the NP luminal wall deposition pattern and its intensity have a significant bearing on the overall drug distribution. Another important observation is that both arterial and plaque heterogeneity modulate drug transport as expected from Hwang and Edelman [30]. Additionally, drug transport is influenced by drug avidity or partitioning into different compositions of the diseased tissue. A direct consequence is the lipid core recruits the hydrophobic drug and retains it such that there is a high local drug concentration within this target region for a reasonably long period of time. This is encouraging from a therapeutic point of view because it may lead to sustained reduction in plaque progression. Furthermore, the time evolution of this local drug distribution within the target region is very closely related to the drug RR profile. Therefore, optimal therapeutic effect may be largely dependent on how well the RR profile can be tailored to achieve peak concentration in synchronization with the time course of desired biological responses. A comparison of the healthy versus diseased patient-specific artery wall drug distribution demonstrates that the existence of vulnerable plaque reduces drug concentration level in the neighboring healthy branches. This indicates that the NPs that are away from the target region can contribute to the overall therapeutic effect facilitated by the highly anisotropic nature of drug diffusion. Observations such as this make patient-specific geometry an essential ingredient in simulating realistic transport forces and flow features that are crucial in drug delivery system design.

The findings of this study provide a research framework for future work. The modular aspect of this computational tool allows incorporation of new functionalities fairly easily that can greatly facilitate design and optimization procedures. For example, one can tailor the existing model to introduce particle shape factor as a parameter in the coupled mass-transport equations. The effects of particle size and shape can be analyzed and validated against existing experimental work [31, 32]. The modular approach will also allow for application of specific boundary condition modification. This will involve more sophisticated mass-transport calculations within the boundary module factoring in fluid dynamic and particle parameters.

## Acknowledgments

This work was partially supported by research Grant No. UTA05-663 from Abbott Vascular Inc. Support of the Texas Advanced Research Program of the Texas Higher Education Coordinating Board through Grant No. 003658-0025-2006 and Portuguese CoLab Grant No. 04A is also gratefully acknowledged.

## References

[1] AHA (2006) *Heart Disease and Stroke Statistics – 2006 Update*, American Heart Association, Dallas, TX.
[2] AHA (2005) *Heart Disease and Stroke Statistics – 2005 Update*, American Heart Association, Dallas, TX.
[3] Lau, J., Kent, D., Tatsioni, A. *et al.* (2004) *Vulnerable Plaques: A Brief Review of the Concept and Proposed Approaches to Diagnosis and Treatment*, US Department of Health and Human Services, Rockville, MD, pp. 1–33.
[4] Falk, E., Shah, P.K., and Fuster, V. (1995) Coronary plaque disruption. *Circulation*, **92** (3), 657–671.
[5] Aikawa, M., Rabkin, E., Sugiyama, S. *et al.* (2001) An HMG-CoA reductase inhibitor, cerivastatin, suppresses growth of macrophages expressing matrix metalloproteinases and tissue factor in vivo and in vitro. *Circulation*, **103** (2), 276–283.
[6] Aikawa, M. and Libby, P. (2000) Lipid lowering reduces proteolytic and prothrombotic potential in rabbit atheroma. *Annals of the New York Academy of Sciences*, **902**, 140–152.
[7] Aikawa, M., Rabkin, E., Okada, Y. *et al.* (1998) Lipid lowering by diet reduces matrix metalloproteinase activity and increases collagen content of rabbit atheroma: a potential mechanism of lesion stabilization. *Circulation*, **97** (24), 2433–2444.
[8] Tonkin, A., Aylward, P., Colquhoun, D. *et al.* (1998) Prevention of cardiovascular events and death with pravastatin in patients with coronary heart disease and a broad range of initial cholesterol levels. *New England Journal of Medicine*, **339** (19), 1349–1357.
[9] Sacks, F.M., Pfeffer, M.A., Moye, L.A. *et al.* (1996) The effect of pravastatin on coronary events after myocardial infarction in patients with average cholesterol levels. *New England Journal of Medicine*, **335** (14), 1001–1009.
[10] Shepherd, J., Cobbe, S.M., Ford, I. *et al.* (1995) Prevention of coronary heart disease with pravastatin in men with hypercholesterolemia. *New England Journal of Medicine*, **333** (20), 1301–1308.
[11] Pedersen, T.R., Kjekshus, J., Berg, K. *et al.* (1994) Randomised trial of cholesterol lowering in 4444 patients with coronary heart disease: the Scandinavian Simvastatin Survival Study (4S). *Lancet*, **344**, 1383–1389.
[12] Creel, C.J., Lovich, M.A., and Edelman, E.R. (2000) Arterial paclitaxel distribution and deposition. *Circulation Research*, **86** (8), 879–884.
[13] Lovich, M.A., Creel, C., Hong, K. *et al.* (2001) Carrier proteins determine local pharmacokinetics and arterial distribution of paclitaxel. *Journal of Pharmaceutical Sciences*, **90** (9), 1324–1335.
[14] Hossainy, S.F.A., Prabhu, S., Hossain, S.S. *et al.* (2008) Mathematical modeling of bi-phasic mixed particle drug release from nanoparticles, in *Proceedings of the 8th World Biomaterials Congress*, Amsterdam.
[15] Hossainy, S.F.A. and Prabhu, S. (2008) A mathematical model for predicting drug release from a biodurable drug-eluting stent coating. *Journal of Biomedical Materials Research Part A*, **87** (2), 487–493.
[16] Hossain, S., Hossainy, S., Bazilevs, Y. *et al.* (2012) Mathematical modeling of coupled drug and drug-encapsulated nanoparticle transport in patient-specific coronary artery walls. *Computational Mechanics*, **49** (2), 213–242.
[17] Saltzman, W.M. (2001) *Drug Delivery*, Oxford University Press, New York, NY.
[18] Hossain, S.S. (2009) Mathematical modeling of coupled drug and drug-encapsulated nanoparticle transport in patient-specific coronary artery walls. Department of Mechanical Engineering, University of Texas at Austin, Austin, TX.
[19] Meyer, G., Merval, R., and Tedgui, A. (1996) Effects of pressure-induced stretch and convection on low-density lipoprotein and albumin uptake in the rabbit aortic wall. *Circulation Research*, **79** (3), 532–540.
[20] Ai, L. and Vafai, K. (2006) A coupling model for macromolecule transport in a stenosed arterial wall. *International Journal of Heat and Mass Transfer*, **49** (9–10), 1568–1591.

[21] Tada, S. and Tarbell, J.M. (2004) Internal elastic lamina affects the distribution of macromolecules in the arterial wall: a computational study. *American Journal of Physiology*, **287**, H905–H913.

[22] Zhang, Y., Bazilevs, Y., Goswami, S. *et al.* (2007) Patient-specific vascular NURBS modeling for isogeometric analysis of blood flow. *Computer Methods in Applied Mechanics and Engineering*, **196** (29–30), 2943–2959.

[23] Hughes, T.J.R., Cottrell, J.A., and Bazilevs, Y. (2005) Isogeometric analysis: CAD, finite elements, NURBS, exact geometry and mesh refinement. *Computer Methods in Applied Mechanics and Engineering*, **194** (39–41), 4135–4195.

[24] Calo, V.M., Brasher, N., Bazilevs, Y., and Hughes, T.J.R. (2008) Multiphysics model for blood flow and drug transport with application to patient-specific coronary artery flow. *Computational Mechanics*, **43** (1), 161–177.

[25] Zarins, C.K., Giddens, D.P., Bharadvaj, B.K. *et al.* (1983) Carotid bifurcation atherosclerosis. Quantitative correlation of plaque localization with flow velocity profiles and wall shear stress. *Circulation Research*, **53** (4), 502–514.

[26] Bazilevs, Y., Calo, V.M., Cottrell, J.A. *et al.* (2007) Variational multiscale residual-based turbulence modeling for large eddy simulation of incompressible flows. *Computer Methods in Applied Mechanics and Engineering*, **197** (1–4), 173–201.

[27] Jansen, K.E., Whiting, C.H., and Hulbert, G.M. (2000) A generalized-$\alpha$ method for integrating the filtered Navier–Stokes equations with a stabilized finite element method. *Computer Methods in Applied Mechanics and Engineering*, **190** (3–4), 305–319.

[28] Bazilevs, Y., Calo, V.M., Tezduyar, T.E., and Hughes, T.J.R. (2007) YZbeta discontinuity capturing for advection-dominated processes with application to arterial drug delivery. *International Journal for Numerical Methods in Fluids*, **54** (6–8), 593–608.

[29] Penn, M.S., Saidel, G.M., and Chisolm, G.M. (1994) Relative significance of endothelium and internal elastic lamina in regulating the entry of macromolecules into arteries in vivo. *Circulation Research*, **74** (1), 74–82.

[30] Hwang, C.-W. and Edelman, E.R. (2002) Arterial ultrastructure influences transport of locally delivered drugs. *Circulation Research*, **90** (7), 826–832.

[31] Gratton, S.E.A., Ropp, P.A., Pohlhaus, P.D. *et al.* (2008) The effect of particle design on cellular internalization pathways. *Proceedings of the National Academy of Sciences of the United States of America*, **105** (33), 11613–11618.

[32] Decuzzi, P., Godin, B., Tanaka, T. *et al.* (2010) Size and shape effects in the biodistribution of intravascularly injected particles. *Journal of Controlled Release*, **141** (3), 320–327.

# 9

# Modeling and Rapid Simulation of High-Frequency Scattering Responses of Cellular Groups

Tarek Ismail Zohdi
*Department of Mechanical Engineering, University of California, Berkeley, USA*

## 9.1 Introduction

In many applications in biophysics, one important desired piece of information is the scattering response of a group of cells to high-frequency waves; for example, the scattering response of red blood cells (RBCs) to optical pulses. The scattering response or "signature" of a group of cells provides information about the absorbance, volume fraction, and morphology, which is important to facilitate noninvasive methods for the detection of hemoglobinapathies (Figure 9.1). In a series of papers, Zohdi [1–4] developed a computational ray-tracing method that was utilized in Zohdi and Kuypers [5] to ascertain the scattering response of RBCs with a high degree of experimental correlation. Because the wavelength of visible light ($3.8 \times 10^{-7} \leq \lambda \leq 7.8 \times 10^{-7}$ m) is approximately at least an order of magnitude smaller than the diameter of a typical RBC scatterer ($d \approx 8 \times 10^{-6}$ m), ray-tracing theory is applicable and can serve to develop fast numerical techniques to quickly ascertain the amount of optical energy, characterized by the Poynting vector, that is reflected and absorbed by multiple RBCs. This chapter describes the basic techniques needed to model and simulate the response of a group of cell-like scatterers using ray-based methods. Specifically, this type of ray-based method is a discretization of the Eikonal equation, which results from an asymptotic analysis of the wave equation (that arises from Maxwell's equations for the specific cases under consideration) in the zero-wavelength limit. Ray-based methods are important since, in the high-frequency regime, nodal or element discretization becomes computationally untenable, because the number of nodes needed to resolve the waves becomes extraordinarily large. Ray tracing is a method that is employed to produce approximate solutions to wave equations for high-frequency/small-wavelength applications. Essentially, ray-tracing methods proceed by initially representing wave fronts by an array of discrete rays. Thereafter, *the problem becomes one of a purely geometric character*, where

*Multiscale Simulations and Mechanics of Biological Materials*, First Edition. Edited by Shaofan Li and Dong Qian.

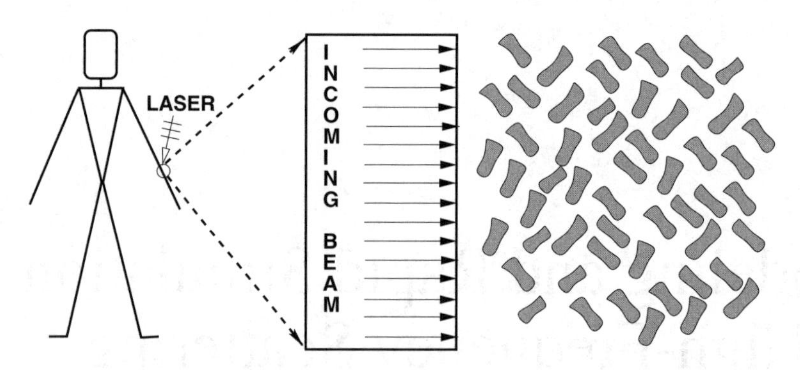


**Figure 9.1** A schematic of the noninvasive test for cell hemoglobinapathies

one tracks the changing trajectories of individual rays which are dictated by the Fresnel conditions (if a ray encounters a material interface). Ray-tracing methods are well suited for computation of scattering in complex systems that are difficult to mesh/discretize (for example, with a procedure such as the finite-difference or finite-element method) and, therefore,they are frequently employed by analysts in such situations. For a review of the state of the art in ray tracing, see Gross [6].

The primary objective of this work is to introduce the reader to the essential ingredients of classical ray-tracing theory as a foundation for rapid computational techniques involving large numbers of scatterers in high-frequency regimes.

**Remark 1** The response of a group of RBCs is important, since changes in the absorbance, volume fraction, and morphology are a sign of blood pathologies. For example, for RBCs, genetic disorders in cytoskeletal proteins (the cell wall "scaffolding") results in RBC pathologies, such as hereditary spherocytosis (spherical RBCs) and hereditary elliptocytosis (elliptical RBCs [7–9]). Deviations in cytosolic and membrane proteins may affect the state of hydration of the cell, and thereby its characteristics. In sickle-cell disease, hemoglobin polymers will distort the shape of the cell. The number of humans that are affected by sickle-cell disease, thalassemia, and other hemoglobinopathies runs in the millions [10, 11], and leads to altered hemoglobin and subsequent blood pathologies.

**Remark 2** Owing to the growing number of applications involving optics and inverse problems, there is renewed interest in ray-tracing methods. The most common physical phenomena associated with "rays" is in optics, although many other wave phenomena (e.g., acoustics) can be described in this manner. Historically, there have been many misconceptions about the nature of light; for example, by Newton, who asserted that rays of light were very small bodies emitted from shining objects. It was Huygens (1629–1695) who first hypothesized wave motion. Maxwell (1831–1879) was the first to convincingly prove that light was one form of electromagnetic energy, which includes radio waves, infrared radiation, X-rays and $\gamma$-rays. It is now commonly accepted that (from quantum theory, pioneered by Planck, Einstein, and Bohr during the early 20th century) electromagnetic energy is quantized; it can only be imparted to or taken away from the electromagnetic field in discrete amounts called photons. Light is said to have a dual nature. Certain phenomena, such as interference, have a wavelike character, while others, like the photoelectric effect, display a particle-like character. Maxwell's theory treats the propagation of light, while quantum theory describes the interaction of light and matter – in particular, the absorption and emission of light. This combined theory is known as quantum electrodynamics. However, despite the well-developed mathematical framework of quantum electrodynamics, the true nature of light remains quite elusive and, for the purposes of this chapter, irrelevant.

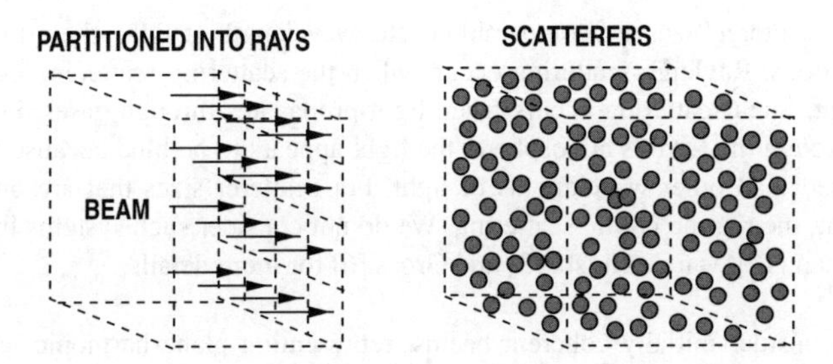


**Figure 9.2** The scattering system considered, comprised of a beam, comprised of multiple rays, incident on a collection of randomly distributed scatterers [1–4]

## 9.2 Ray Theory: Scope of Use and General Remarks

In this work, we assume that the scatterer sizes are much greater than the wavelength of the incident light, thus allowing the use of ray theory. In ray theory, an incident beam of light may be thought to consist of separate rays of light, each of which travel along their own path.[1] Typically, for a scatterer of radius 10 or more times the size of the wavelength of light, it is possible to distinguish quite clearly between the rays incident on the scatterer and rays passing around the scatterer. Furthermore, experimentally speaking, it is possible to distinguish between rays hitting various parts of the scatterer's surface. Thus, the rays may be idealized as being localized (Figure 9.2).

One can think of ray theory, often referred to as "geometrical optics" when dealing with light scattering, as the limiting case of wave optics where the wavelength $\lambda$ tends toward zero, and as being an approximation to Maxwell's equations, in the same spirit as Maxwell's equations being an approximation to quantum mechanics models.In other words, classical mechanics is precisely the same limiting approximation to quantum mechanics as geometrical optics is to wave propagation. Essentially, in ray tracing/geometrical optics, the phase of the wave is considered irrelevant. Thus, for ray tracing to be a valid approach, the wavelengths should be much smaller than those associated with the length scales of the scatterers of the problem at hand (Figure 9.2). Ray theory is a simple and intuitive approximate theory which can provide sufficiently accurate, quantitative information for scattering problems in complex media. It should not be discarded simply because a "more rigorous" theory, namely wave mechanics, is available, just as wave theory should not be abandoned for quantum mechanics theory. A further caveat is that ray theory has nearly ideal characteristics for high-performance numerical simulation.

**Remark 3** As previously noted, the wavelengths of visible light fall approximately within $3.8 \times 10^{-7} \leq \lambda \leq 7.8 \times 10^{-7}$ m. Note that all electromagnetic radiation travels at the speed of light in a vacuum, $c \approx 3 \times 10^{8}$ m/s. A more precise value, given by the National Institute of Standards and Technology, is $c \approx 2.99792458 \times 10^{8} \pm 1.1$ m/s. Generally, it is assumed that the length scale of the surface features of the scatterers is large enough, relative to the electromagnetic wavelength, that the reflections are specular (coherent) and not diffuse, thus allowing ray-tracing theory to be employed. High-frequency applications include: (a) regimes where the scatterers and surface features are larger than visible light rays ($3.8 \times 10^{-7} \leq \lambda \leq 7.8 \times 10^{-7}$ m); (b) regimes where the scatterers and surface features are larger than ultraviolet rays ($10^{-9}$ m $\leq \lambda \leq 10^{-8}$ m); (c) regimes where the scatterers and surface features are larger than X-rays ($10^{-11} \leq \lambda < 10^{-9}$ m); and (d) regimes where the scatterers and surface features are larger than $\gamma$-rays ($10^{-12} \leq \lambda \leq 10^{-11}$ m).

[1]Rays provide an easy way to visualize the progression of light. A ray is a line drawn in space corresponding to the flow of radiant energy. They are perpendicular to the wavefront. They are not physical, just simply a mathematical construct.

**Remark 4** If the scatterer sizes are comparable to the wavelength of light, then it is inappropriate to use ray representations. Rayleigh scattering occurs when the scattering scatterers are smaller than the wavelength of light. Such scattering occurs when light propagates through gases. For example, when sunlight travels through the Earth's atmosphere, the light appears to be blue because blue light is more thoroughly scattered than other wavelengths of light. For scatterer sizes that are on the order of the wavelength of light, the regime is Mie scattering. We do not consider such systems in this chapter. See Bohren and Huffman [12], van de Hulst [13] and Gross [6] for more details.

**Remark 5** We consider initially coherent beams, representing plane harmonic waves (Figure 9.2) composed of multiple collinear rays, where each ray is a vector in the direction of the flow of electromagnetic energy, which, in isotropic media, corresponds to the normal to the wave front. Thus, for isotropic media, the rays are parallel to the wave's propagation vector, denoted $\mathbf{k}$ (see Figure 9.5). Of particular interest is to describe the breakup of initially highly directional coherent beams, which, under normal circumstances, do not spread out into multidirectional rays.

**Remark 6** We will ignore the phenomenon of diffraction, which originally meant, within the field of optics, a small deviation from rectilinear propagation, but which has come to mean a variety of things to different researchers; for example, the generation of a "shadow" behind a scatterer or the "bending around corners" of incident optical (electromagnetic) waves. Although we restrict ourselves to ray-tracing methods in this chapter, it is important to realize that many advanced methods, such as *physical optics*, which are beyond the scope of the present introductory treatment, have ray tracing (geometrical optics) as their starting point.

**Remark 7** In ray-tracing methodology, an incident beam of light, which forms a plane wave front, which is considered "infinite" in extent (in the lateral directions), relative to the wavelength of light, can be thought of as being comprised of separate rays of light, each of which pursues its own path. Thus, it almost goes without saying that the width $w$ of a beam must satisfy $w \gg \lambda$ for the representation as multiple rays to make sense (Figure 9.2). One can consider the representation of a beam by multiple rays as simply taking a large "sampling" of the diffraction by the beam (wave front) over the portion of the scatterer where the beam is incident. The trajectory of harmonic plane waves, and the corresponding ray-representation direction, can actually be derived from Maxwell's equations, which reduce to the classical amplitude and trajectory "eikonal" equations, described shortly. For more details see Bohren and Huffman [12], Elmore and Heald [14], van de Hulst [13] and Gross [6].

**Remark 8** When a beam strikes a surface, the back scatter is referred to as "reflection." $\lambda/2$ is the layer responsible ( Figure 9.3b) for reflecting (vibrating atoms). This is approximately 1000 atomic layers. The larger the wavelength of the incoming light, the larger the interacting layer is. Each atomic layer is less efficient as a scatterer as $\lambda$ gets larger. Thus, transparent media involve a lot of scatterers and mix all wavelengths of light to yield "white light."

**Remark 9** The dependence of the refractive index (or wave speed) on the wavelength is called dispersion. With normal dispersion, the refractive index $n$ is higher at shorter wavelengths $n = c_o/c$, where $c_o$ is the speed of light in a vacuum and $c$ is the speed of light in the medium (Figure 9.3a). Thus, we have $\mathrm{d}n/\mathrm{d}\lambda < 0$ or $\mathrm{d}c/\mathrm{d}\lambda > 0$. This means that shorter wavelength disturbances moves more slowly than longer wavelength disturbances do.

The trigonometric "algebra" that one has to deal with in wave propagation (upcoming) analysis can simplified by using complex exponentials. Recall that a complex number can be written as $z = x + \mathrm{j}y$ ($\mathrm{j} \stackrel{\text{def}}{=} \sqrt{-1}$), or $z = A\,\mathrm{e}^{\mathrm{j}\phi} = A(\cos\phi + \mathrm{j}\sin\phi)$, where $zz^* = A^2$ and $z^* = x - \mathrm{j}y$ is the complex conjugate. Also, $z_1 z_2 = A_1 A_2 = A_1 A_2\,\mathrm{e}^{\mathrm{j}(\phi_1+\phi_2)}$. Thus, if we write, as we shall do later, solutions of

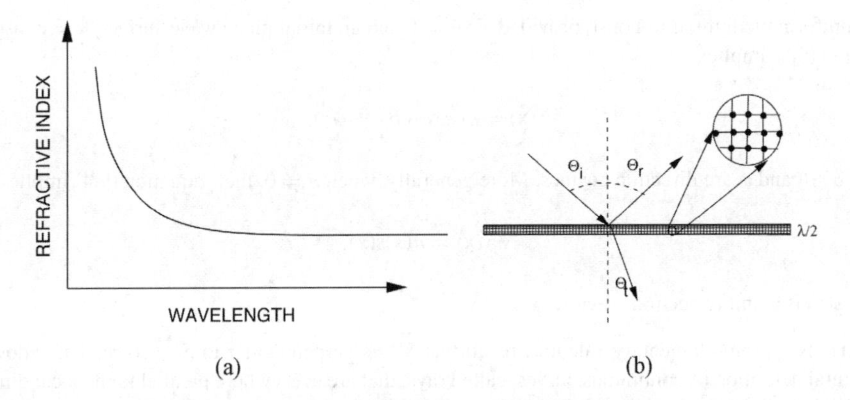


**Figure 9.3** (a) Behavior of the refractive index as a function of wavelength. (b) An interface

the form $\psi(x,t) = A\,e^{j\phi}$, where $\phi = kx - \omega t$, we can manipulate exponentials and get back sine and cosine forms of the wave by taking the real or imaginary part of the answers after calculations have taken place. It is customary to use the real part of the complex portion for the physically meaningful part of the solution.

## 9.3 Ray Theory

Following a somewhat classical analysis found in, for example, Born and Wolf [15], Bohren and Huffman [12], Elmore and Heald [14] and Gross [6], we consider the propagation of a general disturbance $\psi$ governed by a generic wave equation:

$$\nabla^2 \psi = \frac{1}{c^2(\mathbf{x})} \frac{\partial^2 \psi}{\partial t^2}. \tag{9.1}$$

Here, $c(\mathbf{x})$ is a spatially varying wave speed corresponding to a general inhomogeneous medium, where $c(\mathbf{x}) = c_o$ in a homogeneous reference medium and where the refractive index is defined as $n = c_o/c(\mathbf{x})$. Consider a trial solution of the form

$$\psi(\mathbf{x}, t) = A(\mathbf{x})\,e^{j(k_o S(\mathbf{x}) - \omega t)}, \tag{9.2}$$

where $A(\mathbf{x})$ is the amplitude of the disturbance and where $k_o = \omega/c_o = 2\pi/\lambda$ is the wave number in the reference medium. The function $S(\mathbf{x})$ (dimensions of length) is known as the "eikonal," which in Greek means "image." One can interpret a set of waves as simply a family of surfaces for which the values of $k_o S(\mathbf{x})$ differ in increments of $2\pi$. Substituting the trial solution into the wave equation, one obtains

$$k_o^2 A(n^2 - \nabla S \cdot \nabla S) + jk_o(2\nabla A \cdot \nabla S + A\nabla^2 S) + \nabla^2 A = 0. \tag{9.3}$$

There are a variety of arguments to motivate so-called "ray theory." Probably the simplest is to require that, as $k_o \to \infty$, each of the $k_o$ terms, the zeroth-order $k_o$ term, the first-order $k_o$ term, and the second-order $k_o$ term, must vanish. Applying this requirement to the second-order $k_o$ term yields

$$n^2 = \nabla S \cdot \nabla S = \|\nabla S\|^2. \tag{9.4}$$

For a uniform medium, $n = \text{const}$, provided $\nabla^2 A = 0$ and an initial plane wave surface $S = \text{const}$, then Equation (9.3) implies

$$S(\mathbf{x}) = n(\alpha x + \beta y + \phi z), \tag{9.5}$$

where $\alpha$, $\beta$, and $\phi$ are direction cosines. More generally, when $n \neq 0$, then Equation (9.4) implies

$$\nabla S(\mathbf{x}) = n(\mathbf{x})\hat{\mathbf{s}}(\mathbf{x}), \tag{9.6}$$

where $\hat{\mathbf{s}}(\mathbf{x})$ is a unit (direction) vector.

**Remark 10** From elementary calculus, recall that $\nabla S$ is perpendicular to $S = \text{const}$. This allows for the natural definition of continuous curves, called rays, that are everywhere parallel to the local direction $\hat{\mathbf{s}}(\mathbf{x})$. Rearranging the first-order $k_o$ term of Equation (9.3):

$$\frac{1}{A}\nabla A \cdot \nabla S = -\frac{1}{2}\nabla^2 S = -\frac{1}{2}\nabla \cdot (n\hat{\mathbf{s}}). \tag{9.7}$$

Recalling the directional derivative, $\mathrm{d}(\cdot)/\mathrm{d}s \stackrel{\text{def}}{=} \hat{\mathbf{s}} \cdot \nabla(\cdot)$, we have

$$\left(\frac{\nabla S}{\|\nabla S\|}\right) \cdot \nabla A = \left(\frac{\nabla S}{n}\right) \cdot \nabla A = \frac{\mathrm{d}A}{\mathrm{d}s}, \tag{9.8}$$

where $s$ is the arc-length coordinate along the ray. With this definition, once $S(\mathbf{x})$ is known, the component of $\nabla A$ in the $\hat{\mathbf{s}}(\mathbf{x})$ can be found from Equations (9.7) and (9.8):

$$\frac{1}{A}\frac{\mathrm{d}A}{\mathrm{d}s} = -\frac{1}{2n}\nabla \cdot (n\hat{\mathbf{s}}). \tag{9.9}$$

Thus, we are able to determine how the amplitude of the trial solution changes along a ray, but not perpendicular to the trajectory.

**Remark 11** From the previous equations, the general equation for the trajectory of a ray is

$$\frac{\mathrm{d}(n\tilde{\mathbf{s}})}{\mathrm{d}s} = \tilde{\mathbf{s}} \cdot \nabla(n\tilde{\mathbf{s}}) = \nabla n. \tag{9.10}$$

For constant $n$, then $\nabla n = 0$; thus, $\tilde{\mathbf{s}} \cdot \nabla(n\tilde{\mathbf{s}}) = n\tilde{\mathbf{s}} \cdot \nabla\tilde{s} = \mathbf{0}$. This yields $\tilde{\mathbf{s}} = \text{constant}$, until an interface is encountered. Thus, the ray position is simply updated ("time-marched") by

$$\mathbf{r}(t + \Delta t) = \mathbf{r}(t) + \Delta t\mathbf{v}(t), \tag{9.11}$$

with trajectory changes computed when an interface is encountered.[2]

**Remark 12** Ray-tracing deals directly with the ray trajectories, rather than finding them as a by-product of the solution of the wave equation for the eikonal function $S$ and the resulting wave front. To eliminate

[2] Although we do not consider the case of individual materials with nonconstant $n$, which yields $\nabla n \neq 0$, and thus $\tilde{\mathbf{s}} \cdot \nabla(n\tilde{\mathbf{s}}) = \nabla n$, we note that the governing equations can be solved with dynamic relaxation or Newton's method to get the ray trajectory.

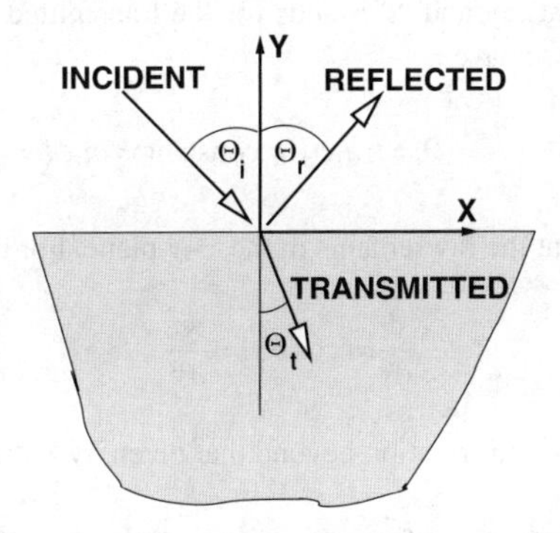


**Figure 9.4** A local coordinate system for a ray reflection

$S$ we need to look at the rate of change of the quantity $n\hat{\mathbf{s}}$ along the ray. Making repeated use of Equation (9.6), we can also derive Equation (9.10) as follows:

$$\frac{\mathrm{d}(n\hat{\mathbf{s}})}{\mathrm{d}s} = \hat{\mathbf{s}} \cdot \nabla(\nabla S) = \frac{\nabla S}{n} \cdot \nabla(\nabla S) = \frac{1}{2n}\nabla(\nabla S \cdot \nabla S) = \frac{1}{2n}\nabla n^2 = \nabla n, \tag{9.12}$$

where $\mathrm{d}(\cdot)/\mathrm{d}s \stackrel{\text{def}}{=} \hat{\mathbf{s}} \cdot \nabla(\cdot)$. The previous equation allows us to find the trajectories of a ray ($\hat{\mathbf{s}}$), given only the refractive index $n(\mathbf{x})$ and the initial direction $\hat{\mathbf{s}}$ of the desired ray.

**Remark 13** If the medium is uniform, then $\mathrm{d}\hat{\mathbf{s}}/\mathrm{d}s = 0$. It follows that $\hat{\mathbf{s}} = \text{const}$. Consequently, because $\hat{\mathbf{s}}$ is a vector, the ray is simply a straight line. Now consider a case where the refractive index varies only in the $y$ direction and that the incident ("i") ray lies initially in the $x$–$y$ plane (Figure 9.4), with a direction given by $\hat{\mathbf{s}}_{\mathrm{i}} = \sin\theta_{\mathrm{i}}\mathbf{e}_x - \cos\theta_{\mathrm{i}}\mathbf{e}_y + \phi_{\mathrm{i}}\mathbf{e}_z$, where $\phi_{\mathrm{i}} = 0$. Breaking Equation (9.12) (which represents the change along the ray direction) into components, we have for the $x$-direction

$$\frac{\mathrm{d}}{\mathrm{d}s}(n\sin\theta) = \frac{\partial n}{\partial x} = 0 \Rightarrow n_{\mathrm{i}}\sin\theta_{\mathrm{i}} = \text{constant} = n_{\mathrm{t}}\sin\theta_{\mathrm{t}}, \tag{9.13}$$

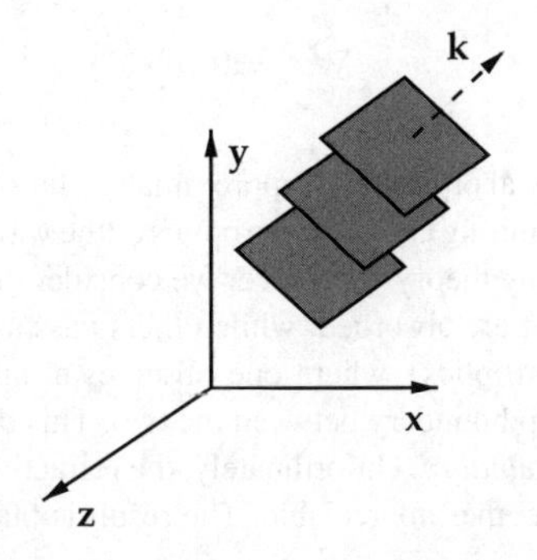


**Figure 9.5** A wave front and propagation vector

where "i" stands for the incident ray and "t" stands for the transmitted ray. Equation (9.13) is simply Snell's law. For the $z$-direction we have

$$\frac{\mathrm{d}}{\mathrm{d}s}(n\phi_\mathrm{i}) = \frac{\partial n}{\partial z} = 0, \Rightarrow n_\mathrm{i}\phi_\mathrm{i} = \text{constant} = n_\mathrm{t}\phi_\mathrm{t} \Rightarrow \phi_\mathrm{t} = 0; \tag{9.14}$$

thus, Equation (9.14) implies that the ray remains in the $x$–$y$ plane. For the $y$-direction we have

$$\frac{\mathrm{d}}{\mathrm{d}s}(n\cos\theta) = \frac{\partial n}{\partial y}. \tag{9.15}$$

Equation (9.15) provides no extra information beyond that given by Equation (9.13).

**Remark 14** A more general derivation of the eikonal equation can be found in, for example, Cerveny *et al.* [16], and starts by assuming a trial solution of the form

$$\psi(\mathbf{x}, t) = A(\mathbf{x})\Phi(t - \Lambda(\mathbf{x})) \tag{9.16}$$

where $\Lambda$ is an eikonal function and the waveform function $\alpha$ is assumed to be of high frequency.[3] This function is then substituted into the wave equation to yield

$$\nabla^2 A\Phi + 2\nabla A \cdot \nabla\Phi + A\nabla^2\Phi = \frac{1}{c^2}A\frac{\partial^2\Phi}{\partial t^2}. \tag{9.17}$$

After using the chain rule of differentiation, this can be written as

$$\frac{\partial^2\Phi}{\partial\Lambda^2}A\left(\nabla\Lambda\cdot\nabla\Lambda - \frac{1}{c^2}\right) + \frac{\partial\Phi}{\partial\Lambda}\left(2\nabla A\cdot\nabla\Lambda + A\nabla^2\Lambda\right) + \Phi\nabla^2 A = 0. \tag{9.18}$$

Analogous to the special case considered before, to motivate so-called "ray theory" one requires that the coefficients of $\partial^2\Phi/\partial\Lambda^2$, $\partial\Phi/\partial\Lambda$, and $\Phi$ are satisfied separately; in other words, the following hold:

$$\nabla\Lambda\cdot\nabla\Lambda - \frac{1}{c^2} = 0, \tag{9.19}$$

$$2\nabla A\cdot\nabla\Lambda + A\nabla^2\Lambda = 0, \tag{9.20}$$

and

$$\nabla^2 A = 0. \tag{9.21}$$

**Remark 15** Ray tracing (geometrical optics) is an approximate solution to the wave equation, neglecting the term $\nabla^2 A$. It is essentially the limiting case of wave optics as the wavelength tends towards zero. Thus, there are limitations to the use of ray theory. Whenever we consider the edge of a scatterer, geometrical ray tracing asserts that rays that hit are absorbed, while others pass as straight lines. This is analogous to everyday experience with light (optics), where one observes a uniformly illuminated region and a dark region, with a somewhat sharp boundary between the two. This description is more or less correct, *away from the illuminated–dark boundary*. Unfortunately, the refractive index does not vary "slowly" at the illuminated edge, and $\nabla^2 A$ becomes appreciable. The result is that there exists a gradual transition

[3]This is a more general case than the one considered in Equation (9.2), where $\Phi(t - \Lambda(\mathbf{x})) = \mathrm{e}^{\mathrm{j}(k_\mathrm{o}S(\mathbf{x})-\omega t)}$.

(boundary layer) from dark to light zones, with a complex structure in the transition region.[4] This effect, which is beyond the scope of our discussion in this chapter, can be explained by appealing to Fresnel diffraction by a knife edge.[5]

**Remark 16** A class of solution methods that extend geometric optics are *physical optics*, which entail using ray methods to estimate the field on a surface and integrating that field over the surface to calculate the transmitted or scattered field. These approaches are not discussed in this chapter.

## 9.4 Plane Harmonic Electromagnetic Waves

### 9.4.1 General Plane Waves

We recall the basic form of the wave equation

$$\frac{\partial^2 A}{\partial x^2} + \frac{\partial^2 A}{\partial y^2} + \frac{\partial^2 A}{\partial z^2} = \frac{1}{v^2}\frac{\partial^2 A}{\partial t^2}, \tag{9.22}$$

where $A$ is a variable and $v$ is the wave speed. We consider time-harmonic plane-wave solutions; that is, those solutions of the form

$$A(\mathbf{r}, t) = A_o \cos(\mathbf{k} \cdot \mathbf{r} - \omega t), \tag{9.23}$$

where $\mathbf{r}$ is an initial position vector to the wave front, where $\mathbf{k}$ is in the direction of propagation. For plane waves, $\mathbf{k} \cdot \mathbf{r} =$ constant. We refer to the phase as

$$\phi = \mathbf{k} \cdot \mathbf{r} - \omega t \tag{9.24}$$

and $\omega = 2\pi/\tau$ as the angular frequency, where $\tau$ is the period. The wave front, over which the phase is constant, is a plane for "plane waves" and is orthogonal to the direction of propagation.

### 9.4.2 Electromagnetic Waves

Following Zohdi and Kuypers [1–5], the propagation of light can be described via an electromagnetic formalism, via Maxwell's equations (in simplified form), in free space:

$$\nabla \times \mathbf{E} = -\mu_o \frac{\partial \mathbf{H}}{\partial t}, \quad \nabla \times \mathbf{H} = \epsilon_o \frac{\partial \mathbf{E}}{\partial t}, \quad \nabla \cdot \mathbf{H} = 0, \quad \text{and} \quad \nabla \cdot \mathbf{E} = 0, \tag{9.25}$$

where $\mathbf{E}$ is the electric field intensity, $\mathbf{H}$ is the magnetic flux intensity, $\epsilon_o$ is the permittivity, and $\mu_o$ is the permeability. Using standard vector identities, one can show that

$$\nabla \times (\nabla \times \mathbf{E}) = -\mu_o \epsilon_o \frac{\partial^2 \mathbf{E}}{\partial t^2} \quad \text{and} \quad \nabla \times (\nabla \times \mathbf{H}) = -\mu_o \epsilon_o \frac{\partial^2 \mathbf{H}}{\partial t^2}, \tag{9.26}$$

and that

$$\nabla^2 \mathbf{E} = \frac{1}{c^2}\frac{\partial^2 \mathbf{E}}{\partial t^2} \quad \text{and} \quad \nabla^2 \mathbf{H} = \frac{1}{c^2}\frac{\partial^2 \mathbf{H}}{\partial t^2}, \tag{9.27}$$

[4] The sharp shadow is the limiting case of vanishing wavelength.

[5] See Elmore and Heald [14] for more detailed discussions.

and that, employing a Cartesian coordinate system,

$$\frac{\partial^2 E_x}{\partial x^2} + \frac{\partial^2 E_x}{\partial y^2} + \frac{\partial^2 E_x}{\partial z^2} = \frac{1}{c^2}\frac{\partial^2 E_x}{\partial t^2}, \tag{9.28}$$

where $c = 1/\sqrt{\epsilon_o \mu_o}$, with identical relations holding for $E_y$, $E_z$, $H_x$, $H_y$, and $H_z$. In the case of plane harmonic waves, for example of the form

$$\mathbf{E} = \mathbf{E}_o \cos(\mathbf{k} \cdot \mathbf{r} - \omega t) \quad \text{and} \quad \mathbf{H} = \mathbf{H}_o \cos(\mathbf{k} \cdot \mathbf{r} - \omega t), \tag{9.29}$$

we have

$$\mathbf{k} \times \mathbf{E} = \mu_o \omega \mathbf{H} \quad \text{and} \quad \mathbf{k} \times \mathbf{H} = -\epsilon_o \omega \mathbf{E} \tag{9.30}$$

and

$$\mathbf{k} \cdot \mathbf{E} = 0 \quad \text{and} \quad \mathbf{k} \cdot \mathbf{H} = 0. \tag{9.31}$$

The three vectors, **k**, **E**, and **H** constitute a mutually orthogonal triad. The direction of ray propagation is given by $\mathbf{E} \times \mathbf{H}/\|\mathbf{E} \times \mathbf{H}\|$.

Recall that since the free-space propagation velocity is given by $c = 1/\sqrt{\epsilon_o \mu_o}$ for an electromagnetic wave in a vacuum and $v = 1/\sqrt{\epsilon \mu}$ for electromagnetic waves in another medium, we can define the index of refraction as

$$n \stackrel{\text{def}}{=} \frac{c}{v} = \sqrt{\frac{\epsilon \mu}{\epsilon_o \mu_o}}. \tag{9.32}$$

### 9.4.3 *Optical Energy Propagation*

Light waves traveling through space carry electromagnetic energy which flows in the direction of wave propagation. The energy per unit area per unit time flowing perpendicularly into a surface in free space is given by the Poynting vector **S**, where

$$\mathbf{S} = \mathbf{E} \times \mathbf{H}. \tag{9.33}$$

Since at optical frequencies **E**, **H**, and **S** oscillate rapidly, it is impractical to measure instantaneous values of **S** directly. Now consider the harmonic representations in Equation (9.29), which leads to

$$\mathbf{S} = \mathbf{E}_o \times \mathbf{H}_o \cos^2(\mathbf{k} \cdot \mathbf{r} - \omega t), \tag{9.34}$$

and, consequently, the average value over a longer (but still quite short) time interval than that of the time scale of rapid random oscillation,

$$\langle \mathbf{S} \rangle_T = \mathbf{E}_o \times \mathbf{H}_o \langle \cos^2(\mathbf{k} \cdot \mathbf{r} - \omega t) \rangle_T = \frac{1}{2}\mathbf{E}_o \times \mathbf{H}_o, \tag{9.35}$$

where $\langle (\cdot) \rangle_T \stackrel{\text{def}}{=} \frac{1}{T}\int_0^T (\cdot)\, dt$. We define the irradiance as

$$I \stackrel{\text{def}}{=} \langle \|\mathbf{S}\| \rangle_T = \frac{1}{2}\|\mathbf{E}_o \times \mathbf{H}_o\| = \frac{1}{2}\sqrt{\frac{\epsilon_o}{\mu_o}}\|\mathbf{E}_o\|^2. \tag{9.36}$$

Thus, the rate of flow of energy is proportional to the square of the amplitude of the electric field. Furthermore, in isotropic media, which we consider for the duration of the work, the direction of energy is in the direction of **S** and in the same direction as **k**. Since $I$ is the energy per unit area per unit time, if we multiply by the "cross-sectional" area of the ray $a_r$, we obtain the energy associated with the ray, denoted as $Ia_r = Ia_b/N_r$, where $a_b$ is the cross-sectional area of a beam (comprising all of the rays) and $N_r$ is the number of rays in the beam (Figure 9.2).

### 9.4.4 *Reflection and Absorption of Energy*

Now we consider a plane harmonic wave incident upon a plane boundary separating two optically different materials, which produces a reflected wave and a transmitted (refracted) wave (Figure 9.6). The space–time dependence of the three waves is given by:

1. $e^{j(\mathbf{k}_i \cdot \mathbf{r} - \omega t)}$ for the incident wave (with propagation vector $\mathbf{k}_i$);
2. $e^{j(\mathbf{k}_r \cdot \mathbf{r} - \omega t)}$ for the reflected wave (with propagation vector $\mathbf{k}_r$); and
3. $e^{j(\mathbf{k}_t \cdot \mathbf{r} - \omega t)}$ for the transmitted wave (with propagation vector $\mathbf{k}_t$).

In order for a time-invariant relation to hold for all points on the boundary, and for all values of $t$, we must have that the arguments of the exponential function be equal on the boundary. Therefore, since the $\omega t$ terms are the same, we have, at the boundary, $\mathbf{k}_i \cdot \mathbf{r} = \mathbf{k}_r \cdot \mathbf{r} = \mathbf{k}_t \cdot \mathbf{r}$, which implies that the waves are coplanar and that their projection onto the plane boundary is equal. We call the plane that contains all three waves the incident plane. Consequently, we have a relation between the propagation constants' magnitudes, $k_i \sin\theta_i = k_r \sin\theta_r = k_t \sin\theta_t$, which, because the reflected and incident medium are the same, implies $\theta_i = \theta_r$. By taking the ratio of the magnitudes of the propagation constants of the transmitted wave and the incident wave, we have

$$\frac{k_t}{k_i} = \frac{\omega/v_t}{\omega/v_i} = \frac{c/v_t}{c/v_i} = \frac{n_t}{n_i} \stackrel{\text{def}}{=} \hat{n}. \tag{9.37}$$

Therefore, we have

$$\frac{\sin\theta_i}{\sin\theta_t} = \hat{n}, \tag{9.38}$$

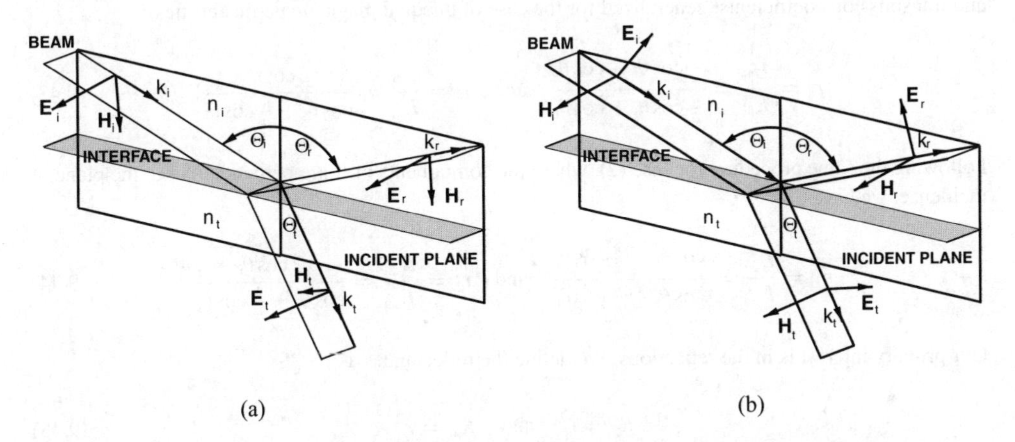


**Figure 9.6** The nomenclature for Fresnel's equations, for the case where the electric field vectors are (a) perpendicular to the plane of incidence and (b) parallel to the plane of incidence [1–5]

which is sometimes referred to as the law of refraction. To compute the amount of energy transmitted (absorbed) and reflected by electromagnetic waves, let $\mathbf{E}_i$ now denote the (vectorial) amplitude of a plane harmonic wave that is incident on a plane boundary separating two materials. Also, let $\mathbf{E}_r$ and $\mathbf{E}_t$ be the amplitudes of the reflected and transmitted waves, respectively. Equations (9.30) and (9.31) collapse, for the incident, reflected, and transmitted magnetic waves, to

$$\mathbf{H}_i = \frac{1}{\mu_i \omega}\mathbf{k}_i \times \mathbf{E}_i, \quad \mathbf{H}_r = \frac{1}{\mu_r \omega}\mathbf{k}_r \times \mathbf{E}_r, \quad \text{and} \quad \mathbf{H}_t = \frac{1}{\mu_t \omega}\mathbf{k}_t \times \mathbf{E}_t. \tag{9.39}$$

respectively. Let us now consider an oblique angle of incidence. Consider two cases for the electric field vector: (1) electric field vectors that are parallel (∥) to the plane of incidence; (2) electric field vectors that are perpendicular (⊥) to the plane of incidence. In either case, the tangential components of the electric and magnetic fields are required to be continuous across the interface. Consider case (1). We have the following general vectorial representations:

$$\mathbf{E}_\parallel = E_\parallel \cos(\mathbf{k}\cdot\mathbf{r} - \omega t)\,\mathbf{e}_1 \quad \text{and} \quad \mathbf{H}_\parallel = H_\parallel \cos(\mathbf{k}\cdot\mathbf{r} - \omega t)\,\mathbf{e}_2, \tag{9.40}$$

where $\mathbf{e}_1$ and $\mathbf{e}_2$ are orthogonal to the propagation direction $\mathbf{k}$, and where $E_\parallel$ and $H_\parallel$ are the amplitudes of the parallel field components. By employing the law of refraction ($n_i \sin\theta_i = n_t \sin\theta_t$) we obtain the following conditions relating the incident, reflected, and transmitted components of the electric field quantities:

$$E_{\parallel i}\cos\theta_i - E_{\parallel r}\cos\theta_r = E_{\parallel t}\cos\theta_t \quad \text{and} \quad H_{\perp i} + H_{\perp r} = H_{\perp t}. \tag{9.41}$$

Since, for plane harmonic waves, the magnetic and electric field amplitudes are related by $H = E/v\mu$, we then have

$$E_{\parallel i} + E_{\parallel r} = \frac{\mu_i}{\mu_t}\frac{v_i}{v_t}E_{\parallel t} = \frac{\mu_i}{\mu_t}\frac{n_t}{n_i}E_{\parallel t} \stackrel{\text{def}}{=} \frac{\hat{n}}{\hat{\mu}}E_{\parallel t}, \tag{9.42}$$

where $\hat{\mu} \stackrel{\text{def}}{=} \mu_t/\mu_i$, $\hat{n} \stackrel{\text{def}}{=} n_t/n_i$, and where $v_i$, $v_r$, and $v_t$ are the values of the velocity in the incident, reflected, and transmitted directions.[6] By again employing the law of refraction, we obtain the reflection and transmission coefficients, generalized for the case of unequal magnetic permeabilities:

$$r_\parallel = \frac{E_{\parallel r}}{E_{\parallel i}} = \frac{\frac{\hat{n}}{\hat{\mu}}\cos\theta_i - \cos\theta_t}{\frac{\hat{n}}{\hat{\mu}}\cos\theta_i + \cos\theta_t} \quad \text{and} \quad t_\parallel = \frac{E_{\parallel t}}{E_{\parallel i}} = \frac{2\cos\theta_i}{\cos\theta_t + \frac{\hat{n}}{\hat{\mu}}\cos\theta_i}. \tag{9.43}$$

Following the same procedure for case (2), where the components of $\mathbf{E}$ are perpendicular to the plane of incidence, we have

$$r_\perp = \frac{E_{\perp r}}{E_{\perp i}} = \frac{\cos\theta_i - \frac{\hat{n}}{\hat{\mu}}\cos\theta_t}{\cos\theta_i + \frac{\hat{n}}{\hat{\mu}}\cos\theta_t} \quad \text{and} \quad t_\perp = \frac{E_{\perp t}}{E_{\perp i}} = \frac{2\cos\theta_i}{\cos\theta_i + \frac{\hat{n}}{\hat{\mu}}\cos\theta_t}. \tag{9.44}$$

Our primary interest is in the reflections. We define the reflectances as

$$R_\parallel \stackrel{\text{def}}{=} r_\parallel^2 \quad \text{and} \quad R_\perp \stackrel{\text{def}}{=} r_\perp^2. \tag{9.45}$$

[6] Throughout the analysis we assume that $\hat{n} \geq 1$.

Particularly convenient forms for the reflections are

$$r_\parallel = \frac{\frac{\hat{n}^2}{\hat{\mu}}\cos\theta_\mathrm{i} - (\hat{n}^2 - \sin^2\theta_\mathrm{i})^{1/2}}{\frac{\hat{n}^2}{\hat{\mu}}\cos\theta_\mathrm{i} + (\hat{n}^2 - \sin^2\theta_\mathrm{i})^{1/2}} \quad \text{and} \quad r_\perp = \frac{\cos\theta_\mathrm{i} - \frac{1}{\hat{\mu}}(\hat{n}^2 - \sin^2\theta_\mathrm{i})^{1/2}}{\cos\theta_\mathrm{i} + \frac{1}{\hat{\mu}}(\hat{n}^2 - \sin^2\theta_\mathrm{i})^{1/2}}. \tag{9.46}$$

Thus, the total energy reflected can be characterized by

$$R \stackrel{\text{def}}{=} \left(\frac{E_\mathrm{r}}{E_\mathrm{i}}\right)^2 = \frac{E_{\perp\mathrm{r}}^2 + E_{\parallel\mathrm{r}}^2}{E_\mathrm{i}^2} = \frac{I_{\parallel\mathrm{r}} + I_{\perp\mathrm{r}}}{I_\mathrm{i}}. \tag{9.47}$$

If the resultant plane of oscillation of the (polarized) wave makes an angle of $\gamma_\mathrm{i}$ with the plane of incidence, then

$$E_{\parallel\mathrm{i}} = E_\mathrm{i}\cos\gamma_\mathrm{i} \quad \text{and} \quad E_{\perp\mathrm{i}} = E_\mathrm{i}\sin\gamma_\mathrm{i}, \tag{9.48}$$

and it follows from the previous definition of $I$ that

$$I_{\parallel\mathrm{i}} = I_\mathrm{i}\cos^2\gamma_\mathrm{i} \quad \text{and} \quad I_{\perp\mathrm{i}} = I_\mathrm{i}\sin^2\gamma_\mathrm{i}. \tag{9.49}$$

Substituting these expression back into the expressions for the reflectances yields

$$R = \frac{I_{\parallel\mathrm{r}}}{I_\mathrm{i}}\cos^2\gamma_\mathrm{i} + \frac{I_{\perp\mathrm{r}}}{I_\mathrm{i}}\sin^2\gamma_\mathrm{i} = R_\parallel\cos^2\gamma_\mathrm{i} + R_\perp\sin^2\gamma_\mathrm{i}. \tag{9.50}$$

For natural or unpolarized light, the angle $\gamma_\mathrm{i}$ varies rapidly in a random manner, as does the field amplitude. Thus, since

$$\langle\cos^2\gamma_\mathrm{i}(t)\rangle_T = \frac{1}{2} \quad \text{and} \quad \langle\sin^2\gamma_\mathrm{i}(t)\rangle_T = \frac{1}{2}, \tag{9.51}$$

and, therefore, for natural light

$$I_{\parallel\mathrm{i}} = \frac{I_\mathrm{i}}{2} \quad \text{and} \quad I_{\perp\mathrm{i}} = \frac{I_\mathrm{i}}{2}; \tag{9.52}$$

therefore,

$$r_\parallel^2 = \left(\frac{E_{\parallel\mathrm{r}}^2}{E_{\parallel\mathrm{i}}^2}\right)^2 = \frac{I_{\parallel\mathrm{r}}}{I_{\parallel\mathrm{i}}} \quad \text{and} \quad r_\perp^2 = \left(\frac{E_{\perp\mathrm{r}}^2}{E_{\perp\mathrm{i}}^2}\right)^2 = \frac{I_{\perp\mathrm{r}}}{I_{\perp\mathrm{i}}}. \tag{9.53}$$

Thus, the total reflectance becomes

$$R = \frac{1}{2}(R_\parallel + R_\perp) = \frac{1}{2}(r_\parallel^2 + r_\perp^2), \tag{9.54}$$

where $0 \le R \le 1$.

**Remark 17** For the cases where $\sin\theta_\mathrm{t} = \sin\theta_\mathrm{i}/\hat{n} > 1$, one may rewrite the reflection relations as

$$r_\parallel = \frac{\frac{\hat{n}^2}{\hat{\mu}}\cos\theta_\mathrm{i} - \mathrm{j}(\sin^2\theta_\mathrm{i} - \hat{n}^2)^{1/2}}{\frac{\hat{n}^2}{\hat{\mu}}\cos\theta_\mathrm{i} + \mathrm{j}(\sin^2\theta_\mathrm{i} - \hat{n}^2)^{1/2}} \quad \text{and} \quad r_\perp = \frac{\cos\theta_\mathrm{i} - \frac{1}{\hat{\mu}}\mathrm{j}(\sin^2\theta_\mathrm{i} - \hat{n}^2)^{1/2}}{\cos\theta_\mathrm{i} + \frac{1}{\hat{\mu}}\mathrm{j}(\sin^2\theta_\mathrm{i} - \hat{n}^2)^{1/2}}, \tag{9.55}$$

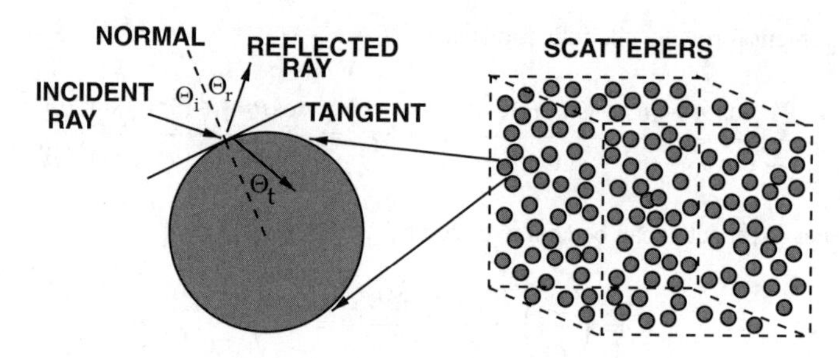


**Figure 9.7** The nomenclature for Fresnel's equations for an incident ray that encounters a scatterer [1–5]

where $\mathrm{j} = \sqrt{-1}$ and in this complex case[7]

$$R_{\parallel} \stackrel{\text{def}}{=} r_{\parallel}\bar{r}_{\parallel} = 1 \quad \text{and} \quad R_{\perp} \stackrel{\text{def}}{=} r_{\perp}\bar{r}_{\perp} = 1, \tag{9.56}$$

where $\bar{r}_{\parallel}$ and $\bar{r}_{\perp}$ are complex conjugates. Thus, for angles larger than the critical angle $\theta_i^*$, all of the energy is reflected.

**Remark 18** Next we shall consider a plane harmonic wave incident upon a plane boundary separating two different optical materials, which produces a reflected wave and a transmitted (refracted) wave (Figure 9.7). The amount of incident electromagnetic energy $I_i$ that is reflected $I_r$ is given by the total reflectance

$$R \stackrel{\text{def}}{=} \frac{I_r}{I_i}, \tag{9.57}$$

where $0 \leq R \leq 1$ and where, explicitly for unpolarized (natural) light,

$$R = \frac{1}{2}\left\{\left[\frac{\frac{\hat{n}^2}{\hat{\mu}}\cos\theta_i - (\hat{n}^2 - \sin^2\theta_i)^{1/2}}{\frac{\hat{n}^2}{\hat{\mu}}\cos\theta_i + (\hat{n}^2 - \sin^2\theta_i)^{1/2}}\right]^2 + \left[\frac{\cos\theta_i - \frac{1}{\hat{\mu}}(\hat{n}^2 - \sin^2\theta_i)^{1/2}}{\cos\theta_i + \frac{1}{\hat{\mu}}(\hat{n}^2 - \sin^2\theta_i)^{1/2}}\right]^2\right\}. \tag{9.58}$$

Here, $\hat{n}$ is the ratio of the refractive indices of the ambient (incident) medium $n_i$ and transmitted (scatterer) medium $n_t$, $\hat{n} = n_t/n_i$, where $\hat{\mu}$ is the ratio of the magnetic permeabilities of the surrounding incident medium $\mu_i$ and transmitted scatterer medium $\mu_t$, $\hat{\mu} = \mu_t/\mu_i$. Consistent with the optical regime, for the remainder of the chapter, we shall take $\hat{\mu} = 1$; that is, $\mu_o = \mu_i \approx \mu_t$.

In the upcoming analysis, the ambient medium is assumed to behave as a vacuum. Thus, there are no energetic losses as the electromagnetic rays pass through it. However, we assume that all electromagnetic energy that is absorbed by a scatterer becomes trapped, not re-emitted, and is converted into heat. The thermal conversion process, and subsequent infrared radiation emmission, is not considered here. Modeling of the thermal coupling in such optical processes can be found in Zohdi [3, 4]. Thus, we ignore the transmission of light through the scatterer, as well as dispersion; that is, the decomposition of light into its component wavelengths (or colors). Dispersion occurs because the index of refraction

[7]The limiting case $\sin\theta_i^*/\hat{n} = 1$ is the critical angle ($\theta_i^*$) case.

of a transparent medium is greater for light of shorter wavelengths. Thus, whenever light is refracted in passing from one medium to the next, the violet and blue light of shorter wavelengths is bent more than the orange and red light of longer wavelengths. Dispersive effects introduce a new level of complexity, primarily because of the refraction of different wavelengths of light, leading to a dramatic growth in the number of rays of varying intensities and color (wavelength).

Also, notice that as $\hat{n} \to 1$ we have complete absorption, while as $\hat{n} \to \infty$ we have complete reflection. The total amount of absorbed power by the scatterers is $(1 - R)I_{\text{i}}$. As mentioned previously, the medium surrounding the scatterers is assumed to behave as a vacuum; that is, there are no energetic losses as the electromagnetic rays pass through it. However, we assume that all electromagnetic energy that is absorbed from a ray by a scattereris converted into heat and that no electromagnetic rays are refracted or dispersed. Heat generation and accompanying infrared radiation emission (with wavelengths in the ranges of $7.1 \times 10^{-7} \leq \lambda \leq 10^{-3}$ m) is addressed later.

**Remark 19** Although not considered here, we note that one can directly relate the irradiance to the photonic energy. The linear momentum of a photon can be obtained from the Einstein mass–energy relation:

$$h\nu = mc^2, \tag{9.59}$$

where $\nu$ is the frequency, $m$ is the mass, $c$ is the speed of light, and $h$ is Planck's constant. Note that a photon must have a zero mass at rest since

$$m = m_o \frac{1}{\sqrt{1 - \frac{u^2}{c^2}}} \tag{9.60}$$

would become infinite at the speed of light. The linear momentum is $p = mc = h\nu/c$ or, alternatively, in terms of the wavelength $p = h/\lambda$. If there are $N$ photons per unit area that hit per second (rate), then the irradiance $I$ of the beam is $I = Nh\nu$.

### 9.4.5 *Computational Algorithm*

The computational algorithm is as follows, starting at pseudo-time $t = 0$ and ending at $t = T$:

1. COMPUTE POSSIBLE RAY REFLECTIONS:
   (a) CHECK IF A RAY HAS ENCOUNTERED A SURFACE
   (b) COMPUTE NEW RAY TRAJECTORY (FIGURE 9.7)
   (c) COMPUTE NEW RAY MAGNITUDE (EQUATION (9.58))
   (d) COMPUTE SCATTERER'S ABSORPTION
2. INCREMENT ALL RAY POSITIONS: $\mathbf{r}_i(t + \Delta t) = \mathbf{r}_i(t) + \Delta t \mathbf{v}_i(t)$, $i = 1, \ldots,$ RAYS
3. GO TO (1) AND REPEAT WITH $t = t + \Delta t$.

The "pseudo-time" step size $\Delta t$ is dictated by the size of the scatterers. A somewhat ad-hoc approach is to scale the time step size according to $\Delta t \propto \xi b/\|\mathbf{v}\|$, where $b$ is the radius of the scatterers, $\|\mathbf{v}\|$ is the magnitude of the velocity of the rays, and $\xi$ is a scaling factor, typically $0.05 \leq \xi \leq 0.1$.

**Example 1** *A single spherical scatterer.* Consider a single reflecting scatterer, with incident rays as shown in Figure 9.8. The algorithm above is straightforward to implement, for example with an initial (overkill) $30 \times 30$ ray grid (Figure 9.8).

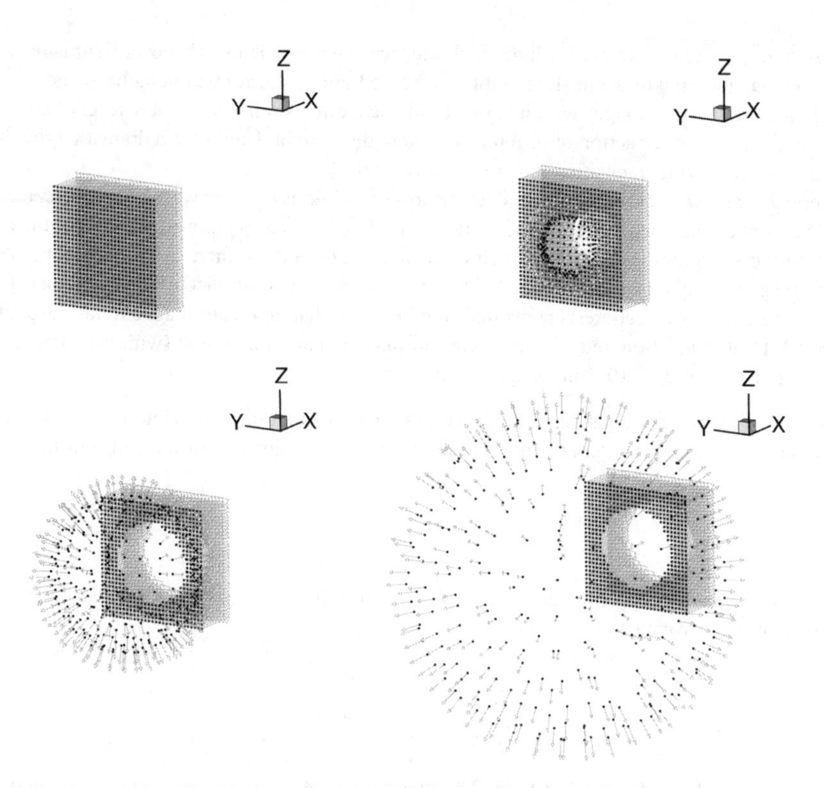


**Figure 9.8** Top to bottom and left to right (single spherical scatterer, $\hat{n} = 4$, not shown), the progressive movement of the Poynting vectors directly from the ray solution

**Example 2** *M*ultiple spherical scatterers. Next we considered a group of $N_s$ randomly dispersed spherical scatterers, of equal size, in a cubical domain of dimensions $D \times D \times D$. The scatterer size and volume fraction were determined by a scatterer/sample size ratio, which was defined via a subvolume size $V \stackrel{\text{def}}{=} (D \times D \times D)/N_s$, where $N_s$ was the number of scatterers in the entire cube. The non-dimensional ratio between the radius $b$ and the subvolume was denoted by $\mathcal{L} \stackrel{\text{def}}{=} b/V^{1/3}$. The volume fraction occupied by the scatterers consequently can be written as $v_s \stackrel{\text{def}}{=} 4\pi\mathcal{L}^3/3$. Thus, the total volume occupied by the scatterers,[8] denoted $\zeta$, can be written as $\zeta = v_s N_s V$. We used $N_s = 1000$ scatterers and $N_r = 1600$ rays, arranged in a square $40 \times 40$ pattern (Figure 9.9). This system provided stable results; that is, increasing the number of rays and/or the number of scatterers beyond these levels resulted in negligibly different overall system responses. The beam was one-half the width of the target sample. The irradiance beam parameter was set to $I_o = \|I(0)\|$, which is the magnitude of the initial irradiance at time $t = 0$. The irradiance for each ray was calculated as $I a_b/N_r$. Owing to the problem's linearity, it is insensitive to the initial magnitude of $I_o$ (it scales out).

[8] For example, if one were to arrange the scatterers in a regular periodic manner, then, at the length scale ratio of $\mathcal{L} = 0.25$, the distance between the centers of the scatterer becomes four scatterer radii. In theoretical studies it is often stated that the critical separation distance between scatterers is approximately three radii to be sufficient to treat the scatterers as independent scatterers, and simply to sum the effects of the individual scatterers to compute the overall response of the aggregate. See Bohren and Huffman [12] or van de Hulst [13] for more details.

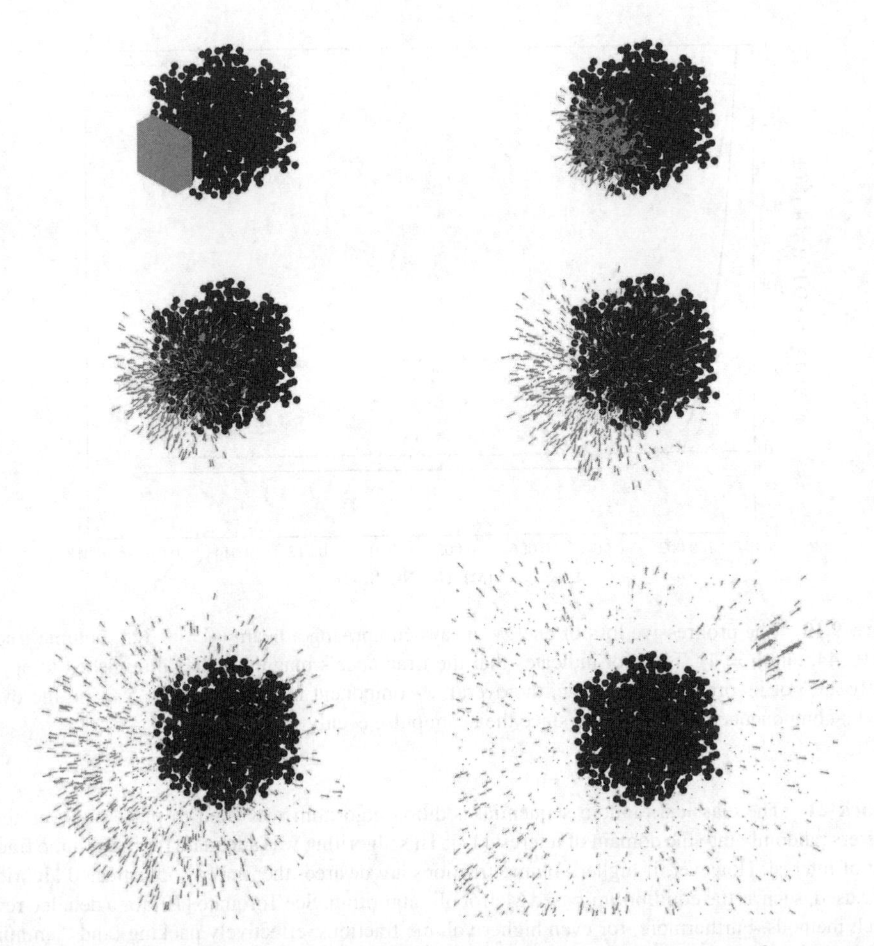

**Figure 9.9** Starting from left to right and top to bottom, the progressive movement of rays comprising a beam ($\mathcal{L} = 0.325$, volume fraction $v_s = 0.144$, and $\hat{n} = 4$) for a set of scatterers. The length of the vectors indicates the irradiance

**Remark 20** We repeatedly refined the "ray grid" up to 100 × 100 rays (10 000 total) and found no significant difference compared with the 40 × 40 result. Therefore, we consider the responses to be, for all practical purposes, independent of the ray-grid density. This scatterer–ray system provided stable results; that is, increasing the number of rays and/or the number of scatterers surrounding the beam resulted in negligibly different overall system responses. Of course, there can be cases where much higher resolution may be absolutely necessary. Thus, it is important to note that a straightforward, natural, algorithmic parallelism is possible with this computational technique. This can be achieved in two possible ways: (1) By assigning each processor its share of the rays, and checking which cells make contact with those rays; or (2) by assigning each processor its share of scatterers and checking which rays make contact with those scatterers. High-performance computational methods for the determination of ray–scatterer intersection can be developed by slightly modifying fast contact-detection algorithms found in, for example, Pöschel and Schwager [17].

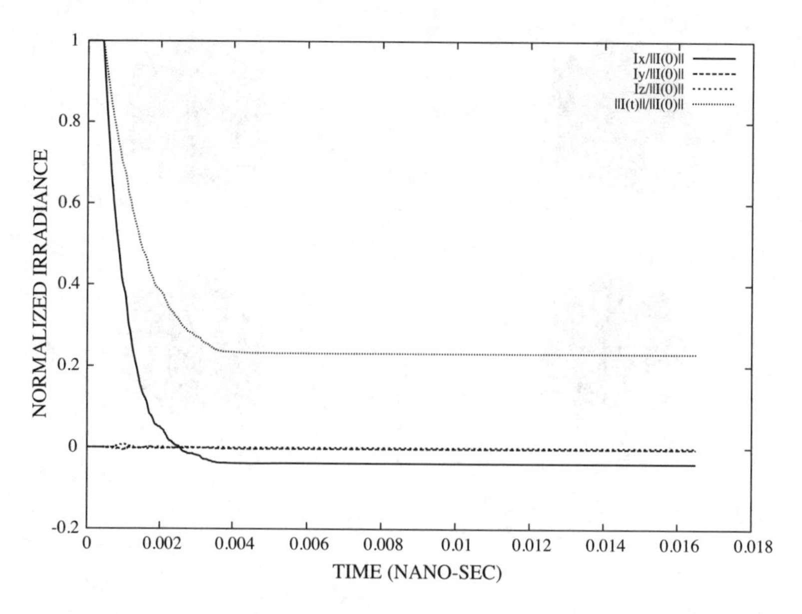


**Figure 9.10** The progressive loss of energy of rays comprising a beam ($\mathcal{L} = 0.325$, volume fraction $v_s = 0.144$, and $\hat{n} = 4$). The plot indicates that the irradiance's magnitude has diminished to approximately 23% of its original magnitude, the overall $x$-component is approximately 5% and the overall $y$- and $z$-components are nearly zero, since the beam/pulse evenly splits

**Remark 21** The classical random sequential addition algorithm was used to place nonoverlapping scatterers randomly into the domain of interest [18]. This algorithm was adequate for the volume fraction range of interest. However, if higher volume fractions are desired, then more sophisticated algorithms can be used, such as the equilibrium-based Metropolis algorithm. See Torquato [19] for a detailed review of such methods. Furthermore, for even higher volume fractions, effectively packing (and "jamming") scatterers, a relatively new class of efficient methods, based on simultaneous scatterer flow and growth, has been developed by Torquato and coworkers [19–23].

**Remark 22** The simulations were run until the rays completely exited the domain, which corresponded to a time scale on the order of $\frac{D}{c}$. The initial velocity vector for all of the initially collinear rays comprising

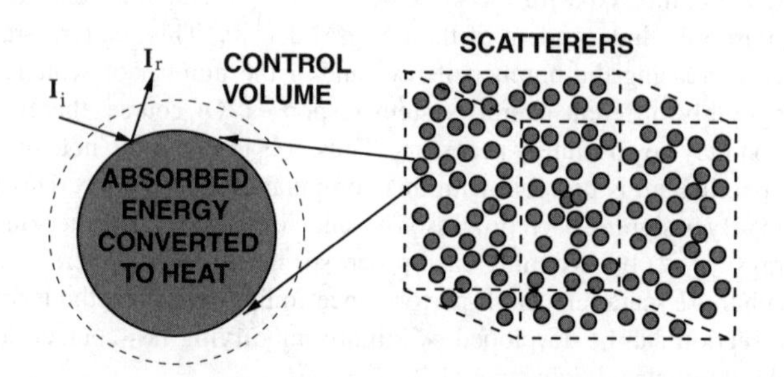


**Figure 9.11** Control volume for heat transfer [3, 4]

the beam was $\mathbf{v} = (c, 0, 0)$. Figure 9.9 shows successive frames of the rays as they move through the system. For more studies on ray propagation through complex media, see Zohdi and Kuypers [1–5]. It is important to emphasize that these calculations were performed *within a few minutes on a single standard laptop*.

**Remark 23** It is important to recognize that one can describe the aggregate ray behavior in an even more detailed manner via a higher moment distributions of rays; for example, by employing the skewness, kurtosis, and so on.

### *9.4.6 Thermal Conversion of Optical Losses*

The conversion of the losses of optical energy into thermal energy are discussed in detail by Zohdi [3, 4], where it was assumed that the scatterers were small enough to consider that the temperature fields are uniform in the scatterers.[9] Zohdi [3, 4] considered an energy balance governing the interconversions of mechanical, thermal, and chemical energy in a system, dictated by the first law of thermodynamics. Accordingly, one requires the time rate of change of the sum of the kinetic energy $\mathcal{K}$ and stored energy $\mathcal{S}$ to be equal to the work rate (power $\mathcal{P}$) and the net heat supplied $\mathcal{H}$:

$$\frac{\mathrm{d}}{\mathrm{d}t}(\mathcal{K} + \mathcal{S}) = \mathcal{P} + \mathcal{H}. \tag{9.61}$$

Here, the stored energy is comprised of a thermal part, $\mathcal{S}(t) = mC\theta(t)$, where $C$ is the heat capacity per unit mass. We assume that the scatterers deform negligibly during the process; thus, $\mathcal{K}$ and $\mathcal{P}$ are insignificant, and we have $\mathrm{d}\mathcal{S}/\mathrm{d}t = \mathcal{H}$. The primary source of heat is due to the incident rays. The energy input from the reflection of a ray is defined as

$$\Delta\mathcal{H}^{\text{rays}} \stackrel{\text{def}}{=} \int_t^{t+\Delta t} \mathcal{H}^{\text{rays}}\,\mathrm{d}t \approx (I_{\mathrm{i}} - I_{\mathrm{r}})a_{\mathrm{r}}\Delta t = (1 - R)I_{\mathrm{i}}a_{\mathrm{r}}\Delta t, \tag{9.62}$$

where $a_{\mathrm{r}}$ is a parameter indicating the "cross-sectional area" of a ray. After an incident ray is reflected, it is assumed that a process of heat transfer transpires. It is assumed that the temperature fields are uniform within the scatterers; thus, conduction within the scatterers is negligible. We remark that the validity of using a lumped thermal model (i.e., ignoring temperature gradients and assuming a uniform temperature within a scatterer) is dictated by the magnitude of the Biot number. A small Biot number (significantly less than unity) indicates that such an approximation is reasonable. The Biot number for spheres scales with the ratio of scatterer volume $V$ to scatterer surface area $a_{\mathrm{s}}$, $V/a_{\mathrm{s}} = b/3$, which indicates that a uniform temperature distribution is appropriate, since the scatterers, by definition, are small.

The first law reads

$$\frac{\mathrm{d}\mathcal{S}}{\mathrm{d}t} = mC\dot{\theta} = \underbrace{-h_{\mathrm{c}}a_{\mathrm{s}}(\theta - \theta_{\mathrm{o}})}_{\text{convective heating}} - \underbrace{\mathcal{B}a_{\mathrm{s}}\varepsilon(\theta^4 - \theta_{\mathrm{s}}^4)}_{\text{thermal radiation}} + \underbrace{\mathcal{H}^{\text{rays}}}_{\text{sources}}, \tag{9.63}$$

[9] Thus, the gradient of the temperature within the scatterer is zero; that is, $\nabla\theta = \mathbf{0}$. Thus, a Fourier-type law for the heat flux will register a zero value, $\boldsymbol{q} = -\mathbb{K}\cdot\nabla\theta = \mathbf{0}$. Furthermore, we assume that the space between the scatterers (i.e., the "ether"), plays no role in the heat transfer process.

where $\theta_o$ is the temperature of the ambient fluid, $\theta_s$ is the temperature of the far field surface (e.g., a container surrounding the flow) with which radiative exchange is made, $\mathcal{B} = 5.67 \times 10^{-8}$ W/(m$^2$K) is the Stefan–Boltzmann constant, $0 \leq \varepsilon \leq 1$ is the emissivity, which indicates how efficiently the surface radiates energy compared with a black-body (an ideal emitter), where $0 \leq h_c$ is the heating due to convection (Newton's law of cooling) into the dilute gas and where $a_s$ is the surface area of a scatterer. It is assumed that the *thermal radiation exchange between the scatterers is negligible*. For the applications considered presently, typically $h_c$ is quite small and plays a small role in the heat transfer processes. From a balance of momentum we have $m\dot{\mathbf{v}} \cdot \mathbf{v} = \mathbf{\Psi}^{tot} \cdot \mathbf{v}$ and Equation (9.63) becomes

$$m C \dot{\theta} = -h_c a_s(\theta - \theta_o) - \mathcal{B} a_s \varepsilon(\theta^4 - \theta_s^4) + \mathcal{H}^{rays} \stackrel{\text{def}}{=} \mathcal{F}(\theta) + \mathcal{H}^{rays}. \tag{9.64}$$

Therefore, after temporal integration with a finite-difference time-step of $\Delta t$ we have, using the trapezoidal rule ($0 \leq \phi \leq 1$),

$$\theta(t + \Delta t) = \theta(t) + \frac{\Delta t}{mC}[\phi \mathcal{F}(\theta(t + \Delta t)) + (1 - \phi)\mathcal{F}(\theta(t)) + \mathcal{H}^{rays}] \tag{9.65}$$

This implicit nonlinear equation for $\theta$, *for each scatterer, is then added into the ray-tracing algorithm.* See Zohdi [3, 4] for details. Employing a staggered fixed-point scheme, we have

$$\theta^{K+1}(t + \Delta t) = \theta(t) + \frac{\Delta t}{mC}[\phi \mathcal{F}(\theta^K(t + \Delta t)) + (1 - \phi)\mathcal{F}(\theta(t)) + \mathcal{H}^{rays}]. \tag{9.66}$$

This iterative procedure is embedded into the overall ray-tracing scheme. The overall algorithm is as follows, starting at $t = 0$ and ending at $t = T$:

1. COMPUTE RAY MAGNITUDES AND ORIENTATIONS AFTER REFLECTION
2. COMPUTE ABSORPTION CONTRIBUTIONS TO THE SCATTERERS: $\Delta\mathcal{H}^{rays}$
3. COMPUTE SCATTERER TEMPERATURE (RECURSIVELY, $K = 1, 2, \ldots$ UNTIL CONVERGENCE):

$$\theta^{K+1}(t + \Delta t) = \theta(t) + \frac{\Delta t}{mC}[\phi \mathcal{F}(\theta^K(t + \Delta t)) + (1 - \phi)\mathcal{F}(\theta(t)) + \mathcal{H}^{rays}]$$

4. INCREMENT ALL RAY POSITIONS: $\mathbf{r}_i(t + \Delta t) = \mathbf{r}_i(t) + \Delta t \mathbf{v}_i(t)$
5. GO TO (1) AND REPEAT ($t = t + \Delta t$).

As before, in order to capture all of the internal reflections that occur when rays enter the scatter systems, the time-step size $\Delta t$ is dictated by the size of the scatterers. A somewhat ad-hoc approach is to scale the time-step size according to $\Delta t = \xi b$, where $b$ is the radius of the scatterers and typically $0.05 \leq \xi \leq 0.1$. Figure 9.12 illustrates the thermal absorption of the scatterers. Other simulation parameters of importance are

- The dimensions of the sample were $10^{-3}$ m $\times$ $10^{-3}$ m $\times$ $10^{-3}$ m.
- The time-scale was set to $3 \times 10^{-3}$ m/$c$, where $c = 3 \times 10^8$ m/s is the speed of light.
- The initial velocity vector for all initially collinear rays comprising the beam was $\mathbf{v} = (c, 0, 0)$.
- The irradiance beam parameter was set to $I = 10^{18}$ Nm/(m$^2$s), where the irradiance for each ray was calculated as $I^{ray}(t = 0)a_r \stackrel{\text{def}}{=} I a_b / N_r$, where $N_r = 40 \times 40 = 1600$ is the number of rays in the beam and $a_b = 0.5 \times 10^{-3}$ m $\times$ $0.5 \times 10^{-3}$ m $= 0.25 \times 10^{-6}$ m$^2$ is the cross-sectional area of the beam.

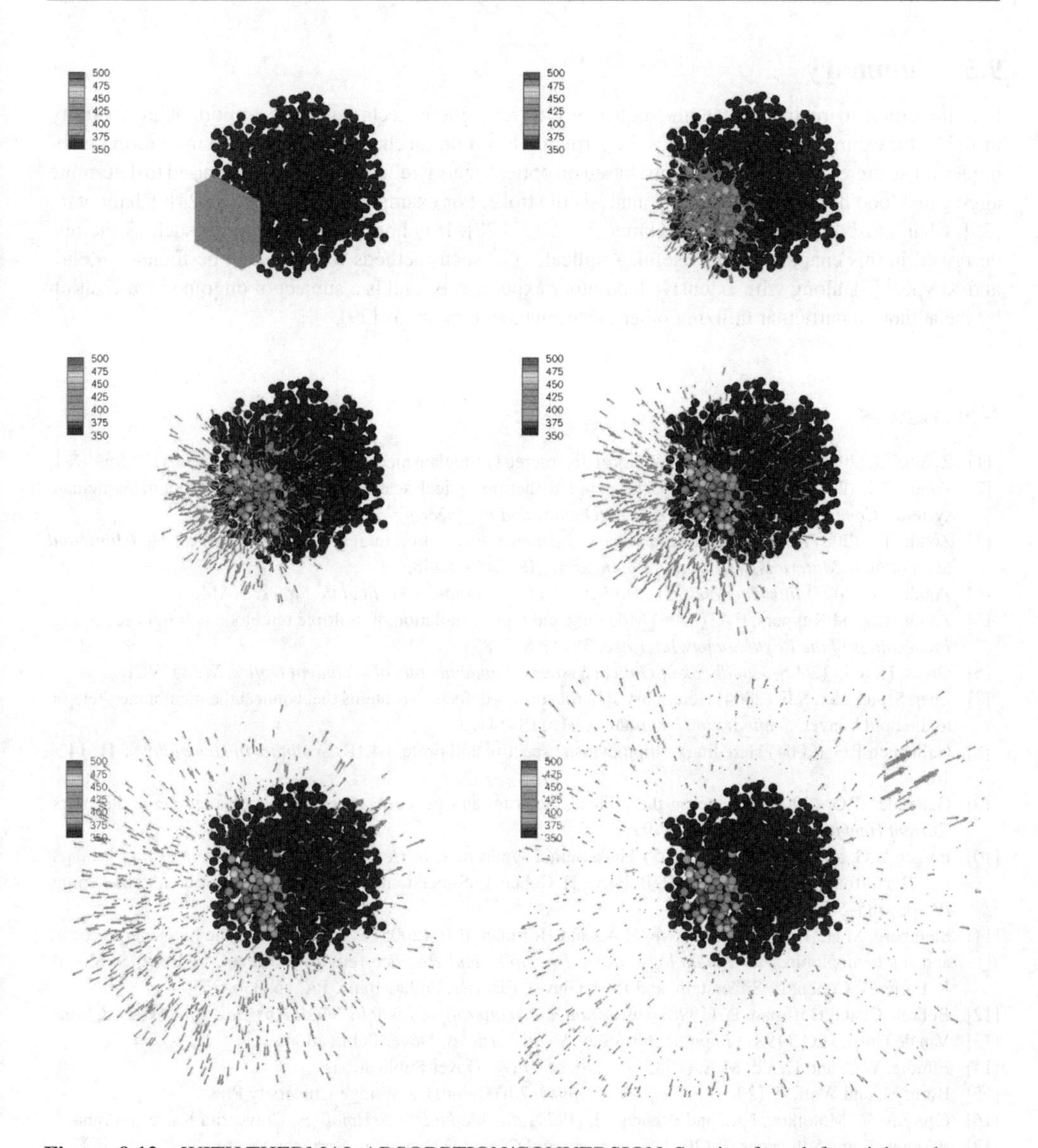


**Figure 9.12** WITH THERMAL ABSORPTION/CONVERSION. Starting from left to right and top to bottom, the progressive movement of rays comprising a beam ($\mathcal{L} = 0.325$, volume fraction $v_s = 0.144$, and $\hat{n} = 4$), for a set of scatterers. The length of the vectors indicates the irradiance $I(t)$ and the legend indicates the temperature in kelvin

One could couple the optical field and thermal fields further by introducing a temperature-dependent refractive index, for example through changes in $\epsilon$ and $\mu$, with virtually no change in the algorithm, although this was not undertaken in this analysis.

**Remark 24** The irradiance beam parameter, which was set to $I = 10^{18}$ N m/(m$^2$ s), was chosen to be artificially high in order to illustrate how the temperature evolves. In reality, the pulses would be much less intense, but repeated over much longer periods of time.

## 9.5 Summary

It is important to realize that high-frequency electromagnetic techniques are important in a variety of fields; for example, fluid mechanics, in particular bio-fluid mechanics. Noninvasive monitoring techniques for scatterer-laden bio-fluid flow, based on optical energy reflectance, are widely used to determine aggregate blood flow properties in the analysis of stroke. For example, see Chau *et al.* [24], Khalil *et al.* [25], Chan *et al.* [26], and Bale-Glickman *et al.* [27, 28]. It is hoped that techniques such as the one described in this chapter will be useful. Application of such methods to RBCs can be found in Zohdi and Kuypers [5], along with extensive laboratory experiments, and is a subject of ongoing investigation by the author, in particular utilizing other electromagnetic methods [29].

## References

[1] Zohdi, T.I. (2006) On the optical thickness of disordered particulate media. *Mechanics of Materials*, **38**, 969–981.
[2] Zohdi, T.I. (2006) Computation of the coupled thermo-optical scattering properties of random particulate systems. *Computer Methods in Applied Mechanics and Engineering*, **195**, 5813–5830.
[3] Zohdi, T.I. (2007) P-wave induced energy and damage distribution in agglomerated granules. *Modelling and Simulation in Materials Science and Engineering*, **15**, S435–S448.
[4] Zohdi, T.I. (2007) *Introduction to the Modeling and Simulation of Particulate Flows*, SIAM.
[5] Zohdi, T.I. and Kuypers, F.A. (2006) Modeling and rapid simulation of multiple red blood cell light scattering. *Proceedings of the Royal Society Interface*, **3** (11), 823–831.
[6] Gross, H. (ed.) (2005) *Handbook of Optical Systems. Fundamentals of Technical Optics*, Wiley–VCH.
[7] Eber, S. and Lux, S.E. (2004) Hereditary spherocytosis – defects in proteins that connect the membrane skeleton to the lipid bilayer. *Seminars in Hematology*, **41**, 118–141.
[8] Gallagher, P.G. (2004) Hereditary elliptocytosis: spectrin and protein 4.1R. *Seminars in Hematology*, **41**, 142–164.
[9] Gallagher, P.G. (2004) Update on the clinical spectrum and genetics of red blood cell membrane disorders. *Current Hematology Reports*, **3**, 85–91.
[10] Forget, B.G. and Cohen, A.R. (2005) Thalassemia syndromes, in *Hematology: Basic Principles and Practice* (eds B. Hoffman, E. Benz, S. Shattil, B. Furie, H. Cohen, L. Silberstein, and P. McGlave), Elsevier, Philadelphia, PA, pp. 485–509.
[11] Steinberg, M.H., Benz, E., Adewoye, H.A., and Hoffman, B.E. (2005) Pathobiology of the human erythrocyte and it's hemoglobins, in *Hematology: Basic Principles and Practice* (eds B. Hoffman, E. Benz, S. Shattil, B. Furie, H. Cohen, L. Silberstein, and P. McGlave), Elsevier, Philadelphia, PA, pp. 356–371.
[12] Bohren, C. and Huffman, D. (1998) *Absorption and Scattering of Light by Small Particles*, John Wiley & Sons.
[13] Van de Hulst, H.C. (1981) *Light Scattering by Small Particles*, Dover Publications.
[14] Elmore, W.C. and Heald, M.A. (1985) *Physics of Waves*, Dover Publications.
[15] Born, M. and Wolf, E. (2003) *Principles of Optics*, 7th edition, Cambridge University Press.
[16] Cerveny, V., Molotkov, I.A., and Psencik., I. (1977) *Ray Methods in Seismology*, Univerzita Karlova, Praha.
[17] Pöschel, T. and Schwager, T. (2004) *Computational Granular Dynamics*, Springer-Verlag.
[18] Widom, B. (1966) Random sequential addition of hard spheres to a volume. *Journal of Chemical Physics*, **44**, 3888–3894.
[19] Torquato, S. (2001) *Random Heterogeneous Materials: Microstructure and Macroscopic Properties*, Springer-Verlag, New York, NY.
[20] Kansaal, A., Torquato, S., and Stillinger, F. (2002) Diversity of order and densities in jammed hard-particle packings. *Physical Review E*, **66**, 041109.
[21] Donev, A., Cisse, I., Sachs, D. *et al.* (2004) Improving the density of jammed disordered packings using ellipsoids. *Science*, **303**, 990–993.
[22] Donev, A., Torquato, S., and Stillinger, F. (2005) Neighbor list collision-driven molecular dynamics simulation for nonspherical hard particles. I. Algorithmic details. *Journal of Computational Physics*, **202**, 737–764.
[23] Donev, A., Torquato, S., and Stillinger, F. (2005) Neighbor list collision-driven molecular dynamics simulation for nonspherical hard particles. II. Application to ellipses and ellipsoids. *Journal of Computational Physics*, **202**, 765–793.

[24] Chau, A.H., Chan, R.C., Shishkov, M. *et al.* (2004) Mechanical analysis of atherosclerotic plaques based on optical coherence tomography. *Annals of Biomedical Engineering*, **32** (11), 1492–1501.
[25] Khalil, A.S., Chan, R.C., Chau, A.H. *et al.* (2005) Tissue elasticity estimation with optical coherence elastography: toward mechanical characterization of in vivo soft tissue. *Annals of Biomedical Engineering*, **33** (11), 1631–1639.
[26] Chan, R.C., Chau, A.H., Karl, W.C. *et al.* (2004) Oct-based arterial elastography: Robust estimation exploiting tissue biomechanics. *Optics Express*, **12** (19), 4558–4572.
[27] Bale-Glickman, J., Selby, K., Saloner, D., and Savas, O. (2003) Physiological flow studies in exact-replica atherosclerotic carotid bifurcation, in *Proceedings of IMECE'03, 2003 ASME International Mechanical Engineering Congress and Exposition, Washington, DC*, pp. 16–21.
[28] Bale-Glickman, J., Selby, K., Saloner, D., and Savas, O. (2003) Experimental flow studies in exact-replica phantoms of atherosclerotic carotid bifurcations under steady input conditions. *Journal of Biomechanical Engineering*, **125**, 38–48.
[29] Zohdi, T.I., Kuypers, F.A., and Lee, W.C. (2010) Estimation of red blood cell volume fraction from overall permittivity measurement. *International Journal of Engineering Science*, **48**, 1681–1691.

# 10

# Electrohydrodynamic Assembly of Nanoparticles for Nanoengineered Biosensors

Jae-Hyun Chung, Hyun-Boo Lee, and Jong-Hoon Kim
*University of Washington, USA*

## 10.1 Introduction for Nanoengineered Biosensors

Controlled manipulation of nanoparticles and biomolecules gains increasing attention in the field of nanoengineered biosensors, disease diagnosis, and drug discovery. Various methods, including centrifugation, magnetic fields, and optical fields, have been investigated for manipulation of particles. Unlike such methods, an electric-field-based method has been applied for precise and parallel manipulation of nanoscale particles. Nanoparticles, such as quantum dots or nanowires, can be assembled on electrodes creating the structures of nanodevices. These nanodevices can be modified through direct manipulation of biomolecules for biosensing and disease diagnostic applications.

For manipulation of nanoparticles, an electric potential is applied through microscale and nanoscale electrodes. Under an electric field, various phenomena are generated by electrostatic interaction, electric polarization, and Joule heating. The electric-field-induced effects are dependent on various electrical and physical parameters; however, selectivity can be achieved by controlling the frequency and amplitude of the electric potential. In this chapter, a theoretical background of electric-field-induced effects is described with the introduction of numerical approaches. The physical insight of the analytical equations is discussed with current experimental results.

## 10.2 Electric-Field-Induced Phenomena

When an electric potential (volts) is applied, an electric field (volts per meter) is generated. Under an electric field, electrophoresis, dielectrophoresis, electroosmosis, and electrothermal flow are induced. Electrophoresis is the motion of a charged particle due to electrostatic interaction under an electric field. Dielectrophoresis is induced by the polarization of a particle in a medium. Electroosmosis is generated

*Multiscale Simulations and Mechanics of Biological Materials*, First Edition. Edited by Shaofan Li and Dong Qian.

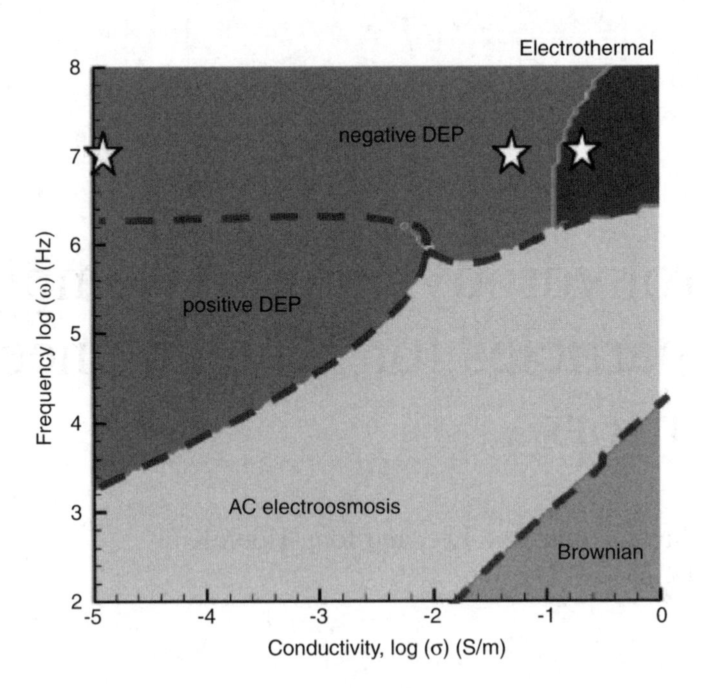


**Figure 10.1** Frequency–conductivity phase diagram [1]. Reprinted with permission from the American Chemical Society © 2008

by ions that are distributed by an electric field. Electrothermal flow is generated by Joule heating in a conductive medium. Electrophoresis and dielectrophoresis describe the motion of a particle within a medium, while electroosmosis and electrothermal flow describe the motion of a medium which generates flow. Such phenomena can be dominant or negligible in various experimental conditions.

Figure 10.1 shows the dominant regions of dielectrophoresis, electroosmosis, and electrothermal flow in the context of electric conductivity and frequency of an electric field [1]. In the graph, electrophoresis is not shown but is typically generated when the frequency is smaller than 1 MHz. To generate electrophoresis, the motion of a particle should be faster than the polarity change of an electric field. Otherwise, a nanoparticle does not respond to the change in a frequency, thus no longer inducing electrophoresis. In this section, the details of electric-field-induced phenomena are discussed with the goal of understanding the underlying physics through analytical equations. Note that the part of the description with analytical equations is taken from the Brown *et al.* [2].

### *10.2.1 Electrophoresis*

Electrophoresis is the motion of a charged particle in a fluid by electrostatic force in the presence of an electric field. An electrophoretic force on a particle is derived from Coulomb's law:

$$\mathbf{F} = \frac{q_1 q_2}{4\pi \varepsilon_0 r^2}\hat{\mathbf{r}}, \tag{10.1}$$

where $\mathbf{F}$ is the force acting between two particles of charges $q_1$ and $q_2$ separated by the distance $r$, $\varepsilon_0$ is the permittivity of vacuum, and $\hat{\mathbf{r}}$ is the unit vector parallel to the line between two particles.

A particle having charge $q$ can create an electric field $\mathbf{E}$ at a point where $r$ is the distance from the particle:

$$\mathbf{E} = \frac{q}{4\pi\varepsilon_0 r^2}\hat{\mathbf{r}}. \tag{10.2}$$

By substituting Equation (10.2) into Equation (10.1), the electrophoretic force $\mathbf{F}_{\mathrm{EP}}$ acting on a particle is

$$\mathbf{F}_{\mathrm{EP}} = q\mathbf{E}. \tag{10.3}$$

The direction of an electrophoretic force is the same as that of an electric field. Electrophoresis is a simple means to manipulate particles for assembly. It should be noted that the charge $q$ in Equation (10.3) does not adequately represent the total charge of a particle in an ionic medium. For electrophoretic motion, an effective charge should be considered for computation because the original charge of a particle is changed by the surrounding ions in a solution.

The effective charge of a particle in solution can be computed by considering the zeta potential $\zeta$ on a particle:

$$\zeta = \frac{q_{\mathrm{o}}}{4\pi\varepsilon r} - \frac{q_{\mathrm{o}}}{4\pi\varepsilon(r+\lambda_{\mathrm{D}})}, \tag{10.4}$$

where $q_{\mathrm{o}}$ is the original charge of a particle, $r$ is the particle radius, and $\lambda_{\mathrm{D}}$ is the Debye length. The first term of the right-hand side in Equation (10.4) is from the original charge of a particle and the second term is from the surrounding ions by distance $\lambda_{\mathrm{D}}$. This can be assumed as the zeta potential on a sphere having radius $r + \lambda_{\mathrm{D}}$ and effective charge $q_{\mathrm{e}}$:

$$\zeta = \frac{q_{\mathrm{e}}}{4\pi\varepsilon(r+\lambda_{\mathrm{D}})}. \tag{10.5}$$

By combining Equations (10.4) and (10.5), the effective charge of a particle is

$$q_{\mathrm{e}} = \frac{\lambda_{\mathrm{D}}}{r}q_{\mathrm{o}}. \tag{10.6}$$

According to Equation (10.6), the magnitude of an electrophoretic force depends on the ratio of the Debye length to the particle radius. Considering the effective charge, the actual behavior of a particle is dependent on the charge of a particle and the ionic medium suspending the particle. Since $\lambda_{\mathrm{D}}$ also depends on surface charges, the surface coating of a particle is important in determining the motion of a particle. When DNA molecules are manipulated under electrophoresis, the effective charge is reduced due to the surrounding counter charge [3]. DNA is strongly negatively charged, which is surrounded by positively charged ions. Thus, the effective charge is reduced by the counter-balance of the ionic solution. The result shows that an effective electrophoretic force can be reduced due to an ionic medium.

### *10.2.2 Dielectrophoresis*

Dielectrophoresis is the movement of a particle in a fluid by an induced dipole moment in an inhomogeneous electric field. The magnitude of force is determined by the polarizability of a particle. One of the methods to calculate dielectrophoretic force is the effective dipole moment (EDM) theory. In this method, the net force on a particle due to an induced dipole can be computed as

$$\mathbf{F} = q\mathbf{E}(\mathbf{r}+\mathbf{d}) - q\mathbf{E}(\mathbf{r}), \tag{10.7}$$

where $\mathbf{r}$ is the position vector of negatively charged pole and $\mathbf{E}(\mathbf{r}+\mathbf{d})$ and $\mathbf{E}(\mathbf{r})$ indicate the electric field on each pole with distance $|\mathbf{d}|$. Using a Taylor series expansion, the first term in the right-hand side can be transformed to

$$\mathbf{F} = q\mathbf{d} \cdot \nabla \mathbf{E}(\mathbf{r}), \tag{10.8}$$

where the higher order terms are neglected.

The dipole moment $\mathbf{p}$ is defined as

$$\mathbf{p} = q\mathbf{d}. \tag{10.9}$$

In an AC field,

$$\mathbf{p} = V\alpha e^{i\omega t}\,\mathbf{E}, \tag{10.10}$$

where $V$ is the particle volume, $\alpha$ is the polarization factor, also called as Clausius–Mossoti factor, and $\omega$ is the angular frequency of an input voltage. The time-averaged dielectrophoretic force on a particle can be derived from Equations (10.8), (10.9), and (10.10):

$$\langle \mathbf{F}_{\mathrm{DEP}} \rangle = \frac{1}{4} V \mathrm{Re}[\alpha] \nabla |\mathbf{E}|^2. \tag{10.11}$$

The magnitude of dielectrophoretic force can be varied with the gradient of an electric field. Another parameter, polarization factor, can affect the magnitude and direction of the force. This factor depends on the electrical properties of a particle and medium (conductivity and permittivity), the geometry of a particle, and the frequency of an electric field.

However, the EDM theory is not accurate enough to calculate a dielectrophoretic force acting on a particle when the particle size is comparable to the characteristic length of an electric field because the EDM theory is based on the assumption that the length of a dipole is small enough to neglect the higher order terms in the calculation of dielectrophoretic force. Maxwell stress tensor (MST) theory has been used to overcome this limitation. In this method, dielectrophoretic force is computed by the surface integral of the Maxwell stress tensor over the particle surface:

$$\langle \sigma_{\mathrm{MST}} \rangle = \frac{1}{4} \varepsilon_{\mathrm{m}} (\mathbf{E} \otimes \mathbf{E}^* + \mathbf{E}^* \otimes \mathbf{E} - |\mathbf{E}|^2\,\mathbf{I}), \tag{10.12}$$

$$\langle \mathbf{F}_{\mathrm{DEP}} \rangle = \oint \langle \sigma_{\mathrm{MST}} \rangle \cdot \mathbf{n}\, \mathrm{d}S, \tag{10.13}$$

where $\sigma_{\mathrm{MST}}$ is the Maxwell stress tensor, $\varepsilon_{\mathrm{m}}$ is the permittivity of medium, $\mathbf{E}^*$ is the complex conjugate of electric field $\mathbf{E}$, and $\otimes$ is the dyadic product.

Figure 10.2 shows the comparison result of dielectrophoretic force computed by the EDM and MST. Two different particle shapes and sizes were considered for computation of dielectrophoretic force. The height from electrode surface to a particle is also changed.

According to the computation results, when a particle size is small in comparison with the characteristic length of an electric field, the EDM theory showed similar results to the MST theory. Also, the EDM theory is similar to the MST when the particle is a prolate shape rather than a rod shape. As the height from the electrode increases, the EDM results become similar to the MST results. Considering the numerical results, the EDM cannot be applicable when the particle shape is complex or variations of electric field gradients near the particle are significant.

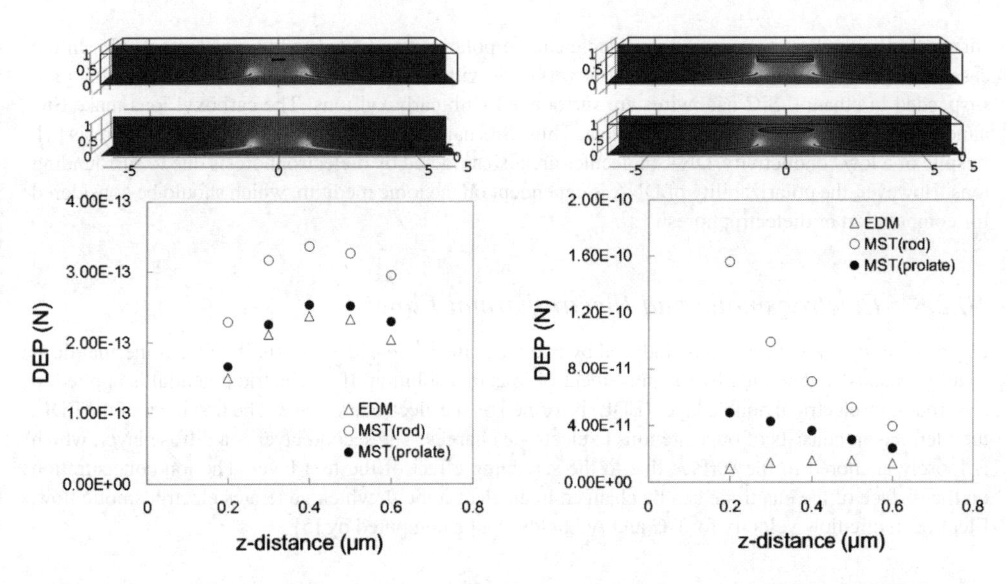


**Figure 10.2** Comparison between dielectrophoretic force calculation by the EDM and MST theories with different particle sizes and positions: (a) particle length is 0.5 μm; (b) particle length is 2.0 μm [2]. Reprinted with permission

The differences between the EDM and the MST theories are summarized in Table 10.1. In the EDM theory, the electric field in the entire domain Ω is solved by the Laplace equation using only the medium properties without considering the frequency input. Dielectrophoretic force is simply calculated from the electric field norm at the particle center. The frequency dependency is applied by the Clausius–Mossoti factor. Note that this factor is dependent on the particle geometry, and thus can be given for simple geometries. On the other hand, the complex Laplace equation in the solid domain $\Omega_s$ and medium domain $\Omega_m$ should be solved for the MST theory. Dielectrophoretic force is computed by the integration of MST over the particle surface. Therefore, dielectrophoresis on any arbitrary shape of particle can be computed by the MST.

Electrically conductive particles are attracted by dielectrophoresis because such particles can be easily polarized. Although a particle is not easily polarized in an electric field, if the particle is coated or

**Table 10.1** Comparison between EDM and MST [2], reprinted with permission

| Method | EDM | MST |
|---|---|---|
| Electric field | $\nabla \cdot (\sigma_m \nabla V) = 0, \quad \text{in } \Omega$<br>$E = -\nabla V$ | $\nabla \cdot (\sigma_m^* \nabla V) = 0, \quad \text{in } \Omega_m$<br>$\nabla \cdot (\sigma_s^* \nabla V) = 0, \quad \text{in } \Omega_s$<br>$E = -\nabla V$ |
| Dielectrophoresis calculation | $\langle \mathbf{F}_{DEP} \rangle = \frac{1}{4} V \mathrm{Re}[\alpha] \nabla \lvert \mathbf{E} \rvert^2$ | $\langle \sigma_{MST} \rangle = \frac{1}{4} \varepsilon_m (\mathbf{E} \otimes \mathbf{E}^* + \mathbf{E}^* \otimes \mathbf{E} - \lvert \mathbf{E} \rvert^2 \mathbf{I})$<br>$\langle \mathbf{F}_{DEP} \rangle = \oint \langle \sigma_{MST} \rangle \cdot \mathbf{n} \, dS$ |
| Shape dependency | Spherical, oblate, and prolate geometries | Arbitrary shape |

surrounded with charges or ions, the particle can be polarized and attracted by an electric field. In the case of SiC nanowires, the conductivity is very low with low permittivity. When SiC nanowires are suspended in ethanol, SiC nanowires are surrounded with carboxyl ions. The carboxyl ions make SiC nanowires more polarizable than the solvent. Thus, SiC nanowires are attracted by dielectrophoresis [4]. In spite of a low conductivity, DNA molecules are also attracted by dielectrophoresis due to surrounding ions. However, the polarizability of DNA is dependent on the ionic medium, which should be considered for computation of dielectrophoresis.

### *10.2.3 Electroosmotic and Electrothermal Flow*

Electrohydrodynamic flow can be induced by two mechanisms: electroosmosis and electrothermal flow. Electroosmosis is generated by the movement of ions in a solution. If an electric potential is applied to electrodes, an electrical double layer (EDL) is formed on the electrode surface. The first layer of the EDL, the Stern layer, consists of opposite ions to electrode charges. The second layer is a diffuse layer, which is loosely anchored to the surface due to the screening effect of the first layer. The ion concentration on the surface of the electrode can be changed in an electric field, which generates electroosmotic flow. Electroosmotic flow velocity for DC and AC fields can be computed by [5]

$$U_{\mathrm{DCEO}} = -\frac{\varepsilon_{\mathrm{m}}\zeta E_{\mathrm{t}}}{\mu}, \tag{10.14}$$

$$U_{\mathrm{ACEO}} = \Lambda \frac{\pi \varepsilon_{\mathrm{m}} V_{\mathrm{o}}^2}{8\mu x} \frac{\Omega^2}{(1+\Omega^2)^2}, \tag{10.15}$$

where $\mu$ is the viscosity of solution, $E_{\mathrm{t}}$ is the tangential directional electric field on the surface of electrode, $\Lambda$ is the relative capacitance of the Stern layer with respect to the overall double layer ($\Lambda = C_{\mathrm{S}}/(C_{\mathrm{S}} + C_{\mathrm{D}})$, where $C_{\mathrm{S}}$ is the capacitance of the Stern layer and $C_{\mathrm{D}}$ is the capacitance of the diffuse layer), $V_{\mathrm{o}}$ is the input voltage, $x$ is the characteristic length (a half length of electrode gap), and $\Omega$ is the nondimensional frequency and is defined as

$$\Omega = \frac{\pi \varepsilon_{\mathrm{m}} \omega x}{2\sigma_{\mathrm{m}} \lambda_{\mathrm{D}}}, \tag{10.16}$$

where $\omega$ is the angular frequency of input voltage and $\sigma_{\mathrm{m}}$ is the conductivity of the medium.

Under an electric field, electroosmotic flow is generated in the form of circulating flow in combination with electrophoresis and dielectrophoresis. Figure 10.3 shows a trajectory of 3 μm diameter spheres in the vicinity of a microtip where an electric potential is applied. Initially, the spheres are located as shown in Figure 10.3a. When an electrophoretic force is applied, the spheres are attracted through the electric field and located on the surface of the microtip. When dielectrophoresis is applied, the spheres are attracted to an edge of the microtip because the spheres are attracted to a higher electric field strength. When electroosmotic flow is applied, the spheres approached the microtip with the convective flow but are carried away with the flow. Therefore, the design of electric fields and the geometry of electrodes are important for a high-yield assembly of biomolecules and nanoparticles.

Electrothermal flow is generated by Joule heating that changes the electric properties of medium, which generates the body force on the fluid. Electrothermal force can be computed by [5]

$$\mathbf{F}_{\mathrm{ET}} = \frac{1}{2}\mathrm{Re}\left[\frac{\sigma_{\mathrm{m}}\varepsilon_{\mathrm{m}}(\alpha-\beta)}{\sigma_{\mathrm{m}} + i\omega\varepsilon_{\mathrm{m}}}(\nabla T \cdot \mathbf{E})\mathbf{E}^* - \frac{1}{2}\varepsilon_{\mathrm{m}}\alpha|\mathbf{E}|\nabla T\right], \tag{10.17}$$

where $T$ is the temperature and $\alpha$ and $\beta$ are defined as $\alpha = (1/\varepsilon_{\mathrm{m}})(\partial\varepsilon_{\mathrm{m}}/\partial T)$ and $\beta = (1/\sigma_{\mathrm{m}})(\partial\sigma_{\mathrm{m}}/\partial T)$, respectively. The first term in the right-hand side of Equation (10.17) is the Coulomb force that is

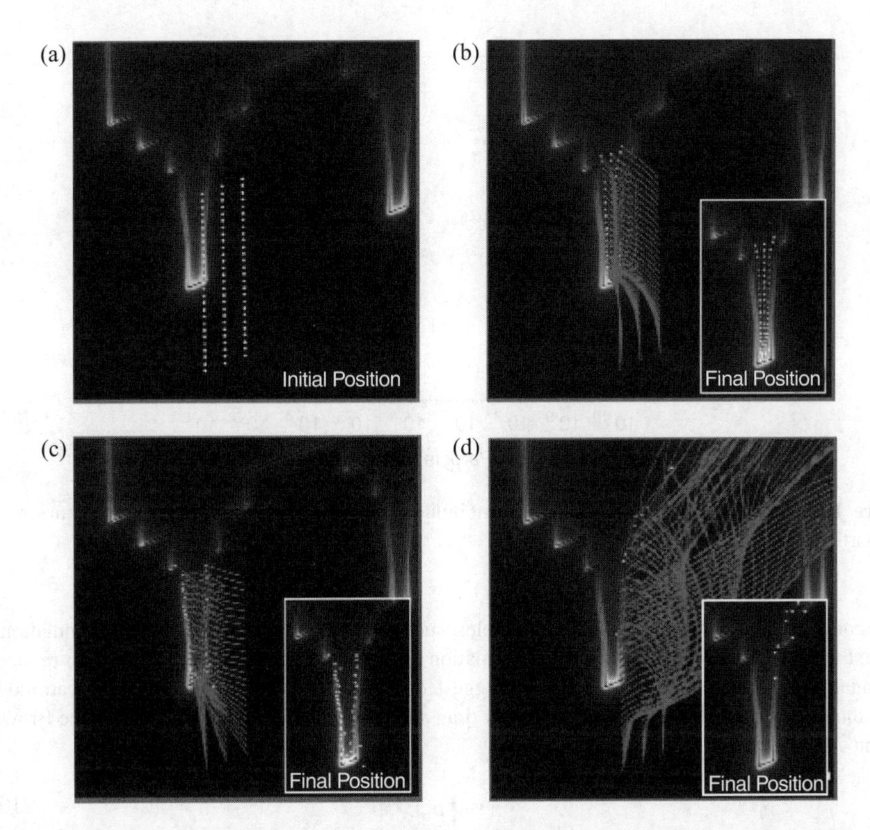


**Figure 10.3** Trajectories of a 3 μm diameter sphere near the microtip according to electrophoresis, dielectrophoresis, and electroosmotic flow. (a) Initial position of particles at 100 μm distance from a microtip. (b) Trajectories of spheres under electrophoresis. The inset image shows the final location of spheres on the microtip surface. (c) Trajectories of spheres under dielectrophoresis. The inset image shows that spheres are attracted at the edge of the microtip. (d) Trajectories of the spheres under electroosmotic flow. In the inset image, spheres are finally located on the microtip surface along an electric field [3]. Reprinted with permission from Springer © 2012

dominant at low frequencies. The second term is the dielectric force that is dominant at high frequencies. These two forces show different flow patterns.

### *10.2.4 Brownian Motion Forces and Drag Forces*

Brownian motion should be considered for nanoscale particles to understand the movement of a particle in a solution. Brownian force is computed as follows [6]:

$$\mathbf{F}_{\mathrm{B}} = \sqrt{\frac{12\pi \mu r k_{\mathrm{B}} T}{\Delta t}}, \tag{10.18}$$

where $k_{\mathrm{B}}$ is the Boltzmann constant and $t$ is the time.

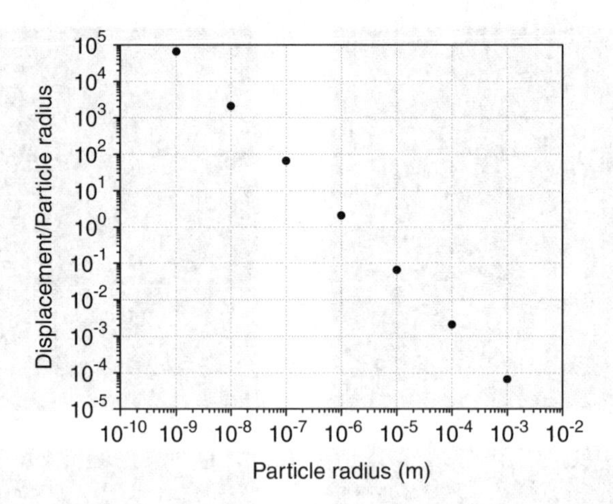


**Figure 10.4** Normalized particle displacement induced by Brownian motion according to various radii of a particle

To compute the motion of deformable particles, such as DNA, a molecular dynamics simulation can be used. However, a molecular dynamics simulation needs high computation resources. To reduce the computation cost, coarse-scale modeling is suggested. A Brownian dynamics simulation can model a DNA molecule as a spring–bead model. If only drag and Brownian forces are considered, the Brownian motion can be derived as

$$\mathbf{F}_{\mathrm{N}} = \mathbf{F}_{\mathrm{D}} + \mathbf{F}_{\mathrm{B}}, \tag{10.19}$$

$$0 = -6\pi\mu r\left(\frac{\Delta\mathbf{x}}{\Delta t} - \mathbf{u}\right) + \mathbf{F}_{\mathrm{B}}, \tag{10.20}$$

$$|\Delta\mathbf{x}| = \sqrt{\frac{2k_{\mathrm{B}}T}{6\pi\mu r}\Delta t}, \tag{10.21}$$

where $\mathbf{F}_{\mathrm{N}}$ is the net force and $\mathbf{F}_{\mathrm{D}}$ is the drag force that is computed from the relative motion of the particle to fluid flow. From Equation (10.20), the net force is assumed as zero at an equilibrium condition. The fluid velocity is set to zero to compute the Brownian motion of particle in Equation (10.21). Figure 10.4 shows the normalized displacement of a particle with various radii due to Brownian motion. According to the computation results, the Brownian-motion-induced force is negligible for a micrometer-size particle.

## 10.3 Geometry Dependency of Dielectrophoresis

As we discussed Section 10.2.2, a dielectrophoretic force is generated from the induced dipole in a particle. The dielectrophoretic force is dependent on the particle geometry, unlike other electric-field-induced forces. As an example of geometric effects on dielectrophoresis, the force acting on a silicon wire of conical shape was examined.

Figure 10.5 shows the simulation result of the force and the torque on a conical-shaped wire positioned under the tip-shaped electrode. Based on the simulation, the conical-shaped wire has a preferential

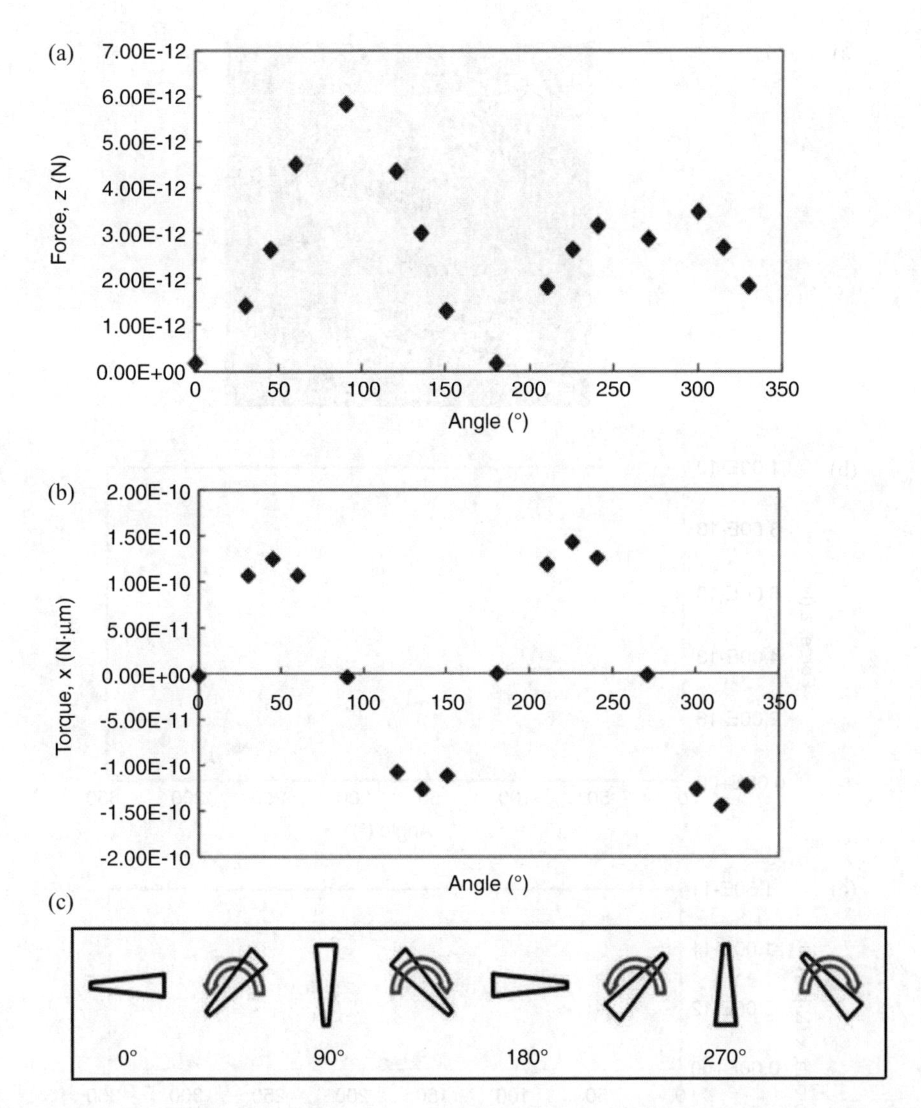


**Figure 10.5** Dielectrophoretic force and torque acting on a conical-shaped particle with different initial angles. (a) Dielectrophoretic force and (b) dielectrophoretic torque on the particle with different orientation; (c) particle orientation and direction of torque by dielectrophoresis

direction by dielectrophoresis. Depending on the initial orientation, the wire can be rotated in clockwise or counter-clockwise directions. The force at each angle varies due to the conical geometry.

Additional analysis is conducted to study a coating effect for dielectrophoresis. The wire has the same conical shape as in Figure 10.5 but is coated with gold on one end. By coating the wire with a high-conductivity layer, the force and the torque show a maximum at a specific angle (Figure 10.6). In a similar way, the dielectrophoretic manipulation of a particle can be controlled by the geometry, a specific coating, and surface modification. The numerical results also show that the effect of surface coating is more important than that of the particle core for dielectrophoretic manipulation.

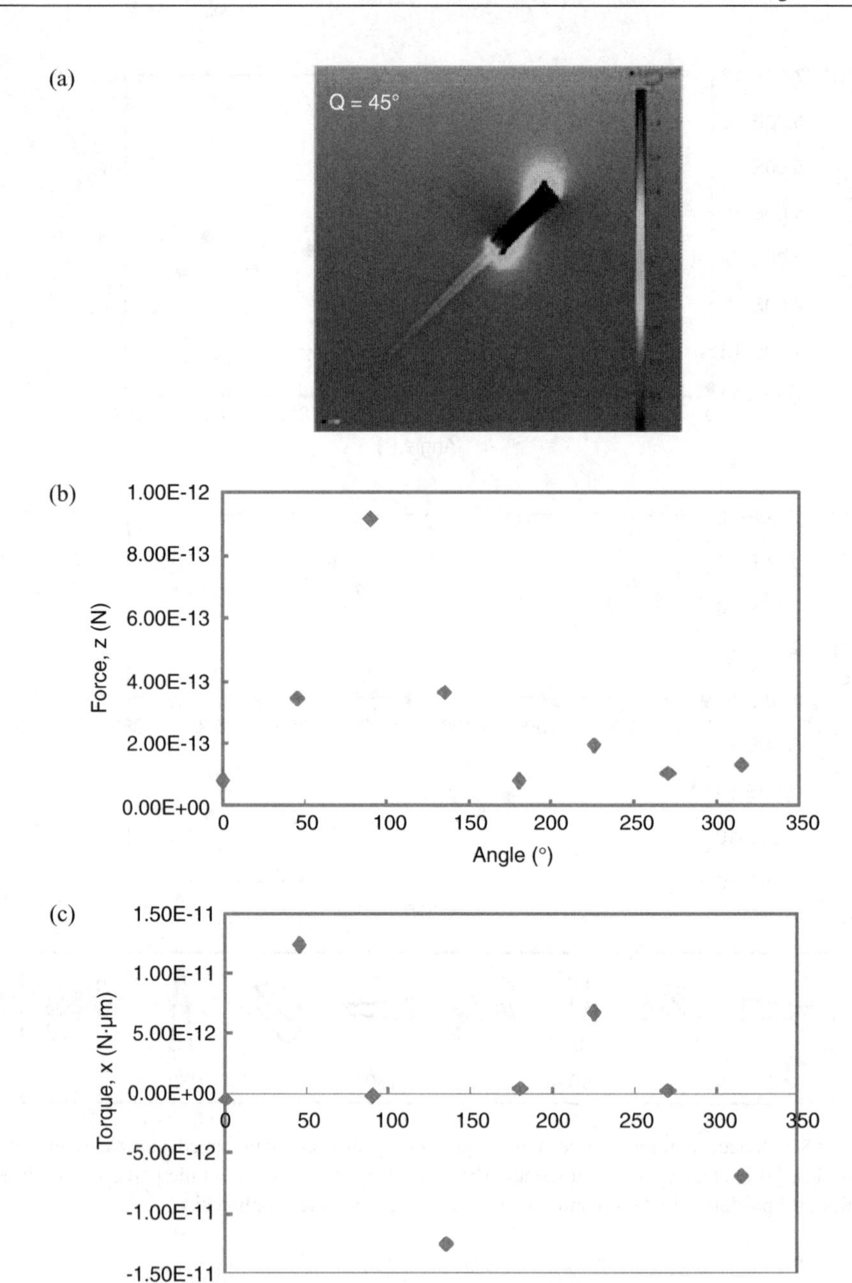


**Figure 10.6** Dielectrophoretic force and torque on a conical-shaped particle coated with gold on one end at different initial orientations: (a) geometry of particle with 45° orientation; (b) dielectrophoretic force; (c) dielectrophoretic torque on the particle

## 10.4 Electric-Field-Guided Assembly of Flexible Molecules in Combination with other Mechanisms

Electric fields can be used to manipulate molecules in a solution in the form of electrophoresis, dielectrophoresis, electroosmosis, and electrothermal flow. Among the electric-field-induced effects, the dielectrophoretic assembly of molecules has become more popular because the particles can be precisely assembled on a substrate. Furthermore, the integration of dielectrophoresis with other mechanisms, such as microfluidic flow, binding affinity, capillary, and viscous forces, can allow enhancement of efficiency for the assembly.

### *10.4.1 Dielectrophoresis in Combination with Fluid Flow*

Dielectrophoretic assembly in the presence of a fluid flow could offer a high-yield placement of nanowires [7]. This approach combines an AC electric field and shear flow in a microfluidic channel for the precise assembly and orientation of nanowires, which is a result of spatial confinement, relatively uniform wire orientation in the channel, and the competition between fluid-induced drag- and dielectrophoretic-forces. In addition, the uniform flow transports nanowires to the vicinity of electrodes, which can compensate the small magnitude of the dielectrophoretic force. This approach has been further developed to have a high deposition yield of 98.5% for Si nanowires [8].

### *10.4.2 Dielectrophoresis in Combination with Binding Affinity*

The specificity of molecular binding between the target and the probe molecule (e.g., between antigen and antibody or between two complementary DNAs) is the principle of affinity assays. Dielectrophoresis in combination with affinity binding could offer a selective assembly of molecules, such as bacteria cells and DNA [9–15]. This technique is useful in a biosensor platform for enhancing the efficiency of concentration and capture. The enhanced efficiency could improve the sensitivity [9–15].

Two approaches have been studied for selective assembly with binding affinity. First, both targets and nontargets are assembled by dielectrophoresis on the electrode surface. Target analytes are attracted and bound to the surface by probe molecules. After the binding, the dielectrophoretic force is released and nontarget molecules are rinsed by fluid flow. As a result, only target molecules are left on the substrate. The other approach is that the dielectrophoretic force is optimized to attract only the target molecules on to substrate under fluid flow. At an optimum frequency, target molecules could be captured on the substrate by binding to probe molecules, while nontargeted molecules are directly transported under fluid flow. Therefore, two functions of dielectrophoresis enhance the capture efficiency: (1) dielectrophoresis increases the number of target analytes that are attracted to the surface of the substrate and (2) dielectrophoresis enhances the binding of targets to probe molecules by steering the targets on to the immobilized probes. Dielectrophoresis can attract the target and change the orientation of probe molecules, which can change the association and dissociation constants for binding, which should be studied further.

### *10.4.3 Dielectrophoresis in Combination with Capillary Action and Viscosity*

Dielectrophoretic force combined with capillary action and viscosity can be used to produce one-dimensional structures and assemble nanoparticles on a wire [16]. The assembled structures are often used as biosensors and chemical sensors [16–18], force transducers [19], different types of beam emitters [17], mass sensors [17, 19], nanomanipulation devices [17, 18], nanoelectromechanical switches

[17, 19], and size-dependent particle separation [16]. Carbon nanotubes (CNTs) are also often assembled onto atomic force microscope tips [17, 18, 20–23].

The common method to assemble particles by dielectrophoresis in conjunction with capillary action is the application of an electric field in between a base electrode and a tip-shaped electrode [18, 22]. The base electrode can hold the solution containing particles. Within the solution, dielectrophoresis attracts the particles in the vicinity of a tip-shaped electrode surface. When the tip-shaped electrode is withdrawn from the solution, the concentrated particles are aligned by capillary and viscous forces, and bound to the terminal end of a tip-shaped electrode by van der Waals forces [24, 25]. In comparison with the assembly method using a nanomanipulator, the tip-based assembly is faster and shows a higher yield [18]. Potentially, other particles in solution can contaminate the surface of the assembled structure [18]. To avoid damage to the electrode surface, an AC electric field instead of a DC electric field can reduce the amount of impurities by eliminating electrophoretic force [17, 22]. Also, choosing the correct electric field properties as well as the curvature of the tip-shaped electrode can significantly increase the yield of this process [22].

## 10.5 Selective Assembly of Nanoparticles

### *10.5.1 Size-Selective Deposition of Nanoparticles*

Nanoparticles and biomolecules can be selectively attracted depending on size. Since the dipole moment is dependent on particle length, larger lengths of DNA molecules could be selectively attracted by dielectrophoresis [13]. Nanoparticles could be sorted by lengths via dielectrophoresis. It was shown that the length of deposited nanowires across a gap was dependent on the gap size (Figure 10.7) [4, 7]. The nanowires, with lengths comparable to the gap size, were selectively deposited across electrodes. Nanowires much smaller or larger than the gap size were not attracted because of lower dielectrophoretic force. The maximum dielectrophoretic force is found when the gap size is 0.8 times the nanowire length. The experimental result [7] verifies the fact that the length of the deposited nanowires is determined

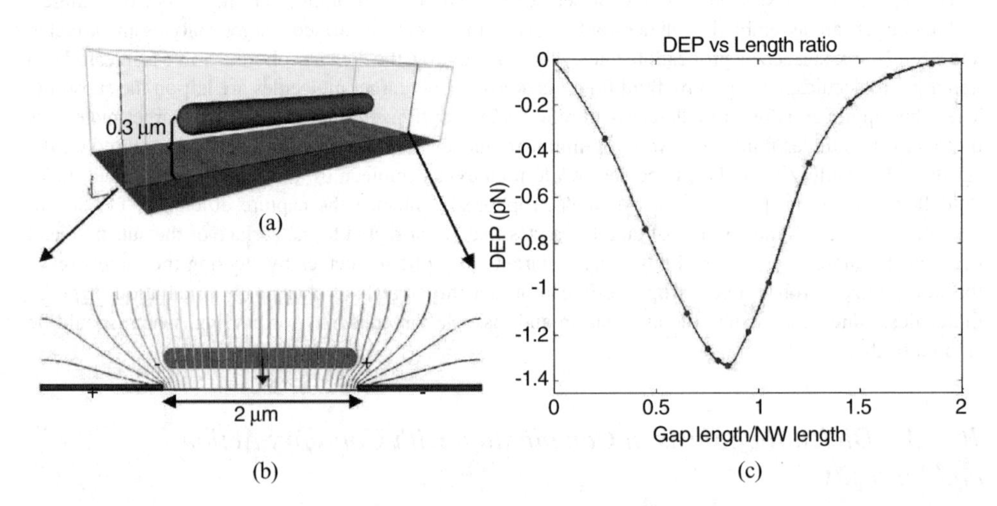


**Figure 10.7** Dielectrophoretic force on a nanowire [4]. (a) A 2.0 μm long nanowire aligns across a gap between electrodes. (b) Cross-sectional view. (c) Dielectrophoretic force on a nanowire at fixed height for various values of the ratio of gap length to nanowire length. Reprinted with permission from American Chemical Society © 2006

by the gap size in dielectrophoretic assembly. This sorting principle can be used for other nanowires and biomolecules that are dominated by dielectrophoresis, which has potential uses in customized nanostructure fabrication, as well as biomolecule assembly and sorting.

### 10.5.2 Electric-Property Sorting of Nanoparticles

Using dielectrophoresis, nanoparticles and biomolecules can be selectively collected due to the difference in electric properties from other particles. The major difference is the permittivity and the conductivity of particles. Although the permittivity is a material property, the polarizability of particles can be controlled by surface modification, as mentioned previously. Since the permittivity in a medium is changed by the surface ions or molecules on a particle, the dielectrophoretic attraction or repulsion can be controlled. Although SiC nanowires are not attracted by dielectrophoresis due to the low conductivity and low permittivity in solvent, SiC nanowires coated with ions in solvent could be attracted by dielectrophoresis [4]. Single-walled CNTs (SWCNTs) can be sorted by their electrical properties [26]. When SWCNTs are attracted by dielectrophoresis, metallic SWCNTs are dominantly attracted by positive dielectrophoresis. Bacterial cells can be attracted when the cells are alive [27]. Dead cells are not attracted because the damaged cells walls are not electrically polarized. Considering the intrinsic electric property, DNA molecules are not attracted by dielectrophoresis. However, DNA molecules in ionic solution are attracted by dielectrophoresis [3]. Therefore, the surface properties can be manipulated and utilized for selective attraction of nanoparticles and bioparticles.

## 10.6 Summary and Applications

Manipulation of nanoparticles and biomolecules has been actively investigated since 1990s. With the fabrication of microscale and nanoscale electrodes, an electric field over 5 kV/cm can easily be generated to manipulate molecules. Using an electric field, electrophoresis, dielectrophoresis, electroosmosis, and electrothermal flow can be generated. Such electric-field-induced phenomena are combined with binding affinity, capillary action, and fluid flow in order to achieve specific binding, size specificity, and high-yield assembly. In the development of electric-field-induced methods, analytical and numerical methods have mainly contributed to enhance the understanding of the physical phenomena and reduce the incubation time for development of devices. Significant effort to understand and manipulate the behavior of biomolecules is likely to greatly improve biomedical applications.

## References

[1] Park, S. and Beskok, A. (2008) Alternating current electrokinetic motion of colloidal particles on interdigitated microelectrodes. *Analytical Chemistry*, **80**, 2832–2841.

[2] Brown, D.A., Kim, J-H., Lee, H.-B. *et al.* (2012) Electric field guided assembly of one-dimensional nanostructures for high performance sensors. *Sensors*, **12**, 5725–5751.

[3] Kalyanasundaram, D., Inoue, S., Kim, J.-H. *et al.* (2012) Electric field-induced concentration and capture of DNA onto microtips. *Microfluidics and Nanofluidics*, **13** (2), 217–225.

[4] Liu, Y., Chung, J.H., Liu, W.K., and Ruoff, R.S. (2006) Dielectrophoretic assembly of nanowires. *Journal of Physical Chemistry B*, **10**, 14098–14106.

[5] Burg, B.R., Bianco, V., Schneider, J., and Poulikakos, D. (2010) Electrokinetic framework of dielectrophoretic deposition devices. *Journal of Applied Physics*, **107**, 124308.

[6] Ramos, A., Morgan, H., Green, N.G., and Castellanos, A. (1998) AC electrokinetics: a review of forces in microelectrode structures. *Journal of Physics D: Applied Physics*, **31**, 2338–2353.

[7] Oh, K., Chung, J.H., Riley, J.J. *et al.* (2007) Fluid flow-assisted dielectrophoretic assembly of nanowires. *Langmuir*, **23**, 11932–11940.

[8] Freer, E.M., Grachev, O., Duan, X.F. *et al.* (2010) High-yield self-limiting single-nanowire assembly with dielectrophoresis. *Nature Nanotechnology*, **5**, 525–530.

[9] Yang, L.J., Banada, P.P., Chatni, M.R. *et al.* (2006) A multifunctional micro-fluidic system for dielectrophoretic concentration coupled with immuno-capture of low numbers of *Listeria monocytogenes*. *Lab on a Chip*, **6**, 896–905.

[10] Yang, L.J. (2012) A review of multifunctions of dielectrophoresis in biosensors and biochips for bacteria detection. *Analytical Letters*, **45**, 187–201.

[11] Yang, L.J. (2009) Dielectrophoresis assisted immuno-capture and detection of foodborne pathogenic bacteria in biochips. *Talanta*, **80**, 551–558.

[12] Gagnon, Z., Senapati, S., Gordon, J., and Chang, H.C. (2008) Dielectrophoretic detection and quantification of hybridized DNA molecules on nano-genetic particles. *Electrophoresis*, **29**, 4808–4812.

[13] Kawabata, T. and Washizu, M. (2001) Dielectrophoretic detection of molecular bindings. *IEEE Transactions on Industry Applications*, **37**, 1625–1633.

[14] Suehiro, J., Ohtsubo, A., Hatano, T., and Hara, M. (2006) Selective detection of bacteria by a dielectrophoretic impedance measurement method using an antibody-immobilized electrode chip. *Sensors and Actuators B: Chemical*, **119**, 319–326.

[15] Kim, J.H., Yeo, W.H., Shu, Z.Q. *et al.* (2012) Immunosensor towards low-cost, rapid diagnosis of tuberculosis. *Lab on a Chip*, **12**, 1437–1440.

[16] Yeo, W.H., Chung, J.H., Liu, Y.L., and Lee, K.H. (2009) Size-specific concentration of DNA to a nanostructured tip using dielectrophoresis and capillary action. *Journal of Physical Chemistry B*, **113**, 10849–10858.

[17] Choi, J.S., Kwak, Y., and Kim, S. (2008) Carbon nanotube samples prepared by an electric-field-assisted assembly method appropriate for the fabrication processes of tip-based nanodevices. *Journal of Micromechanics and Microengineering*, **18**, 035008.

[18] Park, J.-K., Kim, J.-E., and Han, C.-S. (2005) Use of dielectrophoresis in a high-yield fabrication of a carbon nanotube tip. *Japanese Journal of Applied Physics*, **44**, 3235–3239.

[19] Lim, D., Kwon, S., Lee, J. *et al.* (2009) Deterministic fabrication of carbon nanotube probes using the dielectrophoretic assembly and electrical detection. *Review of Scientific Instruments*, **80**, 105103.

[20] Kim, J.-E. and Han, C.-S. (2005) Use of dielectrophoresis in the fabrication of an atomic force microscope tip with a carbon nanotube: a numerical analysis. *Nanotechnology*, **16**, 2245.

[21] Tang, J., Yang, G., Zhang, Q. *et al.* (2009) Rapid and reproducible fabrication of carbon nanotube AFM probes by dielectrophoresis. *Nano Letters*, **5**, 11–14.

[22] Lee, H.W., Kim, S.H., Kwak, Y.K., and Han, C.S. (2004) Nanoscale fabrication of a single multiwalled carbon nanotube attached atomic force microscope tip using an electric field. *Review of Scientific Instruments*, **76**, 046108.

[23] Kuwahara, S., Akita, S., Shirakihara, M. *et al.* (2006) Fabrication and characterization of high-resolution AFM tips with high-quality double-wall carbon nanotubes. *Chemical Physics Letters*, **429**, 581–585.

[24] Wang, M.C.P., Zhang, X., Majidi, E. *et al.* (2010) Electrokinetic assembly of selenium and silver nanowires into macroscopic fibers. *ACS Nano*, **4**, 2607–2614.

[25] Tang, J., Yang, G., Zhang, J. *et al.* (2003) Controlled assembly of carbon nanotube fibrils by dielectrophoresis. *Materials Research Society – Proceedings*, **788**, 539–544.

[26] Krupke, R., Hennrich, F., von Lohneysen, H., and Kappes, M.M. (2003) Separation of metallic from semiconducting single-walled carbon nanotubes. *Science*, **301**, 344–347.

[27] Pohl, H.A. and Hawk, I. (1966) Separation of living and dead cells by dielectrophoresis. *Science*, **152**, 647–649.

# 11

# Advancements in the Immersed Finite-Element Method and Bio-Medical Applications

Lucy Zhang, Xingshi Wang, and Chu Wang

*Department of Mechanical, Aerospace, and Nuclear Engineering, Rensselaer Polytechnic Institute, USA*

## 11.1 Introduction

Numerical simulations of the bio-medical applications often involve in the prediction of the fluid–solid behaviors and dealing with the moving objects and interfaces. In the past few decades, numerous research efforts have been directed to methods development towards efficient and accurate modeling and simulations of fluid–structure interactions.

One of the most well-known and popularly used methods to capture the interactions between a solid structure and its surrounding fluid is the arbitrary Lagrangian–Eulerian method [1, 2]. It allows arbitrary motion of grid/mesh points with respect to their frame of reference by taking the convection of the material points into account. However, for large translations and rotations of the object or inhomogeneous movements of the grid point, fluid elements tend to become severely distorted. It would then require a process of re-meshing or mesh updating, in which the whole domain or part of the domain is spatially re-discretized. This process can be computationally expensive, if frequently used. Most importantly, it can reduce the accuracy at the interface due to the transferring of the solutions from the degenerated mesh to the new mesh.

To avoid constant re-meshing, a nonconforming approach is widely adopted, in which the solid object and the background fluid grid are meshed independently. The object can freely move on top of the fluid grid without deformation. One of the most commonly used numerical methods for simulating biointerfaces is the immersed boundary method [3], which was initially proposed by Peskin to study the blood flow around heart valves [4–6]. It employs a mixture of Eulerian and Lagrangian descriptions

*Multiscale Simulations and Mechanics of Biological Materials*, First Edition. Edited by Shaofan Li and Dong Qian.

for fluid and solid domains, respectively. The interaction is accomplished by interpolating the interfacial forces and velocities between the Eulerian and Lagrangian domains through a smoothed approximation of the Dirac delta function by assuming the fluid enclosed in the boundary has the same properties as the outside and no-slip boundary condition at the interface. One major advantage of nonconforming methods is that the interface is tracked automatically, which circumvents the necessity in using costly mesh-updating algorithms.

The immersed boundary method inspired researchers around the world to further develop and enhance the accuracy and efficiency of the method. In order to achieve a realistic representation of the solid material, an extension of the immersed boundary method, the immersed finite element method (IFEM) [7–12], is developed to represent the background viscous fluid with an unstructured finite-element mesh and finite elements for the immersed deformable solid. Similar to the immersed boundary method, the fluid domain is defined on a fixed Eulerian grid and the solid domain is constructed independently with a Lagrangian mesh. The major difference between these two approaches, however, is that with the IFEM the solid material can be described with a detailed constitutive model such as the linear elastic material, hyperelastic material, or viscoelastic material. It is no longer limited to just a boundary layer; instead, it can occupy a volume space in the entire computational domain. This feature is particularly useful when accurate estimations in deformation or movement of the material must be realized and its affected hydrodynamics are altered through interactions with the surrounding fluid.

The two-way coupled approach (i.e., the interpolation and the distribution of the velocity and the forces between the two domains) is quite robust when the solid behaves very much like the fluid. In fact, one of the premises in the development of the approach is that the soft tissues interacting with the surrounding blood have similar material properties in terms of density and viscosity. However, if large disparities occur in the material properties, the force and the velocity to be interpolated between the two fields can no longer provide consistent convergence, due to the high discontinuity in density as well as other intrinsic parameters in the solid and the fluid.

A semi-implicit algorithm for the IFEM is to alleviate the situations when large disparities in density and material properties are used in the solid and fluid domains. The semi-implicit algorithm enforces the interaction force to be more realistic or more "controlled" within each time step because it is evaluated based on a converged current solution. Therefore, large density differences and large solid stiffness that may result in great discontinuities in stress and force terms can be handled properly.

## 11.2 Formulation

### *11.2.1 The Immersed Finite Element Method*

The IFEM is a finite element based formulation derived from the immersed boundary method. The fluid is assumed to be incompressible and fulfills the entire computational domain, while the solid occupies a volume in the fluid domain and is described with proper constitutive equation. The fact that the solid domain is a separate entity means that it can have a different density and generates different internal stresses in the solid domain that are different from the fluid. In the original derivation of the IFEM, we started with the weak form of the entire system stating that the total work done in both the fluid and the solid must be maintained in equilibrium.

In the IFEM algorithm, the fluid–solid interaction (FSI) force $\mathbf{f}^{FSI,s}$ is computed in the solid domain and then passed onto the fluid domain as an external force. This interaction force is evaluated based on the solid inertial, internal and body forces. However, the existence of the "artificial" fluid in the solid domain must be taken into account and the amount of work it produces must be removed. Thus, the interaction force is a force balance of the solid and its overlapping artificial fluid.

The total work done in the solid domain can be calculated as

$$\underbrace{\int_{\Omega^s} \delta v_i^s \left[ (\rho^s - \rho^f) \frac{dv_i^s}{dt} - (\sigma_{ij,j}^s - \sigma_{ij,j}^f) - f_i^{ext,s} \right] d\Omega}_{\mathbf{f}^{FSI,s}} + \underbrace{\int_{\Omega^s} \delta v_i^s \left( \rho^f \frac{dv_i^s}{dt} - \sigma_{ij,j}^f \right) d\Omega}_{\text{artificial fluid in } \Omega^s} = 0. \tag{11.1}$$

Equation (11.1) contains two terms: the first term is the work done by the solid in the solid domain minus the work done by the artificial fluid. The second term represents the work done by the artificial fluid in the solid domain, which will then be grouped together with the real fluid in $\Omega^f$; together, the artificial and the real fluids are solved together using the Navier–Stokes equations in the entire computational domain $\Omega$. Owing to the fact that the integrations are done in two independent domains (i.e., the solid domain and the entire computational domain filled with fluid), we can further evaluate the strong form and define that the first term in Equation (11.1) be the FSI force $\mathbf{f}^{FSI,s}$:

$$\mathbf{f}^{FSI,s} = -(\rho^s - \rho^f)\ddot{\mathbf{u}}^s + \nabla \cdot \sigma^s - \nabla \cdot \sigma^f + (\rho^s - \rho^f)\mathbf{g} \quad \text{in } \Omega^s, \tag{11.2}$$

where $\rho^s$ and $\rho^f$ are the solid and fluid densities; $\sigma^s$ and $\sigma^f$ are the internal stresses of the solid and fluid, respectively; $\mathbf{u}^s$ is the solid displacement and $\mathbf{g}$ is the body force. The internal stress of the solid $\sigma^s$ is determined by the material constitutive law.

Once $\mathbf{f}^{FSI,s}$ is computed, we may employ an appropriate interpolation procedure to distribute this force from the solid domain onto its surrounding fluid nodes in the computational domain. This distribution of the force can be written as

$$\mathbf{f}^{FSI,f} = \int_{\Omega^s} \mathbf{f}^{FSI,s} \phi(\mathbf{x} - \mathbf{x}^s) \, d\Omega, \tag{11.3}$$

where $\phi(\mathbf{x} - \mathbf{x}^s)$ is the interpolation function, which is a function of the distance between a solid node and its surrounding fluid nodes in its influence domain.

With the appropriate interaction force calculated in the fluid domain, we may solve the fluid using Navier–Stokes equations as the governing equation and the interaction force $\mathbf{f}^{FSI,f}$ as the external force:

$$\nabla \cdot \mathbf{v}^f = 0, \tag{11.4}$$

$$\rho^f(\mathbf{v}^f_{,t} + \mathbf{v}^f \cdot \nabla \mathbf{v}^f) = -\nabla p^f + \mu \nabla^2 \mathbf{v}^f + \mathbf{f}^{FSI,f}, \quad \text{in } \Omega, \tag{11.5}$$

where $\mu$ is the fluid viscosity. The Navier–Stokes equations are solved in the *entire* computational domain, including the volume the solid occupies. Since the effects of the artificial fluid has been accounted for in the calculation of the FSI force, we are *not* double counting that overlapping domain.

Knowing $\mathbf{v}^f$, the solid velocity $\mathbf{v}^s$ is obtained by using the same interpolation function as in Equation (11.3):

$$\mathbf{v}^s = \int_{\Omega} \mathbf{v}^f \phi(\mathbf{x} - \mathbf{x}^s) \, d\Omega. \tag{11.6}$$

Finally, the solid nodal displacement can be updated explicitly in time:

$$\mathbf{u}^{s,n+1} = \mathbf{v}^{s,n+1}\Delta t. \tag{11.7}$$

## 11.2.2 Semi-Implicit Immersed Finite Element Method

Although the force or the work is balanced seamlessly in the strong and weak forms at every time step, the fluid domain is numerically balanced with the FSI force evaluated based on the solid configuration of the *previous* time step.

Performing the Taylor expansion at time step $n$ for $\mathbf{f}^{\mathrm{FSI}}$:

$$\left(\mathbf{f}^{\mathrm{FSI,f}}\right)^{n-1} = \left(\mathbf{f}^{\mathrm{FSI,f}}\right)^{n} - \left(\mathbf{f}^{\mathrm{FSI,f}}_{,t}\right)^{n}\Delta t + O(\Delta t^2). \tag{11.8}$$

The error due to the explicit coupling can be approximated as

$$\mathrm{Error}^{\mathrm{Coupling}} = \frac{1}{\rho^f}\mathbf{f}^{\mathrm{FSI,f}}_{,t}(t)\Delta t + O(\Delta t^2). \tag{11.9}$$

Based on the definition of the FSI force in Equation (11.2), each term on the right-hand side of this FSI force contributes to the accumulative coupling error. These terms are proportional to the density ratio $\rho^s/\rho^f - 1$, the stiffness $K/\rho^f$ and the gravity $(\rho^s/\rho^f - 1)g$, respectively. Here, $K$ is an equivalent Young's modulus of the solid representing the stiffness of the solid material. From this analysis, we can see that if any of these terms are large (i.e., the density ratio or solid stiffness), the resulting error due to the coupling would be large. These large errors often result in instability or divergence of the solution. Therefore, the explicit IFEM algorithm is indeed limited for applications with small density difference between fluid and solid domains and soft solid materials. As a matter of fact, it is the limitation for all non-boundary fitted fluid–structure algorithms when the fluid and solid domains are coupled explicitly. Furthermore, the acceleration and the strain rate of the solid may also have a contribution to the coupling error which makes things much more complicated, especially when we are considering dynamic problems where the solid movement is initiated from a resting position.

Reducing time-step size will always be the natural solution to ensure the stability and accuracy of the coupling between the fluid and solid domains. However, for some cases, the time-step limitation due to the explicit coupling between the fluid and solid domains could easily be several orders of magnitude smaller than the time-step size required by the fluid solver. Such a restriction prohibits the broader range of applications that the algorithm can be applied to because of the impractical required small time-step constraint and encumbered efficiency of the algorithm. Therefore, an improved algorithm is needed to resolve this issue.

The semi-implicit way of computing FSI force should alleviate the restrictions in the time step as well as solution-convergence problem when highly discontinuous material properties are used.

Here, we redefine the interaction force $\mathbf{f}^{\mathrm{FSI,s}}$ in the solid domain, which *only* includes the internal forces for the fluid and solid from the original definition, such that

$$\mathbf{f}^{\mathrm{FSI,s}} = \nabla\cdot\sigma^s - \nabla\cdot\sigma^f \quad \text{in } \Omega^s. \tag{11.10}$$

We then distribute the newly defined $\mathbf{f}^{\mathrm{FSI,s}}$ to the fluid domain using Equation (11.3). The rest of the terms in the original explicit formulation are now incorporated in the fluid equations. The Navier–Stokes equations now must also be redefined as follows:

$$\nabla\cdot\mathbf{v}^f = 0, \tag{11.11}$$

$$\bar{\rho}(\mathbf{v}^f_{,t} + \mathbf{v}^f\cdot\nabla\mathbf{v}^f) = -\nabla p^f + \mu\nabla^2\mathbf{v}^f + \mathbf{f}^{\mathrm{FSI,f}} + \bar{\rho}\mathbf{g} \quad \text{in } \Omega, \tag{11.12}$$

where $\bar{\rho}$ is defined as

$$\bar{\rho} = \rho^{\mathrm{f}} + (\rho^{\mathrm{s}} - \rho^{\mathrm{f}})I(\mathbf{x}). \tag{11.13}$$

Here, the indicator function $I(\mathbf{x})$ is to identify the real fluid region $\Omega^{\mathrm{f}}$, the artificial fluid region or the solid region $\Omega^{\mathrm{s}}$, and the fluid–structure interface $\Gamma^{\mathrm{s}}$, in the computational domain $\Omega$. The value of the indicator function is ranged between zero and one, where it is zero if an entire element belongs to the fluid and one if an entire element belongs to the solid. This newly revised fluid's momentum equation combines the first and the fourth terms in the original FSI force equation.

Comparing with the original IFEM algorithm, the first and the last terms in Equation (11.2) are now been considered in the governing equation and can be evaluated iteratively with the most updated velocity field. This is, therefore, considered as a semi-implicit IFEM algorithm.

In this formulation, the solid internal stress is still evaluated using the velocity from the previous time step because calling the solid solver in every iteration can be extremely time consuming. Even though this term is evaluated explicitly, the semi-implicit scheme still significantly improves the convergence of the solution when the solid material is very stiff.

If we revisit the coupling error equation, Equation (11.9), the magnitude of the coupling error in the internal stress term is proportional to the stiffness ratio $K/\bar{\rho}$ . Noting that, in the semi-implicit form, $\bar{\rho}$ is defined in Equation (11.13), the stiffness ratio here is in fact $K/\rho^{\mathrm{s}}$. For most of the cases, the solid density is larger than the fluid density, which reduces the coupling error compared with $K/\rho^{\mathrm{f}}$ from the original explicit form. Therefore, although the solid internal force is still computed explicitly, the coupling error for the semi-implicit scheme is smaller than the explicit scheme if we consider only the contribution from the solid internal force.

Overall, this semi-implicit scheme relaxes the small time-step requirement and ensures the stability of the FSI force estimation. In particular, this algorithm can handle a much larger range of fluid and solid properties without sacrificing the computational time.

## 11.3 Bio-Medical Applications

The IFEM and semi-implicit IFEM have been used to solve many bio-medical applications [8, 9, 11]. Here, we will show two recent applications: red blood cells (RBC) flowing in bifurcated blood vessels and modeling human vocal folds vibrations during phonation, which is a self-sustained and self-oscillated fluid–structure problem.

### *11.3.1 Red Blood Cell in Bifurcated Vessels*

Understanding the behavior of RBCs flowing in blood vessels, especially when bifurcation happens, is important in estimating the nonuniform hematocrit distribution that would affect the microvasular oxygen distribution, the effective viscosity of blood in microvessels, and the distribution of other metabolites. Learning the behaviors of diseased RBCs that have abnormal rigidity, radius, and shape can be helpful in designing medical therapy. Using the established IFEM method, one can simulate the motion and the deformation of the RBCs within the vessels, and study in detail how the geometry of bifurcation and fluid field affect and direct which daughter branch the RBC flows into.

The geometry of a bifurcated blood vessel is shown in Figure 11.1, where $w_0$ is the diameter of the mother vessel; $w_1$ and $w_2$ are the diameters of the daughter vessels on the top and bottom, respectively; $\beta_1$ and $\beta_2$ are the respective branching angles of the daughter vessels; the branching fillet radii $r_0$, $r_1$, and $r_2$ are given as 3 μm to make the vessel branching transition smooth; $Q_0$, $Q_1$ and $Q_2$ represent the flow rate of each vessel. A RBC is placed near the inlet of the vessel. The radius of the RBC is also given as 2.66 μm. The incoming velocity of the mother vessel is set to be a constant as 0.1 cm/s. The branching

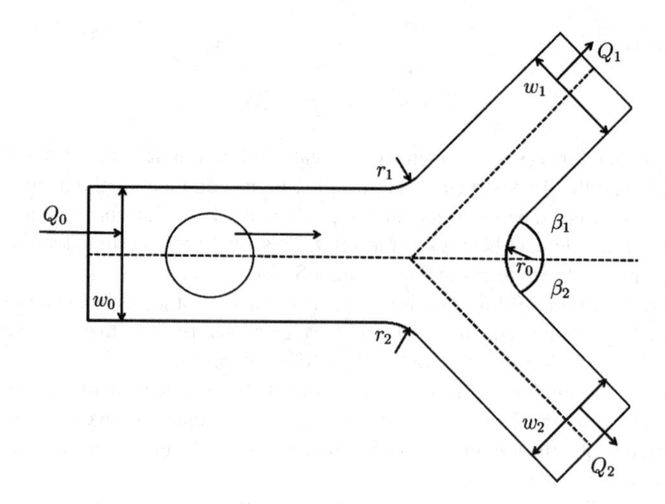


**Figure 11.1** Bifurcation geometry

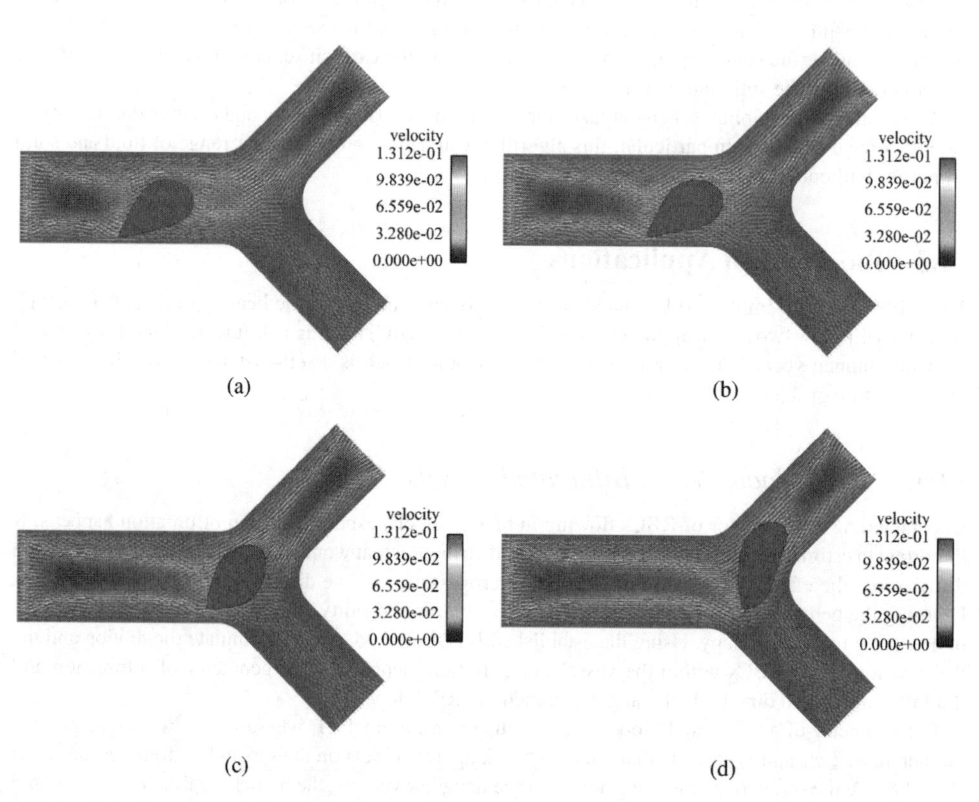


**Figure 11.2** A RBC flowing in a symmetric bifurcated vessel with $Q_1/Q_2 = 3$: (a) $t = 0.008$ s; (b) $t = 0.012$ s; (c) $t = 0.016$ s; (d) $t = 0.020$ s

angles $\beta_1$ and $\beta_2$ are set to be equal and constant, $\beta_1 = \beta_2 = \pi/4$. In this study, we set the diameter of the mother branch to be $w_0 = 8$ μm and consider two sets of diameter ratios, $r_d = w_1/w_2$, to be 1.0 and 1.44. When the diameter ratio is 1.0, it is considered as symmetric bifurcation; when it is not 1.0, then it is considered as asymmetric.

Figures 11.2 and 11.3 represent the blood cell behaviors when encountering bifurcation in symmetric and asymmetric vessels with different original positions and ratios of flow rate of each daughter vessel. Based on Figure 11.2, we note that although the daughter branches are symmetric in the geometry, due to the asymmetric boundary conditions where the ratio of the flow rates for the two daughter branches is $Q_1/Q_2 = 3$, the blood cell tends to move to the daughter branch with a higher mass flow rate. When the daughter branches are asymmetric in geometry but with the same mass flow rate $Q_1/Q_2 = 1$, as shown in Figure 11.3, the blood cell moves to the one with a smaller cross-sectional area, which is due to the higher average velocity in that daughter branch.

A multi-blood cell case is also simulated and the result is shown in Figure 11.4. The cell-to-cell interaction can be thought as two parts: (1) one cell changes the deformation and motion of the other cells by directly contacting each other; (2) one cell affects the other cells indirectly by changing surrounding the fluid field. For this case, instead of specifying the flow rate ratio between daughter vessels, a constant inlet flow rate is given. Therefore, by simulating two RBCs lying in a line profile going through the

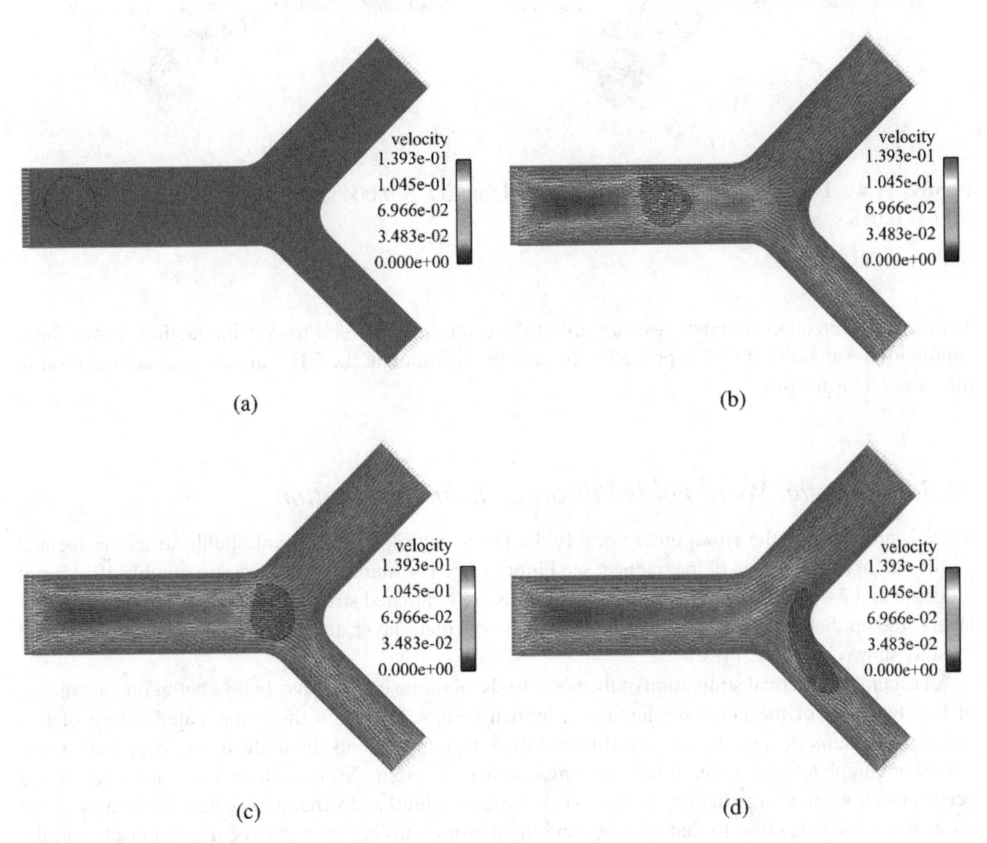


**Figure 11.3** An RBC flowing in an asymmetric bifurcation vessel with $Q_1/Q_2 = 1$: (a) $t = 0.00$ s; (b) $t = 0.01$ s; (c) $t = 0.02$ s; (d) $t = 0.03$ s

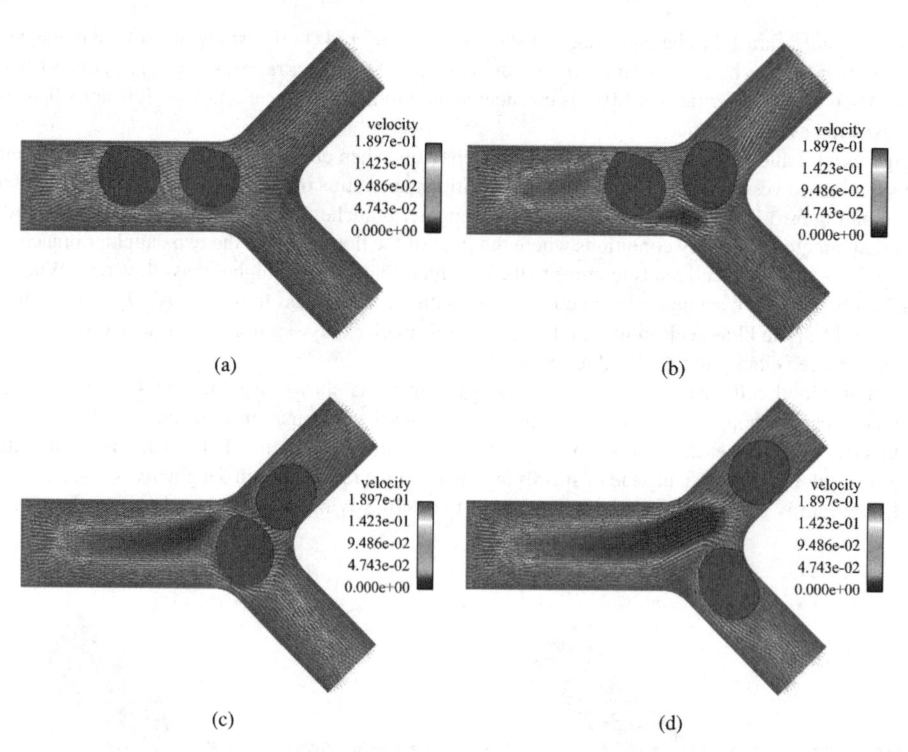


**Figure 11.4** Two RBCs in a symmetric bifurcation: (a) $t = 0.004$ s; (b) $t = 0.008$ s; (c) $t = 0.012$ s; (d) $t = 0.018$ s

symmetric bifurcation together, we are able to see this indirect cell-to-cell interaction. From those simulation results, the IFEM is proved to be suitable to simulate the RBC motion and deformation in microvessel bifurcation.

## *11.3.2 Human Vocal Folds Vibration during Phonation*

Voice is produced by the vibration of vocal folds. The vocal folds are a pair of pliable structures located within the larynx at the top of the trachea; see Figure 11.5. The human vocal folds are roughly 10–15 mm in length and 3–5 mm thick. The human vocal folds are laminated structures composed of five different layers: the epithelium, the superficial layer, the intermediate layer, the deep layer, and thyroarytenoid muscle, shown in Figure 11.5.

An accurate numerical simulation of the vocal folds vibration can help us obtain a better understanding of the dynamics of the voice production in human beings. Owing to the complicated nature of this problem, the numerical model has to fulfill the following requirements. First, the numerical model has to represent completely a coupled fluid–structure interaction system. Second, the numerical model should perform well when a large density ratio exists between the fluid and structure because the density of the vocal fold muscle is close to that of water and the density ratio between the vocal fold muscle and the airflow is about 1000. Third, the motion and deformation of the structure have to be predicted accurately with complicated geometry and material descriptions because the vocal folds have a complex shape and

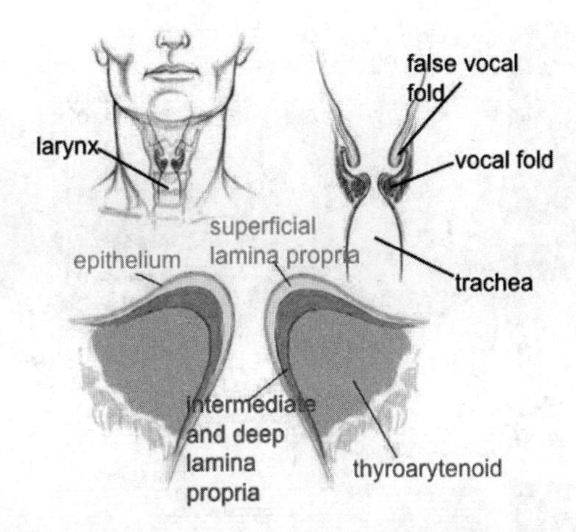


**Figure 11.5** Human vocal folds. (Image is modified from http://haqeeqat.org.pk.)

layer-structures and are viscoelastic materials. The IFEM is a perfect numerical method to perform the simulation of this complex problem.

The geometry of the self-oscillated vocal folds model is shown in Figure 11.6. Since sound is generated by the compression of air, the working fluid is taken as compressible air governed by the ideal gas law at room temperature. The density of the fluid is $\rho^{f} = 1.3 \times 10^{-3}$ g/cm$^3$ and the viscosity of the fluid is $\mu = 1.8 \times 10^{-4}$ g/(cm-s).

The vocal fold muscle is considered an isotropic viscoelastic material. The vocal fold is assumed to have a layered structure: outside cover layer and inside body layer. The cover layer is much softer than the body layer. For the cover layer the Young's modulus is $E = 10$ kPa, whereas $E = 40$ kPa for the body layer. The densities of both the cover and body layers are assumed to be the same: $\rho^{s} = 1.0$ g/cm$^3$. The Poisson ratio is $\nu = 0.3$. Two vocal folds have the exact same geometry and material description, sit in the fluid channel symmetric about the central line. A constant total pressure boundary condition of $P_{in} = 1$ kPa is applied at the channel inlet and the outflow boundary is given at the channel exit. No-slip and no-penetration boundary conditions are applied on the channel walls and on the vocal folds surfaces.

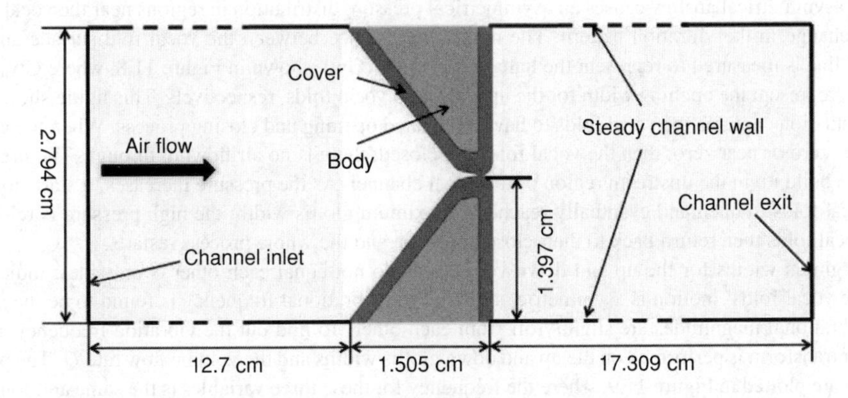


**Figure 11.6** Two-dimensional two-layer self-oscillated vocal folds model

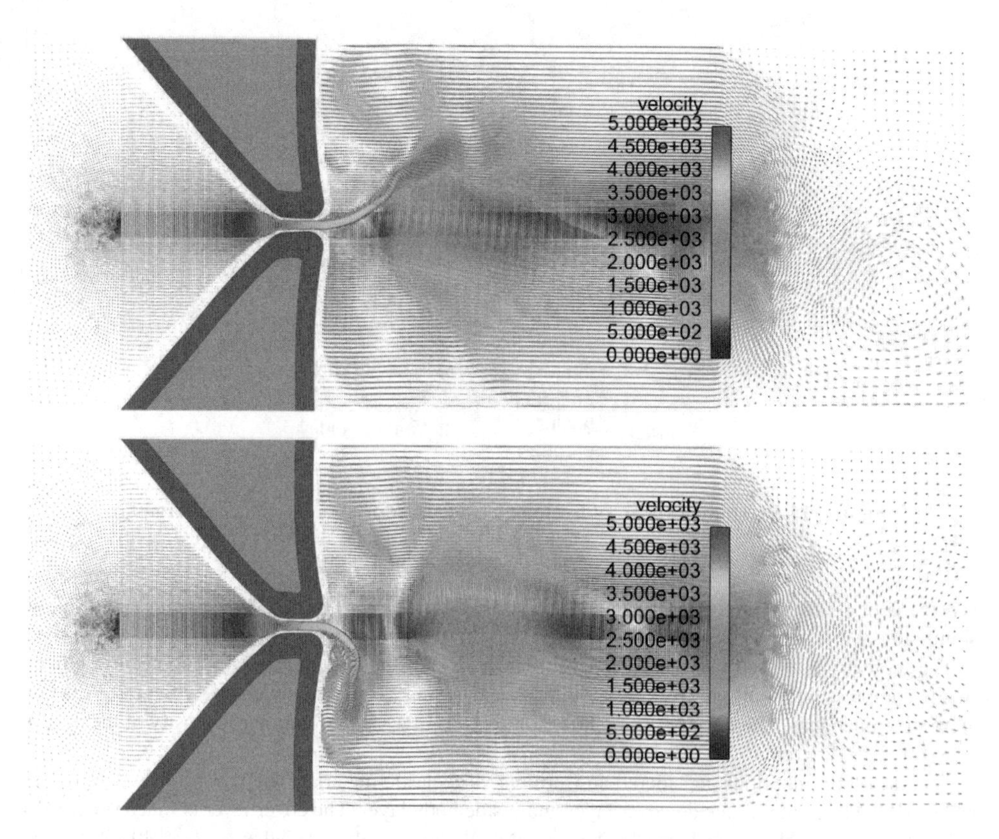


**Figure 11.7** Fluid velocity field at two typical instances during steady vibration

Snapshots of the fluid velocity fields at two typical instances during a steady vibration cycle are shown in Figure 11.7. One can see that the fluid field is not symmetric about the central line during the vibration. The glottal jet tends to attach to one side of the vocal folds randomly, which is the so-called the "Coanda effect" [13, 14].

The asymmetrical airflow causes an asymmetrical pressure distribution in regions near the vocal folds and a change in the vibration pattern. The minimum distance between the vocal fold surface and the central line is measured to represent the half glottis width (Gw), shown in Figure 11.8, where $Gw_{up}$ and $Gw_{down}$ represent the opening width for the up and down vocal folds, respectively. This figure shows that the simulation captures the vocal folds to have a repeated opening and closing process. When the glottis width is zero or near zero, then the vocal folds are closed; there is no air flowing through. The pressure starts to build up in the upstream region of the vocal channel. As the pressure increases, it starts to push the vocal folds to open and eventually reaches a maximum glottis width; the high pressure is released. The vocal folds then return back to their closed position and the whole process restarts.

The glottis widths for the up and down vocal folds do not equal each other over cycles, indicating that the vocal folds' motion is asymmetric, although the vibrational frequency is found to be the same. The vibrational magnitudes are slightly off from each other. To find out the vibration frequency, a fast Fourier transform is performed on the up and down glottis widths and the volume flow rate $Q$. The power spectra are plotted in Figure 11.9, where the frequency for these three variables is the same and found to be 234 Hz, which is in the expected range of a female vocal fold vibrational frequency.

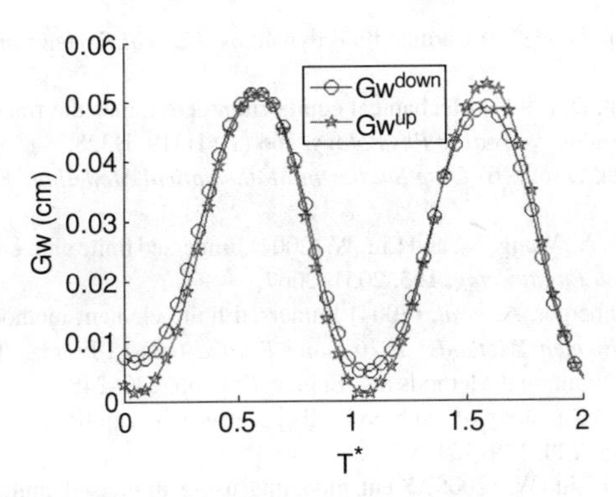


**Figure 11.8** Half glottis width of top and bottom vocal folds

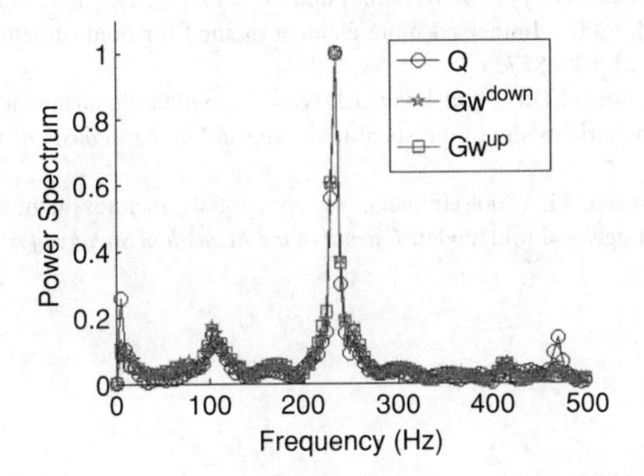


**Figure 11.9** Spectrum plot of flow rate and half glottis width of the top and bottom vocal folds

## 11.4 Conclusions

In this study we showed the recent advancements in the IFEM, from an explicit to semi-implicit fully coupled approach to study fluid–structure interactions. We show that semi-implicit approach can alleviate the time-step constraint and the density ratio constraint that are faced in the original explicit IFEM approach. Two particular biomedical applications are shown. These two examples demonstrate that the IFEM is a powerful technique that can produce realistic and accurate results. The results show that there are no numerical convergence issues when dealing with large disparities in material properties between the fluid and the solid structures.

## References


[1] Hughes, T., Liu, W., and Zimmermann, T. (1981) Lagrangian–Eulerian finite element formulation for incompressible viscous flows. *Computer Methods in Applied Mechanics and Engineering*, **29**, 329–349.

[2] Liu, W., Belytschko, T., and Chang, H. (1986) An arbitrary Lagrangian–Eulerian finite element method for path-dependent materials. *Computer Methods in Applied Mechanics and Engineering*, **58**, 227–245.

[3] Peskin, C. and McQueen, D. (1989) A three-dimensional computational method for blood flow in the heart. I. Immersed elastic fibers in a viscous incompressible fluid. *Journal of Computational Physics*, **81** (2), 372–405.

[4] Peskin, C. and McQueen, D. (1992) Cardiac fluid dynamics. *Critical Reviews in Biomedical Engineering*, **20** (6), 451–459.

[5] Peskin, C. and McQueen, D. (1994) Mechanical equilibrium determines the fractal fiber architecture of aortic heart valve leaflets. *American Journal of Physiology*, **266** (1), H319–H328.

[6] Peskin, C. and McQueen, D. (1996) *Case Studies in Mathematical Modeling: Ecology, Physiology, and Cell Biology*, Prentice-Hall.

[7] Zhang, L., Gerstenberger, A., Wang, X., and Liu, W. (2004) Immersed finite element method. *Computer Methods in Applied Mechanics and Engineering*, **193**, 2051–2067.

[8] Liu, W., Liu, Y., Gerstenberger, A. *et al.* (2004) Immersed finite element method and applications to biological systems, in *Finite Element Methods: 1970's and Beyond* (eds L. Franca, T. Tezduyar, and A. Masud), International Center for Numerical Methods and Engineering, pp. 233–248.

[9] Liu, Y. and Liu, W. (2006) Rheology of red blood cell aggregation in capillary by computer simulation. *Journal of Computational Physics*, **220**, 139–154.

[10] Gay, M., Zhang, L., and Liu, W. (2006) Stent modeling using immersed finite element method. *Computer Methods in Applied Mechanics and Engineering*, **195**, 4358–4370.

[11] Liu, W., Liu, Y., Farrell, D. *et al.* (2006) Immersed finite element method and its applications to biological systems. *Computer Methods in Applied Mechanics and Engineering*, **195**, 1722–1749.

[12] Zhang, L. and Gay, M. (2007) Immersed finite element method for fluid–structuure interactions. *Journal of Fluids and Structures*, **23**, 839–857.

[13] Tao, C., Zhang, Y., Hottinger, D.G., and Jiang, J.J. (2007) Asymmetric airflow and vibration induced by the Coanda effect in a symmetric model of the vocal folds. *Journal of the Acoustical Society of America*, **122** (4), 2270–2278.

[14] Drechsel, J.S. and Thomson, S.L. (2008) Influence of supraglottal structures on the glottal jet exiting a two-layer syngthetic, self-oscillating vocal fold model. *Journal of the Acoustical Society of America*, **123** (6), 4434–4445.

# 12

# Immersed Methods for Compressible Fluid–Solid Interactions

Xiaodong Sheldon Wang
*College of Science and Mathematics, Midwestern State University, USA*

## 12.1 Background and Objectives

Fluid–solid interaction (FSI) plays a very important role in many engineering problems. In the past few decades, numerous research efforts have been directed to method development for modeling of FSI systems [1–4]. Traditionally, staggered iterations are used to link available finite-element codes for solids and structures with computational fluid dynamics codes [5, 6]. Although this iterative procedure is convenient, complex dynamical system behaviors often get lost in the process [7, 8]. In general, mesh updating and remeshing processes such as the arbitrary Lagrangian–Eulerian (ALE) are computationally intensive and often topologically challenging [9–13]. Procedures such as immersed interface methods [14, 15], level set methods [16, 17], front tracking methods [18–20], and other finite-element- and finite-difference-based techniques [21–25] have also been proposed to trace moving interfaces and to eliminate numerical problems associated with large motions of immersed objects.

With the development of computer hardware and software, it is now becoming feasible to employ so-called monolithic approaches to solve FSI systems simultaneously as a whole. As a consequence, reduced-order modeling techniques can be implemented to derive the dominant coupled FSI models and allow more effective and efficient exploration of phase and parametric spaces.

Similar to the fictitious domain method and the extended finite-element method, immersed methods provide a very direct coupling between immersed deformable solids and the surrounding fluid [26–28]. In fact, the same principle can also be used to handle deformable solids immersed in another deformable solid medium.

The immersed boundary method was originally developed by Peskin for the computation of blood flows interacting with heart valves [29–31]. Since its inception, the immersed boundary method has been

*Multiscale Simulations and Mechanics of Biological Materials*, First Edition. Edited by Shaofan Li and Dong Qian.

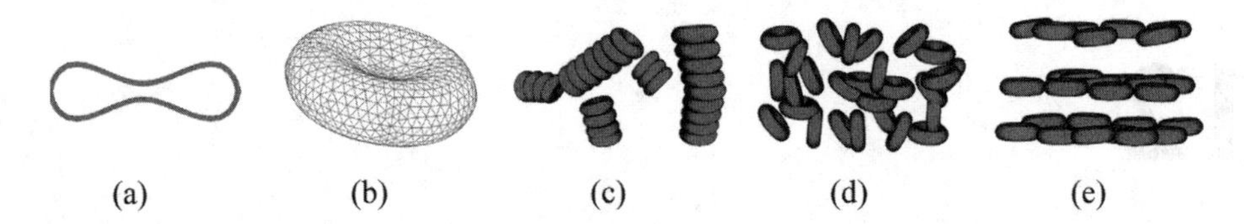

(a) (b) (c) (d) (e)

**Figure 12.1** A three-dimensional finite-element mesh of a single RBC model, and blood microscopic changes under different shear rates: (a) section; (b) surface; (c) low; (d) mid; (e) high

extended to various problems, including swimming motions of marine worms [32], wood pulp fiber dynamics [33], wave propagation in cochleae [34], and biofilm processes [35].

The initial application of immersed methods is for very flexible structures interacting with surrounding viscous fluid. In current versions of immersed methods, complex nonlinear structures or solids can be represented by fiber/beam networks or continuum, often modeled with various finite-element formulations. Preliminary results of the implicit immersed method for compressible media immersed in another compressible medium have also been reported.

In finite-element formulations for immersed methods [36–39], various meshless kernels or radial-based interpolation functions are employed for unstructured background meshes [40]. This mesh-free delta function provides not only a higher order smoothness, but also the ability to handle nonuniform fluid grids. It has been pointed out and confirmed mathematically that, as long as the power input to the surrounding fluid is preserved, the communication between fluid and solid domains can be simply accomplished by finite-element interpolation functions within each element [41–43]. A wonderful comparison of both implicit and explicit approaches for material point methods has also been presented by Sulsky and Kaul [44]. Similar formulations have already been extended to electrokinetics and mass and heat transfers [45–47].

For illustration purpose, a few recent results produced with various immersed methods are included in this chapter [37, 46, 48, 49]; for example, Figure 12.1 shows multiple immersed deformable cells and Figures 12.2 and 12.3 show a study of cell–cell interaction mechanisms in an aqueous environment. In early versions of immersed boundary methods, immersed elastic fibers were employed to construct various objects. As shown in Figure 12.4, torsion, bending, and transverse shear effects have also been added to immersed co-dimension-one structures, namely beams instead of fibers. In addition to the comparison with other numerical or analytical solutions, experimental evidence has also been documented. For instance, through the collaboration with ABIOMED, one experimental validation, as illustrated in Figure 12.5, was performed with the same pulsatile flow and structural conditions as used in the simulation.

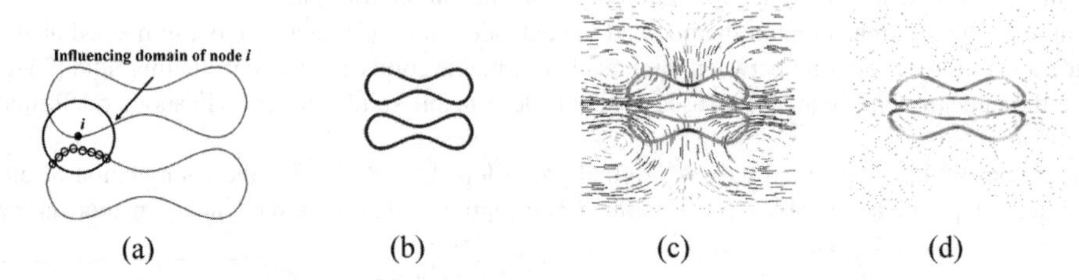


(a) (b) (c) (d)

**Figure 12.2** Illustrations of the initial configuration of the two-RBC model, the local cell–cell contacts, and the deformation in the quiescent fluid under the influence of cell–cell interaction forces: (a) interaction; (b) initial; (c) deformability; (d) equilibrium

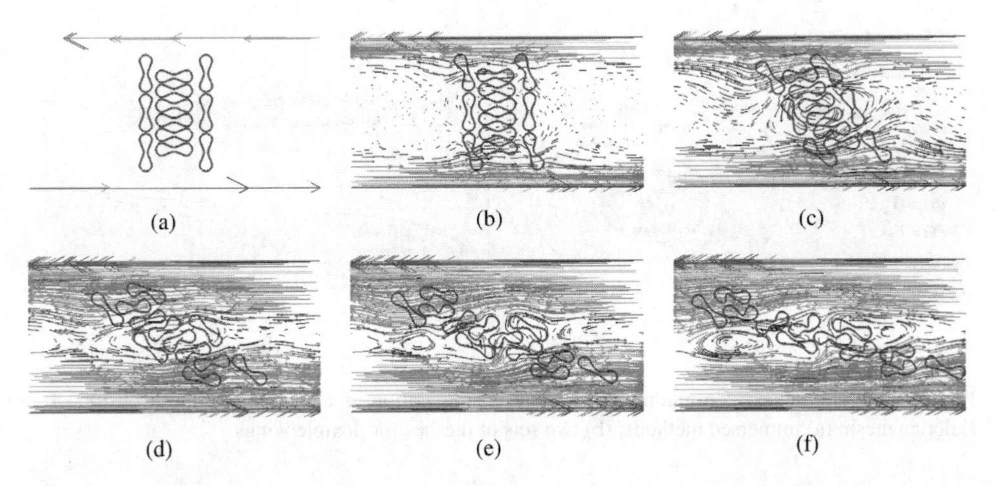


**Figure 12.3** The shear of a multiple-cell aggregate model with a shear rate of 3.0 $s^{-1}$ at (a) $t = 0$, (b) 0.6, (c) 1.2, (d) 1.8, (e) 2.4, and (f) 3 s

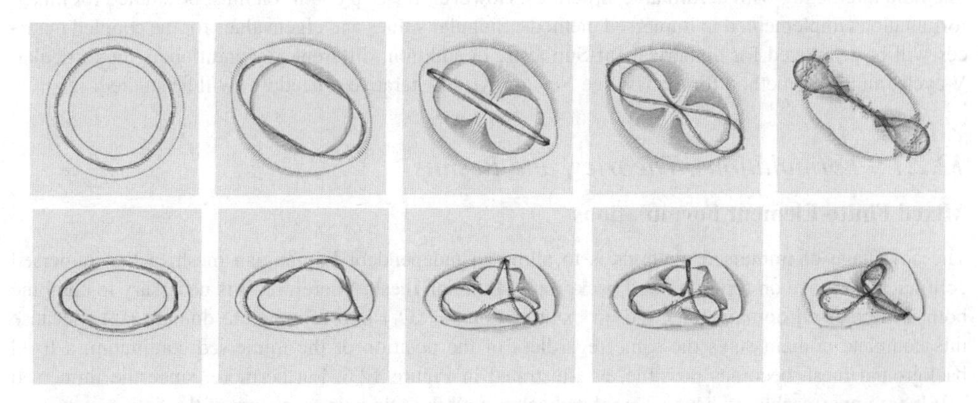

**Figure 12.4** Snapshots showing the ring and vectors of the triads together with fluid markers

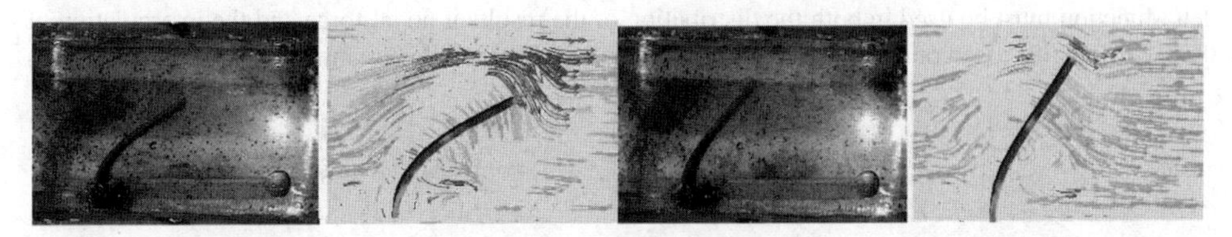

**Figure 12.5** A preliminary comparison between experimental observation and computer simulation of a three-dimensional shell-like structure deflecting in a pulsatile flow through a square cross-sectional channel

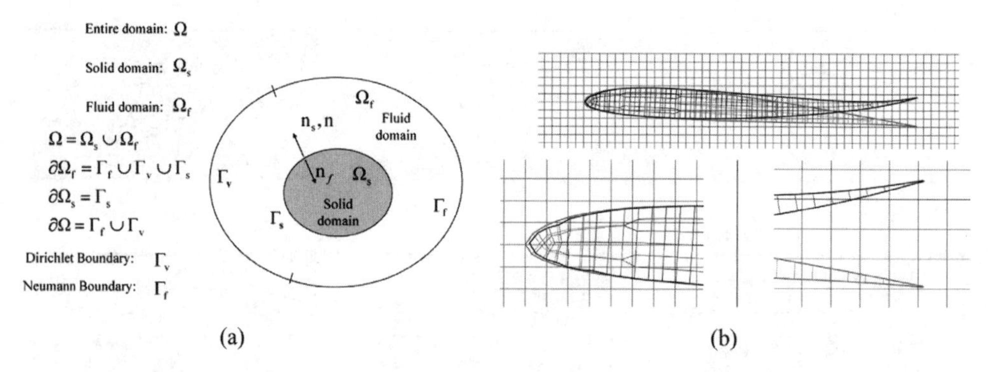


**Figure 12.6** Immersed continuum concept and an illustration of a Lagrangian mesh on top of an Eulerian mesh: (a) immersed methods; (b) two sets of meshes for flexible wings

## 12.2 Results and Challenges

Immersed methods are applicable for compressible FSI systems ranging from acoustic FSIs to compressible fluid interacting with deformable structures. However, Inf-Sup condition must be studied for mixed formulations implemented in immersed methods. Singular values and eigenvalues for the coupled matrices will be employed for numerical Inf-Sup tests. In addition, different preconditioners, in particular, V-cycle, multigrid method for matrix-free Newton-Krylov iterative procedures will be studied.

### *12.2.1 Formulations, Theories, and Results*

#### Mixed Finite-Element Formulations

The advantage of immersed methods is to allow an independent Lagrangian mesh of the immersed continuum to move on top of a fixed background Eulerian mesh. Therefore, it is necessary to combine both an immersed domain ($\Omega_s$) and an exterior domain ($\Omega_f$) into a complete domain ($\Omega$). Because this complete domain stays the same regardless of the position of the immersed continuum, a fixed background mesh becomes possible, as illustrated in Figure 12.6. Furthermore, since the immersed solids are impermeable, to have a complete background fluid domain, concepts of the fictitious domain method [28, 50, 51], as illustrated in Figure 12.7, should be employed.

The key to immersed methods is energy conservation; namely, the energy input to the fluid domain from the immersed solid/structure is the same as that from the equivalent body force. Moreover, the same delta function must be used in both the distribution of the resultant nodal force and the interpolation of the solid velocity based on the surrounding fluid velocities. In fact, this treatment in immersed methods

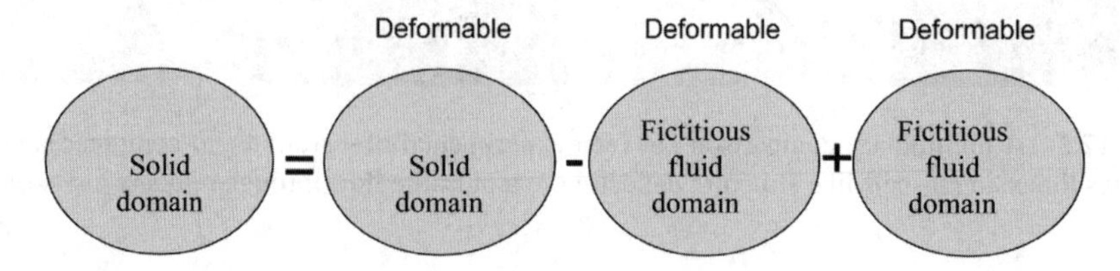


**Figure 12.7** The concept of a fictitious fluid domain versus immersed domain

can be viewed as the synchronization of the fluid motion with the solid motion within the immersed solid domain $\Omega_s$; namely, $\mathbf{v}^s = \mathbf{v}^f$.

This constraint introduces the (distributed) Lagrange multiplier as the equivalent body force. In this case, the equivalent body forces can be directly calculated along with independent fluid and solid velocity vectors [52]. Naturally, if the surrounding fluid is viscous and incompressible, the immersed solid must be incompressible in immersed methods [53–56]. In this study, we consider compressible solids immersed in an compressible fluid medium.

The weak form of the fluid–solid system can be written as

$$\forall \mathbf{w} \in [H^1_{0,\Gamma_v}(\Omega)]^d$$

$$\int_\Omega [w_i \rho_f(\dot{v}_i - g_i) + w_{i,j}\sigma_{ij}]\,\mathrm{d}\Omega - \int_{\Gamma_f} w_i f_i^{\Gamma_f}\,\mathrm{d}\Gamma - \int_{\Omega_s} w_i^s f_i^s\,\mathrm{d}\Omega = 0, \tag{12.1}$$

with

$$\int_{\Omega_s} w_i^s f_i^s\,\mathrm{d}\Omega = -\int_{\Omega_s} [w_i(\rho_s - \rho_f)(\dot{v}_i - g_i) + w_{i,j}(\sigma_{ij}^s - \sigma_{ij}^f)]\,\mathrm{d}\Omega. \tag{12.2}$$

In the Newtonian fluid model, the pressure $p$ is subtracted from the stress components $\sigma_{ij}$ to obtain the deviatoric stress components $\tau_{ij}$:

$$\sigma_{ij} = -p\delta_{ij} + \tau_{ij}, \tag{12.3}$$

with $\tau_{ij} = \mu(v_{j,i} + v_{i,j})$. Moreover, the continuity equation of the compressible viscous fluid is expressed as

$$v_{i,i} + \frac{\dot{p}}{\kappa} = 0, \tag{12.4}$$

where $\kappa$ is the bulk modulus of the fluid.

For typical fluids, the change of the fluid density can be easily updated with the pressure change through the state equation. Likewise, the solid stress tensor is decomposed as a hydrostatic pressure $p^s$ and a deviatoric stress tensor $\tau_{ij}^s$:

$$\sigma_{ij}^s = -p^s\delta_{ij} + \tau_{ij}^s. \tag{12.5}$$

In nonlinear mechanics, the solid deformation gradient

$$F_{ij} = \frac{\partial x_i^s(t)}{\partial x_j^s(0)}$$

has to be introduced, from which the Green–Lagrangian strain $\epsilon_{ij}$ can be derived. Consequently, the energy conjugate stress $S_{ij}$, namely the second Piola–Kirchhoff stress, can be derived from the elastic energy $\bar{W}$, which is often related to the three invariants of the Cauchy–Green deformation tensor $\mathbf{C}$ defined as $\mathbf{F}^T\mathbf{F}$. To better handle the compressibility issues, the additional term $-[p^s + \kappa^s(J_3 - 1)]^2/2\kappa^s$ is added to the elastic energy $\bar{W}$ and the solid unknown pressure $p^s$ is introduced as a separate unknown, matching the treatment of the fluid, in the continuity equation:

$$J_3 - 1 + \frac{p^s}{\kappa^s} = 0, \tag{12.6}$$

where $\kappa^s$ is the solid bulk modulus and $J_3$ stands for the determinant of the deformation gradient.

Of course, to relate to the expression in Equation (12.5), the solid Cauchy stress is converted from the second Piola–Kirchhoff stress:

$$\sigma_{ij}^{\mathrm{s}} = \frac{1}{\det(\mathbf{F})} F_{i,m} S_{mn} F_{j,n}. \tag{12.7}$$

Since the solid displacement is dependent on the fluid velocity, the primary unknowns for the coupled fluid–solid system are the fluid velocity $\mathbf{v}$, the fluid pressure $p$, and the solid pressure $p^{\mathrm{s}}$. Employing the Sobolev spaces, the weak form of governing equations can be expressed as $\forall q \in L^2(\Omega)$, $q^{\mathrm{s}} \in L^2(\Omega_{\mathrm{s}})$, $\mathbf{w} \in [H^1_{0,\Gamma_v}(\Omega)]^d$, which includes $\forall \mathbf{w}^{\mathrm{s}} \in [H^1(\Omega_{\mathrm{s}})]^d$, and find $\mathbf{v}$ and $p \in \Omega$, $p^{\mathrm{s}} \in \Omega_{\mathrm{s}}$, such that

$$\begin{aligned}
&\int_{\Omega} w_i \rho(\dot{v}_i - g_i)\,\mathrm{d}\Omega + \int_{\Omega} (w_{i,j}\tau_{ij} - p w_{i,i})\,\mathrm{d}\Omega - \int_{\Gamma_{\mathrm{f}}} w_i f_i^{\Gamma_{\mathrm{f}}}\,\mathrm{d}\Gamma \\
&+ \int_{\Omega_{\mathrm{s}}} [w_i^{\mathrm{s}}(\rho_{\mathrm{s}} - \rho)(\dot{v}_i - g_i) + w_{i,j}^{\mathrm{s}}(\tau_{ij}^{\mathrm{s}} - \tau_{ij}^{\mathrm{f}}) - (p^{\mathrm{s}} - p)w_{i,i}^{\mathrm{s}}]\,\mathrm{d}\Omega \\
&+ \int_{\Omega} q\left(v_{j,j} + \frac{p_{,t}}{\kappa}\right)\mathrm{d}\Omega + \int_{\Omega_{\mathrm{s}}} q^{\mathrm{s}}\left(J_3 - 1 + \frac{p^{\mathrm{s}}}{\kappa^{\mathrm{s}}}\right)\mathrm{d}\Omega = 0.
\end{aligned} \tag{12.8}$$

For the fluid domain, the following interpolations for the entire domain $\Omega$ are introduced:

$$\mathbf{v}^h = N_I^v \mathbf{v}_I, \quad \mathbf{w}^h = N_I^v \mathbf{w}_I, \quad p^h = N_I^p p_I, \quad q^h = N_I^p q_I, \tag{12.9}$$

where $N_I^v$ and $N_I^p$ stand for the interpolation functions at node $I$ for the velocity vector and the pressure; and $\mathbf{v}_I$, $\mathbf{w}_I$, $p_I$, and $q_I$ are the nodal values of the discretized velocity vector, admissible velocity variation, pressure, and pressure variation, respectively.

Notice that, in general, the interpolation functions for the velocity vector unknowns denoted by the superscript $v$ and the pressure unknowns denoted by $p$ are different. Likewise for the solid domain $\Omega_{\mathrm{s}}$, the discretization is based on the following:

$$\mathbf{u}^{\mathrm{s},h} = N_J^u \mathbf{u}_J^{\mathrm{s}}, \quad \mathbf{w}^{\mathrm{s},h} = N_J^u \mathbf{w}_J^{\mathrm{s}}, \quad p^{\mathrm{s},h} = N_J^{p^{\mathrm{s}}} p_J^{\mathrm{s}}, \quad q^{\mathrm{s},h} = N_J^{p^{\mathrm{s}}} q_J^{\mathrm{s}}, \tag{12.10}$$

where $N_J^u$ and $N_J^{p^{\mathrm{s}}}$ stand for the interpolation functions at node J for the displacement vector and the pressure unknowns; and $\mathbf{u}_J^{\mathrm{s},h}$, $\mathbf{w}_J^{\mathrm{s},h}$, $p_J^{\mathrm{s},h}$, and $q_J^{\mathrm{s},h}$ are the nodal values of the discretized displacement vector, admissible velocity variation, pressure, and pressure variation, respectively.

Substituting both discretizations (12.9) and (12.10) into Equation (12.8), we obtain the following discretized weak form: $\forall q^h \in L^2(\Omega^h)$, $q^{\mathrm{s},h} \in L^2(\Omega_{\mathrm{s}}^h)$, $\mathbf{w}^h \in [H^{1,h}_{0,\Gamma_v^h}(\Omega^h)]^d$, which includes $\forall \mathbf{w}^{\mathrm{s},h} \in [H^{1,h}(\Omega_{\mathrm{s}}^h)]^d$,

$$\begin{aligned}
&\int_{\Omega^h} w_{iI} N_I^v \rho \dot{v}_i^h\,\mathrm{d}\Omega - \int_{\Gamma_{\mathrm{f}}^h} w_{iI} N_I^v f_i^{\Gamma_{\mathrm{f}}^h}\,\mathrm{d}\Gamma + \int_{\Omega^h} (w_{iI} N_{I,j}^v \tau_{ij} - p^h w_{iI} N_{I,i}^v)\,\mathrm{d}\Omega \\
&+ \int_{\Omega_{\mathrm{s}}^h} [w_{iJ}^{\mathrm{s}} N_J^u (\rho_{\mathrm{s}} - \rho)(\dot{v}_i^h - g_i) + w_{iJ}^{\mathrm{s}} N_{J,i}^u (\sigma_{ij}^{\mathrm{s}} - \sigma_{ij}^{\mathrm{f}})]\,\mathrm{d}\Omega \\
&- \int_{\Omega^h} w_{iI} N_I^v \rho g_i\,\mathrm{d}\Omega + \int_{\Omega^h} q_I N_I^p \left(v_{j,j}^h + \frac{p_{,t}^h}{\kappa}\right)\mathrm{d}\Omega \\
&+ \int_{\Omega_{\mathrm{s}}^h} q_J^{\mathrm{s}} N_J^{p^{\mathrm{s}}} \left(J_3 - 1 + \frac{p^{\mathrm{s},h}}{\kappa^{\mathrm{s}}}\right)\mathrm{d}\Omega = 0.
\end{aligned} \tag{12.11}$$

Therefore, for the entire domain $\Omega$, due to the arbitrariness of the variations $w_{iI}$, $q_I$, and $q_J^s$, four equations at each fluid node denoted as $I$ and one equation at each solid node denoted as $J$ are developed:

$$r_{iI}^v = 0, \quad r_I^p = 0, \quad r_J^{p^s} = 0, \tag{12.12}$$

where the residuals are defined as

$$\begin{aligned} r_{iI}^v &= \int_{\Omega^h} N_I^v \rho \dot{v}_i^h \, d\Omega + \int_{\Omega^h} [N_{I,j}^v \tau_{ij} - p^h N_{I,i}^v] \, d\Omega - \int_{\Gamma_f^h} N_I^{v,\Gamma_f^h} f_i^{\Gamma_f^h} \, d\Gamma \\ &\quad + \int_{\Omega_s^h} \tilde{N}[N_J^u(\rho_s - \rho)(\dot{v}_i^h - g_i) + N_{J,i}^u(\sigma_{ij}^s - \sigma_{ij}^f)] \, d\Omega - \int_{\Omega^h} N_I^v \rho g_i \, d\Omega, \\ r_I^p &= \int_{\Omega^h} N_I^p \left( v_{j,j}^h + \frac{p_{,t}^h}{\kappa} \right) d\Omega, \\ r_J^{p^s} &= \int_{\Omega_s^h} N_J^{p^s} \left( J_3 - 1 + \frac{p^{s,h}}{\kappa^s} \right) d\Omega. \end{aligned} \tag{12.13}$$

Note that the nonlinear mapping $\tilde{N}$ from $\mathbf{w}_I$ to $\mathbf{w}_J^s$, namely from the fluid mesh to the solid mesh, depends on the choice of kernel functions, which will be discussed in a separate section.

This isentropic compressible fluid formulation should yield the same results to mixed finite element formulations for acoustic fluid-solid interactions [57, 58]. With sufficiently small time steps, pressure wave propagation and related scattering and radiation phenomena will be captured. The traditional assumption for acoustic FSI systems is small strain and small displacement, namely, the FSI interface does not move at all. It seems that immersed methods bear no advantages over traditional mixed finite element formulations or potential formulations [4, 38, 39, 59]. However, acoustic fluid-solid interaction models provide a perfect venue to validate and to improve immersed methods.

In Figures 12.8, 12.9, and 12.10, typical spurious pressure modes and normal pressure distributions are depicted for couple examples. Naturally, in this work, we must also pay attention to spurious pressure modes. As depicted in Figure 12.11, from the preliminary study of a driven cavity problem with five compressible solids immersed in a compressible fluid, the results are encouraging.

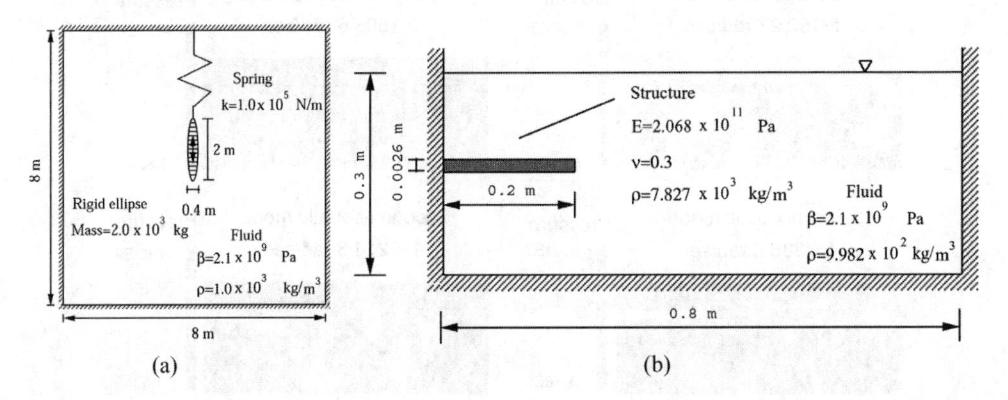


**Figure 12.8** Acoustic FSI systems [4]: (a) immersed ellipse; (b) immersed cantilever beam, reprinted with permission from Elsevier © 2008

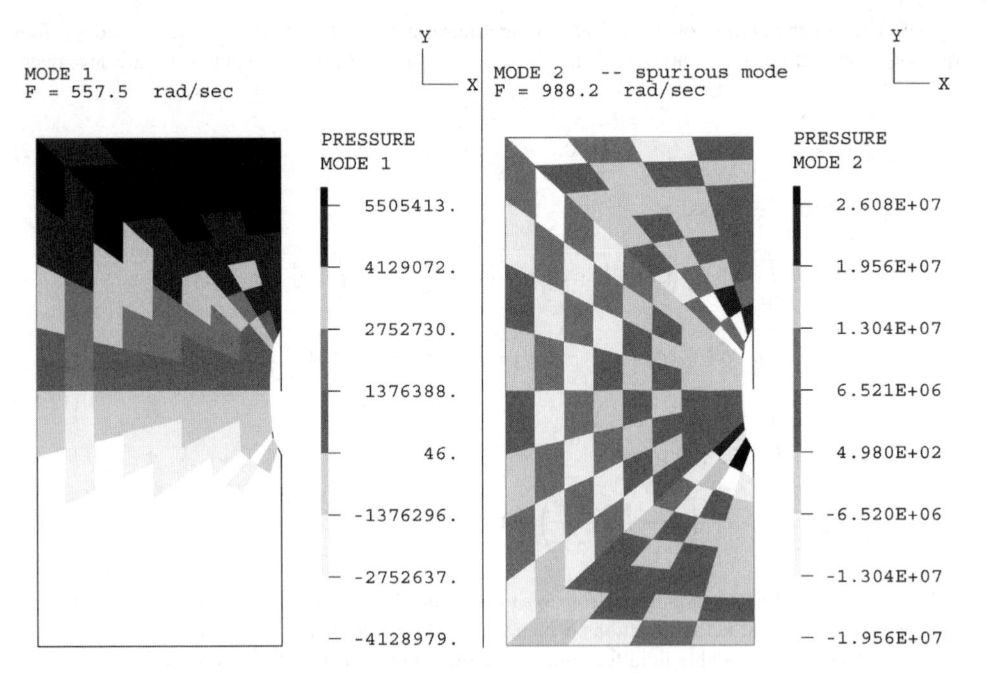


**Figure 12.9** Typical spurious pressure modes in acoustic FSI systems [57], reprinted with permission from John Wiley & Sons © 1997

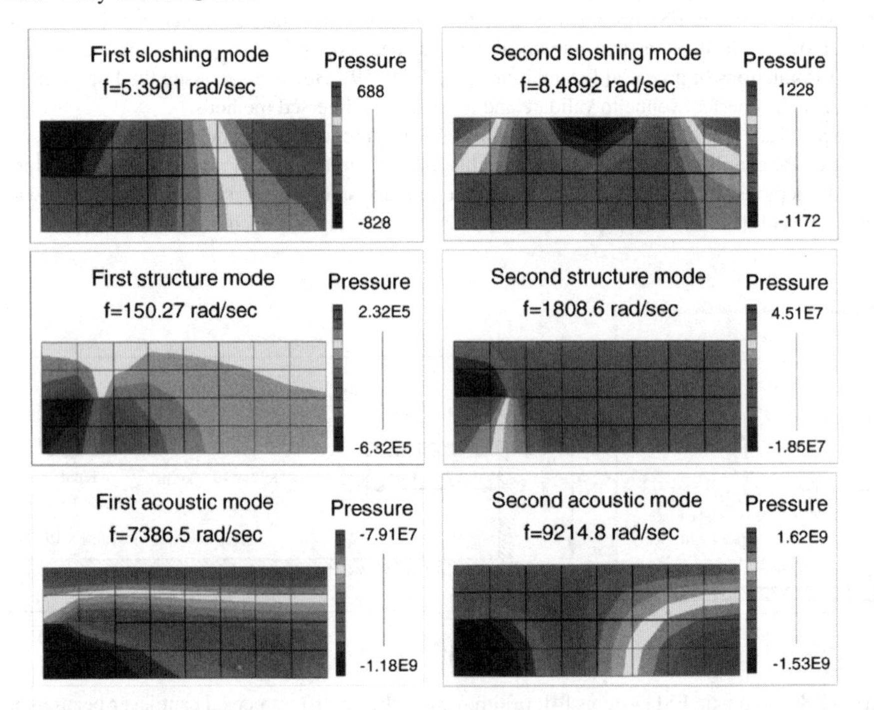


**Figure 12.10** Pressure distribution of an acoustic FSI system [4], reprinted with permission from Elsevier © 2008

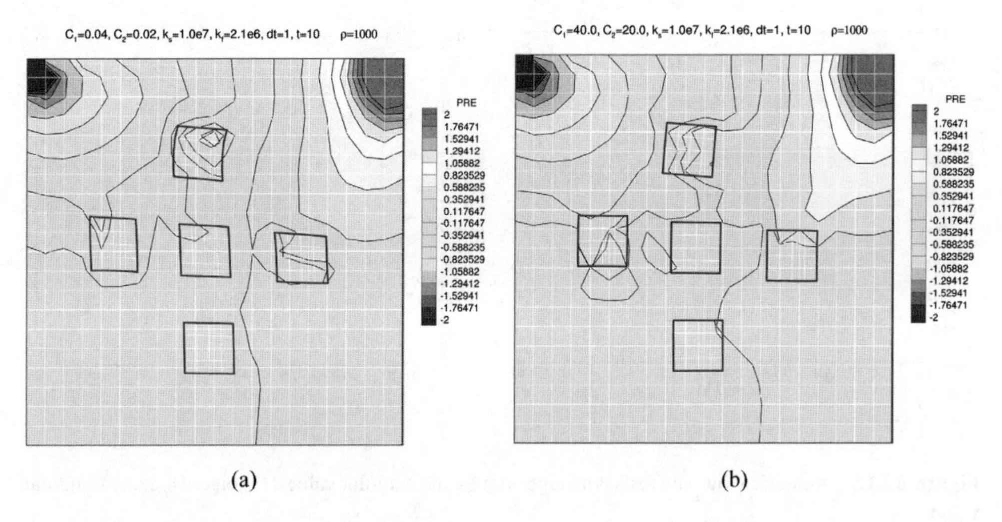


**Figure 12.11** Pressure distribution snapshots ($t = 10$) of an FSI model with five compressible solids immersed in a cavity with compressible fluid: (a) soft; (b) hard

## 12.2.2 *Stability Analysis*

For step-by-step linearized transient analysis, after transformation, the following equation is considered for every time step:

$$\begin{pmatrix} (\mathbf{K}^*_{uu})_h & (\mathbf{K}_{us})_h \\ (\mathbf{K}_{us})^{\mathrm{T}}_h & (\mathbf{K}_{ss})_h \end{pmatrix} \begin{pmatrix} \bar{\mathbf{U}}_h \\ \bar{\mathbf{S}}_h \end{pmatrix} = \begin{pmatrix} \mathbf{R}^*_h \\ \mathbf{0} \end{pmatrix}, \tag{12.14}$$

where $\mathbf{R}^*_h$ is an effective load vector.

The stability and corresponding mixed elements employed in this study are verified through ellipticity and inf–sup conditions, which are necessary and sufficient conditions for well-posedness:

**Ellipticity condition**

$$\bar{\mathbf{V}}^{\mathrm{T}}_h(\mathbf{K}_{uu})^*_h\bar{\mathbf{V}}_h \geq c_1\|\bar{\mathbf{V}}_h\|^2_V \ \forall\ \bar{\mathbf{V}}_h \in \ker[(\mathbf{K}_{us})^{\mathrm{T}}_h], \tag{12.15}$$

where $c_1 > 0$,

$$\|v\|^2_V = \sum_{i,j}\left\|\frac{\partial v_i}{\partial x_j}\right\|^2_{L^2(\mathrm{Vol})},$$

and

$$\ker[(\mathbf{K}_{us})^{\mathrm{T}}_h] = \{\bar{\mathbf{V}}_h | \bar{\mathbf{V}}_h \in R^n,\ (\mathbf{K}_{us})^{\mathrm{T}}_h\bar{\mathbf{V}}_h = 0\}.$$

**Inf–sup condition**

$$\inf_{\bar{\mathbf{S}}_h}\sup_{\bar{\mathbf{U}}_h}\frac{\bar{\mathbf{U}}^{\mathrm{T}}_h(\mathbf{K}_{us})_h\bar{\mathbf{S}}_h}{\|\bar{\mathbf{U}}_h\|\|\bar{\mathbf{S}}_h\|} \geq c_2 > 0, \tag{12.16}$$

where the constant $c_2$ is independent of the mesh size $h$ and the bulk modulus $\beta$.

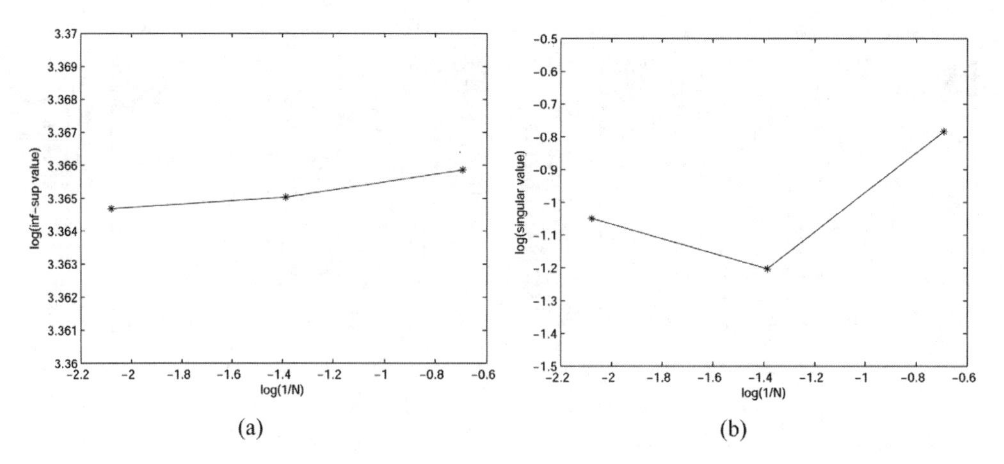


**Figure 12.12** Numerical inf–sup tests with eigenvalues and singular values: (a) eigenvalue; (b) singular value

Although the inf–sup condition for mixed finite-element formulations is elegant and useful, the analytical proof for specific elements and systems can be very difficult. In practice, numerical inf–sup tests using eigenvalues and singular values associated with $\mathbf{K}_{us}$ are often employed [4]. A preliminary set of results is documented in Figure 12.12. It is clear that the lowest singular value of $\mathbf{K}_{us}$ yields the same prediction as the lowest nonzero eigenvalue, as proposed by Bathe [60].

## *12.2.3 Kernel Functions*

The reproducing kernel particle method (RKPM) was originally proposed for meshless finite-element methods [40, 61–64]. The properties of modified window functions can significantly improve the computational efficiency, in particular for problems involving large deformations.

In immersed methods, the key aspect is to employ the same discretized delta function for the distribution of forces and the interpolation of velocities. In this study, we also adopt the modified window function as the discretized delta function. Thus, it is possible to introduce nonuniform meshing in the background fluid domain. Both wavelet and smooth particle hydrodynamics (SPH) methods belong to a class of reproducing kernel methods where the "reproduced" function $u^{\mathrm{R}}(x)$ is derived as

$$u^{\mathrm{R}}(x) = \int_{-\infty}^{+\infty} u(y)\phi(x-y)\,\mathrm{d}y, \tag{12.17}$$

with a projection operator or a window function $\phi(x)$. The reproducing condition requires that up to $n$th-order polynomial can be reproduced:

$$x^n = \int_{-\infty}^{+\infty} y^n\phi(x-y)\,\mathrm{d}y. \tag{12.18}$$

From the convolution theorem, in the Fourier transform domain, Equation (12.17) can be expressed as

$$\hat{u}^{\mathrm{R}}(\omega) = \hat{u}(\omega)\hat{\phi}(\omega), \tag{12.19}$$

where $\omega$ is the natural frequency or wave number of the functions $u(x)$ and $\phi(x)$.

Suppose $\hat{\phi}(\omega)$ is a perfect rectangular window function, a so-called ideal low-pass filter function, which corresponds to the sinc function in the function domain, spatial or temporal, $\hat{u}(\omega)$ contains signals within $|\omega| \leq \omega_c$, where $\omega_c$ is the cut-off or band-limited frequency or wave number defined by the rectangular window function. Nevertheless, it has been recognized that the ideal low-pass filter does not have a compact support in the function domain and, hence, is not useful as an interpolation function in computational mechanics. Therefore, a correction function $C(x; x - y)$ is introduced in the finite computation domain $\Omega$ (i.e., the domain of influence or support):

$$C(x; x - y) = \sum_{k=0}^{n} \beta_k(x)(x - y)^k, \tag{12.20}$$

and the consequent modified window function $\phi$ is constructed as

$$\bar{\phi}(x; x - y) = \sum_{k=0}^{n} \beta_k(x)(x - y)^k \phi(x - y). \tag{12.21}$$

Define $m_k(x)$ as the $k$th moment of the window function $\phi(x)$, with $\mathrm{i} = \sqrt{-1}$, and we have

$$m_k(x) = \int_{\Omega} (x - y)^k \phi(x - y)\,\mathrm{d}y = \mathrm{i}^k \hat{\phi}^{(k)}(0), \quad k = 0, 1, 2, \ldots, n. \tag{12.22}$$

Similarly, define $\bar{m}_k(x)$ as the $k$th moment of $\bar{\phi}(x)$, and we have

$$\bar{m}_k(x) = \int_{\Omega} (x - y)^k \bar{\phi}(x - y)\,\mathrm{d}y = \mathrm{i}^k \hat{\bar{\phi}}^{(k)}(0), \quad k = 0, 1, 2, \ldots, n. \tag{12.23}$$

In general, with a proper construction of the correction function with respect to the selected window functions, such as the scaling function, wavelet, or spline family, we will be able to enforce

$$\hat{\bar{\phi}}(0) = 1, \hat{\bar{\phi}}^{(1)}(0) = 0, \ldots, \hat{\bar{\phi}}^{(n)}(0) = 0. \tag{12.24}$$

In essence, the window function is required to be flatter at $\omega = 0$ in the Fourier domain as the order $n$ of reproducing gets higher in Equation (12.18). This implies that, as $n \to \infty$, $\hat{\bar{\phi}}(\omega)$ becomes flatter at $\omega = 0$ and approaches an ideal filter. Therefore, based on Equations (12.22), (12.23) and (12.24), we can set up the moment equations to solve for $\beta_k$:

$$\begin{bmatrix} m_0(x) & m_1(x) & \cdots & m_n(x) \\ m_1(x) & m_2(x) & \cdots & m_{n+1}(x) \\ \vdots & \vdots & \ddots & \vdots \\ m_n(x) & m_{n+1}(x) & \cdots & m_{2n}(x) \end{bmatrix} \begin{Bmatrix} \beta_0(x) \\ \beta_1(x) \\ \vdots \\ \beta_n(x) \end{Bmatrix} = \begin{Bmatrix} 1 \\ 0 \\ \vdots \\ 0 \end{Bmatrix}. \tag{12.25}$$

Furthermore, we can also define a dilation parameter $a$ and introduce

$$\bar{\phi}_a(x - y) = \frac{1}{a} \bar{\phi}\left(\frac{x - y}{a}\right). \tag{12.26}$$

Thus, by changing the value of the dilation parameter $a$, usually by a factor of 2, a sequence of low-pass filters is defined. Note that the dilation parameter $a$ is directly linked to the mesh density. The difference between two such low-pass filters defines a high-pass filter. Therefore, we can perform a multi-resolution analysis with the projection operator $P_a$ expressed as

$$P_a u(x) = \int_{\Omega} u(y)\bar{\phi}_a(x-y)\,\mathrm{d}y. \tag{12.27}$$

Based on the resolution of the projection operator $P_{a/2^n}$, a hierarchical representation of a function $u(x)$ is defined as

$$u(x) = \lim_{n\to\infty} P_{a/2^n}u(x), \ldots, P_a u(x), \ldots, \lim_{n\to\infty} P_{2^n a}u(x) = \emptyset.$$

Furthermore, we also define a complementary projection operator $Q_{2a}$, such that the higher scale projected solution $P_a u$ can be represented by

$$P_a u(x) = P_{2a}u(x) + Q_{2a}u(x), \tag{12.28}$$

the sum of the low- and high-scale projections, where the $Q_{2a}$ projection, which can be viewed as a *peeled off* scale or the *rate of variation* of $P_a u(x)$, is simply given as a wavelet projection:

$$Q_{2a}u(x) = \int_{\Omega} u(y)\bar{\psi}_{2a}(x-y)\,\mathrm{d}y, \tag{12.29}$$

with $\bar{\psi}_{2a}(x-y) = \bar{\phi}_a(x-y) - \bar{\phi}_{2a}(x-y)$.

Figure 12.13 provides a comparison of various kernel functions in function and spectrum domains. As expected, with the increase of the reproducing order $n$, for the same finite support region, the kernel functions approach more to the ideal low-pass filter.

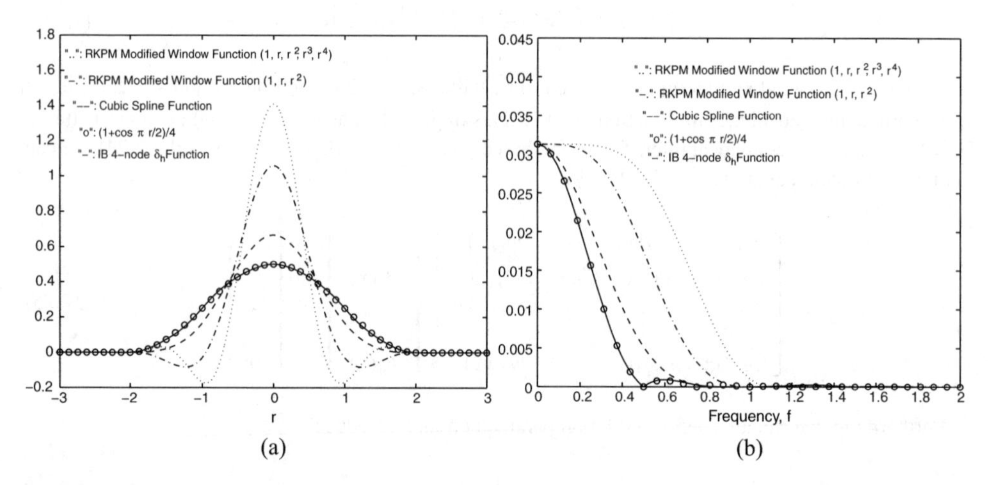


**Figure 12.13** A comparison of various discretized delta functions and their spectra: (a) function domain; (b) spectrum domain

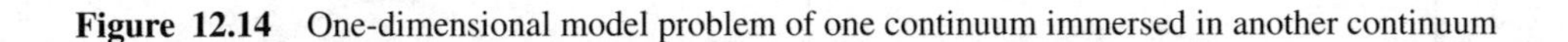

**Figure 12.14** One-dimensional model problem of one continuum immersed in another continuum

## 12.2.4 A Simple Model Problem

We start with a one-dimensional model problem with a continuum immersed in another one-dimensional continuum (Figure 12.14). Assume both domains with a constant cross-section $A$ are subject to the gravitational constant $g$ in $x$ direction the governing equation for regions $(0, L_1) \cup (L_2, L)$ and $(L_1, L_2)$ are written as

$$\frac{\partial}{\partial x}\left(E_{\mathrm{f}}\frac{\partial u}{\partial x}\right) = \rho_{\mathrm{f}}\frac{\partial^2 u}{\partial t^2} - \rho_{\mathrm{f}} g, \qquad \frac{\partial}{\partial x}\left(E_{\mathrm{s}}\frac{\partial u}{\partial x}\right) = \rho_{\mathrm{s}}\frac{\partial^2 u}{\partial t^2} - \rho_{\mathrm{s}} g. \tag{12.30}$$

Equation (12.30) represents second-order partial differential equations for three connected regions. If we consider $L_1$ and $L_2$ variables dependent on time and the state variable $u$, this problem will not be trivial. To start, let us look at the static equation in which Equation (12.30) represents second-order differential equations. Overall, six constants $c_i$ will be introduced for the solution $u(x)$ for all three regions, namely $(0, L_1)$, $(L_2, L)$, and $(L_1, L_2)$, and the solution $u(x)$ can be simply expressed as

$$-\frac{\rho_{\mathrm{f}}}{E_{\mathrm{f}}} g \frac{x^2}{2} + \frac{c_1}{E_{\mathrm{f}}} x + c_4, \quad -\frac{\rho_{\mathrm{f}}}{E_{\mathrm{f}}} g \frac{x^2}{2} + \frac{c_2}{E_{\mathrm{f}}} x + c_5, \quad \text{and} \quad -\frac{\rho_{\mathrm{s}}}{E_{\mathrm{s}}} g \frac{x^2}{2} + \frac{c_3}{E_{\mathrm{s}}} x + c_6. \tag{12.31}$$

Therefore, six constants $c_i$ will be derived based on the following six boundary conditions:

$$\begin{aligned}
u(0) &= 0, & u(L) &= 0, \\
u(L_1^-) &= u(L_1^+), & u(L_2^+) &= u(L_2^-), \\
E_{\mathrm{f}}\frac{\partial u}{\partial x}(L_1^-) &= E_{\mathrm{s}}\frac{\partial u}{\partial x}(L_1^+), & E_{\mathrm{f}}\frac{\partial u}{\partial x}(L_2^+) &= E_{\mathrm{s}}\frac{\partial u}{\partial x}(L_2^-).
\end{aligned} \tag{12.32}$$

To further simplify the study, we assign $L_1 = L/3 = l$, $L_2 = 2L/3 = 2l$, $E_{\mathrm{s}} = aE_{\mathrm{f}} = aE$, and $\rho_{\mathrm{s}} = b\rho_f = b\rho$. In addition, by scaling $c_i$, $i = 1, 2, 3$, by $l/E$ and defining $\beta = \rho g l$, the solution of $\mathbf{c}$ for this simplified problem can be given as

$$\mathbf{c} = \frac{\beta}{2}\langle b+2,\ -b+4,\ 3b,\ 0,\ 3b-3,\ 1+b-2b/a\rangle. \tag{12.33}$$

The comparisons illustrated in Figure 12.15 confirm the validity of immersed methods. In fact, this particular model problem represents a subtle shift to a continuum immersed in another continuum.

## 12.2.5 Compressible Fluid Model for General Grids

In this study we adopt the conservative form for the background compressible fluid domain. Defining the total specific energy as $E$, we have $E = e + v_i v_i/2$. Hence, the pressure $p$ can also be written as $p = \rho(\gamma - 1)e = \rho(\gamma - 1)(E - v_i v_i/2)$. In addition, for isentropic (quasistatic, adiabatic, and reversible) gas, according to the rst law of thermodynamics, namely $de = -p\, d(1/\rho) + dq$, where the reciprocal of the density $\rho$ is in fact the volume and for isentropic gas we have $dq = 0$. Therefore, we have $c_v\, dT = -p\, d(1/\rho)$ and, combined with the ideal gas law, we also have $c_v/R\, d(p/\rho) = -p\, d(1/\rho)$. Considering the universal gas constant $R$ is also written as $c_p - c_v = (\gamma - 1)c_v$ and that $\gamma\, d\rho/\rho = dp/p$,

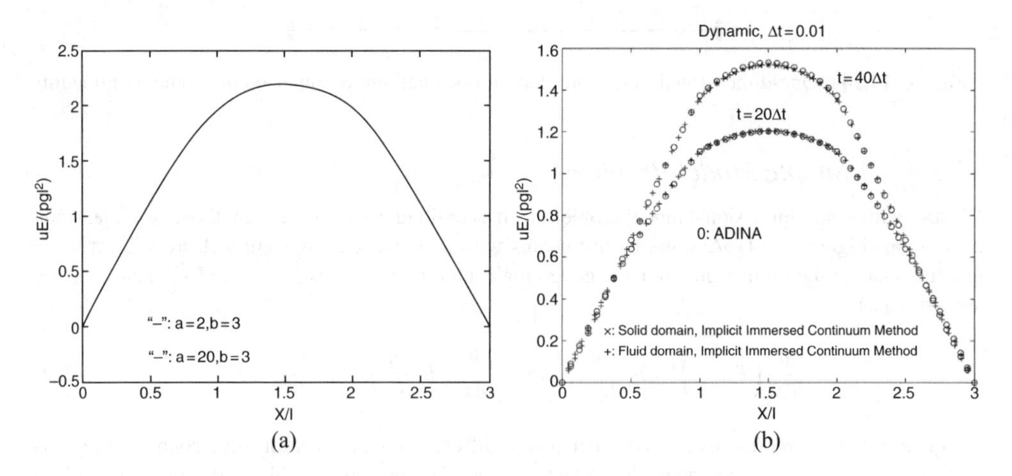


**Figure 12.15** Expected analytical and implicit immersed continuum solution: (a) analytical; (b) numerical

then for an isentropic gas the pressure $p$ is also governed by $p = \rho^{\gamma} C$, where $C$ is a constant; as a consequence, the speed of sound $c$ is defined as $c^2 = \mathrm{d}p/\mathrm{d}\rho = \gamma p/\rho$. Finally, using the specific enthalpy form, $H = E + p/\rho = c^2/(\gamma - 1) + v_i v_i/2$, the conservation of energy is written as

$$\frac{\partial(\rho E)}{\partial t} + \frac{\partial}{\partial x_i}[(\rho E + p)v_i] - \frac{\partial}{\partial x_i}\left(k\frac{\partial T}{\partial x_i}\right) = \frac{\partial}{\partial x_i}(\tau_{ij} v_j) + f_i v_i + q, \tag{12.34}$$

where $q$, different from the state variables $\mathbf{q}$, stands for the heat source per unit volume and time.

Thus, the governing equations are expressed in a unified conservative form:

$$\frac{\partial \mathbf{q}}{\partial t} + \frac{\partial \mathbf{F}_i}{\partial x_i} = \frac{\partial \mathbf{P}_i}{\partial x_i}, \qquad \frac{\partial(J\mathbf{q})}{\partial \tau} + \frac{\partial(B_{ji}\mathbf{F}_j)}{\partial \xi_i} = \frac{\partial(B_{ji}\mathbf{P}_j)}{\partial x_i}, \tag{12.35}$$

where the primary unknown vector $\mathbf{q}$ is defined as a column vector $\langle \rho, \rho v_j, \rho E\rangle$, with $j = 1, 2, 3$, and the flux vectors $\mathbf{F}_i$ for $i = 1, 2, 3$, are defined as a combination of two vectors, namely $\mathbf{F}_i = \langle \rho v_i, \rho v_i v_j, (\rho E + p)v_i\rangle + \langle 0, p\delta_{ij}, 0\rangle$, and the $i$th power vectors $\mathbf{P}_i$ for $i = 1, 2, 3$, are also defined as $\mathbf{P}_i = \langle 0, \tau_{ij}, \tau_{ik} v_k - k\frac{\partial T}{\partial x_i}\rangle$.

## 12.2.6 Multigrid Preconditioner

In the matrix-free Newton–Krylov iteration, the solution vector $\Delta\boldsymbol{\Theta}$, or rather $\Delta\mathbf{V}$, $\Delta\mathbf{P}$, and $\Delta\mathbf{P}^s$ is expressed as

$$\Delta\boldsymbol{\Theta}^{k,n} = \Delta\boldsymbol{\Theta}^{k,0} + \sum_{i=1}^{n} y_i \mathbf{q}^i$$

or

$$\langle \Delta\mathbf{V}^{\mathrm{T}}, \Delta\mathbf{P}^{\mathrm{T}}, \Delta(\mathbf{P}^s)^{\mathrm{T}}\rangle = \Delta\boldsymbol{\Theta}^{k,0} + \sum_{i=1}^{n} y_i \mathbf{q}^i. \tag{12.36}$$

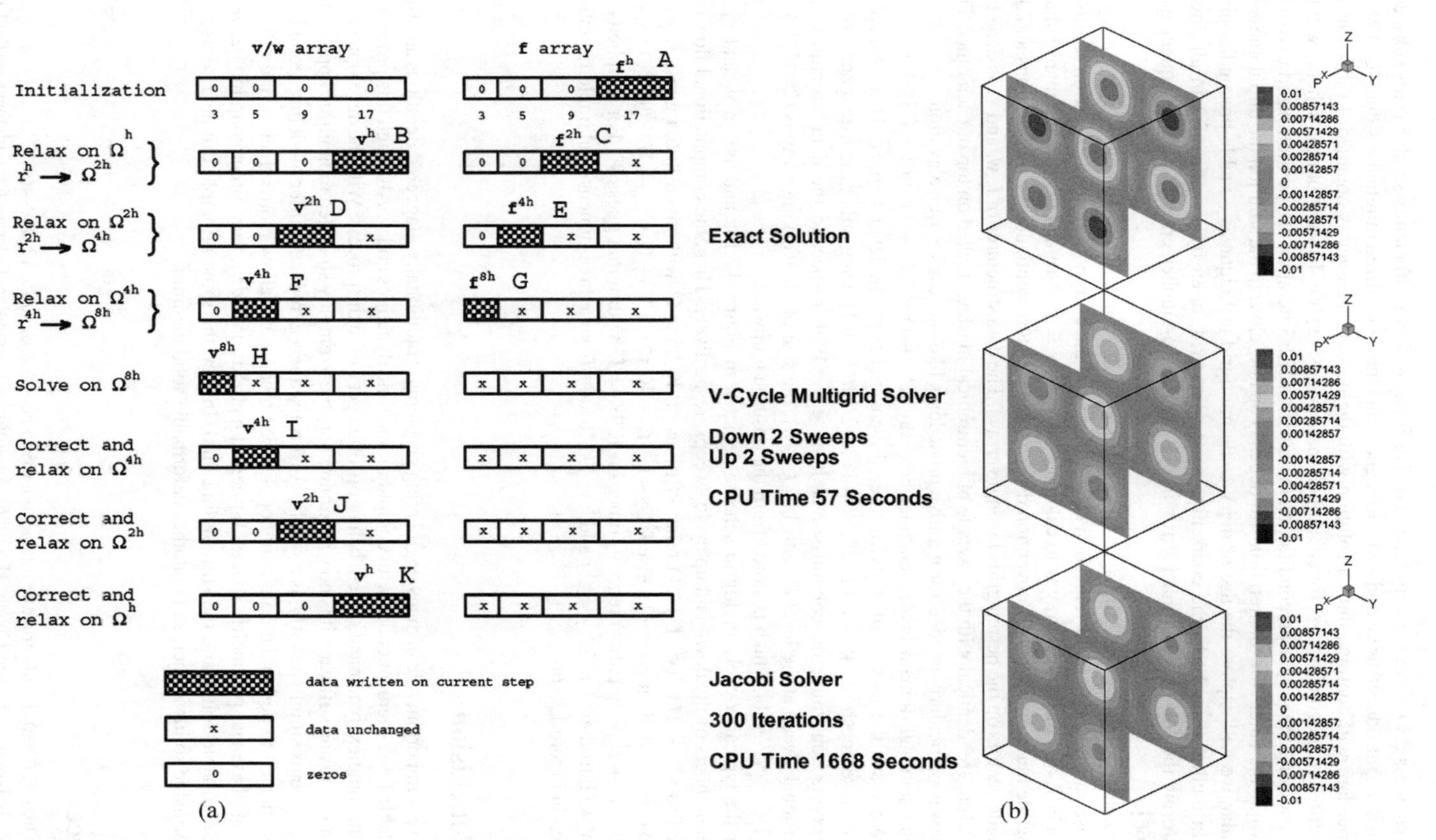


**Figure 12.16** (a) Data structure of V-cycle multigrid procedures and (b) a three-dimensional implementation [4], reprinted with permission from Elsevier © 2008

If the initial guess $\Delta\mathbf{\Theta}^{k,0}$ does not produce a good estimate within a sufficiently small Krylov subspace $\mathbf{K}^n$, $\Delta\mathbf{\Theta}^{k,n}$ will be introduced as an updated initial guess and the generalized minimal residual (GMRES) iteration procedure will continue until a solution with the desired accuracy is obtained. In this part of the project, different preconditioners for an implicit matrix-free Newton–Krylov solution scheme will be studied. By choosing the proper preconditioner, the inner GMRES iteration can be minimized to two or three steps while the outer Newton iteration still depends on how close to the solution the initial guess is.

In particular, we will explore the use of the V-cycle multigrid technique as a viable preconditioning candidate. In multigrid methods, the coarsening of the fine grid, often by a factor of 2 in all directions, transforms the low wave-number error into a high wave-number error at all subsequent grid levels [65, 66].

In the implementation, we store the right-hand side source or boundary terms in the working array $\mathbf{f}$. In order to use various iteration or relaxation schemes, we use the array $\mathbf{v}$ to store the iteration results $\mathbf{u}$ based on $\mathbf{Au} = \mathbf{f}$, and the buffer array $\mathbf{w}$ to calculate the error $\mathbf{r} = \mathbf{f} - \mathbf{Au}$ and to store the intermediate result during the iteration process. Convention dictates that we store contiguously the solutions and the right-hand side vectors on the nested grids in single arrays. The data structures of $\mathbf{f}$, $\mathbf{w}$, and $\mathbf{v}$, as depicted in Figure 12.16a, are identical with the cascade of hierarchical meshes stacked on top of each other. The buffer array $\mathbf{w}$ is particularly needed for iterative procedures like successive overrelaxation.

If the computational domain is a rectangular box, say $(2^{n_1} + 1) \times (2^{n_2} + 1) \times (2^{n_3} + 1)$, with $n_1 \leq n_2 \leq n_3$, we coarsen the mesh uniformly in three dimensions from the level 1 to $n_1$. Then, we solve exactly the $n_1$-level mesh $(2^1 + 1) \times (2^{n_2-n_1+1} + 1) \times (2^{n_3-n_1+1} + 1)$ using either the tridiagonal solver if $n_1 = n_2$ or the nonsymmetric column solver if $n_1 < n_2$. With respect to the data structures, the three-dimensional sweep at the level $k$, with $1 \leq k \leq n_1$, the size of the working array is $(2^{n_1-k+1} + 1)$ $(2^{n_2-k+1} + 1)(2^{n_3-k+1} + 1)$, which is stored from the position $n$ onward.

Consider the typical V-cycle multigrid scheme depicted in Figure 12.16a and assign $\nu_1$ and $\nu_2$ as the iteration numbers of the down and up cycle sweeps, respectively. Then, the computational flow can be illustrated as $\mathbf{v}^h \leftarrow MV^h(\mathbf{v}^h, \mathbf{f}^h)$: (1) relax $\nu_1$ times on $\mathbf{K}^h\mathbf{b}^h = \mathbf{f}^h$ with given initial guess $\mathbf{v}^h = \mathbf{0}$; (2) if $\Omega^h$ coarsest grid, then go to (4), else $\mathbf{f}^{2h} \leftarrow \mathbf{I}_h^{2h}(\mathbf{f}^h - \mathbf{K}^h\mathbf{v}^h)$, $\mathbf{v}^{2h} \leftarrow \mathbf{0}$, $\mathbf{v}^{2h} \leftarrow MV^{2h}(\mathbf{v}^{2h}, \mathbf{f}^{2h})$; (3) convert $\mathbf{v}^h \leftarrow \mathbf{v}^h + \mathbf{I}_{2h}^h\mathbf{v}^{2h}$; and (4) relax $\nu_2$ times on $\mathbf{K}^h\mathbf{b}^h = \mathbf{f}^h$ with initial guess $\mathbf{v}^h$. For simplicity of the notation, we introduce $\bar{i} = 2i$, $\bar{j} = 2j$ and $\bar{k} = 2k$. Results from a three-dimensional implementation are presented in Figure 12.16b.

## 12.3 Conclusion

FSI plays a very important role in many engineering problems. Traditionally, staggered iterations are used to link available finite-element codes with computational fluid dynamics codes. Although this procedure is convenient, complex dynamical system behaviors often get lost in the process. With the development of computer hardware and software, it is now becoming feasible to employ so-called monolithic approaches to solve FS) systems simultaneously as a whole. As a consequence, reduced-order modeling techniques can be implemented to derive the dominant coupled FSI models and allow more effective and efficient exploration of phase and parametric spaces. Immersed methods provide a very direct coupling between immersed deformable solids and the surrounding fluid. In fact, the same principle can also be used to handle deformable solids immersed in another deformable solid medium.

## References

[1] Belytschko, T. (1980) Fluid–structure interaction. *Computers & Structures*, **12**, 459–469.

[2] Liu, W., Belytschko, T., and Chang, H. (1986) An arbitrary Lagrangian–Eulerian finite element method for path-dependent materials. *Computer Methods in Applied Mechanics and Engineering*, **58**, 227–245.

[3] Shyy, W., Udaykumar, H., Rao, M., and Smith, R. (1996) *Computational Fluid Dynamics with Moving Boundaries*, Taylor & Francis.

[4] Wang, X. (2008) *Fundamentals of Fluid–Solid Interactions – Analytical and Computational Approaches*, Elsevier Science.

[5] Bathe, K. and Zhang, H. (2004) Finite element developments for general fluid flows with structural interactions. *International Journal for Numerical Methods in Engineering*, **60**, 213–232.

[6] Bathe, K., Zhang, H., and Wang, M. (1995) Finite element analysis of incompressible and compressible fluid flows with free surfaces and structural interactions. *Computers & Structures*, **56** (2–3), 193–213.

[7] Zhang, J., Childress, S., Libchaber, A., and Shelley, M. (2000) Flexible filaments in a flowing soap film as a model for one-dimensional flags in a two-dimensional wind. *Nature*, **408**, 835–838.

[8] Tornberg, A.K. and Shelley, M. (2004) Simulating the dynamics and interactions of flexible fibers in Stokes flows. *Journal of Computational Physics*, **196**, 8–40.

[9] Tezduyar, T. (1992) Stabilized finite element formulations for incompressible-flow computations. *Advances in Applied Mechanics*, **28**, 1–44.

[10] Tezduyar, T. (2001) Finite element methods for flow problems with moving boundaries and interfaces. *Archives of Computational Methods in Engineering*, **8**, 83–130.

[11] Huerta, A. and Liu, W. (1988) Viscous flow with large free surface motion. *Computer Methods in Applied Mechanics and Engineering*, **69**, 277–324.

[12] Liu, W. and Ma, D. (1982) Computer implementation aspects for fluid–structure interaction problems. *Computer Methods in Applied Mechanics and Engineering*, **31**, 129–148.

[13] Zhang, L., Wagner, G., and Liu, W. (2003) Modeling and simulation of fluid structure interaction by meshfree and FEM. *Communications in Numerical Methods in Engineering*, **19**, 615–621.

[14] Li, Z. and LeVeque, R. (1994) The immersed interface methods for elliptic equations with discontinuous coefficients and singular sources. *SIAM Journal on Scientific Computing*, **31**, 1019–1994.

[15] LeVeque, R. and Li, Z. (1997) Immersed interface methods for Stokes flow with elastic boundaries or surface tension. *SIAM Journal on Scientific Computing*, **18**, 709–735.

[16] J.A. Sethian (1996) *Level Set Methods and Fast Marching Methods*, Cambridge University Press.

[17] Osher, S. and Fedkiw, R. (2002) *Level Set Methods and Dynamic Implicit Surfaces*, Springer.

[18] Glimm, J., Grove, J., Li, X., and Tan, D. (2000) Robust computational algorithms for dynamic interface tracking in three dimensions. *SIAM Journal on Scientific Computing*, **21**, 2240–2256.

[19] Tryggvason, G. (1988) Numerical simulations of the Rayleigh–Taylor instability. *Journal of Computational Physics*, **75**, 253–282.

[20] Li, X., Jin, B., and Glimm, J. (1996) Numerical study for the three-dimensional Rayleigh–Taylor instability through the TVD/AC scheme and parallel computation. *Journal of Computational Physics*, **126**, 343–355.

[21] Farhat, C., Geuzaine, P., and Grandmont, C. (2001) The discrete geometric conservation law and the nonlinear stability of ALE schemes for the solution of flow problems on moving grids. *Journal of Computational Physics*, **174**, 669–694.

[22] Shyy, W., Berg, M., and Ljungqvist, D. (1999) Flapping and flexible wings for biological and micro air vehicles. *Progress in Aerospace Sciences*, **35**, 455–505.

[23] Kamakoti, R. and Shyy, W. (2004) Fluid–structure interaction for aeroelastic applications. *Progress in Aerospace Sciences*, **40**, 535–558.

[24] Lian, Y. and Shyy, W. (2005) Numerical simulations of membrane wing aerodynamics for micro air vehicle applications. *Journal of Aircraft*, **42**, 865–873.

[25] Hughes, T., Liu, W., and Zimmerman, T. (1981) Lagrangian–Eulerian finite element formulations for incompressible viscous flows. *Computer Methods in Applied Mechanics and Engineering*, **29**, 329–349.

[26] Chessa, J. and Belytschko, T. (2003) The extended finite element method for two-phase fluids. *Journal of Applied Mechanics*, **70**, 10–17.

[27] Wagner, G., Moës, N., Liu, W., and Belytschko, T. (2001) The extended finite element method for rigid particles in stokes flow. *International Journal for Numerical Methods in Engineering*, **51**, 293–313.

[28] Glowinski, R., Pan, T., Hesla, T. *et al.* (2001) A fictitious domain approach to the direct numerical simulation of incompressible viscous flow past moving rigid bodies: application to particulate flow. *Journal of Computational Physics*, **169**, 363–426.

[29] Peskin, C. (2002) The immersed boundary method. *Acta Numerica*, **11**, 479–517.

[30] Peskin, C. (1977) Numerical analysis of blood flow in the heart. *Journal of Computational Physics*, **25**, 220–252.

[31] McQueen, D. and Peskin, C. (1985) Computer-assisted design of butterfly bileaflet valves for the mitral position. *Scandinavian Journal of Thoracic and Cardiovascular Surgery*, **19**, 139–148.
[32] Fauci, L. and Peskin, C. (1988) A computational model of aquatic animal locomotion. *Journal of Computational Physics*, **77**, 85–108.
[33] Stockie, J. and Wetton, B. (1999) Analysis of stiffness in the immersed boundary method and implications for time-stepping schemes. *Journal of Computational Physics*, **154**, 41–64.
[34] Beyer, Jr., R. (1992) A computational model of the cochlea using the immersed boundary method. *Journal of Computational Physics*, **98**, 145–162.
[35] Dillon, R., Fauci, L., Fogelson, A., and Gaver, III, D. (1996) Modeling biofilm processes using the immersed boundary method. *Journal of Computational Physics*, **129**, 57–73.
[36] Gay, M., Zhang, L., and Liu, W. (2005) Stent modeling using immersed finite element method. *Computer Methods in Applied Mechanics and Engineering*, **195**, 4358–4370.
[37] Liu, W., Liu, Y., Farrell, D. *et al.* (2006) Immersed finite element method and its applications to biological systems. *Computer Methods in Applied Mechanics and Engineering*, **195**, 1722–1749.
[38] Wang, X. (2006) From immersed boundary method to immersed continuum method. *International Journal for Multiscale Computational Engineering*, **4**, 127–145.
[39] Wang, X. (2007) An iterative matrix-free method in implicit immersed boundary/continuum methods. *Computers & Structures*, **85**, 739–748.
[40] Liu, W., Jun, S., and Zhang, Y. (1995) Reproducing kernel particle methods. *International Journal for Numerical Methods in Fluids*, **20**, 1081–1106.
[41] Boffi, D. and Gastaldi, L. (2003) A finite element approach for the immersed boundary method. *Computers & Structures*, **81**, 491–501.
[42] Newren, E., Fogelson, A., Guy, R., and Kirby, R. (2007) Unconditionally stable discretizations of the immersed boundary equations. *Journal of Computational Physics*, **222**, 702–719.
[43] Mori, Y. and Peskin, C. (2008) Implicit second-order immersed boundary methods with boundary mass. *Computer Methods in Applied Mechanics and Engineering*, **197**, 2049–2067.
[44] Sulsky, D. and Kaul, A. (2004) Implicit dynamics in the material-point method. *Computer Methods in Applied Mechanics and Engineering*, **193**, 1137–1170.
[45] Fogelson, A., Wang, X., and Liu, W. (2008) Special issue: Immersed boundary method and its extensions – Preface. *Computer Methods in Applied Mechanics and Engineering*, **197**, 25–28.
[46] Liu, Y., Liu, W., Belytschko, T. *et al.* (2007) Immersed electrokinetic finite element method. *International Journal for Numerical Methods in Engineering*, **71**, 379–405.
[47] Liu, W., Kim, D., and Tang, S. (2007) Mathematical foundations of the immersed finite element method. *Computational Mechanics*, **39**, 211–222.
[48] Liu, Y., Zhang, L., Wang, X., and Liu, W. (2004) Coupling of Navier–Stokes equations with protein molecular dynamics and its application to hemodynamics. *International Journal for Numerical Methods in Fluids*, **46**, 1237–1252.
[49] Lim, S., Ferent, A., Wang, X., and Peskin, C. (2008) Dynamics of a closed rod with twist and bend in fluid. *SIAM Journal on Scientific Computing*, **31**, 273–302.
[50] Van Loon, R., Anderson, P., Van de Vosse, F., and Sherwin, S. (2006) Comparison of various fluid–structure interaction methods for deformable bodies. *Computers & Structures*, **85**, 833–843.
[51] Yu, Z. (2005) A DLM/FD method for fluid/flexible-body interactions. *Journal of Computational Physics*, **207**, 1–27.
[52] Sulsky, D. and Brackbill, J. (1991) A numerical method for suspension flow. *Journal of Computational Physics*, **96**, 339–368.
[53] Wang, X. and Liu, W. (2004) Extended immersed boundary method using FEM and RKPM. *Computer Methods in Applied Mechanics and Engineering*, **193**, 1305–1321.
[54] Zhang, L., Gerstenberger, A., Wang, X., and Liu, W. (2004) Immersed finite element method. *Computer Methods in Applied Mechanics and Engineering*, **193**, 2051–2067.
[55] Boffi, D., Gastaldi, L., and Heltai, L. (2007) On the CFL condition for the finite element immersed boundary method. *Computers & Structures*, **85**, 775–783.
[56] Boffi, D., Gastaldi, L., Heltai, L., and Peskin, C. (2008) On the hyperelastic formulation of the immersed boundary method. *Computer Methods in Applied Mechanics and Engineering*, **197**, 2210–2231.
[57] Wang, X. and Bathe, K. (1997) Displacement/pressure based finite element formulations for acoustic fluid–structure interaction problems. *International Journal for Numerical Methods in Engineering*, **40**, 2001–2017.

[58] Wang, X. and Bathe, K. (1997) On mixed elements for acoustic fluid–structure interactions. *Mathematical Models & Methods in Applied Sciences*, **7**, 329–343.

[59] Bao, W., Wang, X., and Bathe, K. (2001) On the inf–sup condition of mixed finite element formulations for acoustic fluids. *Mathematical Models & Methods in Applied Sciences*, **11**, 883–901.

[60] Bathe, K. (1996) *Finite Element Procedures*, Prentice Hall, Englewood Cliffs, NJ.

[61] Liu, W., Chen, Y., Chang, C., and Belytschko, T. (1996) Advances in multiple scale kernel particle methods. *Computational Mechanics*, **18**, 73–111.

[62] Liu, W. and Chen, Y. (1995) Wavelet and multiple scale reproducing kernel methods. *International Journal for Numerical Methods in Fluids*, **21**, 901–932.

[63] Li, S. and Liu, W. (1999) Reproducing kernel hierarchical partition of unity, Part I: formulation and theory. *International Journal for Numerical Methods in Engineering*, **45**, 251–288.

[64] Liu, W., Chen, Y., Uras, R., and Chang, C. (1996) Generalized multiple scale reproducing kernel particle methods. *Computer Methods in Applied Mechanics and Engineering*, **139**, 91–157.

[65] McCormick, S. (1987) *Multigrid Methods*, SIAM.

[66] Axelsson, O. (1996) *Iterative Solution Methods*, Cambridge University Press.

# Part III

# From Cellular Structure to Tissues and Organs

In this book, a multiscale approach is adopted in presentation of modeling of biological systems; that is, from cellular and subcellular structure level to tissue and organ level. In this part, we select six chapters to demonstrate various technical approaches at the different scales from cellular level to organ and tissue level. In Chapter 13, Foucard, Espinet, Benet, and Vernerey discuss the role of the cortical membrane in cell mechanics, in which they developed a finite deformation theory of cortical membrane; by combining it with the extended finite-element method they are able to reproduce different equilibrium shapes exhibited by the red blood cell and can handle the very large deformations experienced by the membrane in the endocytosis and blebbing problems. In Chapter 14, Zhang and Zeinali-Davarani study the role of elastin in arterial mechanics, and they have built an accurate constitutive model for elastin that is validated by their experimental study. Coming to the organ and tissue level, in Chapter 15, Professor E. Oñate, a world renowned scholar in computational mechanics, and his co-workers present their recent study on modeling of urinary bladder. In this work they have used the mixture theory of hyperelastic matrix and viscoelastic fibers to build a constitutive model of the bladder and they employ the particle finite-element method (PFEM) to realize geometric modeling of the bladder. In Chapter 16, Liu, Yang, Zou, and Hu discuss the biomechanics and design of vascular stents. One of the main contributions in Wing Liu's career is to develop the reproducing kernel particle method (RKPM). In Chapter 17, Dr. Simkins Jr. presents his recent research on organ modeling. In particular, he has successfully applied RKPM to model and simulate pelvic organs and deformation of the levator ani. The last contribution of this part is Chapter 18, in which Bhamare, Mannava, Kirschman, Vasudevan, and Qian discuss the biomechanics and design of orthopedic implants.

# 13

# The Role of the Cortical Membrane in Cell Mechanics: Model and Simulation

Louis Foucard, Xavier Espinet, Eduard Benet, and Franck J. Vernerey*
*Department of Civil, Environmental and Architectural Engineering, University of Colorado, USA*

## 13.1 Introduction

The lipid bilayer membrane is one of the most fundamental entities in biology. It is omnipresent in organisms, from the nanoscale with the viral envelope [1] and the cell's organelles to the microscale with the cell's membrane itself. Its partitioning abilities allow the cell to separate its interior from its outside environment, making it selectively permeable to molecules and ions. Associated with an underlying mesh of actin/myosin filaments called cortex, it also enables the cell to maintain its shape when subjected to external forces and to mechanically sense its environment. The mechanics of this membrane–cortex complex (MCC) also becomes of critical importance in processes such as cell blebbing, involved in cancer spreading [2–5] or the endocytosis, which is the main mechanism cells use to intake particles and can also be hijacked by viruses to enter the cell [6–9]. A critical challenge in biology, therefore, is to develop a good understanding of the mechanics of MCC, on which rely many cell functions, by elaborating theoretical and numerical models that are biologically and mechanically accurate.

Many theoretical efforts have been directed toward modeling the mechanics of lipid bilayers, which includes the elastic bending and stretching of the lipid bilayer, and its interactions with surrounding fluids, substrates or particles, such as osmotic pressure, viscous friction or adhesion strength. One of the first attempts to describe a the fluid membrane was made by Canham [10], Helfrich [11] and Evans [12]. Focusing on the membrane bending, they considered the lipid bilayer as an incompressible fluid

*Corresponding author: Franck J Vernerey, Department of Civil, Environmental and Architectural Engineering, University of Colorado, 1111 Engineering Drive, 428 UCB, ECOT 422 Boulder, CO 80309-0428 USA. **phone**: 303-492-7165, **e-mail**: Franck.Vernerey@colorado.edu

*Multiscale Simulations and Mechanics of Biological Materials*, First Edition. Edited by Shaofan Li and Dong Qian.

and developed the most widely accepted bending energy, which is a quadratic function of the membrane curvature. By minimizing this bending energy, Jenkins [13] developed equilibrium equations that enabled the prediction of a variety of vesicles equilibrium shapes and the biconcave shape of red cells for example [10]. However, solving these equilibrium equations analytically was far from trivial and could only be done within the constraint of axisymmetry.

The challenge inherent to solving for the vesicles equilibrium shapes in more complicated geometries led to the use of numerical methods such as the finite-element method (FEM) that locally minimizes the bending energy in order to obtain non-axisymmetric equilibrium shapes. The FEM can be used to describe the mechanics of the MCC within three different frames: a fully Lagrangian, a mixed Eulerian–Lagrangian, or a fully Eulerian frame. The Lagrangian formulation was successfully used to compute the different red blood cell equilibrium shapes [14] or complex membrane loadings in non-axisymmetric geometries [15–18], but did not take intracellular fluid motion into account, making it impossible to investigate the effect of a fluid flowing around the cell, for example. The mixed Lagrangian–Eulerian method, also called the immersed boundary method, developed by Peskin [19] aimed at describing the fluid–structure interactions by solving the fluid equations in an Eulerian frame while a Lagrangian frame was used to follow the structure motion. Numerous biological problems, such as blood clotting or cell division, were approached in this manner [20–22]. However, it has been shown [23] that the cell packs reserves of surface area by tightly bundling its membrane, and can therefore experience large increases in its surface area by unfolding its membrane "reserves," making the cell membrane stretchable at a larger scale. The Lagrangian description of the membrane used in the models presented above is not suited for large deformations and leads to severe distortions of the finite-element mesh, which can only be avoided by calling for complex mesh regularization techniques [24]. This constitute another challenge in the immersed cell problem and motivates the need to develop a model that allows for the large deformations of the membrane that can occur in the cell blebbing or the cell endocytosis problems, for example.

A full Eulerian formulation of the fluid–membrane problem that eliminates the issues due to a Lagrangian description of the membrane was presented in Cottet and Maitre [25], where the kinematic quantities of the membrane (such as position or dilation) are accounted for by a level-set function defined and updated on a fixed Eulerian grid. The formulation allows one to successfully compute the equilibrium shape of a red blood cell and can very naturally deal with large deformations. However, it is limited to membrane constitutive relations that only depend on the membrane change of area and cannot account for more general elastic materials. Moreover, one of the assumptions used in this model is the continuity of the fluid velocity across the interface. This does not accommodate cases where fluids have a lower viscosity and a boundary layer thin enough to make the fluid velocity and pressure essentially discontinuous across the membrane at a larger scale.

In this chapter we propose a full Eulerian formulation that is able to describe the rich mechanics and interactions between the cell membrane and cortex and between the membrane and the external environment. The formulation takes into account the mechanics of both the lipid bilayer membrane and the cortex, as well as their interactions. We show that these considerations are of critical importance in situations that lead to debonding between the membrane and the cortex, as is the case in the cell blebbing problem for example. The formulation is then coupled with the extended finite-element method (X-FEM), a numerical method that naturally allows for pressure and velocity discontinuities across the membrane, and the grid-based particle method that allows us to accurately track the membrane geometry and large deformations as it evolves in an Eulerian frame.

The outline of this chapter is as follows: first we provide a biological and mechanical description of the MCC and introduce the main constitutive equations. In Section 13.3, the mathematical formulation used to describe the membrane mechanics, membrane–cortex interactions, and the membrane–fluid interactions in an Eulerian framework is presented, followed by the derivation of the governing equations from energetic considerations. Section 13.4 discusses the numerical implementation in the framework of the X-FEM and the grid based particle method. In Section 13.5, the problems of the red blood cell equilibrium shapes, the cell blebbing and the cell endocytosis are investigated to demonstrate the model's abilities to describe the rich mechanics of the MCC. Finally, Section 13.6 concludes with a discussion of the results and recommendations for future improvements.

## 13.2 The Physics of the Membrane–Cortex Complex and Its Interactions

In many biological problems such as endocytosis, blebbing or the red blood cell biconcavity, the evolution of the shape the cell is driven by the mechanics of the cell membrane and its interaction and adhesion with surrounding fluids or objects. In order to understand the debonding that occurs during cell blebbing between the membrane and the cortex, or the intake by the cell membrane of nanoparticles during endocytosis, one must therefore first investigate the mechanics of each component as well as the forces responsible for their interactions.

### *13.2.1 The Mechanics of the Membrane–Cortex Complex*

In most cells the membrane consists of the juxtaposition of two layers: the lipid bilayer and the cortex, often refereed as the MMC. The lipid bilayer itself is made of a double layer of phospholipids, whose hydrophylic heads point outside of the membrane and hydrophobic tails point in to the core of the membrane. Attached to the lipid bilayer via membrane–cortex attachment proteins is the cortex, composed of a mesh of actin/myosin/spectrin filaments that provides the cell with resistance to deformations and possesses contractile abilities [5] (Figure 13.1). The MMC is generally viewed as an elastic membrane with values of stretching and bending stiffnesses that vary with the type of cell considered. For example, a leukocyte (or white blood cell) has reserves of membrane area stored in micro-folds that reduce its stretching stiffness, whereas the membrane of a erythrocyte (or red blood cell) has no such reserves and is assumed in most studies to be essentially inextensible (maximum area change $\leq 4\%$ [26]). Some biological processes, such as cell blebbing, can lead to debonding between the cortex and the lipid bilayer by rupturing the membrane–cortex attachment proteins. In this context, the cortex and the lipid bilayer

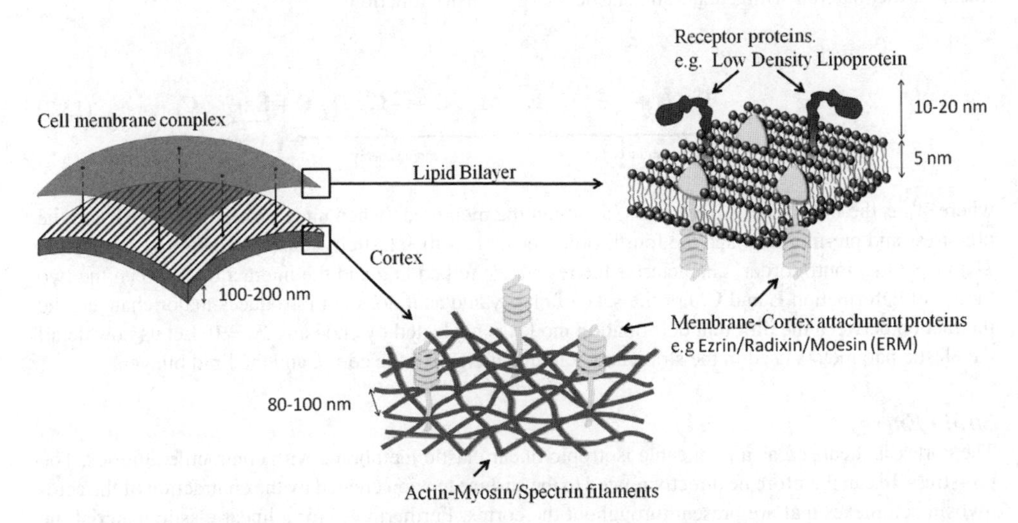


**Figure 13.1** Global scheme of the cell membrane with the detailed parts of the cortex and the lipid bilayer

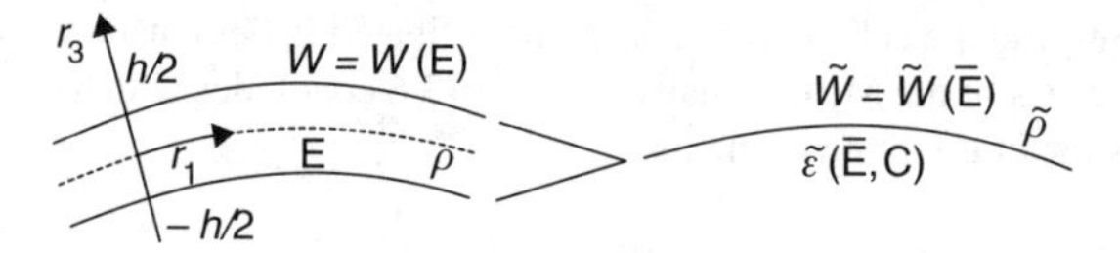


**Figure 13.2** Microscale and mesoscale description of the membrane

become two separate membranes with different parameters of elasticity. The mechanical characteristics of the membrane also vary with the scale of the problem considered: in the case of endocytosis, only the lipid bilayer mechanics are relevant in the uptake of nanoparticles by the cell membrane, since the spacing between the cortical filament (200 nm) is much larger than the size of the particles (10–20 nm). Here, we propose to model the MCC, the cortex, and the lipid bilayer as elastic membranes with different elasticities. In this view, let us consider a generic elastic membrane (that can be the MCC, the cortex, or the lipid bilayer) whose geometry is characterized by a thickness negligible compared with its other dimensions. The analysis of this membrane can then be viewed as a multiscale problem. At the microscale, the membrane is described as a thin layer of thickness $h$ with an elastic energy density $W$, a function of the membrane strain $\mathbf{E}$ that varies along the membrane thickness. At the mesoscale, the membrane is considered as a zero-thickness surface in which the strain $\mathbf{E} = \mathbf{E}(\bar{\mathbf{E}}, \mathbf{C})$ and the elastic energy density $\tilde{W} = \tilde{W}(\bar{\mathbf{E}}, \mathbf{C})$ can be approximated as functions of the membrane in-plane strain $\bar{\mathbf{E}}$ and second fundamental form $\mathbf{C}$ (Figure 13.2). An averaging operation is then introduced in order to reconcile the microscopic and mesoscopic descriptions of the membrane, such that the elastic energy and densities at the two scales are related as follows [27, 28]:

$$\tilde{W} = \frac{1}{h}\int_{-h/2}^{h/2} W \, \mathrm{d}r_3 \quad \text{and} \quad \tilde{\rho} = \frac{1}{h}\int_{-h/2}^{h/2} \rho \, \mathrm{d}r_3. \tag{13.1}$$

Although the formulation allows for more complex strain energy functions, we concentrate here on a linear elastic material with a quadratic elastic energy density function:

$$\begin{aligned} \tilde{W}(\bar{\mathbf{E}}, \mathbf{C}) = {} & W^0 \\ & + \underbrace{\mathbf{T}^0 : \bar{\mathbf{E}} + \frac{1}{2}\bar{\mathbf{E}} : \mathcal{C} : \bar{\mathbf{E}}}_{\text{strain energy}} + \underbrace{\mathbf{M}^0 : \mathbf{C} + \frac{1}{2}\mathbf{C} : \mathcal{D} : \mathbf{C}}_{\text{bending energy}} + \underbrace{\bar{\mathbf{E}} : \mathcal{F} : \mathbf{C}}_{\text{mixed}}, \end{aligned} \tag{13.2}$$

where $W_0$ is the initial energy (per mass) stored in the membrane when undeformed, $\mathbf{T}^0$ and $\mathbf{M}^0$ are the pre-stress and pre-moment, and the fourth-order tensor $\mathcal{C}$ is the elasticity tensor. In addition, the tensors $\mathcal{D}$ and $\mathcal{F}$ (also fourth order) characterize the resistance to bending and the interaction between the two modes of deformation $\bar{\mathbf{E}}$ and $\mathbf{C}$. For the sake of clarity and as it does not introduces major changes, the interaction between the different deformation modes is neglected by choosing $\mathcal{F} = 0$. Let us now detail the elastic parameters used in the strain and bending energy for the cortex and the lipid bilayer.

### *Strain Energy*

The cortex is idealized as a permeable isotropic linear elastic membrane with contractile abilities. The pre-stress $\mathbf{T}^0$ can therefore be directly related to the surface tension created by the contraction of the acto-myosin complexes that are present throughout the cortex. Furthermore, for a linear elastic material the elasticity tensor is written as $\mathcal{C} = \bar{\lambda}\mathbf{g} \otimes \mathbf{g} + 2\bar{\mu}\mathcal{I}$, with $\bar{\lambda}$ and $\bar{\mu}$ the first and second Lamé coefficients and $\mathcal{I}_{ijkl} = \mathbf{g}_{ik}\mathbf{g}_{jl} + \mathbf{g}_{il}\mathbf{g}_{jk}$ [29], where the tensor $\mathbf{g}$ is the membrane's metric and will be defined in Section

13.3.1. On the other hand, the lipid bilayer is usually described as a surface fluid with no resistance to shear forces but high resistance to surface dilation. This behavior can be modeled by choosing an incompressible linear elastic material (but not inextensible, since the lipid bilayer can sustain around 4% of surface dilation) where the first and second Lamé coefficient are respectively equal to the bulk modulus $\bar{\lambda} = \bar{K}$ and $\bar{\mu} = 0$. However, in certain types of cells, such as leukocytes for example, the presence of undulation of the lipid bilayer over the cortex has been observed. These undulations, stabilized by the membrane–cortex bonds, constitute a reserve of lipid bilayer area (that can reach up to 2.6 times the apparent area in some white blood cells) that imparts the membrane with a nonlinear elastic behavior and allows it to endure deformations much larger than 4% in processes such as cell blebbing. A constitutive relation between the membrane tension and the membrane increase of area was introduced by Rawicz *et al.* [26] as

$$\epsilon = \frac{\Delta A}{A_0} = \frac{k_B T}{8\pi k_c} \ln\left(1 + \frac{c\tau_m A}{K_h^{\text{lipid}}}\right) + \frac{\tau_m}{K_A}, \tag{13.3}$$

where $\epsilon = \Delta A/A_0$ is the membrane dilation or relative increase of apparent area, $K_A$ is the elastic modulus for direct stretch, $k_B T$ is the thermal energy, $c = 0.1$ reflects the type of mode used to describe the undulations, and $\tau_m$ the membrane tension. Equation (13.3) has two main regimes: for low and high values of area increase. For low values of area increase, the membrane stiffness remains low and increases logarithmically, which corresponds to the smoothing of the membrane undulations. For higher values of area increase, the lipid bilayer is completely smoothed out and enters a pure stretch regime that is governed by an elastic linear relationship with a much higher elastic modulus, making the lipid bilayer essentially inextensible (Figure 13.3a). From Equation (13.3), one can find the lipid bilayer nonlinear bulk modulus $\bar{\lambda}(\epsilon) = \bar{K}(\epsilon)$ such that $\tau_m = \bar{K}(\epsilon)\epsilon$. Finally, the pre-stress $\mathbf{T}^0$ is taken to be zero in the case of the lipid bilayer since it has not been shown to exhibit contractile abilities.

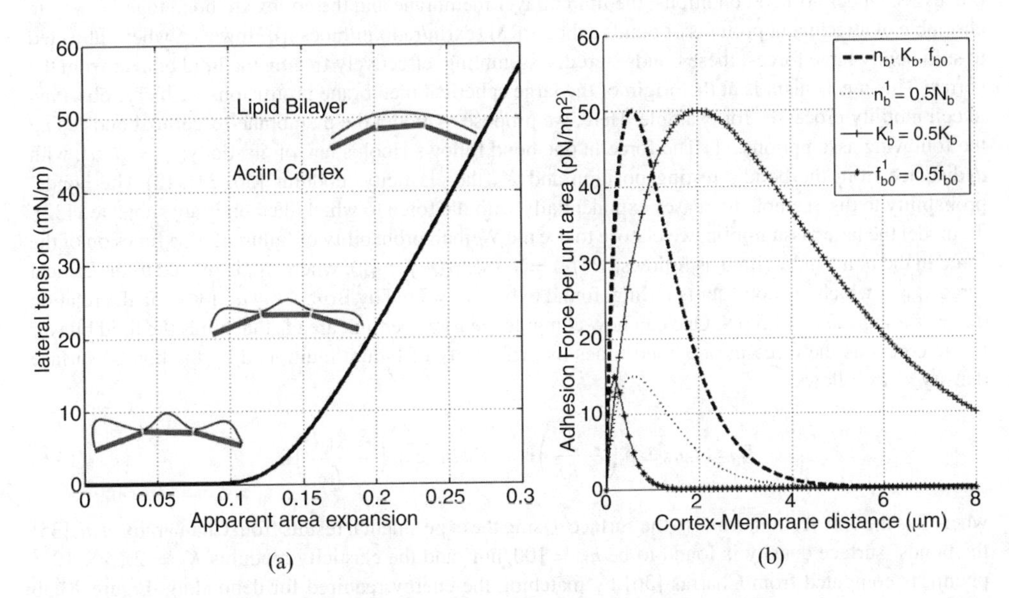


**Figure 13.3** (a) The tension–dilation curve for the unfolding membrane [26]. (b) The behavior of the function $\mathbf{f}_b$ for different values of parameters $K_b$, $f_{b0}$, and $n_b$

### *Bending Energy*

For both the cortex and the lipid bilayer, we choose to use the widely accepted Canham–Helfrich bending energy with different bending stiffnesses, as the cortex offers a greater resistance to bending than the lipid bilayer does. The Canham–Helfrich energy is written $E(\mathbf{C}) = K_h/2(H - C_0)^2 + K_g K$, with $H$ and $K$ the mean and Gauss curvatures, $K_h$ and $K_g$ the bending parameters, and $C_0$ the spontaneous curvature [11]. If the membrane is closed (as is the case in the different problems considered here), the Gaussian curvature energy $K_g K$ has no contribution and can be discarded. This leaves us with the energy $E(\mathbf{C}) = K_h/2(H - C_0)^2$ that is taken into account in $\tilde{W}$ by choosing the different parameters as follows: $W_0 = 1/2K_h C_0^2$, $\mathbf{M}^0 = -2K_h C_0 \mathbf{g}$ and $\mathcal{D} = 1/2K_h \mathbf{g} \otimes \mathbf{g}$.

With the elasticity parameters defined, the elastic energy densities for the cortex and the lipid bilayer now read

$$\text{cortex} \quad \tilde{W}(\bar{\mathbf{E}}, \mathbf{C}) = \underbrace{\mathbf{T}^0 : \bar{\mathbf{E}} + \frac{1}{2}\bar{\mathbf{E}} : \mathcal{C} : \bar{\mathbf{E}}}_{\text{strain energy}} + \underbrace{\frac{K_h^{\text{cortex}}}{2}(H - C_0)^2}_{\text{bending energy}} \tag{13.4}$$

$$\text{lipid bilayer} \quad \tilde{W}(\bar{\mathbf{E}}, \mathbf{C}) = \underbrace{1/2\bar{K}\epsilon^2}_{\text{strain energy}} + \underbrace{\frac{K_h^{\text{lipid}}}{2}(H - C_0)^2}_{\text{bending energy}} \tag{13.5}$$

where $\epsilon = \bar{\mathbf{E}} : \mathbf{g} = \text{tr}\,\bar{\mathbf{E}}$ and $H = \mathbf{C} : \mathbf{g} = \text{tr}\,\mathbf{C}$. Having defined the strain and bending energy densities of both the cortex and the lipid bilayer membrane, let us now turn to the membrane–cortex attachment energy to investigate the mechanical behavior of the MCC, and in particular the debonding between the lipid bilayer membrane and the cortex that occurs during cell blebbing.

### *Membrane–cortex Attachment Energy*

The interaction between the lipid bilayer membrane and the cortex plays a critical role in many cellular processes. Under normal conditions, the lipid bilayer membrane and the cortex are bond together via an ensemble of attachment proteins or bonds called ERM (exrin/radixin/moesin). However, when subjected to sufficiently large forces, these bonds start disassembling, effectively freeing the lipid bilayer from the cortex. This mechanism is at the origin of the large spherical membrane protrusions, or blebs, observed in cell motility processes for example. Here, we propose to model the membrane-to-cortex bonds using the following assumptions. (1) The force in one bond follows Hook's law of elasticity: $f_b = K_b d$, with $d$ distance from the bond's resting position and $K_b$ the elasticity modulus [30, 31]. (2) The bonds' probability to disassemble increases exponentially with the force to which the bonds are subjected [32]. To model the latter assumption, we choose to use the Weibull probability of failure [33], a function of the force in the bond and written as follows: $P(f_b) = 1 - \exp(-f_b/f_{b0})$, where $f_{b0}$ is related to the critical force $f_{\text{crit}}$ at which the bond starts failing, found to be $f_{\text{crit}}$ =5 pN by Brugués *et al.* [34], via the relation $f_{b0} = f_{\text{crit}} \exp(1) = 13.5$ pN. One can now compute the force per unit area $\mathbf{f}_b$ that binds the lipid bilayer to the cortex as the force in one bond times its probability of failure multiplied by the bonds' surface density $n_b$ as follows:

$$\mathbf{f}_b = f_b n_b \exp\left(-\frac{f_b}{f_{b0}}\right)\bar{\mathbf{n}} = K_b d n_b \exp\left(-\frac{K_b d}{f_{b0}}\right)\bar{\mathbf{n}}, \tag{13.6}$$

where $\bar{\mathbf{n}}$ is the unit vector normal to the surface. Using the experimental results from Chamaraux *et al.* [35] the bonds' surface density is found to be $n_b = 100/\text{nm}^2$ and the elasticity modulus $K_b = 2.25 \times 10^{-2}$ pN/nm is computed from Charras [36] by matching the energy required for debonding. Figure 13.3b shows the behavior of the attachment force for different parameter values: reducing the bond elastic modulus $K_b$ increases the total energy required to break the bonds, whereas decreasing the bond density

$n_b$ or the bond critical force $f_{b0}$ has the opposite effect. The membrane-cortex attachment energy (or power) density can then be found as the force $\mathbf{f}_b$ times the membrane–cortex relative displacement (or velocity).

Finally, the MCC total elastic energy is computed as the sum of the lipid bilayer membrane and cortex elastic energy plus the membrane–cortex attachment energy. Let us now turn to the interaction between the lipid bilayer membrane and the external environment, and in particular to the adhesion forces that allow the cell to intake external particles during the endocytosis process.

### *13.2.2 Interaction of the Membrane with the Outer Environment*

An important aspect of the MCC mechanics is to understand how the cell (and particularly the membrane) senses and interacts with its external environment. Membrane adhesion is at the heart of these interactions and plays a primordial role in the intake of external bodies such as vesicles, viruses, or other nanoparticles [37]. This process, known as endocytosis, is one of the mechanisms the cell uses to absorb particles that are too large to simply diffuse through the lipid bilayer membrane. Here, we choose to focus on the receptor-mediated endocytosis that is used by the cell for the uptake of particles with a diameter of the order of tens of nanometers. In this context, the mechanism of endocytosis consists of the wrapping of particles by the lipid bilayer until complete absorption. During this process, two forces are at play and compete against each other: the adhesion between the membrane and the particle and the lipid bilayer bending resistance. If the adhesion that pulls the membrane toward the particle is strong enough to overcome the lipid bilayer bending resistance, the particle becomes completely wrapped and eventually absorbed inside the cell. On the other hand, if the bending resistance is stronger than the adhesion forces, then only partial wrapping is observed and the endocytosis process cannot be achieved.

One of the well-accepted models for cell adhesion was proposed by Bell *et al.* [37], who developed a thermodynamic approach to cell adhesion. In this work, three main variables are at play: the concentration of free ligand on the particle surface, the concentration of free receptors on the membrane surface, and the number of bound ligand–receptor complexes (Figure 13.4a). The system's equilibrium is found by minimizing its total free energy $\Delta G$, given by

$$\begin{aligned}\Delta G = \{(N_r - N_b)[\mu_r(N_r - N_b)] - N_r[\mu_r(N_r)]\} \\ + \{(N_l - N_b)[\mu_l(N_l - N_b)] - N_l[\mu_l(N_l)]\} \\ + N_b[\mu_b(S, N_b)] + \Gamma(d),\end{aligned} \tag{13.7}$$

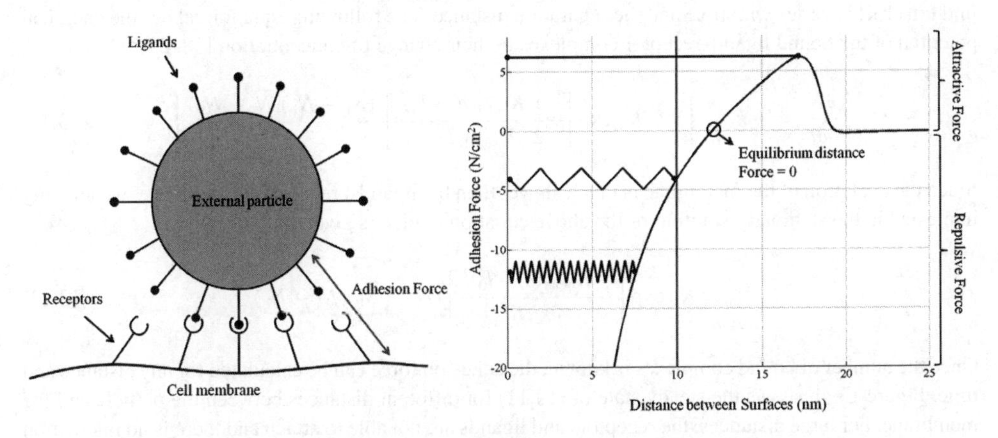


**Figure 13.4** Equilibrium force for different fixed distances

where $N_r$, $N_l$, and $N_b$ are the densities (number per unit surface) of free receptors, ligands, and bonds respectively, and $\mu_r$, $\mu_l$, and $\mu_b$ are their chemical potentials. For dilute solutions, the chemical potentials of the free ligands and receptors are equal to the reference chemical potential plus the entropy of configuration:

$$\mu_\alpha = \mu_\alpha^0 + k_B T \ln(N_\alpha - N_b) \quad \text{with} \quad \alpha = \text{r, l}. \tag{13.8}$$

In the case of the bound ligand–receptor complexes, however, the free energy also includes the elastic energy that comes from stretching these bonds. Considering the bonds as linear elastic "springs" of stiffness $K_{\text{bond}}$ and natural length $l_0$, the bonds' mechano-chemical potential $\mu_b$ is written

$$\mu_b = \mu_b^0 + \frac{1}{2} K_{\text{bond}}(d - l_0)^2 + k_B T \ln(N_b), \tag{13.9}$$

where $d$ denotes the distance between the membrane and the particle. The term $\Gamma(d)$ in (13.7) was introduced to take into account the van der Waals repulsive force between the membrane and particle at small distances. This energy depends on a characteristic decay length $\tau$ and on a strength parameter $\gamma$ as

$$\Gamma(d) = \frac{\gamma}{d} \exp\left(-\frac{d}{\tau}\right). \tag{13.10}$$

The force produced by the attached complexes must overcome this repulsion in order to be able to uptake the external particle. One can then translate the total free energy (13.7) to a tension $\bar{\mathbf{t}}$ applied on the cell membrane by taking its derivative with respect to the separation distance:

$$\bar{\mathbf{t}} = \frac{\partial \Delta G}{\partial \mathbf{d}} = \left[ N_b K_{\text{bond}}(d - l_0) - \frac{\gamma}{d^2 \cdot \tau} \exp\left(\frac{-d}{\tau}\right)(d + \tau) \right] \mathbf{n}, \tag{13.11}$$

where $\mathbf{d} = d\mathbf{n}$ and $\mathbf{n}$ is a unit vector normal to the particle surface. The first term of this equation represents the force associated with the bond complex deformation while the latter part represents the repulsion due to the van der Waals forces. In this chapter, and for the sake of simplicity, we consider both the free ligands and receptors to be abundant enough for their concentration to be considered constant and their diffusion instantaneous. However, the concentration of bound ligand-receptor complexes needed to find $\bar{\mathbf{t}}$ in (13.11) varies with time and the separation distance. The following equation relates the chemical potential of the bound ligand–receptor complexes to their change of concentration [38, 39]:

$$\frac{\partial N_b}{\partial t} = -k_r N_b \left\{ 1 - K_L \exp\left[ -\frac{1}{2} \frac{K_{\text{bond}}(d - l_0)^2}{k_B T} \right] \frac{(N_r - N_b)(N_l - N_b)}{N_b} \right\}. \tag{13.12}$$

Since the evolution of the membrane is much slower then the ligands–receptors binding rate, we are only interested in the stationary solution of the above equation, which is given by

$$N_b = K_L \exp\left[ \frac{-K_{\text{bond}}(d - l_0)^2}{2 k_B T} \right] (N_r - N_b)(N_l - N_b). \tag{13.13}$$

Once the number of bound complexes is known, the adhesion force can be calculated for any distance and time. Figure 13.4b shows the steady state of (13.11) for different distances between the particle and the membrane. For large distances, the receptors and ligands are not able to attach and there is no interaction between the particle and the membrane. As they come closer together, a few receptors are able to attach to the particle's ligands and act as stretched springs, exerting an attractive force between the membrane and the particle. As the process unfolds, the bonds' concentration increases and the membrane wraps around the particle until complete absorption. With the interaction forces $\mathbf{f}_b$ and $\bar{\mathbf{t}}$ defined, let us now derive the equations governing the evolution of the MCC.

## 13.3 Formulation of the Membrane Mechanics and Fluid–Membrane Interaction

The objective of this section is to introduce the reader to a formulation for the elastic membrane–fluid problem that is versatile enough to describe the wide variety of membranes found in biology and their interactions with the surrounding fluids and adhesion forces to external particles. The formulation presented here has the following characteristics: (a) it can describe the dramatically large deformations observed in biology, such as cell blebbing and endocytosis, thanks to an Eulerian description of the problem; (b) it includes discontinuities in pressure and velocity that may appear across the membrane; and (c) it is general enough to accommodate for the description of any hyperelastic membrane. This formulation is then coupled to a novel numerical method that allows for the description of discontinuities in velocity and pressure that arise from the presence of the membrane.

### *13.3.1 Kinematics of Immersed Membrane*

In this section we consider the particular problem of a closed elastic membrane immersed in a three-dimensional domain $\Omega$ filled with fluids that may have different mechanical properties. One of the characteristics of the biological membranes modeled here is the fact that their thickness $h$ is negligible compared with their other dimensions. A cell membrane can therefore be described as a two-dimensional surface $\Gamma$ that splits the fluid domain in two volumes: an enclosed volume $\Omega_1$ and an outer volume $\Omega_2$, such that $\Omega = \Omega_1 \cup \Omega_2$. We choose to express the kinematics of fluids in the embedding three-dimensional Eucledian space $\Omega$ with a fixed Cartesian coordinate system $\{x_1, x_2, x_3\}$ as follows. The motion of a physical particle denoted by point P in Figure 13.5 is expressed by the mapping $\mathbf{x} = \phi(\mathbf{X}, t)$ between its coordinate $\mathbf{X} = \{X_1, X_2, X_3\}$ at a chosen reference time $t = t_0$ and its coordinate $\mathbf{x} = \{x_1, x_2, x_3\}$ at an arbitrary time $t > t_0$. The velocity $\mathbf{v}(\mathbf{x}, t)$ of the particle, located at $\mathbf{x}$ at time $t$, is then written in terms of the material time derivative $\mathbf{v}(\mathbf{x}, t) = D\mathbf{x}/Dt$. An instantaneous measure of deformation can then be provided by the rate of deformation $\mathbf{D} = 1/2(\nabla\mathbf{v} + (\nabla\mathbf{v})^{\mathrm{T}})$, where $\nabla$ is the gradient operator (with respect to the current particle coordinates) in the embedding Euclidean space and the superscript T denotes the transpose operation.

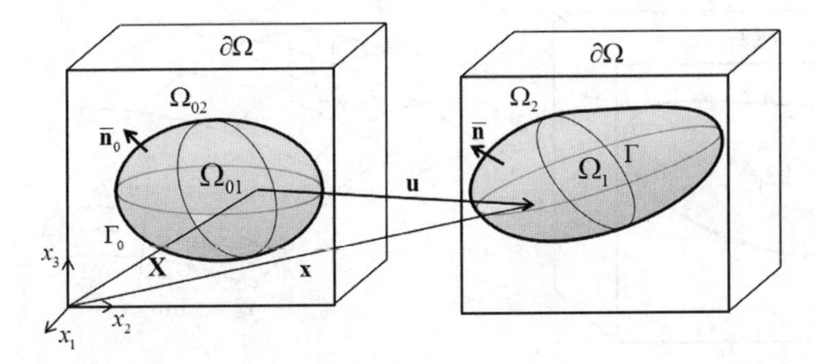


**Figure 13.5** Reference and current configuration of an immersed membrane in a fluid domain $\Omega$

Besides the above kinematic variables that are generally enough to describe the motion of a fluid, one needs to introduce additional quantities to account for the interaction of the fluids with the immersed elastic membrane. As explained earlier, the presence of the membrane generates discontinuities in velocity and stress across $\Gamma$ in the tangential and normal directions. It is of interest, therefore to evaluate the projections of the various continuum fields considered onto the surface $\Gamma$. To do so, we here define the following respective normal and tangential projection operators as [27, 28]

$$\mathbf{P}^{\perp} = \bar{\mathbf{n}} \otimes \bar{\mathbf{n}} \quad \text{and} \quad \mathbf{P}^{\parallel} = \mathbf{I} - \bar{\mathbf{n}} \otimes \bar{\mathbf{n}}, \tag{13.14}$$

where $\bar{\mathbf{n}}$ is a unit vector normal to $\Gamma$, $\mathbf{I}$ is the second-order unit tensor, and $\otimes$ is the dyadic product. Using these projection operators, we can now decompose the velocity field $\mathbf{v}$ into its normal and tangential contributions as follows:

$$\mathbf{v} = \mathbf{v}^{\perp} + \mathbf{v}^{\parallel} \quad \text{with} \quad \mathbf{v}^{\perp} = \mathbf{P}^{\perp} \cdot \mathbf{v} \quad \text{and} \quad \mathbf{v}^{\parallel} = \mathbf{P}^{\parallel} \cdot \mathbf{v}. \tag{13.15}$$

The normal and tangential contributions of any other vectorial or tensorial fields can be obtained in the same fashion [27, 40]. Next, we assign two different values of the fluid tangential velocity $\mathbf{v}^{\parallel +}$ and $\mathbf{v}^{\parallel -}$ on the outer and inner sides of the membrane, respectively (Figure 13.6). With these definitions and noting $\bar{\mathbf{v}}$ the membrane velocity, the jumps of tangential velocity between the fluid and the membrane can be written:

$$[\mathbf{v}]^{\parallel +} = \mathbf{v}^{\parallel +} - \bar{\mathbf{v}}^{\parallel}, \quad [\mathbf{v}]^{\parallel -} = \mathbf{v}^{\parallel -} - \bar{\mathbf{v}}^{\parallel}, \tag{13.16}$$

and

$$[\mathbf{v}]^{\parallel} = \mathbf{v}^{+\parallel} - \mathbf{v}^{-\parallel} = [\mathbf{v}]^{+\parallel} - [\mathbf{v}]^{-\parallel}, \tag{13.17}$$

where $[\mathbf{v}]^{\parallel}$ represents the total tangential velocity jump across $\Gamma$. Additionally, let us denote $\{\bar{\mathbf{v}}^{\parallel}\} = (\bar{\mathbf{v}}^{\parallel +} + \bar{\mathbf{v}}^{\parallel -})/2$ as the mean value of $\bar{\mathbf{v}}^{\parallel}$ across the membrane.

With the kinematics of the fluids and its interaction with the membrane defined, we now turn to deformation measures of the membrane itself. In view of this, one first needs to introduce certain geometrical measures, such as the metric and the second fundamental form, that describe the curvature of the surface $\Gamma$. Since the membrane is represented by a two-dimensional surface, we choose to parameterize it by a local two-dimensional coordinate system $(s^1, s^2) \mapsto \mathbf{x}(s^1, s^2)$ in $\mathbb{R}^3$, as depicted in Figure 13.6. The covariant basis vectors of the plane tangent to the membrane surface $\Gamma$ can be computed

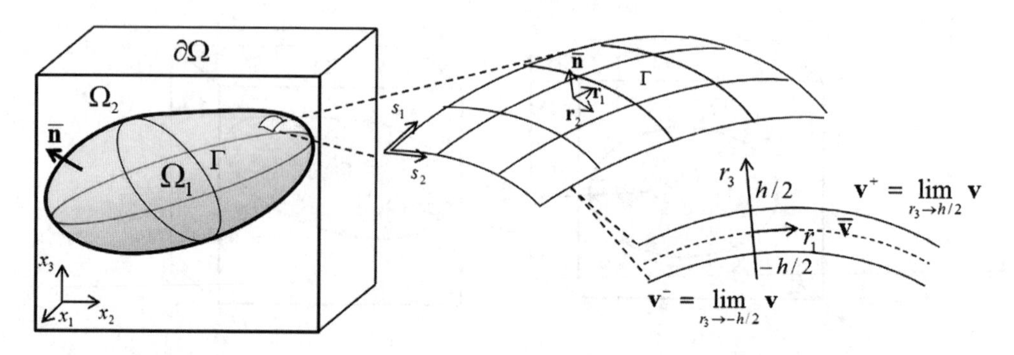


**Figure 13.6** Surface parameterization of the membrane and fields discontinuities

from the derivatives of $\mathbf{x}$ with respect to $(s^1, s^2)$ as $\mathbf{r}_\alpha = \mathbf{x}_{,\alpha}$. The contravariant basis vectors $\mathbf{r}^\alpha$ follow as $\mathbf{r}_\alpha \cdot \mathbf{r}^\beta = \delta_\alpha^\beta$ and the covariant and contravariant metric tensors as $g_{\alpha\beta} = \mathbf{r}_\alpha \cdot \mathbf{r}_\beta$ and $g^{\alpha\beta} = \mathbf{r}^\alpha \cdot \mathbf{r}^\beta$. The vector $\bar{\mathbf{n}}$ normal to the surface and the components of the second fundamental form $\mathbf{C}$ are then written as

$$\bar{\mathbf{n}} = \mathbf{r}_3 = \frac{\mathbf{r}_1 \times \mathbf{r}_2}{|\mathbf{r}_1 \times \mathbf{r}_2|} \quad \text{and} \quad C_{\alpha\beta} = \mathbf{r}_{\alpha,\beta} \cdot \bar{\mathbf{n}}. \tag{13.18}$$

We choose to use the Hencky strain to describe the membrane deformations, as it is suitable for large deformations and adapted to an Eulerian formulation of the problem. In the local coordinate system $\{r_1, r_2, r_3\}$ oriented in the directions $\mathbf{r}_1$, $\mathbf{r}_2$, and $\mathbf{r}_3$, we define the deformation gradient $\bar{F}_{ij}(t) = \partial \bar{x}_i(\bar{\mathbf{X}}, t)/\partial \bar{X}_j$ at time $t$, $\bar{\mathbf{X}}$ and $\bar{\mathbf{x}}$ being the initial and current positions of a material point on the membrane. This second-order tensor represents the local deformation of a material point $\bar{\mathbf{X}}$. Taking advantage of the polar decomposition theorem, we write the deformation tensor $\bar{\mathbf{F}}(t) = \bar{\mathbf{R}}(t)\bar{\mathbf{V}}(t)$ as the product of a symmetric positive-definite stretch tensor $\bar{\mathbf{V}}(t)$ and an orthogonal rotation tensor $\bar{\mathbf{R}}(t)$. The Hencky strain is defined as the logarithm of the left stretch tensor, and the Hencky strain rate is identified with the membrane deformation rate $\bar{\mathbf{D}}$ [41]:

$$\bar{\mathbf{E}}(t) = \ln(\bar{\mathbf{V}}(t)) \quad \text{and} \quad \dot{\bar{\mathbf{E}}} = \bar{\mathbf{D}}, \tag{13.19}$$

where the dot represents the absolute time derivative. The rate of deformation $\bar{\mathbf{D}}$ of the membrane is given by [42]

$$\bar{\mathbf{D}} = \frac{1}{2}[\nabla^{\parallel}\bar{\mathbf{v}}^{\parallel\flat} + (\nabla^{\parallel}\bar{\mathbf{v}}^{\parallel\flat})^{\mathrm{T}}] + (\bar{\mathbf{v}}^{\perp} \cdot \bar{\mathbf{n}})\mathbf{C}, \tag{13.20}$$

where $(\nabla^{\parallel}\bar{\mathbf{v}}^{\parallel\flat})_{ij} = \bar{\mathbf{v}}^{\parallel}_{i|j} = (\partial \bar{v}^{\parallel}_i/\partial r_j) - \Gamma^k_{ij}\bar{v}^{\parallel}_k$ denotes the covariant derivative of $\bar{\mathbf{v}}^{\parallel\flat}$ (the vector $\bar{\mathbf{v}}^{\parallel}$ in the covariant basis), $\Gamma^k_{ij}$ denotes the Christoffel symbols (whose numerical evaluation is described in Section 13.4), and the subscripts $\parallel$ and $\perp$ indicate the tangential and normal contributions of the membrane velocity: $\bar{\mathbf{v}} = \bar{\mathbf{v}}^{\parallel} + \bar{\mathbf{v}}^{\perp}$.

### 13.3.2 Variational Formulation of the Immersed MCC Problem

Let us now focus on deriving the governing equations of the immersed cortex–membrane problem. We choose to do so through the use of the principle of virtual power, as the mathematical formulation of variational calculus it employs will allow us to obtain the final set of governing equations with minimal effort. In this view, we first consider the internal power of the system, considering flows with low Reynolds number (the kinetic energy becomes negligible), when subjected to the rates of deformations $\mathbf{D}$ and $\dot{\bar{\mathbf{E}}}$ of the fluids and the MCC. The total internal energy of the immersed MCC is written as

$$P_{int} = P^{\mathrm{f}}_{\mathrm{int}} + P^{\mathrm{fric}}_{\mathrm{int}} + P^{\mathrm{c}}_{\mathrm{int}} + P^{l}_{\mathrm{int}} + P^{b}_{\mathrm{int}}, \tag{13.21}$$

where $P^{\mathrm{f}}_{\mathrm{int}}$ denotes the internal power of the fluids inside and surrounding the cell, $P^{\mathrm{fric}}_{\mathrm{int}}$ is the dissipation power of frictional forces between the fluids and the lipid bilayer (the cortex being considered as permeable), $P^{\mathrm{c}}_{\mathrm{int}}$ and $P^{\mathrm{l}}_{\mathrm{int}}$ are the elastic power of the cortex and of the lipid bilayer, and $P^{\mathrm{b}}_{\mathrm{int}}$ represents the power generated by the membrane–cortex binding force defined in Equation (13.6). When no debonding occurs between the lipid bilayer and the cortex, the MCC can be viewed as a single elastic membrane. In that case, $P^{\mathrm{m}}_{\mathrm{int}} = P^{\mathrm{c}}_{\mathrm{int}} + P^{\mathrm{l}}_{\mathrm{int}}$ represents the internal power of a single membrane with mechanical properties equivalent to those of the combined lipid bilayer and cortex, and the power $P^{\mathrm{b}}_{\mathrm{int}}$ generated by the membrane–cortex binding force vanishes since the membrane–cortex relative velocity is zero.

If debonding occurs between the lipid bilayer and the cortex (as during the blebbing problem – see Section 13.5), the binding power $P^{\mathrm{b}}_{\mathrm{int}}$ is nonzero, and the cortex and the lipid bilayer are viewed as two separate membranes with velocities $\bar{\mathbf{v}}_{\mathrm{c}}$ and $\bar{\mathbf{v}}_{\mathrm{l}}$, strains $\bar{\mathbf{D}}_{\mathrm{c}}$ and $\bar{\mathbf{D}}_{\mathrm{l}}$, and second fundamental forms $\mathbf{C}_{\mathrm{c}}$ and $\mathbf{C}_{\mathrm{l}}$, respectively. Let us first turn to $P^{\mathrm{f}}_{\mathrm{int}}$, the fluids internal power, and write it as the integral over the volume of fluid $\Omega$ of the fluid power density:

$$P^{\mathrm{f}}_{\mathrm{int}} = \int_{\Omega} \mathbf{T} : \mathbf{D}\, \mathrm{d}\Omega, \tag{13.22}$$

where $\mathbf{T} = \sigma - p\mathbf{I}$ is the fluid stress with $p$ the pressure, $\sigma = \mu\mathbf{D}$ is the viscous stress exerted by the fluid, and $\mu = \mu_1$ in $\Omega_1$ and $\mu = \mu_2$ in $\Omega_2$ are the viscosities of the fluids. Next, the frictional power $P^{\mathrm{fric}}_{\mathrm{int}}$ is written as

$$P^{\mathrm{fric}}_{\mathrm{int}} = \int_{\Gamma_{\mathrm{l}}} \frac{1}{2}(\mathbf{f}^{+} \cdot [\mathbf{v}]^{\|+} + \mathbf{f}^{-} \cdot [\mathbf{v}]^{\|-})\, \mathrm{d}\Gamma_{\mathrm{l}}, \tag{13.23}$$

where $\Gamma_{\mathrm{l}}$ and $\Gamma_{\mathrm{c}}$ respectively denote the lipid bilayer membrane and the cortex and the friction forces are defined as $\mathbf{f}^{+,-} = \mu^{+,-}_{\mathrm{f}}[\mathbf{v}]^{\|+,-}$, where $\mu^{+,-}_{\mathrm{f}}$ are the friction coefficients of the fluids on each side of the membrane [43, 44]. The extreme cases $\mu^{+}_{\mathrm{f}}, \mu^{-}_{\mathrm{f}} = 0$ and $\mu^{+}_{\mathrm{f}}, \mu^{-}_{\mathrm{f}} = \infty$ respectively correspond to the free-slip and no-slip condition between the membrane and the fluids. In case (a), the lipid bilayer $\Gamma_{\mathrm{l}}$ and the cortex $\Gamma_{\mathrm{c}}$ are represented by a single membrane $\Gamma$; hence, the frictional power in this context reads $P^{\mathrm{fric}}_{\mathrm{int}} = \int_{\Gamma} 1/2(\mu^{+}_{\mathrm{f}}[\mathbf{v}]^{\|+} \cdot [\mathbf{v}]^{\|+} + \mu^{-}_{\mathrm{f}}[\mathbf{v}]^{\|-} \cdot [\mathbf{v}]^{\|-})\, \mathrm{d}\Gamma$. Moreover, the power of the binding force $\mathbf{f}_{\mathrm{b}}$ can be computed as the integral over $\Gamma_{\mathrm{c}}$ and $\Gamma_{\mathrm{l}}$ of the force times the velocity as follows:

$$P^{\mathrm{b}}_{\mathrm{int}} = \int_{\Gamma_{\mathrm{c}}} \mathbf{f}_{\mathrm{b}} \cdot \bar{\mathbf{v}}_{\mathrm{c}}\, \mathrm{d}\Gamma_{\mathrm{c}} - \int_{\Gamma_{\mathrm{l}}} \mathbf{f}_{\mathrm{b}} \cdot \bar{\mathbf{v}}_{\mathrm{l}}\, \mathrm{d}\Gamma_{\mathrm{l}}. \tag{13.24}$$

Note that this quantity vanishes when the cortex and the lipid bilayer follow the same motion ($\Gamma_{\mathrm{c}} = \Gamma_{\mathrm{l}}$ and $\bar{\mathbf{v}}_{\mathrm{c}} = \bar{\mathbf{v}}_{\mathrm{l}}$). Let us now turn to the cortex and lipid bilayer membrane elastic power $P^{i}_{\mathrm{int}}$ where $i = \mathrm{c, l}$. We describe here a generic elastic membrane with a velocity $\bar{\mathbf{v}}$, strain $\bar{\mathbf{E}}$, and second fundamental form $\mathbf{C}$ that can be adapted to account for the MCC, the lipid bilayer, or the cortex by choosing the elastic parameter and kinematic variables that correspond to each of those cases. Using the elastic energy density $\tilde{W}$ defined in Equation (13.2), one can find the internal elastic power of the membrane by taking the absolute time derivative of the elastic energy [40]:

$$P^{i}_{\mathrm{int}} = \frac{\mathrm{d}}{\mathrm{d}t} \int_{\Gamma} \bar{\rho}\tilde{W}\, \mathrm{d}\Gamma = \frac{\mathrm{d}}{\mathrm{d}t} \int_{\Gamma} \bar{\rho}\bar{\mathbf{f}} \cdot \bar{\mathbf{v}}\, \mathrm{d}\Gamma, \tag{13.25}$$

where

$$\bar{\mathbf{f}} = \underbrace{(-\bar{\mathbf{T}} : \mathbf{C}\,\bar{\mathbf{n}} - \nabla^{\|} \cdot \bar{\mathbf{T}}^{\sharp})}_{\text{stretching}} + \underbrace{(\nabla^{\|} \cdot (\nabla^{\|} \cdot \bar{\mathbf{m}}) - \bar{\mathbf{m}} : \mathbf{C}^2)\bar{\mathbf{n}}}_{\text{bending}}, \tag{13.26}$$

and where $\bar{\mathbf{T}} = \bar{\rho}\tilde{W}_{,\bar{\mathbf{E}}}$ denotes the membrane stress and $\bar{\mathbf{m}} = \bar{\rho}\tilde{W}_{,\bar{\mathbf{C}}}$ is the membrane moment, respectively defined as the energy conjugate of the membrane strain and curvature. The sharp operator # raises the indices of a tensor (from $\bar{T}_{ij}$ to $\bar{T}^{ij}$, for example). We can now recognize in the above equation the elastic force of the membrane $\mathbf{f}$, where the first term represents the force resulting from stretching the membrane and the second term denotes the force resulting from bending it. By choosing the Helfrich energy for the membrane bending energy, the bending component of the membrane force (13.26) can be rewritten as [40]

$$\bar{\mathbf{f}} = (-\bar{\mathbf{T}} : \mathbf{C}\bar{\mathbf{n}} - \nabla^{\parallel} \cdot \bar{\mathbf{T}}^{\sharp}) + K_{\mathrm{h}}\left[\Delta^{\parallel} H + \frac{H - C_0}{2}(H^2 - 4K + HC_0)\right]\bar{\mathbf{n}}, \tag{13.27}$$

where we recognize the classical Canham–Helfrich bending force [45]. Having defined the internal powers of the embedding fluids and the membranes (cortex or lipid bilayer), one can now write the total internal variational power $\delta P_{\mathrm{int}}$ as

$$\delta P_{\mathrm{int}} = \delta P^{\mathrm{f}}_{\mathrm{int}} + \delta P^{\mathrm{fric}}_{\mathrm{int}} + P^{\mathrm{c}}_{\mathrm{int}} + P^{\mathrm{b}}_{\mathrm{int}} + \delta P^{\mathrm{l}}_{\mathrm{int}}, \tag{13.28}$$

with

$$\delta P^{\mathrm{f}}_{\mathrm{int}} = \int_{\Omega} (\sigma : \delta\mathbf{D} + \nabla \cdot \mathbf{v}\delta p)\, \mathrm{d}\Omega, \tag{13.29}$$

$$\text{friction power} \quad \delta P^{\mathrm{fric}}_{\mathrm{int}} = \int_{\Gamma} (\mathbf{f}^{+}_{\mathrm{f}} \cdot \delta[\mathbf{v}]^{\parallel +} + \mathbf{f}^{-}_{\mathrm{f}} \cdot \delta[\mathbf{v}]^{\parallel -})\, \mathrm{d}\Gamma, \tag{13.30}$$

$$\text{membrane–cortex binding power} \quad \delta P^{\mathrm{b}}_{\mathrm{int}} = \int_{\Gamma_{\mathrm{c}}} \mathbf{f}_{\mathrm{b}} \cdot \delta\bar{\mathbf{v}}_{\mathrm{c}}\, \mathrm{d}\Gamma_{\mathrm{c}} - \int_{\Gamma_{\mathrm{l}}} \mathbf{f}_{\mathrm{b}} \cdot \delta\bar{\mathbf{v}}_{\mathrm{l}}\, \mathrm{d}\Gamma_{\mathrm{l}}, \tag{13.31}$$

$$\text{cortex} \quad \delta P^{\mathrm{c}}_{\mathrm{int}} = \int_{\Gamma_{\mathrm{c}}} \bar{\mathbf{f}}_{\mathrm{c}} \cdot \delta\bar{\mathbf{v}}_{\mathrm{c}}\, \mathrm{d}\Gamma_{\mathrm{c}}, \tag{13.32}$$

$$\text{lipid bilayer} \quad \delta P^{\mathrm{l}}_{\mathrm{int}} = \int_{\Gamma_{\mathrm{l}}} \bar{\mathbf{f}}_{\mathrm{l}} \cdot \delta\bar{\mathbf{v}}_{\mathrm{l}}\, \mathrm{d}\Gamma_{\mathrm{l}}, \tag{13.33}$$

where $\bar{\mathbf{f}}_{\mathrm{c}}$ and $\bar{\mathbf{f}}_{\mathrm{l}}$ denote the cortex and lipid bilayer force and are obtained from $\bar{\mathbf{f}}$ by changing the elastic parameters. This expression will prove useful in applying the principle of virtual power to obtain the governing equations in Section 13.3.3.

### *13.3.3 Principle of Virtual Power and Conservation of Momentum*

To complete the formulation of the cell problem, we now turn to the balance of momentum in the fluids and cell membrane. Although including the inertial terms in the formulation would not present major difficulties, we consider here a quasi-static system where the fluids and membrane move slowly enough for the inertia forces to become negligible. The principle of virtual power states that the change in the power of the system due to a change in fluids and membranes (cortex or lipid bilayer) velocities is $\delta P_{\mathrm{int}} = \delta P_{\mathrm{ext}}$, where $\delta P_{\mathrm{int}}$ was defined in (13.28) and $\delta P_{\mathrm{ext}}$ includes the power of body forces $\mathbf{b}$ on the fluids as well as the power of traction forces $\bar{\mathbf{t}}$ on $\Gamma_{\mathrm{l}}$ (such as the binding force between the vesicle and the lipid bilayer in case (b)) and $\mathbf{t}$ on $\delta\Omega$. The lipid bilayer $\Gamma_{\mathrm{l}}$ is also subjected to Dirichlet-type boundary conditions in order to satisfy the Navier boundary condition on the normal component of velocities: $\bar{\mathbf{v}}^{\perp} = \mathbf{v}^{\perp}$ everywhere on $\Gamma_{\mathrm{l}}$. The subsets $\Gamma^{\mathrm{d}}_{\mathrm{l}}$ and $\delta\Omega^{\mathrm{d}}$ may as well be subjected to fixed velocities $\bar{\mathbf{v}}_{\mathrm{l}} = \bar{\mathbf{v}}^{\mathrm{d}}_{\mathrm{l}}$ and $\mathbf{v} = \mathbf{v}^{\mathrm{d}}$ (Figure 13.7).

Ultimately, the contribution from external forces on the fluids $\Omega$ and the lipid bilayer $\Gamma_{\mathrm{l}}$ reads

$$\delta P_{\mathrm{ext}} = \int_{\Omega} \mathbf{b} \cdot \delta\mathbf{v}\, \mathrm{d}\Omega + \int_{\delta\Omega_{\mathrm{n}}} \mathbf{t} \cdot \delta\mathbf{v}\, \mathrm{d}\delta\Omega_{\mathrm{n}} + \int_{\Gamma^{\mathrm{n}}_{\mathrm{l}}} \bar{\mathbf{t}} \cdot \delta\bar{\mathbf{v}}_{\mathrm{l}}\, \mathrm{d}\Gamma^{\mathrm{d}}_{\mathrm{l}}. \tag{13.34}$$

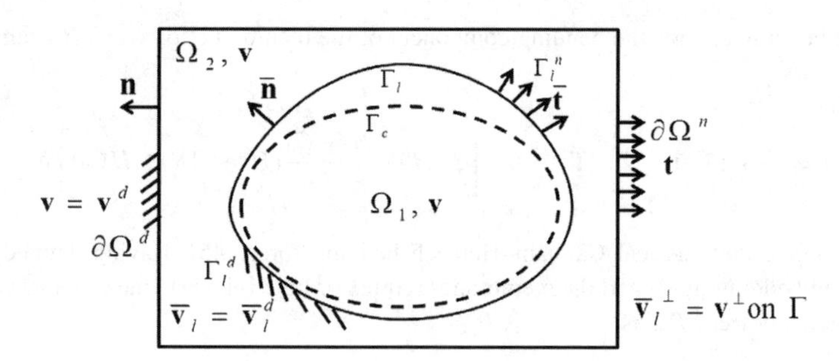


**Figure 13.7** Boundary conditions

To describe the Dirichlet and Neumann boundary conditions, the boundaries $\Gamma$ and $\delta\Omega$ are decomposed into two parts: $\Gamma_\mathrm{l} = \Gamma_\mathrm{l}^\mathrm{d} \cap \Gamma_\mathrm{l}^\mathrm{n}$ and $\delta\Omega = \delta\Omega^\mathrm{d} \cap \delta\Omega^\mathrm{n}$, where the subscript "d" and "n" respectively denote Dirichlet and Neumann boundary-type conditions. To derive the governing equations, we first integrate (13.28) by parts and apply the divergence theorem. This leads to the following expression of the variational internal power:

$$\begin{aligned} \delta P_\mathrm{int} = & -\int_\Omega [(\nabla\cdot\mathbf{T})\cdot\delta\mathbf{v} + \nabla\cdot\mathbf{v}\delta p]\,\mathrm{d}\Omega \\ & + \int_{\Gamma_\mathrm{l}} ([\mathbf{T}\cdot\bar{\mathbf{n}}\cdot\delta\mathbf{v}] + \mathbf{f}_\mathrm{f}^{+}\cdot\delta[\mathbf{v}]_\mathrm{l}^{\parallel+} + \mathbf{f}_\mathrm{f}^{-}\cdot\delta[\mathbf{v}]_l^{\parallel-} + \bar{\mathbf{f}}_l\cdot\delta\bar{\mathbf{v}}_\mathrm{l} - \mathbf{f}_\mathrm{b}\cdot\delta\bar{\mathbf{v}}_\mathrm{l})\,\mathrm{d}\Gamma_\mathrm{l} \\ & + \int_{\Gamma_\mathrm{c}} (\bar{\mathbf{f}}_\mathrm{c}\cdot\delta\bar{\mathbf{v}}_\mathrm{c} + \mathbf{f}_\mathrm{b}\cdot\delta\bar{\mathbf{v}}_\mathrm{c})\,\mathrm{d}\Gamma_\mathrm{c} + \int_{\delta\Omega}\mathbf{T}\cdot\mathbf{n}\cdot\delta\mathbf{v}\,\mathrm{d}\delta\Omega, \end{aligned} \tag{13.35}$$

where $\bar{\mathbf{n}}$ and $\mathbf{n}$ are respectively the unit vectors normal to $\Gamma_\mathrm{l}$ and $\delta\Omega$. Using the fact that $\delta P_\mathrm{int} = \delta P_\mathrm{ext}$ remains true for any arbitrary fields $\delta\mathbf{v}$, $\delta\bar{\mathbf{v}}_\mathrm{l}$, $\delta\bar{\mathbf{v}}_\mathrm{c}$, $\delta\mathbf{v}^{\parallel+}$, $\delta\mathbf{v}^{\parallel-}$, $\delta[\mathbf{v}]$, $\delta p$, and $\delta\bar{p}$ that satisfy

$$\delta\bar{\mathbf{v}}_\mathrm{l} = 0 \text{ on } \Gamma_\mathrm{l}^\mathrm{d} \quad \text{and} \quad \delta\mathbf{v} = 0 \text{ on } \delta\Omega^\mathrm{d}, \tag{13.36}$$

one can derive the following set of governing equations:

$$\text{in the fluids } \Omega \quad \begin{cases} \nabla\cdot\mathbf{T} + \mathbf{b} = 0, & (11.37\text{a}) \\ \nabla\cdot\mathbf{v} = 0, & (11.37\text{b}) \end{cases}$$

$$\text{on the lipid bilayer } \Gamma_\mathrm{l} \quad \begin{cases} [\mathbf{T}\cdot\bar{\mathbf{n}}]^{\parallel} - \mathbf{f}_f^{+} - \mathbf{f}_f^{-} = 0, & (11.38\text{a}) \\ [\mathbf{T}\cdot\bar{\mathbf{n}}]^{\perp} + \bar{\mathbf{f}}_l^{\perp} - \bar{\mathbf{t}}^{\perp} - (\mathbf{f}_\mathrm{b})^{\perp} = 0, & (11.38\text{b}) \\ \mathbf{f}_\mathrm{f}^{+} + \mathbf{f}_\mathrm{f}^{-} + \bar{\mathbf{f}}_\mathrm{l}^{\parallel} - \bar{\mathbf{t}}^{\parallel} - (\mathbf{f}_\mathrm{b})^{\parallel} = 0, & (11.38\text{c}) \\ \bar{\mathbf{v}}_\mathrm{l}^{\perp} = \mathbf{v}^{\perp}, & (11.38\text{d}) \end{cases}$$

$$\text{and on the cortex } \Gamma_\mathrm{c} \quad \mathbf{f}_\mathrm{c} + \mathbf{f}_\mathrm{b} = 0, \tag{13.39}$$

associated with the boundary conditions

$$\mathbf{T}\cdot\mathbf{n} - \mathbf{t} = 0 \text{ on } \delta\Omega^\mathrm{n}, \quad \mathbf{v} = \mathbf{v}_\mathrm{l}^\mathrm{d} \text{ on } \delta\Omega^\mathrm{d}, \quad \text{and} \quad \bar{\mathbf{v}}_\mathrm{l} = \bar{\mathbf{v}}_\mathrm{l}^\mathrm{d} \text{ on } \Gamma_\mathrm{l}^\mathrm{d}. \tag{13.40}$$

Equation (13.37a) represents the force equilibrium and incompressibility in the fluids $\Omega_1$ and $\Omega_2$. The equilibrium between the fluids and the membrane forces is represented by Equations (13.38a)–(13.38d); Equation (13.38a) shows the dependence of the tangential jump of stress $[\mathbf{T}\cdot\bar{\mathbf{n}}]^{\parallel}$ across $\Gamma_1$ on the friction exerted by the lipid bilayer on the fluids. The normal jump of stress $[\mathbf{T}\cdot\bar{\mathbf{n}}]^{\perp}$ across $\Gamma_1$ is shown to result from the normal component of the membrane elastic force in Equation (13.38b). Finally, Equation (13.38c) states that the tangential component of the membrane elastic force reacts to the sum of the frictional forces exerted by the fluids onto the membrane, Equation (13.38d) ensures the continuity of the normal components of the fluids and membrane velocities at the membrane, and Equation (13.39) reflects the equilibrium between the elastic and contractile force of the cortex and the membrane–cortex binding force. Taking into account the tangential velocity and stress discontinuities makes these equations extremely versatile and able to describe a wide variety of problems in an Eulerian framework.

## 13.4 The Extended Finite Element and the Grid-Based Particle Methods

This section describes a numerical strategy to solve governing equations (13.37a)–(13.40) on a structured Eulerian finite-element mesh in the two-dimensional axisymmetric case. The particularity of our system is the presence of a moving membrane $\Gamma$ that separates two fluids, creating discontinuities in the velocity and pressure fields across the membrane. These types of discontinuities cannot be naturally handled by the regular finite-element method. We therefore turn to the X-FEM formalism [46, 47] to solve this issue, since it enables the presence of these discontinuities and, more importantly, makes the description of the membrane $\Gamma$ independent from the spatial discretization at minimal computational cost. The Eulerian domain used for our formulation is subdivided into a mixed formulation mesh with 4/9-node elements to solve for the velocity and pressure. To allow for the existence of discontinuous fields in individual elements, we write the interpolated velocity $\tilde{\mathbf{v}}^e$ and pressure $\tilde{p}^e$ in an element $e$ as the sum of two terms parameterized by nodal values $\mathbf{v}$ and $\hat{\mathbf{v}}$ and by $p$ and $\hat{p}$, respectively, as follows [46, 47]:

$$\tilde{\mathbf{v}}_i^e(\mathbf{x}) = \sum_{I=1}^{n_9} N_I^9(\mathbf{x})\mathbf{v}_i^I + \sum_{J=1}^{m_9} N_J^9(\mathbf{x})(H(\phi(\mathbf{x})) - H(\phi(\mathbf{x}_J)))\hat{\mathbf{v}}_i^J, \tag{13.41}$$

$$\tilde{p}^e(\mathbf{x}) = \sum_{I=1}^{n_4} \mathbf{N}_I^4(\mathbf{x})p_I + \sum_{J=1}^{m_4} \mathbf{N}_J^4(\mathbf{x})(H(\phi(\mathbf{x})) - H(\phi(\mathbf{x}_J)))\hat{p}_J, \tag{13.42}$$

where the lower case $i$ indicates the component of a vector and upper case is used for node number. Furthermore, $n_\alpha$ is the total number of nodes associated with an element $\alpha$ and $m_\alpha$ is the number of enriched nodes (Figure 13.8a). The membrane $\Gamma$ is defined by the intersection of a level-set surface $\phi(\mathbf{x})$ (taken to be the signed distance function when close to $\Gamma$, a positive constant outside of $\Gamma$, and a negative constant inside of $\Gamma$) with a cutting plane as shown in Figure 13.8, and the Heaviside function $H(\phi(\mathbf{x}))$ provides the discontinuity needed to describe the fields. The functions $N_I^\alpha$ are the finite-element shape functions associated with node $I$ of the nine- ($\alpha = 9$) or four- ($\alpha = 4$) node elements and $N_J^\alpha$ the shape functions associated with node $J$ of an element cut by $\Gamma$ (Figure 13.8a). They are constructed from Lagrange polynomials as follows:

$$N_I^\alpha(\xi,\eta) = \prod_{\substack{k=1\\k\neq I}}^{n_{\alpha x}} \frac{\xi - \xi_k}{\xi_I - \xi_k} \prod_{\substack{k=1\\k\neq I}}^{n_{\alpha y}} \frac{\eta - \eta_k}{\eta_I - \eta_k}, \tag{13.43}$$

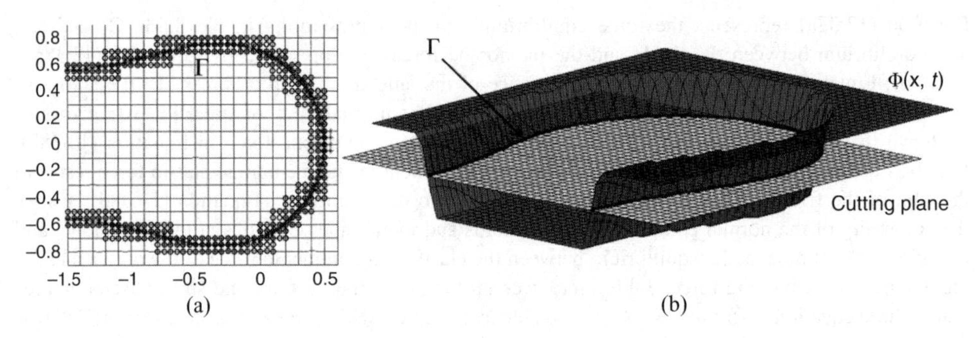


**Figure 13.8** (a) Elements split by membrane Γ and the enriched nodes (circle). The three-dimensional distance function can be seen in (b) and its intersection with the plan $z = 0$ defines Γ, and (c) is the Heaviside function $H$

where $(\xi, \eta)$ are the coordinates in parent elements (that vary from $-1$ to $+1$), and $(n_{\alpha x}, n_{\alpha y})$ the number of nodes in individual elements in the $x$- and $y$-directions. In order to numerically solve the membrane–fluid interaction problem, one also needs to track the membrane position, store and update its Lagrangian information (such as the membrane strain $\bar{\mathbf{E}}$), and compute the high-order derivatives of geometrical information such as the surface Laplacian of the curvature or the Christoffel symbols. The method chosen here is the updated grid-based particle method described by Leung and Zhao [48]. In this work, the interface is represented and tracked by particles (denoted as foot points) associated with grid points on the fixed underlying mesh (Figure 13.9a). To update the position of the membrane, one only needs to interpolate the membrane velocity (found in the previous iteration) at those points and move them accordingly. As the membrane evolves, the grid points associated with the foot points are resampled in order to insure a homogeneous repartition of the foot points on the membrane. Using these particles, one can easily construct polynomials that locally approximate the membrane surface as well as any Lagrangian information (elasticity parameters, strain, etc.) and compute various geometrical quantities: given a point $\mathbf{p}$ on the membrane, the closest foot points are collected to construct a polynomial $s_{(}s_1, s_2)$ (chosen to be of order 2 in this work) that locally approximates the surface at that point. This polynomial is constructed in a local system oriented in the direction normal to the surface $\bar{\mathbf{n}}$ at the closest foot point $(\mathbf{x}_0)$ and is a function of the parameters $(s_1, s_2)$ that run in the directions $(\mathbf{r}_1, \mathbf{r}_2)$ tangent to the surface (Figure 13.9b):

$$\tilde{y}(s_1, s_2) = a_{0,0} + a_{1,0}s_1 + a_{0,1}s_2 + a_{2,0}s_1^2 + a_{0,2}s_2^2, \tag{13.44}$$

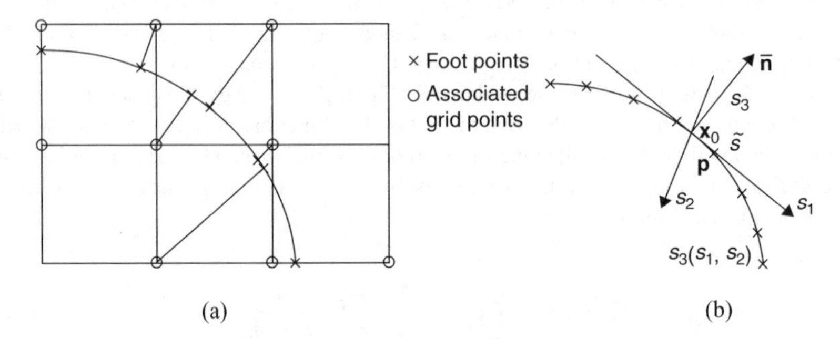


**Figure 13.9** (a) Foot points and associated grid points; (b) local polynomial approximation of the surface (and of any Lagrangian field)

where the coefficient $a_{i,j}$ are found using least square fitting. A point on the membrane is therefore given in local coordinates by $\mathbf{s} = \{s_1, s_2, s_3\}^\top$, and in global coordinate by $\mathbf{x}$, found by operating a rotation and a translation on $\mathbf{s}$:

$$\mathbf{x} = \begin{bmatrix} \bar{\mathbf{n}}_2 & \bar{\mathbf{n}}_1 \\ -\bar{\mathbf{n}}_1 & \bar{\mathbf{n}}_2 \end{bmatrix} \mathbf{s} + \mathbf{x}_0. \tag{13.45}$$

From this function one can extract geometrical information such as the normal $\bar{\mathbf{n}}$ and the metric $\mathbf{g}$ at a point $\mathbf{p}$ [48]:

$$\bar{\mathbf{n}}(\mathbf{p}) = \left.\frac{\mathbf{x}_{,s_1} \times \mathbf{x}_{,s_2}}{|\mathbf{x}_{,s_1} \times \mathbf{x}_{,s_2}|}\right|_{\mathbf{p}}, \quad [g_{ij}] = \begin{bmatrix} 1 + (\mathbf{x}_{,s_1})^2 & \mathbf{x}_{,s_1} \cdot \mathbf{x}_{,s_2} \\ \mathbf{x}_{,s_1} \cdot \mathbf{x}_{,s_2} & 1 + (\mathbf{x}_{,s_2})^2 \end{bmatrix} \tag{13.46}$$

and $[g^{ij}] = [g_{ij}]^{-1}$. One can then easily compute the second fundamental form:

$$\mathbf{C}_{\alpha\beta} = \mathbf{x}_{,s_\alpha s_\beta} \cdot \bar{\mathbf{n}}. \tag{13.47}$$

In order to calculate higher order derivatives of the different geometrical information, one needs to construct polynomials that approximate the second fundamental form and the metric. This is done in the same fashion as for the approximation of the membrane surface, since the value of those fields is known at each foot point. After constructing the polynomial that approximates the metric $\mathbf{g}$ around a point $\mathbf{p}$, the Christoffel symbols at that point (needed to compute the covariant derivative in Equation (13.20)) are given by

$$\Gamma_{ijk} = \frac{1}{2}(\mathbf{g}_{ij,k} + \mathbf{g}_{ik,j} + \mathbf{g}_{jk,i}) \tag{13.48}$$

and the surface Laplacian of the mean curvature $H$ used in Equation (13.27) is [48]

$$\Delta^{\|} H = \sum_{i,j=1}^{2} \frac{1}{\sqrt{g}} \frac{\partial}{\partial s_i} \left( \sqrt{g} g^{ij} \frac{\partial H}{\partial s_j} \right), \tag{13.49}$$

where $\sqrt{g}$ is the square root of the determinant of the metric $[g_{ij}]$. Finally, as for any other Lagrangian fields, the membrane strain is automatically transported with the position update of the foot points as the membrane evolves. The strain at those foot points at time $t$ is then updated using the strain rate at the position of those points at time $t-1$ as follows:

$$\bar{\mathbf{E}}(\mathbf{x}^t) = \bar{\mathbf{E}}(\mathbf{x}^{t-1}) + \dot{\bar{\mathbf{E}}}(\mathbf{x}^{t-1})\,\mathrm{d}t. \tag{13.50}$$

For more details concerning the grid-based particle method, the reader is referred to the work of Leung and Zhao [48].

## 13.5 Examples

### *13.5.1 The Equilibrium Shapes of the Red Blood Cell*

In this section we simulate the different equilibrium shapes observed in red blood cells to validate our formulation of immersed membrane mechanics. The red blood cell equilibrium shapes are obtain by minimizing the Canham–Helfrich bending energy of the lipid bilayer while keeping the total cell surface area constant and reducing its internal volume. We therefore choose to model the red blood cell MCC as an incompressible (but not completely inextensible, as red blood cells can sustain a maximum of 4%

stretch) elastic membrane, with no debonding between the cortex and the lipid bilayer. In that case, the lipid bilayer and the cortex follow the same motion $\Gamma_c = \Gamma_l$, the membrane–cortex binding power $P^{b}_{int}$ vanishes and the total internal power of the system reads

$$P_{int} = P^{f}_{int} + P^{fric}_{int} + P^{m}_{int}, \tag{13.51}$$

where $P^{m}_{int}$ is the MCC elastic power and is computed as given in Equation (13.5):

$$\tilde{W}(\bar{\mathbf{E}}, \mathbf{C}) = 1/2\bar{K}\epsilon^2 + \frac{K_h}{2}(H - C_0)^2, \tag{13.52}$$

while the external power $P_{ext} = 0$ since there are no external forces applied on the membrane of the cell. Using this energy density function and Equation (13.27), one can derive the force exerted by the membrane on the fluid:

$$\bar{\mathbf{f}} = \underbrace{2H\epsilon\bar{K}\bar{\mathbf{n}} - \nabla^{\|}(K\epsilon)}_{\text{stretching}} + \underbrace{K_h\left[\Delta^{\|}H + \frac{H - C_0}{2}(H^2 - 4K + HC_0)\right]\bar{\mathbf{n}}}_{\text{bending}}. \tag{13.53}$$

Numerically, we choose to implement the problem axisymmetrically as shown in Figure 13.10. Initially, the cell has a spherical shape. However, in order to break the central symmetry of the sphere and obtain interesting equilibrium shapes, we choose to give the cell an initially slight ellipsoidal shape. We then

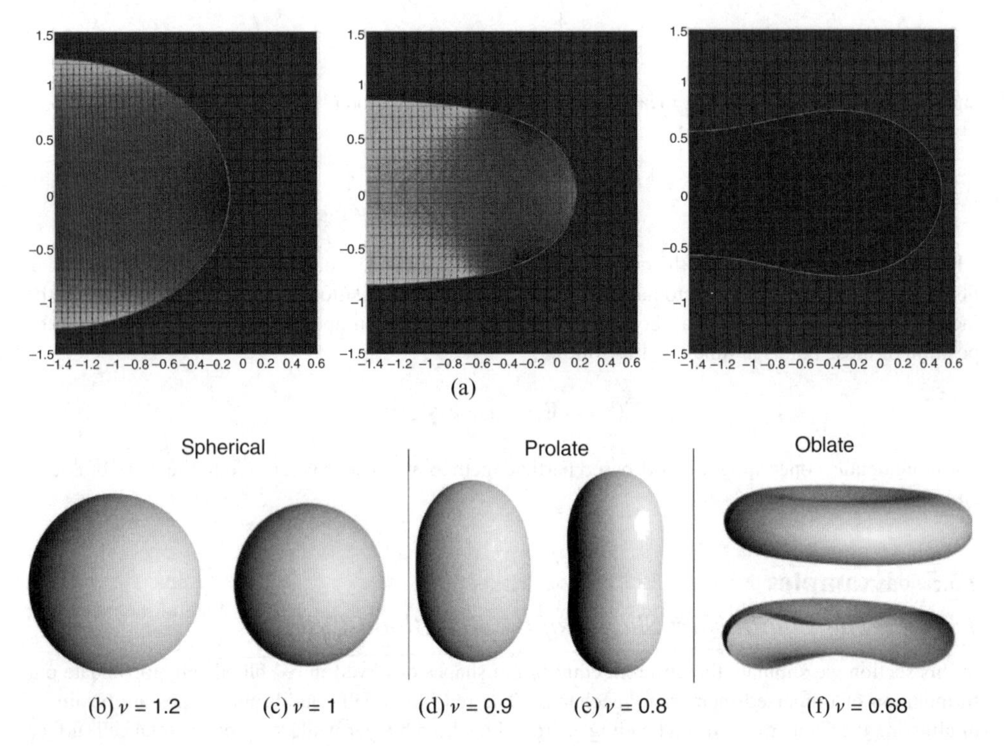


**Figure 13.10** (a) Evolution of the cell as it converges towards its equilibrium shape; (b)–(f) red blood cell equilibrium shapes for different reduced volumes

reduce the internal volume of the cell (but not its surface area) and let the cell find its equilibrium shape. Figure 13.10 shows the evolution of the cell toward its equilibrium position for a reduced volume of $\nu = 0.67V_0$, with $V_0$ the initial volume of the cell as a sphere. We can observe the discontinuity in pressure and velocity allowed by the X-FEM formulation between the inside and outside of the cell. Moreover, the discretization of the cell membrane remains uniform throughout, regardless of the large deformations and displacements the cell exhibits.

Figure 13.10b–f shows the different red blood cell equilibrium shapes for various values of reduced (or augmented) volume. The augmented volume equilibrium shape was obtained by adding a source of fluid at the center of the cell. As expected, the equilibrium shape for augmented volumes remains spherical (Figure 13.10b and c). For reduced volumes $\nu$ between 0.99 and 0.8, the red blood cell takes a prolate equilibrium shape (Figure 13.10c and d), while the cell adopts an oblate (discocyte) shape for $\nu$ between 0.79 and 0.65. The three equilibrium shapes (spherical, prolate, oblate) all constitute local equilibria for any $\nu$ between 1.2 and 0.65, but they successively present a global equilibrium as $\nu$ decreases. The equilibrium shapes found are in very good agreement with the phase diagrams found in Seifert and Berndl [49].

### *13.5.2 Cell Endocytosis*

Cell endocytosis is currently being studied in different areas, such as the creation of new drug delivery tools or in allergy control. Different investigations proved that there is a close dependence between the size and shape of the nanoparticle and the rate of endocytosis, explaining why different particles have more or less affinity to be uptaken. In this section we show that our model is able to represent the endocytosis of a single nanoparticle, and could be used in the future to investigate the effect of different factors, such as particles shape and elasticity.

The case of endocytosis being studied here involves the uptake of a single, rigid particle in the range of tens of nanometers in diameter. At this scale, the cell membrane can be represented as a flat surface, since the cell is orders of magnitude ($\sim$50 μm) larger than the particle. As a result, we impose a constraint on the membrane on the right edge of the (axysimmetric) computational domain to ensure that the membrane remains flat away from the particle. Moreover, both the membrane and the cytosol can flow freely in and out of the bottom and right edges of the computational domain. This choice of boundary conditions is motivated by the fact that a nanoparticle only displaces an infinitesimal portion of the total cell volume, and that the membrane has enough area reserves to wrap around the particle without being stretched (the membrane stays in the unfolding regime). The internal power of the system considered here consists of the elastic power of the lipid bilayer $P^{\mathrm{l}}_{\mathrm{int}}$ (the cortex elastic power and the binding force do not contribute to the internal power here), the fluid dissipation power $P^{\mathrm{f}}_{\mathrm{int}}$, and the fluid–membrane frictional power $P^{\mathrm{fric}}_{\mathrm{int}}$:

$$P_{\mathrm{int}} = P^{\mathrm{f}}_{\mathrm{int}} + P^{\mathrm{fric}}_{\mathrm{int}} + P^{\mathrm{l}}_{\mathrm{int}}, \tag{13.54}$$

while the external power is written $P_{\mathrm{ext}} = \int_{\Gamma_{\mathrm{l}}} \bar{\mathbf{t}} \cdot \bar{\mathbf{v}} \, \mathrm{d}\Gamma_{\mathrm{l}}$ and comes from the adhesion force $\bar{\mathbf{t}}$ between the particle and the membrane defined in Equation (13.11). Figure 13.11 shows the evolution of a cell membrane around a particle of 45 nm in radius. The semicircular line is a level-set function representing the nanoparticle, while the other line is a second level-set representing the cell membrane. The adhesion force represented by the black arrows is shown to be much larger in the leading edge of the membrane as it uptakes the particle.

A critical question in the cell endocytosis problem is to find the minimal particle size for a complete particle uptake. When a spherical particle is considered, most studies agree that the minimum radius for complete wrapping lies between 20 and 25 nm [7, 50–53]. Figure 13.12 shows how the total wrapped area in the steady state increases with the size of the particle. This is explained by the fact that curvature and, therefore, the bending energy decrease with the size of the particle, allowing the adhesion force to

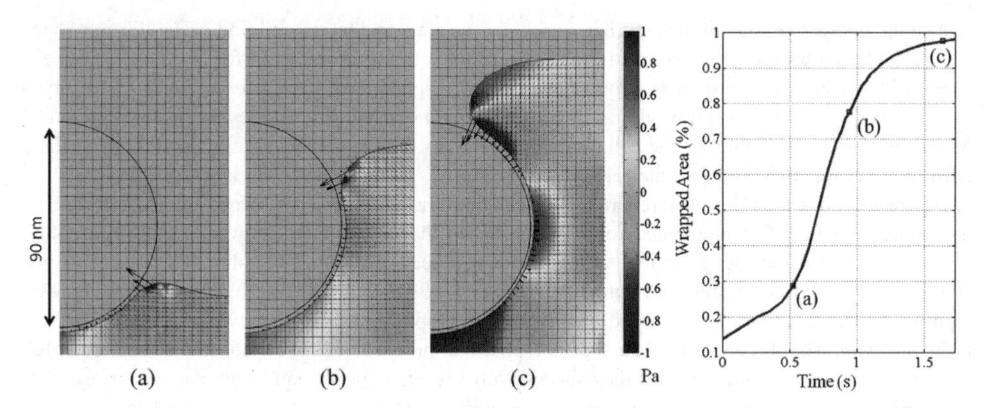


**Figure 13.11** The evolution of the cell membrane around a nanoparticle of 45 nm in radius. (a)–(c) The pressure and velocity field of the fluid at different moments of the endocytosis. The last one is the evolution of the wrapped area with the time. The values of the different parameters are $K_{\text{bond}} = 2 \times 10^{-3}$ N/m, $l_0 = 10$ nm, $N_r = 5 \times 10^{-3}/\text{nm}^2$, $N_l = 5 \times 10^{-3}/\text{nm}^2$, $K_L = 10^9$ nm$^2$, $k_B T = 4.1 \times 10^{-21}$ J, $\tau = 5$ nm, and $\gamma = 1$ pN

push the membrane higher around the particle, until complete wrapping occurs. Our simulations predict that this occurs for a particle radius of around 22–23 nm, which is in good agreement with experiments and other studies such as Decuzzi and Ferrari [6] and Gao *et al.* [7].

### *13.5.3 Cell Blebbing*

The blebbing of a cell has been shown to be involved in the spreading of cancerous cells as an alternative mechanism of motility [3–5]. The typical evolution of a bleb may be described in the following manner [2]. First, the cell cortex generates surface tension by contracting itself via the acto-myosin mechanism

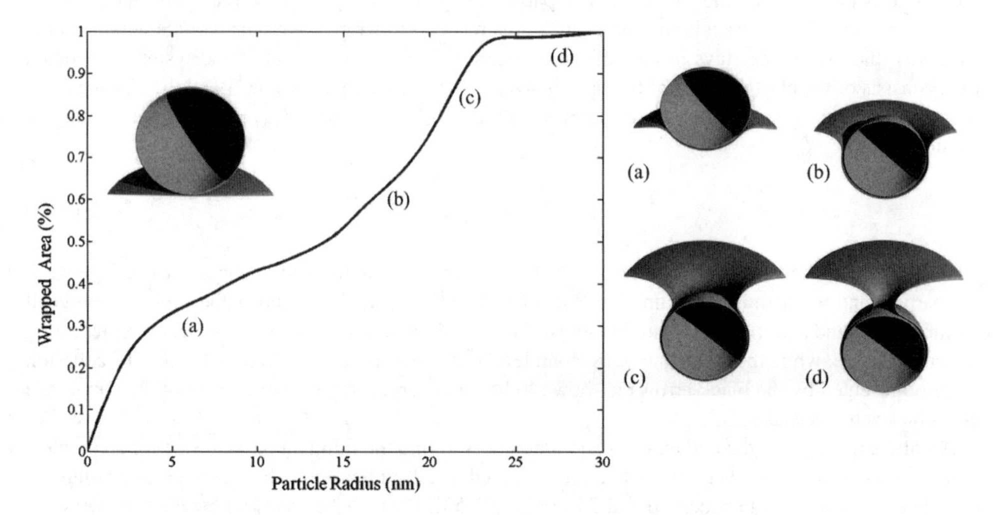


**Figure 13.12** Percentage of particle area wrapped by the membrane as a function of the particle size

that leads to an increase of the internal pressure in the cytosol. This first stage can then be followed by two possible outcomes: the rupture of the actin cortex or delamination of the cell membrane; that is, the debonding between the lipid bilayer and the cortex. Both of these outcomes produce a local decrease of pressure at the rim of the cell which drives cytosol into that region and initiates bleb growth. As the bleb grows (the growth time is on order of a minute), the actin cortex starts reassembling beneath the bleb membrane. The newly formed actin cortex in the bleb region generates tangential contraction ($\mathbf{T}_0$) that locally raises the pressure in the bleb, effectively slowing down the growth and reversing the process. This last stage is known as bleb retraction and ends once the bleb has disappeared [2].

Numerically, this process can be described by introducing two level-set functions associated with the debonding lipid bilayer and cortex. The interactions between the two level-sets are governed by the membrane–cortex binding force $\mathbf{f}_b$ given in Equation (13.6). This force is not homogeneous throughout the cell surface; in fact, its heterogeneity is responsible of the bleb initiation: as the cortex increases the cell's internal pressure, the debonding between the lipid bilayer and the cortex starts in the cell region that has the lowest binder concentration. Here, to describe the debonding of the MCC, the internal power of the system is written as in Equation (13.21):

$$P_{\text{int}} = P_{\text{int}}^{\text{f}} + P_{\text{int}}^{\text{fric}} + P_{\text{int}}^{\text{c}} + P_{\text{int}}^{\text{l}} + P_{\text{int}}^{\text{b}}, \tag{13.55}$$

and the external power $P_{\text{ext}} = 0$ since there are no external forces applied on the membrane here.

Figure 13.13a shows the evolution of the blebbing process from the bleb initiation to its retraction, while Figure 13.13b shows the relationship between the cell cortex contractile tension and the resulting bleb volume. The results compare qualitatively with the experimental data by Tinevez *et al.* [3]; however, the simulations show bigger volumes of blebs. This could be explained by the fact that the experiments of Tinevez *et al.* [3] use a pipette to create the bleb, adding friction energy on the pipette's wall that potentially reduces the bleb sizes. While the bleb is growing, the cell membrane stretches so that its surface increases up to 2.6 times its initial surface [36] thanks to the mechanism of undulations unfolding described in Section 13.2.2. Here, we assume the cell membrane stays in the unfolding regime and, therefore, offers very small resistance to stretching. Let us now turn to the retraction phase of the blebbing process. The reconstruction of the bleb cortex is led by the polymerization of actin monomers (associated with myosin) into contractile filaments described by the following equation [54–56]:

$$C_{\text{p}}(t) = C_{\text{m0}}[1 - \exp(-k_1 C_{\text{m0}} t)], \tag{13.56}$$

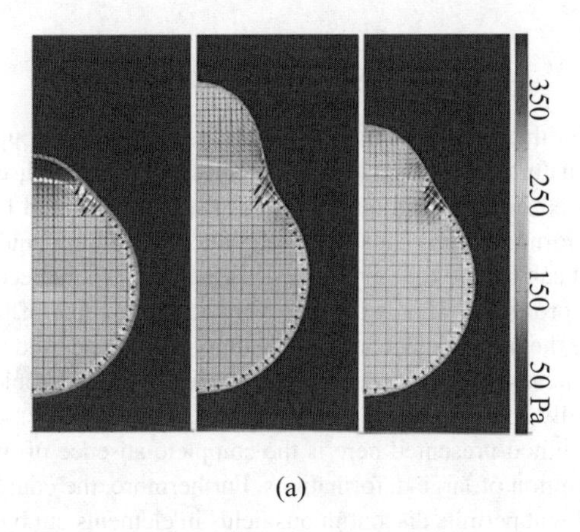


(a)

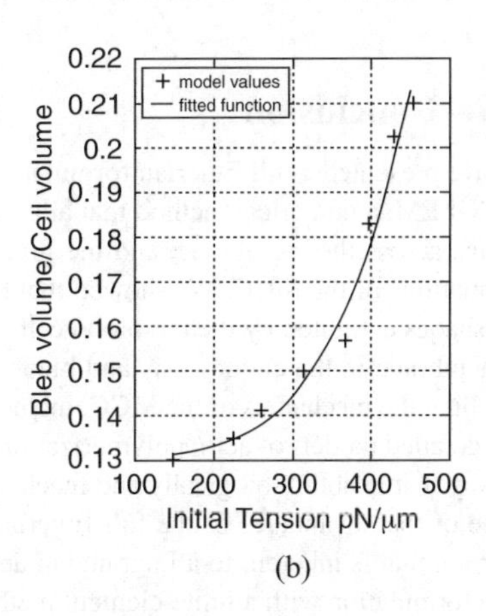


(b)

**Figure 13.13** (a) Evolution of the blebbing process. The arrows denote the adhesion force between the cortex and the membrane. (b) Bleb volume as a function of cortical tension

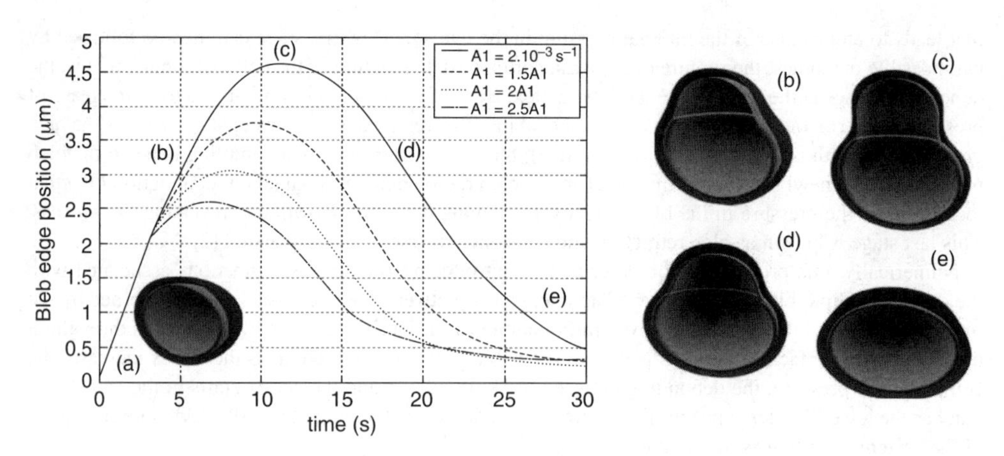


**Figure 13.14** Bleb evolution for different values of actin polymerization rates

where $C_\mathrm{p}$ is the polymerized actin concentration, $C_{\mathrm{m}0} = 0.6\ \mu\mathrm{M}$ is the initial actin monomer concentration, and $k_1 = 12\ \mu\mathrm{M}^{-1}\mathrm{s}^{-1}$ is the polymerization rate [57]. The cortex contraction can then be found as follows:

$$\gamma_\mathrm{c} = \gamma_{\mathrm{c}0}\frac{C_\mathrm{p}}{C_{\mathrm{m}0}}, \tag{13.57}$$

where $\gamma_{\mathrm{c}0} = 1000$ pN/μm is the maximum cortex tension [58]. This equation assumes that the cortex tension is directly proportional to the actin/myosin concentration. Figure 13.14 shows the surface of the bleb with three different values of polymerization rate. As expected, we observe that a higher polymerization rate results in a smaller bleb and a faster retraction. Here, we have shown that the proposed formulation is able to capture the basic stages of bleb growth and retraction, including the large deformations inherent to the cell blebbing problem, based on a simplified model of retraction. However, better descriptions of actin polymerization can later be incorporated, such as a thermodynamical model of actin polymerization that depends on the strains in the bleb [59, 60].

## 13.6 Conclusion

We have presented a full Eulerian formulation for the analysis of the fluid–membrane problem, coupled with X-FEM, a numerical method that allows a natural implementation of velocity and pressure discontinuities across the membrane, and the grid-based particle method that tracks the geometry and large deformations of the MCC. We showed that the formulation is capable of reproducing different equilibrium shapes exhibited by the red blood cell, and can handle the very large deformations experienced by the membrane in the endocytosis and blebbing problems. This is the first formulation capable of fully describing the mechanics of the MCC, including the membrane–cortex debonding, and can be used with more detailed models of actin polymerization or adhesion forces in order to describe biological problems in a way that is both biologically and mechanically accurate.

One of the advantages of the full Eulerian method presented here is the complete absence of mesh distortion that is inherent to a Lagrangian description of large deformations. Furthermore, the coupling of the formulation with a finite-element method that permits discontinuous fields in elements cut by the membrane enables an easy enforcement of a wide variety of boundary conditions on the membrane. In the present state, the finite-element mesh size is limited by the maximum curvature along the membrane. Using a curvature-based refinement algorithm (easy to implement on a fixed structured mesh, and naturally compatible with the grid-based particle method) could greatly improve the method's performance by coarsening or refining the mesh in regions of low or high curvature respectively.

## Acknowledgments

FJV gratefully acknowledges the University of Colorado CRCW Seed Grant and NIH grant number 1R21AR061011 in support of this work.

## References

[1] Lucas, W. (2002) Viral capsids and envelopes: structure and function. *Life Sciences*, 1–7, http://dx.doi.org/10.1002/9780470015902.a0001091.pub2.

[2] Charras, G. and Paluch, E. (2008) Blebs lead the way: how to migrate without lamellipodia. *Nature Reviews. Molecular Cell Bbiology*, **9** (9), 730–736, doi: 10.1038/nrm2453, http://www.ncbi.nlm.nih.gov/pubmed/18628785.

[3] Tinevez, J.Y., Schulze, U., Salbreux, G. *et al.* (2009) Role of cortical tension in bleb growth. *Proceedings of the National Academy of Sciences of the United States of America*, **106** (44), 18581–18586, doi:10.1073/pnas.0903353106, http://www.pubmedcentral.nih.gov/articlerender.fcgi?artid=2765453\&tool=pmcentrez\&rendertype=abstract.

[4] Fackler, O.T. and Grosse, R. (2008) Cell motility through plasma membrane blebbing. *Journal of Cell Biology*, **181** (6), 879–884, doi:10.1083/jcb.200802081, http://www.pubmedcentral.nih.gov/articlerender.fcgi?artid=2426937\&tool=pmcentrez\&rendertype=abstract.

[5] Charras, G.T., Hu, C.K., Coughlin, M., and Mitchison, T.J. (2006) Reassembly of contractile actin cortex in cell blebs. *Journal of Cell Biology*, **175** (3), 477–490, doi:10.1083/jcb.200602085, http://www.pubmedcentral.nih.gov/articlerender.fcgi?artid=2064524\&tool=pmcentrez\&rendertype=abstract.

[6] Decuzzi, P. and Ferrari, M. (2007) The role of specific and non-specific interactions in receptor-mediated endocytosis of nanoparticles. *Biomaterials*, **28** (18), 2915–2922, doi:10.1016/j.biomaterials.2007.02.013.

[7] Gao, H., Shi, W., and Freund, L.B. (2005) Mechanics of receptor-mediated endocytosis. *Proceedings of the National Academy of Sciences of the United States of America*, **102** (27), 9469–9474, doi:10.1073/pnas.0503879102.

[8] Steinman, R.M., Mellman, I.S., Muller, W.A., and Cohn, Z.A. (1983) Endocytosis and the recycling of plasma membrane. *Journal of Cell Biology*, **96** (1), 1–27.

[9] Wileman, T., Harding, C., and Stahl, P. (1985) Receptor-mediated endocytosis. *Biochemical Journal*, **232** (1), 1–14.

[10] Canham, P.B. (1970) The minimum energy of bending as a possible explanation of the biconcave shape of the human red blood cell. *Journal of Theoretical Biology*, **26** (1), 61–81, http://www.ncbi.nlm.nih.gov/pubmed/5411112.

[11] Helfrich, W. (1973) Elastic properties of lipid bilayers: theory and possible experiments. *Zeitschrift fur Naturforschung Teil C: Biochemie, Biophysik, Biologie, Virologie*, **28** (11), 693–703, http://ludfc39.u-strasbg.fr/pdflib/membranes/Elasticity/1973_Helfrich_b.pdf.

[12] Evans, E.A. (1974) Bending resistance and chemically induced moments in membrane bilayers. *Biophysical Journal*, **14** (12), 923–931, http://linkinghub.elsevier.com/retrieve/pii/S000634957485959X.

[13] Jenkins, J.T. (1977) Static equilibrium configurations of a model red blood cell. *Journal of Mathematical Biology*, **4** (2), 149–169, http://www.ncbi.nlm.nih.gov/pubmed/886227.

[14] Feng, F. and Klug, W. (2006) Finite element modeling of lipid bilayer membranes. *Journal of Computational Physics*, **220** (1), 394–408, doi:10.1016/j.jcp.2006.05.023, http://linkinghub.elsevier.com/retrieve/pii/S0021999106002476.

[15] Farsad, M., Vernerey, F., and Park, H. (2012) An XFEM-based numerical strategy to model mechanical interactions between biological cells and a deformable substrate. *International Journal of Numerical Methods in Engineering*, **92** (3), 238–267, doi:10.1002/nme.4335.

[16] Farsad, M. and Vernerey, F. (2010) An extended finite element/level set method to study surface effects on the mechanical behavior and properties of nanomaterials. *International Journal of Numerical Methods in Engineering*, **84** (12), 1466–1489.

[17] Vernerey, F.J., Foucard, L., and Farsad, M. (2011) Bridging the scales to explore cellular adaptation and remodeling. *BioNanoScience*, **1** (3), 110–115, doi:10.1007/s12668-011-0013-6, http://www.springerlink.com/index/10.1007/s12668-011-0013-6.
[18] Vernerey, F. and Farsad, M. (2011) An Eulerian/XFEM formulation for the large deformation of cortical cell membrane. *Computer Methods in Biomechanics and Biomedical Engineering*, **14** (5), 433–445.
[19] Peskin, C. (1972) Flow patterns around heart valves: a numerical method. *Journal of Computational Physics*, **10** (2), 252–271, http://dx.doi.org/10.1016/0021-9991(72)90065-4.
[20] Eggleton, C.D. and Popel, A.S. (1998) Large deformation of red blood cell ghosts in a simple shear flow. *Physics of Fluids*, **10** (8), 1834–1845, doi:10.1063/1.869703, http://link.aip.org/link/PHFLE6/v10/i8/p1834/s1&\Agg=doi.
[21] Dillon, R., Owen, M., and Painter, K. (2008) A single-cell-based model of multicellular growth using the immersed boundary method. *AMS Contemporary Mathematics*, **466**, 1–15.
[22] Li, Y., Yun, A., and Kim, J. (2012) An immersed boundary method for simulating a single axisymmetric cell growth and division. *Journal of Mathematical Biology*, **65** (4), 653–675, doi:10.1007/s00285-011-0476-7, http://www.ncbi.nlm.nih.gov/pubmed/21987086.
[23] Groulx, N., Boudreault, F., Orlov, S.N., and Grygorczyk, R. (2006) Membrane reserves and hypotonic cell swelling. *Journal of Membrane Biology*, **214** (1), 43–56, http://www.ncbi.nlm.nih.gov/pubmed/17598067.
[24] Ma, L. and Klug, W.S. (2008) Viscous regularization and r-adaptive remeshing for finite element analysis of lipid membrane mechanics. *Methods*, **227**, 5816–5835, doi:10.1016/j.jcp.2008.02.019.
[25] Cottet, G.H. and Maitre, E. (2006) A level set method for fluid–structure interactions with immersed surfaces. *Mathematical Models and Methods in Applied Sciences*, **16**, 415–438, http://ljk.imag.fr/membres/Georges-Henri.Cottet/ref27.pdf.
[26] Rawicz, W., Olbrich, K.C., McIntosh, T. *et al.* (2000) Effect of chain length and unsaturation on elasticity of lipid bilayers. *Biophysical Journal*, **79** (1), 328–339, doi:10.1016/S0006-3495(00)76295-3.
[27] Vernerey, F. (2012) The effective permeability of cracks and interfaces in porous media. *Transport in Porious Media*, **93** (3), 815–829.
[28] Vernerey, F. (2011) A theoretical treatment on the mechanics of interfaces in deformable porous media. *International Journal of Solids and Structures*, **48** (22–23), 3129–3141.
[29] Lecieux, Y. and Bouzidi, R. (2012) Numerical wrinkling prediction of thin hyperelastic structures by direct energy minimization. *Advances in Engineering Software*, **50**, 57–68, doi:10.1016/j.advengsoft.2012.02.010, http://linkinghub.elsevier.com/retrieve/pii/S0965997812000415.
[30] Evans, E., Heinrich, V., Leung, A., and Kinoshita, K. (2005) Nano- to microscale dynamics of P-selectin detachment from leukocyte interfaces. I. Membrane separation from the cytoskeleton. *Biophysical Journal*, **88** (3), 2288–2298, doi:10.1529/biophysj.104.051698.
[31] Young, J. and Mitran, S. (2010) A numerical model of cellular blebbing: a volume-conserving, fluid–structure interaction model of the entire cell. *Journal of Biomechanics*, **43** (2), 210–220, doi:10.1016/j.jbiomech.2009.09.025.
[32] Bell, G.I. (1978) Models for the specific adhesion of cells to cells. *Science (New York, NY)*, **200** (4342), 618–627.
[33] Weibull, W. (1951) A statistical distribution function of wide applicability. *Journal of Applied Mechanics*, **18**, 283–297.
[34] Brugués, J., Maugis, B., Casademunt, J. *et al.* (2010) Dynamical organization of the cytoskeletal cortex probed by micropipette aspiration. *Proceedings of the National Academy of Sciences of the United States of America*, **107** (35), 15415–15420, doi:10.1073/pnas.0913669107, http://www.pubmedcentral.nih.gov/articlerender.fcgi?artid=2932608\&tool=pmcentrez\&rendertype=abstract.
[35] Chamaraux, F., Ali, O., Keller, S. *et al.* (2008) Physical model for membrane protrusions during spreading. *Physical Biology*, **5** (3), 036009, doi:10.1088/1478-3975/5/3/036009.
[36] Charras, G.T. (2008) A short history of blebbing. *Journal of Microscopy*, **231** (3), 466–478, doi:10.1111/j.1365-2818.2008.02059.x, http://www.ncbi.nlm.nih.gov/pubmed/18755002.
[37] Bell, G.I., Dembo, M., and Bongrand, P. (1984) Cell adhesion. Competition between nonspecific repulsion and specific bonding. *Biophysical Journal*, **45** (6), 1051–1064, doi:10.1016/S0006-3495(84)84252-6.
[38] Olberding, J.E., Thouless, M.D., Arruda, E.M., and Garikipati, K. (2010) A theoretical study of the thermodynamics and kinetics of focal adhesion dynamics, in *IUTAM Symposium on Cellular, Molecular and Tissue Mechanics* (eds K. Garikipati and E.M. Arruda), vol. 16 of *IUTAM Bookseries*, Springer, Dordrecht, pp. 181–192, doi:10.1007/978-90-481-3348-2.

[39] Dembo, M., Torney, D.C., Saxman, K., and Hammer, D. (1988) The reaction-limited kinetics of membrane-to-surface adhesion and detachment. *Proceedings of the Royal Society B: Biological Sciences*, **234** (1274), 55–83, doi:10.1098/rspb.1988.0038.

[40] Foucard, L. and Vernerey, F. (2012) A general theoretical and numerical formulation for immersed membranes undergoing large morphological changes. *Journal of Computational Physics*, under review.

[41] Meyers, A., Schiebe, P., and Bruhns, O.T. (2000) Some comments on objective rates of symmetric Eulerian tensors with application to Eulerian strain rates. *Acta Mechanica*, **103**, 91–103.

[42] Kadianakis, N. (2009) Evolution of surfaces and the kinematics of membranes. *Journal of Elasticity*, **99** (1), 1–17, doi:10.1007/s10659-009-9226-0, http://www.springerlink.com/index/10.1007/s10659-009-9226-0.

[43] Clopeau, T., Mikelic, A., and Robert, R. (1998) On the vanishing viscosity limit for the 2D incompressible Navier–Stokes equations with the friction type boundary conditions. *Nonlinearity*, **11** (6), 1625–1636, doi:10.1088/0951-7715/11/6/011, http://iopscience.iop.org/0951-7715/11/6/011.

[44] Lopes Filho, M.C., Nussenzveig Lopes, H.J., and Planas, G.V. (2005) On the inviscid limit for 2D incompressible flow with Navier friction condition. *SIAM Journal on Mathematical Analysis*, **36**, 1130–1141, http://arxiv.org/abs/math/0307295.

[45] Arroyo, M. and DeSimone, A. (2009) Relaxation dynamics of fluid membranes. *Physical Review E*, **79** (3 Pt 1), 031915, doi:10.1103/PhysRevE.79.031915, http://link.aps.org/doi/10.1103/PhysRevE.79.031915.

[46] Moës, N., Dolbow, J., and Belytschko, T. (1999) A finite element method for crack growth without remeshing. *International Journal for Numerical Methods in Engineering*, **46** (1), 131–150, doi:10.1002/(SICI)1097-0207(19990910)46:1<131::AID-NME726>3.0.CO;2-J, http://doi.wiley.com/10.1002/(SICI)1097-0207(19990910)46:1<131::AID-NME726>3.0.CO;2-J.

[47] Belytschko, T., Parimi, C., Moës, N. *et al.* (2003) Structured extended finite element methods for solids defined by implicit surfaces. *International Journal for Numerical Methods in Engineering*, **56** (4), 609–635, doi:10.1002/nme.686, http://doi.wiley.com/10.1002/nme.686.

[48] Leung, S. and Zhao, H. (2009) A grid based particle method for moving interface problems. *Journal of Computational Physics*, **228** (8), 2993–3024, doi:10.1016/j.jcp.2009.01.005, http://linkinghub.elsevier.com/retrieve/pii/S0021999109000138.

[49] Seifert, U. and Berndl, K. (1991) Shape transformations of vesicles: phase diagram for spontaneous-curvature and bilayer-coupling models. *Physical Review A*, **44** (2), 1182–1202, http://pra.aps.org/abstract/PRA/v44/i2/p1182_1.

[50] Chaudhuri, A., Battaglia, G., and Golestanian, R. (2011) The effect of interactions on the cellular uptake of nanoparticles. *Physical Biology*, **8** (4), 046002, doi:10.1088/1478-3975/8/4/046002.

[51] Zhang, S., Li, J., Lykotrafitis, G. *et al.* (2009) Size-dependent endocytosis of nanoparticles. *Advanced Materials*, **21** (4), 419–424, doi:10.1002/adma.200801393.

[52] Yuan, H., Huang, C., and Zhang, S. (2010) Virus-inspired design principles of nanoparticle-based bioagents. *PloS ONE*, **5** (10), e13495, doi:10.1371/journal.pone.0013495.

[53] Yuan, H. and Zhang, S. (2010) Effects of particle size and ligand density on the kinetics of receptor-mediated endocytosis of nanoparticles. *Applied Physics Letters*, **96** (3), 033704, doi:10.1063/1.3293303.

[54] Korn, E.D. (1982) Actin polymerization and its regulation by proteins from nonmuscle cells. *Physiological Reviews*, **62** (2), 672–737.

[55] Wegner, A. (1976) Head to tail polymerization of actin. *Journal of Molecular Biology*, **108** (1), 139–150.

[56] Brenner, S.L. and Korn, E.D. (1983) On the mechanism of actin monomer–polymer subunit exchange at steady state. *Journal of Biological Chemistry*, **258** (8), 5013–5020.

[57] Van der Gucht, J. and Sykes, C. (2009) Physical model of cellular symmetry breaking. *Cold Spring Harbor Perspectives in Biology*, **1** (1), a001909, doi:10.1101/cshperspect.a001909.

[58] Thoumine, O., Cardoso, O., and Meister, J.J. (1999) Changes in the mechanical properties of fibroblasts during spreading: a micromanipulation study. *European Biophysics Journal EBJ*, **28** (3), 222–234, doi:10.1111/j.1463-1326.2006.00665.x, http://www.ncbi.nlm.nih.gov/pubmed/10192936.

[59] Foucard, L. and Vernerey, F.J. (2012) A thermodynamical model for stress-fiber organization in contractile cells. *Applied Physics Letters*, **100** (1), 13702, doi:10.1063/1.3673551, http://www.ncbi.nlm.nih.gov/pubmed/22271931.

[60] Foucard, L. and Vernerey, F. (2012) On the dynamics of stress fibers turnover in contractile cells. *Journal of Engineering Mechanics*, **138** (10), 1282–1287.

# 14

# Role of Elastin in Arterial Mechanics

Yanhang Zhang[a,b] and Shahrokh Zeinali-Davarani[b]
[a]*Department of Mechanical Engineering, Boston University, USA*
[b]*Department of Biomedical Engineering, Boston University, USA*

## 14.1 Introduction

Blood vessels are complex organs with hierarchical ultra-structures of extracellular matrix (ECM) and smooth muscle cells (SMCs) that bear the majority of wall stress and determine the mechanical and biological responses of the vessel wall. Elastic fibers, principal components of ECM assemblies that endow blood vessels with flexibility and extensibility [1], are essential to accommodate deformations encountered during the cardiac cycle in order to provide a smooth flow of blood. These resilient networks are made primarily of a protein called elastin. Elastic fibers consist of an inner elastin core surrounded by a mantle of fibrillin-rich microfibrils [2]. Elastin molecules are cross-linked together via linkages called desmosines and isodesmosines to form rubberlike elastic fibers [1], which are extremely stable with a very low turnover rate. When subjected to stretch, each elastin molecule uncoils into a more extended conformation and the cross-links establish the load transfer within the elastic fiber.

In elastic arteries, such as aorta, elastic fibers form thick concentric fenestrated layers of elastic lamellae with inter-laminar connecting fibers distributed radially through the vessel wall [3]. Each elastic lamella alternates with a layer of SMCs; together, they organize into a lamellar unit which is considered as the functional unit of the vessel wall [4, 5].

In order to study the structure and function of elastin in vascular mechanics, isolated elastin was obtained with the removal of cells, collagen fibers, and other ECM components [6, 7]. A scanning electron microscope image of isolated elastin reveals a regularized concentric layer of elastin sheets (Figure 14.1b), which are less obvious with the presence of collagen fibers and smooth muscle cells in the intact aorta (Figure 14.1a). Histology images demonstrate the apparent difference between the microstructure of aorta and its corresponding elastin. Elastin samples are empty of cells compared with

*Multiscale Simulations and Mechanics of Biological Materials*, First Edition. Edited by Shaofan Li and Dong Qian.

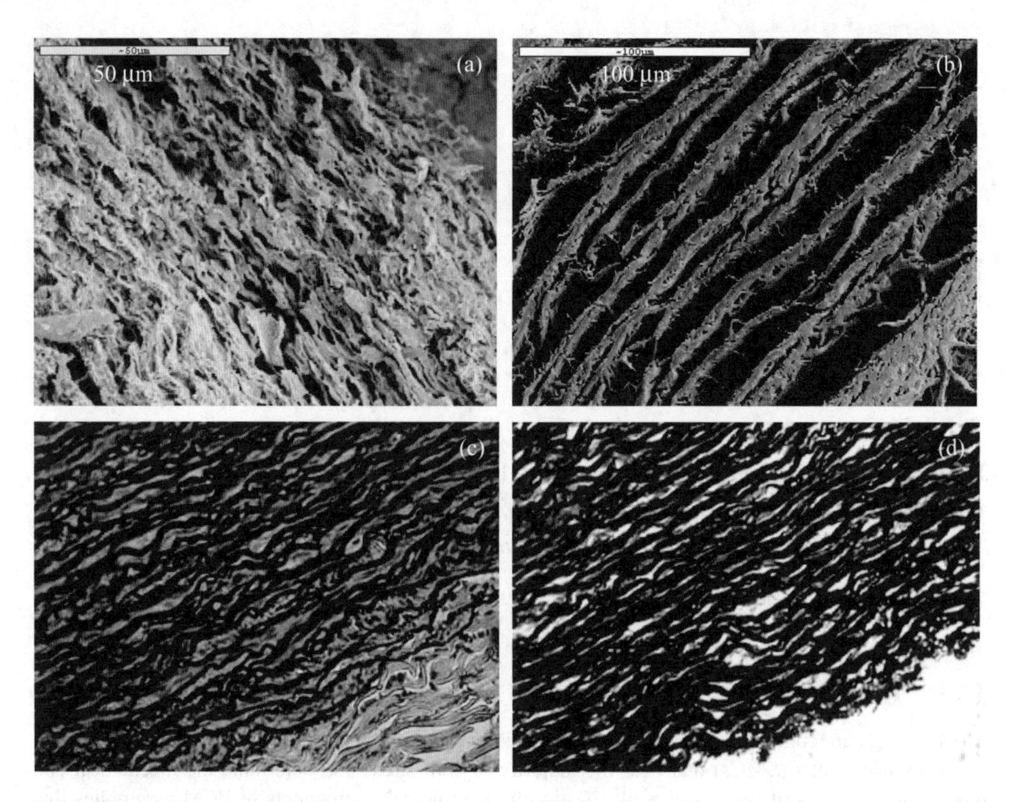


**Figure 14.1** Scanning electron microscopy images of (a) intact porcine aorta and (b) isolated elastin. Histology images of (c) intact aorta and (d) isolated elastin

intact aorta samples; also, elastin samples have a lower density of three-dimensional (3D) ECM network (Figure 14.1d). The aorta samples show cells embedded in a cross-linked network of elastin and collagen fibers (Figure 14.1c), while the same stain of elastin samples shows that only elastic fibers are present in the isolated elastin.

## 14.2 The Role of Elastin in Vascular Diseases

Elastin is a critical component of the arterial wall, providing not only the structural resilience, but also the biological signaling essential in vascular morphogenesis and maintenance of the mechanical homeostasis [4, 8, 9]. Elastin helps vascular SMCs to maintain their contractile state and prevents their migration [10–12] and modulation to synthetic phenotype [13, 14] that are common in many diseases. Pathogenesis of vascular diseases, including hypertension, atherosclerosis, stenosis, and aneurysms, has been associated with elastin disorders in structure and function and its altered interaction with other arterial constituents. For example, abdominal aortic aneurysms (AAAs) have been associated with substantial loss of the elastin content [15, 16] and significant increase in the arterial wall stiffness – see reviews by Vorp [17] and Humphrey and Holzapfel [18]. Hypertension has been associated with significant structural remodeling leading to increased wall thickness and stiffness [19, 20]. Although the elastin content does not seem to change significantly in experimental hypertension [19, 21], the crosslinking of the elastin fibers may increase [22]. A significant portion of the elevated stiffness in pulmonary hypertension is attributed to the increase in elastin stiffness [23]. In atherosclerosis, an increase in synthesis of elastin is reported by some experimental studies [24], mainly in intima [25], while the media contains disorganized and immature elastin [26]. Also, accumulation of lipid has been suggested to alter the functional role of elastin in preventing SMC migration as well as its overall contribution to tissue elasticity [27–30]. It is speculated that the mechanical properties of the newly synthesized elastic fibers under pathological conditions may not be the same as the one produced during development [20, 31].

These findings show a close association between vascular disease progression and the impaired elastin function/structure. From the biomechanical viewpoint, abnormalities in elastin can alter mechanical homeostasis and promote SMCs proliferation, migration and synthesis of ECM, and eventually contribute to the cycle of diseases. It is important to have a better understanding of the mechanical properties of elastin and its alteration in diseases, as well as the mechanisms by which elastin is degraded/damaged, influencing the arterial remodeling.

## 14.3 Mechanical Behavior of Elastin

### *14.3.1 Orthotropic Hyperelasticity in Arterial Elastin*

It is well known that blood vessels possess anisotropic material properties; that is, the mechanical response in the circumferential and longitudinal directions is not the same. In a few studies, mechanical testing of the elastic and viscoelastic properties of the isolated elastin has been investigated [32–34]. However, uniaxial loadings were used in these studies and are insufficient to elaborate the mechanical anisotropy of elastin.

The biaxial tensile testing device is known to be first redesigned for soft tissues by Y.C. Fung's group [35] and ever since improved [36, 37] and used by many research groups on biological tissues; for example, engineered and native heart valve [38, 39], healthy or diseased arteries [40–42], and intraluminal thrombus layer in AAAs [43]. These studies have demonstrated the importance of rigorous experimental approaches to elucidate the structure–function relationship of tissues and its underlying evolutions in diseases. Gundiah *et al.* [44] performed uniaxial and equi-biaxial tensile tests on isolated elastin. In their equi-biaxial tensile tests the strain was up to about 4%, within which the elastin appears to be isotropic, manifested by the equivalent tangent modulus in the axial and circumferential directions.

On the other hand, more recently, Zou and Zhang [6] demonstrated an anisotropic response of elastin isolated from the bovine thoracic aorta even. As shown in Figure 14.2, elastin shows strong anisotropy that is comparable to the intact arteries, with the circumferential direction being stiffer than the longitudinal direction. This trend was found to be consistent along the thoracic aorta with varying thickness but insignificant variation in elastin content. Elastin isolated from thinner sections was found to be stiffer than thicker sections of the artery in both circumferential and longitudinal directions, consistent with Lillie and Gosline [33]. In a study of circumferential heterogeneity of aortic wall, Kim and Baek [45] also showed that thinner (posterior) regions of aorta were stiffer than thicker (anterior) regions. Together, these results may be explained by the vascular adaptation process and mechanical homeostasis [8, 46], reflecting the important role of elastin in mechanobiology.

The discrepancy between the observed isotropic and anisotropic mechanical responses may be explained through the mechanical tests being performed on arteries of different species and different anatomical locations. According to Lillie and Gosline [33], elastin seems to be isotropic in proximal aorta and its anisotropy increases distally. The differences in elastin isolation methods and storage can also be a contributing factor. In the study by Gundiah *et al.* [44], elastin was isolated by autoclave and hot alkali methods from porcine aorta, and then stored in 80% ethanol for about a week. The elastin

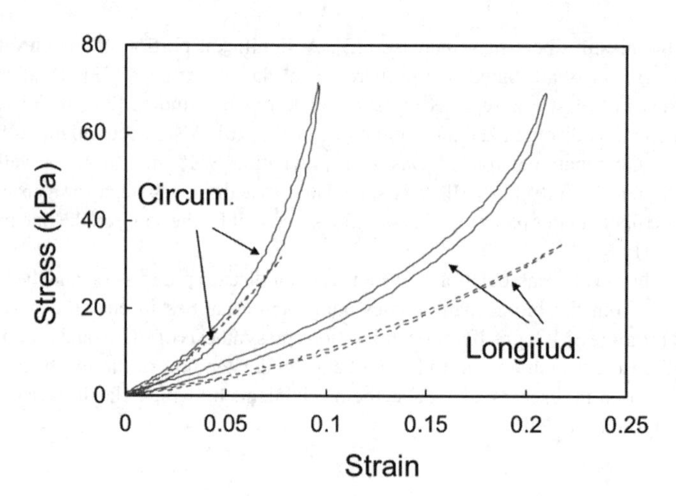


**Figure 14.2** Cauchy stress versus logarithmic strain in circumferential and longitudinal directions for the intact bovine thoracic aorta sample and its isolated elastin from the medium section of the bovine thoracic aorta under equi-biaxial loading. Solid lines correspond to intact aorta and dashed lines correspond to elastin (adapted from Zou and Zhang [6])

samples were then washed and rehydrated in distilled water before testing. Recent research showed that the use of 70% ethanol can cause elastin tissue to swell [47]. Moreover, the mechanics of elastin have been shown to correlate strongly with its water content [48]. Studies considering these conditions are necessary in order to elucidate the cause of the aforementioned discrepancies.

It also appears that elastin is mainly responsible for the linear elastic response of blood vessels under lower strains. Whereas, at higher strains the aorta becomes much stiffer in both directions due to the strain stiffening and the involvement of collagen fibers, and the stress–strain responses of intact aorta are more nonlinear compared with elastin (Figure 14.2). The aorta and elastin have similar tangent moduli in the circumferential direction when strain is below 5%. The initial tangent modulus is about 100 kPa and it increases almost linearly to 200 kPa at 5% strain. As the strain increases further, the tangent modulus of aorta increases dramatically and results in a much higher modulus. The tangent modulus in the circumferential direction at physiological strains is consistent with previous *in vivo* and *in vitro* studies on aortic mechanics [49–51], in which the tangent moduli at low tensile strains were shown to vary from 100 kPa to 350 kPa. The tangent modulus in the longitudinal direction is significantly lower than that in the circumferential direction. Elastin has an initial tangent modulus of about 30 kPa. The tangent modulus of aorta is about twice as that of elastin below 15% strain, and it increases significantly at higher strains. These results further proves that elastin contributes to the aortic mechanics in the physiological range. The physiological strains of aorta in the circumferential and longitudinal directions have been reported to be about 20% from *in vitro* studies [52–54].

Comparison between the equi-biaxial mechanics of elastin and aorta sheds light on understanding the role of the microstructural components in the mechanical behavior of the arterial wall, and making functional biological scaffolds with desired biomechanical properties. Previous studies on scaffolds with different ratios of collagen and elastin [34, 55, 56] have shown that scaffolds with higher collagen content had a higher tensile strength, whereas addition of elastin increased elasticity and distensibility. While the results in Figure 14.2 provide quantitative information on the mechanical properties of elastin compared with intact artery, the interconnections between elastin and other ECM components and cells are important in the functionality of arterial wall as well.

### 14.3.2 Viscoelastic Behavior

Arterial wall and its constituents undergo cyclic hemodynamic loads and it is crucial to understand their time-dependent properties. The relevance of the changes in viscoelastic responses of the arterial tissue and, in particular, elastin to the progression of vascular diseases may lead to better diagnostic and therapeutic approaches. The viscoelastic properties of elastin are highly associated with its microstructure, mechanical and chemical environments, temperature, and hydration level [32, 48, 57, 58]. The viscoelastic mechanical behavior of elastin, as of most materials, is characterized by its creep and stress relaxation responses. The relaxation time, associated with the time for molecular rearrangement, is related to the available free volume at the molecular level [59, 60].

At physiological stress levels, intact aortic tissue was shown to relax more rapidly and exhibit more obvious stress relaxation than purified elastin [7], mainly due to the presence of the SMCs and collagenous content in the intact aorta (Figure 14.3). This is consistent with Nagatomi *et al.* [61], who found a correlation between collagen content and stress relaxation of bladder tissue. Repeating the relaxation tests at different initial stress levels shows that, as the initial stress increases, the rate of relaxation increases linearly for the aortic tissue, presumably because of the engagement of collagen fibers at higher stress levels [7]. In contrast, the rate of relaxation for elastin increases only until a certain limit of stress level (far below physiological range) is reached (Figure 14.4). Other studies have demonstrated the dependence of the magnitude of stress relaxation on initial stress–strain levels for ligament [62, 63], showing that the rate of stress relaxation decreases with higher stress or strain level. They suggested that the decrease in the relaxation rate with increasing strain could be due to greater water loss that makes the tissue more elastic and less viscous. A study by Nagatomi *et al.* [64] on the bladder wall reported similar phenomena. The discrepancy between these findings could be due to the differences in the composition of biological tissues. Aorta is an elastic artery, composed of elastin up to 30% of the total weight [8], unlike ligaments that are mainly composed of collagen. It is widely known that elastin carries the most loads at lower levels of loading and collagen fibers engage in load bearing at higher levels. The complex interactions and the shift in mechanical role may be contributing to the discrepancy [7]. However, with regard to elastin, apparently at physiologically relevant stress levels, the rate of relaxation is almost insensitive to the initial stress within purified elastin, consistent with

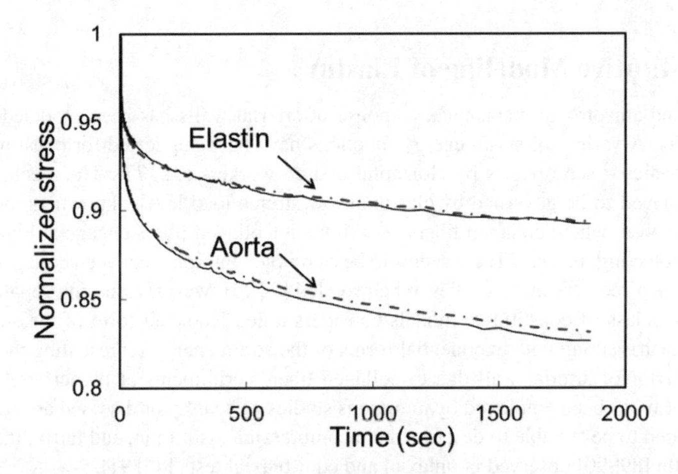


**Figure 14.3** Representative normalized stress relaxation curves of elastin and intact aorta. Solid and dotted lines correspond to circumferential and longitudinal directions, respectively (adapted from Zou and Zhang [7])

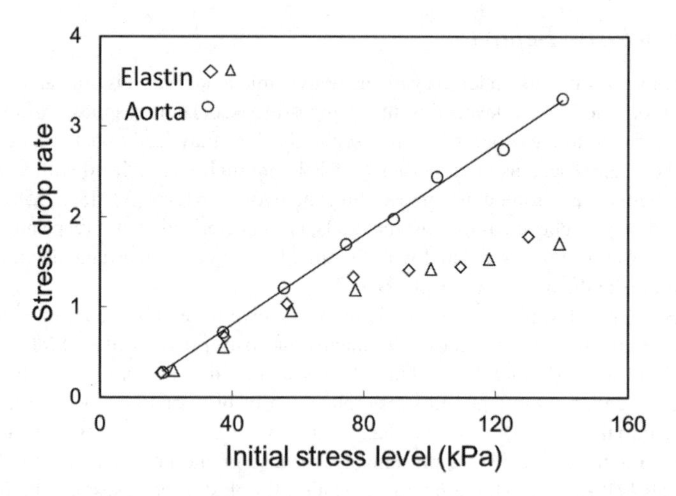


**Figure 14.4** Stress relaxation (drop) rate as a function of initial stress levels for intact aorta and elastin samples with added linear trend line. The stress drop rates of two elastin samples were presented to better observe the trend (adapted from Zou and Zhang [7])

other studies that indicate much less relaxation for elastin compared with smooth muscle cells or collagen [65].

Interestingly, the creep behavior of elastin differs from the stress relaxation. The isolated aortic elastin exhibits negligible creep response compared with its prominent stress relaxation response [7]. Lack of creep response is uniquely identified in soft biological tissues and appears to be the consequence of a functionally independent mechanism from that of stress relaxation [63, 64, 66–70]. Although the underlying mechanisms that differentiate the creep response from stress relaxation in arterial wall remain to be elucidated, some studies have pointed out the gradual recruitment of crimped (collagen) fibers as a protective mechanism that determines the creep response in ligaments [71].

## 14.4 Constitutive Modeling of Elastin

The nonlinear and anisotropic mechanical response of arterial walls has been modeled in the context of finite elasticity. A variety of strain energy functions have been suggested for constitutive modeling of arterial mechanics – see reviews by Holzapfel and co-workers [72, 73]. The mechanical response of arteries is believed to be governed by elastin at low stretch/load levels, known as the toe region of stress–strain response, where collagen fibers are crimped. Collagen fibers engage at higher load levels, stiffening and protecting the wall. There seems to be a complex interplay between collagen and elastin in bearing loads which requires more studies to be revealed [74]. However, assuming separate mechanical roles, a major subclass of constitutive models considers a neo-Hookean form of the strain energy for the amorphous matrix along with exponential forms of the strain energy representing the stiffening and anisotropic behavior of arterial wall due to collagen fiber recruitments at higher stretches [75]. The neo-Hookean form has been employed by numerous studies of healthy or diseased arterial wall [76–88] and has been found to be suitable to describe the incompressible, isotropic, and fairly linear response of the arterial elastin [89, 90] observed in uniaxial and equi-biaxial tests [44, 91].

Anisotropic behavior of elastin in arteries has also been reported and there have been efforts to model this direction-dependent mechanical response of elastin. The strain energy function for elastin presented by Rezakhaniha and Stergiopulos [92] has two parts: a neo-Hookean component and a component accounting for longitudinally oriented fibers that, together, resemble a transversely isotropic neo-Hookean model [93]. A similar but orthotropic form has been employed by Kao *et al.* [94] considering two orthogonal fiber families reinforcing the neo-Hookean matrix. Zou and Zhang [6] observed significant anisotropic response of the arterial elastin and took a statistical mechanic-based approach to describe the orthotropic response of elastin in relation to its microstructure.

In the study by Zou and Zhang [6], based on the analogy between the entangled long molecular chains and the structural protein framework seen in the elastin network, the eight-chain element is selected as the basic unit while every single chain is modeled as a freely jointed chain [95]. That is, different dimensions of the basic unit element in the three orthogonal material directions induce the orthotropic response. The final homogenized form of the strain energy function for the material response is given as

$$w = w_0 + \frac{nk\Theta}{4}\left[N\sum_{i=1}^{4}\left(\frac{\rho^{(i)}}{N}\beta_\rho^{(i)} + \ln\frac{\beta_\rho^{(i)}}{\sinh\beta_\rho^{(i)}}\right) - \frac{\beta_p}{\sqrt{N}}\ln(\lambda_1^{a^2}\lambda_2^{b^2}\lambda_3^{c^2})\right] + B[\cosh(J-1)-1], \tag{14.1}$$

where $w$ is the overall energy, $w_0$ is a constant, $B$ is a parameter that controls the bulk compressibility, and $J$ is the determinant of deformation gradient tensor. Parameters $a$, $b$, and $c$ are normalized dimensions along the principal material directions, and $\lambda_1$, $\lambda_2$, and $\lambda_3$ are stretches along these directions. $N = (a^2 + b^2 + c^2)/4$ is the number of links within each individual chain. $\rho^{(i)}$ is the normalized deformed lengths of the constituent chains in the unit element. $\beta_\rho^{(i)} = \ell^{-1}(\rho^{(i)}/N)$ is the inverse Langevin function and $p = \frac{1}{2}\sqrt{a^2 + b^2 + c^2}$ is the initial normalized length of each chain. Parameter $n$ is the chain density per unit volume, $k$ is Boltzmann's constant, and $\Theta$ is absolute temperature (see Zou and Zhang [6] for more details).

Zou and Zhang [6] demonstrated the potential of a statistical-mechanics-based method for the study of elastin mechanics. The constitutive model (14.1) using an eight-chain unit element is derived based on an idealization of the ECM structure. It is useful to relate the input material parameters in the model with the microstructure and the mechanical behavior of the elastin. Four independent material parameters in the model, $a$, $b$, $c$, and $n$, relate the microstructure to the mechanical behavior of the elastin. $a$ and $b$ are dimensions along the longitudinal and circumferential directions of the arterial wall, respectively. As shown in the representative stress–strain curves in Figure 14.5, the relative values $a$, $b$, and $c$ provide information on the fiber orientation and the degree of orthotropy that exists in the network. The dependent parameter $N$ is related to the chain length between the two cross-links at the micro-level,and the extensibility of the material at the macro-level. A less extensible material will be expected to be more cross-linked and thus have a smaller $N$. The chain density per unit volume $n$ is related to the elastin content and the initial tangent modulus of elastin.

Figure 14.6 shows the finite-element simulation results of the equi- and nonequi-biaxial tests of elastin. Experimental results were also plotted for the purpose of comparison. Material parameters were obtained by fitting the simulation results to the equi-biaxial test data and then used to predict the mechanical responses under nonequi-biaxial loading conditions and to compare with experimental results. Good agreement between data and model prediction suggests that material parameters obtained from the equi-biaxial test were able to predict the stress–strain responses of elastin under arbitrary nonequi-biaxial loading conditions and confirms that the microstructural model is suitable for the study of the anisotropic hyperelastic mechanical behavior of arterial elastin.

Modeling studies on elastin have also been extended to its viscoelastic properties based on the theory of quasi-linear viscoelasticity (QLV) originally introduced to biomechanical studies by Fung [65]. This model is characterized by an instantaneous elastic stress response and a relaxation function. Owing to its relative ease to implement and the limited number of material parameters required to model the viscoelastic response of soft tissue, the theory of QLV has been widely used to model the viscoelastic

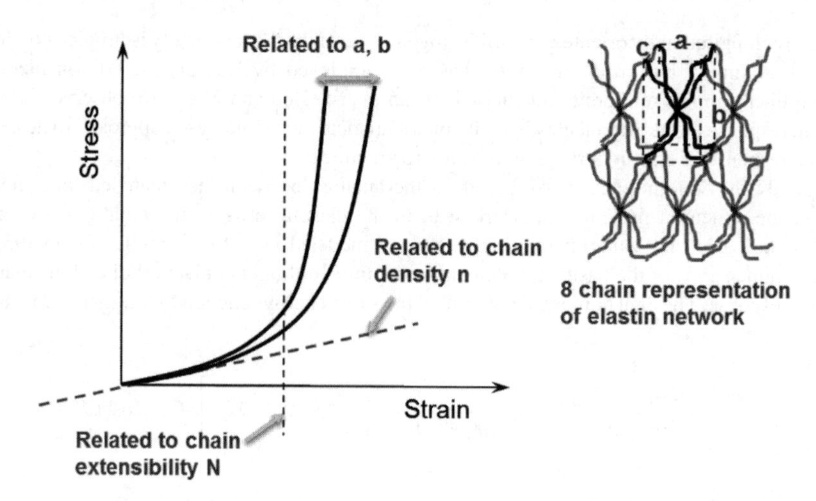


**Figure 14.5** Typical hyperelastic orthotropic stress–strain response and model parameters associated with the eight-chain orthotropic unit element (adapted from Zou and Zhang [6])

response of soft biological tissues [96–99]. Accordingly, the general form of the second Piola–Kirchhoff stress $\mathbf{S}(t)$ is given as

$$\mathbf{S}(t) = \int_{0}^{t} G(t-\tau)\frac{\partial \mathbf{S}^{\mathrm{e}}}{\partial \tau}\,\mathrm{d}\tau, \tag{14.2}$$

where $\mathbf{S}^{\mathrm{e}}$ is the elastic stress function and $G(t)$, the reduced relaxation function, is given as

$$G(t) = \frac{1 + C[E_1(t/\tau_2) - E_1(t/\tau_1)]}{1 + C\ln(\tau_2/\tau_1)}, \tag{14.3}$$

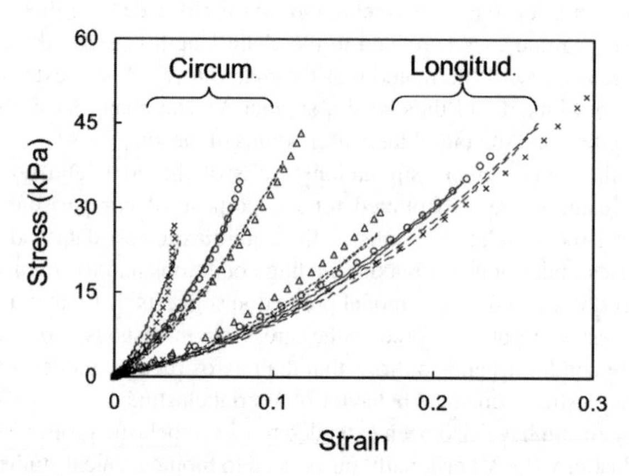


**Figure 14.6** Cauchy stress versus logarithmic strain for elastin from bovine thoracic aorta under equi-biaxial (circles) and nonequi-biaxial (other symbols) loading in circumferential and longitudinal directions (adapted Zou and Zhang [6])

where $C$, $\tau_1$, and $\tau_2$ are material parameters and $E_1(z)$ is an exponential integral function:

$$E_1(z) = \int_z^{\infty} \frac{e^{-t}}{t}\, dt \quad (|\arg z| < \pi).$$

The QLV model at fiber level has been used to model the tissue-level orthotropic viscoelastic properties. For example, Bischoff [100] considered a chain of freely jointed viscoelastic fibers and incorporated it into the eight-chain unit elements to model the macroscopic viscoelastic response of tissues. A similar concept has been followed by Zou and Zhang [7] to model the orthotropic viscoelastic response of arterial elastin. That is, the elastic stress tensor $\mathbf{S}^e$ is derived from the eight-chain orthotropic hyperelastic model (14.1) as

$$\mathbf{S}^e = \frac{\partial W}{\partial \mathbf{E}} = \frac{n}{4}\left[\sum_{i=1}^{4} \mathbf{S}^e_{\text{chain}} - k\Theta \frac{\beta_p}{\sqrt{N}}\left(\frac{a^2}{\lambda_a^2}\mathbf{a}\otimes\mathbf{a} + \frac{b^2}{\lambda_b^2}\mathbf{b}\otimes\mathbf{b} + \frac{c^2}{\lambda_c^2}\mathbf{c}\otimes\mathbf{c}\right)\right] + BJ\mathbf{C}^{-1}\sinh(J-1), \tag{14.4}$$

where $\mathbf{C}$ is the right Cauchy–Green tensor and $\mathbf{C} = \mathbf{F}^T\mathbf{F}$, $\mathbf{E} = 1/2(\mathbf{C} - \mathbf{I})$ is the Green–Lagrange strain tensor, and $\mathbf{F}$ is the deformation gradient. $\mathbf{S}^e_{\text{chain}}$ is the elastic stress of a single freely jointed chain, given by

$$\mathbf{S}^e_{\text{chain}} = k\Theta \frac{\mathbf{p}^i \otimes \mathbf{p}^i}{\rho^i}\beta_{\rho^i}, \tag{14.5}$$

where $\mathbf{p}^i$ is the normalized chain vector describing the undeformed chain. The resulting viscoelastic stress tensor $\mathbf{S}$ in the macroscopic continuum model is then given as

$$\mathbf{S} = \frac{n}{4}\left[\sum_{i=1}^{4}\int_0^t G(t-\tau)\frac{\partial(\mathbf{S}^e_{\text{chain}})}{\partial\tau}\, d\tau - k\Theta\frac{\beta_p}{\sqrt{N}}\left(\frac{a^2}{\lambda_a^2}\mathbf{a}\otimes\mathbf{a} + \frac{b^2}{\lambda_b^2}\mathbf{b}\otimes\mathbf{b} + \frac{c^2}{\lambda_c^2}\mathbf{c}\otimes\mathbf{c}\right)\right]$$
$$+ BJ\mathbf{C}^{-1}\sinh(J-1). \tag{14.6}$$

The reduced relaxation function is approximated by a series of exponential functions, and can be written as [101]

$$G(t) = G_e + \frac{G_0 - G_e}{N_d + 1}\sum_{I=0}^{N_d}\exp(-t/10^{I+I_0}), \tag{14.7}$$

where $G_e$, $G_0$, $N_d$, and $I_0$ are material parameters that will be obtained by fitting the simulation results to the experimental stress relaxation data.

There are two groups of material parameters: a group of four independent parameters ($a$, $b$, $c$, and $n$) describing the orthotropic hyperelastic behavior of elastin, as described earlier, and an additional group of four parameters ($G_e$, $G_0$, $N_d$, and $I_0$) describing the viscoelastic response. The two groups of parameters are, respectively, estimated by fitting the equi-biaxial tensile test data and stress relaxation test data. As shown in Figure 14.7, the results demonstrate that the orthotropic viscoelastic constitutive model incorporating fiber-level viscoelasticity into the eight-chain statistical-mechanics-based microstructural model is suitable to describe the stress relaxation behavior of elastin.

Although the simulation results fit the experimental data reasonably well, there are limitations associated with using the fiber-level QLV model with the orthotropic hyperelastic constitutive model. The QLV model considers separate strain and time dependency for the stress response and, therefore, it is not able to predict the initial stress dependency of the stress relaxation, limiting the usage of the QLV

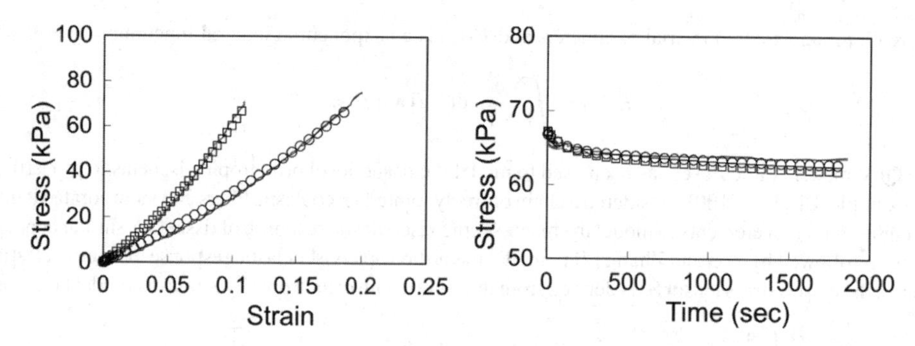


**Figure 14.7** Simulation results of elastin under equi-biaxial tensile test (left) and stress relaxation test (right). Circles and squares represent longitudinal and circumferential directions, respectively (adapted from Zou and Zhang [7])

model in studies of viscoelastic properties of materials which show an obvious stress dependency, such as aorta. That being said, as shown in Figure 14.4, there is a small dependence of the rate of elastin stress relaxation on the initial stress level at physiological load levels and the material parameters fitted from one test can be used to describe the stress relaxation of elastin under different initial stress levels. Also, the QLV model is found to be insufficient to describe the creep behavior due to its intrinsic assumptions of separable strain and time dependency of the stress response. Despite the deficiencies of the QLV model, the negligible creep behavior and minimal initial stress dependency of stress relaxation of elastin make this model adequate enough to investigate the mechanics of elastin.

## 14.5 Conclusions

Changes in the biomechanical function of elastin have important medical and physiological consequences. It is necessary to better understand the inherent structure and mechanical behavior of elastin, its contribution to the whole arterial mechanics, and its interaction with other vascular wall constituents. Such knowledge would benefit tissue engineering, lead to a better understanding of the pathophysiology of vascular diseases, and facilitate the development of more accurate mathematical models that predict the mechanical response of tissues and its evolution in diseases.

Anisotropic response of the arterial elastin may have important structure–function and mechanobiological implications. The lack of creep response suggests different mechanisms involved in the stress relaxation and creep of elastin. Although elastin shows less stress relaxation behavior than intact arteries, its viscoelastic properties in conjunction with degradation during the lifetime of the cardiovascular system can deteriorate, contributing to diseases, and hence needs to be investigated. For example, biochemical damage results in changes in the microstructure of elastin, and thus in the mechanical functionalities of tissues. The alteration of elastin orthotropic and viscoelastic properties due to glycation was shown by Zou and Zhang [102] and needs to be understood in cardiovascular diseases and aging. The current stage of experimental and modeling results sets a platform to launch further and more detailed studies on the biomechanical and biological roles of elastin in arterial wall.

## Acknowledgments

This work was made possible through research funding from the National Heart, Lung, and Blood Institute grant HL-098028 and National Science Foundation grants CMMI 1100791 and CAREER 0954825 to Y. Zhang.

## References

[1] Kielty, C.M., Sherratt, M.J., and Shuttleworth, C.A. (2002) Elastic fibers. *Journal of Cell Science*, **115**, 2817–2828.

[2] Mecham, R.P. (2008) Methods in elastic tissue biology: elastin isolation and purification. *Methods*, **45**, 32–41.

[3] O'Connell, M.K., Murthy, S., Phan, S. *et al.* (2008) The three dimensional micro- and nanostructure of the aortic medial lamellar unit measured using 3D confocal and electron microscopy imaging. *Matrix Biology*, **27**, 171–181.

[4] Brooke, B.S., Bayes-Genis, A., and Li, D.Y. (2003) New insights into elastin and vascular disease. *Trends in Cardiovascular Medicine*, **13**, 176–181.

[5] Wolinski, H. and Glagov, S.A. (1967) Lamellar unit of aortic medial structure and function in mammals. *Circulation Research*, **20**, 99–111.

[6] Zou, Y. and Zhang, Y. (2009) An experimental and theoretical study on the anisotropy of elastin network. *Annals of Biomedical Engineering*, **37**, 1572–1583.

[7] Zou, Y. and Zhang, Y. (2011) The orthotropic viscoelastic behavior of aortic elastin. *Biomechanics and Modeling in Mechanobiology*, **10**, 613–625.

[8] Humphrey, J.D. (2002) *Cardiovascular Solid Mechanics; Cells, Tissues, and Organs*, Springer, New York, NY.

[9] Humphrey, J.D. (2008) Mechanisms of arterial remodeling in hypertension: coupled roles of wall shear and intramural stress. *Hypertension*, **52**, 195–200.

[10] Ito, S., Ishimaru, S., and Wilson, S.E. (1997) Inhibitory effect of type 1 collagen gel containing alpha-elastinon proliferation and migration of vascular smooth muscle and endothelial cells. *Cardiovascular Surgery*, **5**, 176–183.

[11] Li, D.Y., Brooke, B.S., Davis, E.C. *et al.* (1998) Elastin is an essential determinant of arterial morphogenesis. *Nature*, **393**, 276–280.

[12] Karnik, S.K., Brooke, B.S., Bayes-Genis, A. *et al.* (2003) A critical role for elastin signaling in vascular morphogenesis and disease. *Development*, **130**, 411–423.

[13] Ailawadi, G., Moehle, C.W., Pei, H. *et al.* (2009) Smooth muscle phenotypic modulation is an early event in aortic aneurysms. *Journal of Thoracic and Cardiovascular Surgery*, **138**, 1392–1399.

[14] Yamamoto, M., Yamamoto, K., and Noumura, T. (1993) Type I collagen promotes modulation of cultured arterial smooth muscle cells from a contractile to a synthetic phenotype. *Experimental Cell Research*, **204**, 121–129.

[15] Campa, J.S., Greenhalgh, R.M., and Powell, J.T. (1987) Elastin degradation in abdominal aortic aneurysm. *Atherosclersis*, **65**, 13–21.

[16] Menashi, S., Campa, J.S., Greenhalgh, R.M., and Powell, J.T. (1987) Collagen in abdominal aortic aneurysm: typing, content, and degradation. *Journal of Vascular Surgery*, **6**, 578–582.

[17] Vorp, D.A. (2007) Biomechanics of abdominal aortic aneurysm. *Journal of Biomechanics*, **40**, 1887–1902.

[18] Humphrey, J.D. and Holzapfel, G.A. (2012) Mechanics, mechanobiology, and modeling of human abdominal aorta and aneurysms. *Journal of Biomechanics*, **45**, 805–814.

[19] Kobs, R.W., Muvarak, N.E., Eickhoff, J.C., and Chesler, N.C. (2005) Linked mechanical and biological aspects of remodeling in mouse pulmonary arteries with hypoxia-induced hypertension. *American Journal of Physiology – Heart and Circulatory Physiology*, **288**, H1209–H1217.

[20] Arribas, S.M., Hinek, A., and Gonzalez, M.C. (2006) Elastic fibres and vascular structure in hypertension. *Pharmacology & Therapeutics*, **111**, 771–791.

[21] Boumaza, S., Arribas, S.M., Osborne-Pellegrin, M. *et al.* (2001) Fenestrations of the carotid internal elastic lamina and structural adaptation in stroke-prone spontaneously hypertensive rats. *Hypertension*, **37**, 1101–1107.

[22] Chow, M.J., Zou, Y., Huamei, H. *et al.* (2012) Obstruction-induced pulmonary vascular remodeling. *Journal of Biomechanical Engineering*, **133**, 111009.

[23] Lammers, S.R., Kao, P.H., Qi, H.J. *et al.* (2008) Changes in the structure–function relationship of elastin and its impact on the proximal pulmonary arterial mechanics of hypertensive calves. *American Journal of Physiology – Heart and Circulatory Physiology*, **295**, H1451–H1459.

[24] Fischer, G.M., Cherian, K., and Swain, M.L. (1981) Increased synthesis of aortic collagen and elastin in experimental atherosclerosis. *Atherosclerosis*, **39**, 463–467.

[25] Nakatake, J. and Yamamoto, T. (1987) Three-dimensional architecture of elastic tissue in athero-arteriosclerotic lesions of the rat aorta. *Atherosclerosis*, **164**, 191–200.
[26] Krettek, A., Sukhova, G.K., and Libby, P. (2003) Elastogenesis in human arterial disease: a role for macrophages in disordered elastin. *Arteriosclerosis, Thrombosis, and Vascular Biology*, **23**, 582–587.
[27] Grande, J., Davis, H.R., Bates, S. *et al.* (1987) Effect of an elastic growth substrate on cholesteryl ester synthesis and foam cell formation by cultured aortic smooth muscle cells. *Atherosclerosis*, **68**, 87–93.
[28] Noma, A., Takahashi, T., and Wada, T. (1981) Elastin–lipid interaction in the arterial wall. Part 2. In vitro binding of lipoprotein–lipid to arterial elastin and the inhibitory effect of high of high density lipoproteins on the process. *Atherosclerosis*, **38**, 373–382.
[29] Seyama, Y., Hayashi, M., Usami, E. *et al.* (1990) Basic study on nondelipidemic fractionation of aortic connective tissue of human and experimental atherosclerosis. *Japanese Journal of Clinical Chemistry*, **19**, 53–61.
[30] Wachi, H. (2011) Role of elastic fibers on cardiovascular disease. *Journal of Health Science*, **57**, 449–457.
[31] Jacob, M.P. (2003) Extracellular matrix remodeling and matrix metalloproteinases in the vascular wall during ageing and in pathological conditions. *Biomedical & Pharmacology Journal*, **57**, 195–200.
[32] Lillie, M.A. and Gosline, J.M. (2002) The viscoelastic basis for the tensile strength of elastin. *International Journal of Biological Macromolecules*, **30**, 119–127.
[33] Lillie, M.A. and Gosline, J.M. (2007) Mechanical properties of elastin along the thoracic aorta in the pig. *Journal of Biomechanics*, **10**, 2214–2221.
[34] Lu, Q., Ganesan, K., Simionescu, D.T., and Vyavahare, N.R. (2004) Novel porous aortic elastin and collagen scaffolds for tissue engineering. *Biomaterials*, **25**, 5227–5237.
[35] Lanir, Y. and Fung, Y.C. (1974) Two-dimensional mechanical properties of rabbit skin. II. Experimental results. *Journal of Biomechanics*, **7**, 171–174.
[36] Humphrey, J.D., Vawter, D.L., and Vito, R.P. (1986) Mechanical behavior of excised canine visceral pleura. *Annals of Biomedical Engineering*, **14**, 451–466.
[37] Vito, R.P. (1980) The mechanical properties of soft tissues. I. A mechanical system for biaxial testing. *Journal of Biomechanics*, **13**, 947–950.
[38] Adamczyk, M.M., Lee, T.C., and Vesely, I. (2000) Biaxial strain properties of elastase-digested porcine aortic valves. *Journal of Heart Valve Disease*, **9**, 445–453.
[39] Sacks, M.S. and Sun, W. (2003) Multiaxial mechanical behavior of biological materials. *Annual Review of Biomedical Engineering*, **5**, 251–284.
[40] Lally, C., Reid, A.J., and Prendergast, P.J. (2004) Elastic behavior of porcine coronary artery tissue under uniaxial and equibiaxial tension. *Annals of Biomedical Engineering*, **23**, 1355–1364.
[41] Vande Geest, J.P., Sacks, M.S., and Vorp, D.A. (2004) Age dependency of the biaxial biomechanical behavior of human abdominal aorta. *Journal of Biomechanical Engineering*, **126**, 815–822.
[42] Vande Geest, J.P., Sacks, M.S., and Vorp, D.A. (2006) The effects of aneurysm on the biaxial mechanical behavior of human abdominal aorta. *Journal of Biomechanics*, **39**, 1324–1334.
[43] Vande Geest, J.P., Sacks, M.S., and Vorp, D.A. (2006) A planar biaxial constitutive relation for the luminal layer of intra-luminal thrombus in abdominal aortic aneurysms. *Journal of Biomechanics*, **39**, 2347–2354.
[44] Gundiah, N., Ratcliffe, M.B., and Pruitt, L.A. (2007) Determination of strain energy function for arterial elastin: experiments using histology and mechanical tests. *Journal of Biomechanics*, **40**, 586–594.
[45] Kim, J. and Baek, S. (2011) Circumferential variations of mechanical behavior of the porcine thoracic aorta during the inflation test. *Journal of Biomechanics*, **44**, 1941–1947.
[46] Humphrey, J.D. (2008) Vascular adaptation and mechanical homeostasis at tissue, cellular, and sub-cellular levels. *Cell Biochemistry and Biophysics*, **50**, 53–78.
[47] Lillie, M.A. and Gosline, J.M. (2007) Limits to the durability of arterial elastin tissue. *Biomaterials*, **28**, 2021–2031.
[48] Lillie, M.A. and Gosline, J.M. (1990) The effects of hydration on the dynamic mechanical properties of elastin. *Biopolymers*, **29**, 1147–1160.
[49] Wells, S.M., Langille, B.L., and Adamson, S.L. (1998) In vivo and in vitro mechanical properties of the sheep thoracic aorta in the perinatal period and adulthood. *American Journal of Physiology*, **274**, H1749–H1760.

[50] Wells, S.M., Langille, B.L., Lee, J.M., and Adamson, S.L. (1999) Determinants of mechanical properties in the developing ovine thoracic aorta. *American Journal of Physiology*, **277**, 1385–1391.
[51] Zhang, Y., Dunn, M.L., Drexler, E.S. *et al.* (2005) A microstructural hyperelastic model of pulmonary arteries under normo- and hypertensive conditions. *Annals of Biomedical Engineering*, **33**, 1042–1052.
[52] Guo, X., and Kassab, G.S. (2004) Distribution of stress and strain along the porcine aorta and coronary arterial tree. *American Journal of Physiology-Heart & Circulatory Physiology*, **286**, H2361–H2368.
[53] Stergiopulos, N., Vulliemoz, S., Rachev, A. *et al.* (2001) Assessing the homogeneity of the elastic properties and composition of the pig aortic media. *Journal of Vascular Research*, **38**, 237–246.
[54] Han, H.-C. and Fung Y.-C. (1995) Longitudinal strain of canine and porcine aortas. *Journal of Biomechanics*, **28**, 637–641.
[55] Daamen, W.F., van Moerkerk, H.T.B., Hafmans, T. *et al.* (2003) Preparation and evaluation of molecularly-defined collagen elastin glycosaminoglycan scaffolds for tissue engineering. *Biomaterials*, **24**, 4001–4009.
[56] Black, L.D., Allen, P.G., Morris, S.M. *et al.* (2008) Mechanical and failure properties of extracellular matrix sheets as a function of structural protein composition. *Biophysics*, **94**, 1916–1929.
[57] Spina, M., Friso, A., Ewins, A.R. *et al.* (1999) Physicochemical properties of arterial elastin and its associated glycoproteins. *Biopolymers*, **49**, 255–265.
[58] Weinberg, P.D., Winlove, C.P., and Parker, K.H. (1995) The distribution of water in arterial elastin: effects of mechanical stress, osmotic pressure, and temperature. *Biopolymers*, **35**, 161–169.
[59] Knauss, W.G. and Emri, I.J. (1981) Non-linear viscoelasticity based on free volume consideration. *Computers & Structures*, **13**, 123–128.
[60] Lillie, M.A. and Gosline, J.M. (1996) Swelling and viscoelastic properties of osmotically stressed elastin. *Biopolymers*, **39**, 641–693.
[61] Nagatomi, J., Gloeckner, D.C., Chancellor, M.B. *et al.* (2004) Changes in the biaxial viscoelastic response of the urinary bladder following spinal cord injury. *Annals of Biomedical Engineering*, **32**, 1409–1419.
[62] Hingorani, R.V., Provenzano, P.P., Lakes, R.S. *et al.* (2004) Nonlinear viscoelasticity in rabbit medial collateral ligament. *Annals of Biomedical Engineering*, **32**, 306–312.
[63] Provenzano, P., Lakes, R., Keenan, T., and Vanderby, R. Jr. (2001) Nonlinear ligament viscoelasticity. *Annals of Biomedical Engineering*, **29**, 908–914.
[64] Nagatomi, J., Toosi, K.K., Chancellor, M.B., and Sacks, M.S. (2008) Contribution of the extracellular matrix to the viscoelastic behavior of the urinary bladder wall. *Biomechanics and Modeling in Mechanobiology*, **7**, 395–404.
[65] Fung, Y.C. (1993) *Biomechanics: Mechanical Properties of Living Tissues*, Springer, New York, NY.
[66] Boyce, B.L., Jones, R.E., Nguyen, T.D., and Grazier, J.M. (2007) Stress-controlled viscoelastic tensile response of bovine cornea. *Journal of Biomechanics*, **40**, 2367–2376.
[67] Grashow, J.S., Sacks, M.S., Liao, J., and Yoganathan, A.P. (2006) Planar biaxial creep and stress relaxation of the mitral valve anterior leaflet. *Annals of Biomedical Engineering*, **34**, 1509–1518.
[68] Liao, J., Yang, L., Grashow, J., and Sacks, M.S. (2007) The relationship between collagen fibril kinematics and mechanical properties in the mitral valve anterior leaflet. *Journal of Biomechanical Engineering*, **129**, 78–87.
[69] Stella, J.A., Liao, J., and Sacks, M.S. (2007) Time-dependent biaxial mechanical behavior of the aortic heart valve leaflet. *Journal of Biomechanics*, **40**, 3169–3177.
[70] Thornton, G.M., Oliynyk, A., Frank, C.B., and Shrive, N.G. (1997) Ligament creep cannot be predicted from stress relaxation at low stress: a biomechanical study of the rabbit medial collateral ligament. *Journal of Orthopaedic Research*, **15**, 652–656.
[71] Thornton, G.M., Frank, C.B., and Shrive, N.G. (2001) Ligament creep behavior can be predicted from stress relaxation by incorporating fiber recruitment. *Journal of Rheology*, **45**, 493–507.
[72] Holzapfel, G.A., Gasser, T.C., and Ogden, R.W. (2000) A new constitutive framework for arterial wall mechanics and a comparative study of material models. *Journal of Elasticity*, **61**, 1–48.
[73] Holzapfel, G.A. and Ogden, R.W. (2010) Constitutive modelling of arteries. *Proceedings of the Royal Society A*, **466**, 1551–1597.
[74] Chow, M.J., Mondonedo, J.R., Johnson, V., and Zhang, Y. (2012) Progressive structural and biomechanical changes in elastin degraded aorta. *Biomechanics and Modeling in Mechanobiology*, doi:10.1007/s10237-012-0404-9.

[75] Holzapfel, G.A. and Weizsacker, H.W. (1998) Biomechanical behavior of the arterial wall and its numerical characterization. *Computers in Biology and Medicine*, **28**, 377–392.

[76] Baek, S., Gleason, R.L., Rajagopal, K.R., and Humphrey, J.D. (2007) Theory of small on large: potential utility in computations of fluidsolid interactions in arteries. *Computer Methods in Applied Mechanics and Engineering*, **196**, 3070–3078.

[77] Cardamone, L., Valentin, A., Eberth, J.F., and Humphrey, J.D. (2009) Origin of axial prestretch and residual stress in arteries. *Biomechanics and Modeling in Mechanobiology*, **8**, 431–446.

[78] Driessen, N.J.B., Cox, M.A.J., Bouten, C.V.C., and Baaijens, F.P.T. (2008) Remodelling of the angular collagen fiber distribution in cardiovascular tissues. *Biomechanics and Modeling in Mechanobiology*, **7**, 93–103.

[79] Eberth, J.F., Cardamone, L., and Humphrey, J.D. (2011) Evolving biaxial mechanical properties of mouse carotid arteries in hypertension. *Journal of Biomechanics*, **44**, 2532–2537.

[80] Ferruzzi, J., Vorp, D.A., and Humphrey, J.D. (2011) On constitutive descriptors of the biaxial mechanical behavior of human abdominal aorta and aneurysms. *Journal of the Royal Society Interface*, **8**, 435–450.

[81] Figueroa, C.A., Baek, S., Taylor, C.A., and Humphrey, J.D. (2009) A computational framework for fluid–solid-growth modeling in cardiovascular simulations. *Computer Methods in Applied Mechanics and Engineering*, **198**, 3583–3602.

[82] Gasser, T.C., Ogden, R.W., and Holzapfel, G.A. (2006) Hyperelastic modelling of arterial layers with distributed collagen fibre orientations. *Journal of the Royal Society Interface*, **3**, 15–35.

[83] Hansen, L., Wan, W., and Gleason, R.L. (2009) Microstructurally motivated constitutive modeling of mouse arteries cultured under altered axial stretch. *Journal of Biomechanical Engineering*, **131**, 101015.

[84] Haskett, D., Johnson, G., Zhou, A. *et al.* (2010) Microstructural and biomechanical alterations of the human aorta as a function of age and location. *Biomechanics and Modeling in Mechanobiology*, **9**, 725–736.

[85] Wagner, H.P. and Humphrey, J.D. (2011) Differential passive and active biaxial mechanical behaviors of muscular and elastic arteries: basilar versus common carotid. *Journal of Biomechanical Engineering*, **133**, 051009.

[86] Watton, P.N., Hill, N.A., and Heil, M. (2004) A mathematical model for the growth of the abdominal aortic aneurysm. *Biomechanics and Modeling in Mechanobiology*, **3**, 98–113.

[87] Wicker, B.K., Hutchens, H.P., Wu, Q. *et al.* (2008) Normal basilar artery structure and biaxial mechanical behaviour. *Computer Methods in Biomechanics and Biomedical Engineering*, **11**, 539–551.

[88] Zeinali-Davarani, S., Choi, J., and Baek, S. (2009) On parameter estimation for biaxial mechanical behavior of arteries. *Journal of Biomechanics*, **42**, 524–530.

[89] Aaron, B.B. and Gosline, J.M. (1981) Elastin as a random-network elastomer – a mechanical and optical analysis of single elastin fibers. *Biopolymers*, **20**, 1247–1260.

[90] Watton, P.N., Yiannis, V., and Holzapfel, G.A. (2009) Modelling the mechanical response of elastin for arterial tissue. *Journal of Biomechanics*, **42**, 1320–1325.

[91] Gundiah, N., Ratcliffe, M.B., and Pruitt, L.A. (2009) The biomechanics of arterial elastin. *Journal of the Mechanical Behavior of Biomedical Materials*, **2**, 288–296.

[92] Rezakhaniha, R. and Stergiopulos, N. (2008) A structural model of the venous wall considering elastin anisotropy. *Journal of Biomechanical Engineering*, **130**, 031017.

[93] DeBotton, G., Hariton, I., and Socolsky, E.A. (2006) Neo-Hookean fiber-reinforced composites in finite elasticity. *Journal of the Mechanics and Physics of Solids*, **54**, 533–559.

[94] Kao, P.H., Lammers, S.R., Tian, L. *et al.* (2011) A microstructurally driven model for pulmonary artery tissue. *Journal of Biomechanical Engineering*, **133**, 051002.

[95] Bischoff, J.E., Arruda, E.A., and Grosh, K. (2002) A microstructurally based orthotropic hyperelastic constitutive law. *Journal of Applied Mechanics*, **69**, 570–579.

[96] Doehring, T.C., Carew, E.O., and Vesely, I. (2004) The effect of strain rate on the viscoelastic response of aortic valve tissue: a direct-fit approach. *Annals of Biomedical Engineering*, **32**, 223–232.

[97] Giles, J.M., Black, A.E., and Bischoff, J.E. (2007) Anomalous rate dependence of the preconditioned response of soft tissue during load controlled deformation. *Journal of Biomechanics*, **40**, 777–785.

[98] Kwan, M.K., Li, T.H.D., and Woo, S.L.Y. (1993) On the viscoelastic properties of the anteromedial bundle of the anterior cruciate ligament. *Journal of Biomechanics*, **26**, 447–452.

[99] Sverdlik, A. and Lanir, Y. (2002) Time-dependent mechanical behavior of sheep digital tendons, including the effects of preconditioning. *Journal of Biomechanical Engineering*, **124**, 78–84.

[100] Bischoff, J.E. (2006) Reduced parameter formulation for incorporating fiber level viscoelasticity into tissue level biomechanical models. *Annals of Biomedical Engineering*, **34**, 1164–1172.

[101] Puso, M.A. and Weiss, J.A. (1998) Finite element implementation of anisotropic quasi-linear viscoelasticity using a discrete spectrum approximation. *Journal of Biomechanical Engineering*, **120**, 62–70.

[102] Zou, Z. and Zhang, Y. (2012) The biomechanical function of arterial elastin in solutes. *Journal of Biomechanical Engineering*, **134**, 071002.

# 15

# Characterization of Mechanical Properties of Biological Tissue: Application to the FEM Analysis of the Urinary Bladder

Eugenio Oñate[a], Facundo J. Bellomo[b], Virginia Monteiro[a], Sergio Oller[a], and Liz G. Nallim[b]

[a] *International Center for Numerical Method in Engineering (CIMNE), Technical University of Catalonia, Spain*

[b] *INIQUI (CONICET), Faculty of Engineering, National University of Salta, Argentina*

## 15.1 Introduction

This chapter presents an approach for the mechanical behavior of soft biological tissue using the finite-elements method (FEM) and a general constitutive model. Specifically, we analyze the mechanical behavior of a urinary bladder starting from a procedure for obtaining the mechanical characterization of the biological tissue. The difficulty in this study lies not only in modeling the mechanical behavior of the bladder subjected to inflation under the presence of an internal fluid, but also in the difficulty encountered in determining the biomechanical properties of the biological tissue that forms the bladder. Bladder tissue is modeled as a composite material formed by soft matrix reinforced with preferentially oriented fibers. In the first part of the chapter we present a procedure for identifying the mechanical properties of biological tissue's main constituents by an inverse method. Then this information is used for the numerical simulation of the mechanical behavior of the bladder within the FEM.

The formulation can be applied to various types of biological tissues, both in the field of material characterization and in the numerical simulation of the tissue's biomechanical behavior. The approach presented in this chapter has been applied to the study of the arterial tissue behavior [1].

*Multiscale Simulations and Mechanics of Biological Materials*, First Edition. Edited by Shaofan Li and Dong Qian.

## 15.2 Inverse Approach for the Material Characterization of Biological Soft Tissues via a Generalized Rule of Mixtures

There are few experimental data related to the mechanical properties of the main load-bearing constituents of soft biological tissues (collagen, elastin). Veseley [2] estimated the relative contribution of elastin to the mechanics of the porcine aortic valve. Roeder *et al.* [3] studied the structural–mechanical relationship of three-dimensional type I collagen matrices prepared in vitro. Holzapfel [4] presented a complete review of the biomechanical role of collagen in arterial walls. There are many difficulties associated with the isolation and testing of the individual components of soft tissues. For these reasons, experimental studies on mechanical properties of collagen are mainly related to tendons and ligaments [5–8]. On the other hand, several experimental studies had been carried out regarding the mechanical properties of extracellular matrix (ECM) scaffolds [9–16], derived mostly from porcine small intestinal submucosa and the urinary bladder. These scaffolds have become widely studied recently for a number of applications, including repair of the urinary bladder, esophagus, and myocardium [17]. These studies provide a very useful insight into the mechanics, fiber alignment, and kinematics of the ECM material.

A comprehensive study of soft tissues requires an adequate knowledge of the mechanical properties of their components. In this chapter a constitutive model based on a generalized rule of mixtures is proposed. The tissue is modeled as a biological composite material reinforced by families of collagen fibers. The material parameters of the model are identified by an inverse method. The application of this methodology leads to the mechanical characterization of matrix and collagen reinforcement. The model described in the following sections has been implemented in a finite-element formulation and applied to the mechanical analysis of the urinary bladder.

### 15.2.1 *Constitutive Model for Material Characterization*

The constitutive model presented in this chapter is an extension of the generalized rule of mixtures proposed by Car *et al.* [18]. This theory allows the study of the behavior of composite materials as a combination of individual components with their own constitutive law, each one satisfying appropriate serial–parallel compatibility equations. These equations establish the inter-material kinematic conditions.

#### Generalized Rule of Mixtures for Finite Strains

The free energy in the reference configuration is given by

$$m\Psi = \sum_{c=1}^{n} m_c k_c \Psi_c, \tag{15.1}$$

where $m$ and $m_c$ are, respectively, the composite mass and the $c$th component mass and $\Psi$ and $\Psi_c$ are, respectively, the composite and the $c$th component free energy. The Green–Lagrange strain tensor of the $c$th component on the reference configuration is given by [19]

$$\boldsymbol{E}_c = [(1 - \chi_c)\boldsymbol{I}_4 + \chi_c \boldsymbol{\phi}_c] : \boldsymbol{E}, \tag{15.2}$$

where $\boldsymbol{E}$ is the composite Green–Lagrange strain tensor and $(\phi_{ijkl})_c$ is the serial behavior tensor in the reference configuration given by

$$(\phi_{ijkl})_c = \left(\frac{\partial^2 \Psi_c}{\partial (E_{ij})_c \partial (E_{rs})_c}\right)^{-1} : \left[\sum_{c=1}^{n} k_c \left(\frac{\partial^2 \Psi_c}{\partial (E_{rs})_c \partial (E_{kl})_c}\right)^{-1}\right]^{-1}, \tag{15.3}$$

where $(\partial^2\Psi_c/\partial(E_{ij})_c\partial(E_{rs})_c)^{-1}$ is the constitutive tensor of the $c$th component and $[\sum_{c=1}^{n} k_c(\partial^2\Psi_c/\partial(E_{rs})_c\partial(E_{kl})_c)^{-1}]^{-1}$ is the composite serial constitutive tensor, both expressed in the reference configuration. The fourth-order tensor $\boldsymbol{\phi}_c$ maps the strain to its serial counterpart in the reference configuration ensuring the serial equilibrium constraint.

The second Piola–Kirchhoff stress tensor can be obtained as follows:

$$\boldsymbol{S} = m\frac{\partial\Psi}{\partial\mathbf{E}} = \sum_{c=1}^{n} k_c m_c \frac{\partial\Psi_c}{\partial\boldsymbol{E}_c}\frac{\partial\boldsymbol{E}_c}{\partial\boldsymbol{E}} = \sum_{c=1}^{n} k_c\mathbf{S}_c\frac{\partial\boldsymbol{E}_c}{\partial\boldsymbol{E}}. \tag{15.4}$$

For further applications it is useful to write the Cauchy stress via a push-forward operation

$$\sigma = \frac{1}{J}\mathbf{F}\cdot\mathbf{S}\cdot\mathbf{F}^{\mathrm{T}}, \tag{15.5}$$

where $\mathbf{F}$ is the deformation gradient.

In order to consider near incompressibility in soft issues, the energy is split into its isochoric and volumetric parts as follows:

$$\Psi = \Psi_{\mathrm{vol}}(J) + \Psi_{\mathrm{iso}}(\bar{\mathbf{C}}), \tag{15.6}$$

where $\bar{\mathbf{C}}$ is the deviatoric part of the Cauchy–Green rigth tensor $\mathbf{C}$. The volumetric part of the free energy accounts for the tissue fluids' contribution, which is mainly responsible for the tissue quasi-incompressibility. The interstitial tissue pressure is considered to be the same for all the solid phases, while the serial-parallel behavior affects the solid components only. The stress expression is obtained by

$$S = S_{\mathrm{vol}} + S_{\mathrm{iso}} = 2\frac{\partial\Psi_{\mathrm{vol}}(J)}{\partial C} + \sum_{c=1}^{n} k_c[(1-\chi_c)\mathbf{I}_4 + \chi_c(\boldsymbol{\phi})_c]\cdot\bar{\mathbf{S}}_c, \tag{15.7}$$

where $\mathbf{S}_{\mathrm{vol}}$ is the volumetric part of the stress tensor, $\mathbf{S}_{\mathrm{iso}}$ stands for the isochoric stresses, and $\overline{\mathbf{S}}_c$ are the deviatoric stresses for the $c$th component.

## Mixing Rule for the Uniaxial Case: A Discrete and Incremental Approach

The inverse method proposed in this chapter employs results from uniaxial tests [20]. The uniaxial stress field is deduced from the general expressions (15.4) and (15.7) as

$$\mathbf{S}^{\theta} = \mathbf{S}_{\mathrm{vol}} + \sum_{c=1}^{n} k_c\bar{\mathbf{S}}_c\frac{\partial\mathbf{E}_c}{\partial\mathbf{E}} = \mathbf{S}_{\mathrm{vol}} + \sum_{c=1}^{n} k_c\bar{\mathbf{S}}_c\frac{\partial}{\partial\mathbf{E}}\{[(1-\chi_c)\mathbf{I}_4 + \chi_c\boldsymbol{\phi}_c]\mathbf{E}\}, \tag{15.8}$$

where all tensorial quantities are now reduced to scalar ones.

The fibers induce anisotropy in the composite behavior and, accordingly, the stresses are also functions of the stretching direction, characterized by the $\theta$ angle between the fibers and the stretching direction. Superscript $\theta$ in the second Piola–Kirchhoff stress $\mathbf{S}^{\theta}$ indicates the corresponding fiber direction.

An incremental form of Equation (15.8) suitable for solving the inverse problem can be obtained by considering a linearized stress–strain relationship over a finite time step; that is,

$$\Delta\mathbf{S}^{\theta} = \Delta\mathbf{S}_{\mathrm{vol}} + \sum_{c=1}^{n} k_c\Delta\bar{\mathbf{S}}_c[(1-\chi_c)\mathbf{I}_4 + \chi_c\boldsymbol{\phi}_c]. \tag{15.9}$$

Similarly, the linearization of Equation (15.3) yields [20]

$$\boldsymbol{\phi}_c = \left(\frac{\Delta\bar{\mathbf{S}}_c}{\Delta\mathbf{E}_c}\right)^{-1} : \left[\sum_{c=1}^{n} k_c \left(\frac{\Delta\bar{\mathbf{S}}_c}{\Delta\mathbf{E}_c}\right)^{-1}\right]^{-1}. \tag{15.10}$$

The serial–parallel coupling parameter $\chi_c$ ranges between zero and one. If the stretch is applied in the fiber direction then the components exhibit a pure parallel behavior and $\chi_c = 0$, while a pure serial case occurs when the stretch is perpendicular to the fiber direction, giving $\chi_c = 1$. When the angle between the fibers and the stretching direction $\theta$ is different from 0° or 90°, the composite response is a combination of serial and parallel behaviors. The value of $\chi_c$ in these intermediate cases is obtained as the ratio between the stiffness contribution of the fibers in the $\theta$ direction and the stiffness contribution of the fibers working in parallel.

A soft tissue considered as a biological composite material can be modeled as an isotropic matrix reinforced with long collagen fibers with preferred orientations [21]. Thereby, soft tissue can be considered in a simplified way as a collagen-reinforced composite. The amount of serial and parallel behavior of the whole composite depends on the angle between the fibers direction and the stretching direction. During a test, the fibers rotate; consequently, the corresponding angle is updated according to the change of configuration. The updated angle is obtained by means of a push-forward operation as

$$\theta' = \arctan\left[\frac{\tan(\theta)}{\sqrt{\lambda^3}}\right], \tag{15.11}$$

where $\lambda$ is the uniaxial stretch.

In this model, only the main orientation of the fibers is considered. However, it is possible to consider as many "families" of fibers as desired, each characterized by its mean orientation and volumetric fraction.

### *15.2.2 Definition of the Objective Function and Materials Characterization Procedure*

As previously stated, soft tissue can be considered in a simplified way as a collagen-reinforced composite. Experimental tests of the main individual components of soft tissues are difficult. Therefore, the development of an analytical–numerical approach for the mechanical characterization of the main tissue components is very useful. In this chapter a constrained optimization problem is proposed and solved using experimental results.

The matrix main load-bearing components are elastin and a small amount of collagen fibers with random orientations. Owing to the presence of these collagen fibers, the matrix also exhibits a strong nonlinear behavior.

A hyperelastic neo-Hookean model is considered for the matrix and for the elastic contribution of the fibers. The goal of the optimization problem is to find the material parameters that best fit experimental data. The minimum experimental data necessary to carry out the numerical computations are two uniaxial tests of the tissue. Angle $\theta$ between the collagen fibers and the stretching direction must be different for each test, thus allowing one to asses the influence of the reinforcing orientation on the composite stiffness. The two tests are identified by means of these two angles termed $\theta_1$ and $\theta_2$.

The estimation of the material parameters requires the minimization of an objective function. The goal is the simultaneous minimization of the differences between the experimentally measured and the theoretically determined Cauchy stresses $\sigma$ for the two stretching directions $\theta_1$ and $\theta_2$ at each equilibrium configuration $k$ [21]. The objective function $R$ is defined as [1]

$$R = \sum_{k=1}^{m} \left[ \left( \sigma_{\theta_1}|_{\rm nu} - \sigma_{\theta_1}|_{\rm ex} \right)_k^2 + \left( \sigma_{\theta_2}|_{\rm nu} - \sigma_{\theta_2}|_{\rm ex} \right)_k^2 \right] \tag{15.12}$$

where subscripts "ex" and "nu" denote experimental and numerical values, respectively, and $m$ is the number of equilibrium points considered.

The values of the material parameters are obtained from Equation (15.12) using a subspace trust region method based on the interior-reflective Newton method [22]. Upper and lower bounds for the material parameters are set for each constitutive model following Humphrey [21].

### *15.2.3 Validation of the Inverse Model for Urinary Bladder Tissue Characterization*

The validation of the proposed model for characterization of the bladder tissue is presented next. The biological tissue is composed of a matrix of a mix of elastin, muscle, aqueous ground substance, and collagen fibers and it is considered to have an isotropic behavior. The matrix is reinforced by families of long collagen fibers with preferred orientations along the longitudinal direction of the bladder (i.e., oriented from the apex to the neck). The active behavior of the soft muscle is not considered, with collagen and elastin being the principal load-bearing materials.

Experimental data has been obtained from two different sources. The mechanical properties of the fiber considering its orientation and the anisotropy of the tissue are taken from mechanical data of Gilbert *et al.* [17] using ECM scaffolds derived from the porcine urinary bladder. The second data source is the results of uniaxial tests of intact human bladder tissue published by Martins *et al.* [23]. These data are used to obtain the matrix properties of bladder tissue.

From the analysis of the fiber alignment mapping from Gilbert *et al.* [17] an average mean orientation of 8.5° with respect to the longitudinal direction of the bladder is adopted. The experimental stress–strain curves along the circumferential and longitudinal directions derived from Gilbert *et al.* [17] are shown in Figure 15.1. In these samples the fibers are more closely aligned with the longitudinal direction; consequently, the tissue stiffness is larger in that direction. The stress–strain relationships of the fibers are obtained by the application of the inverse method proposed using the ECM data shown in Figure 15.2. Most of the matrix components are removed during the scaffold preparation procedure. Consequently, a fiber participation ratio of 0.9 is adopted and the rest of the matrix takes the remaining volume. In order to simplify the resolution of the inverse problem, a fully imcompresible neo-Hookean model is considered for the fibers. The stress–strain relationship for the fiber obtained using this model is shown in Figure 15.2.

The fit of the ECM matrix obtained cannot be considered representative of the intact tissue matrix, and for this reason a second application of the inverse method is necessary. To this end, the data from the uniaxial test of intact urinary bladder tissue along its longitudinal direction are used (Figure 15.3). The data plotted are obtained by averaging the results corresponding to the uniaxial tests of human female bladder published by Martins *et al.* [23]. These tests correspond to 13 samples from female cadavers without observable clinical pelvic floor dysfunction with ages between 18 and 65 years. Consequently, they can be considered representative results for human bladder tissue.

In this second step the fiber stress–strain relationship is adopted from the results previously obtained and only the matrix model parameters are adjusted to fit the data from Figure 15.3. The stress–strain relationship of the human bladder matrix obtained is plotted in Figure 15.4.

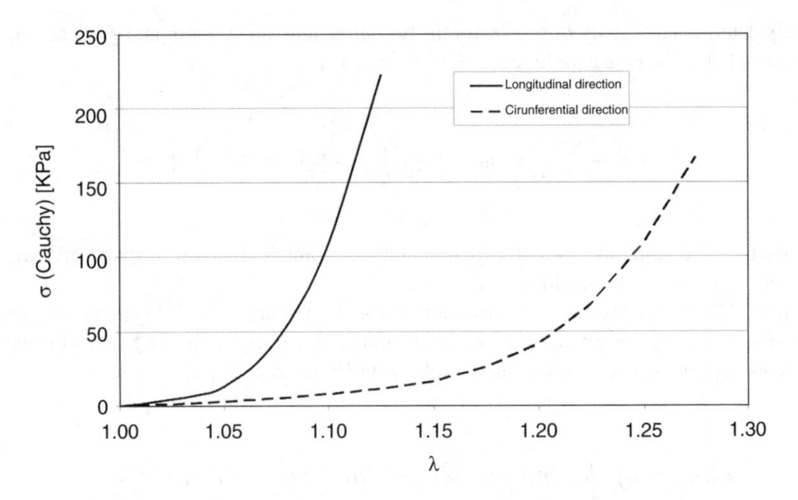


**Figure 15.1** Average stress–strain relationship for the longitudinal and circumferential directions of ECM scaffolds derived from Gilbert *et al.* [17], with permission from Elsevier © 2008

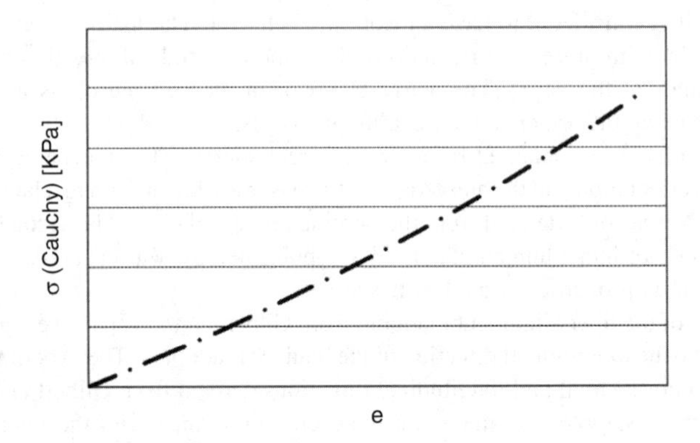


**Figure 15.2** Stress–strain relationship obtained for the fiber fitting of experimental data from Gilbert *et al.* [17], with permission from Elsevier © 2008

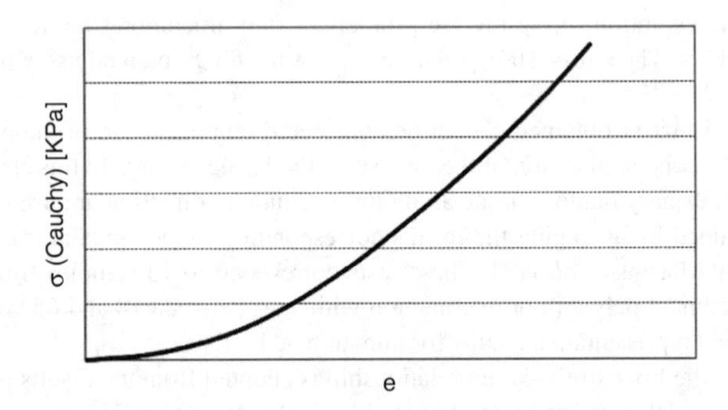


**Figure 15.3** Averaged results of uniaxial tests along the longitudinal axis of human bladder tissue derived from Martins *et al.* [23], with permission from Springer © 2011

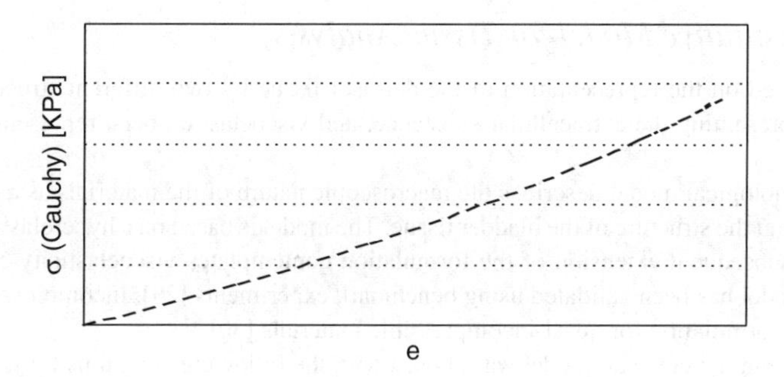


**Figure 15.4** Stress–strain relationship obtained for the matrix fitting using experimental data from Martins *et al.* [23], with permission from Springer © 2011

This validation example shows that the procedure is suitable for the characterization of soft biological tissues modeled as fiber-reinforced composites. The methodology is flexible enough to be applied combining different sets of data, which is especially useful given the scarcity of experimental data regarding the mechanical behavior of soft-tissue components. In this first example, two components are just employed in the rule of mixtures; one component is the matrix, which is a mix of different materials, and the second component is constituted by the collagen fibers reinforcement. However, the rule of mixtures allows us to work with any number of components. Hence, a larger number of constituents can be considered if there is quantitative information regarding the constituent's morphology, volume participation ratios, preferred orientation, and so on.

## 15.3 FEM Analysis of the Urinary Bladder

The material properties characterized in the previous sections have been implemented in an FEM code for studying the mechanical behavior of the bladder. Namely, we have studied the detrusor. This is a smooth muscle which is responsible for maintaining an almost constant pressure inside the bladder during the filling and the storage of the urine [24].

Smooth muscles are known by their nonlinear behavior, and in the specific case of the detrusor the change in the stiffness is due not only to its mechanical properties but also to chemical reactions. The detrusor is innervated by an autonomic nervous system that allows the muscle to be partly contracted, maintaining tonus for prolonged periods with low energy consumption.

Given the complexity of biological materials and their multiscale hierarchy, some simplifications were made in the classical nonlinear continuum mechanics theory. The proposed model is based on the three-dimensional representation of the detrusor tissue by two different structures: a hyperelastic matrix, representing the extracellular substance, and viscoelastic fibers, representing the passive fibers [25]. For the sake of simplicity chemical reactions were not taken into account. Homogenization theory is used to represent the integration of these two main structures, whose resultant mechanical properties have been obtained from the rule of mixtures approach described in previous sections.

To represent the mechanics of the bladder, the constitutive equations and the kinematics are expressed in a total Lagragian description [26, 27].

The model considers the implementation of a hyperelastic–viscoelastic constitutive law into a finite-element formulation accounting for fluid–structure interaction (FSI) [28] effects for the simulation of the deformation of the bladder wall as this is filled with urine. The resulting nonlinear FEM problem is solved by the Newton–Raphson method with a Bossak scheme.

### *15.3.1 Constitutive Model for Tissue Analysis*

The model is based on the representation of the detrusor tissue by two different structures: a hyperelastic matrix, representing the extracellular substance, and viscoelastic fibers, representing the passive fibers [29].

The phenomenological model describes the macroscopic nature of the materials as a continuum, and intends to represent the structure of the bladder tissue. The model is based on a hyperelastic neo-Hookean formulation. A subsequent extension of the formulation contemplates viscoelasticity effects and fiber behavior. The model has been validated using benchmark experiments [29]. Incompressibility is treated using the Ogden formulation for quasi-incompressible materials [30].

The neo-Hookean hyperelastic model was chosen with the following equations for the isochoric part of the strain energy potential and the stress respectively:

$$\begin{aligned} \Psi_{\text{iso}} &= \frac{1}{2}\mu(J^{-2/3}C_{kk} - 3), \\ S_{ij\,\text{iso}} &= \frac{\partial \Psi_{\text{iso}}}{\partial E_{ij}} = \mu J^{-2/3}(\delta_{ij} - \tfrac{1}{3}C_{kk}C_{ij}^{-1}). \end{aligned} \tag{15.13}$$

The hyperelastic constitutive tensor matrix is obtained from

$$\mathbf{D} = \frac{\partial \mathbf{S}}{\partial \mathbf{E}} = 2\frac{\partial \mathbf{S}}{\partial \mathbf{C}} = 4\frac{\partial^2 \Psi}{\partial \mathbf{C}\partial \mathbf{C}}. \tag{15.14}$$

The isochoric part of the elasticity tensor is given by

$$D_{ij\,\text{iso}} = -\frac{1}{3}\mu J^{-2/3} I C_{ij}^{-1} + \left(\frac{1}{9}\mu J^{-2/3}\right) C_{ij}^{-1} C_{ij}^{-1} - \frac{1}{3}\mu J^{-2/3}\left(-\frac{1}{2}C_{ik}^{-1}C_{jl}^{-1} + C_{il}^{-1}C_{jk}^{-1}\right). \tag{15.15}$$

The Ogden model formulation for quasi-incompressible, rubber-like materials has been implemented. A shear modulus high enough to approach the condition of quasi-incompressibility has been considered.

The volumetric part of the strain energy is given by

$$\Psi_{\text{vol}} = kG(J) \quad \text{with} \quad G = \beta^{-2}(\beta \cdot \ln J + J^{-\beta} - 1) \quad \text{and} \quad \beta = 9. \tag{15.16}$$

The volumetric part of the stress is computed as

$$\mathbf{S}_{\text{vol}} = 2\frac{\partial \Psi_{\text{vol}}(J)}{\partial \mathbf{C}} = Jp\mathbf{C}^{-1}. \tag{15.17}$$

The hydrostatic pressure $p$ and the volumetric part of the stress are obtained as

$$p = \frac{\mathrm{d}\Psi_{\text{vol}}(J)}{\mathrm{d}\mathbf{C}} = k\frac{\mathrm{d}G(J)}{\mathrm{d}J}, \qquad \mathbf{S}_{\text{vol}} = k\frac{1}{\beta}\left(1 - \frac{1}{J^{\beta}}\right)\mathbf{C}^{-1} \tag{15.18}$$

and the volumetric part of the elasticity tensor as:

$$\mathbf{D}_{\text{vol}} = \frac{\partial \mathbf{S}_{\text{vol}}}{\partial \mathbf{E}} = 2\frac{\partial \mathbf{S}_{\text{vol}}}{\partial \mathbf{C}} = 4\frac{\partial^2 \Psi_{\text{vol}}}{\partial \mathbf{C}\partial \mathbf{C}}. \tag{15.19}$$

After some algebra we obtain

$$D_{ij\,\mathrm{vol}} = \frac{k}{2}\left(\frac{11}{2}J^{-5/2} - J^{-9}\right)C_{ij}^{-1}C_{ij}^{-1} + \frac{k}{2}(1 - J^{-9})\left(-\frac{1}{2}C_{ik}^{-1}C_{jl}^{-1} + C_{il}^{-1}C_{jk}^{-1}\right). \tag{15.20}$$

The final form of the elasticity tensor is given by Equation (15.15) and Equation (15.19):

$$\mathbf{D} = \mathbf{D}_{\mathrm{iso}} + \mathbf{D}_{\mathrm{vol}},$$

$$D_{ij} = -\frac{1}{3}\mu J^{-2/3} I C_{ij}^{-1} + \left[\frac{1}{9}\mu J^{-2/3} + \frac{k}{2}\left(\frac{11}{2}J^{-5/2} - J^{-9}\right)\right]C_{ij}^{-1}C_{ij}^{-1}$$
$$+\left[-\frac{1}{3}\mu J^{-2/3} + \frac{k}{2}(1 - J^{-9})\right]\left(-\frac{1}{2}C_{ik}^{-1}C_{jl}^{-1} + C_{il}^{-1}C_{jk}^{-1}\right). \tag{15.21}$$

The previous elastic model has also been extended to account for viscoelastic effects. The viscoelastic model is based on the generalized Maxwell model [29, 31]. The kinematics of the viscoelastic body are represented by a superposition of a purely elastic body with a fixed reference configuration $\Omega_0$ and a Maxwell body with a moving reference configuration $\Omega_0^v$.

The thermodynamic process of the body is described by the deformation measures $J(\mathbf{X}, t)$, $\bar{\mathbf{C}}(\mathbf{X}, t)$, and $\bar{\mathbf{C}}^v(\mathbf{X}_v, t)$ and the second-order structural tensor $\mathbf{A}(\mathbf{X})$.

For incompressible material, viscous effects only affect the isochoric part of the free energy, split into a thermodynamic part (pure hyperelastic deformation) and a dissipative part (viscoelastic response). Consequently, the second Piola–Kirchhoff stress is written as

$$\mathbf{S} = 2\frac{\partial \Psi(J(t), \bar{\mathbf{C}}(t), \bar{\mathbf{C}}^v(t), \mathbf{A})}{\partial \mathbf{C}(t)} = \mathbf{S}_{\mathrm{vol}}^{\infty} + \mathbf{S}_{\mathrm{iso}}^{\infty} + \mathbf{S}_{\mathrm{iso}\,0}^{v} \tag{15.22}$$

where $\mathbf{S}_{\mathrm{vol}}^{\infty}$ corresponds to the volumetric part of the second Piola–Kirchhoff stress, $\mathbf{S}_{\mathrm{iso}}^{\infty}$ are the hyperelastic stresses, and $\mathbf{S}_{\mathrm{iso}\,0}^{v}$ represents the overstress of the viscoelastic solid accounting for time-dependent effects governed by the dissipative potential $\Upsilon_{\mathrm{iso}}$.

The different components of the second Piola–Kirchhoff stresses are computed as

$$\mathbf{S}_{\mathrm{vol}}^{\infty} = 2\frac{\partial \Psi_{\mathrm{vol}}^{\infty}(J(t))}{\partial \mathbf{C}(t)}, \qquad \mathbf{S}_{\mathrm{iso}}^{\infty} = 2\frac{\partial \Psi_{\mathrm{vol}}^{\infty}(\bar{\mathbf{C}}(t), \mathbf{A})}{\partial \mathbf{C}(t)}, \qquad \mathbf{S}_{\mathrm{iso}\,0}^{v} = 2\frac{\partial \Upsilon_{\mathrm{iso}}(\bar{\mathbf{C}}(t), \bar{\mathbf{C}}^v(t), \mathbf{A})}{\partial \mathbf{C}(t)}. \tag{15.23}$$

The overstress of the viscoelastic solid reads

$$\mathbf{S}_{\mathrm{iso}\,0_{n+1}}^{v} = \xi_{n+1}^{2}\mathbf{S}_{\mathrm{iso}\,0_n}^{v} + \beta\xi_{n+1}\left(\mathbf{S}_{\mathrm{iso}_{n+1}}^{\infty} - \mathbf{S}_{\mathrm{iso}_n}^{\infty}\right). \tag{15.24}$$

A more refined model for the detrusor was also developed by assuming a hyperelastic matrix reinforced with viscoelastic fibers. A random orientation of the fibers was considered. The model intends to represent the phenomenological behavior of the microstructure of the detrusor tissue, where the matrix consists of a neo-Hookean hyperelastic material and the collagen fibers are represented by two classes of perpendicular viscoelastic fibers at the element level. A similar model was proposed by Kondo *et al.* [32] using experiments to obtain the hyperelastic and viscoelastic parameters.

In order to account for the effect of fibers, the phenomenological model considers the second Piola–Kirchhoff stress as a summation of the following tensors:

$$\mathbf{S} = \mathbf{S}_{\mathrm{vol}} + \mathbf{S}_{\mathrm{iso}}^{\infty} + \mathbf{S}_{\mathrm{f}} \tag{15.25}$$

where $\mathbf{S}_{vol}$ and $\mathbf{S}_{iso}^{\infty}$ are respectively the volumetric and the hyperelastic contributions of the stress, computed as described previously, and $\mathbf{S}_f$ is the fiber contribution to the stress tensor, defined as

$$\mathbf{S}_f = \mathbf{S}_f^{\infty} + \mathbf{S}_f^{v}, \tag{15.26}$$

where $\mathbf{S}_f^{\infty}$ corresponds to the stress contribution to the elastic part of the Maxwell model (the hyperelastic response described in Equation (15.22)) and $\mathbf{S}_f^{v}$ corresponds to the viscous contribution of the Maxwell model.

Collagen fibers are considered by introducing the unit fiber direction $\mathbf{M}$ denoting the referential/local orientation of the fibers. The local structure of the material in the reference configuration $\Omega_0$ is defined by a symmetric second-order structural tensor $\mathbf{A} = \mathbf{M} \otimes \mathbf{M}$. The expression of this vector in the current configuration is $\mathbf{a} = \mathbf{FAF}^{\mathrm{T}}$.

Once fibers are accounted for, the material becomes anisotropic. We thus introduce a fourth invariant tensor $\mathbf{I}_4$ to represent the fibers orientation, defined as

$$\mathbf{I}_4 = \bar{\mathbf{C}} : \mathbf{A}. \tag{15.27}$$

The stored energy function reads

$$\Psi = \frac{k_1}{2k_2} \exp[k_2(I_4 - 1)^2 - 1], \tag{15.28}$$

where $k_1$ and $k_2$ are fiber parameters. The first derivative of the stored energy function with respect to the fourth invariant is

$$\frac{\partial \Psi}{\partial \mathbf{I}_4} = k_1(\mathbf{I}_4 - 1) \exp[k_2(\mathbf{I}_4 - 1)^2]. \tag{15.29}$$

Finally, the second Piola–Kircchoff stress contribution of the fiber is

$$\mathbf{S}_f = 2\frac{\partial \Psi(\mathbf{I}_4)}{\partial \mathbf{C}} = 2\frac{\partial \Psi}{\partial \mathbf{I}_4}\frac{\partial \mathbf{I}_4}{\partial \mathbf{C}} = 2J^{-2/3}\frac{\partial \Psi}{\partial \mathbf{I}_4}\left[\mathbf{A} - \frac{1}{3}(\mathbf{A} : \bar{\mathbf{C}})\bar{\mathbf{C}}^{-1}\right]. \tag{15.30}$$

The previous constitutive model was implemented into a standard finite-element formulation using both three-noded membrane triangles, rotation-free shell triangles, and stabilized four-noded tetrahedra to model the bladder wall. For details, see Monteiro [29].

## 15.3.2 *Validation. Test Inflation of a Quasi-incompressible Rubber Sphere*

The hyperelastic neo-Hookean model and the finite-element formulation were validated in the study of the inflation of a quasi-incompressible rubber sphere [29, 33, 34]. For the sake of simplicity, only a half-sphere was considered, accounting from symmetry (Figure 15.5).

The inflation pressure is a function of Cauchy stress and the radius and thickness of the sphere as

$$p_i = 2\frac{h}{r}\sigma. \tag{15.31}$$

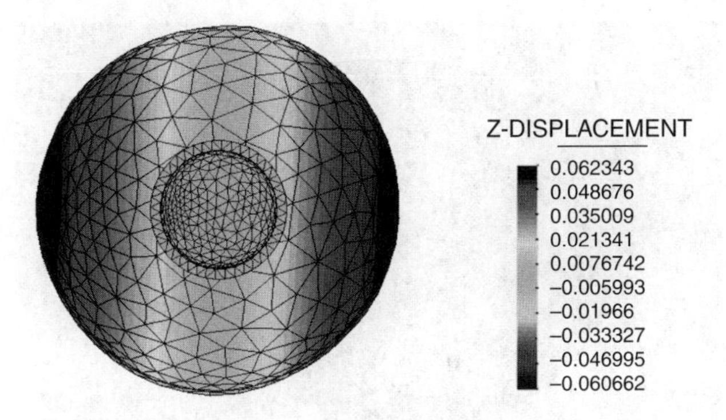


**Figure 15.5** Results for the rubber sphere inflation. The internal half-sphere shows the initial geometry

Taking into account the incompressibility of the material, and using the constitutive equation, we can obtain the relation between the stretch in the principal direction (or circumferential stretch) and the associated circumferential Cauchy stress as

$$\sigma = \mu_p(\lambda^{\alpha_p} - \lambda^{-2\alpha_p}), \tag{15.32}$$

where $\alpha_p = 2$ and $\mu = \mu_1 = 624.0$ kPa is the shear modulus, previously represented by $C_1$ for the neo-Hookean model.

The geometrical data considered for the validation test are initial internal radius 0.03 m and initial thickness 0.005 m. The finite-element mesh consists of 3278 four-noded tetrahedra and 675 nodes.

Figure 15.6 shows the results for evolution of the circumferential Cauchy stress versus the stretch $\lambda$, considering a bulk modulus of 10 000 kPa. The analytical result for incompressible rubber is also shown.

The inflation of a half-sphere under internal pressure provides a first approximation of bladder inflation.

### 15.3.3 Mechanical Simulation of Human Urinary Bladder

Our goal is to simulate the human bladder under filling conditions. For this purpose, we start with the geometry from a simplified high-resolution three-dimensional model of the human bladder of a

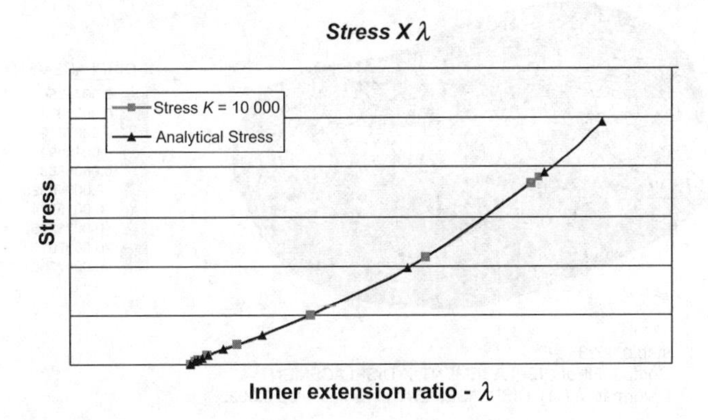


**Figure 15.6** Circumferential Cauchy stress versus stretch $\lambda$

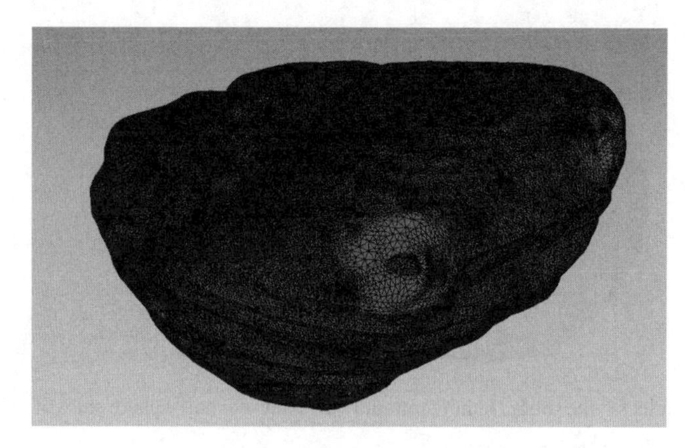

**Figure 15.7** Three-dimensional model of the urinary bladder meshed with four-noded tetrahedra

39-year-old man with no known urological diseases [29]. The morphology was obtained from the Visible Human Project (VH) and the input geometry from the work of Tonar *et al.* [35]. The geometry was treated with the pre–post processor GID (www.gidhome.com). The initial mesh size discretizing the internal volume is of the order of 1 million four-noded tetrahedra (Figure 15.7).

In order to get a faster convergence and a reduced number of elements, the discretization of the bladder wall was simplified using three-noded membrane triangles (Figure 15.8).

The membrane formulation for the bladder wall was used in combination with the particle FEM (PFEM) [36–40] simulating the flow of urine during filling accounting for the bladder and urine interactions.

To simulate the filling of the bladder with urine, we start with a simple geometry considering only part of one ureter connected to a reservoir. Before proceeding to the simulation of bladder filling, boundary conditions are applied based on the study of the physiology and dynamics of the urinary apparatus and pelvic region. In the urinary bladder, the region corresponding to the trigone is considered to be fixed and zero displacements are imposed on this region (Figure 15.9). The surrounding lower part of the bladder also has restricted mobility to represent the contact with the pelvic musculature (Figure 15.10).

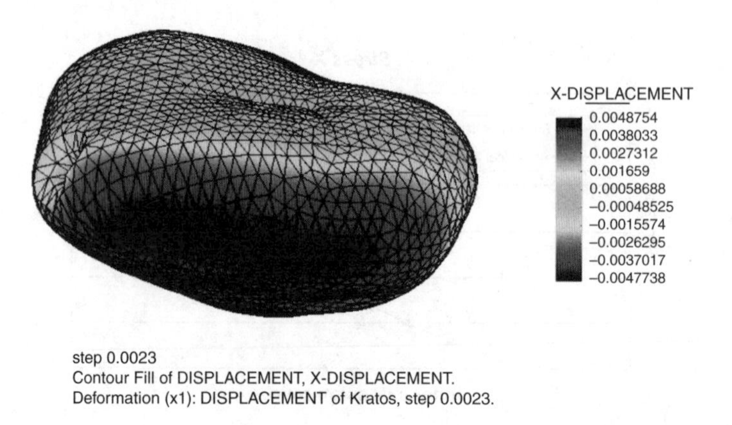


**Figure 15.8** Urinary bladder wall meshed with three-noded membrane triangles

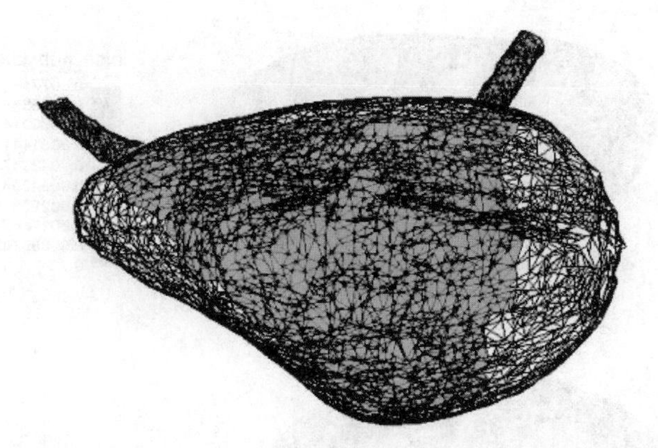

**Figure 15.9** Zero displacement imposed on the trigone area (shaded)

The nonlinear viscoelastic constitutive model was tested first for the geometry of the bladder meshed with three-noded quasi-incompressible membrane elements. The simplified condition of zero displacement in the junction with the ureters was imposed. The initial internal pressure of 67 Pa corresponds to the maximum hydrostatic pressure within the 50 ml bladder, computed with the PFEM. The pressure was then increased up to a maximum of 1000.0 Pa, (or 10 cm $H_2O$), which corresponds to the expected pressure of urine within the bladder for an average adult during filling conditions [38].

Figure 15.11 shows the deformed shape of the bladder at two filling instants.

## *15.3.4 Study of Urine–Bladder Interaction*

The entrance of urine in the bladder was simulated by injecting fluid nodes at the extremity of each ureter. Figure 15.12 shows the mesh used for the bladder filling study accounting for urine–bladder interaction. In this simulation, the initial volume of the bladder, also known as residual volume, is of the order of 50 ml. Figure 15.13 shows the geometry of the bladder at two filling instants.

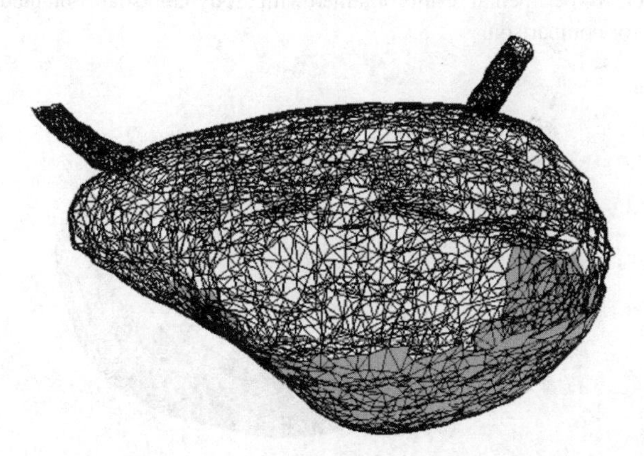

**Figure 15.10** Restricted displacement imposed on the area in contact with pelvic musculature (shaded)

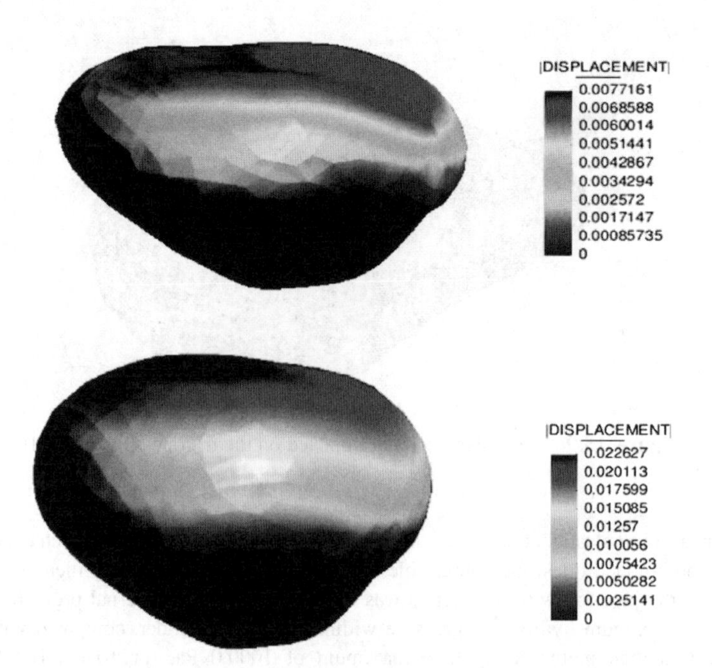


**Figure 15.11** Bladder displacements with different volumes: 70 ml (top) and 100 ml (bottom)

Figure 15.14 shows the results for the Cauchy stresses at different points of the bladder wall for the numerical analysis considering the constitutive model with fibers. Results are compared with approximated expected values computed with the Laplace equation:

$$T = P_{\text{ves}} \frac{R}{2d} \tag{15.33}$$

where $P_{\text{ves}}$ is the internal pressure, $d$ is current thickness, and $R$ is the current radius.

Figure 15.15 shows the results for the pressure-filling volume curve obtained accounting for urine–wall interaction effects. Experimental results obtained with urodynamic data obtained in cystometry tests [24] are also shown for comparison.

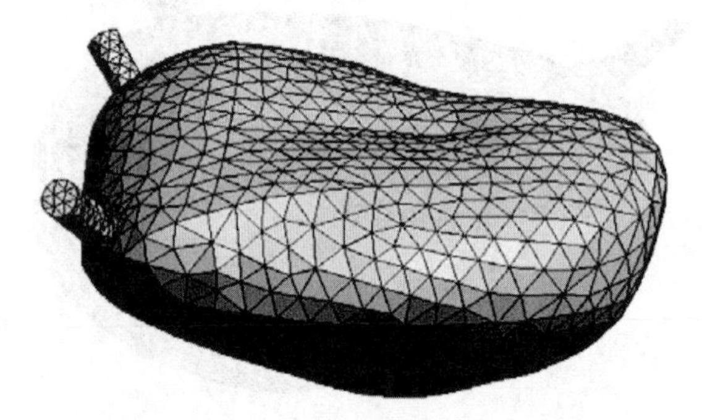

**Figure 15.12** Mesh for bladder filling study

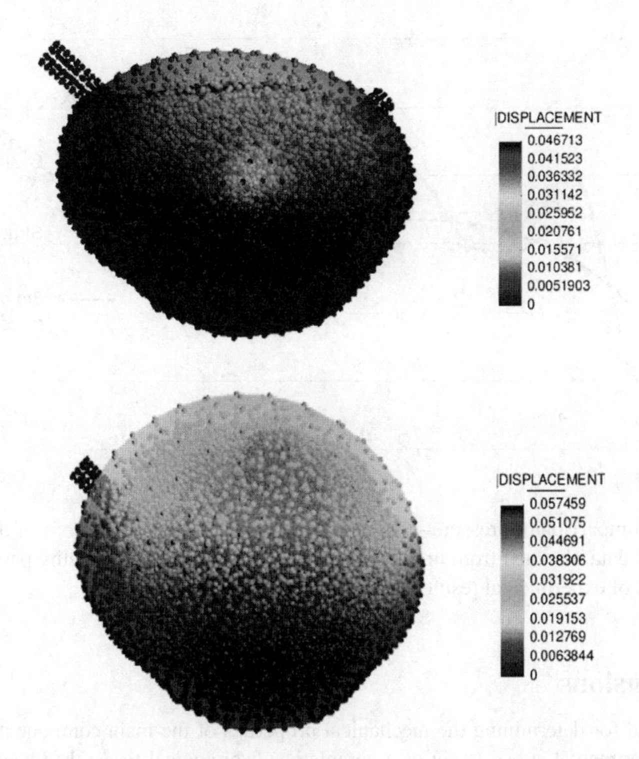


**Figure 15.13** Displacement distribution and fluid nodes for filling volumes of 65 ml (top) and 113 ml (bottom)

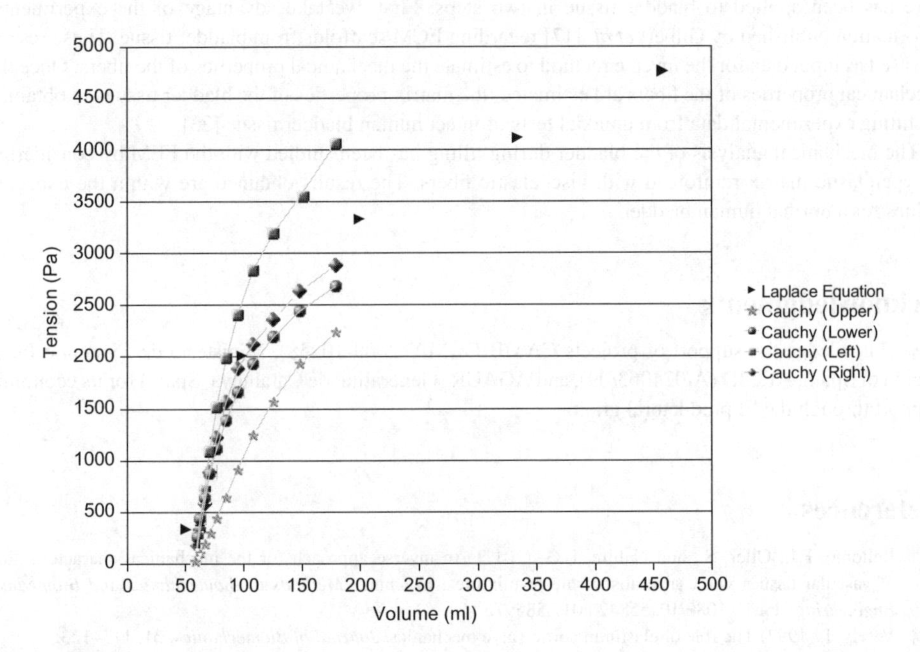


**Figure 15.14** Principal Cauchy stresses versus volume (homogenized model, shear modulus 5 kPa)

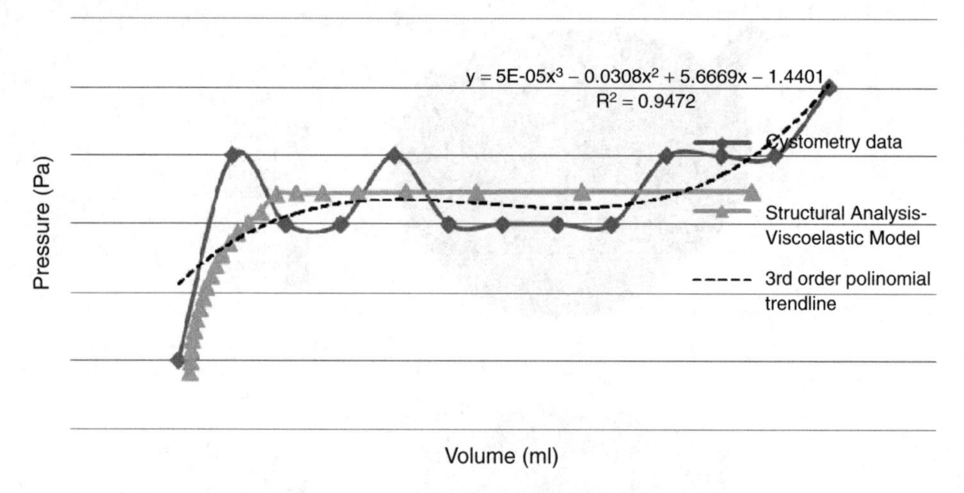


**Figure 15.15** Comparison of pressure–volume results for the structural analysis of the bladder filling with experimental data obtained from urodynamic tests (cystometry) on a healthy patient. Dashed line shows the average of experimental results

## 15.4 Conclusions

An inverse method for determining the mechanical properties of the main components of soft tissues, starting from experimental stress–strain of a complete soft biological tissue, has been presented. The determination of the mechanical properties of tissue constituents is a necessary step for successfully modeling the behavior of the tissue as a reinforced composite material. The inverse problem proposed here has been applied to bladder tissue in two steps. First, we take advantage of the experimental information published by Gilbert *et al.* [17] regarding ECM scaffolds from bladder tissue. These results are used as input data for the inverse method to estimate the mechanical properties of the fibers. Once the mechanical properties of the fibers are estimated, the matrix properties of the bladder tissue are obtained by fitting experimental data from uniaxial tests of intact human bladder tissue [23].

The mechanical analysis of the bladder during filling has been studied with the FEM by considering a hyperelastic matrix reinforced with viscoelastic fibers. The results obtained are within the expected values for a normal human bladder.

## Acknowledgments

We acknowledge the support of projects CAMBIO (MAT2009-10258), Ministerio de Ciencia e Innovación of Spain, AECID (A/024063/09), and AGAUR (Generalitat de Catalunya, Spain) for its economic support through the FI predoctoral grant.

## References


[1] Bellomo, F.J., Oller, S., and Nallim, L.G. (2011) An inverse approach for the mechanical characterisation of vascular tissues via a generalised rule of mixtures. *Computer Methods in Biomechanics and Biomedical Engineering*, doi:10.1080/10255842.2011.585976.

[2] Vesely, I. (1997) The role of elastin in aortic valve mechanics. *Journal of Biomechanics*, **31**, 115–123.

[3] Roeder, B.A., Kokini, K., Sturgis, J.E. *et al.* (2002) Tensile mechanical properties of three-dimensional type I collagen extracellular matrices with varied microstructure. *Journal of Biomechanical Engineering*, **124**, 214–222.
[4] Holzapfel, G.A. (2008) Collagen in arterial walls: biomechanical aspects, in *Collagen, Structure and Mechanics* (ed. P. Fratzl), Springer-Verlag, Heidelberg, pp. 285–324.
[5] Fratzl, P., Misof, K., Zizak, I. *et al.* (1997) Fibrillar structure and mechanical properties of collagen. *Journal of Structural Biology*, **122**, 119–122.
[6] Puxkand, R., Zizak, I., Paris, O. *et al.* (2002) Viscoelastic properties of collagen: synchrotron radiation investigations and structural model. *Philosophical Transactions of the Royal Society B: Biological Sciences*, **357**, 191–197.
[7] Eppell, S.J., Smith, B.N., Kahn, H., and Ballarini, R. (2006) Nano measurements with micro-devices: mechanical properties of hydrated collagen fibrils. *Journal of the Royal Society Interface*, **3**, 117–121.
[8] Shen, Z.L., Dodge, M.R., Kahn, H. *et al.* (2008) Stress–strain experiments on individual collagen fibrils. *Biophysical Journal*, **95**, 3956–3963.
[9] Sacks, M.S. and Gloeckner, D.C. (1999) Quantification of the fiber architecture and biaxial mechanical behavior of porcine intestinal submucosa. *Journal of Biomedical Materials Research*, **46** (1), 1–10.
[10] Gloeckner, D.C., Sacks, M.S., Billiar, K.L., and Bachrach, N. (2000) Mechanical evaluation and design of a multilayered collagenous repair biomaterial. *Journal of Biomedical Materials Research*, **52** (2), 365–373.
[11] Gilbert, T.W., Sacks, M.S., Grashow, J.S. *et al.* (2006) Fiber kinematics of small intestinal submucosa under biaxial and uniaxial stretch. *Journal of Biomechanical Engineering*, **12** (6), 890–898.
[12] Atala, A., Bauer, S.B., Soker, S. *et al.* (2006) Tissue-engineered autologous bladders for patients needing cystoplasty. *Lancet*, **367**, 1241–1246.
[13] Badylak, S.F., Vorp, D.A., Spievack, A.R. *et al.* (2005) Esophageal reconstruction with ECM and muscle tissue in a dog model. *Journal of Surgical Research*, **128** (1), 87–97.
[14] Bolland, F., Korossis, S., Wilshaw, S.P. *et al.* (2007) Development and characterisation of a full-thickness acellular porcine bladder matrix for tissue engineering. *Biomaterials*, **28** (6), 1061–1070.
[15] Brown, A.L., Farhat, W., Merguerian, P.A. *et al.* (2002) 22 week assessment of bladder acellular matrix as a bladder augmentation material in a porcine model. *Biomaterials*, **23** (10), 2179–2190.
[16] Kochupura, P.V., Azeloglu, E.U., Kelly, D.J. *et al.* (2005) Tissue-engineered myocardial patch derived from extracellular matrix provides regional mechanical function. *Circulation*, **30** (112, Suppl. 9), I144–I149.
[17] Gilbert, T.W., Wognum, S., Joyce, E.M. *et al.* (2008). Collagen fiber alignment and biaxial mechanical behavior of porcine urinary bladder derived extracellular matrix. *Biomaterials*, **29** (36), 4775–4782.
[18] Car, E., Oller, S., and Oñate, E. (2000) An anisotropic elasto-plastic constitutive model for large strain analysis of fiber reinforced composite material. *Computer Methods in Applied Mechanics and Engineering*, **185** (2–4), 245–277.
[19] Bellomo, F.J., Oller, S., Armero, F., and Nallim, L.G. (2011) A general constitutive model for vascular tissue considering stress driven growth and biological availability. *CMES – Computer Modeling in Engineering and Sciences*, **80** (1), 1–21.
[20] Bellomo, J.F. (2012) Numerical simulation of mechanical behavior of in vivo soft biological tissue. PhD thesis, Faculty of Engineering of National University of Salta, Argentina (in Spanish).
[21] Humphrey, J.D. (2002) *Cardiovascular Solid Mechanics, Cells, Tissues and Organs*, Springer.
[22] Coleman, T.F. and Li, Y. (1996). An interior, trust region approach for nonlinear minimization subject to bounds. *SIAM Journal on Optimization*, **6**, 418–445.
[23] Martins, P.A., Filho, A.L., Fonseca, A.M. *et al.* (2011) Uniaxial mechanical behavior of the human female bladder. *International Urogynecology Journal*, **22** (8), 991–995.
[24] Wein, A., Kavoussi, L., Novick, A. *et al.* (2007). *Campbell–Walsh Urology*, 9th edition, Saunders.
[25] Cimrman, R. (2005) Mathematical modelling of biological tissues. PhD thesis, University of West Bohemia.
[26] Holzapfel, G.A. (2000) *Nonlinear Solid Mechanics: A Continuum Approach for Engineering*, John Wiley & Sons.
[27] Belytschko, T., Liu, W.K., and Moran, B. (2000) *Nonlinear Finite Elements for Continua and Structures*, John Wiley & Sons.
[28] Idelsohn, S., Del Pin, F., Rossi, R., and Oñate, E. (2009) Fluid–structure interaction problems with strong added mass effect. *International Journal for Numerical Methods in Engineering*, **80** (10), 1261–1294.
[29] Monteiro, V.S.A. (2012) Computational model of the human urinary bladder. PhD thesis, Department of Strength of Material and Structures of Technical University of Catalonia (Barcelona Tech. Spain).

[30] Ogden, R.W. (1984) *Non-Linear Elastic Deformations*, Ellis Horwood Ltd.

[31] Holzapfel, G. and Gasser, C. (2001) A viscoelastic model for fiber-reinforced composites at finite strains: continuum basis, computational aspects and applications. *Computer Methods in Applied Mechanics and Engineering*, **190**, 4379–4403.

[32] Kondo, A., Susset, J., and Lefaivre, J. (1972) Viscoelastic properties of bladder. I. Mechanical model and its mathematical analysis. *Investigative Urology*, **10** (2), 154–163.

[33] Treloar, L.R.G. (1944) Stress–strain data for vulcanised rubber under various types of deformation. *Transactions of the Faraday Society*, **40**, 59–70.

[34] Alexander, H. (1971) Tensile instability of initially spherical balloons. *International Journal of Engineering Sciences*, **9**, 151–162.

[35] Tonar, Z., Zátura, F., and Grill, R. (2004) Surface morphology of kidney, ureters and urinary bladder models based on data from the visible human male. *Biomedical Papers of the Medical Faculty of the University of Palacky Olomuc Czech Republic*, **148** (2), 249–251.

[36] Oñate, E., Idelsohn, S.R., Del Pin.F., and Aubry, R. (2004) The particle finite element method, an overview. *International Journal of Computational Methods*, **1** (2), 267–307.

[37] Oñate, E., Idelsohn, S.R., Celigueta, M.A., and Rossi, R. (2008) Advances in the particle finite element method for the analysis of fluid–multibody interaction and bed erosion in free surface flows. *Computer Methods in Applied Mechanics and Engineering*, **197**, 1777–1800.

[38] Chapple, C.R. and MacDiarmid, S.A. (2000) *Urodynamics Made Easy*, 2nd edition, Churchill Livingstone, Philadelphia, PA.

[39] Rossi, R. (2005). Light-weight structures: structural analysis and coupling issues. PhD thesis, University of Padova.

[40] Rossi, R., Ryzhakov, P.B., and Oñate, E. (2009) A monolithic FE formulation for the analysis of membranes in fluids. *International Journal of Spatial Structures*, **24** (4), 205–210.

# 16

# Structure Design of Vascular Stents

Yaling Liu[a,b], Jie Yang[a], Yihua Zhou[a], and Jia Hu[a,c]
[a]*Department of Mechanical Engineering and Mechanics, Lehigh University, USA*
[b]*Bioengineering Program, Lehigh University, USA*
[c]*School of Mechanics and Engineering, Southwest Jiaotong University, Republic of China*

## 16.1 Introduction

Vascular stents, especially drug-eluting stents, are a splendid creation of the twentieth century. With continuous developments of stents design, vascular stents have become the first choice for the treatment of vascular stenosis, including coronary heart diseases, for their advantages of having good clinical efficiency, being minimal invasive for patients, and so on [1]. The further request for stents from clinics has stimulated the innovation of stent design. Currently, more than 100 different designs of stents have been in use and there are a growing number of new stent designs coming out yearly [2, 3]. Some important parameters and properties of commercial stents are listed in Table 16.1, though these data needs constant updating with the rapid appearance of new stent designs. From Table 16.1 it can be seen that most vascular stents have similarities, such as the material selection of 316L stainless steel, the structure form of a tube, and the $<20\%$ metal surface coverage.

Research has shown that the stent design has a great influence on the therapeutic effect, especially on the treatment of restenosis and the acute thrombosis [4, 5]. Rittersma *et al.* found that thickness of stents was related to the endothelialization after stent implantation and that the thickness should not exceed a specific value [6]. Hara *et al.* summarized the relations of the stent design and in-stent restenosis. It is revealed that such stent design parameters, as material selection, structure form, metal coverage, strut shape, and drug properties, has a subtle influence on treatment efficacy after stent implantation [7].

The stent design affects the therapeutic efficacy of stenting. Thus, almost all stent designers are pursuing an ideal design for the best curative effect. Then, what properties should the ideal design have? What main parameters should the ideal design determine? What fast, effective method should be used to design the idealized stent? The answers to those questions have important significance in the design of better stents and further expansion of stent clinical application.

*Multiscale Simulations and Mechanics of Biological Materials*, First Edition. Edited by Shaofan Li and Dong Qian.

**Table 16.1** Main properties and parameters of vascular stents (by permission from Elsevier, *Journal of the American College of Cardiology*, copyright (2002))

| | NIR (BOSTN) | RADIUS (BOSTN) | WALLSTENT (BOSTN) | BX-VELOCITY | MULTILINK (GUIDANT) | BESTENT (METRONIC) | S660 (METRONIC) | S670 (METRONIC) |
|---|---|---|---|---|---|---|---|---|
| Expansion method | Balloon-expandable | Self-expanding | Self-expanding | Balloon-expandable | Balloon-expandable | Balloon-expandable | Balloon-expandable | Balloon-expandable |
| Material | Stainless steel/golden coating | Ni–Ti alloy | Platinum/cobalt | Stainless steel | Stainless steel | Stainless steel | Stainless steel | Stainless steel |
| Geometric pattern | Closed cell/V-linker | Multi-segments/oblique jointed | Circle-wire | Sinusoidal ring/S-shape linker | Sinusoidal ring/straight linker | Rotational locked linker | Sinusoidal ring/ellipse-filleted rectangular meshed | Sinusoidal ring/ellipse-filleted rectangular meshed |
| Thickness (mm) | 0.1 | 0.11 | 0.08 | 0.14 | 0.09–0.13 | 0.085 | 0.13–0.15 | 0.13–0.15 |
| Metal coverage rate (%) | 12–16 | 20 | 14 | 15–20 | 14–9 | 12–17 | 20 | 17–23 |
| Ratio of axial shrink (%) | <5 | <5 | 15–20 | <1.5 mm | 2.7 | 0 | 1.5 | 3.5 |
| Ratio of radial recoil (%) | <0.5 | 0 | 0 | <0.5 | <2 | <2.2 | <2 | <1.5 |
| Outer diameter during transportation (mm) | 1.1–1.2 | 1.42 | 1.53 | 1–1.2 | 1–1.42 | 1.0 | 1.0 | 1.1 |
| Axial length (mm) | 9, 12, 15, 18, 25, 31 | 14, 20, 31 | 17–48 | 8, 13, 18, 23, 28, 33 | 8, 13, 18, 28, 33, 38 | 9, 12, 15, 18, 24, 30 | 9, 12, 15, 18, 24 | 9, 12, 15, 18, 24, 30 |
| Minimal guiding tubular radius (mm) | 0.064 | 0.066 | 0.064 | 0.067 | 0.056–0.075 | 0.064 | 0.064 | 0.064 |
| Expanded diameter (mm) | 2.5–4.0 | 3.0–4.0 | 3.5–6.0 | 2.25–5.0 | 2.5–5.5 | 3.0–4.5 | 2.5 | 3.0–4.0 |
| Releasing pressure (atm) | 8 | – | | 10–16 | 8–9 | 8 | 8 | 8 |
| Exposure pressure | 16 | – | | 16 | 14–16 | 16 (3.0, 3.5 mm) 15 (4.0 mm) | 16 | 16 15 (4.0 mm) |

## 16.2 Ideal Vascular Stents

The properties and design of stents are key factors in the clinical efficacy after stents are implanted. Optimal stents should be fixed on the balloon, the delivery system should be flexible and easy to transport and to be traced in order to settle the stent to the lesion region, minimizing the injury to the vascular wall caused by the stent during the implanting process. Expanded stents should offer strong support along the radial direction, have low foreshortening, and small elastic recoil. Key parameters of properties for optimal stents are listed in detail as follows:

1. *Dimension limitation* The inner radius of vascular stents is the principal parameter for the choice of stents. Implanted stents under common conditions have an *inserting* action on the vascular wall. It will promote endothelialization, decrease thrombosis, and reduce the resistance and vortex of blood flow. Usually, it is expected that, at the cost of the smallest damage to the vascular wall, patients could get their blocked vessel as large as possible. The lumen diameter, which is smaller than the normal one, will cause a higher rate of restenosis; on the contrary, too large a diameter leads to an increase in vascular damage. The thickness of vascular stents is also an important parameter for evaluating the loss of lumen diameter. Comparison of stents with different thicknesses showed that relatively thin struts have much less late luminal loss than thick struts do [8].
2. *Biocompatibility* The common stent materials are metal and polymer. The material of stents implanted in human bodies requires good resistance to corrosion, thrombosis, and blood coagulation. Those properties depend on the stent materials, and now material surface modification is a popular approach to obtain a better biocompatibility.
3. *Flexibility* Flexibility is the degree to which the stent can be bent along the axial direction after it is implanted into the blood vessel, which depends strongly on the design of the bridged struts of the stents. The flexibility of stents can generally be judged by the rules between the transverse loads (or bending moments) of the structure and curvatures of bending.
4. *Expansion* Expansion is how large a diameter in the lesion. This shows the size of blood vessels to which the stent can apply. However, the stress of the stent will increase as the expansion of the stent increases. If the expansion exceeds the critical value and goes beyond the material's ultimate strength, local concentration of strain will be caused easily and, thus, the support strength and longevity of the stent will decrease.
5. *Radial force* The support force is the load the stent can sustain before permanent deformation occurs or the stent is completely destroyed. In other words, it is the force which the stent provides to keep the blocked lesion open in the corresponding diameter. Generally speaking, the releasing pressure of the balloon to expand the stent can be regarded as a measure index for the support force. But the support force should not be too large. On the one hand, a larger support will lead to a larger releasing pressure, decreasing the safety of stents during the expansion period; on the other hand, the support force depends on the structure and thickness of the stents. A higher releasing pressure requires a thicker strut, which leads to a conflict with the design requirement of minimum thickness. Meanwhile, the stent's support force is against the flexibility in some circumstances; a strong support always sacrifices some flexibility. In summary, the current best design strategy is to choose stents based on the lesion condition of vascular vessels. For example, it is advisable to choose stents with larger support for the lesion with a relatively hard plaque.
6. *The metal coverage rate* The metal coverage rate is the ratio of area covered by metal to the whole stent surface. The thrombosis of the stent surface is highly related to the metal coverage rate. Studies have found that a higher metal coverage rate is regarded as easier for the initiation of thrombosis. That is, the less area the metal covers, the smaller the possibility is that thrombosis exits [9].
7. *Axial shrinking ratio* The shrinking ratio of vascular stents is the ratio of stent length before expansion and after fully expanded into the lesion. The majority of stents are meshed structures which are cut by laser from the deformed tube and then pressed onto the balloon and packed into a

slender catheter. Therefore, a nonlinear axial shrinking will be caused unavoidably during the stent expansion process. When the stent is fully expanded, its shrinking ratio reaches the maximum value and the acceptable clinical shrinking ratio of stents is between 0 nd 20% [10]; of course, it will be better to be smaller.

8. *Elastic recoil ratio* The property of elastic recoil ratio is specifically used to express balloon-expansion stents. It is the percentage of the stent's diameter when the balloon is expanded to the maximal level with respect to the diameter when the balloon is released. There are two reasons causing elastic recoil: one is the recoil of the stent itself and the other is the compression of the blood vessel toward the stent. A stent with low elastic recoil can reduce blood vessel injury and decrease the rate of restenosis.
9. *Fatigue resistance* According to the design requirement, vascular stents should endure 4 billion cycles in an accelerated fatigue test in order to maintain their structural integrity in their 10-year life cycle [11]. This allow evaluation of whether the implant will lose efficacy in vivo and whether the stent will maintain its size as well as structural integrity in the long run.
10. *Transportation* The property of transportation is the ability to deliver stents to the lesion during the interventional operation. Good transportation requires that the stent can be pressed to the balloon firmly and will not detach from the balloon, and that the stent should have good flexibility to pass the curved lumen of blood vessels as well. Moreover, the stent should be visible under X-rays to allow the operation to carry on during the transportation process. Transportation is the supreme factor in the design of stents, and is related to the safety of the interventional operation.
11. *Apposition property* Apposition is the ability of stents to fully cover and support the blood vessel. It is related to the support force and the uniformity of expansion of the stents. Stent collapse, migration, and in-stent restenosis are more likely to happen after interventional therapy if the stent possesses a poor apposition property.
12. *Expansion uniformity* Expansion uniformity means vascular stents can be expanded uniformly during the expansion process. If the deformation is nonuniform, extremely large plastic deformation will occur locally in the stent, leading to stent fracture starting from that region. Meanwhile, more injury will be caused to the vascular wall by the extra large local deformation. The issues raised by nonuniform expansion of stents demand the stent design to be symmetrical in the circumferential direction.
13. *Hemodynamic property* Vascular stents can change the hemodynamic properties of the lesion domain after implantation. Usually, blood flow is likely to cause vortices and detention near stents, leading to a high possibility of thrombus. Meanwhile, stents will reduce the shear stress of the vascular wall near the lesion domain after implantation, and this may be a key factor in in-stent restenosis [12]. This is related to the stent structure, section property, section dimensions, and other design parameters.

Currently, there are no perfect stents; every stent has its own advantages and drawbacks. Harmonious stent structure usually offers several good properties but cannot cover all biomechanical properties. For example, flexibility conflicts with the support force and axial shrink and expansion should be balanced to reach a better comprehensive performance during the design. Consequently, optimization of stent design involves evaluation of the biomechanical properties that should be regarded as a guide for the optimization.

## 16.3 Design Parameters that Affect the Properties of Stents

The design of stents is related to such multiple disciplines as materials, mechanics, medicine, biochemistry, and so on. Before the design, the influence of design parameters of stent properties like the working mechanism, material forms, and geometry should be considered. A reasonable choice of design parameters is the core of design. This needs the accumulation of experience over years. The aim of choosing parameters is to meet the design indexes of ideal stents, including expansion method, structure form of stents, section shape and dimensions of main struts, height of struts, linking forms, allocation of linkers, materials selection, metal coverage rate, and so on.

### 16.3.1 Expansion Method

Stents can be specified as self-expanding and balloon-expanded according to their expansion method. Self-expanding stents use their own elastic force to open the narrowed blood vessels. This kind of stent is usually made of shape memory alloy. Self-expanding stents cause little injury toward the wall of blood vessels, but their support is weak. Balloon-expanded stents use a balloon to deploy the stent and cause it to plastically deform and finally hold the blocked vessel open by its deformed shape. This stent mostly uses 316L stainless steel as its skeleton.

### 16.3.2 Stent Materials

From the point view of stent materials, currently used materials include 316L stainless steel, Ni–Ti alloy, tantalum–iridium alloy, polymer, magnesium alloy, and pure iron. The dominant material is 316L stainless steel, and the popularity of polymer material, magnesium alloy, and pure iron is growing.

The determination of stent materials is an important and basic process of design. The material should first have good biocompatibility, then excellent radiographic properties, namely high resolution under X-rays to be convenient for the operation. Finally, the material should meet certain mechanical properties to guarantee the support, recoil, shrinking rate, and other key design indexes. It is the key process of the design. Commonly used materials in the clinic include:

1. *Non-biodegradable materials* Metal materials mainly include 316L stainless steel, Ni–Ti alloy, tantalum alloy, cobalt alloy, and so on. 316L stainless steel is more commonly used because it is cheap, easily processed, and has satisfying hardness and strength after the heating process. Since it contains chromium and easily forms an oxidized chromium surface, it has good support, biocompatibility, and corrosion resistance; this led to it being widely used during the early period of stent applications. In 2001, the US ASTMF association announced it as the standard material for surgical implants [6].

   Ni–Ti memory alloy is also widely used in the medical field owing to its extreme elasticity and shape memory effect. In practical application, 90% of these materials make use of the elasticity and 10% of them use the shape memory effect. Ni–Ti alloy is usually applied for fabricating self-expanding vascular stents because of its good MRI compatibility, biocompatibility, and corrosion resistance; it is also widely used for coronary artery stents, peripheral vascular stents, intracranial stents, esophageal stents, biliary stents, tracheal stents, urethral stents, and so on.

   Cobalt-based alloys mainly contain chromium, nickel, and manganese. The widely used types of this alloy are L605, MP35N, Phyno, and Elgiloy. It is usually used in the fabrication of hip joint equipment. The density, tensile modulus of elasticity, and mechanical properties of cobalt alloy are all higher than those of 316L stainless steel [7]. Furthermore, the high density and non-ferromagnetic property of cobalt alloy make it easier for X-ray detection and guarantee a better MRI compatibility. Owing to its admirable material characteristics, advanced cobalt alloy represents the main developing trend of stents. Tantalum alloy belongs to rare metals, its high density also offers good X-ray visibility and better MRI compatibility, biocompatibility, and high corrosion resistance, but it has drawbacks like poor support along the radial direction and ease of collapse, so it has not been widely used.
2. *Biodegradable materials* Nowadays, biodegradable materials can be sorted as degradable polymer materials and degradable metals. Degradable polymer materials include PGA, PLA, PLLA, PGLA, and PHBV. PGLA and PLLA are more widely used, and these two materials have been approved as biomaterial implants by the US Food and Drug Administration (FDA) [8]. Degradable metals mainly mean magnesium alloy and pure iron. Magnesium alloy's safety has been verified, but it has some extent of elastic recoil and noeintimal hyperplasia. The axial support force is smaller than that of the

**Figure 16.1** Palmaz–Schatz stents. Reproduced from the *Journal of Clinical Pathology* (2005), with permission from BMJ Publishing Group Ltd © 2005

iron stent. In addition, the density of magnesium is lower, so the visibility under X-ray is poorer, requiring ultrasound orientation and increasing the surgery's difficulty. Magnesium stents have not been widely used so far and their clinical efficacy still needs to be verified by long term observation and data.

### *16.3.3 Structure of Stents*

From the point of structure form, early stent structure can be sorted as slotted tubes and twining wire. The representative of the former type is Palmaz–Schat (PS) stents (Figure 16.1) and the latter is Giantuco–Roubin stents (Figure 16.2). The tubular stents have stronger support along the radial direction but poor flexibility and compliance, while the wired stents hold the opposite properties.

Later stent structures are mostly a combination of the two forms. Currently, stents can be divided into three forms: tubular structure, wired structure, and the combined structure.

Wired stents are manufactured by twining or knitting wires. The twining type has good flexibility along the axial direction, but is easy to expand nonuniformly during the expansion process. For the knitting type, though it has good flexibility along the axial direction, the shrinking along the axial direction is large. This structure is usually used to fabricate self-expanding stents.

Stents of tube structure are usually made of metal tubes processed by laser cutting or electrochemical etching. This type of stent has strong support along the radial direction and low recoil. After continuous development, this structure has reached an outstanding combination of support along the radial direction and flexibility along the axial direction. Nowadays, this form of stent occupies 70% of the stent market. The structure commonly contains circular Z-shaped ring struts and different forms and numbers of linkers. Based on the position of strut heads on different segments, stents can be sorted as along-head type and opposite-head type (Figure 16.3 and Figure 16.4). Usually, the linkers can be located easily

**Figure 16.2** Giantuco–Roubin stents. Reproduced from the *Journal of Clinical Pathology* (2005), with permission from BMJ Publishing Group Ltd © 2005

**Figure 16.3** "Along-head" type of stent

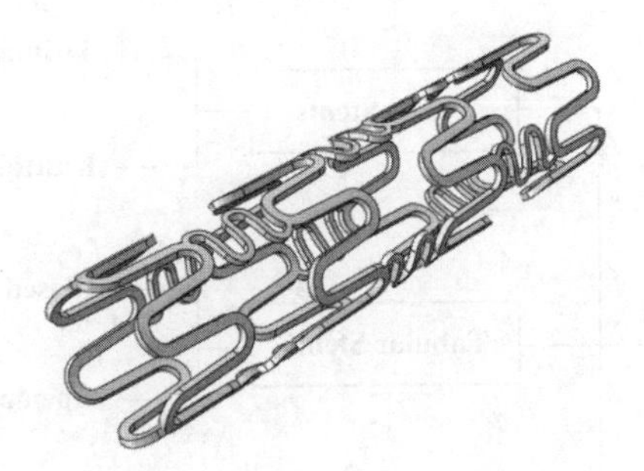

**Figure 16.4** "Opposite-head" type of stent

**Figure 16.5** Closed-cell stents, adapted from Hara *et al.* [7], with permission from Elsevier, *Advanced Drug Delivery Reviews*, © 2006

in the opposite-head type, but the shrinking rate is larger than the along-head type. Along-head types of stents are more likely to have poor axial flexibility and compliance, if the linkers are not correctly located. In order to solve this problem, S-shaped, V-shaped, and U-shaped linkers appeared to enhance the axial flexibility and compliance.

According to the cell region enclosed by the struts and linkers, stents can be sorted as closed-cell and open-cell (Figure 16.5 and Figure 16.6) [7]. Usually the closed-cell stent has stronger support than the open-cell one, but a little poorer axial flexibility and compliance than the open-cell ones.

Combined structures of stents are more likely used as aneurysm stents and nonvascular stents, and will not be discussed much here. Figure 16.7 summarizes the different stent structure forms and their representative products [2, 3].

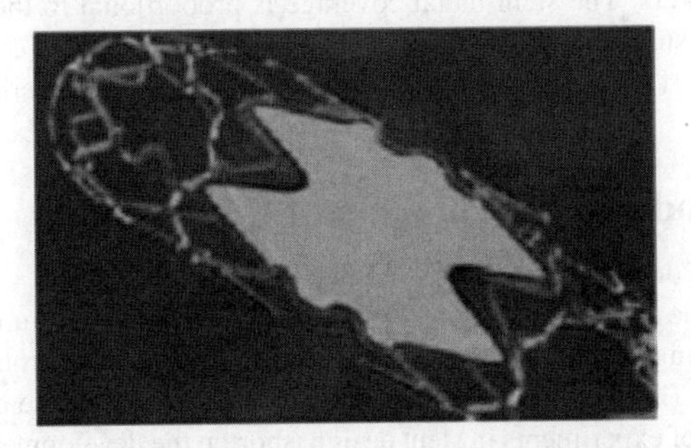

**Figure 16.6** Open-cell stents, adapted from Hara *et al.* [7], with permission from Elsevier, *Advanced Drug Delivery Reviews*, © 2006

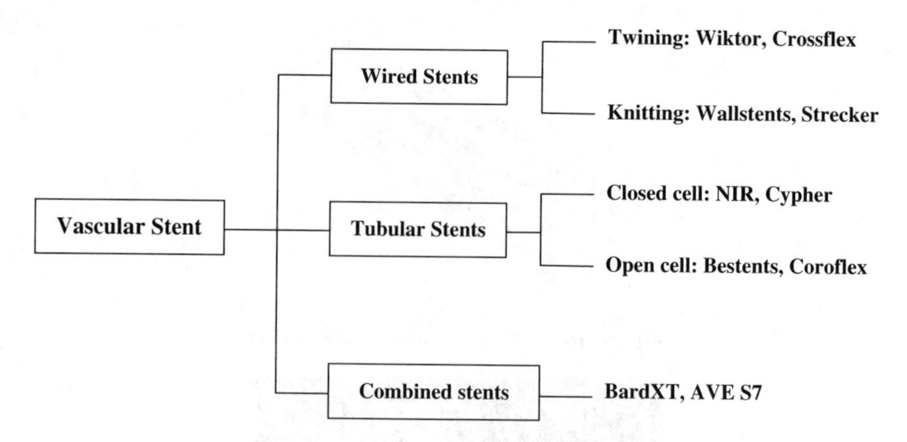


**Figure 16.7** Structure forms of vascular stents

### *16.3.4 Effect of Design Parameters on Stent Properties*

The selected materials should have good biocompatibility, good ductility, high yield stress, good X-ray penetrability, and some other properties. Among them, yield stress has a larger impact on stent support force, recoil rate, and axial shrinkage [6]. Choosing the kind of materials should be according to the specific design of the stent, processing techniques, and other factors.

The support force primarily depends on the design of the main struts. Generally, stent support force is proportional to the material yield strength and width and height of the strut cross-section, but inversely proportional to the square of the main strut height.

The stent compression design is related to the cross-section dimensions of the main struts and the dimensions of the linkers. It is more difficult to meet the compression requirements if the width of the main struts and linkers becomes larger. In addition, the arrangement of linkers should be uniform, otherwise the stent is easy to expand highly nonuniformly.

The recoil rate mainly depends on the struts design, and it is inversely proportional to the material yield strength and elastic modulus. The larger the strut width is, the smaller the recoil rate will be. If the main strut height becomes larger, the recoil rate will become larger. In the design, the typical curve of the support load and radial displacement is determined by the Finite-Element Method. If it does not meet the design requirements, we need to modify the design parameters according to the various rules.

Metal coverage is a geometric problem; it can be calculated easily by width and number of struts and the dimensions of the linkers. The stent metal coverage is proportional to the width and length of the main struts and linkers. Usually, the design of metal coverage can be achieved by correcting the design parameters after the support force and other important indexes meet the criterion.

## 16.4 Main Methods for Vascular Stent Design

A complete vascular stent development process is shown in Figure 16.8.

Parameter selection is the core of the stent design process. It needs a large number of design experiences and profound understanding of the factors which affect the mechanical properties of vascular stents. Finite-element analysis (FEA) is an important and necessary method in the design. On the one hand, FEA can reduce the scale of experiments in stent design, shorten the development cycle, and decrease the design costs. On the other hand, most of the structure sizes of the vascular stents are at the 0.1 mm level, and many of the mechanical properties are difficult to obtain purely relying on experiments. Therefore,

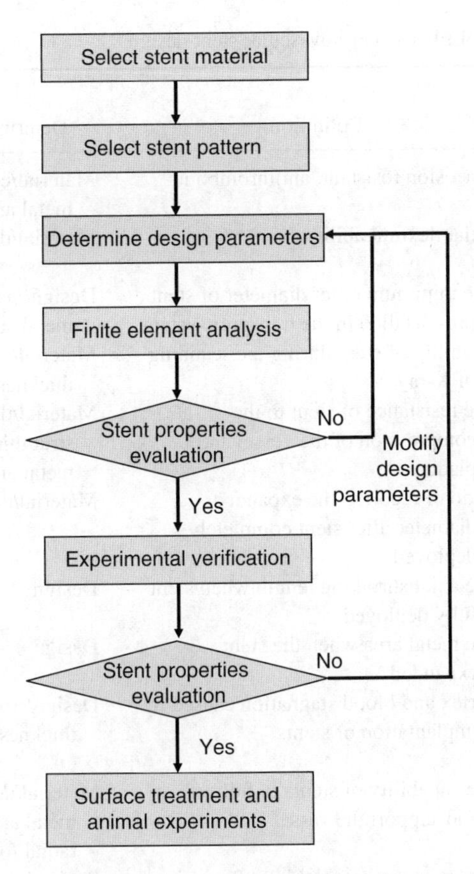


**Figure 16.8** Stent design flow chart

the US FDA demands that every commercial vascular stent product in the market must be accompanied by a detailed FEA.

Currently, FEA has been widely used in the development process of vascular stents and other biomedical engineering products. Table 16.2 summarizes the determining factors of the main design index in the vascular stent den process and the information of whether FEA can be applied in this field [2, 3].

FEA in the vascular stent design process is broadly divided into solid analysis and fluid analysis (Figure 16.9). Solid analysis is for vascular stent mechanical property analysis and mutual coupling analysis of blood vessel and stent after the stent is implanted into the coronary arteries. The mechanical property analysis of the stent itself generally uses a single stent model. It is a basic solid mechanics analysis in the design process and some design indexes can be obtained through it, such as the support force, compression property, recoil ratio, compliance, and axial contraction. Then the analysis models can be developed. Currently, vascular stent solid mechanics analysis models mainly include: a stent–balloon coupling analysis model, which can further analyze the uniformity of stent expansion and the "dog bone" phenomenon; a stent–balloon–lesion coupling analysis model, which combines with the ideal lesions model and is able to analyze the interaction of vessels and plaques after the vascular stent is implanted into the body; a mechanical coupling model based on medical imaging analysis, which details the vascular lesions information, using the lesion model reconstructed from medical CT images [13–19].

**Table 16.2** Application of FEA in cardiovascular stent design

| Characteristics of optimal stents | Definition | Determinants | Can FEA be applied? |
|---|---|---|---|
| Excellent biocompatibility | Corrosion resistant, antithrombotic | Material/coating/ metal area | No |
| Excellent flexibility | Axial flexural ability | Material/design | Yes |
| Small compressed size | The minimum outer diameter of stent pre-installed in the transport system | Design/material/ metal area | Yes |
| High X-ray visibility | Visibility of stent during the scanning of X-ray | Material/strut thickness | No |
| Suitable radial force | The resistance of stent to the compression of the vessel and plaque | Material/design/ strut thickness/ metal area | Yes |
| Low rate of recoil | Ratio of recoil to the expanded diameter after stent completely deployed | Material/design | Yes |
| Low axial shrink | The axial shrinking length when stent fully deployed | Design | Yes |
| Suitable metal covering area | The metal area when the stent expanded | Design | No |
| Excellent hemodynamic characteristics | Vortex and blood stagnation caused by implantation of stents | Design/strut thickness | Yes |
| Excellent stent apposition | The capability of stents to fully cover and support the vessel wall | Material/design/ metal area/ radial force | Yes |
| Good blood liquidity in bifurcation | Impact of the branch blood flow after stent implantation | Design/metal area | Yes |
| Uniform deformation | Uniform expansion of stents | Design | Yes |
| Excellent fixity | Security and stability of the stent attached to the transport system | Material/design/ minimum compressed size | Yes |
| Excellent anti-fatigue capability | The ability of stent to resist the continual heart pulsating load | Material/design | Yes |

The mechanical analysis based on CT images can provide more details of the impact of the vascular stent on blood vessels and plaques after the stent has been implanted into the lesion. At the same time, it makes the mechanical properties analysis more accurate when stents serve in human bodies. On the one hand, these analyses provide evidence for optimization of the vascular stent design. On the other hand, they support pathological research, such as in-stent restenosis and stent fracture.

Vascular stent fluid mechanics analyses mainly focus on the changes in the vascular lesions' hemodynamic properties after the newly designed stent is implanted into the body. Studies show that vascular stent implantation can reduce the shear stress of the vascular wall and it may be an important factor in

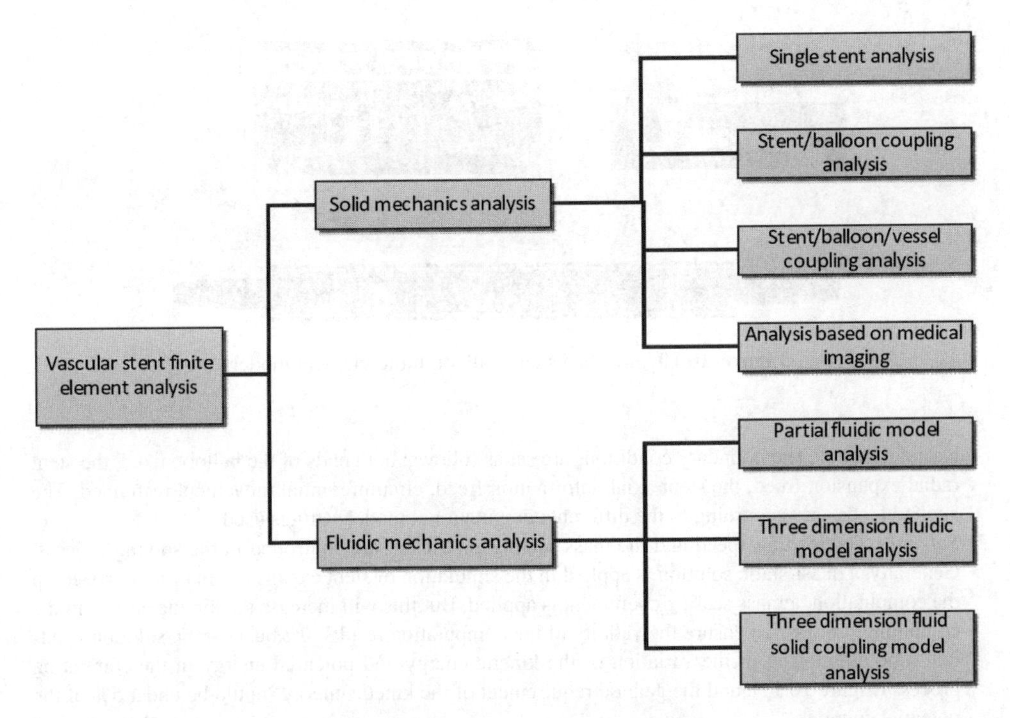


**Figure 16.9** Vascular stent FEA model

in-stent restenosis [20]. The fluid mechanical analysis model goes through a partial model, selecting a few stent cross-sections as the object to analyze the flow field properties. Then it goes to a three-dimensional model, which is capable of simulating the whole flow field after vascular stents are implanted into the blood vessel, and finally to a three-dimensional fluid–solid coupling analysis, considering the vascular wall elastic factor. Those analyses not only help the optimization of stent design, but also provide a basis for the research of the complications of stenting.

The above solid mechanics analysis models are for the highly nonlinear problem which is difficult to solve, coupling of the material nonlinearity, large geometric deformation, and state nonlinearity. Currently, most solutions use an explicit integral quasi-static computing method. The analysis software includes ABAQUS/EXPLICT, ANSYS, and MARC.

The fluid analysis software used frequently is, for example, Fluent and CFX.

The main steps of FEA can be divided as follows:

1. *Modeling* The stent geometric model is established according to the designed stent parameters, usually using geometry modeling software like PRO/E to model the vascular stent. After obtaining the geometry model, the model is meshed in the finite-element software. It is better to choose a hexahedral element, which can provide the best computational efficiency and accuracy.

   Balloon models mostly employ the membrane element, and a few use the entity element. Before computation, the balloon model needs to be deflated and folded, and this process can deal with imposing a negative pressure on the balloon. Figure 16.10 shows a meshed stent–balloon finite-element model.

   After completing the mesh, the material and boundary conditions need to be set. Setting the materials based on the actual material properties, here the balloon material can be superelastic material or general

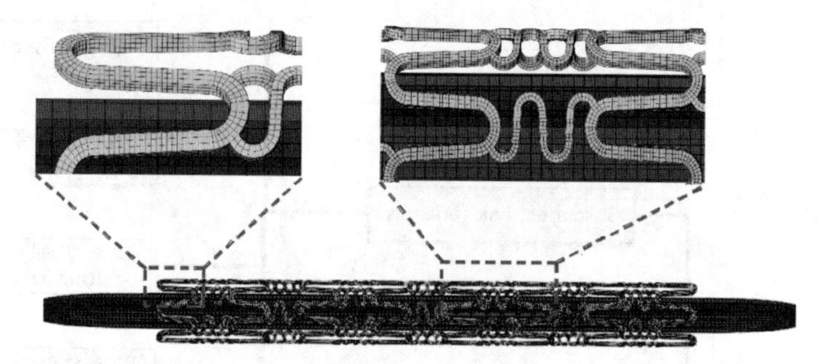

**Figure 16.10** Meshed stent–balloon finite-element model

elastic material. The boundary conditions are set as follows: both ends of the balloon fixed, the stent radial expansion freed, the stent axial deformation freed, circumferential movement restrained. The applied loads vary according to the different computational model settings used.

2. *Solving* The loading speed and the mass scaling coefficient are controlled in the solving process. Generally, a quasi-static solution is applied in the simulation of stent expansion. In order to speed up the computation, a mass scaling coefficient is applied. But this will increase the kinetic energy in the computing process. To ensure the validity of the computation results, the quasi-static solution needs real-time monitoring of the variation of the kinetic energy and potential energy in the computing process (Figure 16.11), and the general requirement of the kinetic energy should be under 5% of the potential energy.
3. *Post-processing* For different design indexes of vascular stent there would be different finite-element models and methods of dealing with the results.
   (i) *Stent supporting behavior research* The expansion loads can be applied on the inner surface of struts or by balloon, considering the loading and unloading processes. The boundary conditions are the same as discussed above. The radial displacement–pressure curve, as shown in Figure 16.12, is a typical result handling the supporting index of stents. From Figure 16.12 we can get the stent support force, recoil rate, stiffness, and other important mechanical properties. Figure 16.13

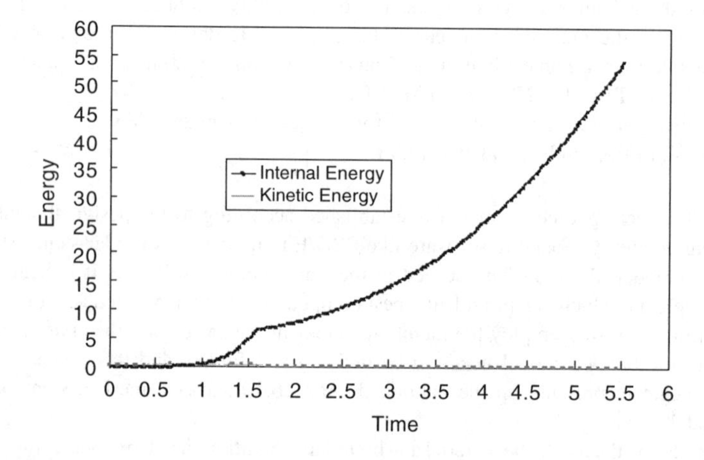


**Figure 16.11** Variation of kinetic energy and potential energy in the simulation

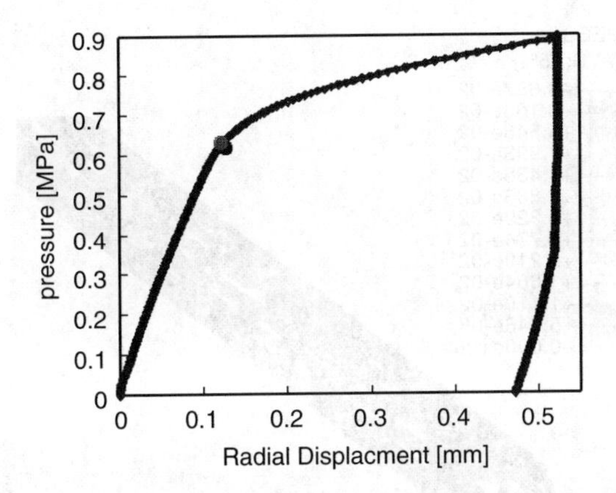


**Figure 16.12** Stent pressure–radial displacement curve, with permission from BMJ Publishing Group Ltd. *Journal of Clinical Pathology*, © 2005

shows the plastic strain distribution after the stent is deployed, and through analysis we can obtain some important design indexes, such as the uniformity of stent expansion and the stent safety.

(ii) *Stent compression property analysis* Generally, this analysis imposes radial displacement on the outside surface of the stent, to compress the stent to reach a diameter of about 1.0 mm. Considering the unloading recoil, we can similarly get a stent compression–recoil curve and the stress–strain distribution after compression. Figure 16.14 shows the strain distribution after stent compression. This figure clearly reflects the deformation of the stent compressed to an outside diameter of 1.0 mm, characterization of the stent compression property, the stent safety after compression, and other properties.

(iii) *Stent flexibility analysis* The loading method of three-point bending can be used in the flexibility analysis. Two ends of the stent are constrained and the middle part of the stent is subjected to a

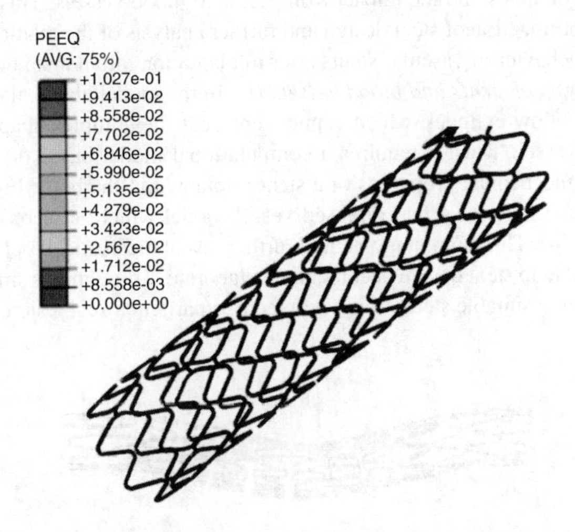


**Figure 16.13** Plastic strain distribution after stent deployed

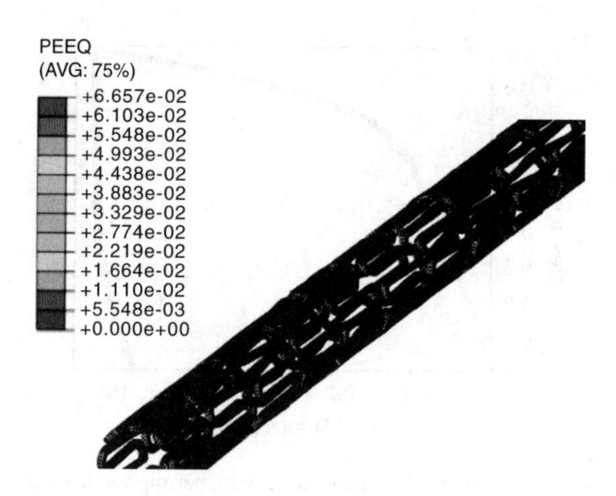


**Figure 16.14** Plastic strain distribution after stent compressed

concentrated force. Through this analysis the relation between the stent bending deformation and the load magnitude can be obtained, which means the flexibility index of the stent design.

(iv) *Nonuniform expansion analysis* For different design indexes, FEA can establish the corresponding computational models. In the analysis of stent nonuniform expansion, Yang and co-workers built up folded-balloon modelling [21, 22], studied the mechanism of nonuniform expansion, showed the phenomenon of nonuniform expansion through the finite-element method and its agreement with experimental results. This method can be a proper option for assessing the expanding uniformity of a newly designed stent.

(v) *Interaction analysis of stents and blood vessels I* For the final clinical application of vascular stents, the analysis of stent mechanical properties in diseased vessels is an important part of the research on stent properties and the optimal design method. The finite-element method technique is widely used in that research; Figure 16.17 shows the expansion effect of stents in a lesion with different design [23]. This technique can provide the mechanical properties of stenting, such as stress and strain of stents and the impact from stents to blood vessels. This mechanical message is vital for the optimization of stent design and further analysis of the relation between lesion and the mechanical behavior of vascular stents after implantation into human bodies.

(vi) *Interaction analysis of stents and blood vessels II* In the model above, the diseased vessel was simplified as a hollow cylinder, which cannot represent the complex shape and lesion level of a human blood vessel. Thus, it requires a computational model based on the clinical lesion to analyze the real mechanical properties of a stent implanted in the body. Figure 16.18 shows the FEA results on the basis of a real diseased vessel model which is reconstructed from clinical medical images [24]. This computation could further reveal mechanical properties of stents in real blood vessels and help stent design become more elaborate. Through this simulation, the designer is able to work out a suitable stent design pointed to a particular real lesion.

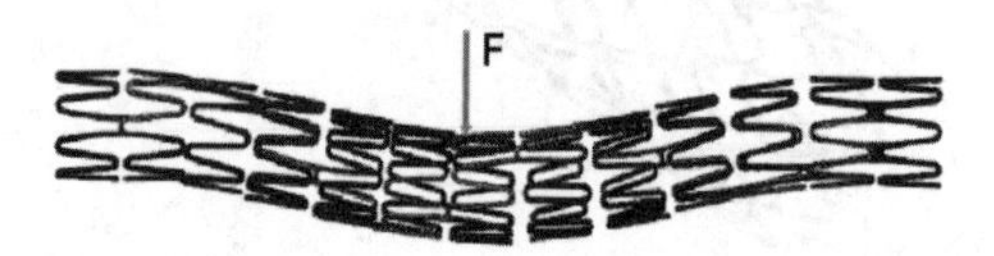


**Figure 16.15** Stent flexibility computation

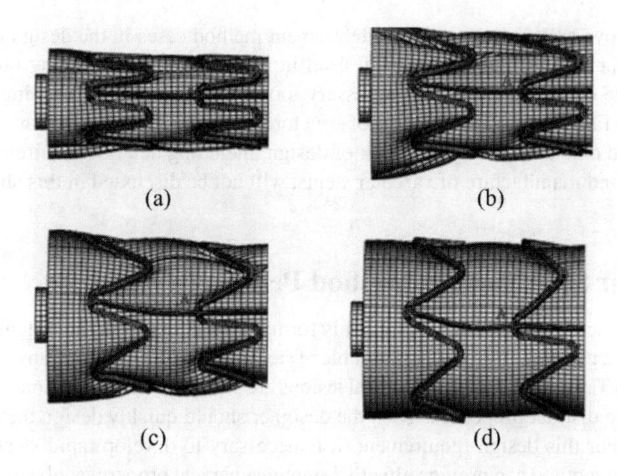


**Figure 16.16** Nonuniform expansion of stents

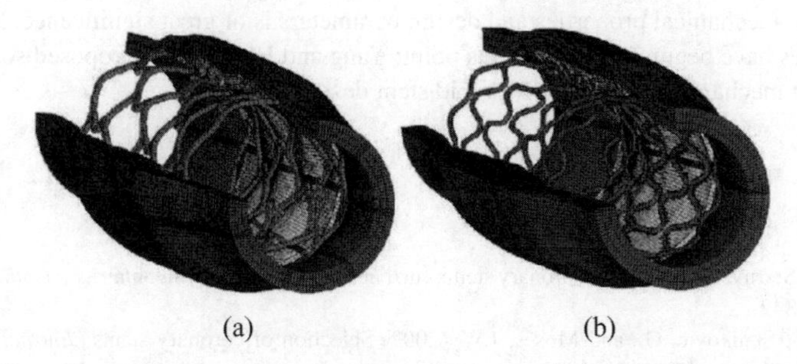


**Figure 16.17** The coupling model of stent and vessel I, with permission from Elsevier, *Journal of Biomechanics*, © 2005

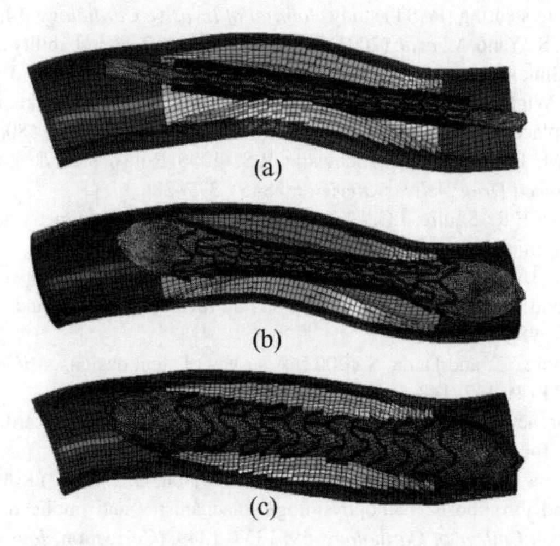


**Figure 16.18** The coupling model of stent and vessel II, with permission from Elsevier, *Journal of Biomechanics*, © 2010

The examples above are just part of the finite-element method cases in the design of vascular stents; the finite-element method technique is widely used in designing and optimizing other stent indexes. At present, the finite-element method is a necessary tool in the process of designing stent mechanical properties. Through FEA and the modification of structure parameters, the preliminary design of vascular stents can be figured out. The subsequent work of design, including detailed requirements and methods in the experiments and manufacture of vascular stents, will not be discussed in this chapter.

## 16.5 Vascular Stent Design Method Perspective

Currently existing vascular stent design is uniformly formatted. To choose a proper stent product for a specific lesion in a clinic, generally the optional variable of stent from a different company is merely the length of stent, as shown in Table 16.1. However, cinical lesions are vastly diverse and in some specific situations the stents need some distinct properties. Then the designer should quickly design the stent according to the actual lesions. For this design requirement, it is necessary to develop rapid stent design methods. While finite-element method can provide effective stent mechanical property analysis, it still needs some time in the whole design process. As a result, studying the rapid stent design method is an important direction of vascular stent design in the future. In the rapid stent design, the theoretical relationship between the vascular stent mechanical properties and design parameters is of great significance. Now, a growing number of studies have begun to focus on this point; Yang and Huang have proposed some mathematic formulas of stent mechanical properties for rapid stent design [25, 26].

## References

[1] Garg, S. and Serruys, P.W. (2010) Coronary stents current status. *Journal of the American College of Cardiology*, **56** (10), S1–S42.

[2] Colombo, A., Stankovic, G., and Moses, J.W. (2002) Selection of coronary stents. *Journal of the American College of Cardiology*, **40**, 1021–1033.

[3] Wang, W. (2005) Finite element analysis of the mechanical behavior of coronary stent and its structure optimization. Dalian University of Technology, Dalian.

[4] Gurble, P.A., Callahan, K.P., Malinin, A.I. *et al.* (2002) Could stent design affect platelet activation? Results of the platelet activation in stenting (PAST) study. *Journal of Invasive Cardiology*, **14**, 584–596.

[5] Yoshitomi, Y. Kojima, S., Yano, M. *et al.* (2001) Does stent design affect probability of restenosis? A randomized trial comparing Multilink stents with GFX stents. *American Heart Journal*, **142** (3), 445–451.

[6] Rittersma, S.Z.H., de Winter, R.J., Koch, K.T. *et al.* (2004) Impact of strut thickness on late luminal loss after coronary artery stent placement. *American Journal of Cardiology*, **93** (4), 477–480.

[7] Hara, H., Nakamura, M., Palmaz, J.C., and Schwartz, R.S. (2006) Role of stent design and coatings on restenosis and thrombosis. *Advanced Drug Delivery Reviews*, **58** (3), 377–386.

[8] Garasic, J.M., Edelman, E.R., Squire, J.C. *et al.* (2000) Stent and artery geometry determine intimal thickening independent of arterial injury. *Circulation*, **101** (7), 812–818.

[9] Schampaert, E. Moses, J.W., Schofer, J. *et al.* (2006) Sirolimus-eluting stents at two years: a pooled analysis of SIRIUS, E-SIRIUS, and C-SIRIUS with emphasis on late revascularizations and stent thromboses. *American Journal of Cardiology*, **98** (1), 36–41.

[10] Stoeckel, D., Bonsignore, C., and Duda, S. (2002) A survey of stent designs. *Minimally Invasive Therapy and Allied Technologies*, **11** (4), 137–147.

[11] YY/T0663–2008, Non-active surgical implants – Particular requirements for cardiac and vascular implants – Specific requirements for arterial stents.

[12] Koskinas, K., Chatzizisis, Y.S., Antoniadis, A.P., and Giannoglou, G.D. (2012) Role of endothelial shear stress in stent restenosis and thrombosis: pathophysiologic mechanisms and implications for clinical translation. *Journal of the American College of Cardiology*, **59**, 1337–1349. (Correction. *Journal of the American College of Cardiology*, **59** (21), 1919.)

[13] Yang, J., Liang, M.B., Huang, N., and Liu, Y.L. (2010) Studying the non-uniform expansion of a stent influenced by the balloon. *Journal of Medical Engineering and Technology*, **35**, 301–305.

[14] Gervaso, F., Capelli, C., Petrini, L. *et al.* (2008) On the effects of different strategies in modelling balloon-expandable stenting by means of finite element method. *Journal of Biomechanics*, **41** (6), 1206–1212.

[15] Etave, F., Finet, G., Boivin, M. *et al.* (2001) Mechanical properties of coronary stents determined by using finite element analysis. *Journal of Biomechanics*, **34** (8), 1065–1075.

[16] Holzapfel, G.A. and Gasser, T.C. (2007) Computational stress-deformation analysis of arterial walls including high-pressure response. *International Journal of Cardiology*, **116** (1), 78–85.

[17] De Beule, M., Mortier, P., Carlier, S.G. *et al.* (2008) Realistic finite element-based stent design: the impact of balloon folding. *Journal of Biomechanics*, **41** (2), 383–389.

[18] Gastaldi, D., Morlacchi, S., Nichetti, R. *et al.* (2010) Modelling of the provisional side-branch stenting approach for the treatment of atherosclerotic coronary bifurcations: effects of stent positioning. *Biomechanics and Modeling in Mechanobiology*, **9** (5), 551–561.

[19] Mortier, P., Holzapfel, G.A., De Beule, M. *et al.* (2010) A novel simulation strategy for stent insertion and deployment in curved coronary bifurcations: comparison of three drug-eluting stents. *Annals of Biomedical Engineering*, **38** (1), 88–99.

[20] Gay, M., Zhang, L., and Liu, W.K. (2006) Stent modeling using immersed finite element method. *Computer Methods in Applied Mechanics and Engineering*, **195** (33–36), 4358–4370.

[21] Yang, J., Huang, N., and Du, Q.X. (2009) A non-uniform expansion mechanical safety model of the stent. *Journal of Medical Engineering & Technology*, **33**, 525–531.

[22] Yang, J., Liang, M.-B., Huang, N. *et al.* (2009) Simulation of stent expansion by finite element method, in *ICBBE 2009. 3rd International Conference on Bioinformatics and Biomedical Engineering*, IEEE CFP0929C-CDR, pp. 1–4.

[23] Lally, C., Dolan, F., and Prendergast, P.J. (2005) Cardiovascular stent design and vessel stresses: a finite element analysis. *Journal of Biomechanics*, **38** (8), 1574–1581.

[24] Zahedmanesh, H., Kelly, D.J., and Lally, C. (2010) Simulation of a balloon expandable stent in a realistic coronary artery – determination of the optimum modelling strategy. *Journal of Biomechanics*, **43** (11), 2126–2132.

[25] Yang, J. and Huang, N. (2009) Mechanical formula for the plastic limit pressure of stent during expansion. *Acta Mechanica Sinica*, **25** (6), 795–801.

[26] Yang, J. and Huang, N. (2010) Formula for elastic radial stiffness of the tubular vascular stent. *IFMBE Proceedings*, **31**, 1435–1438.

# 17

# Applications of Meshfree Methods in Explicit Fracture and Medical Modeling

Daniel C. Simkins, Jr.
*University of South Florida, USA*

## 17.1 Introduction

Various meshfree methods have been proposed over the years as an alternative to mesh-based methods, such as the finite-element method (FEM), for solving partial differential equations. In this chapter, two application areas that can leverage the unique characteristics of meshfree methods will be presented: explicit representation of cracks and automated generation of analytical models from discrete point sets. In the former application, the explicit crack is easily modeled by using a visibility condition to modify the local support of individual nodes, thereby introducing an abrupt discontinuity into the shape functions and explicitly cutting the material. In the latter application, ability of meshfree methods to construct functions spaces without the need for a mesh that meets quality constraints is exploited to convert dense point clouds directly into analytical models.

## 17.2 Explicit Crack Representation

The study of fracture and the subsequent need to model the formation of holes that grow within a solid body is certainly not new. To make this discussion clear, it is beneficial to define precisely what is meant by an explicit crack. For this definition, we turn to the branch of mathematics called topology. The points of the problem domain form a *topological space*. A topological space is *connected* if any two points in the space can be connected by a continuous path consisting of points lying in the space. A space is said to be *1-connected*, or *simply connected* if any path connecting two points can be continuously deformed into any other path connecting the same two points, as shown in Figure 17.1.

*Multiscale Simulations and Mechanics of Biological Materials*, First Edition. Edited by Shaofan Li and Dong Qian.

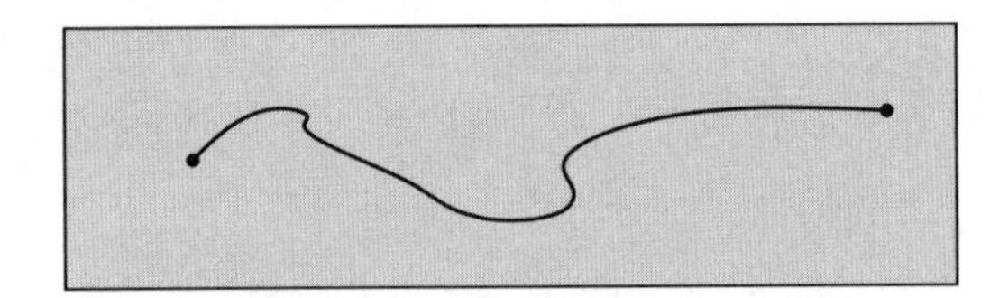

**Figure 17.1** A simply connected space

On the other hand, a multiply connected space is one where one cannot continuously deform one path into another. Figure 17.2 shows such a case. There is no way to deform the solid path into the dashed path in a continuous way.

These examples demonstrate what we mean by an explicit crack: it is one that may change the domain topology, more particularly one that may change the connectedness of the underlying domain.

The method described in this section explicitly represents cracks by cutting the material and changing the topology of the problem domain. An implicit representation, then, is one in which a model is used to say where the crack would be, if the domain was in fact discretized to explicitly represent the cracks. It is a relevant question as to whether an explicit or implicit representation is to be preferred, but that will not be addressed here.

### *17.2.1 Two-Dimensional Cracks*

We will now describe the basic crack algorithm in two-dimensional problems, which can be found in Simkins and Li [1]. The basic building block of the algorithm is the concept that crack tips are always coincident with a meshfree node. Let us carefully consider the diagram in Figure 17.3, as it contains all the necessary concepts underpinning the algorithm. For the remainder of this discussion, assume a total Lagrangian formulation. The problem domain is discretized by nodes, or particles, depicted by open circles. The filled squares are nodes lying on the created crack surface, and filled circles are current crack-tip nodes. The open squares represent former crack-tip nodes. Looking at the figure, along the crack center line immediately preceding the current crack tip is an open square that was the previous crack tip. Imagine the previous state where this was the current crack tip. The dashed circle centered on the old crack tip represents the support radius for that node – the nodes within that circle all interact with the old crack tip node. At a given point in the simulation, the material model indicates that the material at the node has failed, and the crack should propagate. Material model considerations will be discussed later; for now, let us accept that the material model can provide such information in the form of a continuous scalar damage variable, assumed to be in the unit interval with unity indicating failure. The algorithm then searches all the support particles evaluating how close they are to failure, and selects the node with the highest damage level to be the new crack tip. The relative position vector between old crack tip and new crack tip provides a propagation direction. The old crack-tip node is now split into

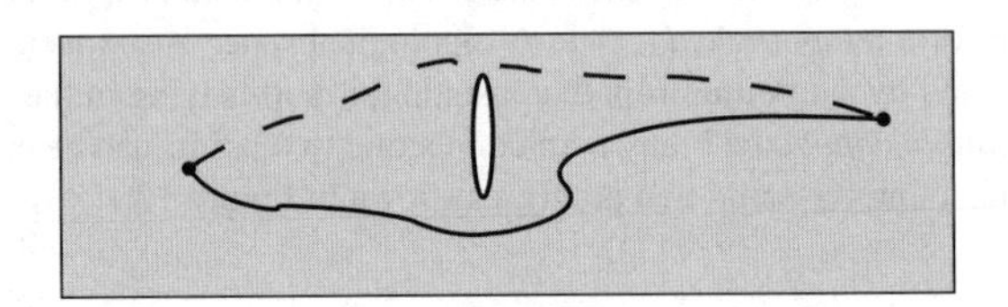

**Figure 17.2** A multiply connected space

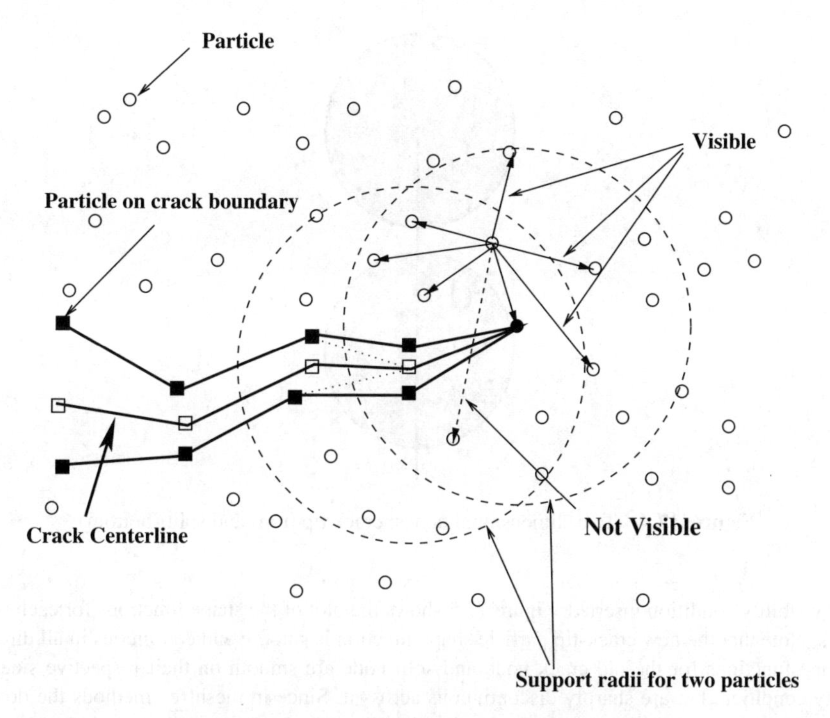


**Figure 17.3** Two-dimensional crack algorithm

two nodes, each node carrying half the volume and mass of the unsplit particle. Other state variables are distributed in a logical fashion; for example, temperature is assumed to be unaffected by the splitting. In the reference frame, the new and old nodes are separated by a small amount perpendicular to the crack direction, a choice of 1% of the incremental crack length seems to work well.

Once the new crack tip has been established and the old tip node split, a *visibility* condition is inserted. The original concept of a visibility condition was first proposed by Belytschko and co-workers [2, 3], but in the context of defining boundaries for meshfree domains. The use of a visibility condition for modeling cracks was first introduced by Simkins and Li [1], and was later applied by Rabczuk *et al.* [4]. In the two-dimensional case, the visibility condition is simply a line segment, shown as the crack center line in the figure, that is used to modify the support node set. Taking the union of the set of support particles for the old tip and the new tip, these represent all the particles whose support connectivity could be affected by the crack opening. Figure 17.3 shows a node with arrows to various support particles. For each pairwise interaction, a line-of-sight check is performed to see whether a visibility line segment is intersected. If there is an intersection, then the two nodes are prevented from interacting. After each propagation, the local support node set is updated and the shape functions recomputed.

The enforcement of the visibility condition is what effects the topology changes in the domain, as we will now show. Recall that the visibility condition terminates at the crack tip, and passing between the old crack-tip nodes and those split from them. The set of support nodes at the crack tip are unaffected by the visibility condition, except that the newly created split particle is added to the support set. Thus, the crack-tip shape function is virtually the same as it was before the crack propagated. On the other hand, the old crack tip and the split particle are affected quite a bit. Figure 17.4 shows three nodes in the reference configuration: one at the new crack tip, one at the old crack tip, and one that has been split

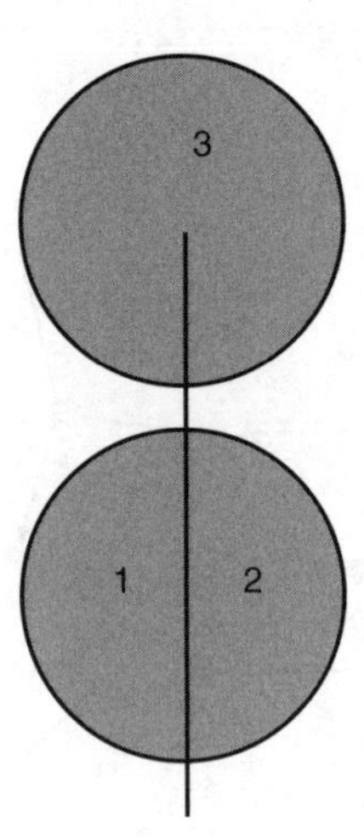


**Figure 17.4** Two-dimensional nodes: crack tip (top) and split (bottom)

and the visibility condition inserted. Figure 17.5 shows the plot of the shape functions for each of these particles. Note that the new crack-tip particle shape function is smooth and continuous in all directions. The shape functions for the old crack node and split node are smooth on their respective side of the visibility condition, but are sharply discontinuous across it. Since in meshfree methods the domain is assumed to exist where the shape functions exist, the discontinuity in the split nodes' shape functions have defined a hole within the domain, potentially changing its connectedness (topology), and hence effectively cutting the material.

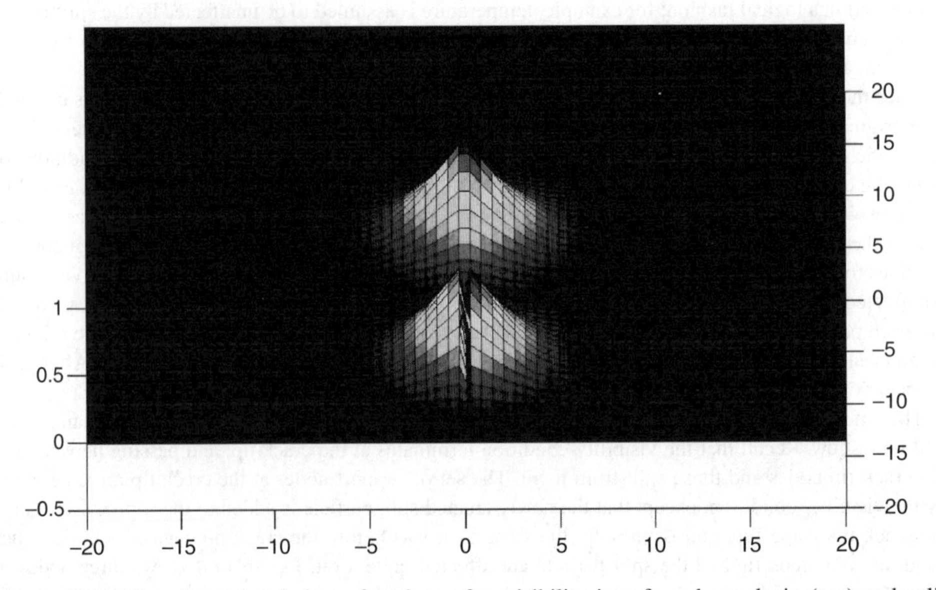


**Figure 17.5** Two-dimensional shape functions after visibility is enforced: crack tip (top) and split (bottom)

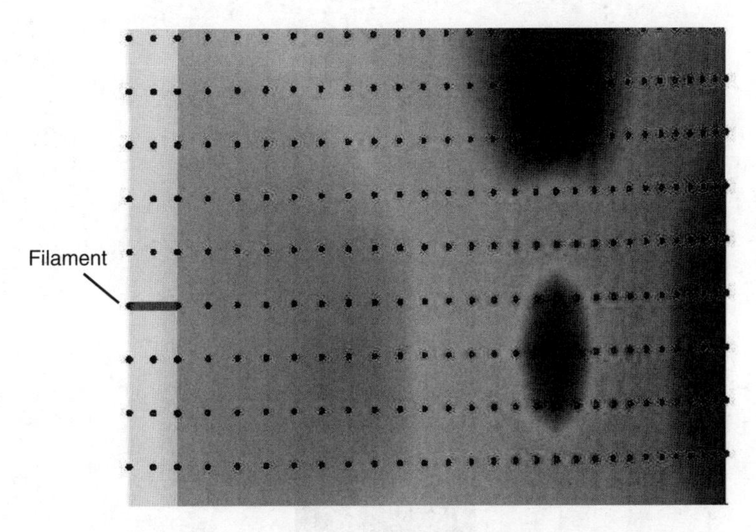


**Figure 17.6** Filament shown in a three-dimensional shell

### *17.2.2 Three-Dimensional Cracks in Thin Shells*

With an understanding of the two-dimensional algorithm, it is easy to extend to three-dimensional thin-shell-type structures. This is accomplished by discretizing the shell through the thickness in a self-similar pattern parallel to the local surface outward normal. For shells that are thin enough, it is reasonable to assume that all cracks are through cracks. For a given nodal location on the surface, the collection of nodes through the thickness act together and are referred to as a *filament*. The entire shell is now modeled as a collection of filaments and, owing to the through-crack assumption, cracks propagate from filament to filament and the two-dimensional algorithm can be applied to filaments. A thin shell with a filament is shown in Figure 17.6. Conceptually, the application of the algorithm to shell structures is straightforward. Some applications, however, require a little more attention to the details of the geometry.

### *17.2.3 Material Model Requirements*

We would now like to elaborate on the requirements for material modeling. The cracking algorithm only requires two things from the material model: (i) a binary determination of whether a material point has failed; (ii) an ability to predict among unfailed points the next point to fail. One suitable way to meet these requirements is for the material model to provide a continuous scalar damage measure and a threshold. One such material model suitable for ductile materials is the Johnson–Cook model [5, 6]. This model provides a continuous scalar damage value. When the damage is greater than unity, material failure is assumed to have occurred. Clearly, the damage value provides a simple ordering on undamaged material points.

Another interesting material model, suitable for glassy polymers in laminated composites, is the Boeing onset theory, formerly called the strain invariant failure theory, and is described in [7–9]. This is the material model used to drive the examples provided in Section 17.2.4 related to laminated composites.

### *17.2.4 Crack Examples*

In this section, two examples of the crack algorithm will be presented, each demonstrating different aspects of the algorithm. The first example is the ductile failure of a pressurized metal cylinder. The

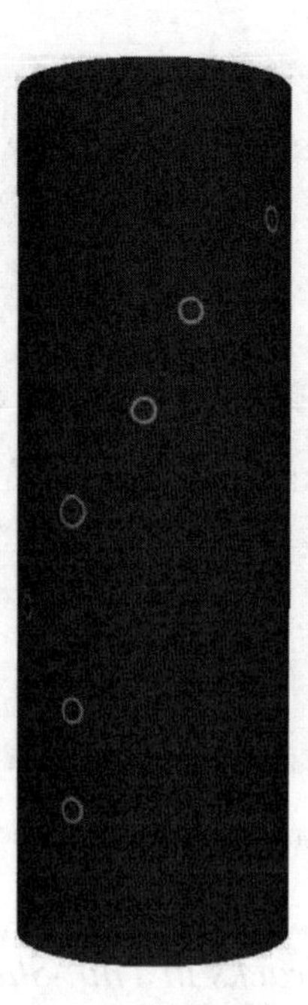

**Figure 17.7** Heated pressure cylinder

cylinder is 1.5 m tall, 0.5 m in diameter with a wall thickness of 3 mm. It is spot heated in several locations, shown as the bright red spots in Figure 17.7. To prevent the material from melting, a thermostat is used to prevent the material heating above 50% of the melt temperature. The Johnson–Cook material model is used in a total Lagrangian coupled thermo-mechanical formulation. Figures 17.8–17.10 show the final configuration after the crack has initiated and propagated to simulation end. Contours are shown for the

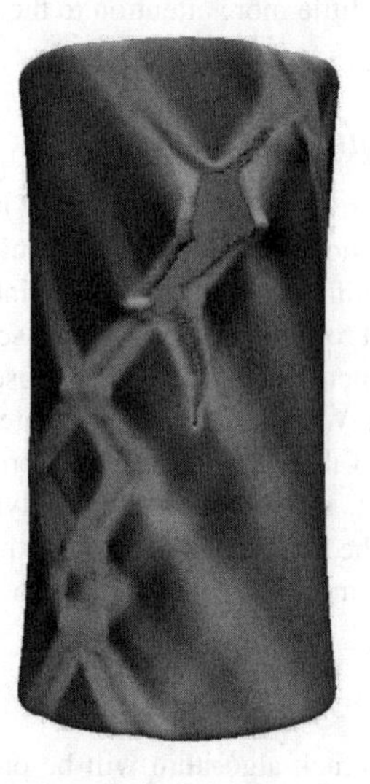

**Figure 17.8** Damage

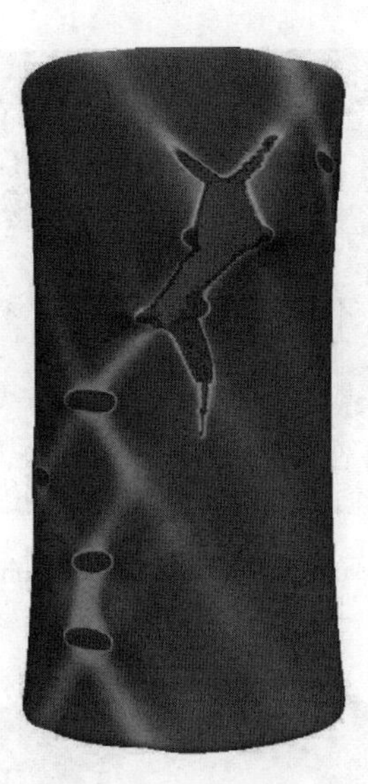

**Figure 17.9** Temperature

scalar damage measure from the Johnson–Cook model, the temperature, and the hoop stress. Note that in Figure 17.8 the crack follows a contour of high damage. Since the model is coupled thermo-mechanical, plastic work is converted to heat; this is shown in Figure 17.9, where the opened crack surfaces are at an elevated temperature. Further, unloading on the crack faces is shown in Figure 17.10. We leave this example with a close-up view of the crack-tip region in Figure 17.11 colored by damage.

**Figure 17.10** Hoop stress

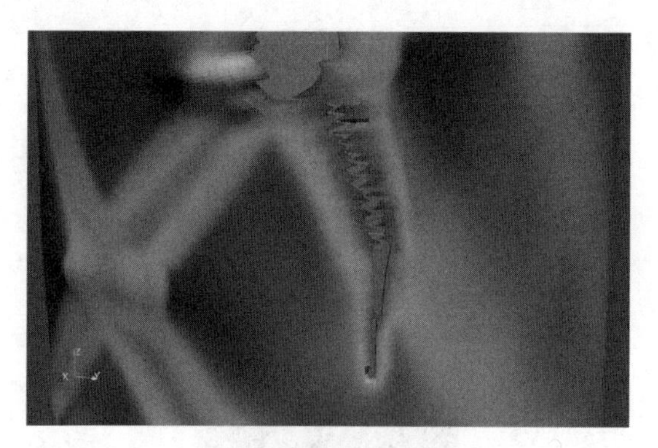

**Figure 17.11** Close-up view of crack-tip region with damage profile

**Figure 17.12** Laminated arch with fiber angle: 0° (dark gray), 90° (mid gray). Final deformed shape in light gray

The three-dimensional shell crack algorithm can be extended to model fracture in laminated composites. This is achieved by modeling each ply as a thin shell with through cracks coupled with a delamination model. Figure 17.12 shows the initial configuration of a laminated arch colored by the local fiber angle. The blue plies have fibers oriented in the local 0° direction, which is tangential to the curvature. The red plies are 90° plies with the fibers oriented coming out of the page. In gray, the final displaced shape is shown superimposed on the initial configuration. The arch is loaded psuedo-statically with applied displacements at the bottom inside corners of the arch pulling it in a wishbone fashion. At some loading, this part begins to delaminate, as shown in Figure 17.13. Generally speaking, delaminations are preceded

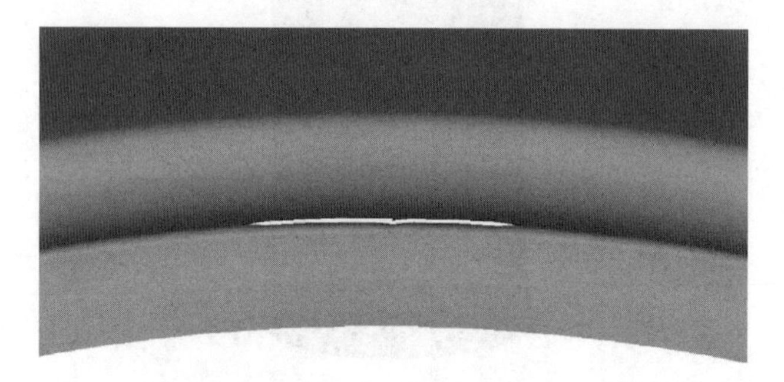

**Figure 17.13** Initial failure in delamination

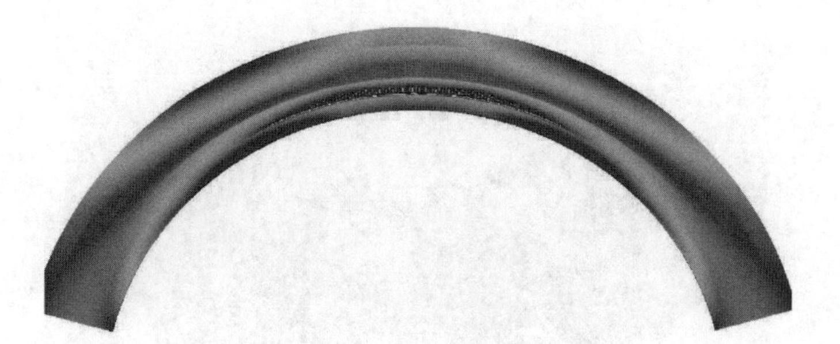

**Figure 17.14** Delamination transition to intra-ply cracks

by intraply cracks that then join from adjacent plies leading to the delamination. In this case, though, the unique loading and geometry develop a large stress pulling the inner ply away from the rest of the beam. Large bending strains are developed in the innermost ply, but this is a 0° ply, so the fibers are oriented to resist this load. Since there are no fibers in the direction of the pull-off force, only the relatively weaker polymer matrix is available to resist the load. This then leads to the initial delamination. With increasing applied displacements, the innermost ply peels away, resulting in two independent beams in bending: one the single 0° ply that has delaminated and the remaining intact plies. Once the delamination has occurred, a 90° ply is now exposed to large bending strains, but the fibers are not oriented to resist the deformation. Thus, transverse intraply cracks develop, as shown in Figure 17.14 and a close-up view in Figure 17.15.

## 17.3 Meshfree Modeling in Medicine

Engineering analysis has been increasingly applied to problems in the life sciences, in particular medicine. While finite-element analysis of artificial devices for medical purposes is commonplace, practitioners now want to be able to apply analysis techniques to individual biological structures; that is, bones and organs. This latter application represents a significant challenge. Man-made devices are generally designed with a computer-aided design (CAD) program, and thus the mathematical representation of the geometry is known. Conversion of a CAD model into analytical models has years of research behind it, and continues to be researched. Biological objects, on the other hand, do not come to the analyst in a

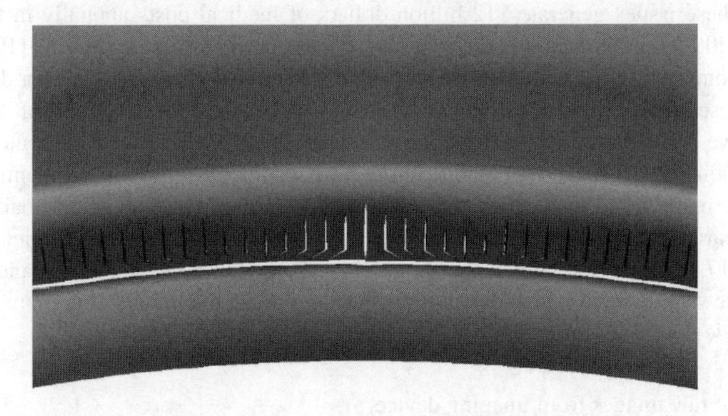

**Figure 17.15** Close-up of center of arch

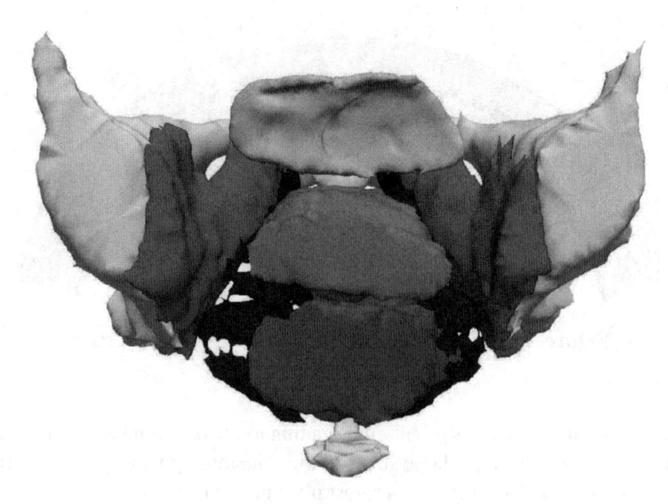

**Figure 17.16** Pelvic organs in a human female

mathematical form; rather, they are generally defined in the form of a set of three-dimensional points in space; for example, a point cloud. The point clouds are generally obtained from a diagnostic technique, such as magnetic resonance imaging (MRI) or computed tomography (CT). The challenge is to infer from this point set a suitable analytic representation so that traditional engineering analyses can be performed. Easily obtained programs, such as the 3D Slicer, can read the data generated by the imaging machines, convert the images to point clouds, and generate meshes. This is not quite a satisfactory result, because the resulting images, while visually appealing, do not stand up under the scrutiny required for analysis. In particular, the meshes generally have duplicate points, edges, and faces; holes and discontinuities; and topological variations such as intersecting faces. The first of these problems can easily be rectified, but the other problems generally require manual intervention to guide software in repairing the mesh. Unfortunately, this latter process can be quite tedious and time consuming.

Meshfree Galerkin methods build their function spaces directly from point sets, and thus are not plagued by the same issues as mesh-based methods. This feature makes meshfree methods attractive for life science applications. Omitting the details, we will now present an example important in the area of women's health.

In the medical field of urogynecology, significant issues involving the female pelvic floor are well correlated with the delivery of a child vaginally. Examples include organ prolapse and urinary and fecal incontinence. These issues generate \$12 billion dollars of medical costs annually in the USA alone. Approximately 40% of all women will develop one of these problems in her lifetime [10]. It is believed by some practitioners that the mechanical deformation of the tissues in the pelvic floor during childbirth are a leading cause of pelvic floor dysfunction. Clearly, experimental investigations of these hypotheses are very sensitive and difficult to perform. Therefore, we seek to develop a computer simulation to perform such studies in a virtual environment. Further, we would like to perform epidemiological studies to investigate the relationship to genetics and other possible factors. Ultimately, the goal is to provide an automatic, or nearly automatic, process to perform engineering analysis on objects defined by medical images. Figure 17.16 shows the human female pelvic region with the major organs and bones. For the present study, the vagina, rectum, bladder, urethra, etc. are not the main focus. In Figure 17.17, we show the primary pelvic floor muscles and pelvis. The basic approach we follow is:

1. Acquire set of raw images from imaging device.
2. Segment images into label maps.

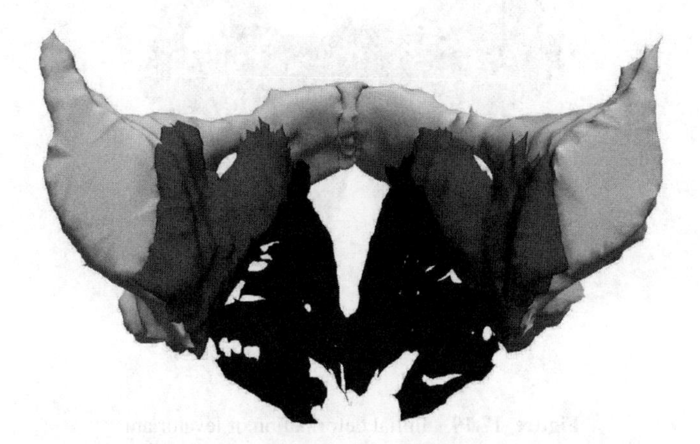

**Figure 17.17** Pelvic floor muscles: obturator (dark gray), levator ani (black), and pelvis (light gray)

3. Process sequence of label maps separating each object and converting from voxels to $x$, $y$, $z$ coordinates. This yields a dense three-dimensional point set of each object.
4. Decimate the dense point set into a set of suitable meshfree points.
5. Locate connections between objects.
6. Reassemble the individual objects into a single model.
7. Construct meshfree function space on the model.
8. Locate fetal head path.
9. Simulate birth as a 9 cm sphere whose center travels along fetal head path.

Our analysis procedure begins with step 3 and is completely automated through step 9. We illustrate step 4 in Figure 17.18. On the left is the dense point set computed in step 3 and on the right is the resulting thinned meshfree point set. In this case, the number of points is reduced from 34 189 in the dense set to 6860 meshfree nodes. Once the entire model is built, the birthing simulation consists of a sphere representing the fetal head that is then pushed through the birth canal, displacing the muscles as it progresses. Figures 17.19–17.21 show the deformation of the levator ani at three different instances of the simulation.

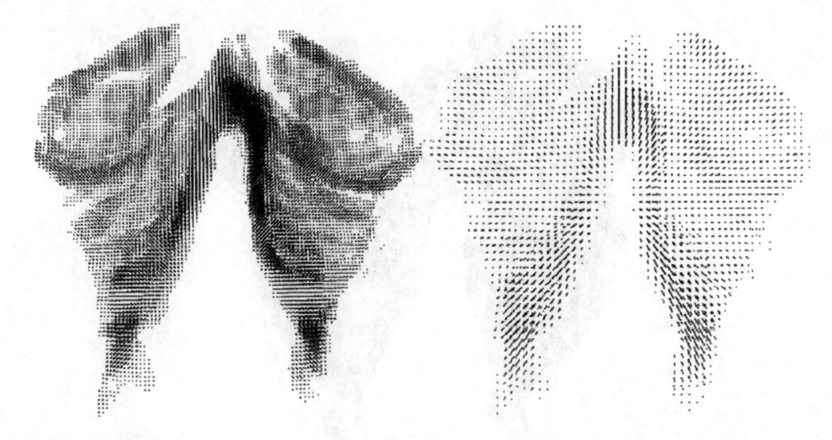

**Figure 17.18** Levator ani: dense voxel points (left) and thinned to meshfree nodal locations (right)

**Figure 17.19** Initial deformation of levator ani

**Figure 17.20** Intermediate deformation of levator ani

**Figure 17.21** Final deformation of levator ani

## Acknowledgments

We would like to acknowledge the University of South Florida's Research Computing for providing the computing resources used to perform the pressure cylinder simulation.

## References

[1] Simkins, D. and Li, S. (2006) Meshfree simulations of thermo-mechanical ductile fracture. *Computational Mechanics*, **38**, 235–249.

[2] Belytschko, T., Krongauz, Y., Organ, D. *et al.* (1996) Meshless methods: an overview and recent developments. *Computer Methods in Applied Mechanics and Engineering*, **139**, 3–47.

[3] Belytschko, T., Y, K., M, F. *et al.* (1996) Smoothing and accelerated computations in the element free Galerkin method. *Journal of Computational and Applied Mathematics*, **74**, 111–126.

[4] Rabczuk, T., Zi, G., Bordas, S., and Nguyen-Xuan, H. (2010) A simple and robust three-dimensional cracking-particle method without enrichment. *Computer Methods in Applied Mechanics and Engineering*, **199** (37–40), 2437–2455.

[5] Johnson, G. and Cook, W. (1983) A constitutive model and data for metals subjected to large strains, high strain rates and high temperature, in *Proceedings of the 7th International Symposium on Ballistics, The Hague, The Netherlands*, pp. 1–7.

[6] Johnson, G. and Cook, W. (1985) Fracture characteristics of three metals subjected to various strains, strain rates, temperatures and pressures. *Engineering Fracture Mechanics*, **21**, 31–48.

[7] Gosse, J. and Christenson, S. (2001) Strain invariant failure criteria for polymers in compostie materials, in *Proceedings of the 42nd AIAA/ASME/ASCE/AHS/ASC Structures, Structural Dynamics, and Materials Conference and Exhibit, Seattle, Washington, 16–19 April, 2001*, American Institute of Aeronautics and Astronautics, AIAA-2001-1184.

[8] Buchanan, D., Gosse, J., Wollschlager, J. *et al.* (2009) Micromechanical enhancement of the macroscopic strain state for advanced composite materials. *Composites Science and Technology*, **69** (11–12), 1974–1978.

[9] Tay, T., Tan, S., Tan, V., and Gosse, J. (2005) Damage progression by the element-failure method (EFM) and strain invariant failure theory (SIFT). *Composites Science and Technology*, **65**, 935–944.

[10] Hoyte, L., Damaser, M., Warfield, S. *et al.* (2008) Quantity and distribution of levator ani stretch during simulated vaginal childbirth. *American Jjournal of Obstetrics and Gynecology*, **199**, 198.e1–198.e5.

# 18

# Design of Dynamic and Fatigue-Strength-Enhanced Orthopedic Implants

Sagar Bhamare[a], Seetha Ramaiah Mannava[a], Leonora Felon[b], David Kirschman[b], Vijay Vasudevan[a], and Dong Qian[c,1]

[a]*School of Dynamic Systems College of Engineering and Applied Science, University of Cincinnati, USA*

[b]*X-spine Systems, Inc., USA*

[c]*Department of Mechanical Engineering, University of Texas at Dallas, USA*

## 18.1 Introduction

Thoracolumbar spine-deformation-related diseases such as scoliosis, kyphosis, cervical spinal disorders, and lower-back-pain-related issues have led to an increasing worldwide demand for spinal implant products, which, in turn, have led to the design of advanced spinal instrumentation systems [1, 2]. Spinal instrumentation is recommended to hold the spine immovable in place after the surgery, till complete healing of the spinal fusion takes place. Spinal instrumentation provides immediate fixation that allows early mobilization of the patient with use of less external support [3]. Traditionally, spinal implant products have been rigid constructs, which allow an irreversible biological fusion of the vertebrae together. Such constructs constitute a set of rigid rods and pedicle screws. Constituents (rods and pedicle screws) are typically manufactured from titanium alloy (Ti–6Al–4V ELI) [4], which is the medical grade of the Ti–6Al–4V, a popular material used in aerospace applications. The spinal instrumentation system used to correct a lumbar spinal disorder is schematically presented in Figure 18.1. Titanium alloy is used because of its good mechanical properties and biocompatibility. Spinal instrumentation is designed considering its biomechanical functions. At the same time it must undergo static and fatigue tests according to the ASTM standard F1717 [5] for design approval from the FDA. The FDA has specific guidance document and product code information to direct the industry to the types of tests

[1]Part of this work was performed by Dong Qian while at the University of Cincinnati.

*Multiscale Simulations and Mechanics of Biological Materials*, First Edition. Edited by Shaofan Li and Dong Qian.

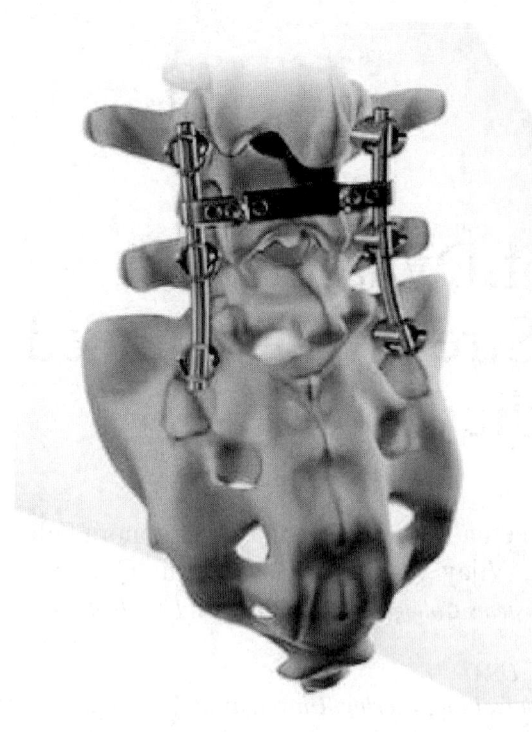

**Figure 18.1** Spinal instrumentation system for lumbar area

and standards that the product must be evaluated to. It is up to the manufacturer to develop acceptance criteria for their product based on the biomechanical needs of the target area. However, the new product must be compared with and seen as equivalent or superior to the current FDA-approved devices. The static and fatigue load-carrying capacities of instrumentation may vary based on the specific applications. However, any instrumentation should withstand at least 5 million cycles of fatigue load on construct set-up at load ratio $R = (\text{Load}_{\text{min}}/\text{Load}_{\text{max}}) = 10$ [5] and maximum load equivalent to 50% of the static load-carrying capacity of the corresponding instrumentation.

Design of the implant rod in the instrumentation system is very critical, as it affects the fatigue and static load-carrying capacities of the instrumentation as well as the global and segmental behavior of the spine and its biomechanics [6, 7]. The current industry standard is to have a rigid rod of diameter 5.5–6.35 mm, which satisfies the FDA requirements. A standard titanium alloy rod is highly stiff, restraining the post-operative movement of the patient. Rigid rods may experience failures due to the sudden overload in case of obese patients. Several workers have indicated the problems with rigid fixation, such as stress shielding of the fusion mass [8, 9], loss of bone-mineral content [10], and failure of the implant itself due to fatigue [11]. These problems may be countered by introducing flexibility in the instrumentation system. Such flexible devices are also called dynamic devices, which allow motion to some degree after surgery. These devices are also called motion-preserving devices, which is the latest trend in the orthopedic implant industry. Dynamic devices permit greater loads on the graft, which further enhance the fusion process [12]. In addition, such a dynamic device reduces the risk of mineral bone loss and disc degeneration [6]. Considering the benefits of the dynamic instrumentation, a number of workers [13, 14] have performed in vitro studies to assess the stability of the dynamic devices. These studies indicate that dynamic devices can be as equally stable as their rigid counterparts.

Dynamic devices can be designed by modifying the rigid rod itself. By reducing the intrinsic stiffness of the rigid rod, flexibility of the overall spinal instrumentation system is increased. Examples are the N-Flex system (Synthes) using a reduced rod diameter and the Accuflex system (Globus Medical) using a spring–rod modification. Unfortunately, these systems have demonstrated premature fatigue failure and are no longer actively marketed in the USA. Introduction of the flexibility by reducing the cross-section reduces the fatigue strength of the implant rod. Additionally, as a dynamic device, the rod will need to undergo higher cycle loading than a rigid device. Rigid devices are shielded from stresses over time as the fusion takes place. After the fusion has occurred (typically within 1 year), the loads placed upon the device essentially fall to zero. A dynamic rod, however, is subject to cyclic loading indefinitely for the life of the patient. As such, the need for increased flexibility and high-cycle fatigue (HCF) strength present significant challenges to dynamic implant devices development. In this chapter, we have developed the robust surface treatment called laser shock peening (LSP) of the dynamic implant rods to tackle this problem.

The LSP process [15, 16] is an advanced surface treatment process used to generate deep compressive stresses in Ti- and Ni-based alloys. The mean and maximum stresses of the applied fatigue loading cycle are affected by the residual stresses. If the residual stresses are compressive, they help reduce the mean stress, thus improving the total fatigue life of the component. The LSP process has been used by the aerospace industry to improve the fatigue life of fan and compressor blades made of titanium alloy Ti–6Al–4V [17, 18]. Some studies have shown tremendous increase in the fatigue strength and resistance to crack propagation of Ti–6Al–4V due to LSP [19, 20]. The material used in most of the orthopedic implants is Ti–6Al–4V ELI, which is the medical grade of Ti–6Al–4V. Thus, it is possible to apply LSP on orthopedic implants to improve the fatigue strength and life in the HCF region. Motivated by the need for dynamic instrumentation with enhanced fatigue life, the purpose of this chapter is to present a novel design philosophy for dynamic orthopedic implants. As a part of the proposed framework, a dynamic implant rod is designed by machining longitudinal grooves, thus introducing the flexibility in the instrumentation. The LSP process is applied to induce a deep layer of compressive residual stresses. Finally, the fatigue performance of untreated and LSP-treated dynamic implant rods is evaluated by performing bending fatigue tests. The effects of LSP are further investigated by performing fractography analysis on fatigue-tested implant rods. A US patent application for the proposed design philosophy has been filed by X-spine Systems, Inc. and the University of Cincinnati on March 24, 2009 (serial #61/162,697) entitled "An implant and a system and method for processing, designing and manufacturing and improved orthopedic implant."

Based on the discussion above, the rest of this chapter is organized as follows: In Section 18.2, we present the fatigue life analysis methods, specifically for orthopedic implant applications and incorporation of surface treatment effects in lifing. A basic introduction to the LSP process is given in Section 18.3. The modeling and numerical simulation methodology for the LSP process on orthopedic implants is further discussed in Section 18.4. A comprehensive example of a dynamic implant rod design is presented in Section 18.5 and the effects of surface treatments on fatigue life are thoroughly discussed. Finally, a brief discussion on the implications of this novel design philosophy is presented in Section 18.6.

## 18.2 Fatigue Life Analysis of Orthopedic Implants

### *18.2.1 Fatigue Life Testing for Implants*

As mentioned in Section 18.1, spinal instrumentation must be tested under fatigue loading on construct set-up as per ASTM standard F1717 [5]. A complete instrumentation system is primarily subjected to the bending loads and corresponding cyclic bending stresses. The focus of this work is to study the changes in fatigue behavior of spinal instrumentation due to added flexibility and residual stresses induced by surface treatments. The implant rod is the main constituent of the spinal instrumentation, which dominates the overall performance. Thus, the fatigue behavior of implant rods can be studied on an individual basis to reflect the changes in the fatigue behavior of spinal instrumentation. The fatigue

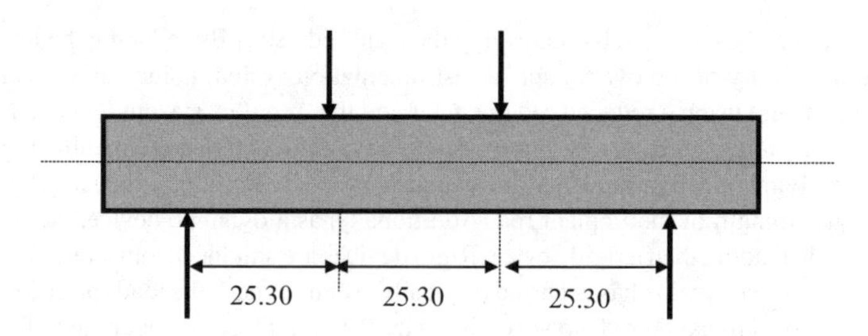


**Figure 18.2** Schematic of four-point bending set-up as per ASTM F2193

characteristics of a dynamic implant rod can be established effectively by performing four-point bending fatigue tests as per ASTM standard 2193 [21], which supplies guidelines for static and fatigue testing of individual components of spinal instrumentation. Based on the geometric dimensions (especially length) of the implant rods, the set-up can be adjusted. A schematic of the set-up used for the current work is shown in Figure 18.2. All dimensions are in millimeters. Fatigue tests must be carried out at applied load ratio $R = 10$ to replicate the fatigue testing required for spinal instrumentation, which essentially represents the cyclic loads on implants while in service. The frequency of the fatigue testing should be maintained between 2 and 30 Hz.

Fatigue tests are performed at various load levels on a given implant to obtain the cycles to failure curve plotted against the applied load (i.e., applied stress or strain), which is also called the *S–N* (stress–life) curve. Typical *S–N* and $\varepsilon$–*N* (strain–life) curves are shown in Figure 18.3. These curves are established to determine the safe level of operation of the implant for the required life. In addition, they are used for designing new implants of the same material under similar loading conditions.

As far as the FDA guidelines are concerned, an implant should have 5 million cycles life under bending fatigue loading at $R = 10$ with the maximum load at least 50% of the static bending-load capacity of the implant. This approach of designing for fatigue life is termed the safe-life approach. The safe-life approach got its name from the fact that a design criterion for this approach is *infinite life* (5 million cycles for the current case) [22]. Implants are designed based on test data using empirical models to last for a certain number of cycles and verified by performing fatigue tests. The presence of structural flaws, such as initial cracks, is ignored in this approach. Although this approach is referred to as an *old philosophy* of design, it is still widely used in industry. The safe-life approach is further divided into two categories called stress–life and strain–life methods. In stress–life approaches, *S–N* curves are established. Mild steels and alloys which harden due to strain-ageing show a plateau in the *S–N* curve.

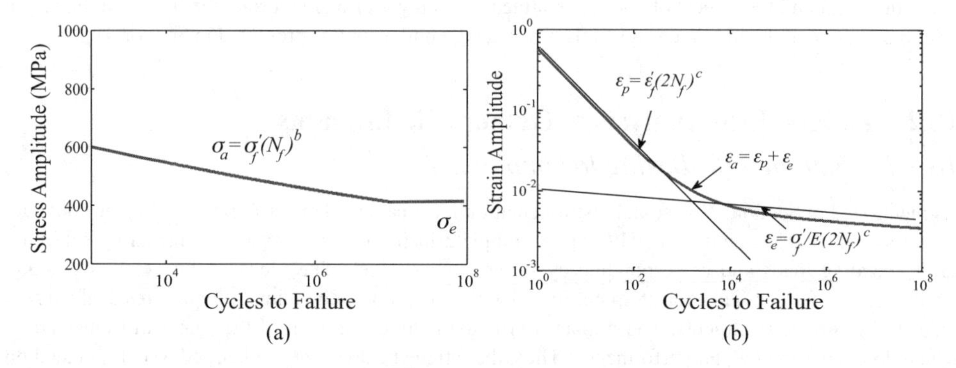


**Figure 18.3** Typical (a) stress–life and (b) strain–life curves obtained from axial fatigue tests

The stress value at which the plateau starts is often called the *endurance limit* $\sigma_e$. The stress–life approach is useful when the loading stress amplitudes are lower than the yield stress, i.e., HCF applications. For the low-cycle fatigue (LCF) and LCF–HCF applications, where components have some constraints and might be subjected to plastic strains, Manson [23] and Coffin [24] in the 1950s independently proposed a law relating the plastic strain amplitude to the fatigue life. Their law combined with HCF behavior from the Basquin [25] relationship is popularly knows as the strain–life ($\varepsilon$–$N$) relationship. Both methods can be employed for predicting the fatigue test results of orthopedic implants.

## *18.2.2 Fatigue Life Prediction*

Fatigue testing of implant rods must be performed at low frequency, which usually ranges from 2 to 30 Hz [5, 21]. Even for a small change in the implant rod design and surface treatment parameters, one needs to go through the rigorous fatigue testing exercise, which eventually increases the product design cycle time. Fatigue life prediction is performed on a routine basis in the aerospace and ground-vehicle industries. However, it is not a common practice in the orthopedic implant industry. In recent years, implant life prediction has gained popularity due to obvious advantages. Wierszycki *et al.* [26, 27] developed a nonlinear finite-element analysis model in ABAQUS [28] to simulate the contact conditions in dental implants and used the fe-safe [29] life prediction program to predict the possible fatigue damage sites and corresponding fatigue lives. The material of this dental implant was Ti–6Al–4V ELI. Their analysis could predict the fatigue crack initiation location with further verification by comparison with experiments. Ploeg *et al.* [30] applied standard strain–life prediction models with notch correction procedures to predict the HCF life of Ti–6Al–4V ELI hip stem implants. A good agreement was observed between predictions and experiments. As far as the spinal implants are concerned, a lot of modeling effort has been devoted to simulating the biomechanical effects of the implants on vertebral system [6, 12]. However, the fatigue behavior is not taken into account. A robust life prediction model for spinal implants can support the rigorous testing required for FDA approval. Within the safe-life methods, the strain–life method is versatile owing to its ability to model both the LCF and HCF regimes. Also, the strain–life methodology is accurate in handling multiaxial loading conditions. Cyclic material behavior is effectively accounted using the hardening models in the strain–life approach, which is not possible in the stress–life method. Thus, the strain–life method seems an apt choice for the life predictions of orthopedic implants. The strain-life equation, which is fit to the experimental fatigue data, is given as

$$\left(\frac{\Delta\varepsilon}{2}\right)_{\text{total}} = \varepsilon_a = \frac{\sigma_f'}{E}(2N_f)^b + \varepsilon_f'(2N_f)^c, \tag{18.1}$$

where $\sigma_f'$ is the fatigue strength coefficient, $b$ is the fatigue strength exponent, $\varepsilon_f'$ is the fatigue ductility coefficient, $c$ is the fatigue ductility exponent (which are obtained by performing strain-controlled axial fatigue tests and fitting experimental data to Equation (18.1)), $N_f$ is fatigue life, and $\varepsilon_a$ is the applied strain amplitude. Ploeg *et al.* [30] recently reported significant cyclic softening behavior for Ti–6Al–4V ELI. Cyclic softening behavior of the implant material can be accounted for in the life prediction algorithm by using the analytical method of notch correction proposed by Glinka [31], called the equivalent strain energy density method (ESED). The Glinka method uses energy concepts to transfer the elastic stress–strains to the cyclic stress–strains. Using the Massing hypothesis in conjunction with the ESED method, the nominal elastic stresses $\sigma_n$ are related to the local stresses $\sigma$ by Equations (18.2) and (18.3):

$$\frac{(K_t\sigma_n)^2}{2E} = \frac{\sigma^2}{2E} + \frac{\sigma}{n'+1}\left(\frac{\sigma}{K'}\right)^{1/n'}, \tag{18.2}$$

$$\frac{(K_t\Delta\sigma_n)^2}{4E} = \frac{\Delta\sigma^2}{4E} + \frac{\Delta\sigma}{n'+1}\left(\frac{\Delta\sigma}{2K'}\right)^{1/n'}, \tag{18.3}$$

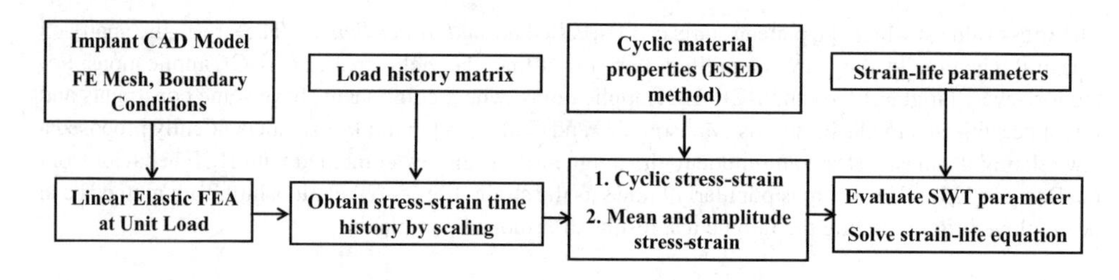


**Figure 18.4** Strain–life algorithm for fatigue life prediction of spinal implants

where $K_t$ is the theoretical stress concentration factor, $E$ is the Young's modulus, $K'$ is the cyclic strength coefficient, and $n'$ is the cyclic hardening exponent. Details of the notch correction methods can be found in Molski and Glinka [31]. Fatigue testing of spinal instrumentation and individual components must be performed at $R \neq -1$; that is, nonzero mean stresses. To account for the mean stresses in the strain–life method, the mean stress correction model by Smith *et al.* [32] (SWT) is implemented in the current work. The SWT parameter $\sigma_{\max}(\Delta\varepsilon/2)$ is obtained by multiplying the strain–life equation by the maximum stress $\sigma_{\max}$ in the given loading cycle. Thus, we have

$$\sigma_{\max}\frac{\Delta\varepsilon}{2} = \frac{(\sigma_f')^2}{E}(2N_f)^{2b} + \sigma_f'\varepsilon_f'(2N_f)^{b+c}. \tag{18.4}$$

The SWT model is used widely in strain–life prediction algorithms and is proven to be effective in a variety of conditions in the published literature. Combining the notch correction method and the SWT model we have the strain–life-based life prediction program, which is shown in flowchart format in Figure 18.4.

Finally, the effect of LSP-induced residual stresses must be accounted for in the life prediction program to predict the fatigue behavior of dynamic orthopedic implants. In safe-life methods, residual stresses are superimposed onto the loading stresses, which essentially change the mean stress of the loading. The relaxation rate of the residual stresses can be included to change the mean stress per cycle. In the current work the strain–life methodology is applied. Thus, residual stresses induced by LSP are superimposed onto stresses induced by the fatigue loading before subjecting to the cyclic correction. The resultant stress–strain data are then used for strain–life prediction, which reflects the effect of residual stresses.

## 18.3 LSP Process

A schematic of LSP process is shown in Figure 18.5. During the LSP process, the target surface or the component is usually covered with the ablation medium, which could either be a thin layer of absorbing tape (vinyl or aluminum) or black paint. The target is then exposed to a high-energy pulsed laser beam with intensity on the order of $10^9$ W cm$^{-2}$. A variety of lasers ranging from short-pulse lasers to solid-state (Nd:YAG and Nd:Glass) lasers can be used for LSP processing [33]. The high-energy laser beam causes evaporation of the ablation layer in a very short period of time (typically a few nanoseconds), which leads to the generation of a plasma. To confine this explosive plasma, a layer of transparent material (typically water) is used on top of the ablation layer. The confined overlay helps to increase the pressure of the plasma, which further induces powerful shock waves in the target material. This type of LSP processing is known as confined-ablation mode [15]. The peak magnitudes of the pressure pulses generated during the LSP process can be in the range 1–10 GPa. If the magnitude of the pressure wave exceeds the Hugoniot elastic limit (HEL) of the target material, then plastic deformation occurs, producing residual stresses on the surface and through the depth of the target material. The magnitude, depth, and uniformity

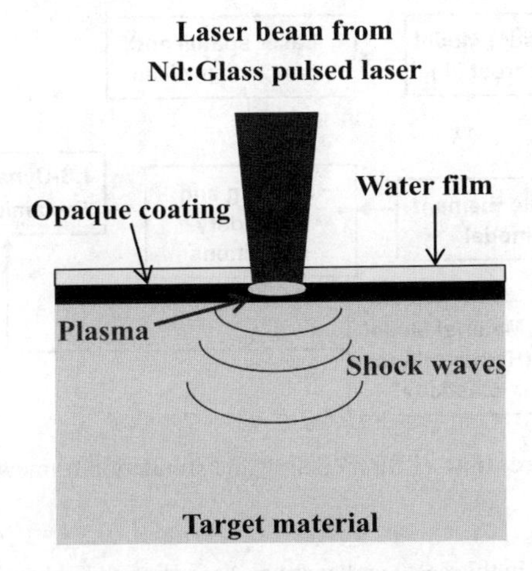


**Figure 18.5** Schematic of LSP process

of the residual stresses (plastic deformation) induced by the LSP process depend on the laser parameters (i.e., laser energy, spot diameter, and pulse width). In addition, processing techniques (i.e., single-sided peening, dual-sided peening, overlap percentage between adjacent laser shots, and, most importantly, the sequence of laser shots) affect the residual stresses. For a given processing technique, the residual stress field is a function of the laser peak power density (PPD = energy/pulse width/spot area). The effect of various parameters on the final distribution has been studied through experiments and simulation by various researchers in the published literature [34–36]. Based on the application requirement (i.e., nature of the fatigue loading and component geometry), LSP parameters are optimized to achieve the required residual stress distribution.

## 18.4 LSP Modeling and Simulation

The LSP process is extremely transient, where final residual stress distribution is function of a large number of parameters. Modeling and simulations play an important role in developing the process. The effects of various parameters on the final residual stress distribution can be effectively studied before process deployment and component fatigue testing. LSP simulation can be effectively performed using the finite-element method (FEM). The basic framework for LSP modeling and simulation is shown in Figure 18.6.

### *18.4.1 Pressure Pulse Model*

In the LSP process, laser energy is converted into mechanical pressure through a complex interaction between laser beam, ablation medium (tape or paint), and confinement medium (water). The first step in the LSP process modeling is to obtain the mechanical pressure as a function of spot diameter, pulse width, and energy of the laser. Fabbro *et al.* [37] performed a detailed analytical and experimental investigation of plasma pressure produced during the confined LSP process and proposed a one-dimensional pressure pulse model. A uniform distribution of laser intensity in the spatial domain was assumed in Fabbro *et al.*'s pressure model, treating wave propagation in the confining medium and the target material as one-dimensional. During the LSP process the laser beam interacts with the coated material to form a plasma. The thickness and the pressure of the plasma at the interface are controlled by the impedances of

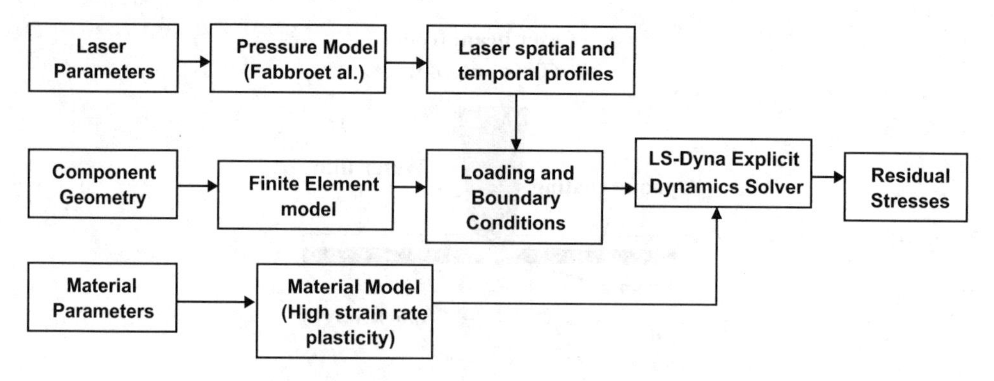


**Figure 18.6** LSP modeling and simulation framework

the confining medium (water in this case) and coating. According to Fabbro *et al.* [37], plasma thickness $L$ at given instant $t$ is related to the shock wave pressure $P$ by

$$\frac{\mathrm{d}L(t)}{\mathrm{d}t} = \frac{2}{Z}P(t), \tag{18.5}$$

where $Z$ is the effective impedance of the interface calculated using Equation (18.6) as a function of impedance confining medium $Z_1$ and coating $Z_2$, respectively:

$$\frac{2}{Z} = \frac{1}{Z_1} + \frac{1}{Z_2}, \tag{18.6}$$

$$I(t) = P(t)\frac{\mathrm{d}L}{\mathrm{d}t} + \frac{3}{2\alpha}\frac{\mathrm{d}}{\mathrm{d}t}[P(t)L(t)]. \tag{18.7}$$

Finally, plasma pressure history is related to the incident laser energy using Equation (18.7). The first part of the right-hand side of the Equation (18.7) gives the work needed for expansion of the plasma, whereas the second part gives the change in internal energy of the plasma. The ratio between the plasma internal energy and thermal energy is given by the correction factor $\alpha$. Equations (18.5) and (18.7) give a nonlinear differential equation, which can be solved using a numerical method to obtain the plasma thickness and pressure pulse history for given laser parameters. Fabbro *et al.*'s model has been successfully used in several LSP simulation studies [38, 39]. From the given laser parameters, peak power density (PPD) is calculated, which is used to fit a short rise time (SRT)-Gaussian laser intensity profile. This profile is then used as an input for the pressure model to obtain the pressure pulse temporal profile.

## *18.4.2 Constitutive Model*

During the LSP process, materials are subjected to very high strain rates (on the order of $10^6$ s$^{-1}$). At such high strain rates, materials show rate-dependent strain hardening and thermal softening due to heat generated during plastic deformation. In the literature, various models have been proposed for modeling such high strain-rate behavior of the materials. These models include Johnson–Cook (JC) [40], Zerilli–Armstrong (ZA) [41], modified ZA [42], mechanical threshold stress (MTS) [43], and Khan–Huang–Liang (KHL) [44]. Meyer [42] recommended use of the ZA model for titanium alloys owing to its ability to model stress saturation in Ti–6Al–4V at high strains. In addition, the ZA model incorporates several key physical mechanism of the dynamic plasticity. However, the standard ZA model does not consider the dependence of the plastic response on pressure and temperature, which is important in the case of high strain-rate response. Considering this, the ZA model combined with a pressure- and temperature-dependence model given by Steinberg *et al.* [45] is used here:

$$\bar{\sigma} = [A + Be^{-(\beta_0 - \beta_1 \ln \dot{\bar{\varepsilon}})T} + C\bar{\varepsilon}^n]\frac{G(P,T)}{G_0}, \tag{18.8}$$

where $\bar{\sigma}$ is the effective stress, $A$, $B$, $C$, $n$, $\beta_0$ and $\beta_1$ are material parameters, $T$ is temperature, $\dot{\bar{\varepsilon}}$ is the effective visco-plastic strain rate, $\bar{\varepsilon}$ is the effective visco-plastic strain, $G_0$ is the shear modulus at room temperature, and $P$ is the pressure. $G(P,T)/G_0$ gives the variation in shear modulus as a function of temperature and pressure, as shown by

$$\frac{G(P,T)}{G_0} = 1 + \left(\frac{G'_p}{G_0}\right)\frac{P}{\eta^{1/3}} + \left(\frac{G'_T}{G_0}\right)(T - 300), \tag{18.9}$$

where $G_0$ is shear modulus at room temperature, $G'_p$ is the derivative of the shear modulus with respect to pressure, $G'_T$ is the derivative of the shear modulus with respect to temperature, $P$ is pressure, and $\eta$ is compressibility. Material parameters are based on Ref. [46] for Ti–6Al–4V, given as $A = 810$ MPa, $B = 1800$ MPa, $C = 250$ MPa, $\beta_0 = 0.009$, $\beta_1 = 0.0005$, and $n = 0.5$. The model parameters related to pressure are based on Ref. [45], which are given as $G_0 = 43.4$ GPa, $G'_p/G_0 = 11.5\ \text{TPa}^{-1}$, and $G'_T/G_0 = 0.62\ \text{kK}^{-1}$.

### *18.4.3 Solution Procedure*

Time-scale in LSP process is very short i.e., few nanoseconds. During the process, high strain rate plastic deformation is experienced by the material and heat is generated due to plastic dissipation. Due to short time scale this process can be considered as adiabatic. Governing equation for the process is then the momentum equation, which can be written in discrete form using the standard FEM interpolation as

$$\mathbf{M}\ddot{\mathbf{u}}^{\mathrm{h}} + \mathbf{f}^{\mathrm{int}} = \mathbf{f}^{\mathrm{ext}}, \tag{18.10}$$

where $\mathbf{f}^{\mathrm{int}}$ and $\mathbf{f}^{\mathrm{ext}}$ are the internal and external nodal force vectors, $\mathbf{u}^{\mathrm{h}}$ is the displacement vector and $\mathbf{M}$ is the mass matrix obtained after finite-element discretization. The energy equation is not considered due to the adiabatic nature of the process. To capture the high-strain-rate nonlinear deformation, the momentum equation may then be solved using the explicit dynamics framework available in commercial FE software or any wave propagation code. Here, LS-DYNA software has been used; a user-defined code was written to implement the proposed ZA model with pressure dependence.

LSP simulation can be separated into two stages. In the first stage, the extremely transient response of the material under the action of LSP-induced intensive shock pressure needs to be captured. Interaction of the shock waves is effectively captured by the explicit dynamics scheme. This stage is executed up to a point where additional load will not produce any plastic deformation. In the second stage, travelling elastic waves in a component are allowed to dissipate using the dynamic relaxation scheme. Artificial damping is introduced to obtain a steady-state solution. After this simulation, residual stresses are extracted at the desired location. For an LSP process involving multiple laser shots in a certain pattern, two stages of the simulation can be repeated using the restart capability of the LS-DYNA software. The pressure pulse needs to be effectively transferred onto the finite-element model as per the sequence of the LSP patch.

## 18.5 Application Example

### *18.5.1 Implant Rod Design*

Following the proposed design methodology, the implant rod was designed by machining longitudinal grooves. Design goals were established for the novel implant rod based on the biomechanical requirements and FDA guidelines. The increase in flexibility was limited to within 7–12%, which essentially means an equivalent decrease in the bending stiffness. The reduction in static 0.2% yield load was limited to 10–15%. The shape of the dynamic rod was conceptualized based on the manufacturing, biomedical, and LSP process consideration. To design the shape and geometric dimensions of the grooves, a parametric study was carried out by performing four-point bending simulations using a three-dimensional nonlinear FEM. Analysis was carried out in the commercial FE software ANSYS. A bilinear hardening model was used to represent the elastic–plastic material behavior observed in the tests. Simulations replicated the tests recommended in ASTM standard F2193 [21] (Figure 18.2). The parametric study yielded a design satisfying the requirements on flexibility and yield load, which is schematically shown in Figure 18.7. To verify the data obtained from finite-element analysis parametric study, the dynamic rod was fabricated and tested on a four-point bending set-up on an MTS machine with displacement-controlled loading. The strain on the outer surface was measured using strain gauges. The current rigid rod was tested to establish the baseline data and evaluate the relative flexibility and yield load data. Results for the relative load ratio (applied load/yield load of rigid rod) against the percentage strain are shown in Figure 18.7. A very close agreement is observed between experiments and simulation predictions. Based on the test data, the flexibility of the dynamic implant rod is increased by 9.78% and the 0.2% yield load is reduced by 10.53% relative to the current rigid rod, thus satisfying the design goals. Residual stresses and fatigue life results are discussed next.

### *18.5.2 Residual Stresses*

The LSP process was first simulated to optimize the laser parameters and shot sequence for obtaining the required compressive residual stress distribution eventually contributing to bending fatigue life improvement. The LSP simulation process as described earlier using the FEM was followed. Three laser intensities (3, 4, and 5 GW $cm^{-2}$) were chosen for simulation based on the previous experiences of LSP on Ti–6Al–4V. An LSP patch of 10 mm length was simulated to save computational time. The LSP shot sequence used for the simulation was a two-step process. In the first step, a ring of shots

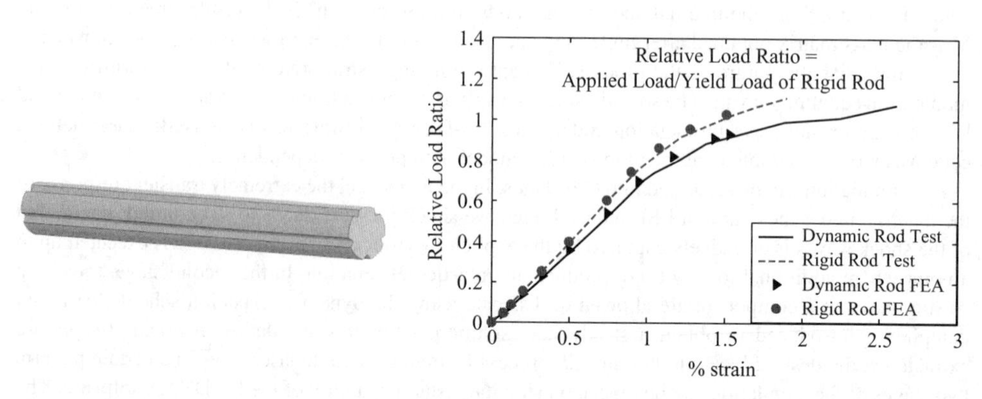


**Figure 18.7** Schematic of the dynamic implant rod and corresponding four-point bending quasi-static FEA and simulation results

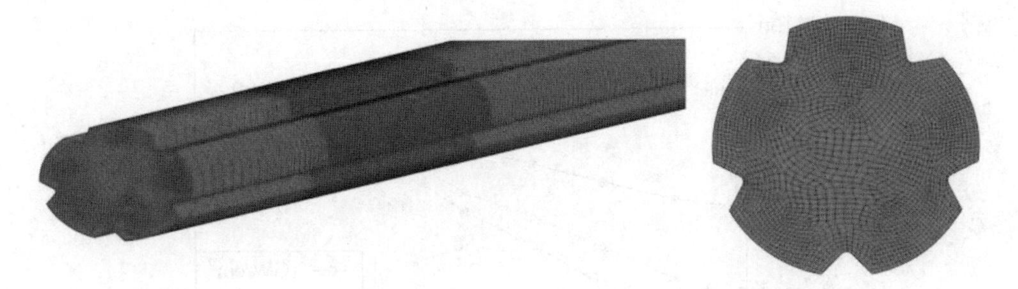

**Figure 18.8** Three-dimensional finite-element mesh of novel implant rod

enveloping the complete circumference of the rod with 50% overlap was processed. The ring of shots was then traversed along the length of the rod, maintaining the 50% overlap with the previous ring. A three-dimensional finite-element model of the dynamic implant rod used for the LSP simulation is shown in Figure 18.8. With the proposed sequence, a simulation of 126 shots was completed. A special algorithm was developed to apply pressure loading on the cylindrical surface for the circular laser spots.

Residual stresses were extracted at the end of each simulation by averaging over the area of 1 mm diameter on the land portion of the dynamic implant rod. A contour plot of the residual stresses obtained after the simulation is shown in Figure 18.9. Axial residual stresses, which are important to resist the bending loads, are extracted and plotted against the depth in Figure 18.10 for three laser intensities. Based on the simulation results, a laser intensity of 5 GW cm$^{-2}$ was chosen for further experimental evaluation. Dynamic implant rods were LSP processed at the University of Cincinnati using a Gen I Nd:Glass set-up. A black tape was used as ablation medium. With the proposed sequence of LSP spots, taped processing takes a very long time, making it commercially ineffective. Therefore, the use of ablation medium was stopped after several trials of taped LSP processing. Instead, the laser was directly applied on the surface. This process is also known as naked LSP processing. The process induces a very thin *recast layer* on the surface due to melting and solidification of alloy during the processing. To remove this recast layer, dynamic implant rods were further processed using a glass-bead peening process.

At the end of the process, residual stresses were measured using the X-ray diffraction technique. Comparison of the simulated and measured residual stresses is shown in Figure 18.11. A reasonable agreement between experimental and simulated residual stress values is observed except near the surface. A sharp gradient obtained in the experimental residual stress profile is due to the glass-bead peening process, which is not accounted for in the simulation. Differences in the stress magnitudes through the

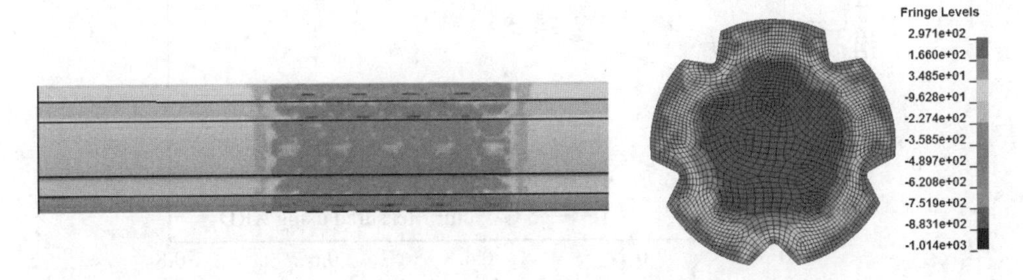


**Figure 18.9** Contour plot of residual stresses along the length and cross-section of the dynamic implant rod

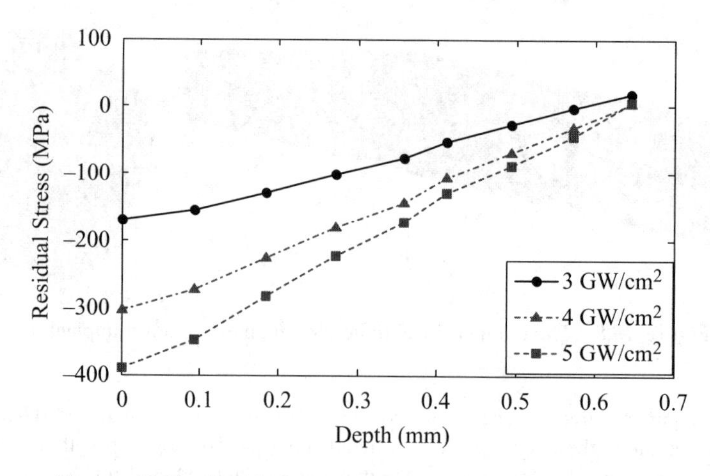


**Figure 18.10** Residual stresses obtained using LSP simulation

depths are attributed to the naked LSP processing. In the simulations, taped processing is assumed where pressure is higher than the naked LSP processing due to the shock impedance mismatch added due to the tape. The goal of LSP simulation was to verify the uniformity of the residual stress field and predict the magnitudes of residual stresses, which are achieved with reasonable accuracy, as shown in Figure 18.9 and Figure 18.11. It is observed that a significant amount of residual stress is introduced in Ti–6Al–4V ELI for a laser intensity of 5 GW $cm^{-2}$.

### *18.5.3 Fatigue Tests and Life Predictions*

To study the effect of flexibility and LSP on fatigue behavior of dynamic implant rods (grooved rods of 120 mm length), four-point bending fatigue tests were performed (see Figure 18.2 for set-up). Current rigid rods (5.5 mm diameter and 120 mm length) were also tested under similar loading conditions to establish the baseline data for comparison. Several rigid and dynamic implant rods were fabricated at

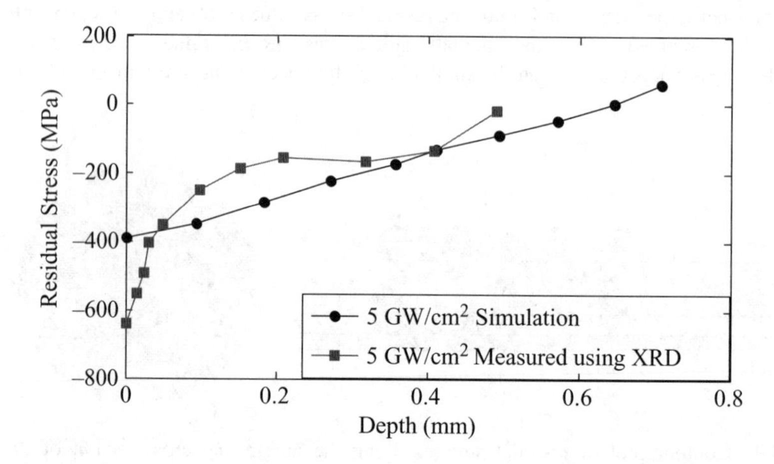


**Figure 18.11** Comparison of the simulated and measured axial residual stresses

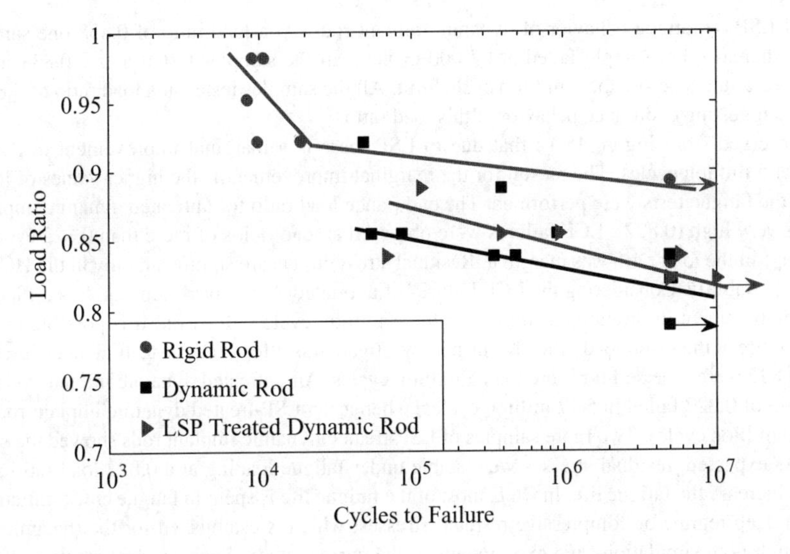


**Figure 18.12** Four-point bending fatigue test results of rigid and dynamic implant rods

X-Spine Inc. Surface roughness of as-received rods was $R_a = 0.4775$ μm. Based on the results obtained from the LSP simulations and experiments, several untreated dynamic implant rods were LSP treated at the University of Cincinnati. The LSP parameters and shot sequence were kept the same as used in the LSP experiments (Section 18.5.2). However, the length of the LSP patch was increased to cover a significant portion of the area that will experience tensile loading stresses during the four-point bending fatigue loading. Surface treatments do not change the yield load of the component, which was verified by conducting the four-point bending quasi-static test on the LSP-treated dynamic implant rod. All the implant rods were tested under load-controlled fatigue at $R = 10$ and 15 Hz frequency. Figure 18.12 shows the results of the fatigue tests where cycles to failure are plotted against the *load ratio*, which is the ratio between the maximum applied load and the 0.2% yield load of the corresponding implant rod.

A very sharp transition is observed in the fatigue test results going from the LCF to HCF region for rigid implant rods. Tests were repeated at critical load ratios to finally determine the run-out load ratio, which turned out to be 0.89 for the current rigid rods. A 5 million cycle life was considered as run-out. A few tests were performed at lower load ratios to check for the repeatability. All of them were run-out, as seen in Figure 18.12. Similar to the rigid rods, a sharp transition is observed in the fatigue tests of untreated dynamic implant rods starting from about 50 000 cycles. Both types of rods had the same surface conditions, which is reflected in their fatigue behavior under bending loading conditions. Several constant-amplitude fatigue tests were conducted at various load ratios to determine the critical load ratio at the 5 million cycle life. At a load ratio of 0.838, scatter was observed where samples failed at 350 000 cycles, 477 200 cycles, and one sample was run-out. All the samples tested above 0.838 load ratio experienced failures before the 5 million cycle limit. Three samples tested at load ratio of 0.822 were run-outs. Also, tests performed at lower load ratios were run-out, suggesting an endurance load ratio of 0.822 for the dynamic implant rods. A sharp transition in the *S–N* curve at such high load ratios is attributed to the surface conditions and the glass-bead peening done on the as-received implant rods. The gradient effect may play a role in bending fatigue, which is further enhanced with pre-existing residual stresses due to the glass-bead peening process. A combination of these factors might arrest the short crack growth resulting in a plateau from LCF–HCF to HCF region. Several constant-amplitude fatigue tests were then conducted at various load ratios to determine the endurance load ratio and evaluate the effects of LSP on fatigue behavior of dynamic implant rods. At a load ratio of 0.838 one sample was run-out, whereas other sample failed at 67 000 cycles. All the samples tested above 0.838 load ratio experienced failures before the 5 million cycle limit. All the samples tested at a load ratio of 0.822 were run-outs, suggesting endurance behavior at this load ratio.

It is observed from Figure 18.12 that due to LSP there is a marginal improvement in the fatigue strength at 5 million cycles. The reason for the marginal improvement is the higher values of load ratio at which the fatigue tests were performed. The endurance load ratio for untreated dynamic implant rods itself was very high (0.822). LCF failures were observed at load ratios of more than 0.822 even though the increase in the load ratio was marginal. Residual stress effects are significant only in the HCF region due to their stability. Considering the LCF failures of untreated/as-received samples, it is difficult to get a large improvement in the fatigue strength at the 5 million cycles. However, it is possible to increase the service life at the given load ratio. To further investigate this effect, fatigue tests at an endurance load ratio of 0.822 were repeated for more than 5 million cycles. An untreated dynamic implant rod tested at a load ratio of 0.822 failed in 5.17 million cycles, whereas an LSP-treated dynamic implant rod did not fail till 10 million cycles. Two more samples of LSP-treated dynamic implant rods showed the same test results. As expected, residual stresses were stable under fatigue loading at a 0.822 load ratio and they helped to increase the fatigue life. In HCF, most of the fatigue life is spent in fatigue crack initiation. LSP induces a deep regime of compressive residual stresses, which is established for the dynamic implant rods through both simulations and experiments in the current study. Such residual stresses slow down the rate of short fatigue crack growth, thus improving the total fatigue life.

Fractography analysis was performed on LSP-treated and untreated dynamic implant rods to study the effect of LSP on short crack growth. The untreated dynamic implant rod that failed at 5.17 million cycles at a load ratio of 0.822 was compared with the LSP-treated implant rod that was run-out (10 million cycles) at the same load ratio. The LSP-treated rod was tested at a higher load ratio to complete failure. Both the samples were then investigated using a scanning electron microscope (SEM) to obtain micrographs, as shown in Figure 18.13. Striation spacing measurements were taken at the same location (360 μmm from the tensile side) on fractured surfaces of both the samples. The average striation spacing rods was measured to be 217.11 nm in untreated rods and 113.3 nm in LSP-treated rods, suggesting short crack growth in LSP-treated rod is much slower than in the untreated rod for the same load (i.e., applied stress

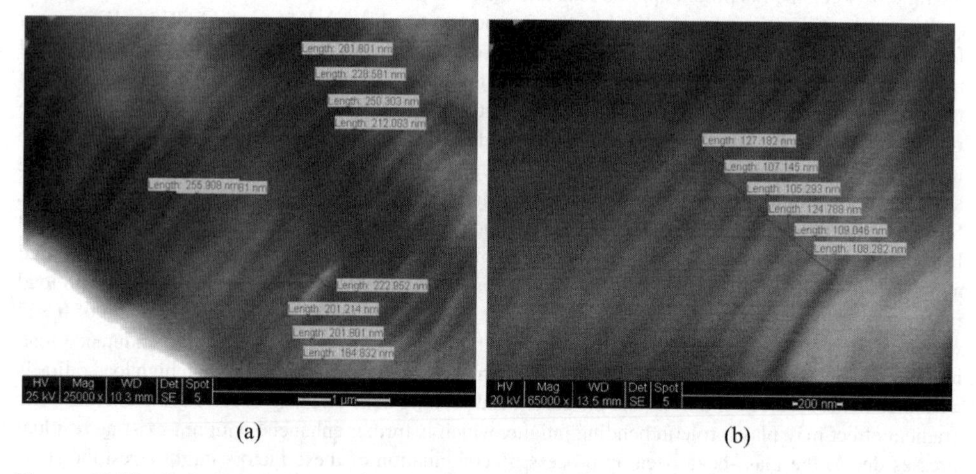


**Figure 18.13** Striation spacing measurements using SEM images on fractured samples of (a) untreated and (b) LSP-treated dynamic implant rods

intensity factor range). This analysis validates the claim that LSP-treated rods have more fatigue life for the same applied load ratio compared with the untreated rods.

The strain–life methodology as discussed in Section 18.2 can then be applied to test its effectiveness in predicting the fatigue behavior of dynamic implant rods. For the strain–life model, $\sigma_f'$, $\varepsilon_f'$, $b$, and $c$ parameters are required. Ploeg *et al.* [30] reported the strain–life parameters for Ti–6Al–4V ELI by performing strain-controlled axial fatigue tests as $\sigma_f' = 819$ MPa, $\varepsilon_f' = 0.886$, $b = -0.0407$, and $c = -0.648$. In their study, coupons were heat treated at 900°C, air cooled, and rough-grit blasted before strain-controlled cycling. As-received implant rods considered in this study had a different preprocessing and surface condition. Also, fatigue testing in the current study is under bending loads. The goal of the life prediction exercise for orthopedic implants is to be able to predict the fatigue strength of implants for the fatigue life limit recommended by FDA guidelines. In order to achieve the goal, strain–life parameters required for the SWT model were calibrated based on the bending fatigue test results for rigid implant rods and subsequently used to predict the fatigue behavior of dynamic implant rods. The calibrated parameters are $\sigma_f' = 1180$ MPa, $\varepsilon_f' = 0.45$, $b = -0.0137$, and $c = -1.25$. Life prediction analysis was then performed for untreated and LSP-treated dynamic implant rods at various load ratios used in fatigue testing to generate the *S–N* curves, which are compared with the fatigue test results in Figure 18.14.

A very good agreement is observed between the fatigue test results and predictions made using the strain–life approach for untreated dynamic rods. At 5 million cycle life, predictions for the fatigue strength of the dynamic implant rods made using the calibrated SWT model parameters agree with the test results of fatigue strength at 5 million cycles within 1.22%. The error percentage is calculated based on the predicted load ratio and the average load ratio observed at 5 million cycles. There is also reasonable agreement between life predictions at other load ratios. For a load ratio of 0.838, a lot of scatter is observed in the test data and predicted life falls approximately at the mean life value from the tests. Thus, it can be concluded that calibrated SWT model parameters are able to predict the bending fatigue behavior of untreated dynamic implant rods. For life prediction analysis of LSP-treated dynamic rods, measured residual stresses (Figure 18.11) were incorporated in the strain–life algorithm. The superposition principle was followed to determine the effective loading stresses. At 5 million cycle life, predictions for the fatigue strength of the LSP-treated dynamic implant rods made using the calibrated SWT model parameters and superimposed residual stresses agree with the test results

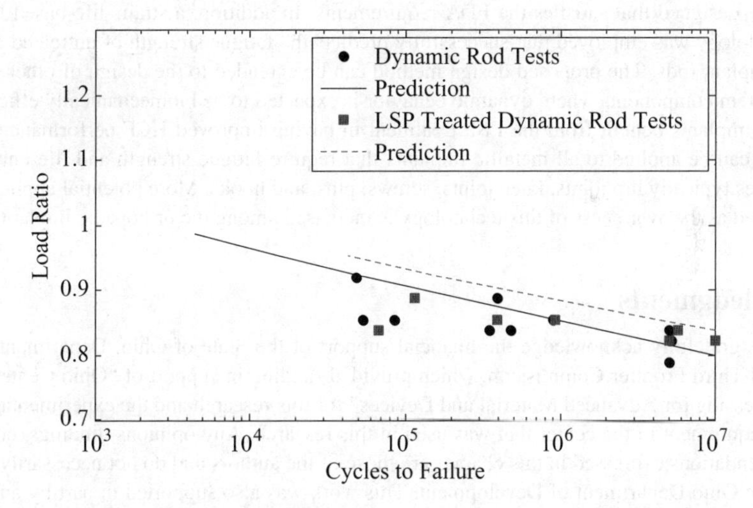


**Figure 18.14** Four-point bending fatigue test results and corresponding strain–life predictions for untreated and LSP-treated dynamic implant rods

of fatigue strength at 5 million cycles within 2.44%. Predictions at other load ratios are in reasonable agreement with the test results. However, predictions are nonconservative for the LSP-treated rods, whereas predictions for untreated rods were conservative. The reason for nonconservative predictions is attributed to the residual stress relaxation, which is expected at the loading levels close to the yield loads. Several workers have reported the residual stress relaxation in titanium alloys under fatigue loading. Nevertheless, predictions are in reasonable agreement with the test results. The strain–life model is able to predict the fatigue strength values for a variety of tests performed under the current study. For the LSP-treated rods, the significant improvement is in fatigue life rather than the fatigue strength, which is attributed to the short crack growth behavior under residual stresses. The strain–life model cannot capture this behavior effectively. A robust short crack growth prediction model can be further employed to predict the improvement in fatigue life of LSP-treated dynamic implant rods at given load ratios.

## 18.6 Summary

Motivated by the need for dynamic spinal instrumentation, a novel design philosophy for dynamic orthopedic implant rods is proposed in this chapter and subsequently applied for the design of dynamic spinal implant rods. The dynamic nature is achieved by introducing flexibility into the implant system. Specifically, for the spinal instrumentation system, flexibility is increased by machining longitudinal grooves in the rigid implant rods. An additional requirement in terms of fatigue life of spinal instrumentation was satisfied by applying an advanced surface treatment called LSP, which is the key element of the proposed design philosophy. A significant amount of compressive residual stresses is introduced in implant rods of Ti–6Al–4V ELI material. To support the experimental investigation, the LSP process is modeled and simulated using the FEM. The LSP process parameters are optimized using simulations; experimental measurements of residual stresses further validate the simulation methodology. Extensive fatigue testing was performed on rigid, untreated dynamic and LSP-treated dynamic implant rods as per FDA guidelines. Dynamic implant rods successfully passed the FDA requirements on fatigue life; that is, the load ratio at 5 million cycles is more than 0.5 (0.822 for the design proposed in this work). The effect of LSP on fatigue life improvement of the dynamic implant rods is clearly evident in the fatigue tests conducted in the HCF regime, where a $2\times$ improvement is obtained at a load ratio of 0.822. Thus, with the proposed design method, a fatigue-strength-enhanced dynamic spinal implant rod was successfully designed that satisfies the FDA requirements. In addition, a strain–life-based life prediction methodology was employed that successfully predicts the fatigue strength of untreated and treated dynamic implant rods. The proposed design method can be extended to the design of other orthopedic implant system components where dynamic behavior is expected to be biomechanically efficient. Also, the current implants benefit from the LSP treatment in having improved HCF performance. The LSP technology can be applied to all metallic implants that require fatigue strength and life enhancement. This includes typically hip joints, knee joints, screws, pins, and hooks. More potential applications will be discovered as the awareness of this technology is increased among the orthopedic implant industry.

## Acknowledgments

The authors gratefully acknowledge the financial support of the State of Ohio, Department of Development and Third Frontier Commission, which provided funding in support of "Ohio Center for Laser Shock Processing for Advanced Material and Devices" for this research and the experimental and computational equipment in the center that was used in this research. Any opinions, findings, conclusions, or recommendations expressed in this chapter are those of the authors and do not necessarily reflect the views of the Ohio Department of Development. This work was also supported in part by an allocation of computing time from the Ohio Supercomputer Center. DQ also acknowledges the support of start-up fund from the University of Texas at Dallas.

## References

[1] Knowledge Enterprise Inc. (2008) The Orthopedic Industry Annual Report, Chagrin Falls.

[2] Boos, N. and Aebi, M. (eds) (2008) *Spinal Disorders. Fundamentals of Diagnosis and Treatment*, Springer-Verlag, Berlin.

[3] Shono, Y., Kaneda, K., Abumi, K. *et al.* (1998) Stability of posterior spinal instrumentation and its effects on adjacent motion segments in the lumbosacral spine. *Spine*, **23** (14), 1550–1558.

[4] Akahori, T. and Niinomi, M. (1998) Fracture characteristics of fatigued Ti–6Al–4V ELI as an implant material. *Materials Science and Engineering A*, **243** (1–2), 237–243.

[5] ASTM (2004) *Standard Test Methods for Spinal Implant Constructs in a Vertebrectomy Model.*

[6] Xu, H.Z., Wang, X.Y., Chi, Y.L. *et al.* (2006) Biomechanical evaluation of a dynamic pedicle screw fixation device. *Clinical Biomechanics*, **21** (4), 330–336.

[7] Korovessis, P., Papazisis, Z., Koureas, G., and Lambiris, E. (2004) Rigid, semirigid versus dynamic instrumentation for degenerative lumbar spinal stenosis: a correlative radiological and clinical analysis of short-term results. *Spine*, **29** (7), 735–742.

[8] McAfee, P.C., Farey, I.D., Sutterlin, C.E. *et al.* (1989) 1989 Volvo Award in Basic Science: Device-related osteoporosis with spinal instrumentation. *Spine*, **14** (9), 919–926.

[9] Johnston, C.E. II, Ashman, R.B., Baird, A.M., and Allard, R.N. (1990) Effect of spinal construct stiffness on early fusion mass incorporation: experimental study. *Spine*, **15** (9), 908–912.

[10] Smith, K.R., Hunt, T.R., Asher, M.A. *et al.* (1991) The effect of a stiff spinal implant on the bone-mineral content of the lumbar spine in dogs. *Journal of Bone and Joint Surgery – Series A*, **73** (1), 115–123.

[11] McLain, R.F., Sparling, E., and Benson, D.R. (1993) Early failure of short-segment pedicle instrumentation for thoracolumbar fractures. A preliminary report. *Journal of Bone and Joint Surgery – Series A*, **75** (2), 162–167.

[12] Goel, V.K. (2001) Hinged-dynamic posterior device permits greater loads on the graft and similar stability as compared with its equivalent rigid device: a three-dimensional finite element assessment. *Journal of Prosthetics and Orthotics*, **13** (1), 17–20.

[13] Asazuma, T., Yamugishi, M., Sato, M. *et al.* (2004) Posterior spinal fusion for lumbar degenerative diseases using the Crock–Yamagishi (C–Y) spinal fixation system. *Journal of Spinal Disorders and Techniques*, **17** (3), 174–177.

[14] Templier, A., Denninger, L., Mazel, C. *et al.* (1998) Comparison between two different concepts of lumbar posterior osteosynthesis implants: a finite-element analysis. *European Journal of Orthopaedic Surgery and Traumatology*, **8** (1), 27–36.

[15] Clauer, A.H. and Lahrman, D.F. (2001) Laser shock processing as a surface enhancement process. *Key Engineering Materials*, **197**, 121–144.

[16] Peyre, P. and Fabbro, R. (1995) Laser shock processing: a review of the physics and applications. *Optical and Quantum Electronics*, **27** (12), 1213–1229.

[17] Clauer, A.H. (1996) Laser shock peening for fatigue resistance, in *Surface Performance of Titanium* (eds J.K. Gregory, H.J. Rack, and D. Eylon), TMS, Warrendale, PA, pp. 217–230.

[18] Mannava, S.R. and Rockstroh, R.J. (2005) Fourth generation of laser shock peening to enhance damage tolerance of aircraft engine components, in *ASME/JSME PVP Conference*.

[19] Ruschau, J.J., John, R., Thompson, S.R., and Nicholas, T. (1999) Fatigue crack nucleation and growth rate behavior of laser shock peened titanium. *International Journal of Fatigue*, **21**, S199–S209.

[20] Nalla, R.K., Altenberger, I., Noster, U. *et al.* (2003) On the influence of mechanical surface treatments – deep rolling and laser shock peening – on the fatigue behavior of Ti–6Al–4V at ambient and elevated temperatures. *Materials Science and Engineering A*, **355**, 216–230.

[21] ASTM (2007) *Standard Specifications and Test Methods for Components Used in the Surgical Fixation of the Spinal Skeletal System.*

[22] Suresh, S. (2001) *Fatigue of Materials*, 2nd edition, Cambridge University Press, Cambridge.

[23] Manson, S.S. (1965) Fatigue: a complex subject – some simple approximations. *Experimental Mechanics*, **5** (7), 193–226.

[24] Coffin, L.F. (1960) The stability of metals under cyclic plastic strain. *ASME Journal of Basic Engineering*, **82**, 671–682.

[25] Basquin, O.H. (1910) The exponential law of endurance tests. *Proceedings – American Society for Testing Materials*, **10**, 625–630.

[26] Wierszycki, M., Kąkol, W., and Łodygowski, T. (2006) The screw loosening and fatigue analyses of three dimensional dental implant model, in *ABAQUS Users' Conference*, Boston, MA.
[27] Wierszycki, M., Kąkol, W., and Łodygowski, T. (2006) Fatigue algorithm for dental implant. *Foundations of Civil and Environmental Engineering*, **7**, 363–380.
[28] *ABAQUS Analysis User's Manual* (2005) ABAQUS, Inc., Pawtucket.
[29] *feSafe User's Manual* (2005) Safe Technology Ltd, Sheffield.
[30] Ploeg, H.-L., Bürgi, M., and Wyss, U.P. (2009) Hip stem fatigue test prediction. *International Journal of Fatigue*, **31** (5), 894–905.
[31] Molski, K. and Glinka, G. (1981) A method of elastic–plastic stress and strain calculation at a notch root. *Materials Science and Engineering*, **50** (1), 93–100.
[32] Smith, K.N., Watson, P., and Topper, T.H. (1970) A stress–strain function for the fatigue of metals. *Journal of Materials*, **15**, 767–778.
[33] Thorslund, T., Kahlen, F.-J., and Kar, A. (2003) Temperatures, pressures and stresses during laser shock processing. *Optics and Lasers in Engineering*, **39** (1), 51–71.
[34] Arif, A.F.M. (2003) Numerical prediction of plastic deformation and residual stresses induced by laser shock processing. *Journal of Materials Processing Technology*, **136** (1–3), 120–138.
[35] Yang, C., Hodgson, P.D., Liu, Q., and Ye, L. (2008) Geometrical effects on residual stresses in 7050-T7451 aluminum alloy rods subject to laser shock peening. *Journal of Materials Processing Technology*, **201** (1–3), 303–309.
[36] Warren, A.W., Guo, Y.B., and Chen, S.C. (2008) Massive parallel laser shock peening: simulation, analysis, and validation. *International Journal of Fatigue*, **30** (1), 188–197.
[37] Fabbro, R., Fournier, J., Ballard, P. *et al.* (1990) Physical study of laser-produced plasma in confined geometry. *Journal of Applied Physics*, **68** (2), 775–784.
[38] Hu, Y. and Yao, Z. (2008) Overlapping rate effect on laser shock processing of 1045 steel by small spots with Nd:YAG pulsed laser. *Surface and Coatings Technology*, **202** (8), 1517–1525.
[39] Zhou, Z., Gill, A.S., Qian, D. *et al.* (2011) A finite element study of thermal relaxation of residual stress in laser shock peened IN718 superalloy. *International Journal of Impact Engineering*, **38** (7), 590–596.
[40] Johnson, G.R. and Cook, W.H. (1983) A constitutive model and data for metals subjected to large strains, high strain rates and high temperatures, in *7th International Symposium on Ballistics*, The Hague, The Netherlands.
[41] Zerilli, F.J. and Armstrong, R.W. (1987) Dislocation-mechanics-based constitutive relations for material dynamics calculations. *Journal of Applied Physics*, **61** (5), 1816–1825.
[42] Meyer, H.W. (2006) A modified Zerilli–Armstrong constitutive model describing the strength and localizing behavior of Ti–6Al–4V. Report ARL-CR-0578, Dynamic Science, Inc., Aberdeen, MD.
[43] Follansbee, P.S. and Kocks, U.F. (1988) A constitutive description of the deformation of copper based on the use of the mechanical threshold stress as an internal state variable. *Acta Metallurgica*, **36** (1), 81–93.
[44] Khan, A.S., Suh, Y.S., and Kazmi, R. (2004) Quasi-static and dynamic loading responses and constitutive modeling of titanium alloys. *International Journal of Plasticity*, **20** (12), 2233–2248.
[45] Steinberg, D.J., Cochran, S.G., and Guinan, M.W. (1979) A constitutive model for metals applicable at high-strain rate. *Journal of Applied Physics*, **51** (3), 1498–1504.
[46] Macdougall, D.A.S. and Harding, J. (1999) A constitutive relation and failure criterion for Ti6Al4V alloy at impact rates of strain. *Journal of the Mechanics and Physics of Solids*, **47** (5), 1157–1185.

# Part IV

# Bio-mechanics and Materials of Bones and Collagens

In this part we present five chapters focusing on the study of the mechanical properties and microstructures of bones. The first contribution of this part is Chapter 19, in which Elkhodary, Greene and O'Connor apply the so-called archetype blending continuum (ABC) theory to study material properties and microstructure of bones. The ABC theory is a recently developed multiscale and multiphysics theory by Wing Liu with his co-workers, and Elkhodary, Greene and O'Connor are the main contributors of the ABC theory. This chapter will be among the first publications of ABC theory in the literature. In Chapter 20, Yang, Chi, and Chen present a multiscale analysis of porous bone structure and materials. They used an image-based micromechanics homogenization technique in their study. Chapter 21 is another study of material properties of bone; however, in this case, Zamiri and De adopt a combined experimental and analytical approach to study the local mechanical strength of bone, namely the nonlinear plasticity of the bone. In Chapter 22, Kim, Kim, and Lee present a systematic study of cellular structure of bones, and they have applied their research results to study the fracture of bone. The last contribution of this book is from Adnan, Sarker, and Ferdous, who present their recent study on mineralized collagens. The chapter is a detailed multiscale study of mineralized collagens or bones, and the authors have examined mineralized collagen materials from their chemical content, molecular structure to mesoscale and macroscale structures and properties.

# 19

# Archetype Blending Continuum Theory and Compact Bone Mechanics

Khalil I. Elkhodary[a], Michael Steven Greene[b], and Devin O'Connor[a]
[a]*Department of Mechanical Engineering, Northwestern University, USA*
[b]*Theoretical & Applied Mechanics, Northwestern University, USA*

## 19.1 Introduction

In this chapter we explore the connection between a structurally complex biomaterial (i.e., bone) and a new theory of generalized continuum mechanics that accounts for microstructural complexity without explicitly modeling it. The hierarchical structure of bone [1, 2] is well suited for applications in generalized continua, which are theories that add higher displacement gradients or extra degrees of freedom to the classical equations in order to approximate fluctuations in deformation occurring at fine scales. The connection has been drawn before by Lakes [3] and Yoon and Katz [4], where the mechanical behavior of bone was compared with features of a Cosserat solid [5] and experimental methods for determination of the model constants were developed.

The interest in bone mechanics and theories of generalized continua stems from their function as mechanical materials that both drive locomotion of the body and protect critical organs. Thus, a natural trade-off exists between weight, which when high slows movement, and strength, which when high better protects. Bone's impressive hierarchical structure strikes a balance in this trade-off between weight and strength, two typically competing features of structural materials, while retaining remarkable properties like high fracture resistance [6–8]. A better understanding of bone behavior under deformation, therefore, will improve patient diagnosis and therapy for osteoporosis, guide bone tissue regenerative engineering [9], enable design of prosthetic limbs and protective gear for military soldiers or professional athletes, explain and avoid bone breaks or brain concussions, and even inform the design of new synthetic materials that mimic the high-strength, high-toughness performance of bone through material hierarchies. To get at this understanding, traditional continuum methods do not suffice, as they are prohibitively restrictive

*Multiscale Simulations and Mechanics of Biological Materials*, First Edition. Edited by Shaofan Li and Dong Qian.

in their information content: How can one constitutive law be expected to describe the five or so levels of structural hierarchy in a typical bone tissue?

The microstructure and mechanical properties of bone have been well studied [2, 10, 11], though we provide a brief summary here so the reader may be familiar with the terminology we adopt throughout the chapter. Following the summary of bone's structure, we will turn our discussion to generalized continuum mechanics and the proposed archetype-blending continuum (ABC) theory. We do this so we may drive the formulation of ABC with concreteness furnished by the definitions from bone mechanics.

### *19.1.1 A Short Look at the Hierarchical Structure of Bone*

Before describing the hierarchical structure of bone it is important to note that there are two main types of bone: *spongy* (also referred to, more precisely, as "trabecular," though "spongy" is more descriptive) and compact (also referred to as "cortical"). As their names indicate, spongy bone is highly porous, whereas compact bone is far denser. The focus of this chapter will be on characteristics of stress, strain, and rotation fields in compact bone at its mesoscale. However, we note that below a certain length scale we refer to as the *mesoscale*, bone tissue is comprised of largely the same substructures; thus, we start from the bottom up in its description. An image of the scale hierarchy in bone is shown in Figure 19.1 and described immediately.

At the *nanoscale*, the constituent phases of bone are collagen proteins (80–90% of all protein), impure minerals whose exact crystal structure is not known, non-collagenous organic material (10–20% of all protein), water, and bone cells. As is explained by Rho *et al.* [2], the in-situ mechanical properties of both collagen and mineral are not well known. The exact volume fraction of each material is highly variable and depends on species, individuals within species, bone type, and location within a single bone. Liu [9], however, gives an approximate value of 60% mineral and 40% collagen. Individual collagen, a structural protein, forms a tropocollagen macromolecule that comprises three individual polypeptides of collagen. The individual collagen chains are of the same length, approximately 300 nm long [2, 11], and are held together by hydrogen bonds that form a triple helix. The minerals at the bone nanoscale are assumed to be stiffer than the surrounding collagen proteins and inorganic matter by two orders of magnitude [12]

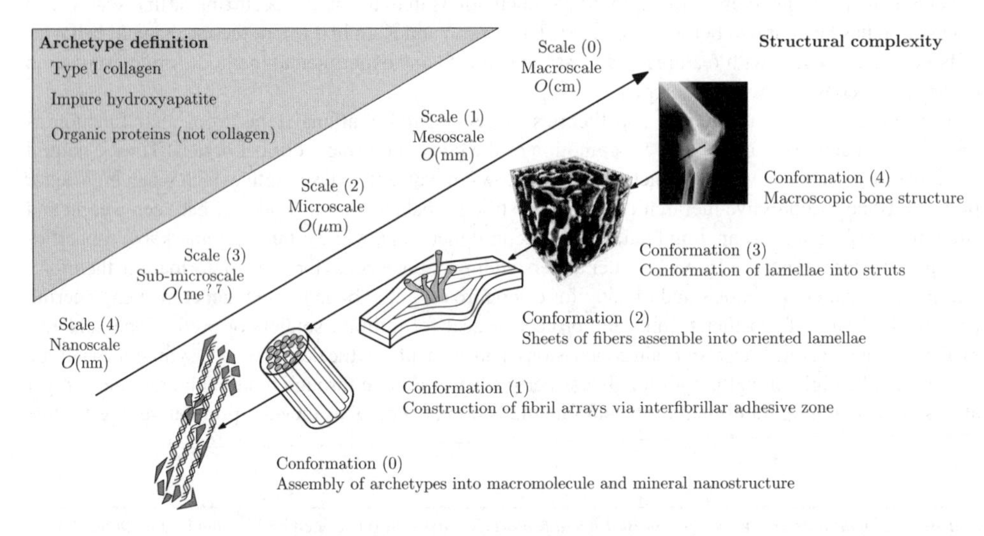


**Figure 19.1** The hierarchical structure of bone

and form into plate-shaped structures $50 \times 25 \times 3$ nm$^3$ in size [2, 11, 13]. Minerals exist in the gaps between, adjacent to, and embedded in collagen macromolecules and the associated microstructure they form [11, 14].

At the next level, which we label the *sub-microscale*, tropocollagen macromolecules pack up side by side through bonding to molecules in neighboring tropocollagen to form stabilized structures termed microfibrils, which in turn aggregate to form fibrils approximately 2–3 μm thick for lamellar bone. Minerals exist around, within, and between (micro)-fibrils in bone, though the extent of the mineral network is unclear. Following the sub-microscale is the microscale, where fibrils assemble in oriented sheets called lamellae. A *lamella* (plural lamellae) is a collection of collagen fibrils arranged in sheets where the fibrils are approximately aligned in the same direction. Lamellae are around 3–7 μm thick [2, 11] and stack on top of one another. The manner in which adjacent lamellae stack is debated in the literature, with some claiming the orientation of adjacent lamellae is that of twisted plywood (gradual change in fibril orientation) [15], and others claiming the change between layers is much sharper. Past the microscale, the hierarchical structure at the mesoscale varies widely between spongy and compact bone. Experimental images of sub-micro and micro bone can be found in Refs. [7, 13, 16].

The primary structural unit of compact bone at the mesoscale is termed the *osteon*, which is a set of lamellae stacked on top of one another and concentrically wrapped to form a roughly cylindrical unit 10–500 μm in diameter. This unit is the osteon. In the middle of developed osteons, a hollowed (Haversian) canal houses a blood vessel about 10–70 μm in diameter [17] that provides nutrients to the surrounding material. Compact bone is solid bone with very little porosity arising from the aforementioned blood vessels, bone cells, and the small canals that transport nutrients and information for the bone cells. The three-dimensional canal network within osteon macrostructure is visualized in Cooper *et al.* [18] and discussed in a review paper on bone and poroelasticity by Cowin [19]. Of importance in this level of bone structure are the *cement sheaths*, also called cement lines, which surround osteons and are roughly 1–50 μm in thickness [20, 21]. Cement sheaths have been described by Lakes [12] as one of the mechanisms causing viscoelasticity in bone, yet Currey [11, see p. 16 and references therein] gives a good account of how little is known about their properties. Cement lines have also been targeted as one component of the compact bone mesoscale that contributes to its fracture resistance, as the cement lines are the site for microcrack nucleation, delamination barriers, and crack-path deflection obstacles [6, 22].

### 19.1.2 A Background of Generalized Continuum Mechanics

Continuum methods have been around for centuries, with the basic idea [23] positing that material points in a continuous, homogeneous medium induce short-range line forces between adjacent material point centroids for a given deformation. In the traditional continuum mechanics, three orthogonal traction vectors are sufficient to uniquely describe the state of stress at a point by Cauchy's famous tetrahedron. In this picture of a continuum, with forces acting on lines of action between material point centroids and force couples at a point vanishing in the infinitesimal limit, linear momentum balance provides the governing partial differential equation of material motion (Newton's second law) while angular momentum balance proves the symmetry of the stress tensor.

However, this picture is simply an approximation of the complex atomic, or for multi-component materials, microstructural interaction occurring within a single material point, which works well when these interactions occur at length scales infinitesimal, or approximately so, relative to the structural size of interest. Clearly these interactions cannot be ignored in all cases. In natural or contemporary synthetic materials, like bone, which contain substructures and several phases that are inseparable from a length-scale standpoint, such mesoscopic fluctuations impact the global behavior of the solid. In this picture, long-range interactions are engendered and force couples mobilize, which has been the physical basis [24] for introducing couple stresses into continuum formulations as power conjugate [25] to second displacement gradient kinematic quantities. Theories which include these quantities, couple stresses and strain gradients, are called high-grade continua [26]. Couple stresses made their first appearance in the early 1900s [5], but have been developed both mathematically (e.g., [27–31]) and applied to many materials since: most notably elasticity in the mid 1900s (e.g., [27, 28, 30, 32]), nonlinear elasticity [33, 34], and plasticity in the last two decades [31, 35–38].

In tandem with high-grade theories, those theories of high-order continua have been both developed and applied to the same purpose: approximating microstructural contributions to global energetics of a continuum structure. These theories, instead of using higher displacement gradients, introduce extra degrees of freedom to describe the deformability of each material point in the continuum. In this way, using the language of Eringen [39], material points are not points but themselves particles described by a continuum that contains the microstructure of the solid invisible at the macroscale. High-order theories derive from a first-order Taylor series expansion about material particle centroids so that an extra macroscopic field is introduced, something like a "micro-deformation," and the variational power is assumed to depend on microstrains and their first gradients [25].

The early work in generalized continua focused on one additional scale described by the high gradients or extra degrees of freedom, though recently the multiresolution theory has developed to include an arbitrary number of $N$ discrete scales [40]. Computational methods have been developed to calibrate scale-specific, potentially nonlinear and complex, constitutive laws [41]. The multiresolution formulation assumes nested subscales are separable (that is, the higher scales include all scales below), with distinct velocity gradients describing the deformability of each scale at which microstructure exists in the same manner as Mindlin [42] described the microdomain. These theories of generalized continua act both to improve predictive capability in situations where size effects exist (e.g., [36, 43–48]) and improve numerical properties of the governing equations by alleviating mesh sensitivity to local phenomena [41, 49, 50]. High-order theories are particularly useful when the scales of stress variation and mesostructure intersect, with application in high-frequency wave propagation [51], localization and failure (shear bands, fracture modes) in metals [40, 41, 52–56], granular materials [50], brittle composites [57], and filled/porous elastomers [47, 48]. Since the high-order theories are complex from an implementation standpoint, the interested reader is referred to Refs. [40, 41, 58] for implementation details of quasi-static high-order theories and dynamic high-order elasticity [51].

The situations where classical continuum mechanics fails to work well are size effects, post localization, and wave dispersion [51, 59], yet these are all phenomena observed in bone. Buechner and Lakes [44] have measured the size effects in viscoelasticity of bone and suggested Cosserat theory to model it. Benecke *et al.* [60] have showed the localization of inelastic strains through digital image correlation experiments on fibrolamellar bone specimens subjected to tension, and Thurner *et al.* [61] showed the presence of local buckling in trabecular struts of spongy bone specimens subjected to uniaxial compression. Wave dispersion has also been captured experimentally in bone by different studies [62–68]. These experimentally observed phenomena in bone justify the use of generalized continuum theories for their prediction, as these are distinct, well-documented cases where classical continuum mechanics breaks down. Section 19.2 covers the high-order theory used to describe bone tissue motion, namely the archetype blending continuum (ABC) theory, which we use due to the shortcomings of the traditional multiresolution philosophy.

### *19.1.3 Notes on the Archetype Blending Continuum Theory*

Before explaining the ABC theory directly, we first offer definitions to ensure consistency of language throughout this chapter. Heterogeneous materials, without question, are aggregates of individual components, which we collectively refer to hereafter as *multicomponent* materials. These components act as material building blocks, termed *archetypes* (see Epstein and Elżanowski [69] for other usage), which by different synthesis and processing techniques self-assemble to form a complex mesostructure or *conformation*. Archetypes and their conformations, therefore, exist in any multicomponent material. In general, archetypes contain their own substructures, meaning they are *submorphic*, and their conformation generates new mesoscale structures defining material *mesomorphism*. An analogy may be found in Lego building blocks: an archetype corresponds to an individual brick whose shape and connectors describe its submorphism. An arrangement of interlocking bricks would correspond to a conformation and its structural properties to mesomorphism. The mesoscopic intersection of archetypes is where modern mechanics research opportunities abound: the scale is too large for explicit descriptions, but too small for traditional continuum mechanics.

Take, for example, the smallest archetypes of bone shown in Figure 19.1: collagen, mineral, and non-collagenous organic material. The pronounced structural hierarchy that exists in bone tissue discussed in Section 19.1.1 arises from the subsequent assembly of tropocollagen macromolecules, impure hydroxyapatite minerals, and non-collagenous organic material to form the structural levels: microfibrils, fibril arrays, oriented lamellar structures similar to fiber-reinforced composites with a pseudo-random stacking structure, and porous spongy networks or Haversian systems (depending on the type of bone tissue) that underlie the macroscopic bone. These complex mesostructures are what give rise to a bone that is simultaneously light and tough [6]. We are careful to note here, however, that for any material system an archetype is an abstract concept. It is the building block of the material below which we explicitly model. Thus, it need not be the smallest space-filling unit of a material, like collagen proteins and mineral crystal unit cells. It can be any building block at a higher level: for instance, a lamella or osteon, depending on the application. We will in fact adopt oriented lamellae as archetypes in the sample analysis we conduct in this chapter.

The ABC theory transforms this abstract picture of materials – archetypes and their conformations – to a mathematically rigorous but tractable generalized continuum mechanics theory suitable for many applications in science and engineering. ABC does so by connecting generalized continuum mechanics, as dicussed in Section 19.1.2, with micromechanics [70], a field that essays to predict apparent properties of microstructured solids. This is a simple but vital concept.

Micromechanics methods make apparent property predictions in a bottom-up fashion through analysis and *blending* (i.e., homogenization) of its components and their interfaces or discontinuities. The reader is referred to Refs. [70–73] for more extended discussions of micromechanical methods. A commonality among micromechanics theories is that they condense all mesostructural information into a single constitutive law, which is variationally conjugate to a single mesoscale velocity field representing the blend. Micromechanics thus limits its predictive reach for multicomponent materials by assigning a single mesoscale kinetic energy for the homogenized mesostructure, thereby losing the ability to track energy distributions within the deforming blend. Simply put, there are too few degrees of freedom to represent a complex mesoscale for any detailed dynamic analysis. More general continuum theories are thus of interest.

Though generalized continuum theories mean to overcome micromechanics limitations by introducing additional degrees of freedom, a disconnect remains between the two fields. As mentioned in Section 19.1.2, all theories to date assume substructures are nested and orders of magnitude apart, a concept referred to as "scale separation." With scale separation, archetypes are not separated from others; rather, constitutive laws are specified for conformations alone, which are assumed to exist at length scales orders of magnitudes apart. This description is limiting for two reasons. First, specifying a constitutive law for a conformation is more difficult than specifying it for an archetype, which is implicitly assumed to be better understood. Second, modern materials, with bone as a pinnacle example, do not have conformations at scales orders of magnitude apart. One need look no further than individual lamellae (cf. Figure 19.1) that are 3–7 μm thick and comprise the osteons which are 10–500 μm wide in their transverse direction. Scale separation is not possible with these realities. In the past multiresolution methods, a philosophical inconsistency exists: an apparent macroscopic constitutive law must be posited for the same material whose global properties are yet to be predicted. Further, an inherent "double-counting" of energy damages the formulation: as any mechanism must be active in every scale enclosing it, constitutive laws at any scale must account for all their subscale mechanisms. Overall, multicomponent materials that exhibit extensive horizontal complexity like bone are inconsistent with these theories.

The ABC theory used here does not have those theoretical deficiencies. Instead, constitutive laws are posited in a modular fashion for (1) individual archetypes, (2) the manner in which they are blended, and (3) archetype interaction terms. ABC theory is itself a generalized theory of the high-order type and enhanced with high gradients; thus, in its full form it requires $\mathscr{C}^1$ continuous interpolation functions. In ABC, the limitations of micromechanics and previous generalized continua are overcome so that energies from individual archetypes and interactions may be better tracked during macroscopic deformation, while constitutive modeling becomes more modular and meaningful. Instead of assuming scale separation, we mathematically expand the true velocity field in the mesoscale by application of the fundamental theorem of calculus (FTC) and require the simple body to satisfy power equivalence to the mesostructure. These concepts are punctuated in Section 19.2 on ABC formulation. In the theory, extra degrees of freedom reflect the contribution of different regions or *partitions*, manifested by sampling points, within the homogenized mesostructure, rather than nested length scales in the generalized theories of old. The difference in ideas among ABC, classical micromechanics, and previous generalized continuum mechanics is summarized in Elkhodary *et al.* [74].

## 19.2 ABC Formulation

### 19.2.1 Physical Postulates and the Resulting Kinematics

Recognizing that phenomenologies governing macroscale deformation and fracture stem from the mesoscale, we propose that all multicomponent material systems can be modeled by developing them through three structural stages: (1) internal structures of mesoscale components (archetypes), (2) the mesostructural conformation of archetypes, and (3) a corresponding simple body or, as we will use interchangeably, the elemental or simulation domain.

In the following, superscript $M$ denotes fields defined on the simple body ($\mathscr{M}$; i.e., in the model or element domain), while superscript $C$ indicates fields defined on the mesostructure ($\mathscr{C}$). We can always map a point from the elemental domain $\xi \in \Omega^M$ to some expansion point $\boldsymbol{p}(\xi) \in \Omega^C$ in the mesostructural neighborhood. See Figure 19.2 for a depiction. We then search for some target point in the neighborhood, usually one beyond which nonlocal effects become negligible. To construct the velocity field, we sweep the neighborhood along an arbitrary smooth path ($\boldsymbol{\alpha}$-curve) between expansion

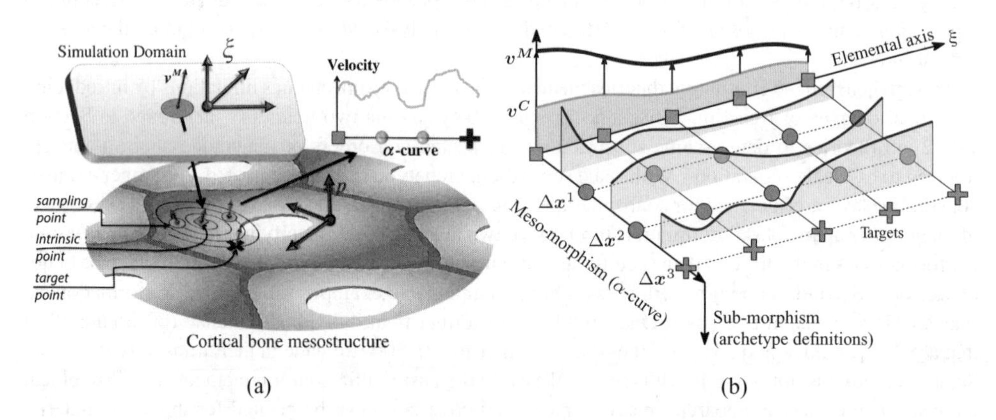


**Figure 19.2** Kinematic description of an ABC continuum point: (a) relating mesostructure to single elemental point; (b) continuous quantities across elemental domain

and target points, sampling all velocities on our way. We can thus describe the velocity $\boldsymbol{v}^M$ in the simple body at a point $\xi$ as a continuous and differentiable function in two sets of variables: $\boldsymbol{v}^M(\xi, l^*)$, where $\xi$ is the position of the point in the element domain and $l^*$ is the target point distance from the expansion point. Let us proceed to construct these velocity components, whose directions will be designated by the lowercase indices within index notation. Throughout the chapter, upper-case sub/superscripts are not indices but part of the variable name, and lower-case sub/superscripts are indices. Bold font indicates a vector or tensor.

Applying the FTC along the curve $\boldsymbol{\alpha}$ in the mesostructure we obtain the following velocity expansion:

$$\boldsymbol{v}^M(\xi, l^*) = \boldsymbol{v}(\boldsymbol{p}(\boldsymbol{\xi}), \boldsymbol{\alpha}(0)) + \int_{\alpha(0)}^{\alpha(l*)} \mathrm{d}\boldsymbol{v} = \hat{\boldsymbol{v}} + \int_{\alpha(0)}^{\alpha(l*)} \partial_\alpha \boldsymbol{v} \cdot \mathrm{d}\alpha. \tag{19.1}$$

To make the notation more compact, we define any quantity $\hat{\boldsymbol{q}} \equiv \boldsymbol{q}(\boldsymbol{p}(\boldsymbol{\xi}), \boldsymbol{\alpha}(0))$ and call it the *intrinsic* solution. We can next discretize the velocity field in Equation (19.1) by splitting the integral into as many subintervals as desired to collect multiple velocity samples from the mesostructural neighborhood. If we further assume that each relative velocity originates from a finite velocity "difference quotient" $\boldsymbol{\Lambda}^{*n}$, defined between any two finitely spaced sampling points on the curve $\boldsymbol{\alpha}$, we can rewrite the integral as

$$\boldsymbol{v}^M \equiv \hat{\boldsymbol{v}} + \boldsymbol{\Lambda}^{*1} \cdot \Delta\boldsymbol{x}^1 + \boldsymbol{\Lambda}^{*2} \cdot \Delta\boldsymbol{x}^2 + \cdots + \boldsymbol{\Lambda}^{*N} \cdot \Delta\boldsymbol{x}^N, \tag{19.2}$$

where $\Delta\boldsymbol{x}^n$ is the vector connecting sampling point $n$ to $n-1$. $\Delta\boldsymbol{x}^n$ may be thought of as a length vector containing "nonlocal" information (due to its finiteness) about the complex mesostructure. $\boldsymbol{\Lambda}^*$ will be supposed to vary smoothly over the element domain, as indicated in Figure 19.2b. For a more detailed discussion, see Elkhodary *et al.* [74]. The velocity gradient in the simple body is defined as $L_{ij}^M(\boldsymbol{\xi}, l^*) \equiv \partial_j(v_i^M)$, which then admits a multirate decomposition of the form

$$\boldsymbol{L}^M(\boldsymbol{\xi}, l^*) = \hat{\boldsymbol{L}} + (\boldsymbol{\Lambda}^{*1}\nabla)\overset{2}{\cdot}\Delta\boldsymbol{x}^1 + \cdots + (\boldsymbol{\Lambda}^{*N}\nabla)\overset{2}{\cdot}\Delta\boldsymbol{x}^N, \quad \text{or} \tag{19.3a}$$

$$\boldsymbol{L}^M(\boldsymbol{\xi}, l^*) \equiv \hat{\boldsymbol{L}} + \boldsymbol{\Lambda}^1 + \cdots + \boldsymbol{\Lambda}^N, \tag{19.3b}$$

where operator $(\overset{2}{\cdot})$ indicates contraction on the second index of a tensor. With these quantities defined, the velocity field at each material point is defined in terms of the degrees of freedom $\{\hat{\boldsymbol{v}}, \boldsymbol{\Lambda}^{*n}\}$.

## *19.2.2 ABC Variational Formulation*

The equations of motion of heterogeneous materials can be derived from the principle of virtual power,

$$\delta\dot{E}_{\text{kin}} = \delta\dot{E}_{\text{imp}} - \delta\dot{E}_{\text{def}}, \tag{19.4}$$

where $\delta\dot{E}_{\text{kin}}$ is the kinetic power of the motion, $\delta\dot{E}_{\text{imp}}$ is the impressed power due to external and body forces, and $\delta\dot{E}_{\text{def}}$ is the deformational power that generates forces internal to the simple body.

The ABC theory aims to satisfy this variational principle, Equation (19.4), by varying all degrees of freedom $\{\hat{\boldsymbol{v}}, \boldsymbol{\Lambda}^{*n}\}$ independently, cf. Figure 19.3b, and stipulating a *power equivalence* between the simple body (continuum) and mesostructure. As will be seen in Section 19.3 on constitutive modeling, these degrees of freedom will correspond to partitions of a material point, with each partition represented by a sampling point. ABC theory thus acts in an expanded configuration space to compute forces conjugate to the expanded velocities $\{\hat{\boldsymbol{v}}, \boldsymbol{\Lambda}^{*n}\}$ at multiple sampling points within each continuum material point to represent the mesostructural deformations more accurately. Once force balance is achieved in an ABC simulation, the assembly of expanded forces will be power equivalent to a resultant force vector $\boldsymbol{f}^M$

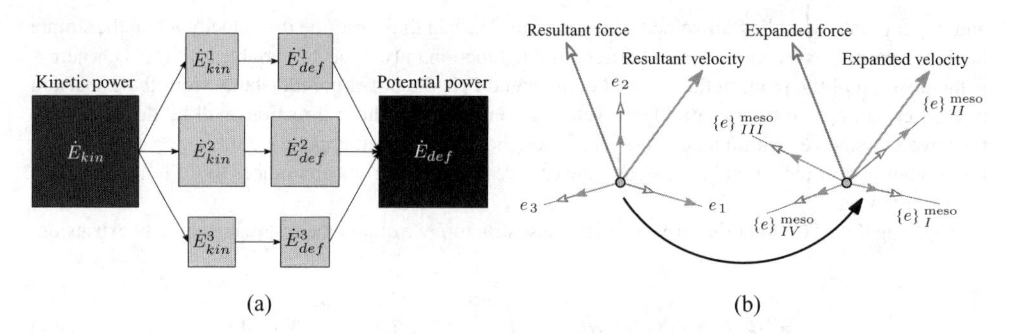


(a) (b)

**Figure 19.3** Using an expanded space for an enhanced variational statement: (a) breaking down kinetic power to solve for components of potential power; (b) forces and velocities in original and expanded spaces

conjugate to $\boldsymbol{v}^M$ (target point velocity) in ordinary configuration space. For more details, see Elkhodary *et al.* [74].

## Kinetic Power

Following the procedure in Germain [25] and Vernerey *et al.* [40], the total kinetic energy variation is

$$\delta E_{\mathrm{kin}} = \int_{\Omega} \rho^M \boldsymbol{v}^M \cdot \delta \boldsymbol{v}^M \,\mathrm{d}V = \int_{\Omega} \rho^M [\hat{\boldsymbol{v}} + \boldsymbol{\Lambda}^{*n} \cdot \Delta \boldsymbol{x}^n] \cdot \delta[\hat{\boldsymbol{v}} + \boldsymbol{\Lambda}^{*n} \cdot \Delta \boldsymbol{x}^n] \,\mathrm{d}V, \tag{19.5}$$

where $\rho^M$ is the material point density, and the latter equality came from plugging in Equation (19.2). The summation convention on two indices (e.g., $n = 1, 2, \ldots, N$) is also assumed in this chapter (unless they are enclosed in parentheses). Taking the material time derivative of Equation (19.5), the variation of kinetic power is written as

$$\delta \dot{E}_{\mathrm{kin}} = \int_{\Omega} [\dot{\boldsymbol{\gamma}} \cdot \delta \hat{\boldsymbol{v}} + \dot{\boldsymbol{\Gamma}}^n : \delta \boldsymbol{\Lambda}^{*n}] \,\mathrm{d}V, \tag{19.6}$$

where the interia terms $\dot{\boldsymbol{\gamma}}$ and $\dot{\boldsymbol{\Gamma}}$ are given as [74]

$$\dot{\boldsymbol{\gamma}} = \rho^M (\dot{\hat{\boldsymbol{v}}} + \dot{\boldsymbol{\Lambda}}^{*n} \cdot \Delta \boldsymbol{x}^n) \quad \text{and} \quad \dot{\boldsymbol{\Gamma}}^n = \dot{\boldsymbol{\gamma}} \otimes \Delta \boldsymbol{x}^n. \tag{19.7}$$

In this formulation, partitions of the material point are coupled kinetically, meaning off-diagonal terms will exist in the discrete approximation of the mass matrix.

## Impressed Power

The impressed power is divided into a surface and body term such that $\delta \dot{E}_{\mathrm{imp}} = \delta \dot{E}^{\Gamma}_{\mathrm{imp}} + \delta \dot{E}^{\Omega}_{\mathrm{imp}}$, which from the kinematic description given in Section 19.2.1 admit the following forms:

$$\delta \dot{E}^{\Gamma}_{\mathrm{imp}} = \int_{\Gamma_t} [\boldsymbol{t} \cdot \delta \hat{\boldsymbol{v}} + \boldsymbol{T}^n : \delta \boldsymbol{\Lambda}^{*n}] \,\mathrm{d}S, \tag{19.8a}$$

$$\delta \dot{E}^{\Omega}_{\mathrm{imp}} = \int_{\Omega} [\boldsymbol{b} \cdot \delta \hat{\boldsymbol{v}} + \boldsymbol{B}^n : \delta \boldsymbol{\Lambda}^{*n}] \,\mathrm{d}V, \tag{19.8b}$$

where $\boldsymbol{t}$, $\boldsymbol{T}^n$ are specific surface tractions and doube tractions respectively, and $\boldsymbol{b}$, $\boldsymbol{B}^n$ are specific body forces and double forces.

**Deformational Power**

The deformational power (internal power) in the elemental domain is a sum of the powers due to strains in Equation (19.3a) and their first gradients (when curvature effects are of interest). Taking the first variation produces

$$\delta \dot{E}_{\mathrm{def}} = \int_{\Omega} (\boldsymbol{\sigma} : \delta\hat{\boldsymbol{L}} + \boldsymbol{s}^n : \delta\boldsymbol{\Lambda}^n + \boldsymbol{\sigma\sigma} \,\vdots\, \delta\hat{\boldsymbol{L}}\nabla + \boldsymbol{ss}^n \,\vdots\, \delta\boldsymbol{\Lambda}^n\nabla)\,\mathrm{d}V, \tag{19.9}$$

Note that second-order tensors $\boldsymbol{\sigma}$ and $\boldsymbol{s}$ and third-order tensors $\boldsymbol{\sigma\sigma}$ and $\boldsymbol{ss}$ take on the meaning of stresses and double stresses conjugate to the velocity gradients and second gradients respectively.

## 19.3 Constitutive Modeling in ABC

### *19.3.1 General Concept*

In this section we present the ABC constitutive modeling strategy. The basic idea is to partition a continuum material point into a number of partitions, Figure 19.4, in a manner that resembles “domain decomposition.” Depending on the mesostructure that a material point represents, different partitioning strategies may be adopted. For instance, in Figure 19.4a an alloy’s mesostructure may be represented by a material point decomposed into the matrix, inclusion (or any heterogeneity), and an interphase. Alternatively, for large lamellar mesostructures, for instance, it may be preferable to decompose the material point into general partitions, according to orientation of lamellae, as in Figure 19.4b, which is the approach we will take in this chapter. The objective is always to better capture the expected variation of a velocity field across the mesostructural span that the material point represents, Figure 19.4c, so that associated kinetic energies may capture strain inhomogeneities from the relative motions (velocities) within a material point of the ABC formulation. Note that from Figure 19.4 the sampling points may represent unequal spans of mesostructure, so that corresponding weighting of their contribution to effective material properties would be needed, as noted by Elkhodary *et al.* [74].

Generally, the ABC formulation requires three elements to construct a constitutive law at a material point: (a) component-level stress–strain laws for the archetypes, and for their interactions in each partition; (b) micromechanics blending laws to generate effective stresses and double stresses for each partition, using mesostructural descriptors that parameterize the blend laws (e.g., volume fractions), as explained

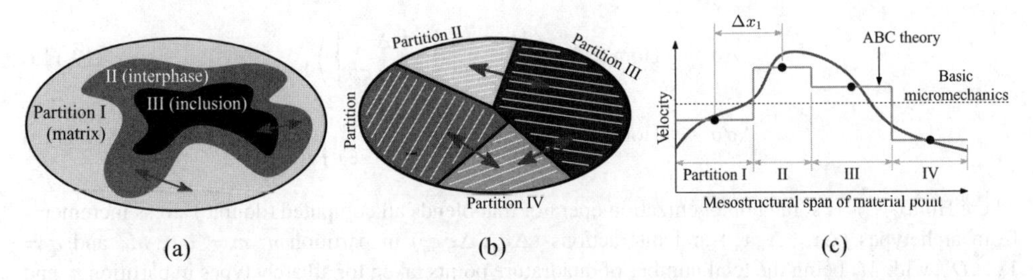


**Figure 19.4** Decomposing a material point into partitions to better capture velocity variation: (a) three concentric partitions; (b) four general partitions; (c) mesostructural velocity field

by Elkhodary *et al.* [74]; and (c) coupling laws across sampling points (represented by the arrows in Figure 19.4).

Coupling between sampling points is needed, just as with any "domain decomposition" approach, since partitions behave collectively as a unit (i.e., a material point). For example, enforcing certain continuity conditions across partition boundaries may represent possible coupling laws. In this chapter we will assume only simple unidirectional coupling from sampling point $n-1$ to $n$ for illustration. We thus define the stresses $\boldsymbol{s}$ and double stress $\boldsymbol{ss}$ at a sampling point $n$, conjugate to the relative velocity gradients and second gradients in Equation (19.9), respectively, via the first-order update equations

$$(\boldsymbol{s}^n)_{t+\Delta t} = (\boldsymbol{s}^n)_t + \alpha \mathbb{C}^{n-\frac{1}{2}} \cdot (\boldsymbol{\Lambda}^{\mathbf{e}}_{\mathbf{sym}})^n \Delta t + (1-\alpha)(\Delta\boldsymbol{\sigma}^n - \Delta\boldsymbol{\sigma}^{n-1}), \tag{19.10a}$$

$$(\boldsymbol{ss}^n)_{t+\Delta t} = (\boldsymbol{ss}^n)_t + \alpha \mathbb{CC}^{n-\frac{1}{2}} \cdot (\boldsymbol{\Lambda}^{\mathbf{e}}_{\mathbf{sym}}\nabla)^n \Delta t + (1-\alpha)(\Delta\boldsymbol{\sigma\sigma}^n - \Delta\boldsymbol{\sigma\sigma}^{n-1}), \tag{19.10b}$$

where

$$\begin{aligned} \mathbb{C}^{n-\frac{1}{2}} &\equiv \mathfrak{f}(\Delta\boldsymbol{\sigma}^n, \Delta\boldsymbol{\sigma}^{n-1}, \ldots) \\ &= \left[\frac{\Delta\boldsymbol{\sigma}^n + \Delta\boldsymbol{\sigma}^{n-1}}{2\Delta t}\right]\left[\frac{(\boldsymbol{D}^{\mathrm{e}})^n + (\boldsymbol{D}^{\mathrm{e}})^{n-1}}{2}\right]^{-1}, \end{aligned} \tag{19.10c}$$

$$\begin{aligned} \mathbb{CC}^{n-\frac{1}{2}} &\equiv \mathfrak{f}(\Delta\boldsymbol{\sigma\sigma}^n, \Delta\boldsymbol{\sigma\sigma}^{n-1}, \ldots) \\ &= \left[\frac{\Delta\boldsymbol{\sigma\sigma}^n + \Delta\boldsymbol{\sigma\sigma}^{n-1}}{2\Delta t}\right]_* \left[\frac{(\boldsymbol{D}^{\mathrm{e}}\nabla)^n + (\boldsymbol{D}^{\mathrm{e}}\nabla)^{n-1}}{2}\right]_*^{-1}. \end{aligned} \tag{19.10d}$$

Tensors $\mathbb{C}^{n-\frac{1}{2}}$ and $\mathbb{CC}^{n-\frac{1}{2}}$ designate an apparent resistance, or "pseudo-stiffness," of the material lying in between (and coupling) sampling points $n$ and $n-1$. $\mathbb{C}^{n-\frac{1}{2}}$ may be visualized as a spring which resists first-order deformations (e.g., shear and tension), connecting two sampling points. $\mathbb{CC}^{n-\frac{1}{2}}$ may be visualized as a spring resisting curvature (i.e., bending). The stiffnesses are computed as an average from sampling point (double) stress and strain (gradient) values. These functional forms in Equation (19.10) could be revised in future research to better account for deformation constraints and/or conditions of interest across sampling points. Nonetheless, this current form premits in addition to the "average" term a "jump" term in the stress increment, which possibly arises due to relative motion between neighboring partitions. For simplicity, both terms are combined by a single weight factor $\alpha$, and we assume $\alpha = 0.5$. The asterisk subscript on brackets, $[\,]_*$, indicates the third-order tensor needs to be imbedded as a block diagonal in a square matrix to obtain the required inverses.

We may write the stress increments in Equation (19.10) for a sampling point representing the $n$th partition of mesostructure as [74]

$$\Delta\boldsymbol{\sigma}^n = \mathrm{Hom}_{M_n+Q_n}\left(\begin{Bmatrix} \varpi \circ \psi \\ \omega \circ \Psi \end{Bmatrix}, \begin{Bmatrix} \Delta\tau_m \\ \Delta\varsigma_q \end{Bmatrix}\right), \tag{19.11a}$$

$$\Delta\boldsymbol{\sigma\sigma}^n = \mathrm{Hom}_{M_n+Q_n}\left(\begin{Bmatrix} \varpi \circ \psi \\ \omega \circ \Psi \end{Bmatrix}, \begin{Bmatrix} \Delta\tau\tau_m \\ \Delta\varsigma\varsigma_q \end{Bmatrix}\right), \tag{19.11b}$$

where $\mathrm{Hom}_{M_n+Q_n}$ is some homogenization operator that blends all computed (double) stress increments from archetypes $(\Delta\tau_m, \Delta\tau\tau_m)$ and interactions $(\Delta\varsigma_q, \Delta\varsigma\varsigma_q)$ in partition $n$. $m = 1\ldots M_n$ and $q = 1\ldots Q_n$, with $M_n$ being the total number of quadrature points taken for all archetypes in partition $n$, and $Q_n$ being the total number of quadrature points taken for the interactions in the same partition. The set $\{\psi, \Psi\}$ defines the mesostructural properties of the $n$th partition, based off experimental images, statistical descriptions of the mesostructure, or literature. $\{\psi, \Psi\}$ specifies how the archetype and interaction mechanisms are currently distributed over the partition to be blended (e.g., volume fractions, in the simplest case). The set $\{\varpi, \omega\}$ defines weighting factors for the archetypes and interactions with respect to a studied macroscopic response of the heterogeneous material.

In this chapter we propose special forms for stress blending laws in cortical bone. These are: (a) for the compartments, an elastic, lamellar composite law, with radial symmetry centered about the Haversian canals; (b) for the cement lines, an isotropic viscoelastic law. For the double stress, we (a) again propose a lamellar composite law for the compartments and (b) select an isotropic elastic law for the cement lines. With these models developed, we investigate the role of mesostructure on dynamic deformation and strain inhomogeneities in cortical bone within the ABC framework.

## *19.3.2 Blending Laws for Cortical Bone Modeling*

### Elasticity of the Osteons

An elastic lamellar model is proposed for the osteons. In the principal frame, stiffness is written as

$$C_\mathrm{p} = \begin{bmatrix} \begin{bmatrix} 1/E_{xx} & -\nu_{yx}/E_{yy} & -\nu_{xz}/E_{xx} \\ -\nu_{xy}/E_{xx} & 1/E_{yy} & -\nu_{zy}/E_{zz} \\ -\nu_{zx}/E_{zz} & -\nu_{yz}/E_{yy} & 1/E_{zz} \end{bmatrix}^{-1} & & 0 & \\ & 2G_{xy} & & \\ 0 & & 2G_{yz} & \\ & & & 2G_{zx} \end{bmatrix}, \tag{19.12}$$

so that the rotated stiffness may be given as [73]

$$C_\mathrm{r} = Q^{-1} C_\mathrm{p} Q, \tag{19.13}$$

where the rotation matrix $Q$ is expressed as

$$Q = \begin{bmatrix} m^2 & n^2 & & & & 2mn \\ n^2 & m^2 & & & & -2mn \\ & & 1 & & & \\ & & & m & -n & \\ & & & n & m & \\ -mn & mn & & & & m^2 - n^2 \end{bmatrix}, \tag{19.14}$$

with $m = \cos(\theta)$, $n = \sin(\theta)$, and $\theta$ being the angle formed between the lamellae and the horizontal. It is assumed that lamellae are aligned normal to the radial position vector as shown in Figure 19.5a, measured from the center of the Haversian canal for each osteon. This radial position will not be updated during the simulation for simplicity, in spite of the evolving deformation. Elastic moduli are shown in Figure 19.6, and compared with an isotropic model. Radial distance from the Haversian canal center is also assumed to influence the stiffness of lamellae, such that material points nearer the Haversian canal exhibit a lower stiffness, consistent with experimental measurements of hardness (e.g., see Ref. [75]). The proposed relation is (no sum on $i$)

$$E_{ii} = E_{ii}\left(1 - f * \frac{R_\mathrm{max} - \|r\|}{R_\mathrm{max} - R_\mathrm{min}}\right), \quad i \in \{x, y, z\}, \tag{19.15}$$

where $f \equiv 0.4$, consistent with experimental trends [75], $R_\mathrm{max}$ is the maximum radius of the osteon, $R_\mathrm{min}$ is the radius of the Haversian canal, and $r$, with $R_\mathrm{min} \le \|r\| \le R_\mathrm{max}$, is the radial position of the

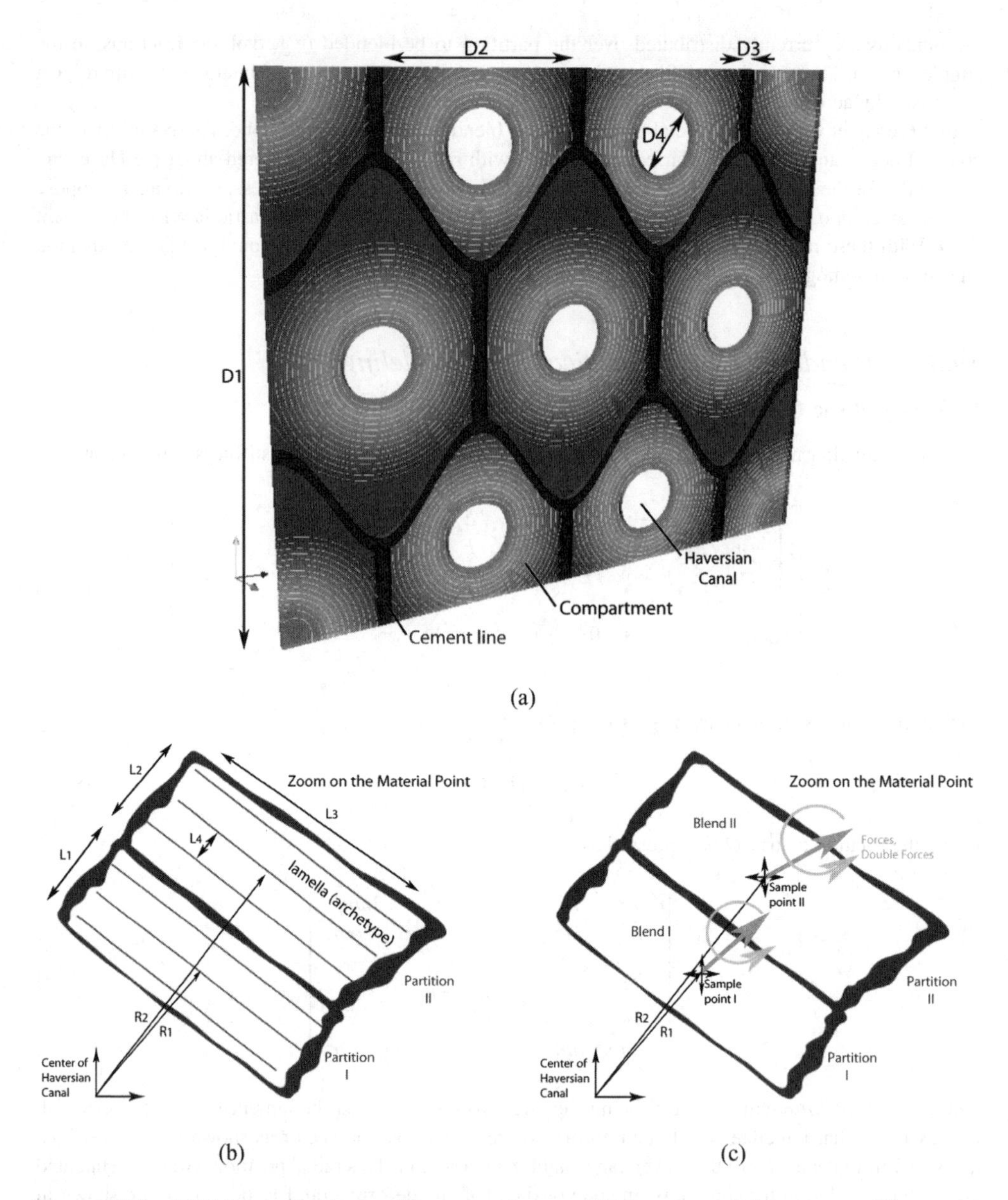


**Figure 19.5** ABC for cortical bones: (a) compact bone exhibits osteon radial symmetry; (b) material point definition; (c) forces and double forces at a material point

material point measured from the center of the Haversian canal. The resulting variation in stiffness can be seen from the contours in Figure 19.5a.

For gradients in strain we define corresponding stiffness within a partition as

$$CC_{\tau} = CC_{\tau}^{*} + \Delta r \otimes \left(Q^{-1} C_{\mathrm{p}} Q\right) \otimes \Delta r, \tag{19.16}$$

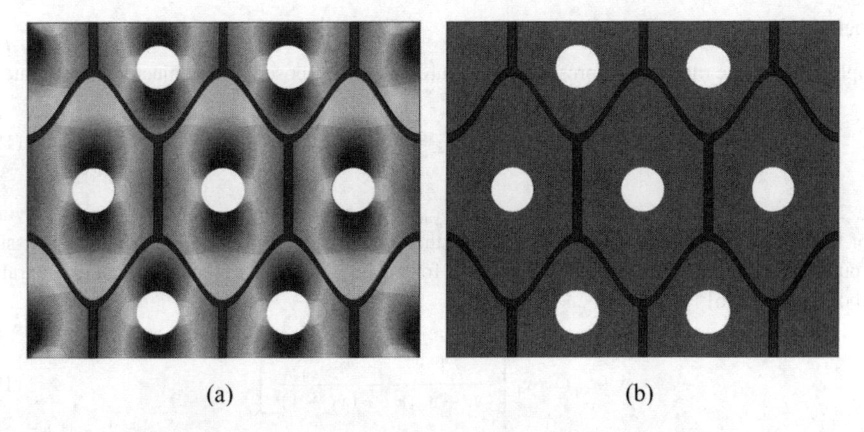


**Figure 19.6** Cortical bone moduli of elasticity. (a) Case I: lamellar. Filled contours are $E_{yy}$; line contours are $E_{xx}$ (max: 30 GPa; min: 19 GPa). (b) Case II: isotropic, spatially homogeneous. $E_{xx} = E_{yy} = 24.5$ GPa

where $\Delta r = L_1\hat{r}$ for the first partition and $\Delta r = L_2\hat{r}$ for the second partition (cf. Figure 19.5a). $CC_{\tau}^{*}$ represents a basic resistance of lamellae to bending, which we here assume is negligible; that is, $CC_{\tau}^{*} = 0$. This is reasonable, as radii of curvature are much larger than lamellar thickness. Thus, double forces will arise due to the finite span of partitions, as shown schematically in Figure 19.7. As each partition stacks more lamellae, it resists bending more. Hence, $\Delta r$ in Equation (19.16) scales with the span of a partition, cf. Figure 19.5b.

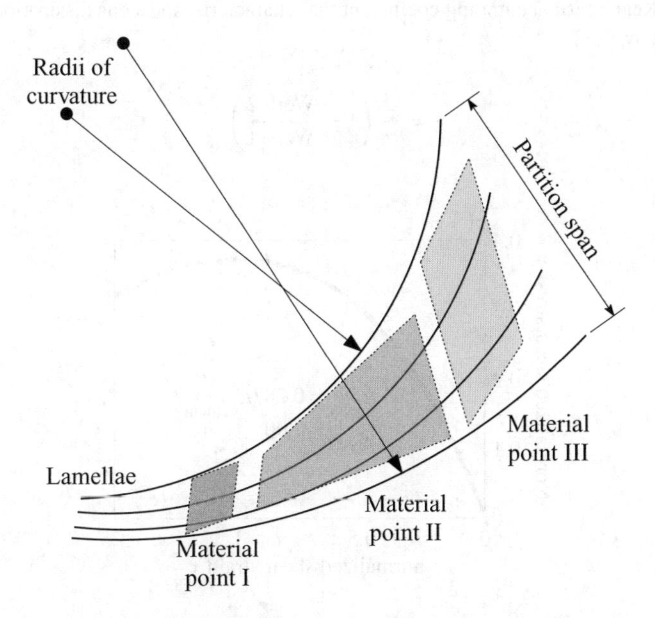


**Figure 19.7** Strain gradients in lamellae

## Viscoelasticity in Cement Lines

A simple viscoelastic model, as represented in Figure 19.8, is proposed. The (objective) stress rate is

$$\sigma_{ij}^{\nabla J} = (2G_{\text{cem}})^* D_{ij}^{\text{dev}} + \kappa_{\text{cem}} D_{kk} = (y' \cdot 2G_{\text{cem}}) D_{ij}^{\text{dev}} + \kappa_{\text{cem}} D_{kk}, \tag{19.17}$$

where $2G_{\text{cem}}$ is the second Lamé parameter and $\kappa_{\text{cem}}$ is the bulk modulus. Thus, $D_{ij}^{\text{dev}}$ is the deviatoric part of the rate of deformation tensor and $D_{kk}$ is the volumetric part. Viscous dissipation is assumed to be purely distortional, so viscosity is accounted for via coefficient $y'$ obtained from a two-parameter hyperbolic model [76]:

$$y'(x_{\text{eq}}) = y_1^2 \left[ \frac{1}{(y_1 + x_{\text{eq}})^2} - \frac{x_{\text{eq}}^m}{(y_1 + 1)^2} \right], \tag{19.18}$$

where

$$x_{\text{eq}} = \frac{\varepsilon_{\text{eq}}}{\gamma_{\max}}, \quad \text{with} \quad \varepsilon_{\text{eq}} = \sum_{\text{inc}} \sqrt{\frac{2}{3} D^{\text{dev}} : D^{\text{dev}}} \Delta t, \tag{19.19a}$$

$$y_1 = \frac{y_{\max}}{1 - 2y_{\max}}, \quad \text{with} \quad y_{\max} = \frac{\tau_{\max}}{G_{\text{cem}} \gamma_{\max}}, \quad \text{and} \tag{19.19b}$$

$$m = \frac{y_{\max}}{1 - y_{\max}}, \; y_{\max} < 0.5 \quad \text{or} \quad 1.1 \left( \frac{y_{\max}}{1 - y_{\max}} \right), \; y_{\max} > 0.5. \tag{19.19c}$$

The summation in Equation (19.19a) spans the time increments ($\Delta t$) in a simulation. In the model, parameters ($\tau_{\max}, \gamma_{\max}$) represent maximum shear stress and maximum shear strain, respectively. Following Prevost and Keane [76], a damping coefficient that characterizes viscous dissipation is phenomenologically defined by

$$\xi = \frac{2}{\pi} \left( 1 - \frac{\text{Work 2}}{\text{Work 1}} \right), \tag{19.20}$$

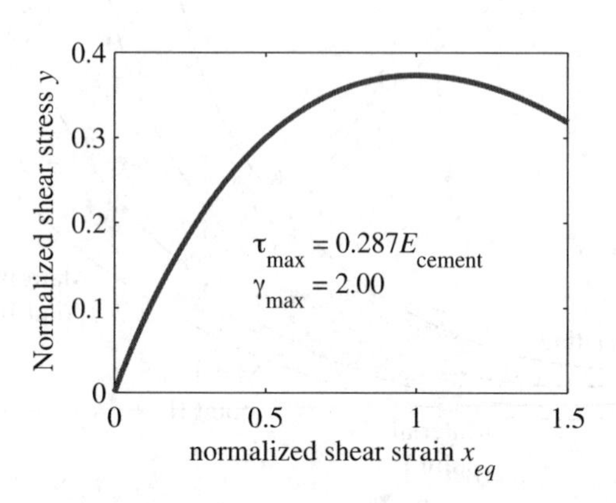


**Figure 19.8** Viscoelastic material law

where

$$\text{Work } 2 = \frac{1}{2} y(x_{eq}) \cdot x_{eq}, \quad \text{Work } 1 = \sum_{inc} \left( y(x_{eq}) \cdot \dot{x}_{eq} \right) \Delta t, \quad \text{for } x_{eq} \leq 1. \tag{19.21}$$

As such, damping is fundamentally seen as a deviation from linearity in cement line stress–strain response.

For the gradients in strain we define a corresponding stiffness of a partition as

$$CC_{cem} = CC^*_{cem} + \Delta a \otimes (C_{cem}) \otimes \Delta a, \tag{19.22}$$

where $C_{cem}$ is the stiffness assuming isotropic elasticity, $\Delta a = L_1^{cem} \hat{1}$ for the first partition, $\Delta a = L_2^{cem} \hat{1}$ for the second partition, and $\hat{1} = \frac{1}{\sqrt{3}}[111]$. Again we assume $CC^*_{cem} = 0$ for the viscous cement line. In this model, all curvature effects are assumed elastic and arise due to the finiteness of the material point partitions.

## 19.4 The ABC Computational Model

The model for cortical bone mesostructure used in this chapter is shown in Figure 19.5a, with material points as defined in Figure 19.5b. The basic dimensions are $D_1 = 75$ μm, $D_2 = 28$ μm, $D_3 = 2$ μm, and $D_4 = 10$ μm. The thickness is taken as $t = 1$ μm. For the material point partitions, the length scales are here assumed as $L_1 = L_2 = L_3 = 3L_4 = 0.3$ μm. In this chapter, only a comparative analysis between lamellar and isotropic models is pursued, so that these dimensional assumptions are not expected to be of direct consequence. The nondimensionalization scheme in Table 19.1 was used, along with the material properties listed. The elastic modulus of the cement line was taken as $E_{cem} = 0.3E_{osteon}$, which is consistent with experimental trends suggesting their lower stiffness (cf. [75]). The viscous properties of cement lines are characterized by the two parameters shown in Figure 19.8. Sampling points will be spaced as $\frac{1}{2}(L_1 + L_2) = 0.3$ nm apart in the radial direction for the lamellar case, and in the [111] direction for the isotropic case.

The mesh used contains approximately 20 600 brick elements. The element formulation is eight-node, iso-parametric and trilinear. A 2 × 2 × 2 Gauss quadrature scheme was used. The element was implemented in LSDYNA as a user element Fortran subroutine (`dyn21b.f`) with 12 degrees of freedom per node: three degrees of freedom for the velocities ($\hat{v}$) of sampling point I and nine for the velocity difference quotient ($\Lambda^*$) at sampling point II. LSDYNA default hourglass control and artificial damping were used to stabilize the dynamic simulation. Also, the large value for the density in Table 19.1 was taken to speed up the dynamic calculations. A nominal strain rate of 5000 $s^{-1}$ was applied by pulling on the top and bottom surfaces. For a plane-strain approximation, all velocity components with a $z$-direction were set to zero ($v_3 = \Lambda^*_{31} = \Lambda^*_{32} = \Lambda^*_{13} = \Lambda^*_{23} = \Lambda^*_{33} = 0$) across the model. Moreover, to excite strain inhomogeneities more visibly across the mesostructure, $\Lambda^*_{22} = 0.1v_2$ was applied along the

**Table 19.1** The nondimensionalization scheme

| Dimensional quantity | Nondimensional quantity | Normalizing quantities |
|---|---|---|
| Length $x$ | $x' = x/l_{ref}$ | $l_{ref} = 300$ nm |
| Time $t$ | $t' = tc_{ref}/l_{ref}$, | $c_{ref} = 4900$ nm/ns |
| Mass density $\rho$ | $\rho' = \rho c^2_{ref}/E_{osteon}$ | $E_{osteon} = 24.5$ GPa |
| Stress $\sigma$ | $\sigma' = \sigma/E_{osteon}$ | $\rho_{ref} = 30 \times 10^3$ kg/m$^3$ |

top and bottom faces (Figure 19.5a), while setting all other $\Lambda^*_{ij} = 0$ on these boundaries. Notice that this loading condition is essentially antisymmetric.

## 19.5 Results and Discussion

In this section we discuss a comparative analysis of cortical bone with lamellar osteons on the one hand versus isotropic osteons on the other hand.

### *19.5.1 Propagating Strain Inhomogeneities across Osteons*

Figure 19.9 predicts relative shear $\Lambda^*_{21}$ evolving between partitions of material points. Relative shear expresses a strain inhomogeneity propagating across the mesostructure. The alternating pattern from positive (red or R) to negative (blue or B) shear predicted in the figure appears to be the result of the global loading condition which enforces net vertical motion of the mesostructure. An immediate comparison of the wave front between case I, accounting for the lamellar structure of osteons, and case II, using an isotropic model, can be made. At time point $t' = 90$, the lamellar structure bends the propagating inhomogeneity wave away from the Haversian canal where the stiffness is lower (cf. Equation (19.15)), whereas in the isotropic case the front remains planar. At time point $t' = 250$, four important comparisons between the lamellar and isotropic case can be made from the middle column of Figure 19.9. First, there appears to be a phase shift between the two cases. In the isotropic case the pattern predicted for $\Lambda^*_{21}$ from the bottom surface is R–B–R–B, whereas for the lamellar case it is R–B–R. The reverse pattern holds moving from the top surface. This result suggests less frequent spatial alterations in

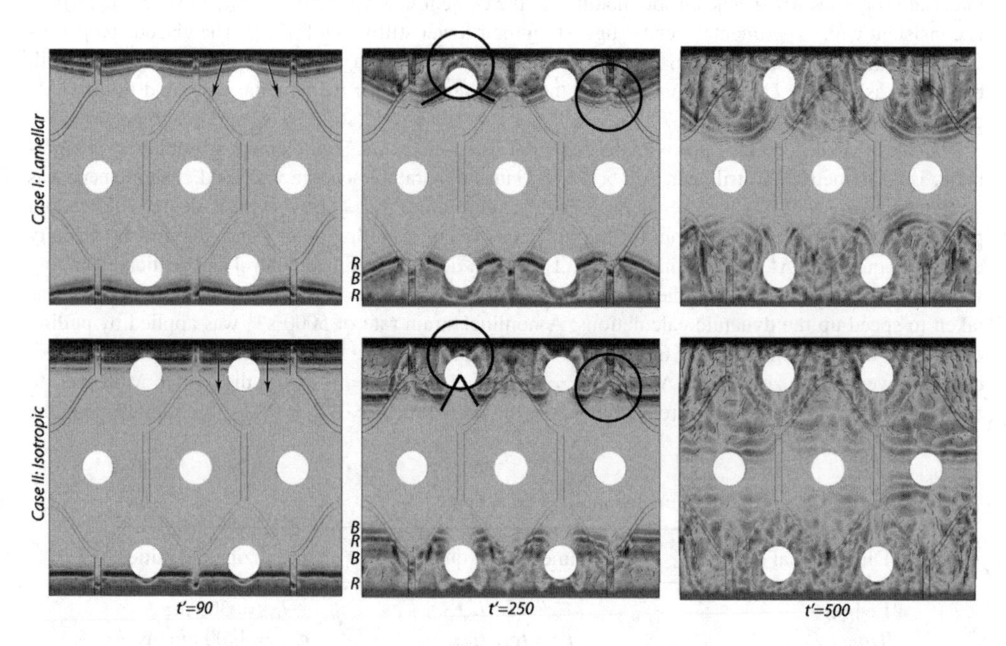


**Figure 19.9** The $\Lambda^*_{21}$ strain inhomogeneity evolution. Color contours vary linearly from $-5 \times 10^{-5} c_{\mathrm{ref}} L^{-1}_{\mathrm{ref}}$ to $5 \times 10^{-5} c_{\mathrm{ref}} L^{-1}_{\mathrm{ref}}$. Max $= 1.79 \times 10^{-4} c_{\mathrm{ref}} L^{-1}_{\mathrm{ref}}$ (lamellar), Max $= 1.9 \times 10^{-4} c_{\mathrm{ref}} L^{-1}_{\mathrm{ref}}$ (isotropic)

evolving strain inhomogeneity patterns across the lamellar structure, so that larger swaths of the material deform in the same way under the given dynamic load. Second, the disruption of a wave traversing a Haversian canal is lessened by the lamellar structure, which can be seen from the broken lines superposed on the figure. For the lamellar case the angle formed by the line is nearer 180° than the isotropic case. It seems, therefore, that the concentric wrapping of the lamellae acts to better distribute the load and deformation away from canals, such that they interfere less with stress flow across the mesostructure. Third, inhomogeneity waves reflected off the canals are predicted to form patterns with a lower spatial frequency (at the given time) in the lamellar structure. This result is seen by looking in the top-left circles superposed on the figure. For the lamellar case the wave forms a single large loop behind the canal, whereas in the isotropic case two large loops with a small loop in between are already formed. Note, however, that as reflected waves continue to interact they form patterns of yet high spatial frequency, so the given mesh does not resolve them perfectly, which leads to some noise in the predictions, as may be seen from the figure. The fourth prediction is that deformation is shared among osteons more readily with a lamellar structure. This result is seen from the circles on the right in the figure. The inhomogeneity wave leads in the osteon in the middle for the lamellar case, whereas it trails its neighbors for the isotropic case. As waves exhibit a faster propagation in the outer lamellae of osteons, due to their higher stiffness, it is thus predicted that, before a single osteon is fully loaded to its interior, neighboring osteons will participate in the deformation for the lamellar case. Finally, this prediction is strengthened by looking at time point $t' = 500$, where it is clear that shear strain inhomogeneities propagate more slowly toward osteon cores for the lamellar case. Thus, under dynamic load, the lamellar structure of osteons gives cortical bone more time to respond to disturbances, as well as better deformation-sharing properties among osteons.

### *19.5.2 Normal and Shear Stresses in Osteons*

Figure 19.10 shows the contours for normal stress component $\sigma_{22}^{n=2}$. The inhomogeneities evolving in Figure 19.9 are too small in magnitude to induce visible differences between the stress contours in the first and second partitions; that is, $\sigma_{22}^{n=2} \simeq \sigma_{22}^{n=1}$. The clearest differences between the lamellar and isotropic cases are three. First, the lamellar case shows the front of stress curving to avoid the Haversian canal, whereas for the isotropic case it is predicted that the Haversian canal acts as a stress concentration site.

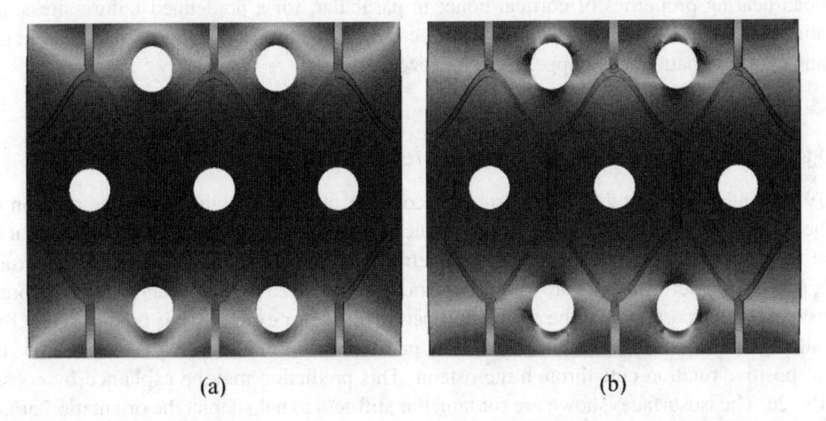

**Figure 19.10** The $\sigma_{22}^{n=2}$ normal stress field. Contours vary linearly from 0.0 to $0.14E_{\text{osteon}}$ on both sub-figures. Max $= 0.1776E_{\text{osteon}}$ (lamellar), Max $= 0.1941E_{\text{osteon}}$ (isotropic). Time point $t' = 820$. (a) Case I: lamellar. (b) Case II: isotropic

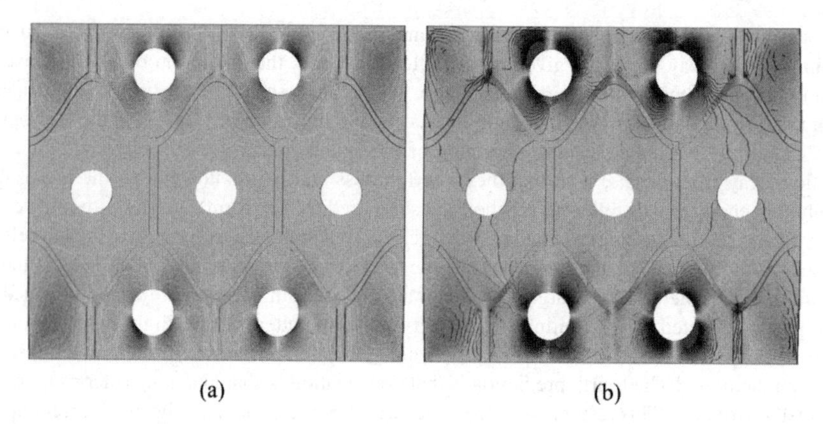

Figure 19.11 The $\sigma_{12}^{n=2}$ shear stress field. Contours vary linearly from $-0.03E_{\text{osteon}}$ to $0.03E_{\text{osteon}}$ on both sub-figures. Max = $0.0342E_{\text{osteon}}$ (lamellar), Max = $0.0835E_{\text{osteon}}$ (isotropic). Time point $t' = 820$. (a) Case I: lamellar. (b) Case II: isotropic

This result is consistent with the physical property of osteons that their natural load-bearing direction is along their lamellae, which wrap around the Haversian canal so that stress flows *around* it, rather than *toward* it. Second, the stress wave front is predicted to propagate much farther in the isotropic case at the given time point, which is again consistent with the results of the previous section that indicate the lamellar structure of osteons slows the propagation of disturbances toward the osteon core, giving cortical bone more time to respond to the energetic excitation. Finally, the third prediction is that the stress magnitude in the lamellar case builds up 8.5% less than for the isotropic case at the given time point, as can be inferred from the caption of Figure 19.10. Though the isotropic case uses an average value for the stiffness, so many material points in the lamellar structure have a normal stiffness that exceeds it, the stress patterns are predicted to evolve in a manner that avoids localization at the Haversian canal, and the structure experiences a smaller maximum. This difference becomes even more emphatic on comparing shear stresses; cf. Figure 19.11. The maximum shear stress in the lamellar case is 59% less than the isotropic case at the given time point. It appears, therefore, that the lamellar structure improves overall load-bearing properties of cortical bone; in particular, for a predefined failure stress, a larger deformation should be endured by the lamellar structure, since normal and shear stresses build up more slowly and evolve in patterns that appear less localized.

### *19.5.3 Rotation and Displacement Fields in Osteons*

Figure 19.12a and b compare the element rotation component $\beta_{12}^{n=1}$. As with stress, the rotation $\beta_{12}^{n=1} \simeq \beta_{12}^{n=2}$. The prediction is that for the lamellar mesostructure the rotations evolve to a maximum that is 7.6% less than the isotropic case. Thus, evolving geometric nonlinearities are smaller when the structure is lamellar, and overall deformations are more uniform. Figure 19.12a highlights an interesting prediction for the osteons on the corners. For the osteon outlined in black, for example, it is predicted that rotations are not all negative when the lamellar structure is accounted for, unlike in the isotropic case. Instead, a band of positive rotation cuts through the osteon. This prediction may be explained by considering Figure 19.12c. The isosurfaces shown are for lamellar stiffness to help depict the orientation of lamellae as deformation proceeds. As such, when the bottom surface is pulled on, the lamellae reorient themselves. As the bottom-right corner is freer to move laterally, it displaces both downward and to the left. This motion induces a negative (clockwise) rotation $\beta$ on the right surface, as indicated by arrow 2 on the

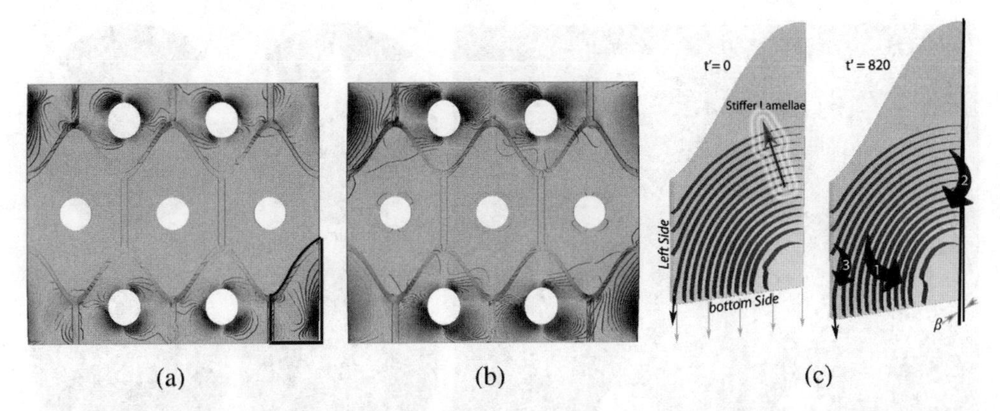


**Figure 19.12** The $\beta_{12}^{n=1}$ rotation field. Contours vary linearly from $-4°$ to $4°$ on sub-figures (a) and (b). Max = 7.3° (lamellar), Max = 7.9° (isotropic). Time point $t' = 820$. (a) Case I: lamellar. (b) Case II: isotropic. (c) Influence of lamellae on rotation

figure. On the other hand, as the bottom surface is pulled down, the lamellae tend to become more elliptical, and, as can be inferred from the small black arrow at the bottom-left of the osteon, a positive (counter-clockwise) rotation ensues (labeled arrow 1 on the figure). For the few lamellae, however, that do not end at the bottom surface but end at the left side of the osteon in this model, their higher stiffness (due to radial position) with respect to the lower part of the osteon promotes a negative rotation as indicated by arrow 3. The lamellar structure, therefore, appears to accommodate a more complex rotation field than the isotropic case. It appears that accommodating such complexity across the entire osteon hinders the localization of rotations at the Haversian canals and accounts for the predicted smaller maxima in the lamellar case. Moreover, by inspecting the lateral displacements for the two cases, Figure 19.13, it is seen that the lamellar structure suppresses the amount of "barreling" of the overall mesostructure, decreasing maximum lateral displacement by 36.5%. Therefore, it may be concluded that the lamellar structure decreases global geometric nonlinearities in deforming cortical bone; that is, the lamellae

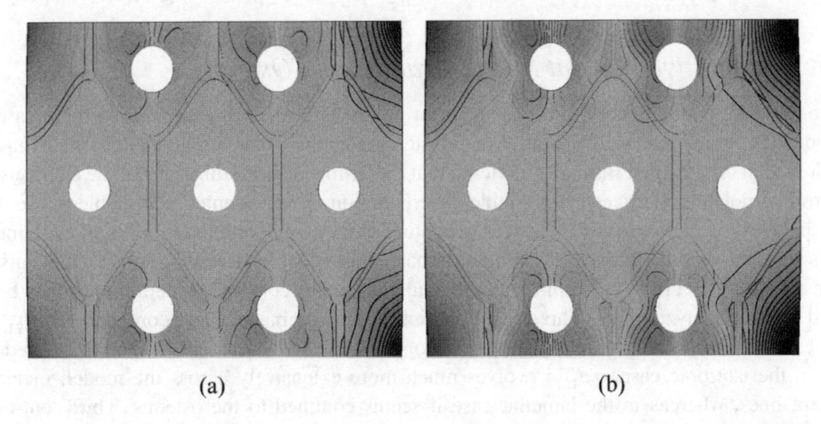


**Figure 19.13** The $u_1^{n=1}$ displacement field. Contours vary linearly from $-1L_{\text{ref}}$ to $+1L_{\text{ref}}$ on both sub-figures. Max = $1.22L_{\text{ref}}$ (lamellar), Max = $1.92L_{\text{ref}}$ (isotropic). Time point $t' = 820$. (a) Case I: lamellar. (b) Case II: isotropic

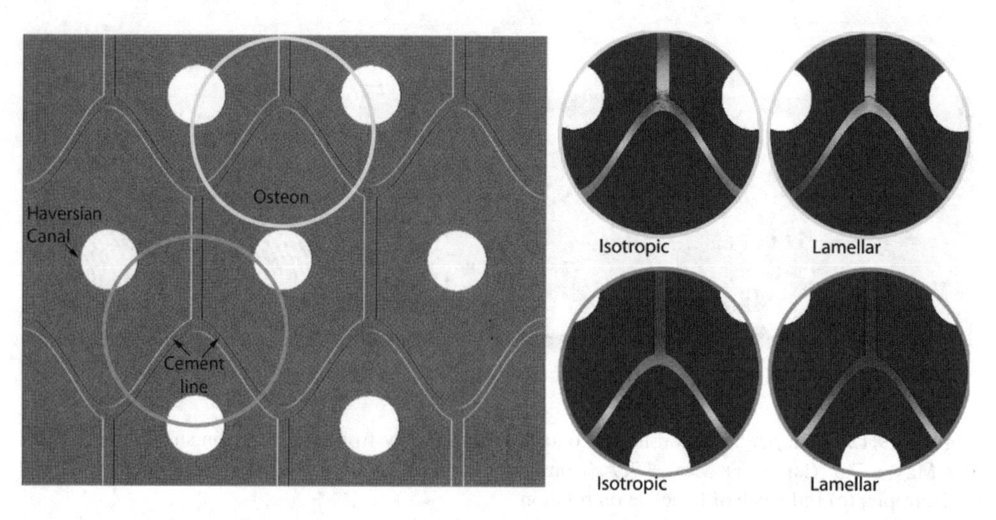


**Figure 19.14** The $\xi^{n=1}$ damping coefficient. Color contours vary linearly from 0.0 to 0.01. Max = 0.014 371 5 (isotropic), Max = 0.011 824 (lamellar). Time point $t' = 820$

promote globally uniform deformation, potentially by accommodating spread-out and complex rotation patterns across osteons.

### *19.5.4 Damping in Cement Lines*

Figure 19.14 compares the plots for damping coefficient $\xi^{n=1}$ for the lamellar and isotropic cases. As with stress, rotation, and displacement, $\xi^{n=1} \simeq \xi^{n=2}$. Two main predictions can be made. First, the lamellar case exhibits less viscous dissipation than the isotropic case by 17.7%. Second, viscous dissipation in the cement lines at the meet of three osteons does not reach the recorded maximum in the model, unlike in the isotropic case. These results again confirm that the lamellar structure of osteons appears to significantly delay the attainment of maximal and/or threshold values (e.g., maximum viscous dissipation) for a given strain, and diminishes mesostructural susceptibility to localization of fields at geometric imperfections.

### *19.5.5 Qualitative Look at Strain Gradients in Osteons*

Here, we compare qualitatively the evolving strain gradients in the lamellar and isotropic cases. The limitation of the comparison to a qualitative nature stems from the rather simplified numerical implementation adopted in this chapter. Basically, noticing that the tri-linear shape functions of the elements do not yield zero interpolation matrices upon double differentiation, it was deemed reasonable to use them as (underpopulated) second-derivative matrices to capture order-of-magnitude estimates of evolving strain gradients across the model. Figure 19.15 thus compares three double-stress ($\sigma\sigma^{n=2}$) components. Here, we again have $\sigma\sigma^{n=2} \simeq \sigma\sigma^{n=1}$. In spite of numerical noise in the contours, general trends may be easily identified and are analyzed here. Three main predictions can be made. First, component $\sigma\sigma_{121}^{n=2}$ in the lamellar case has a maximum smaller than the isotropic case; 7.8% lower according to these predictions. Second, in the isotropic case, $\sigma\sigma_{121}^{n=2}$ evolves much more extensively across the model, including in the cement lines, whereas in the lamellar case it seems confined to the osteons. Third, out-of-plane components of double stress ($\sigma\sigma_{321}^{n=2}, \sigma\sigma_{133}^{n=2}$) become active in the lamellar case, but not in the isotropic case, as can be inferred from the second and third columns of Figure 19.15 (the values reported for

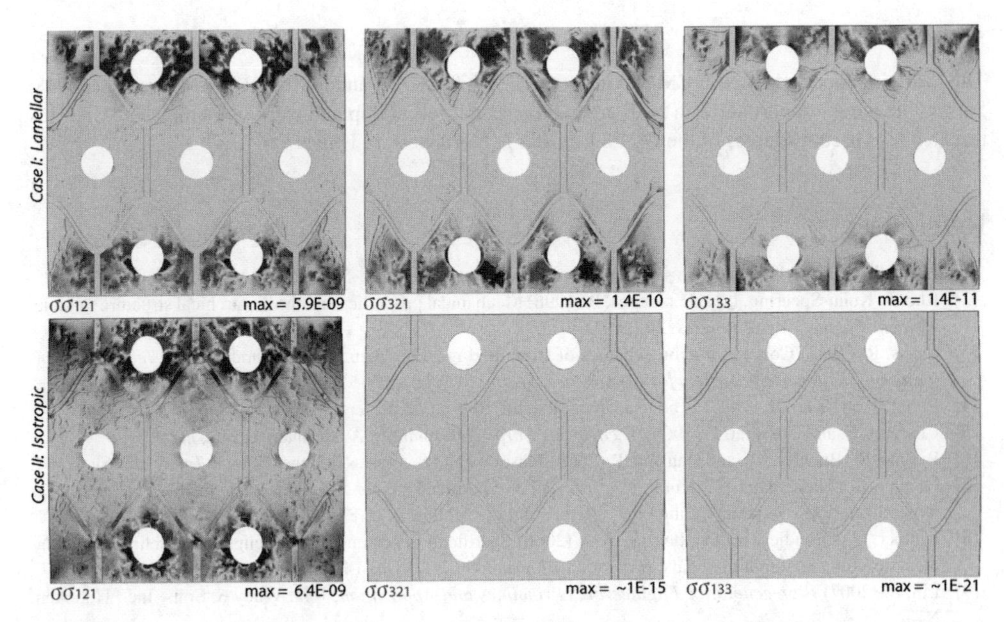


**Figure 19.15** $\{\sigma\sigma^{n=2}\}$ components comparison. Color contours vary linearly from $-16.8 \times 10^{-11} E_{\text{osteon}} (L_{\text{ref}})^2$ to $16.8 \times 10^{-11} E_{\text{osteon}} (L_{\text{ref}})^2$ on all sub-figures. Time point $t' = 820$

the isotropic case in fact give indication of the magnitude of numerical errors for these components in the computational model). Altogether, these predictions confirm the role of the lamellar structure in: (a) decreasing component-wise curvature effects in the cement line and osteons; (b) spreading out strain gradients over multiple directions (components), which collectively encourage less localized deformation of cortical bone.

## 19.6 Conclusion

In this chapter we have presented the ABC theory, with special application to compact bone mechanics. Corresponding constitutive laws for the osteons and cement lines were developed and implemented in the ABC computational model representing the mesostructure of cortical bone. Two partitions per material point were introduced to investigate the propagation of strain inhomogeneities across the mesostructure in a comparative study between lamellar and isotropic mesostructures. Deformation of the osteons with geometric nonlinearities was modeled by using a hypo-elastic constitutive model. First-order and curvature effects were accounted for in osteons by considering the lamellar structure wrapping around the Haversian canal and a radial dependence of stiffness. Viscoelasticity of the cement lines was captured by modifying the second Lamé parameter of an isotropic hypo-elasticity model according to the slope of a hyperbolic material law with two phenomenological parameters. Results were presented for dynamic simulations investigating the propagation of shear strain inhomogeneity, normal and shear stress, rotation and displacement fields, viscous damping, and double-stress components. Results indicate that the lamellar structure of osteons: (a) improves load bearing capacity; (b) reduces localization and geometric nonlinearity; and (c) allows more response time to excitation energies by slowing disturbance propagation toward osteon cores, when compared with an isotropic mesostructure.

## Acknowledgments

This work is supported by NSF CMMI Grants 0823327, 0928320, and NSF IDR CMM Grant 1130948. Steven Greene is supported by an NSF graduate research fellowship and warmly thanks the NSF. This research also used resources of the QUEST cluster at Northwestern University.

## References

[1] Lakes, R. (1993) Materials with structural hierarchy. *Nature*, **361**, 511–515.
[2] Rho, J., Kuhn-Spearing, L., and Zioupos, P. (1998) Mechanical properties and the hierarchical structure of bone. *Medical Engineering & Physics*, **20** (2), 92–102.
[3] Lakes, R. (1988) Cosserat micromechanics of structured media experimental methods, in *3rd Technical Conference of the American Society for Composites, Seattle, WA*, pp. 505–516.
[4] Yoon, H. and Katz, J. (1983) Is bone a Cosserat solid? *Journal of Materials Science*, **18** (5), 1297–1305.
[5] Cosserat, E. and Cosserat, F. (1909) *Theorie des Corps Déformables*, A. Hermann et Fils, Paris.
[6] Ritchie, R., Buehler, M., and Hansma, P. (2009) Plasticity and toughness in bone. *Physics Today*, **62** (6), 41–47.
[7] Fantner, G., Hassenkam, T., Kindt, J. *et al.* (2005) Sacrificial bonds and hidden length dissipate energy as mineralized fibrils separate during bone fracture. *Nature Materials*, **4** (8), 612–616.
[8] Fantner, G., Oroudjev, E., Schitter, G. *et al.* (2006) Sacrificial bonds and hidden length: unraveling molecular mesostructures in tough materials. *Biophysical Journal*, **90** (4), 1411–1418.
[9] Liu, S. (2007) *Bioregenerative Engineering: Principles and Application*, John Wiley & Sons, Inc., Hoboken, NJ.
[10] Vincent, J. (1990) *Structural Biomaterials*, Princeton University Press, Princeton, NJ.
[11] Currey, J. (2002) *Bones: Structure and Mechanics*, Princeton University Press, Princeton, NJ.
[12] Lakes, R. and Katz, J. (1979) Viscoelastic properties of wet cortical bone – II. Relaxation mechanisms. *Journal of Biomechanics*, **12** (9), 679–68.
[13] Weiner, S. and Traub, W. (1992) Bone structure: from angstroms to microns. *FASEB Journal*, **6** (3), 879–885.
[14] Weiner, S. and Price, P. (1986) Disaggregation of bone into crystals. *Calcified Tissue International*, **39**, 365–375.
[15] Giraud-Guille, M. (1988) Twisted plywood architecture of collagen fibrils in human compact bone osteons. *Calcified Tissue International*, **42**, 167–180.
[16] Woodard, J., Hilldore, A., Lan, S. *et al.* (2007) The mechanical properties and osteoconductivity of hydroxyapatite bone scaffolds with multi-scale porosity. *Biomaterials*, **28** (1), 45–54.
[17] Buechner, P. and Lakes, R. (2003) Size effects in the elasticity and viscoelasticity of bone. *Biomechanics and Modeling in Mechanobiology*, **1**, 295–301.
[18] Cooper, D., Turinsky, A., Sensen, C., and Hallgrimsson, B. (2003) Quantitative 3D analysis of the canal network in cortical bone by micro-computed tomography. *The Anatomical Record*, **274B** (1), 169–179.
[19] Cowin, S. (1999) Bone poroelasticity. *Journal of Biomechanics*, **32** (3), 217–238.
[20] Schaffler, M., Burr, D., and Frederickson, R. (1987) Morphology of the osteonal cement line in human bone. *The Anatomical Record*, **217** (3), 223–228.
[21] Burr, D., Schaffler, M., and Frederickson, R. (1988) Composition of the cement line and its possible mechanical role as a local interface in human compact bone. *Journal of Biomechanics*, **21** (11), 939–941, 943–945.
[22] Nalla, R., Kinney, J., and Ritchie, R. (2003) Mechanistic fracture criteria for the failure of human cortical bone. *Nature Materials*, **2** (3), 164–168.
[23] Malvern, L. (1969) *Introduction to the Mechanics of a Continuous Medium*, Prentice-Hall, Upper Saddle River, NJ.
[24] Kroner, E. (1963) On the physical reality of torque stresses in continuum mechanics. *International Journal of Engineering Science*, **1** (2), 261–278.
[25] Germain, P. (1973) The method of virtual power in continuum mechanics. Part 2: microstructure. *SIAM Journal on Applied Mathematics*, **25** (3), 556–575.
[26] Fish, J. and Kuznetsov, S. (2010) Computational continua. *International Journal for Numerical Methods in Engineering*, **84** (7), 774–802.
[27] Aero, E. and Kuvshinkii, E. (1960) Fundamental equations of the theory of elastic media with rotationally interacting particles. *Soviet Physics Solid State*, **2**, 1272–1281.

[28] Toupin, R. (1964) Theories of elasticity with couple-stress. *Archive for Rational Mechanics and Analysis*, **17** (2), 85–112.

[29] Koiter, W. (1964) Couple stress in the theory of elasticity, I and II. *Proceedings of the Koninklijke Nederlandse Akademie van Wetenschappen. Series B: Physical Sciences*, **67**, 17–44.

[30] Mindlin, R. (1965) Second gradient of strain and surface-tension in linear elasticity. *International Journal of Solids and Structures*, **1** (4), 417–438.

[31] Fleck, N. and Hutchinson, J. (1997) Strain gradient plasticity. *Advances in Applied Mechanics*, **30**, 295–361.

[32] Mindlin, R. and Tiersten, H. (1962) Effects of couple-stresses in linear elasticity. *Archive for Rational Mechanics and Analysis*, **11** (1), 415–448.

[33] Triantafyllidis, N. and Aifantis, E. (1986) A gradient approach to localization of deformation. I. Hyperelastic materials. *Journal of Elasticity*, **16** (3), 225–237.

[34] Triantafyllidis, N. and Bardenhagen, S. (1993) On higher order gradient continuum theories in 1-D nonlinear elasticity. Derivation from and comparison to the corresponding discrete models. *Journal of Elasticity*, **33** (3), 259–293.

[35] Aifantis, E. (1984) On the microstructural origin of certain inelastic models. *Journal of Engineering Materials and Technology*, **106** (4), 326–330.

[36] Fleck, N., Muller, G., Ashby, M., and Hutchinson, J. (1994) Strain gradient plasticity: theory and experiment. *Acta Metallurgica et Materialia*, **42** (2), 475–487.

[37] Nix, W. and Gao, H. (1998) Indentation size effects in crystalline materials: a law for strain gradient plasticity. *Journal of the Mechanics and Physics of Solids*, **46** (3), 411–425.

[38] Gao, H., Huang, Y., Nix, W., and Hutchinson, J. (1999) Mechanism-based strain gradient plasticity I. Theory. *Journal of the Mechanics and Physics of Solids*, **47** (6), 1239–1263.

[39] Eringen, A. (1999) *Microcontinuum Field Theories I: Foundations and Solids*, Springer, New York.

[40] Vernerey, F., Liu, W., and Moran, B. (2007) Multi-scale micromorphic theory for hierarchical materials. *Journal of the Mechanics and Physics of Solids*, **55** (12), 2603–2651.

[41] McVeigh, C. and Liu, W. (2008) Linking microstructure and properties through a predictive multiresolution continuum. *Computer Methods in Applied Mechanics and Engineering*, **197** (41–42), 3268–3290.

[42] Mindlin, R. (1964) Micro-structure in linear elasticity. *Archive for Rational Mechanics and Analysis*, **16** (1), 51–78.

[43] Park, H. and Lakes, R. (1986) Cosserat micromechanics of human bone: strain redistribution by a hydration sensitive constituent. *Journal of Biomechanics*, **19** (5), 385–397.

[44] Buechner, P. and Lakes, R. (2003) Size effects in the elasticity and viscoelasticity of bone. *Biomechanics and Modeling in Mechanobiology*, **1** (4), 295–301.

[45] Forest, S., Barbe, F., and Cailletaud, G. (2000) Cosserat modelling of size effects in the mechanical behaviour of polycrystals and multi-phase materials. *International Journal of Solids and Structures*, **37** (46–47), 7105–7126.

[46] De Borst, R. and Remmers, J. (2006) Computational modelling of delamination. *Composites Science and Technology*, **66** (6), 713–722.

[47] Tang, S., Greene, M., and Liu, W. (2011) Two-scale mechanism-based theory of nonlinear viscoelasticity. *Journal of the Mechanics and Physics of Solids*, **60** (2), 199–226.

[48] Tang, S., Greene, M., and Liu, W. (2012) A renormalization approach to model interaction in microstructured solids: application to porous elastomer. *Computer Methods in Applied Mechanics and Engineering*, **217–220**, 213–225.

[49] Schellekens, J. and de Borst, R. (1993) A non-linear finite element approach for the analysis of mode-I free edge delamination in composites. *International Journal of Solids and Structures*, **30** (9), 1239–1253.

[50] Liu, W. and McVeigh, C. (2008) Predictive multiscale theory for design of heterogeneous materials. *Computational Mechanics*, **42** (2), 147–170.

[51] Gonella, S., Greene, M., and Liu, W. (2011) Characterization of heterogeneous solids via wave methods in computational microelasticity. *Journal of the Mechanics and Physics of Solids*, **59** (5), 959–974.

[52] McVeigh, C., Vernerey, F., Liu, W. *et al.* (2007) An interactive micro-void shear localization mechanism in high strength steels. *Journal of the Mechanics and Physics of Solids*, **55** (2), 225–244.

[53] Vernerey, F., Liu, W., Moran, B., and Olson, G. (2008) A micromorphic model for the multiple scale failure of heterogeneous materials. *Journal of the Mechanics and Physics of Solids*, **56** (4), 1320–1347.

[54] Vernerey, F., Liu, W., Moran, B., and Olson, G. (2009) Multi-length scale micromorphic process zone model. *Computational Mechanics*, **44** (3), 433–445.

[55] Tian, R., Chan, S., Tang, S. *et al.* (2010) A multiresolution continuum simulation of the ductile fracture process. *Journal of the Mechanics and Physics of Solids*, **58** (10), 1681–1700.

[56] McVeigh, C. and Liu, W. (2010) Multiresolution continuum modeling of micro-void assisted dynamic adiabatic shear band propagation. *Journal of the Mechanics and Physics of Solids*, **58** (2), 187–205.

[57] McVeigh, C. and Liu, W. (2009) Multiresolution modeling of ductile reinforced brittle composites. *Journal of the Mechanics and Physics of Solids*, **57**, 244–267.

[58] Zervos, A. (2008) Finite elements for elasticity with microstructure and gradient elasticity. *International Journal for Numerical Methods in Engineering*, **73** (4), 564–595.

[59] Erofeyev, V. (2003) *Wave Processes in Solids with Microstructure*, vol. 8 of *Stability, Vibration, and Control of Systems*, World Scientific, Singapore.

[60] Benecke, G., Kerschnitzki, M., Fratzl, P., and Gupta, H. (2009) Digital image correlation shows localized deformation bands in inelastic loading of fibrolamellar bone. *Journal of Materials Research*, **24** (2), 421–429.

[61] Thurner, P., Erickson, B., Jungmann, R. *et al.* (2007) High-speed photography of compressed human trabecular bone correlates whitening to microscopic damage. *Engineering Fracture Mechanics*, **74** (12), 1928–1941.

[62] Yoon, H. and Katz, J. (1976) Dispersion of the ultrasonic velocities in human cortical bone, in *IEEE Ultrasonic Symposium Proceedings* (eds J. de Klerk and B. McAvoy), pp. 48–50.

[63] Lakes, R., Yoon, H., and Katz, J. (1983) Slow compressional wave propagation in wet human and bovine cortical bone. *Science*, **220** (4596), 513–515.

[64] Droin, P., Berger, G., and Laugier, P. (1998) Velocity dispersion of acoustic waves in cancellous bone. *IEEE Transactions on Ultrasonics, Ferroelectrics and Frequency Control*, **45** (3), 581–592.

[65] Wear, K. (2000) Measurements of phase velocity and group velocity in human calcaneus. *Ultrasound in Medicine & Biology*, **26** (4), 641–646.

[66] Wear, K. (2001) A stratified model to predict dispersion in trabecular bone. *IEEE Transactions on Ultrasonics, Ferroelectrics and Frequency Control*, **48** (4), 1079–1083.

[67] Marutyan, K., Holland, M., and Miller, J. (2006) Anomalous negative dispersion in bone can result from the interference of fast and slow waves. *Journal of the Acoustical Society of America*, **120** (5), EL55–EL61.

[68] Vavva, M., Protopappas, V., Gergidis, L. *et al.* (2009) Velocity dispersion of guided waves propagating in a free gradient elastic plate: application to cortical bone. *Journal of the Acoustical Society of America*, **125** (5), 3414–3427.

[69] Epstein, M. and Elżanowski, M. (2007) *Material Inhomogeneities and their Evolution: A Geometric Approach*, Interaction of Mechanics and Mathematics, Springer, New York, NY.

[70] Nemat-Nasser, S. and Hori, M. (1999) *Micromechanics: Overall Properties of Heterogeneous Materials*, Elsevier, New York, NY.

[71] Mura, T. (1987) *Micromechanics of Defects in Solids*, vol. 3 of *Mechanics of Elastic and Inelastic Solids*, 2nd edition, Kluwer Academic Publishers, Norwell, MA.

[72] Ostoja-Starzewski, M. (2002) Lattice models in micromechanics. *Applied Mechanics Reviews*, **55** (1), 35–60.

[73] Daniel, I. and Ishai, O. (2006) *Engineering Mechanics of Composite Materials*, 2nd edition, Oxford University Press, New York, NY.

[74] Elkhodary, K.I., Greene, M.S., Tang, S. *et al.* (2012) Archetype-blending continuum theory. *Computer Methods in Applied Mechanics and Engineering*, in press.

[75] Hogan, H. (1992) Micromechanics modeling of Haversian cortical bone properties. *Journal of Biomechanics*, **25** (5), 549–556.

[76] Prevost, J. and Keane, C. (1990) Shear stress–strain curve generation from simple material parameters. *Journal of Geotechnical Engineering*, **116** (8), 1255–1263.

# 20

# Image-Based Multiscale Modeling of Porous Bone Materials

Judy P. Yang[a], Sheng-Wei Chi[b], and Jiun-Shyan Chen[c]
[a]*Department of Civil & Environmental Engineering, National Chiao Tung University, Taiwan*
[b]*Department of Civil and Environmental Engineering, University of Illinos at Chicago, USA*
[c]*Department of Civil and Environmental Engineering, University of California, Los Angeles, USA*

## 20.1 Overview

Bones comprise about one-fifth of an individual body weight, with the main functions in supporting and protecting organs, performing movements, and producing the blood cells, among others. Based on the microstructural composition, bones can be classified as cortical bone (compact bone) and trabecular bone (cancellous bone or spongy bone). With reference to Figure 20.1a for a femur long bone, the cortical bone constitutes about 80% of the human skeleton mass and forms an outer layer of bones, while the trabecular bone fills the interior with a porous and cancellous structure; see the microstructure of the femur long bone shown in Figure 20.1b. In general, the cortical bone is stiffer, harder, and denser than the trabecular bone so that it has the ability to protect organs and support the body movement, in addition to its ability to transmit chemical components. In contrast, the trabecular bone has a larger surface area, lower density and stiffness, and higher porosity than those of the cortical bone. Owing to the porous nature of the trabecular bone, there exists space for the blood vessels and bone marrow to flow inside the spongy structure, as can be seen from the computed tomography (CT) scan of the trabecular bone microstructure in Figure 20.2. As such, the bone materials can be characterized as porous media of solid skeleton with the pores filled with fluid. Consequently, the fluid-saturated porous material has been introduced to describe the constitutive behavior of bone materials.

Homogenization methods have been introduced to provide a multiscale paradigm for analysis of poroelastic materials, and yield macroscopic balance laws that resemble the Biot's theory [1] while embedding microscopic features [2–4]. This work is motivated by the fact that as high-resolution digital imaging techniques emerge, such as the advancement of micro-CT and micro-magnetic resonance imaging (micro-MRI), the investigation of mechanical properties of cortical and trabecular bones can

*Multiscale Simulations and Mechanics of Biological Materials*, First Edition. Edited by Shaofan Li and Dong Qian.

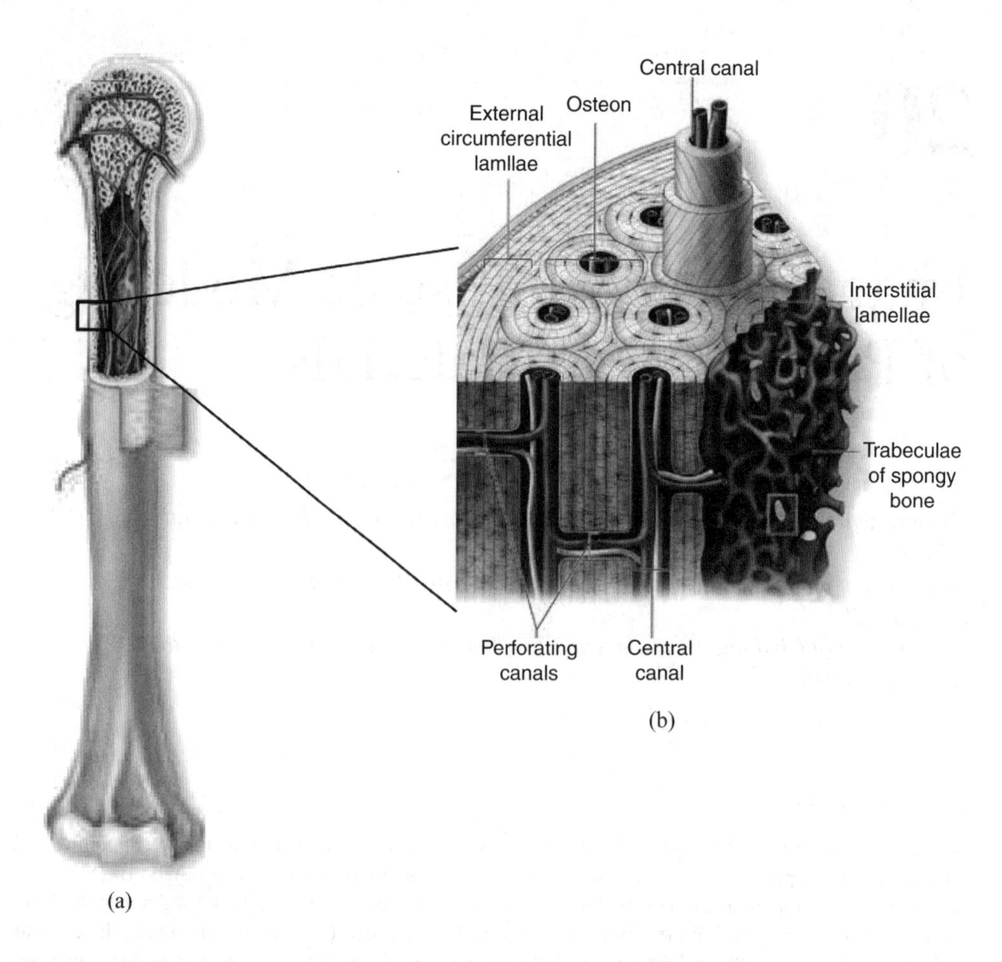


**Figure 20.1** Hierarchical structure of bone: (a) macrostructure of a femur long bone; (b) microstructures of cortical and trabecular bones (adapted from http://academic.kellogg.edu/herbrandsonc/bio201_mckinley/skeletal.htm)

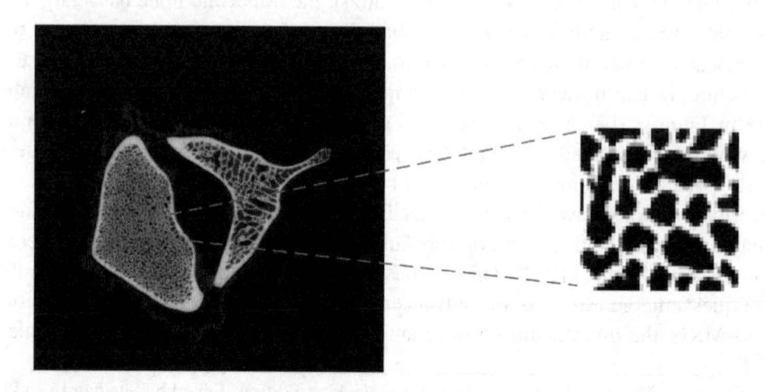

**Figure 20.2** Microstructure of bone, imaged by micro-CT (adapted from http://www.scanco.ch/)

be estimated more precisely, though with challenges remaining to be resolved. For instance, the jagged interface and sharp corner arise when converting the image pixels into $C^0$ finite elements, leading to artificial localized responses such as stress concentration [5, 6]. In other words, avoiding distorted elements in the process of image reconstruction and mesh generation is a tedious task [7–10]. Another issue arising from the model reconstruction process based on high-resolution medical images is the background noise and blurred objects, which could lead to incorrect numerical predictions and diagnoses due to the geometry errors in the simulation models [5, 6, 10–12]. A computational framework that can effectively model biological materials at various length scales is of an urgent need for advancement of bioscience. This work introduces an image-based strong form collocation method in conjunction with image segmentation for multiscale modeling of bone materials. This chapter is arranged as follows. The asymptotic expansion-based homogenization of porous microstructures and the resulting generalized balanced laws are presented in Section 20.2. The level set method for image segmentation and a collocation method for solving the level set equation are introduced in Section 20.3. In Section 20.4, the scale coupling characteristic functions defining the microscopic cell problems will be solved by an image-based gradient-reproducing kernel collocation method (G-RKCM). Application of the proposed computational framework to the modeling of a trabecular bone microstructure is demonstrated in Section 20.5. Concluding remarks are given in Section 20.6.

## 20.2 Homogenization of Porous Microstructures

The mathematical theory of homogenization offers an effective and rigorous way to describe the gross mechanical behaviors of media having a high degree of heterogeneity. For fluid-saturated poroelastic materials composed of solid skeleton and fluid in the pores, the homogenization can be obtained by considering the explicit field and constitutive equations of each constituent and the associated geometric details. The theory of homogenization with consideration of microstructural effects was popularized in the 1970s. The main interest was to determine the effective mechanical properties of composite materials by linking the local features of microstructures to the macroscopic responses. The asymptotic expansion theory was termed the homogenization method and investigated from the mathematical aspect of approximation by Babuska [13]. Bensoussan and his co-workers [14] introduced the periodic structures, the so-called microscopic cells, to the asymptotic analysis. Sanchez-Palencia [15] applied the mathematical framework of homogenization to many physics-related problems, including nonhomogeneous media and vibration theory. Lions [16] investigated homogenization analysis from the viewpoint of mathematical convergence. Guedes and Kikuchi [17] introduced the weak form formulation to the homogenization method for numerical analysis.

Extension of the homogenization method was made to porous materials leading to the generalized Darcy's law that describes the behavior of a flux flowing through a porous medium [18, 19]. Specifically, fluid viscosity is the key component affecting the interaction between the solid and fluid in the microstructures and the scaling to the macroscopic poroelastic behavior. Different choices of the scaling parameters in the asymptotic homogenization method have been discussed by Sanchez-Palencia [15] and Hornung [18]. The procedure of asymptotic expansion-based homogenization for a quasi-static poroelastic medium with fluid of low viscosity will be detailed in the following [19].

### *20.2.1 Basic Equations of Two-Phase Media*

To describe a two-phase medium, we consider the following boundary value problem with reference to Figure 20.3. We use superposed "S" to denote the solid phase and superposed "F" denoting the fluid phase. The solid phase equilibrium equation and boundary conditions are given as

$$\begin{aligned} \sigma_{ij,j}^{S} + \rho^{S} b_i^{S} &= 0 && \text{in } \Omega_S, \\ u_i &= \bar{u}_i && \text{on } \Gamma_{Su}, \\ \sigma_{ij}^{S} n_j^{S} &= \bar{t}_i^{S} && \text{on } \Gamma_{St}, \end{aligned} \tag{20.1}$$

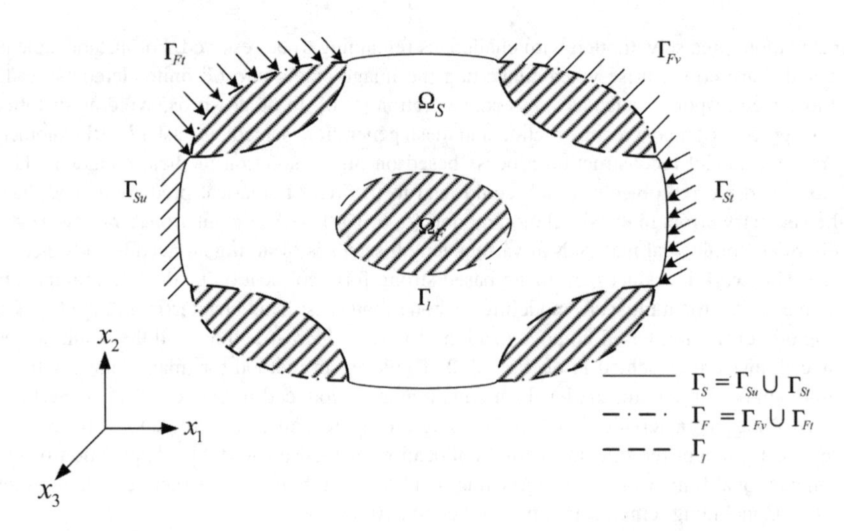


**Figure 20.3** A two-phase porous medium with boundary and interface conditions

where $\sigma_{ij}^{\mathrm{S}}$ is the solid stress, $\rho^{\mathrm{S}}$ is the solid density, $\rho^{\mathrm{S}} b_i^{\mathrm{S}}$ is the solid body force, $u_i$ is the solid displacement with the prescribed displacement $\bar{u}_i$ on $\Gamma_{\mathrm{S}u}$, and $\bar{t}_i^{\mathrm{S}}$ is the surface traction on $\Gamma_{\mathrm{S}t}$. The fluid phase Stokes equation and boundary conditions are expressed as

$$\begin{aligned} &\sigma_{ij,j}^{\mathrm{F}} + \rho^{\mathrm{F}} b_i^{\mathrm{F}} = 0 && \text{in } \Omega_{\mathrm{F}}, \\ &v_{i,i} = 0 && \text{in } \Omega_{\mathrm{F}}, \\ &v_i = \bar{v}_i && \text{on } \Gamma_{\mathrm{F}v}, \\ &\sigma_{ij}^{\mathrm{F}} n_j^{\mathrm{F}} = \bar{t}_i^{\mathrm{F}} && \text{on } \Gamma_{\mathrm{F}t}, \end{aligned} \tag{20.2}$$

where $\sigma_{ij}^{\mathrm{F}}$ is the fluid stress, $\rho^{\mathrm{F}}$ is the fluid density, $\rho^{\mathrm{F}} b_i^{\mathrm{F}}$ is the fluid body force, $v_i$ is the fluid velocity with the prescribed velocity $\bar{v}_i$ on $\Gamma_{\mathrm{F}v}$, and $\bar{t}_i^{F}$ is the surface traction on $\Gamma_{\mathrm{F}t}$. Letting $\Gamma_{\mathrm{I}}$ denote the solid-fluid interface, the corresponding interface equilibrium conditions are

$$\begin{aligned} &v_i = \frac{\partial u_i}{\partial t} && \text{on } \Gamma_{\mathrm{I}}, \\ &\sigma_{ij}^{\mathrm{F}} n_j^{\mathrm{F}} = \sigma_{ij}^{\mathrm{S}} n_j^{\mathrm{S}} && \text{on } \Gamma_{\mathrm{I}}, \end{aligned} \tag{20.3}$$

where $n_j^{\mathrm{F}} = -n_j^{\mathrm{S}}$ are unit normals on $\Gamma_{\mathrm{I}}$.

A porous medium is composed of interconnected canals filled with incompressible fluid of low viscosity. The solid phase is assumed linear elastic with the stress given by

$$\sigma_{ij}^{S} = C_{ijkl}\varepsilon_{kl} = C_{ijkl} u_{(k,l)}, \tag{20.4}$$

where $u_{(i,j)} \equiv (u_{i,j} + u_{j,i})/2$. The stress for the Newtonian fluid is expressed as

$$\sigma_{ij}^{\mathrm{F}} = -p\delta_{ij} + 2\mu D_{ij} + \kappa D_{kk}\delta_{ij}, \tag{20.5}$$

where $p$ is the fluid pressure; $\kappa$ and $\mu$ are bulk viscosity and dynamic viscosity, respectively, and $D_{ij} = (\partial v_i/\partial x_j + \partial v_j/\partial x_i)/2 \equiv v_{(i,j)}$ is the rate of deformation. Assuming that the fluid is incompressible, the stress in fluid reduces to

$$\sigma_{ij}^{\mathrm{F}} = -p\delta_{ij} + 2\mu v_{(i,j)}. \tag{20.6}$$

The weak form of the two-phase problem is then stated as: find $u_i \in H^1$, $u_i = \bar{u}_i$ on $\Gamma_{\mathrm{S}u}$, $v_i \in H^1$, $v_i = \bar{v}_i$ on $\Gamma_{\mathrm{F}v}$, $p \in H^0$, for all $\varpi_i \in H^1$, $\varpi_i = 0$ on $\Gamma_{\mathrm{S}u}$, $w_i \in H^1$, $w_i = 0$ on $\Gamma_{\mathrm{F}v}$, and $\omega \in H^0$, such that

$$\int_{\Omega_{\mathrm{S}}} C_{ijkl} u_{(k,l)} \varpi_{(i,j)} \,\mathrm{d}\Omega = \int_{\Omega_{\mathrm{S}}} \rho^{\mathrm{S}} b_i^{\mathrm{S}} \varpi_i \,\mathrm{d}\Omega + \int_{\Gamma_{\mathrm{S}t}} \bar{t}_i^{\mathrm{S}} \varpi_i \,\mathrm{d}\Gamma + \int_{\Gamma_{\mathrm{I}}} \sigma_{ij}^{\mathrm{F}} n_j^{\mathrm{F}} \varpi_i \,\mathrm{d}\Gamma, \tag{20.7}$$

$$-\int_{\Omega_{\mathrm{F}}} p w_{i,i} \,\mathrm{d}\Omega + \int_{\Omega_{\mathrm{F}}} 2\mu v_{(i,j)} w_{i,j} \,\mathrm{d}\Omega = \int_{\Omega_{\mathrm{F}}} \rho^{\mathrm{F}} b_i^{\mathrm{F}} w_i \,\mathrm{d}\Omega + \int_{\Gamma_{\mathrm{F}t}} \bar{t}_i^{\mathrm{F}} w_i \,\mathrm{d}\Gamma + \int_{\Gamma_{\mathrm{I}}} \sigma_{ij}^{\mathrm{S}} n_j^{\mathrm{S}} w_i \,\mathrm{d}\Gamma, \tag{20.8}$$

$$\int_{\Omega_{\mathrm{F}}} v_{i,i} \omega \,\mathrm{d}\Omega = 0. \tag{20.9}$$

The coupling effect between the solid and fluid phases has been introduced on the interface $\Gamma_{\mathrm{I}}$ in (20.7) and (20.8).

## 20.2.2 Asymptotic Expansion of Two-Phase Medium

As shown in Figure 20.4, the heterogeneous medium is assumed to be assembled from spatially repeated microscopic cells $Y$. The macroscopic coordinate $\mathbf{x}$ and the microscopic coordinate $\mathbf{y}$ are related by a scale parameter $\lambda$ as

$$\mathbf{x} = \lambda \mathbf{y}. \tag{20.10}$$

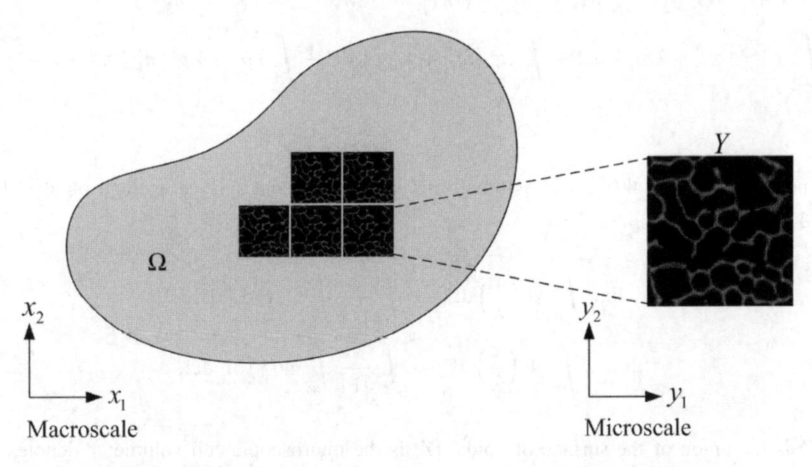


**Figure 20.4** Macroscopic and microscopic coordinate systems

To present the heterogeneous system, $\Omega^\lambda$ is used to denote the total domain of the medium considering heterogeneity, and the field variables with superscript $\lambda$ denoting the total scale of the fields. Rewrite the weak forms in (20.7)–(20.9) with heterogeneity considered as

$$\int_{\Omega_S^\lambda} C_{ijkl} u^\lambda_{(k,l)} \varpi_{(i,j)} \, d\Omega = \int_{\Omega_S^\lambda} \rho^S b_i^S \varpi_i \, d\Omega + \int_{\Gamma_{St}} \bar{t}_i^S \varpi_i \, d\Gamma + \int_{\Gamma_I} \sigma_{ij}^F n_j^F \varpi_i \, d\Gamma, \tag{20.11}$$

$$-\int_{\Omega_F^\lambda} p^\lambda w_{i,i} \, d\Omega + \int_{\Omega_F^\lambda} 2\mu v^\lambda_{(i,j)} w_{i,j} \, d\Omega = \int_{\Omega_F^\lambda} \rho^F b_i^F w_i \, d\Omega + \int_{\Gamma_{Ft}} \bar{t}_i^F w_i \, d\Gamma + \int_{\Gamma_I} \sigma_{ij}^S n_j^S w_i \, d\Gamma, \tag{20.12}$$

$$\int_{\Omega_F^\lambda} v^\lambda_{i,i} \omega \, d\Omega = 0. \tag{20.13}$$

Consider the asymptotic expansion of field and state variables with $Y$-periodicity as

$$u_i^\lambda(\mathbf{x}, t) = u_i^0(\mathbf{x}, \mathbf{y}, t) + \lambda u_i^1(\mathbf{x}, \mathbf{y}, t), \tag{20.14}$$

$$v_i^\lambda(\mathbf{x}, t) = v_i^0(\mathbf{x}, \mathbf{y}, t) + \lambda v_i^1(\mathbf{x}, \mathbf{y}, t), \tag{20.15}$$

$$p^\lambda(\mathbf{x}, t) = p^0(\mathbf{x}, \mathbf{y}, t) + \lambda p^1(\mathbf{x}, \mathbf{y}, t), \tag{20.16}$$

where variables with superscript "0" and "1" are coarse- and fine-scale components of the variables with superscript "$\lambda$", respectively.

## Two-Scale Decomposition of Solid Phase Equilibrium Equation

Owing to the assumption of low viscosity, we have $\sigma_{ij}^F n_j^F = p\delta_{ij} n_j^F$ on the interface. Therefore, the interface traction in (20.11) can be written as $\int_{\Gamma_I} p^\lambda n_j^F \varpi_i \, d\Gamma$. By applying the asymptotic expansion given in (20.14) and taking the derivative following $\partial\Phi(\mathbf{x}_i, \mathbf{y}_i)/\partial x_i = \partial\Phi/\partial x_i + (1/\lambda)\partial\Phi/\partial y_i$, one obtains from (20.11) the following equation:

$$\begin{aligned} &\int_{\Omega_S^\lambda} C_{ijkl} \left( \frac{\partial u_k^0}{\partial x_l} + \frac{1}{\lambda}\frac{\partial u_k^0}{\partial y_l} + \lambda \frac{\partial u_k^1}{\partial x_l} + \frac{\partial u_k^1}{\partial y_l} \right) \left( \frac{\partial \varpi_i^0}{\partial x_j} + \frac{1}{\lambda}\frac{\partial \varpi_i^0}{\partial y_j} + \lambda \frac{\partial \varpi_i^1}{\partial x_j} + \frac{\partial \varpi_i^1}{\partial y_j} \right) d\Omega \\ &= \int_{\Omega_S^\lambda} \rho^S b_i^S \left(\varpi_i^0 + \lambda \varpi_i^1\right) d\Omega + \int_{\Gamma_{St}} \bar{t}_i^S \left(\varpi_i^0 + \lambda \varpi_i^1\right) d\Gamma + \int_{\Gamma_I} \left(p^0 + \lambda p^1\right) n_i^F \left(\varpi_i^0 + \lambda \varpi_i^1\right) d\Gamma. \end{aligned} \tag{20.17}$$

For a $Y$-periodic function $\Psi(y)$, it is assumed that the volume and surface average of $\Psi(y)$ hold the following relations:

$$\lim_{\lambda \to 0^+} \int_{\Omega^\lambda} \Psi\left(\frac{x}{\lambda}\right) d\Omega \to \int_\Omega \frac{1}{|Y|} \int_Y \Psi(y) \, dY \, d\Omega, \tag{20.18}$$

$$\lim_{\lambda \to 0^+} \lambda \int_{\Gamma^\lambda} \Psi\left(\frac{x}{\lambda}\right) d\Gamma \to \int_\Omega \frac{1}{|Y|} \int_{\partial Y} \Psi(y) \, d\Gamma \, d\Omega, \tag{20.19}$$

where $\Gamma^\lambda$ is the union of the surface of voids; $|Y|$ is the microscopic cell volume; $Y$ denotes the cell domain, and $\partial Y$ denotes the union of the surface of voids on a cell boundary, respectively.

Applying the averaging process over the microscopic cell (20.19) to the last term on the right hand side of (20.17), and considering $\varpi_i^0$ being divergence free in $Y$, $p^1 = 0$ can be deduced [15]. Consequently, the following three leading order equations can be obtained from (20.17):

$$\int_{\Omega_S^\lambda} C_{ijkl} \frac{\partial u_k^0}{\partial y_l} \frac{\partial \varpi_i^0}{\partial y_j} \, d\Omega = 0, \tag{20.20}$$

$$\int_{\Omega_S^\lambda} C_{ijkl} \left( \frac{\partial u_k^0}{\partial x_l} \frac{\partial \varpi_i^0}{\partial y_j} + \frac{\partial u_k^1}{\partial y_l} \frac{\partial \varpi_i^0}{\partial y_j} + \frac{\partial u_k^0}{\partial y_l} \frac{\partial \varpi_i^0}{\partial x_j} + \frac{\partial u_k^0}{\partial y_l} \frac{\partial \varpi_i^1}{\partial y_j} \right) d\Omega = 0, \tag{20.21}$$

$$\begin{aligned} &\int_{\Omega_S^\lambda} C_{ijkl} \left( \frac{\partial u_k^1}{\partial x_l} \frac{\partial \varpi_i^0}{\partial y_j} + \frac{\partial u_k^0}{\partial x_l} \frac{\partial \varpi_i^0}{\partial x_j} + \frac{\partial u_k^1}{\partial y_l} \frac{\partial \varpi_i^0}{\partial x_j} + \frac{\partial u_k^0}{\partial y_l} \frac{\partial \varpi_i^1}{\partial x_j} + \frac{\partial u_k^0}{\partial x_l} \frac{\partial \varpi_i^1}{\partial y_j} + \frac{\partial u_k^1}{\partial y_l} \frac{\partial \varpi_i^1}{\partial y_j} \right) d\Omega \\ &= \int_{\Omega_S^\lambda} \rho^S b_i^S \varpi_i^0 \, d\Omega + \int_{\Gamma_{St}} \bar{t}_i^S \varpi_i^0 \, d\Gamma - \int_\Omega \frac{1}{|Y|} \int_{Y_S} p^0 \frac{\partial \varpi_i^1}{\partial y_i} \, dY \, d\Omega. \end{aligned} \tag{20.22}$$

By introducing (20.18) to average over the microscopic cell in (20.20), and considering $\varpi_i^0 = \varpi_i^0(\mathbf{y}, t)$, we have $u_i^0 = u_i^0(\mathbf{x}, t)$. By applying the averaging process (20.18) together with $u_i^0 = u_i^0(\mathbf{x}, t)$, $\varpi_i^0 = \varpi_i^0(\mathbf{x})$, and considering the interface traction, Equation (20.22) can be separated into

$$\int_\Omega \frac{1}{|Y|} \int_{Y_S} C_{ijkl} \left( \frac{\partial u_k^0}{\partial x_l} + \frac{\partial u_k^1}{\partial y_l} \right) \frac{\partial \varpi_i^0}{\partial x_j} \, dY \, d\Omega = \int_\Omega \frac{1}{|Y|} \int_{Y_S} \rho^S b_i^S \varpi_i^0 \, dY \, d\Omega + \int_{\Gamma_{St}} \bar{t}_i^S \varpi_i^0 \, d\Gamma \tag{20.23}$$

and

$$\int_\Omega \frac{1}{|Y|} \int_{Y_S} C_{ijkl} \left( \frac{\partial u_k^0}{\partial x_l} + \frac{\partial u_k^1}{\partial y_l} \right) \frac{\partial \varpi_i^1}{\partial y_j} \, dY \, d\Omega = - \int_\Omega \frac{1}{|Y|} \int_{Y_S} p^0 \frac{\partial \varpi_i^1}{\partial y_i} \, dY \, d\Omega. \tag{20.24}$$

Further introducing the following scale-coupling relation [18, 19]:

$$u_i^1(\mathbf{x}, \mathbf{y}, t) = -\chi_i^{kl}(\mathbf{y}) \frac{\partial u_k^0(\mathbf{x}, t)}{\partial x_l} - \eta_i(\mathbf{y}) p^0(\mathbf{x}, t), \tag{20.25}$$

where $\chi_i^{kl}(\mathbf{y})$ and $\eta_i(\mathbf{y})$ are the characteristic functions, namely, the scale-coupling functions. Introducing (20.25) to the microscopic equation in (20.24) leads to two microscopic cell problems for the scale-coupling functions:

$$\int_{Y_S} C_{ijmn} \frac{\partial \chi_m^{kl}}{\partial y_n} \frac{\partial \varpi_i^1}{\partial y_j} \, dY = \int_{Y_S} C_{ijkl} \frac{\partial \varpi_i^1}{\partial y_j} \, dY, \tag{20.26}$$

$$\int_{Y_S} C_{ijkl} \frac{\partial \eta_k}{\partial y_l} \frac{\partial \varpi_i^1}{\partial y_j} \, dY = \int_{Y_S} \delta_{ij} \frac{\partial \varpi_i^1}{\partial y_j} \, dY. \tag{20.27}$$

Substituting (20.25) into the macroscopic equation in (20.23) gives rise to

$$\begin{aligned} &\int_\Omega \frac{1}{|Y|} \int_{Y_S} \left( C_{ijkl} - C_{ijmn} \frac{\partial \chi_m^{kl}}{\partial y_n} \right) dY \frac{\partial u_k^0}{\partial x_l} \frac{\partial \varpi_i^0}{\partial x_j} - \int_\Omega \frac{1}{|Y|} \int_{Y_S} C_{ijkl} \frac{\partial \eta_k}{\partial y_l} \, dY \, p^0 \frac{\partial \varpi_i^0}{\partial x_j} \, d\Omega \\ &= \int_\Omega \frac{1}{|Y|} \int_{Y_S} \rho^S b_i^S \varpi_i^0 \, dY \, d\Omega + \int_{\Gamma_{St}} \bar{t}_i^S \varpi_i^0 \, d\Gamma. \end{aligned} \tag{20.28}$$

Based on (20.28), the homogenized elasticity tensor $\bar{C}_{ijkl}$, the homogenized effective stress coefficient tensor $\bar{\alpha}_{ij}$, and the average body force of the solid phase $\bar{f}_i^{\mathrm{S}}$ are defined as follows:

$$\bar{C}_{ijkl} = \frac{1}{|Y|}\int_{Y_{\mathrm{S}}}\left(\mathrm{C}_{\mathrm{ijkl}} - C_{ijmn}\frac{\partial \chi_m^{kl}(\mathbf{y})}{\partial y_n}\right)\mathrm{d}Y, \tag{20.29}$$

$$\bar{\alpha}_{ij} = \frac{1}{|Y|}\int_{Y_{\mathrm{S}}} C_{ijkl}\frac{\partial \eta_k(\mathbf{y})}{\partial y_l}\,\mathrm{d}Y, \tag{20.30}$$

$$\bar{f}_i^{\mathrm{S}} = \frac{1}{|Y|}\int_{Y_{\mathrm{S}}} \rho^{\mathrm{S}} b_i^{\mathrm{S}}\,\mathrm{d}Y = \frac{|Y_{\mathrm{S}}|}{|Y|}\rho^{\mathrm{S}} b_i^{\mathrm{S}}. \tag{20.31}$$

With the definitions of the homogenized material parameters in (20.29)–(20.31), the homogenized macroscopic equilibrium equation (20.28) can be recast as

$$\int_{\Omega} \bar{C}_{ijkl}\frac{\partial u_k^0}{\partial x_l}\frac{\partial \varpi_i^0}{\partial x_j}\,\mathrm{d}\Omega - \int_{\Omega}\bar{\alpha}_{ij}p^0\frac{\partial \varpi_i^0}{\partial x_j}\,\mathrm{d}\Omega = \int_{\Omega}\bar{f}_i^{\mathrm{S}}\varpi_i^0\,\mathrm{d}\Omega + \int_{\Gamma_{\mathrm{S}t}}\bar{t}_i^{\mathrm{S}}\varpi_i^0\,\mathrm{d}\Gamma. \tag{20.32}$$

## Two-Scale Decomposition of Fluid Phase Equilibrium Equation

Following the same procedures in the last section, and considering a scale factor $\lambda^2$ for the fluid viscosity $\mu$ to account for the viscous effect [15, 18, 19] as well as negligible fluid shear stress due to low viscosity when imposing equilibrium with the solid traction on the interface, the fluid phase equation (20.12) can be processed to obtain the following leading order equations:

$$-\int_{\Omega_{\mathrm{F}}^{\lambda}} p^0\frac{\partial w_i^0}{\partial y_i}\,\mathrm{d}\Omega = \int_{\Omega}\frac{1}{|Y|}\int_{\partial Y_{\mathrm{I}}} C_{ijkl}\frac{\partial u_k^0}{\partial y_l}n_j^{\mathrm{S}}w_i^1\,\mathrm{d}\Gamma\,\mathrm{d}\Omega, \tag{20.33}$$

$$\begin{aligned}
&\int_{\Omega_{\mathrm{F}}^{\lambda}}\left[-\left(p^0\frac{\partial w_i^0}{\partial x_i} + p^0\frac{\partial w_i^1}{\partial y_i} + p^1\frac{\partial w_i^0}{\partial y_i}\right) + 2\mu\frac{\partial v_i^0}{\partial y_j}\frac{\partial w_i^0}{\partial y_j}\right]\mathrm{d}\Omega = \int_{\Omega_{\mathrm{F}}^{\lambda}}\rho^{\mathrm{F}}b_i^{\mathrm{F}}w_i^0\,\mathrm{d}\Omega + \int_{\Gamma_{\mathrm{F}t}}\bar{t}_i^{\mathrm{F}}w_i^0\,\mathrm{d}\Gamma\\
&\quad + \int_{\Omega}\frac{1}{|Y|}\int_{\partial Y_{\mathrm{I}}} C_{ijkl}\frac{\partial u_k^0}{\partial x_l}n_j^{\mathrm{S}}w_i^1\,\mathrm{d}\Gamma\,\mathrm{d}\Omega + \int_{\Omega}\frac{1}{|Y|}\int_{\partial Y_{\mathrm{I}}} C_{ijkl}\frac{\partial u_k^1}{\partial x_l}n_j^{\mathrm{S}}w_i^0\,\mathrm{d}\Gamma\,\mathrm{d}\Omega\\
&\quad + \int_{\Omega}\frac{1}{|Y|}\int_{\partial Y_{\mathrm{I}}} C_{ijkl}\frac{\partial u_k^1}{\partial y_l}n_j^{\mathrm{S}}w_i^1\,\mathrm{d}\Gamma\,\mathrm{d}\Omega.
\end{aligned} \tag{20.34}$$

Applying the averaging process over the microscopic cell to (20.33), and using the fact $u_i^0 = u_i^0(\mathbf{x}, t)$, we have $p^0 = p^0(\mathbf{x}, t)$. Further, considering $p^1 = 0$, we obtain the following macroscopic and microscopic equations:

$$\int_{\Omega_{\mathrm{F}}^{\lambda}}\left(-p^0\frac{\partial w_i^0}{\partial x_i} + 2\mu\frac{\partial v_i^0}{\partial y_j}\frac{\partial w_i^0}{\partial y_j}\right)\mathrm{d}\Omega = \int_{\Omega_{\mathrm{F}}^{\lambda}}\rho^{\mathrm{F}}b_i^{\mathrm{F}}w_i^0\,\mathrm{d}\Omega + \int_{\Gamma_{\mathrm{F}t}}\bar{t}_i^{\mathrm{F}}w_i^0\,\mathrm{d}\Gamma + \int_{\Omega}\frac{1}{|Y|}\int_{\partial Y_{\mathrm{I}}} C_{ijkl}\frac{\partial u_k^1}{\partial x_l}n_j^{\mathrm{S}}w_i^0\,\mathrm{d}\Gamma\,\mathrm{d}\Omega \tag{20.35}$$

and

$$-\int_{\Omega_{\mathrm{F}}^{\lambda}} p^0\frac{\partial w_i^1}{\partial y_i}\,\mathrm{d}\Omega = \int_{\Omega}\frac{1}{|Y|}\int_{\partial Y_{\mathrm{I}}} C_{ijkl}\frac{\partial u_k^0}{\partial x_l}n_j^{\mathrm{S}}w_i^1\,\mathrm{d}\Gamma\,\mathrm{d}\Omega + \int_{\Omega}\frac{1}{|Y|}\int_{\partial Y_{\mathrm{I}}} C_{ijkl}\frac{\partial u_k^1}{\partial y_l}n_j^{\mathrm{S}}w_i^1\,\mathrm{d}\Gamma\,\mathrm{d}\Omega. \tag{20.36}$$

Applying the averaging process over the microscopic cell to (20.35), considering the far field traction prescribed on the heterogeneous medium vanishing in the microscopic cell, and letting $w_i^0 = w_i^0(\mathbf{y})$ with the divergence-free condition in $Y_{\mathrm{F}}$, we have

$$\begin{aligned}\int_\Omega \frac{1}{|Y|}\int_{Y_{\mathrm{F}}} 2\mu \frac{\partial v_i^0}{\partial y_j}\frac{\partial w_i^0}{\partial y_j}\,\mathrm{d}Y\,\mathrm{d}\Omega &= \int_\Omega \frac{1}{|Y|}\int_{Y_{\mathrm{F}}} \rho^{\mathrm{F}} b_i^{\mathrm{F}} w_i^0\,\mathrm{d}Y\,\mathrm{d}\Omega \\ &+ \int_\Omega \frac{1}{|Y|}\int_{Y_{\mathrm{F}}} C_{ijkl}\left(\frac{\partial \chi_k^{mn}(\mathbf{y})}{\partial y_j}\frac{\partial u_m^0}{\partial x_n \partial x_l} + \frac{\partial \eta_k(\mathbf{y})}{\partial y_j}\frac{\partial p^0}{\partial x_l}\right) w_i^0\,\mathrm{d}\Gamma\,\mathrm{d}\Omega,\end{aligned} \tag{20.37}$$

where we assume that the last force term in (20.37) can be mainly attributed to the pressure gradient in the fluid domain. According to Terada *et al.* [19], the solution of the following form is assumed:

$$v_i^0(\mathbf{x},\mathbf{y},t) = \frac{\partial u_i^0(\mathbf{x},t)}{\partial t} + \left(\rho^{\mathrm{F}} b_j^{\mathrm{F}} - \frac{\partial p^0}{\partial x_j}\right)\kappa_{ij}(\mathbf{y}), \tag{20.38}$$

where $\kappa_{ij}(\mathbf{y})$ is the characteristic function related to the hydraulic permeability. Recalling the interface condition $v_i = \partial u_i/\partial t$ in (20.3), the second term on the right hand side of (20.38) can be viewed as the relative velocity of fluid with respect to the solid phase, which is zero on the interface and in solid. Substituting (20.38) into (20.37) leads to the following microscopic cell problem:

$$\begin{aligned} 2\mu \frac{\partial \kappa_{ij}(\mathbf{y})}{\partial y_k \partial y_k} + \delta_{ij} = 0 \quad &\text{in } Y_{\mathrm{F}}, \\ \kappa_{ij}(\mathbf{y}) = 0 \quad &\text{on } \partial Y, \end{aligned} \tag{20.39}$$

where $\kappa_{ij}(\mathbf{y})$ is a function characterizing the steady-state Stokes flow within the cell, which is $Y$-periodic and divergence free. By averaging over the microscopic cell on the relative velocity in (20.38), the homogenized macroscopic velocity is given by the generalized Darcy's law

$$\bar{v}_i = \frac{\partial u_i^0(\mathbf{x},t)}{\partial t} - \bar{K}_{ij}\frac{\partial \bar{P}}{\partial x_j}, \tag{20.40}$$

where the macroscopic permeability tensor $\bar{K}_{ij}$ and the generalized pressure gradient $\partial \bar{P}/\partial x_j$ are defined as follows:

$$\bar{K}_{ij} = \frac{1}{|Y|}\int_{Y_{\mathrm{F}}} \kappa_{ij}(\mathbf{y})\,\mathrm{d}Y, \tag{20.41}$$

$$\frac{\partial \bar{P}}{\partial x_j} = -\left(\rho^{\mathrm{F}} b_j^{\mathrm{F}} - \frac{\partial p^0(\mathbf{x})}{\partial x_j}\right). \tag{20.42}$$

### Two-Scale Decomposition of Continuity Equation

Introducing the asymptotic expansion of the fluid velocity in (20.15) and the multiscale decomposition to (20.13) yields

$$\int_{\Omega_{\mathrm{F}}^{\lambda}} \left(\frac{\partial v_i^0}{\partial x_i} + \frac{1}{\lambda}\frac{\partial v_i^0}{\partial y_i} + \lambda\frac{\partial v_i^1}{\partial x_i} + \frac{\partial v_i^1}{\partial y_i}\right)\omega\,\mathrm{d}\Omega = 0. \tag{20.43}$$

Substituting $v_i^0$ in (20.38) into (20.43), considering the divergence free condition of $v_i^0$, and taking the limit as $\lambda \to 0^+$, we have the macroscopic continuity equation:

$$\int_{\Omega_F^\lambda} \frac{\partial}{\partial x_i}\left(\frac{\partial u_i^0(\mathbf{x},t)}{\partial t}\right)\omega\,\mathrm{d}\Omega + \int_{\Omega_F^\lambda} \frac{\partial}{\partial x_i}\left[\left(\rho^F b_j^F - \frac{\partial p^0(\mathbf{x})}{\partial x_j}\right)\kappa_{ij}(\mathbf{y})\right]\omega\,\mathrm{d}\Omega + \int_{\Omega_F^\lambda} \frac{\partial v_i^1}{\partial y_i}\omega\,\mathrm{d}\Omega = 0. \tag{20.44}$$

Applying the averaging process over the microscopic cell and assuming $\omega = \omega(\mathbf{x})$ lead to

$$\int_{\Omega} \frac{\partial}{\partial x_i}\left(\frac{\partial u_i^0(\mathbf{x},t)}{\partial t}\right)\omega\,\mathrm{d}\Omega + \int_{\Omega} \frac{1}{|Y|}\int_{Y_F} \frac{\partial}{\partial x_i}\left(\rho^F b_j^F - \frac{\partial p^0(\mathbf{x})}{\partial x_j}\right)\kappa_{ij}(\mathbf{y})\,\mathrm{d}Y\,\omega\,\mathrm{d}\Omega = 0, \tag{20.45}$$

where the last term in (20.44) can be shown to be zero by periodicity. Based on the definitions of the permeability tensor $\bar{K}_{ij}$ and the generalized pressure gradient $\partial\bar{P}/\partial x_j$ in (20.41) and (20.42), respectively, the final form of the homogenized macroscopic continuity equation (20.45) is

$$\int_{\Omega} \frac{\partial}{\partial x_i}\left(\frac{\partial u_i^0(\mathbf{x},t)}{\partial t}\right)\omega\,\mathrm{d}\Omega - \int_{\Omega} \frac{\partial}{\partial x_i}\left(\frac{\partial \bar{P}}{\partial x_j}\right)\bar{K}_{ij}\omega\,\mathrm{d}\Omega = 0. \tag{20.46}$$

## 20.2.3 Homogenized Porous Media

A poroelastic medium composed of an elastic solid and incompressible Newtonian fluid of low viscosity has been analyzed by the two-scale asymptotic expansion method discussed in Sections 20.2.1 and 20.2.2. The homogenized governing equations of poroelastic materials are summarized as follows:

- homogenized macroscopic equilibrium equation

$$\int_{\Omega} \bar{C}_{ijkl}\frac{\partial u_k^0}{\partial x_l}\frac{\partial \varpi_i^0}{\partial x_j}\,\mathrm{d}\Omega - \int_{\Omega} \bar{\alpha}_{ij}p^0\frac{\partial \varpi_i^0}{\partial x_j}\,\mathrm{d}\Omega = \int_{\Omega} \bar{f}_i^S\varpi_i^0\,\mathrm{d}\Omega + \int_{\Gamma_{S_t}} \bar{t}_i^S\varpi_i^0\,\mathrm{d}\Gamma; \tag{20.47}$$

- homogenized macroscopic continuity equation

$$\int_{\Omega} \frac{\partial}{\partial x_i}\left(\frac{\partial u_i^0(\mathbf{x},t)}{\partial t}\right)\omega\,\mathrm{d}\Omega - \int_{\Omega} \frac{\partial}{\partial x_i}\left(\rho^F b_j^F - \frac{\partial p^0(\mathbf{x})}{\partial x_j}\right)\bar{K}_{ij}\omega\,\mathrm{d}\Omega = 0; \tag{20.48}$$

- generalized Darcy's law

$$v_i = \frac{\partial u_i^0(\mathbf{x},t)}{\partial t} - \bar{K}_{ij}\frac{\partial \bar{P}}{\partial x_j}; \tag{20.49}$$

- homogenized macroscopic stress

$$\bar{\sigma}_{ij} = \bar{C}_{ijkl}u_{(k,l)}^0 - \bar{\alpha}_{ij}p^0. \tag{20.50}$$

Equations (20.47)–(20.49) reassemble the governing equations in poroelasticity [1], with embedded microstructural information. The homogenized material parameters with overbars in (20.47)–(20.49) are defined in (20.29)–(20.31), (20.41), and (20.42).

## 20.3 Level Set Method for Image Segmentation

The level set method originally devised by Osher and Sethian [20, 21] was intended to track topological changes such as merging and breaking, which has been extensively applied in practical research areas including computer graphics, image processing, optimization, and computational fluid dynamics. The essential idea of this method is to track the evolution of a surface in $N-1$ dimensions presented by the level set function with $N$ dimensions in space. Consequently, the motion of a boundary is governed by the evolution of the level set function. One advantage of the level set method is the implicit representation of the curve through the level set function, which facilitates the control of the moving interfaces with changing topology effectively in the process of evolution. Here, we introduce the active contour model proposed by Chan and Vese [22, 23] for image segmentation and interface identification based on the Mumford–Shah functional with a level set formulation. This technique can detect objects from images with or without gradients; that is, images with sharp or blurred interfaces.

The level set equation in the active contour model is traditionally solved by the finite difference method (FDM), where the inputs of the image, known as pixels, are discretized uniformly on the grid points. Nevertheless, for problems of objects whose boundaries evolve in response to the physical process, such as contact and penetration problems, fixed grid discretization leads to large approximation errors in the interface representation. Solving level set equations using a Lagrangian grid has been proposed based on the meshfree method for contact problems [24, 25] In this chapter, the strong form collocation method is introduced to solve the level set equation for image segmentation.

### 20.3.1 Variational Level Set Formulation

Given an image enclosed by an open region $\Omega$ with the boundary $\partial\Omega$ in $R^n$ and the associated color code $C(\mathbf{x})$ as illustrated in Figure 20.5, and consider a closed evolving interface $\Gamma$ as the trial of the image interface. Define a level set function $\phi(\mathbf{x}, t)$, which is Lipschitz continuous and satisfies the following conditions:

$$\begin{aligned} &\phi(\mathbf{x}, t) > 0 && \text{if } \mathbf{x} \text{ is inside } \Gamma(t), \\ &\phi(\mathbf{x}, t) = 0 && \text{if } \mathbf{x} \text{ is on } \Gamma(t), \\ &\phi(\mathbf{x}, t) < 0 && \text{if } \mathbf{x} \text{ is outside } \Gamma(t). \end{aligned} \tag{20.51}$$

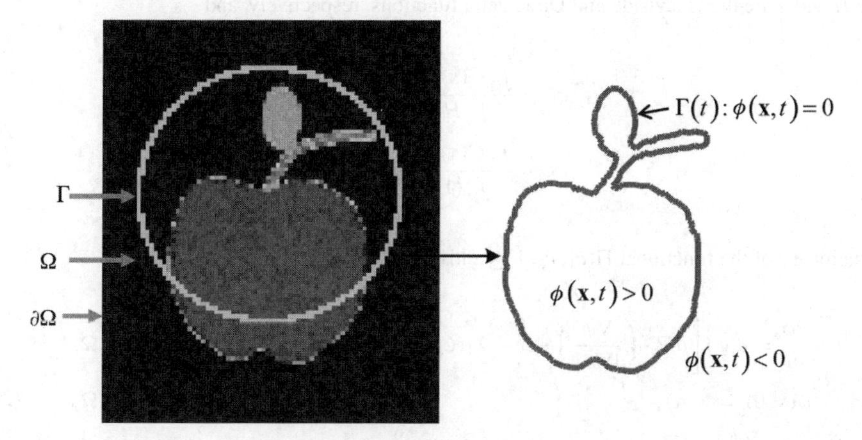


**Figure 20.5** Level set function

Assume that the image contains two regions, i.e., the object to be detected and the remaining region outside the object in the image. The object, an apple, occupies $\Omega^i$ and the rest of the area of the image occupies $\Omega^0$. Define a least-squares functional:

$$\Pi(\Gamma) = \int_{\text{inside }\Gamma} (C(\mathbf{x}) - \bar{c}_1)^2 \, d\mathbf{x} + \int_{\text{outside }\Gamma} (C(\mathbf{x}) - \bar{c}_2)^2 \, d\mathbf{x}, \tag{20.52}$$

where $\Gamma$ is the trial boundary of the object, and constants $\bar{c}_1$ and $\bar{c}_2$ are the averages of $C(\mathbf{x})$ inside and outside $\Gamma$, respectively. For a given initial trial boundary $\Gamma^0$, the minimization of (20.52) drives $\Gamma^0$ toward the true boundary of the object $\partial\Omega^i$.

Based on the Mumford–Shah functional [26] for image segmentation, two terms, called the regularization terms, including the length of $\Gamma$ and the area inside $\Gamma$, have been added to (20.52) to yield the following functional $\Pi(\bar{c}_1, \bar{c}_2, \Gamma)$ [22, 23]:

$$\begin{aligned} \Pi(\bar{c}_1, \bar{c}_2, \Gamma) &= \mu \cdot \text{Length}(\Gamma) + \nu \cdot \text{Area}(\text{inside }\Gamma) \\ &\quad + \lambda_1 \int_{\text{inside }\Gamma} (C(\mathbf{x}) - \bar{c}_1)^2 \, d\mathbf{x} + \lambda_2 \int_{\text{outside }\Gamma} (C(\mathbf{x}) - \bar{c}_2)^2 \, d\mathbf{x}, \end{aligned} \tag{20.53}$$

where $\mu \geq 0, \nu \geq 0, \lambda_1, \lambda_2 > 0$. The first two regularization terms control the smoothness of the detected boundary while the last two constraints measure the errors of the trial interface.

The implicit geometric representation of the moving interface $\Gamma(t)$ in terms of the level set function $\phi(\mathbf{x}, t)$, including the unit outward normal $\mathbf{n} = -\nabla\phi/|\nabla\phi|$, the mean curvature $\kappa = -\nabla \cdot (\nabla\phi/|\nabla\phi|)$, the area $A^+$ inside $\Gamma(t)$, $A^+ = \int_\Omega H(\phi(\mathbf{x}, t)) \, d\mathbf{x}$, the area $A^-$ outside $\Gamma(t)$, $A^- = \int_\Omega [1 - H(\phi(\mathbf{x}, t))] \, d\mathbf{x}$, and the length of $\Gamma(t)$, $\text{Length}(\Gamma) = \int_\Omega \delta(\phi(\mathbf{x}, t))|\nabla\phi(\mathbf{x}, t)| \, d\mathbf{x}$, is introduced in (20.53) to yield

$$\begin{aligned} \Pi(\bar{c}_1, \bar{c}_2, \Gamma) &= \mu \int_\Omega \delta(\phi)|\nabla\phi| \, d\mathbf{x} + \nu \int_\Omega H(\phi) \, d\mathbf{x} + \lambda_1 \int_\Omega (C(\mathbf{x}) - \bar{c}_1)^2 H(\phi) \, d\mathbf{x} \\ &\quad + \lambda_2 \int_\Omega (C(\mathbf{x}) - \bar{c}_1)^2 (1 - H(\phi)) \, d\mathbf{x}, \end{aligned} \tag{20.54}$$

where $H$ and $\delta$ are the Heaviside and Dirac delta functions, respectively, and

$$\begin{aligned} \bar{c}_1 &= \frac{\int_\Omega C(\mathbf{x}) H(\phi) \, d\mathbf{x}}{\int_\Omega H(\phi) \, d\mathbf{x}}, \\ \bar{c}_2 &= \frac{\int_\Omega C(\mathbf{x})(1 - H(\phi)) \, d\mathbf{x}}{\int_\Omega (1 - H(\phi)) \, d\mathbf{x}}. \end{aligned} \tag{20.55}$$

The stationary of the functional $\Pi(\bar{c}_1, \bar{c}_2, \Gamma)$ yields

$$\begin{aligned} &\frac{\partial\phi}{\partial t} = \delta(\phi)\left[\mu\nabla \cdot \left(\frac{\nabla\phi}{|\nabla\phi|}\right) - \nu - \lambda_1 (C(\mathbf{x}) - \bar{c}_1)^2 + \lambda_2 (C(\mathbf{x}) - \bar{c}_2)^2\right] && \text{in } \Omega, \\ &\phi(\mathbf{x}, 0) = \phi_0(\mathbf{x}) && \text{in } \Omega, \\ &\delta(\phi)\frac{\nabla\phi}{|\nabla\phi|} \cdot \mathbf{n} = 0 && \text{on } \partial\Omega, \end{aligned} \tag{20.56}$$

where $\mathbf{n}$ is the outward normal to the boundary $\partial\Omega$. A regularization function $H_\varepsilon$ is introduced to the Heaviside function $H(\cdot)$ with the following form:

$$H_\varepsilon(z) = \frac{1}{2}\left[1 + \frac{2}{\pi}\arctan\left(\frac{z}{\varepsilon}\right)\right], \tag{20.57}$$

where $\varepsilon$ is a positive parameter. The corresponding Delta function $\delta(\cdot)$ becomes

$$\delta_\varepsilon(z) = \frac{1}{\pi\varepsilon}\left[1 + \left(\frac{z}{\varepsilon}\right)^2\right]^{-1}. \tag{20.58}$$

The FDM is usually employed to solve (20.56) numerically for image segmentation. The interested reader is referred to Chan and Vese [23].

### *20.3.2 Strong Form Collocation Methods for Active Contour Model*

Consider the general strong form of the level set equation as follows:

$$\begin{aligned} \frac{\partial\phi}{\partial t} &= L\phi + f \quad \text{in } \Omega, \\ \phi(\mathbf{x}, 0) &= \phi_0(\mathbf{x}) \quad \text{in } \Omega, \\ B\phi &= h \quad \text{on } \partial\Omega, \end{aligned} \tag{20.59}$$

where $L$ and $B$ are the differential operators in the domain $\Omega$ and on the boundary $\partial\Omega$, respectively, and $\phi_0$ is the initial condition. In the image segmentation for boundary and interface identification, the main purpose is to identify the boundary and interface of the object represented by the zero level set function $\phi = 0$ rather than to find $\phi$ in the entire domain. Thus, the real boundary condition on $\partial\Omega$ has an insignificant influence on the solution $\phi$ close to $\phi = 0$, and it can be neglected in the numerical scheme for computational efficiency.

By introducing the collocation method to the level set equation in (20.59), the level set function $\phi$ is approximated by $\phi^h$ as

$$\phi \approx \phi^h = \sum_{I=1}^{N_s} \Psi_I(\mathbf{x})\mathbf{a}_I(t) =: \mathbf{\Psi}^{\mathrm{T}}(\mathbf{x})\mathbf{a}(t), \quad \forall\mathbf{x} \in \Omega, \tag{20.60}$$

where $\mathbf{\Psi}^{\mathrm{T}} = [\Psi_1, \ldots, \Psi_{N_s}]$, $\mathbf{a}^{\mathrm{T}} = [a_1, \ldots, a_{N_s}]$, $N_s$ is the number of source points, $\Psi_I(\mathbf{x})$ is the approximation function with compact support, such as the reproducing kernel (RK) function, and $a_I$ is the generalized coefficient. By substituting the approximation of the level set function and enforcing (20.59) at the collocation points $\mathbf{p}_\ell \in \Omega$ with $\ell = 1, 2, \ldots, N_c$, the semi-discrete collocation system is derived as follows:

$$\begin{aligned} \mathbf{N}\dot{\mathbf{a}}(t) &= \mathbf{L} + \mathbf{f}, \\ \mathbf{N}\mathbf{a}(0) &= \varphi_0, \end{aligned} \tag{20.61}$$

where

$$
\begin{aligned}
\mathbf{L} &= [L\phi^h(\mathbf{p}_1), L\phi^h(\mathbf{p}_2), \ldots, L\phi^h(\mathbf{p}_{N_c})]^{\mathrm{T}}, \\
\mathbf{f} &= [f(\mathbf{p}_1), f(\mathbf{p}_2), \ldots, f(\mathbf{p}_{N_c})]^{\mathrm{T}}, \\
\mathbf{N} &= [\boldsymbol{\Psi}(\mathbf{p}_1), \boldsymbol{\Psi}(\mathbf{p}_2), \ldots, \boldsymbol{\Psi}(\mathbf{p}_{N_c})]^{\mathrm{T}}, \\
\varphi_0 &= [\phi_0(\mathbf{p}_1), \phi_0(\mathbf{p}_2), \ldots, \phi_0(\mathbf{p}_{N_c})]^{\mathrm{T}}.
\end{aligned} \tag{20.62}
$$

Denote the level set function evaluated at the collocation point $\mathbf{p}_\ell$ by $\phi_I^h = \Psi^{\mathrm{T}}(\mathbf{p}_I)\mathbf{a}$, the explicit form of the component $L_I$ in matrix $\mathbf{L}$ is given by

$$
L_I = \delta_\varepsilon(\phi_I^h)\left[\mu\,\Xi\left(\phi_I^h\right) - \nu - \lambda_1(C(\mathbf{p}_I) - \bar{c}_1)^2 + \lambda_2(C(\mathbf{p}_I) - \bar{c}_2)^2\right], \tag{20.63}
$$

where

$$
\begin{aligned}
\Xi\left(\phi_I^h\right) = {} & \left(\phi_{I,xx}^h + \phi_{I,yy}^h + \phi_{I,zz}^h\right)\left(\phi_{I,x}^h\phi_{I,x}^h + \phi_{I,y}^h\phi_{I,y}^h + \phi_{I,z}^h\phi_{I,z}^h\right)^{-1/2} - \left(\phi_{I,x}^h\phi_{I,x}^h + \phi_{I,y}^h\phi_{I,y}^h + \phi_{I,z}^h\phi_{I,z}^h\right)^{-3/2} \\
& \times \left(\phi_{I,xx}^h\phi_{I,x}^h\phi_{I,x}^h + \phi_{I,yy}^h\phi_{I,y}^h\phi_{I,y}^h + \phi_{I,zz}^h\phi_{I,z}^h\phi_{I,z}^h + 2\phi_{I,xy}^h\phi_{I,x}^h\phi_{I,y}^h + 2\phi_{I,yz}^h\phi_{I,y}^h\phi_{I,z}^h + 2\phi_{I,zx}^h\phi_{I,z}^h\phi_{I,x}^h\right)
\end{aligned} \tag{20.64}
$$

and

$$
\begin{aligned}
\bar{c}_1 &= \frac{\sum_{I=1}^{N_c} C(\mathbf{p}_I)H_\varepsilon(\phi_I^h)\,\mathrm{d}A_I}{\sum_{I=1}^{N_c} H_\varepsilon(\phi_I^h)\,\mathrm{d}A_I}, \\
\bar{c}_2 &= \frac{\sum_{I=1}^{N_c} C(\mathbf{p}_I)(1 - H_\varepsilon(\phi_I^h))\,\mathrm{d}A_I}{\sum_{I=1}^{N_c} (1 - H_\varepsilon(\phi_I^h))\,\mathrm{d}A_I},
\end{aligned} \tag{20.65}
$$

where $\mathrm{d}A_I$ is the integration weight associated with point $\mathbf{p}_I$.

By employing the forward Euler's method, the collocation equations in (20.61) become

$$
\mathbf{a}^{n+1} = \mathbf{a}^n + \Delta t \mathbf{N}^{-1}\mathbf{L}^n, \tag{20.66}
$$

$$
\mathbf{a}^0 = \mathbf{N}^{-1}\varphi_0, \tag{20.67}
$$

in which the superscript denotes the $n$th time step and $\Delta t$ is the time step size.

When solving a second-order partial differential equation by the RK approximation with strong form collocation, one requires the shape functions to satisfy the second-order consistency to ensure convergence [27]. However, in solving the level set equation for image processing, the zero level set is our main interest, and we consider an uncorrected kernel function in Equation (20.60) for enhanced computational efficiency.

A numerical example in Figure 20.6 demonstrates the effectiveness of the direct collocation method for image segmentation. The microstructure in Figure 20.6a is presented by 75 × 75 pixels. The parameters for solving the level set equation are given as follows: $\mu = 1000$, $\nu = 0$, $\lambda_1 = \lambda_2 = 3500$, $\mathrm{d}t = 0.1$, and $\varepsilon$ is set to be twice the RK support size. The collocation points are taken the same as the source points, and the residual is set to be $5 \times 10^{-6}$. The quadratic B-spline kernel function is used as the shape function, and the kernel support size is 1.2. To compute the gradient of a level set function in the regularization term when evaluating in the center of the kernel for nonzero $\mu$, a small shifting of the center point is applied. The evolution process is shown in Figure 20.6a–d, where the interface is almost identified in the second step. With a large parameter $\lambda$, the level set function reaches the steady state quickly.

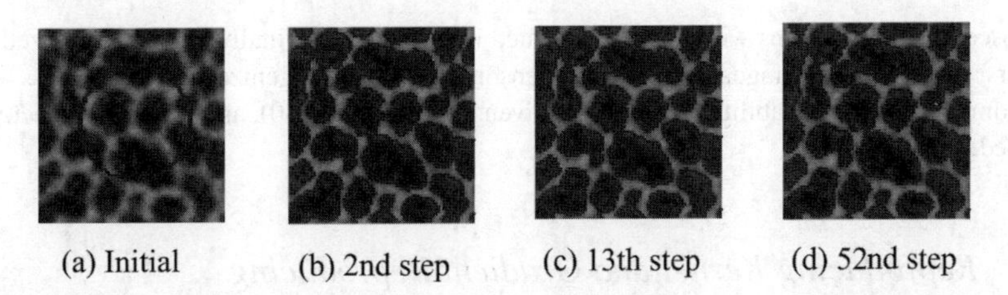


**Figure 20.6** Interface identification of a microstructure by a direct collocation method

## 20.4 Image-Based Microscopic Cell Modeling

Based on the framework of asymptotic-based homogenization, we have formulated the microscopic cell problems for poroelastic materials composed of an elastic solid and Newtonian fluid of low viscosity. With the level set technique for image segmentation, the microstructural models can be constructed directly from images. In this section, the developed image-based strong form collocation method will be employed to solve the microscopic cell problems.

### *20.4.1 Solution of Microscopic Cell Problems*

The strong form of the microscopic cell problem for the $Y$-periodic characteristic functions $\chi$ and $\eta$ in the solid domain are summarized as follows:

$$\left(C_{ijmn}\frac{\partial \chi_m^{kl}}{\partial y_n}\right)_{,j} = 0 \quad \text{in } Y_{\mathrm{S}}, \tag{20.68}$$

$$\left(C_{ijmn}\frac{\partial \chi_m^{kl}}{\partial y_n} - C_{ijkl}\right) n_j^{\mathrm{S}} = 0 \quad \text{on } \partial Y_{\mathrm{S}}, \tag{20.69}$$

and

$$\left(C_{ijkl}\frac{\partial \eta_k}{\partial y_l}\right)_{,j} = 0 \quad \text{in } Y_{\mathrm{S}}, \tag{20.70}$$

$$\left(C_{ijkl}\frac{\partial \eta_k}{\partial y_l} - \delta_{ij}\right) n_j^{\mathrm{S}} = 0 \quad \text{on } \partial Y_{\mathrm{S}}, \tag{20.71}$$

where $C_{ijkl}$ is the elastic tensor of the solid skeleton and $n_j^{\mathrm{S}}$ is the unit normal on the inner boundary of the cell. The strong form of the microscopic cell problem for the $Y$-periodic and divergence-free $\boldsymbol{\kappa}$ in the fluid domain is

$$2\mu\frac{\partial^2 \kappa_{ij}}{\partial y_k \partial y_k} + \delta_{ij} = 0 \quad \text{in } Y_{\mathrm{F}}, \tag{20.72}$$

$$\kappa_{ij} = 0 \quad \text{on } \partial Y_{\mathrm{F}}, \tag{20.73}$$

where $\mu$ is the fluid viscosity. The inner boundary conditions of microscopic cells on $\partial Y_{\mathrm{S}}$ and on $\partial Y_{\mathrm{F}}$ are stated in (20.69), (20.71) and (20.73), respectively, while the periodic boundary conditions are imposed on the outer boundaries of microscopic cells. Upon constructing microscopic cell models directly from CT or MRI, we introduce the G-RKCM [28] to solve the characteristic functions defined in the microscopic cell problems with details described in this section. Finally, the homogenized material parameters, namely, the homogenized elasticity tensor $\bar{C}_{ijkl}$, the homogenized effective stress tensor $\bar{\alpha}_{ij}$, and the homogenized permeability tensor $\bar{K}_{ij}$ as given in (60.29), (60.30), and (60.41), respectively, will be obtained.

## *20.4.2 Reproducing Kernel and Gradient Reproducing Kernel Approximations*

In the RK approximation [29, 30], the unknown $\mathbf{u}$ is approximated by $\mathbf{v}$ as follows:

$$\mathbf{u}(\mathbf{x}) \approx \mathbf{v}(\mathbf{x}) = \sum_{I=1}^{N_s} \Psi_I(\mathbf{x})\mathbf{a}_I, \tag{20.74}$$

where $N_s$ is the number of source points and $\Psi_I(\mathbf{x})$ is the RK shape function satisfying partition of unity and consistency condition expressed as

$$\Psi_I(\mathbf{x}) = \mathbf{H}^T(\mathbf{0})\mathbf{M}^{-1}(\mathbf{x})\mathbf{H}(\mathbf{x}-\mathbf{x}_I)\varphi_a(\mathbf{x}-\mathbf{x}_I), \tag{20.75}$$

with

$$\mathbf{M}(\mathbf{x}) = \sum_{I=1}^{N_s} \mathbf{H}(\mathbf{x}-\mathbf{x}_I)\mathbf{H}^T(\mathbf{x}-\mathbf{x}_I)\varphi_a(\mathbf{x}-\mathbf{x}_I), \tag{20.76}$$

where $\varphi_a(\mathbf{x}-\mathbf{x}_I)$ is the kernel function and $\mathbf{H}(\mathbf{x}-\mathbf{x}_I)$ is the monomial basis vector.

Strong form collocation for second-order differential equations requires taking second-order differentiation on the RK shape functions of (20.75), which is time consuming, especially in calculating higher order derivatives of $\mathbf{M}^{-1}(\mathbf{x})$ at every evaluation point $\mathbf{x}$. As such, we introduce the approximation of $\mathbf{u}_{,\beta}$ [31–33] as

$$\mathbf{u}_{,\beta} \approx \mathbf{w}_\beta = \sum_{I=1}^{N_s} \Psi_I^\beta(\mathbf{x})\mathbf{a}_I, \tag{20.77}$$

where the gradient RK shape function is expressed as

$$\Psi_I^\beta(\mathbf{x}) = C^\beta(\mathbf{x};\mathbf{x}-\mathbf{x}_I)\varphi_a(\mathbf{x}-\mathbf{x}_I), \tag{20.78}$$

with $\beta = (\beta_1, \beta_2, \ldots, \beta_d)$ and $|\beta| = \sum_{i=1}^{d} \beta_i \leq k$ for $\Psi_I^\beta \in C^k$; $C^\beta(\mathbf{x};\mathbf{x}-\mathbf{x}_I)$ is the correction function expressed as

$$\begin{aligned} C^\beta(\mathbf{x};\mathbf{x}-\mathbf{x}_I) &= \sum_{|\alpha|=0}^{q} b_\alpha^\beta(\mathbf{x})(\mathbf{x}-\mathbf{x}_I)^\alpha, \quad q \geq 0 \\ &=: \mathbf{H}^T(\mathbf{x}-\mathbf{x}_I)\mathbf{b}^\beta(\mathbf{x}), \end{aligned} \tag{20.79}$$

where $q$ is degree of the monomial bases. The coefficients $b_{\alpha}^{\beta}$ in the correction function are obtained by the following gradient reproducing conditions:

$$\sum_{I=1}^{N_s} \Psi_I^{\beta} \mathbf{x}_I^{\alpha} = D^{\beta} \mathbf{x}^{\alpha}, \quad 0 \leq |\alpha| \leq q. \tag{20.80}$$

Finally, the gradient RK shape function is obtained as

$$\Psi_I^{\beta}(\mathbf{x}) = (-1)^{|\beta|} D^{\beta} \mathbf{H}^{\mathrm{T}}(0) \mathbf{M}^{-1}(\mathbf{x}) \mathbf{H}(\mathbf{x} - \mathbf{x}_I) \varphi_a(\mathbf{x} - \mathbf{x}_I), \tag{20.81}$$

where $D^{\beta} \equiv \partial^{\beta_1}/\partial^{\beta_1} x_1 \cdot \partial^{\beta_2}/\partial^{\beta_2} x_2 \cdots \partial^{\beta_d}/\partial^{\beta_d} x_d$.

### 20.4.3 Gradient Reproducing Kernel Collocation Method

To introduce gradient RK approximation in the discretization of the strong form, consider the following boundary value problem:

$$\begin{aligned} \mathbf{L}^1 \mathbf{u}_{,x} + \mathbf{L}^2 \mathbf{u}_{,y} &= \mathbf{f} \quad \text{in } \Omega, \\ \mathbf{B}_h^1 \mathbf{u}_{,x} + \mathbf{B}_h^2 \mathbf{u}_{,y} &= \mathbf{h} \quad \text{on } \partial\Omega_h, \\ \mathbf{B}_g \mathbf{u} &= \mathbf{g} \quad \text{on } \partial\Omega_g, \end{aligned} \tag{20.82}$$

where $\mathbf{L}^1$ and $\mathbf{L}^2$ are the differential operators in $\Omega$, $\mathbf{B}_h^1$ and $\mathbf{B}_h^2$ are the boundary operators on $\partial\Omega_h$, and $\mathbf{B}_g$ is the boundary operator on $\partial\Omega_g$. The approximations of $\mathbf{u}$, $\mathbf{u}_{,x}$, and $\mathbf{u}_{,y}$ are given as

$$\mathbf{u} \approx \mathbf{v} = \boldsymbol{\Psi}^{\mathrm{T}} \mathbf{a}, \quad \mathbf{u}_{,x} \approx \mathbf{w}_x = \boldsymbol{\Psi}^{x\mathrm{T}} \mathbf{a}, \quad \mathbf{u}_{,y} \approx \mathbf{w}_y = \boldsymbol{\Psi}^{y\mathrm{T}} \mathbf{a}, \tag{20.83}$$

where $\boldsymbol{\Psi}$, $\boldsymbol{\Psi}^x$, $\boldsymbol{\Psi}^y$, and $\mathbf{a}$ are the vector forms of $\{\Psi_I\}_{I=1}^{N_s}$, $\{\Psi_I^x\}_{I=1}^{N_s}$, $\{\Psi_I^y\}_{I=1}^{N_s}$, and $\mathbf{a}_I$, respectively. Using the equivalence between the least-squares minimization and quadrature rule by gradient RK approximation, the weighted least-squares approximation of the linear system is reached as

$$\mathbf{A}\mathbf{a} = \begin{pmatrix} \mathbf{A}^1 + \mathbf{A}^2 \\ \sqrt{\alpha_h}(\mathbf{A}^3 + \mathbf{A}^4) \\ \sqrt{\alpha_g}\mathbf{A}^5 \end{pmatrix} \mathbf{a} = \begin{pmatrix} \mathbf{b}^1 \\ \sqrt{\alpha_h}\mathbf{b}^2 \\ \sqrt{\alpha_g}\mathbf{b}^3 \end{pmatrix} = \mathbf{b}, \tag{20.84}$$

where $\mathbf{A}$ constitutes matrices associated with the differential operators and $\mathbf{b}$ constitutes vectors related to the source terms, respectively, and $\sqrt{\alpha_h}$ is the weight on $\partial\Omega_h$ and $\sqrt{\alpha_g}$ is the weight on $\partial\Omega_g$, respectively [34, 35].

**Remark 1** From the convergence theory, the corresponding weights $\sqrt{\alpha_h}$ on $\partial\Omega_h$ and $\sqrt{\alpha_g}$ on $\partial\Omega_g$ have been shown to be $\sqrt{\alpha_h} \approx O(1)$ and $\sqrt{\alpha_g} \approx O(\kappa a^{q-p-1})$, respectively, for elasticity problems [28], where $\kappa = \max\{\lambda, \mu\}$, $a$ is the compact support of the RK shape function, $p$ and $q$ are the order of bases in approximating the unknown and its derivative, respectively. For details, see Chi *et al.* [28].

**Remark 2** The convergence of G-RKCM is only dependent on the polynomial degree $q$ in the approximation of $\mathbf{u}_{,x}$ and $\mathbf{u}_{,y}$, and is independent of the polynomial degree $p$ in the approximation of $\boldsymbol{u}$. Further, $q \geq 2$ is needed for convergence. For details, see Chi *et al.* [28].

**Remark 3** The collocation points in the strong form collocation method play a similar role as the quadrature points in the least-squares method. In G-RKCM, it requires only first-order differentiation of the approximation functions for optimal convergence and thus allows the use of the same collocation and source points for sufficient accuracy in the solution process. For details, see Chi *et al.* [28].

For the $\chi$-microscopic cell problem given in (20.72) and (20.73), the gradient RK approximation in two dimensions is introduced as

$$\chi = \sum_{I=1}^{N_s} \mathbf{\Psi}_I \mathbf{a}_I, \quad \chi_{,x} = \sum_{I=1}^{N_s} \mathbf{\Psi}_I^x \mathbf{a}_I, \quad \chi_{,y} = \sum_{I=1}^{N_s} \mathbf{\Psi}_I^y \mathbf{a}_I, \tag{20.85}$$

in which $\mathbf{\Psi}_I$ and $\mathbf{a}_I$ are the matrix and vector composed of the RK shape functions and the generalized coefficients, respectively.

The corresponding matrices in $\mathbf{A}$ and $\mathbf{b}$ for solving the $\chi$-microscopic cell problem are given as

$$\mathbf{A}_{IJ}^1 = \mathbf{L}^1 \mathbf{\Psi}_J^{x^{\mathrm{T}}}(\mathbf{p}_I) = \begin{bmatrix} a_1 & 0 & 0 & a_3 & 0 & 0 \\ 0 & a_1 & 0 & 0 & a_3 & 0 \\ 0 & 0 & a_1 & 0 & 0 & a_3 \\ a_4 & 0 & 0 & a_2 & 0 & 0 \\ 0 & a_4 & 0 & 0 & a_2 & 0 \\ 0 & 0 & a_4 & 0 & 0 & a_2 \end{bmatrix}, \quad \begin{aligned} a_1 &= (\lambda + 2\mu)\Psi_{J,x}^x(\mathbf{p}_J), \\ a_2 &= \mu \Psi_{J,x}^x(\mathbf{p}_J), \\ a_3 &= \mu \Psi_{J,y}^x(\mathbf{p}_J), \\ a_4 &= \lambda \Psi_{J,y}^x(\mathbf{p}_J), \end{aligned} \tag{20.86}$$

$$\mathbf{A}_{IJ}^2 = \mathbf{L}^2 \mathbf{\Psi}_J^{y^{\mathrm{T}}}(\mathbf{p}_I) = \begin{bmatrix} a_5 & 0 & 0 & a_7 & 0 & 0 \\ 0 & a_5 & 0 & 0 & a_7 & 0 \\ 0 & 0 & a_5 & 0 & 0 & a_7 \\ a_8 & 0 & 0 & a_6 & 0 & 0 \\ 0 & a_8 & 0 & 0 & a_6 & 0 \\ 0 & 0 & a_8 & 0 & 0 & a_6 \end{bmatrix}, \quad \begin{aligned} a_5 &= \mu \Psi_{J,y}^y(\mathbf{p}_J), \\ a_6 &= (\lambda + 2\mu)\Psi_{J,y}^y(\mathbf{p}_J), \\ a_7 &= \lambda \Psi_{J,x}^y(\mathbf{p}_J), \\ a_8 &= \mu \Psi_{J,x}^y(\mathbf{p}_J), \end{aligned} \tag{20.87}$$

$$\mathbf{A}_{IJ}^3 = \mathbf{B}_h^1 \mathbf{\Psi}_J^{x^{\mathrm{T}}}(\mathbf{q}_I) = \begin{bmatrix} a_9 & 0 & 0 & a_{11} & 0 & 0 \\ 0 & a_9 & 0 & 0 & a_{11} & 0 \\ 0 & 0 & a_9 & 0 & 0 & a_{11} \\ a_{12} & 0 & 0 & a_{10} & 0 & 0 \\ 0 & a_{12} & 0 & 0 & a_{10} & 0 \\ 0 & 0 & a_{12} & 0 & 0 & a_{10} \end{bmatrix}, \quad \begin{aligned} a_9 &= (\lambda + 2\mu)\Psi_J^x(\mathbf{q}_I) n_1^s, \\ a_{10} &= \mu \Psi_J^x(\mathbf{q}_I) n_1^s, \\ a_{11} &= \mu \Psi_J^x(\mathbf{q}_I) n_2^s, \\ a_{12} &= \lambda \Psi_J^x(\mathbf{q}_I) n_2^s, \end{aligned} \tag{20.88}$$

$$\mathbf{A}_{IJ}^4 = \mathbf{B}_h^2 \mathbf{\Psi}_J^{y^{\mathrm{T}}}(\mathbf{q}_I) = \begin{bmatrix} a_{13} & 0 & 0 & a_{15} & 0 & 0 \\ 0 & a_{13} & 0 & 0 & a_{15} & 0 \\ 0 & 0 & a_{13} & 0 & 0 & a_{15} \\ a_{16} & 0 & 0 & a_{14} & 0 & 0 \\ 0 & a_{16} & 0 & 0 & a_{14} & 0 \\ 0 & 0 & a_{16} & 0 & 0 & a_{14} \end{bmatrix}, \quad \begin{aligned} a_{13} &= \mu \Psi_J^y(\mathbf{q}_I) n_2^s, \\ a_{14} &= (\lambda + 2\mu)\Psi_J^y(\mathbf{q}_I) n_2^s, \\ a_{15} &= \lambda \Psi_J^y(\mathbf{q}_I) n_1^s, \\ a_{16} &= \mu \Psi_J^y(\mathbf{q}_I) n_1^s, \end{aligned} \tag{20.89}$$

and

$$\begin{aligned} \mathbf{b}^1 &= [0 \quad 0 \quad 0 \quad 0 \quad 0 \quad 0]^{\mathrm{T}}, \\ \mathbf{b}^2 &= [(\lambda + 2\mu)n_1^s \quad \lambda n_1^s \quad \mu n_2^s \quad \lambda n_2^s \quad (\lambda + 2\mu)n_2^s \quad \mu n_1^s]^{\mathrm{T}}. \end{aligned} \tag{20.90}$$

Here, we use the property $\chi_{kmn} = \chi_{knm}$, and $\lambda$ and $\mu$ are Lamé constants.

Similar to (20.85), RK and G-RK approximations are introduced for $\boldsymbol{\eta}$ and its derivatives in the microscopic cell. The corresponding matrices in **A** and **b** for solving the $\boldsymbol{\eta}$-microscopic cell problem are as follows:

$$\begin{aligned}
\mathbf{A}_{IJ}^{1} &= \mathbf{L}^{1}\boldsymbol{\Psi}_{J}^{x^{\mathrm{T}}}(\mathbf{p}_I) = \begin{bmatrix} (\lambda+2\mu)\Psi_{J,x}^{x}(\mathbf{p}_I) & \mu\Psi_{J,y}^{x}(\mathbf{p}_I) \\ \lambda\Psi_{J,y}^{x}(\mathbf{p}_I) & \mu\Psi_{J,x}^{x}(\mathbf{p}_I) \end{bmatrix}, \\
\mathbf{A}_{IJ}^{2} &= \mathbf{L}^{2}\boldsymbol{\Psi}_{J}^{y^{\mathrm{T}}}(\mathbf{p}_I) = \begin{bmatrix} \mu\Psi_{J,y}^{y}(\mathbf{p}_I) & \lambda\Psi_{J,x}^{y}(\mathbf{p}_I) \\ \mu\Psi_{J,x}^{y}(\mathbf{p}_I) & (\lambda+2\mu)\Psi_{J,y}^{y}(\mathbf{p}_I) \end{bmatrix}, \\
\mathbf{A}_{IJ}^{3} &= \mathbf{B}_{h}^{1}\boldsymbol{\Psi}_{J}^{x^{\mathrm{T}}}(\mathbf{q}_I) = \begin{bmatrix} (\lambda+2\mu)\Psi_{J}^{x}(\mathbf{q}_I)n_1^{\mathrm{s}} & \mu\Psi_{J}^{x}(\mathbf{q}_I)n_2^{\mathrm{s}} \\ \lambda\Psi_{J}^{x}(\mathbf{q}_I)n_2^{\mathrm{s}} & \mu\Psi_{J}^{x}(\mathbf{q}_I)n_1^{\mathrm{s}} \end{bmatrix}, \\
\mathbf{A}_{IJ}^{4} &= \mathbf{B}_{h}^{2}\boldsymbol{\Psi}_{J}^{y^{\mathrm{T}}}(\mathbf{q}_I) = \begin{bmatrix} \mu\Psi_{J}^{y}(\mathbf{q}_I)n_2^{\mathrm{s}} & \lambda\Psi_{J}^{y}(\mathbf{q}_I)n_1^{\mathrm{s}} \\ \mu\Psi_{J}^{y}(\mathbf{q}_I)n_1^{\mathrm{s}} & (\lambda+2\mu)\Psi_{J}^{y}(\mathbf{q}_I)n_2^{\mathrm{s}} \end{bmatrix}, \\
\mathbf{b}^{1} &= [0 \quad 0]^{\mathrm{T}}, \quad \mathbf{b}^{2} = [n_1^{\mathrm{s}} \quad n_2^{\mathrm{s}}]^{\mathrm{T}}.
\end{aligned} \tag{20.91}$$

Similarly, for the $\boldsymbol{\kappa}$-microscopic cell problem, the associated matrices in **A** and **b** are given by

$$\begin{aligned}
\mathbf{A}_{IJ}^{1} &= \mathbf{L}^{1}\boldsymbol{\Psi}_{J}^{x^{\mathrm{T}}}(\mathbf{p}_I) = 2\mu_{\mathrm{v}} \begin{bmatrix} \Psi_{J,x}^{x}(\mathbf{p}_I) & 0 & 0 \\ 0 & \Psi_{J,x}^{x}(\mathbf{p}_I) & 0 \\ 0 & 0 & \Psi_{J,x}^{x}(\mathbf{p}_I) \end{bmatrix}, \\
\mathbf{A}_{IJ}^{2} &= \mathbf{L}^{2}\boldsymbol{\Psi}_{J}^{y^{\mathrm{T}}}(\mathbf{p}_I) = 2\mu_{\mathrm{v}} \begin{bmatrix} \Psi_{J,y}^{y}(\mathbf{p}_I) & 0 & 0 \\ 0 & \Psi_{J,y}^{y}(\mathbf{p}_I) & 0 \\ 0 & 0 & \Psi_{J,y}^{y}(\mathbf{p}_I) \end{bmatrix},
\end{aligned} \tag{20.92}$$

$$\begin{aligned}
\mathbf{A}_{IJ}^{3} &= \mathbf{B}_{g}\Psi_{J}^{\mathrm{T}}(\mathbf{q}_I) = \begin{bmatrix} \Psi_{J}(\mathbf{q}_I) & 0 & 0 \\ 0 & \Psi_{J}(\mathbf{q}_I) & 0 \\ 0 & 0 & \Psi_{J}(\mathbf{q}_I) \end{bmatrix}, \\
\mathbf{b}^{1} &= [-1 \quad 0 \quad -1]^{\mathrm{T}}, \quad \mathbf{b}^{3}[0 \quad 0 \quad 0]^{\mathrm{T}},
\end{aligned} \tag{20.93}$$

where $\mu_{\mathrm{v}}$ is the fluid viscosity.

## 20.5 Trabecular Bone Modeling

The image-based numerical framework for microstructure construction and for solving microscopic cell models will be employed to investigate the trabecular bone mechanical properties. As shown in Figure 20.7, the trabecular bone image of a sheep vertebra (1024 × 973 pixels), adapted from the website of SCANCO Medical (http://www.scanco.ch/), is obtained by Xtreme-CT (a high-resolution

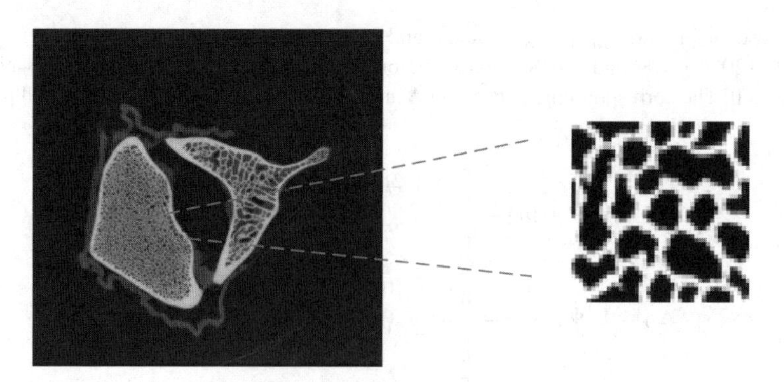

**Figure 20.7** Image of a vertebra with a specified microscopic cell discretized by 45 × 45 pixels

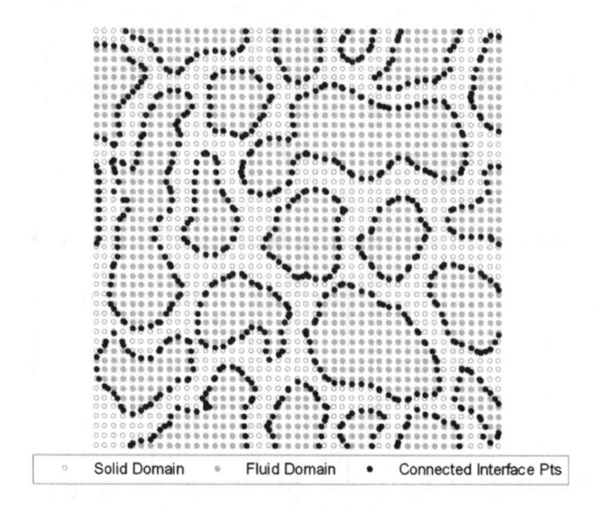


**Figure 20.8** Connected interface and partitioned domain for the microscopic cell

peripheral quantitative-computed tomography, HR-pQCT) with nominal resolution of 41 μm and a maximum scan size 126 mm × 150 mm in the plane. The corresponding microscopic cell, presented by 45 × 45 pixels, is specified by the square, with microscopic cell size 10 mm × 10 mm as shown in Figure 20.8. The material properties of the trabecular ovine bone are given as follows: the Young's modulus is 1192 MPa [36], Poisson's ratio is 0.3 [37], and the blood viscosity in the adult sheep is $4.373 \times 10^{-3}$ Pa.s (N.s/m$^2$) [38].

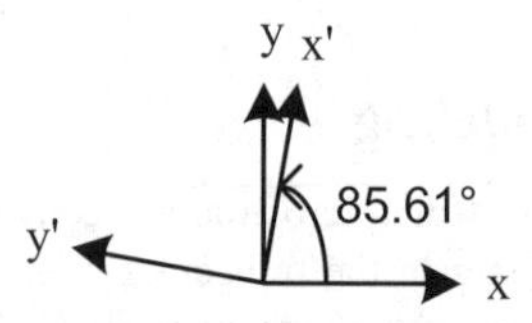


**Figure 20.9** Principal direction of the trabecular bone

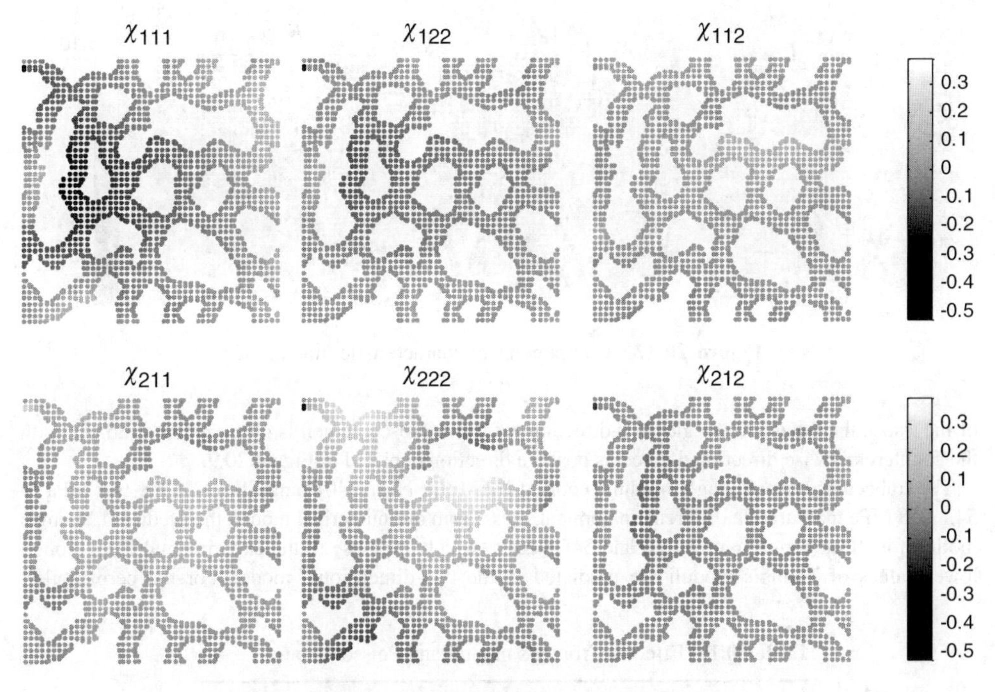


**Figure 20.10** Components of characteristic function $\chi$

The contour plots of the characteristic functions $\chi$, $\eta$, and $\kappa$, the solutions of microscopic cell problems in (20.68)–(20.73), are shown in Figure 20.10, Figure 20.11, and Figure 20.12, respectively. The effective Young's moduli and Poisson's ratio, the effective stress coefficient tensor, and the homogenized permeability tensor obtained by G-RKCM are summarized in Table 20.1, Table 20.2, and Table 20.3, respectively. Following the original work [39], the homogenized elasticity tensor is transformed to the principal planes where trabecular bones exhibit orthotropy. The difference between the components of homogenized material constants in global $x$ and $y$ directions indicates the material anisotropy of the trabecular bone. It is noted that for a two-dimensional image, the angle between the principal direction

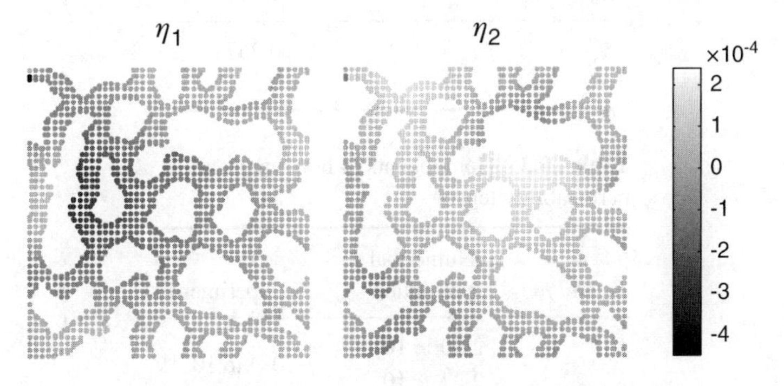


**Figure 20.11** Components of characteristic function $\eta$

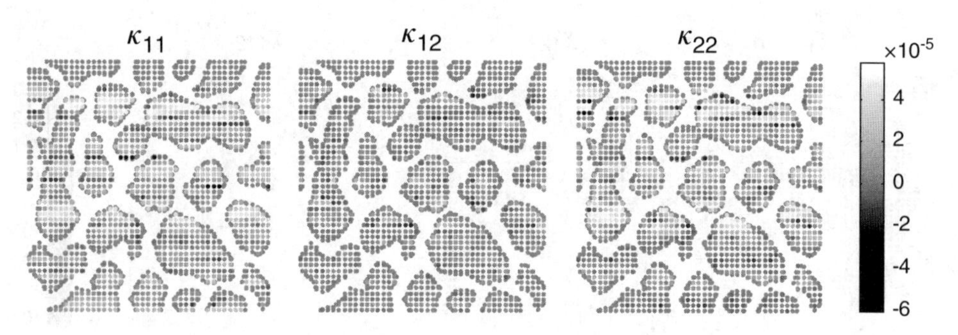


**Figure 20.12** Components of characteristic function **κ**

of the trabecular bone and the global $x$-direction for the microscopic cell is observed to be 85.61° with the counterclockwise direction defined as positive direction depicted in Figure 20.9.

The trabecular bone Young's moduli reported by Mittra *et al.* [40] range from 359 ±104 MPa to 542 ±78 MPa measured in different anatomical sites. From our numerical model, the predicted Young's moduli for sheep vertebra are 318.6 and 342.8 MPa, which are close to the experimental data, though lower values of Young's moduli are predicted in the two-dimensional model. For the permeability

**Table 20.1** Effective Young's moduli and Poisson's ratio

| Material constants | Numerical prediction | Experiment |
|---|---|---|
| $\bar{E}_{11}$ (MPa) | 318.6 | 359 ± 104 to |
| $\bar{E}_{22}$ (MPa) | 342.8 | 542 ± 78 |
| $\bar{\nu}_{12}$ | 0.1714 | – |
| $\bar{\nu}_{21}$ | 0.1844 | |

**Table 20.2** Components of homogenized effective stress coefficient tensor

| $\bar{\alpha}_{ij}$ | Numerical prediction |
|---|---|
| $\bar{\alpha}_{11}$ | 0.2571 |
| $\bar{\alpha}_{22}$ | 0.2545 |

**Table 20.3** Components of homogenized permeability tensor

| $\bar{K}_{ij}$ (m$^2$) | Numerical prediction | Experiment |
|---|---|---|
| $\bar{K}_{11}$ | $2.80 \times 10^{-8}$ | $10^{-8}$ to $10^{-14}$ |
| $\bar{K}_{22}$ | $2.27 \times 10^{-8}$ | |

measured in the trabecular bones, the reported range of bone permeability in literature [41] has a wide span from $10^{-8}$ to $10^{-14}$ $m^2$ due to flow direction, bone porosity, and microstructural morphology such as trabecular architecture and separation for various anatomical sites. The predicted permeability from our numerical simulation is within range $10^{-8}$ $m^2$, which shows agreement with the reported data. On the other hand, the porosity for the trabecular bone microstructure is 0.5620, which lies within the typical porosity range from 0.5 to 0.9 for trabecular bones [42–44], though a two-dimensional model might underestimate the interconnection of pores.

## 20.6 Conclusions

We introduced an asymptotic-based multiscale homogenization method to correlate the macro- and micro-mechanical behaviors of poroelastic materials, where an elastic solid and Newtonian fluid of low viscosity are considered. Through this homogenization process, the generalized Darcy's law, homogenized macroscopic continuity equation, and homogenized macroscopic equilibrium equation were obtained, where the homogenized macroscopic continuity and equilibrium equations reassemble the governing equations in Biot's theory. To effectively construct microstructures with multiple phases from medical images, we introduced the active contour model based on variational level set formulation for interface identification and boundary segmentation without the need of CAD procedures in mesh-based methods. Inspired by the images with pixel point discretization, we introduced the collocation method to solve the level set equation. When using RKCM to solve the level set equation in the image segmentation, it is shown that the consistency conditions imposed in construction of the RK shape function can be relaxed. Specifically, the kernel function can be directly used as the shape function in the strong form collocation method for computational efficiency. The direct collocation method has been introduced to solve the level set equation without regularization term, and the transformation matrix is no longer needed in RKCM for solving the active contour model, thereby making the numerical algorithm very efficient. In the solution process of microscopic cells, we introduced the G-RKCM to solve the microscopic cell problems in determined linear systems, while no second-order differentiation is needed. A multiscale modeling of trabecular bone microstructure is presented herein. The ability to solve the microstructure with complex geometry by the proposed image-based strong form collocation method is demonstrated.

## Acknowledgment

The support of US Army Engineer Research Development Center under the contract W912HZ110027 to UCLA and National Institute of Health under the contract AR053343 to UCSD with subcontract 31928 FDP-UCSD to UCLA are greatly acknowledged.

## References

[1] Biot, M.A. (1941) General theory of three-dimensional consolidation. *Journal of Applied Physics*, **12**, 155–164.
[2] Williams, J.L. and Lewis, J.L. (1982) Properties and an anisotropic model of cancellous bone from the proximal tibia epiphysis. *Journal of Biomechanical Engineering*, **104**, 50–56.
[3] Gibson, L.J. (1985) The mechanical behavior of cancellous bone. *Journal of Biomechanics*, **18**, 317–328.
[4] Hollister, S.J., Fyhrie, D.P., Jepsen, K.J., and Goldstein, S.A. (1991) Application of homogenization theory to the study of trabecular bone mechanics. *Journal of Biomechanics*, **24**, 825–839.
[5] Hollister, S.J. and Riemer, B.A. (1993) Digital-image-based finite element analysis for bone microstructure using conjugate gradient and Gaussian filter techniques, in *Mathematical Methods in Medical Imaging 11, Proceedings SPIE–The International Society for Optical Engineering*, Society of Photo-Optical Instrumentation Engineers, Bellingham, WA, pp. 95–106.
[6] Hollister, S.J. and Kikuchi, N. (1994) Homogenization theory and digital imaging: a basis for studying the mechanics and design principles of bone tissue. *Biotechnology and Bioengineering*, **43**, 586–596.

[7] Hoppe, H., DeRose, T., Duchamp, T. *et al.* (1993) Mesh optimization, in *Proceedings of the 20th Annual Conference on Computer Graphics and Interactive Techniques*, ACM, Anaheim, CA, pp. 19–26.
[8] Ulrich, D., van Rietbergen, B., Weinans, H., and Ruegsegger, P. (1998) Finite element analysis of trabecular bone structure: a comparison of image-based meshing techniques. *Journal of Biomechanics*, **31**, 1187–1192.
[9] Podshivalov, L., Holdstein, Y., Fischer, A., and Bar-Yoseph, P.Z. (2009) Towards a multi-scale computerized bone diagnostic system: 2D micro-scale finite element analysis. *Communications in Numerical Methods in Engineering*, **25**, 733–749.
[10] Podshivalov, L., Fischer, A., and Bar-Yoseph, P.Z. (2011) Multiscale FE method for analysis of bone micro-structures. *Journal of the Mechanical Behavior of Biomedical Materials*, **4**, 888–899.
[11] Charras, G.T. and Guldberg, R.E. (2000) Improving the local solution accuracyof large-scale digital image-based finite element analyses. *Journal of Biomechanics*, **33**, 255–259.
[12] Hara, T., Tanck, E., Homminga, J., and Huiskes, R. (2002) The influence of microcomputed tomography threshold variations on the assessment of structural and mechanical trabecular bone properties. *Bone*, **31**, 107–109.
[13] Babuska, I. (1976) Homogenization approach in engineering, in *Computing Methods in Applied Sciences and Engineering* (eds J.-L. Lions and R. Glowinski), vol. 134 of Lecture Notes in Economics and Mathematical Systems, Springer-Verlag, Berlin, pp. 137–153.
[14] Bensousson, A., Lions, J.-L., and Papanicolaou, G. (1978) *Asymptotic Analysis for Periodic Structures*, North-Holland, Amsterdam.
[15] Sanchez-Palencia, E. (1980) Non-homogeneous media and vibration theory, in *Monte Carlo and Quasi-Monte Carlo Methods 1996* (eds H. Niederreiter, P. Hellekalek, G. Larcher, and P. Zinterhof), vol. 127 of *Lecture Notes in Physics*, Springer-Verlag, Berlin.
[16] Lions, J.-L. (1981) *Some Methods in the Mathematical Analyses of Systems and their Control*, Science Press/Gordon and Breach, Beijing/New York, NY.
[17] Guedes, J.M. and Kikuchi, N. (1990) Preprocessing and postprocessing for materials based on the homogenization method with adaptive finite element methods. *Computer Methods in Applied Mechanics and Engineering*, **83**, 143–198.
[18] Hornung, U. (1997) *Homogenization and Porous Media*, Springer, New York, NY.
[19] Terada, K., Ito, T., and Kikuchi, N. (1998) Characterization of the mechanical behaviors of solid–fluid mixture by the homogenization method. *Computer Methods in Applied Mechanics and Engineering*, **153**, 223–257.
[20] Osher, S. and Sethain, J.A. (1988) Fronts propagation with curvature-dependent speed: algorithms based on Hamilton–Jacobi formulation. *Journal of Computational Physics*, **79**, 12–49.
[21] Osher, S.J. and Fedkiw, R.P. (2002) *Level Set Methods and Dynamic Implicit Surfaces*, Springer Verlag, New York, NY.
[22] Chan, T. and Vese, L. (1999) An active contour model without edges, in *Scale-Space Theories in Computer Vision* (eds M. Neilsen, P. Johansen, O.F. Olsen and J. Weickert), vol. 1687 of Lecture Notes in Computer Science, Springer, Berlin, pp. 141–151.
[23] Chan, T. and Vese, L. (2001) Active contours without edges. *IEEE Transactions on Image Processing*, **10** (2), 266–277.
[24] Guan, P.C. (2008) Adaptive coupling of FEM and RKPM formulations for contact and impact problems, Civil and Environmental Engineering Department, University of California, Los Angeles, CA.
[25] Guan, P.C., Chi, S.W., Chen, J.S. *et al.* (2011) Semi-Lagrangian reproducing kernel particle method for fragment-impact problems. *International Journal of Impact Engineering*, **38**, 1033–1047.
[26] Mumford, D. and Shah, J. (1989) Optimal approximation by piecewise smooth functions and associated variational problems. *Communications on Pure and Applied Mathematics*, **42**, 577–685.
[27] Hu, H.Y., Chen, J.S., and Hu, W. (2011) Error analysis of collocation method based on reproducing kernel approximation. *Numerical Methods for Partial Differential Equations*, **27**, 554–580.
[28] Chi, S.W., Chen, J.S., Hu, H.Y., and Yang, J.P. (2012) A gradient reproducing kernel collocation method for boundary value problems. *International Journal for Numerical Methods in Engineering*, DOI: 10.1002/nme.4432.
[29] Liu, W.K., Jun, S., and Zhang, Y.F. (1995) Reproducing kernel particle methods. *International Journal for Numerical Methods in Fluids*, **20**, 1081–1106.
[30] Chen, J.S., Pan, C., Wu, C.T., and Liu, W.K. (1996) Reproducing kernel particle methods for large deformation analysis of nonlinear structures. *Computer Methods in Applied Mechanics and Engineering*, **139**, 195–227.

[31] Li, S. and Liu, W.K. (1998) Synchronized reproducing kernel interpolant via multiple wavelet expansion. *Computational Mechanics*, **21**, 28–47.
[32] Li, S. and Liu, W.K. (1999) Reproducing kernel hierarchical partition of unity. Part I: Formulation and theory. *International Journal for Numerical Methods in Engineering*, **45**, 251–288.
[33] Li, S. and Liu, W.K. (1999) Reproducing kernel hierarchical partition of unity. Part II: Applications. *International Journal for Numerical Methods in Engineering*, **45**, 289–317.
[34] Zhang, X., Liu, X.H., Song, K.Z., and Lu, M.W. (2001) Least-squares collocation meshless method. *International Journal for Numerical Methods in Engineering*, **51**, 1089–1100.
[35] Hu, H.Y., Chen, J.S., and Hu, W. (2007) Weighted radial basis collocation method for boundary value problems. *International Journal for Numerical Methods in Engineering*, **69**, 2736–2757.
[36] Nafei, A., Danielsen, C.C., Linde, F., and Hvid, I. (2000) Properties of growing trabecular ovine bone. Part I: Mechanical and physical properties. *Journal of Bone & Joint Surgery, British Volume*, **82**, 910–920.
[37] Brown, T.D., Pedersen, D.R., Radin, E.L., and Rose, R.M. (1988) Global mechanical consequences of reduced cement/bone coupling rigidity in proximal femoral arthroplasty: a three-dimensional finite element analysis. *Journal of Biomechanics*, **21**, 115–129.
[38] Windberger, U., Bartholovitsch, A., Plasenzotti, R. *et al.* (2003) Whole blood viscosity, plasma viscosity and erythrocyte aggregation in nine mammalian species: reference values and comparison of data. *Experimental Physiology*, **88**, 431–440.
[39] Cowin, S.C. and Mehrabadi, M.M. (1987) On the identification of material symmetry for anisotropic elastic materials. *Quarterly Journal of Mechanics and Applied Mathematics*, **40**, 451–476.
[40] Mittra, E., Rubin, C., and Qin, Y.-X. (2005) Interrelationship of trabecular mechanical and microstructural properties in sheep trabecular bone. *Journal of Biomechanics*, **38**, 1229–1237.
[41] Nauman, E.A., Fong, K.E., and Keaveny, T.M. (1999) Dependence of intertrabecular permeability on flow direction and anatomic site. *Annals of Biomedical Engineering*, **27**, 517–524.
[42] Gibson, L.J. (1985) The mechanical behavior of cancellous bone. *Journal of Biomechanics*, **18**, 317–328.
[43] Kuhn, J.L., Goldstein, S.A., Feldkamp, L.A. *et al.* (1990) Evaluation of a microcomouted tomography system to study trabecular bone structure. *Journal of Orthopaedic Research*, **8**, 833–842.
[44] Cowin, S.C. (1999) Bone poroelasticity. *Journal of Biomechanics*, **32**, 217–238.

# 21

# Modeling Nonlinear Plasticity of Bone Mineral from Nanoindentation Data

Amir Reza Zamiri and Suvranu De

*Department of Mechanical, Aerospace, and Nuclear Engineering, Rensselaer Polytechnic Institute, New York, USA*

## 21.1 Introduction

Bone is a hierarchical biological composite that includes inorganic minerals within an organic collagen matrix [1, 2]. At the nanometer length scale, bone is composed of self-assembled collagen fibrils and inorganic hydroxyapatite nanocrystals [3, 4]. There is much interest in characterizing the mechanical properties of hydroxyapatite not only for its importance in the overall mechanical behavior of healthy and diseased bone, but also for its wide-scale application in biomaterials, regenerated hard tissue, and medicine [5, 6]. For instance, hydroxyapatite crystals may play a key role in metabolic activity of bone [7] and may induce cytotoxicity of certain cancer cells [8].

Like most crystalline solids, the mechanical properties of hydroxyapatite are strongly influenced by defects such as dislocations through slip-induced plastic deformation. They also have an important role in *in-vivo* dissolution processes and in crystal maturation of biological apatites [9]. Studies using microscale compression tests have shown that the hexagonal lattice of hydroxyapatite crystals exhibits highly anisotropic plastic deformation at high temperatures [10]. Based on these experiments, it is now recognized that the mechanical deformation of hydroxyapatite single crystals is based on crystallographic slip with $\{1010\}\langle 0001\rangle$-type slip systems. Indentation experiments have been performed to investigate the mechanical properties of hydroxyapatite single crystals [11, 12] which exhibit mechanical behavior that is scale dependent. At the microscale, hydroxyapatite crystals are very brittle and crack during microindentation [11]. However, plastic response is observed during nanoindentation [11].

Hydroxyapatite single crystals exhibit anisotropic deformation behavior with the crystal being softer and morc ductile along the [1010] direction [12]. As mentioned above, the mechanical properties of hydroxyapatite crystals are a function of crystal size [13]. At the nanoscale, mechanical properties of crystals become insensitive to the defects and reach their theoretical maximum values. A study of the

*Multiscale Simulations and Mechanics of Biological Materials*, First Edition. Edited by Shaofan Li and Dong Qian.

mechanical response of hydroxyapatite single crystal using atomistic simulations shows that the yield stress and hardness of the crystal significantly increase when the defects in the crystal do not contribute to the deformation [14].

The instrumented indentation studies provide valuable qualitative information about the overall mechanical properties of hydroxyapatite single crystals. However, there is dearth of quantitative data, especially in the nonlinear regime. A number of studies have been focused to find the anisotropic elastic tensor of hydroxyapatite [15–23]. Atomistic simulations have been performed to determine the elastic tensor of hydroxyapatite crystal [14]. However, to develop a nonlinear mechanical model of the elastic–plastic response, more detailed analysis of the indentation data along multiple crystallographic directions must be performed.

Indentation data have been used extensively to determine mechanical response of materials, including Young's modulus, hardness, yield stress, and flow properties [24–26]. However, such studies use finite-element simulations extensively. In this chapter, we present a two-step method in which we first use the hardness and Young's modulus obtained through indentation experiments to compute yield stress and the information to find flow properties.

The chapter is organized as follows. In Section 21.2 a reverse analysis algorithm is introduced to identify the nonlinear elastic–plastic mechanical properties of crystalline materials such as bone mineral. In Section 21.3 the model is used for elastic–plastic mechanical property identification of hydroxyapatite single crystal from nanoindentation data.

## 21.2 Methods

The test specimen is indented with various indenter geometries during instrumented indentation and the force–depth response is measured during the loading and unloading cycle (Figure 21.1a). The hardness $H$ or the average contact pressure $P_{\text{avg}}$ can be extracted from such experiments as

$$H = P_{\text{avg}} = \frac{P_{\text{m}}}{A_{\text{m}}}, \tag{21.1}$$

where $A_{\text{m}}$ is the projected contact area at maximum load $P_{\text{m}}$. For a sharp indenter, the loading curve is assumed to follow the following relationship:

$$P = Ch^2, \tag{21.2}$$

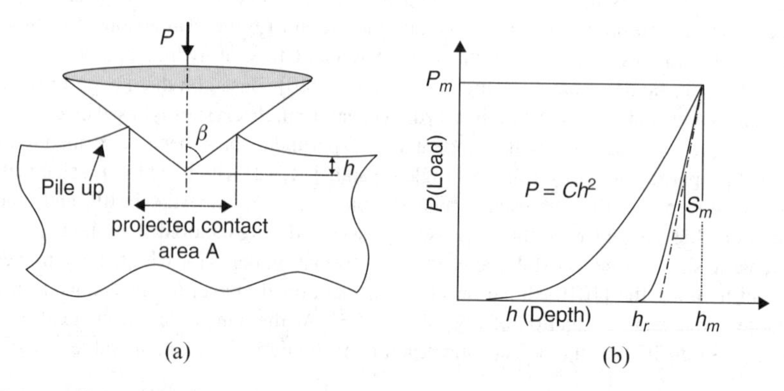


**Figure 21.1** (a) Schematic illustration of instrumented indentation of a homogeneous, isotropic semi-infinite substrate with a sharp conical indenter. (b) Typical load–displacement curves obtained from an indentation experiment

where $P$ is the force of indentation, $h$ is the corresponding depth, and $C$ is the loading curvature. Further analysis is usually performed to process the indentation data to determine elastic–plastic properties, including the loading curvature $C$, and unloading slope $S_m$ at maximum depth $h_m$ and residual indentation depth $h_r$ (Figure 21.1b).

We compute these parameters in a two-step process. In the first step, we compute the yield stress from the hardness and Young's modulus data, while in the second step we determine the hardening parameters, as explained below.

## *Step 1: Determination of Yield Stress*

According to the slip-line field theory for rigid, perfectly plastic materials, the hardness of the material, obtained through indentation, is directly related to the yield stress $H \cong 3\sigma_y$ [27]. However, in the expanding cavity models (ECMs), the hardness is not only related to the yield stress $\sigma_y$ but also to the Young's modulus $E$ of the indented material [28]. Further analyses showed that the hardness of elastic–plastic materials is also a function of the inclination of the indenter face $\beta$ [29, 30]; that is:

$$R_Y = \frac{H}{\sigma_y} = \Psi_1\left(\frac{E}{\sigma_y}, \beta\right). \tag{21.3}$$

For a wedge-type indenter and incompressible material, the functional form of Equation (21.3) may be chosen as [30]

$$R_Y = \frac{H}{\sigma_y} = \frac{\xi}{\sqrt{3}}\left[1 + \ln\left(\frac{4\sigma_y}{3\pi E}\tan\beta\right)\right], \tag{21.4}$$

where $\xi$ is a parameter that depends on the ratio of the yield stress and Young's modulus of the indented material. Equation (21.3) may, therefore, be used to compute the yield stress, provided hardness and Young's modulus are known. Based on observations, for hard metallic alloys and ceramics, $\xi$ may be chosen as $5\sqrt{2}/9$ and 1, respectively.

## *Step 2: Determination of Hardening Parameters*

The stress–strain response of most materials may be expressed as

$$\sigma = \begin{cases} E\varepsilon, & \varepsilon \le \varepsilon_y, \\ K\varepsilon^n, & \varepsilon \ge \varepsilon_y, \end{cases} \tag{21.5}$$

where $\varepsilon_y = \sigma_y/E$ is the yield strain and $n$ and $K$ are the work-hardening exponent and work-hardening rate, respectively. Cheng and Cheng [24, 25] proposed that the load curvature $C$ in Equation (21.2) has the following functional form:

$$C = \frac{P}{h^2} = \bar{E}\Psi_2\left(\frac{\sigma_y}{\bar{E}}, n, \beta\right), \tag{21.6}$$

where $\bar{E}$ is the reduced Young's modulus with the following expression:

$$\bar{E} = \left(\frac{1-\nu^2}{E} + \frac{1-\nu_i^2}{E_i}\right)^2, \tag{21.7}$$

with $(\nu, E)$ and $(\nu_i, E_i)$ being the Poisson's ratio and Young's modulus pairs of the indented specimen and the indenter, respectively. Dao *et al.* [26] proposed an alternative expression of the load curvature:

$$C = \frac{P}{h^2} = \sigma_r \Psi_2 \left( \frac{\bar{E}}{\sigma_r}, n \right), \tag{21.8}$$

where $\sigma_r$ is a representative stress corresponding to a characteristic strain of the indented material.

Since $n$, $\sigma_r$ and $\bar{E}$ are unknown in Equation (21.8), a more useful expression is [26]

$$C = \frac{P}{h^2} = \sigma_r \Psi_3 \left( \frac{\bar{E}}{\sigma_r}, \frac{\sigma_y}{\sigma_r} \right), \tag{21.9}$$

in which only $\sigma_r$ and $\bar{E}$ are unknown.

The following functional form of $\Psi_3$ has been proposed [31–34] based on extensive finite-element analysis:

$$C = \frac{P}{h^2} = A\sigma_{0.29} \left[ 1 + \frac{\sigma_y}{\sigma_{0.29}} \right] \left[ B + \ln \left( \frac{\bar{E}}{\sigma_y} \right) \right], \tag{21.10}$$

where $A$ and $B$ are constants which depend on indenter geometry. $\sigma_{0.29}$ is the stress which corresponds to the characteristic strain $\varepsilon_r = 0.29$ for the indented material in uniaxial compression. Previous finite-element analyses for sharp indenters (e.g., Berkovich and Vickers) suggest that $\varepsilon_r = 0.29$ [32].

Computation of the reduced elastic constant $\bar{E}$ using Equation (21.7) is challenging, as the indenter elastic constant is not easily determined. An easier way is to use the unloading curve. If the unloading force is $P_u$, then the slope of the unloading portion of the indentation curve is

$$S = \frac{dP_u}{dh} = \frac{dP_u}{dh}(h, h_m, \bar{E}, \sigma_r, n). \tag{21.11}$$

Dimensional analysis gives [26]

$$S = \frac{dP_u}{dh} = \bar{E} h \Psi_4 \left( \frac{h_m}{h}, \frac{\sigma_r}{\bar{E}}, n \right). \tag{21.12}$$

At $h = h_m$

$$S_m = \left. \frac{dP_u}{dh} \right|_{h=h_m} = \bar{E} h_m \Psi_4 \left( \frac{\sigma_r}{\bar{E}}, n \right). \tag{21.13}$$

However, the function $\Psi_4$ has three unknowns ($n$, $\sigma_r$, and $\bar{E}$). Based on previous analysis [35], we use a simpler dimensionless function:

$$S_m = \left. \frac{dP_u}{dh} \right|_{h=h_m} = \bar{E} h_m c^* \frac{\sqrt{A_m}}{h_m} \tag{21.14}$$

where $c^*$ is a computationally driven parameter which depends on the geometry of the indenter. Hence,

$$\bar{E} = \frac{1}{c^* \sqrt{A_m}} \left. \frac{dP}{dh} \right|_{h=h_m}. \tag{21.15}$$

The work hardening exponent $n$ can be computed, using $\sigma_{0.29}$ and $\sigma_y$, as

$$n = \frac{\ln(\sigma_{0.29}) - \ln(\sigma_y)}{\ln(0.29) - \ln(\varepsilon_y)}. \tag{21.16}$$

The work hardening rate $K$ is obtained using Equation (21.6).

## 21.3 Results

We apply the analysis algorithm presented above to extract the flow properties of single crystal hydroxyapatite based on nanoindentation data (Table 21.1) for indentation on the prism (1010) and basal (0001) facets [12]. The results are presented in Table 21.2.

**Table 21.1** Indentation data for hydroxyapatite single crystal [12]

| Indented facet | $E$ (GPa) | $H$ (GPa) | $P$ (N) | $C$ (GPa) | $A_m$ (m$^2$) | $S_m$ (N/m) |
|---|---|---|---|---|---|---|
| (0001) | 150.38 | 7.06 | $5 \times 10^{-3}$ | $2.32 \times 10^{11}$ | $7.08 \times 10^{-13}$ | $1.20 \times 10^{5}$ |
| (1010) | 143.56 | 6.41 | $7.5 \times 10^{-3}$ | $2.03 \times 10^{11}$ | $1.17 \times 10^{-12}$ | $1.65 \times 10^{5}$ |

**Table 21.2** Flow properties of hydroxyapatite single crystal

| Facet | $\sigma_y$ (GPa) | $K$ (GPa) | $n$ |
|---|---|---|---|
| (0001) | 2.42 | 18.82 | 0.49 |
| (1010) | 2.17 | 14.47 | 0.45 |

The yield stress is seen to be lower along the [0001] direction than along the [1010] direction. The work hardening rate coefficient and exponent are also higher along the [0001] direction. The corresponding stress–strain curves are plotted in Figure 21.2.

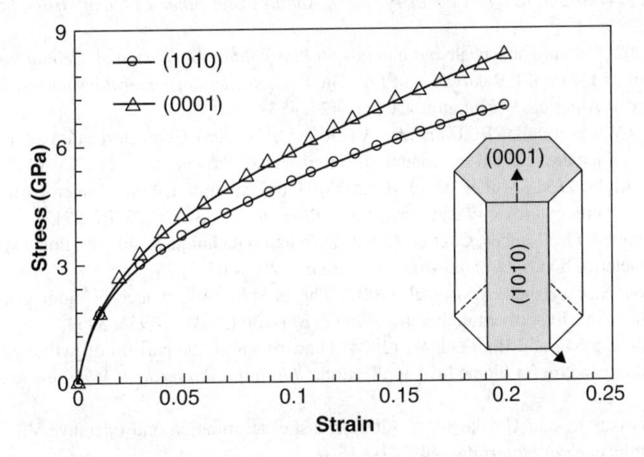


**Figure 21.2** Stress–strain response of the hydroxyapatite single crystal extracted from nanoindentation data

Unlike other hexagonal close-packed crystals, only the $\{1010\}\langle 0001\rangle$ slip systems get activated [10] in hydroxyapatite, which explains its lower yield point along the [1010] direction. Such behavior is not surprising, as hydroxyapatite is a molecular crystalline material and experiences more constraints to dislocation motion along certain crystallographic directions than atomic crystals do.

## 21.4 Conclusions

We presented a method to compute the anisotropic elasto-plastic mechanical response of hydroxyapatite single crystals from nanoindentation data. Since these are molecular crystals, in addition to the effects of dislocation-induced plastic flow, changes in molecular orientation and debonding may be important. However, we have not considered such effects in this study. Based on *ab initio* simulations of defect-free hydroxyapatite single crystals it has been shown [14] that uniaxial loading along the [0001] direction exhibits a more brittle response than along the [1010] direction. During tensile straining it is observed that the deformation behavior of defect-free hydroxyapatite single crystal is mainly due to the rotation of the $PO_4$ tetrahedra with associated motion of both columnar and axial $Ca^{2+}$ ions. However, such *ab initio* simulations cannot consider the effects of dislocations owing to limitations on unit cell size.

The technique presented here may be used for investigating the mechanical properties of other naturally occurring crystalline materials.

## Acknowledgments

We gratefully acknowledge the support of this work through Office of Naval Research grants N000140510686 and N000140810462 with Dr. Clifford Bradford as the cognizant Program Manager. We also acknowledge the help of Rahul in manuscript preparation.

## References

[1] Mann, S. (1993) Molecular tectonics in biomineralization and biomimetic materials chemistry. *Nature*, **365**, 499–505.

[2] Currey, J.D. (2005) Hierarchies in biomineral structures. *Science*, **309**, 253–254.

[3] Elliott, J.C. (1994) *Structure and Chemistry of the Apatites and other Calcium Orthophosphates*, Elsevier, Amsterdam.

[4] Elliott, J.C. (2002) Calcium phosphate biominerals, in *Phosphates: Geochemical, Geobiological and Materials Importance* (eds M.J. Kohn, J. Rakovan, and J.M. Hughes), *Reviews in Mineralogy and Geochemistry*, vol. 48, Mineral Society of America, Washington, DC, pp. 427–453.

[5] Rodrigues, C.V.M., Serricella, P., Linhares, A.B.R. *et al.* (2003) Characterization of a bovine collagen–hydroxyapatite composite scaffold for bone tissue engineering. *Biomaterials*, **24** (27), 4987–4997.

[6] Motskin, M., Wright, D.M., Muller, K. *et al.* (2009) Hydroxyapatite nano and microparticles: correlation of particle properties with cytotoxicity and biostability. *Biomaterials*, **30** (19), 3307–3317.

[7] Bazin, D., Chappard, C., Combes, C. *et al.* (2009) Diffraction techniques and vibrational spectroscopy opportunities to characterise bones. *Osteoporosis International*, **20**, 1065–1075.

[8] Hou, C.H., Hou, S.M., Hsueh, Y.S. *et al.* (2009) The in vivo performance of biomagnetic hydroxyapatite nanoparticles in cancer hyperthermia therapy. *Biomaterials*, **30** (23–24), 3956–3960.

[9] Porter, A.E., Best, S.M., and Bonfield, W. (2004) Ultrastructural comparison of hydroxyapatite and silicon-substituted hydroxyapatite for biomedical applications. *Journal of Biomedical Materials Research A*, **68**, 133–141.

[10] Nakano, T., Awazu, T., and Umakoshi, Y. (2001) Plastic deformation and operative slip system in mineral fluorapatite single. *Scripta Materialia*, **44**, 811–815.

[11] Viswanath, B., Raghavan, R., Ramamurty, U., and Ravishankar, N. (2007) Mechanical properties and anisotropy in hydroxyapatite single crystals. *Scripta Materialia*, **57**, 361–364.

[12] Saber-Samandari, S. and Gross, K.A. (2009) Micromechanical properties of single crystal hydroxyapatite by nanoindentation. *Acta Biomaterialia*, **5**, 2206–2212.

[13] Gao, H.J., Ji, B.H., Jager, I.L. *et al.* (2003) Materials become insensitive to flaws at nanoscale: lessons from nature. *Proceedings of the National Academy of Science of the United States of America*, **100** (10), 5597–5600.

[14] Ching, W.Y., Rulis, P., and Misra, A. (2009) Ab initio elastic properties and tensile strength of crystalline hydroxyapatite. *Acta Biomaterialia*, **5**, 3067–3075.

[15] Guy, J.L., Mann, A.B., Kivi, K.J. *et al.* (2002) Nanoindentation mapping of the mechanical properties of human molar tooth enamel. *Archives of Oral Biology*, **47** (4), 281–291.

[16] Hearmon, R.F.S. (1961) *An Introduction to Applied Anisotropic Elasticity*, Oxford University Press, London.

[17] Katz, J.L. and Ukrainck, K. (1971) Anisotropic elastic properties of hydroxyapatite. *Journal of Biomechanics*, **4** (3), 221–227.

[18] Lees, S.R.F. (1972) Anisotropy in hard dental tissues. *Journal of Biomechanics*, **5**, 557–566.

[19] Gilmore, R.S. and Katz, J.L. (1982) Elastic properties of apatites. *Journal of Materials Science*, **17** (4), 1131–1141.

[20] Lopes, M.A., Silva, R.F., Monteiro, F.J., and Santos, J.D. (2000) Microstructural dependence of Young's and shear moduli of $P_2O_5$ glass reinforced hydroxyapatite for biomedical applications. *Biomaterials*, **21** (7), 749–754.

[21] Nelea, V., Pelletie, H., Iliescu, M. *et al.* (2002) Calcium phosphate thin film processing by paulsed laser deposition. *Journal of Materials Science: Materials in Medicine*, **13**, 1167–1173.

[22] Nieh, T.G., Jankowsk, A.F., and Koike, J. (2001) Processing and characterization of hydroxyapatite coating on titanium produced by magnetron sputting. *Materials Research*, **16** (11), 3238–3245.

[23] Snyders, R., Music, D., Sigumonrong, D. *et al.* (2007) Experimental and ab initio study of the mechanical properties of hydroxyapatite. *Applied Physics Letters*, **90**, 193902.

[24] Cheng, Y.T. and Cheng, C.M. (1998) Scaling approach to conical indentation in elastic–plastic solids with work hardening. *Journal of Applied Physics*, **84**, 1284–1291.

[25] Cheng, Y.T. and Cheng, C.M. (2004) Scaling, dimensional analysis, and indentation measurements. *Materials Science and Engineering: R*, **44**, 91–149.

[26] Dao, M., Chollacoop, N., VanVliet, K.J. *et al.* (2001) Computational modeling of the forward and reverse problems in instrumented sharp indentation. *Acta Materialia*, **49**, 3899–3918.

[27] Tabor, D. (1951) *Hardness of Metals*, Clarendon Press, Oxford.

[28] Marsh, D.M. (1964) Plastic flow in glass. *Proceedings of the Royal Society of. London, Series A*, **279**, 420–435.

[29] Hirst, W. and Howse, M.G.J.W. (1969) The indentation of materials by wedges. *Proceedings of the Royal Society of. London, Series A*, **311**, 429–444.

[30] Johnson, K.L. (1970) The correlation of indentation experiments. *Journal of the Mechanics and Physics of Solids*, **18**, 115–126.

[31] Giannakopoulos, A.E., Larsson, P.L., and Vestergaard, R. (1994) Analysis of Vickers indentation. *International Journal of Solids and Structures*, **31**, 2679–2708.

[32] Giannakopoulos, A.E. and Suresh, S. (1999) Determination of elastoplastic properties by instrumented sharp indentation. *Scripta Materialia*, **40**, 1191–1198.

[33] Larsson, P.L., Giannakopoulos, A.E., Soderlund, E. *et al.* (1996) Analysis of Berkovich indentation. *International Journal of Solids and Structures*, **33** (2), 221–248.

[34] Venkatesh, T.A., Van Vliet, K.J., Giannakopoulos, A.E., and Suresh, S. (2000) Determination of elastoplastic properties by instrumented sharp indentation: guidelines for property extraction. *Scripta Materialia*, **42**, 833–839.

[35] King, R.B. (1987) Elastic analysis of some punch problems for a layered medium. *International Journal of Solids and Structures*, **23** (12), 1657–1664.

# 22

# Mechanics of Cellular Materials and its Applications

Ji Hoon Kim[a], Daeyong Kim[a], and Myoung-Gyu Lee[b]
[a]*Korea Institute of Materials Science, South Korea*
[b]*Pohang University of Science and Technology, South Korea*

## 22.1 Biological Cellular Materials

There are many kinds of natural cellular materials that you can meet in your everyday lives. Among them, we limit our focus on bones that make up the body of a human or animal.

### *22.1.1 Structure of Bone*

Bone is classified into two different types depending on the porosity (or density): the cortical bone and trabecular bone. The cortical bone has a more compact density than the trabecular bone and has a low porosity of less than 10%. The tissue of the cortical bone is laid down in lamellae within which the collagen fibers are aligned in parallel. The cortical bone has a cylindrical structure with a diameter of approximately 200 μm and a length of 10 mm along the longitudinal axis of the bone. On the other hand, the trabecular bone consists of a matrix which has a three-dimensional (3D) porous (or cellular) network connected with each other. The space inside the cellular structure is filled with bone marrow, providing the bone with a vascular environment. Trabecular bone or cancellous bone is usually found at the end of long bones in the vertebral bodies such as ribs and scapula. Figure 22.1 shows the hierarchical bone structure at various levels of scale [1, 2]. Figure 22.2 shows the detailed 3D structure of a trabecular bone image obtained by the 3D micro-computed tomography (μCT) technique. As shown in this figure, the trabeculae are organized as a network of closed plates and rods (Figure 22.2a). As the bone is aged, it is known that the density of the trabecular bone becomes lower and the types of trabeculae change from plate type to thin rod type, as shown in Figure 22.2b.

### *22.1.2 Mechanical Properties of Bone*

The bone has been considered as a highly anisotropic elastic material like fiber-reinforced composites on the microscopic scale. The structural alignment of lamellae along the longitudinal direction results

*Multiscale Simulations and Mechanics of Biological Materials*, First Edition. Edited by Shaofan Li and Dong Qian.

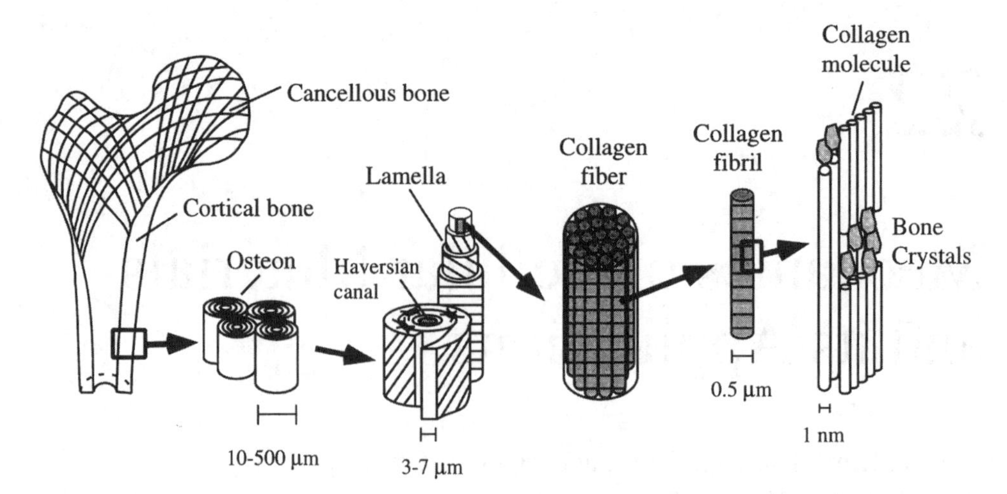


**Figure 22.1** Hierarchical structure of bone [1]. Reproduced with permission from Elsevier

in significant anisotropic properties. In addition to this anisotropy, the cortical bone also has strong asymmetry between tension and compression. For example, as reported by Reilly and Burstein [4], the elastic moduli of cortical bone in the longitudinal and transverse directions are 17 GPa and 11.5 GPa, respectively, and the ultimate tensile strengths in those directions are 133 MPa and 51 MPa, respectively. Moreover, the compressive ultimate strength is much larger than that of the tensile direction; that is, 51 MPa in tension and 133 MPa in compression in the transverse direction, for example. The anisotropic and asymmetric mechanical properties in cortical bone can be thought as naturally optimized because the loading conditions in daily life are compressive in the longitudinal direction.

The structure of trabecular bone is similar to the foam or cellular solid materials and its mechanical properties are highly dependent on the apparent density. Owing to its high porosity compared with

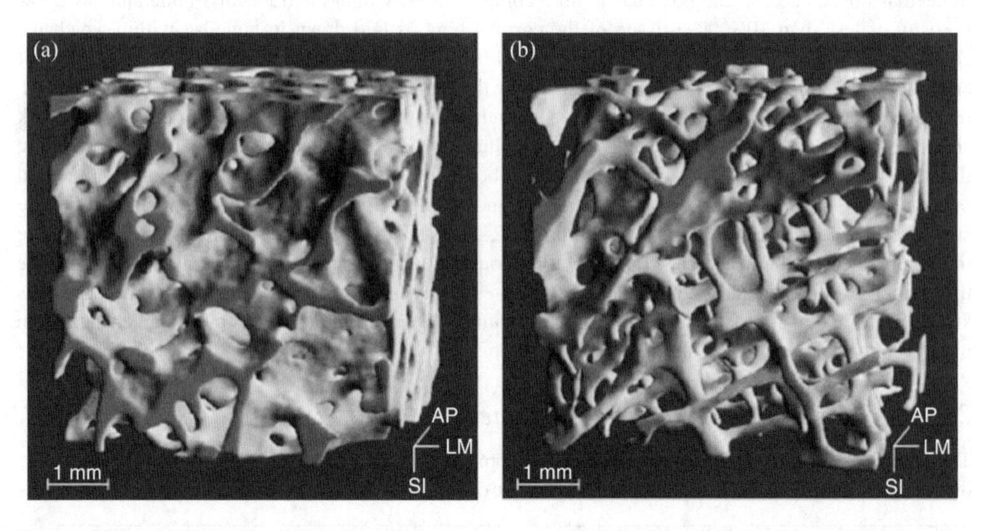


**Figure 22.2** μCT showing 3D trabecular bone structure [3]: (a) of the trabecular bone from the iliac crest of the pelvis of a 37 year-old man; (b) of a 73 year-old woman with osteoporosis. Reproduced with permission from Elsevier

cortical bone, the typical uniaxial stress–strain curve in compression can be equivalently explained by the characteristic structure of trabecular bone. When the trabecular bone is under compressive loading, the bone behaves as a linear elastic solid in the small strain range, but the response becomes nonlinear as the stress continues to increase. Since trabecular bone consists of lots of trabecular struts which are oriented to the direction of the principal stress [5], the start of nonlinear response may correspond to when the trabecular structures are collapsed (or elastically buckled). The stress at this point is usually termed the yield strength in classical plasticity theory. During the progressive collapsing of the trabecular structure, the stress–strain response shows a long anelastic region until the collapsed rods and plates contact others. Then, the stress acting on the trabecular bone sharply increases again. Note that this unique stress–strain behavior of trabecular bone structure is very similar to general cellular solid materials even with elastic–plastic material properties.

Since the mechanical properties of trabecular bone are highly dependent on the apparent density, many researches have been worked on finding out an accurate relationship between the uniaxial mechanical properties of trabecular bone and density. Carter and Hayes [6] provided an empirical relationship between the density and the elastic modulus or compressive strength of trabecular bone. In their analysis, the compressive elastic modulus has a cubic relationship with the density, while the ultimate compressive strength has a quadratic relationship. Based on the cellular solid structure of trabecular bone, the hypothesis stating that the compressive yield strain is positively correlated with the apparent density due to the mechanism of buckling (or collapse of rod structure) has been frequently accepted [7–9]. A simple axial strut cellular solid unit cell model can be constructed with the variation of strut thickness, vertical and horizontal strut lengths, as shown in Figure 22.3.

A more thorough study was conducted by Kopperdahl and Keaveny [10], where the relationship between the apparent density and the compressive or tensile strains for human vertebral trabecular bone

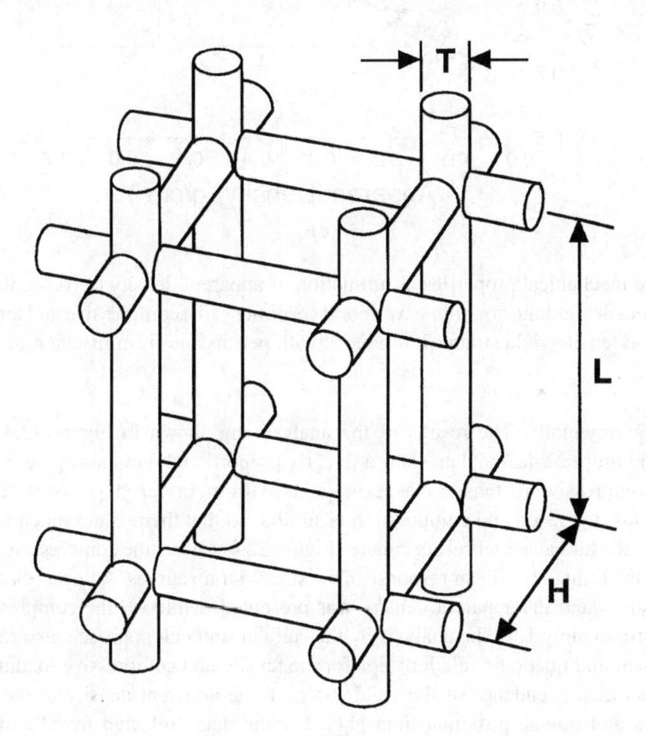


**Figure 22.3** A unit cell of an axial strut cellular solid model [10]. Reproduced with permission from Elsevier

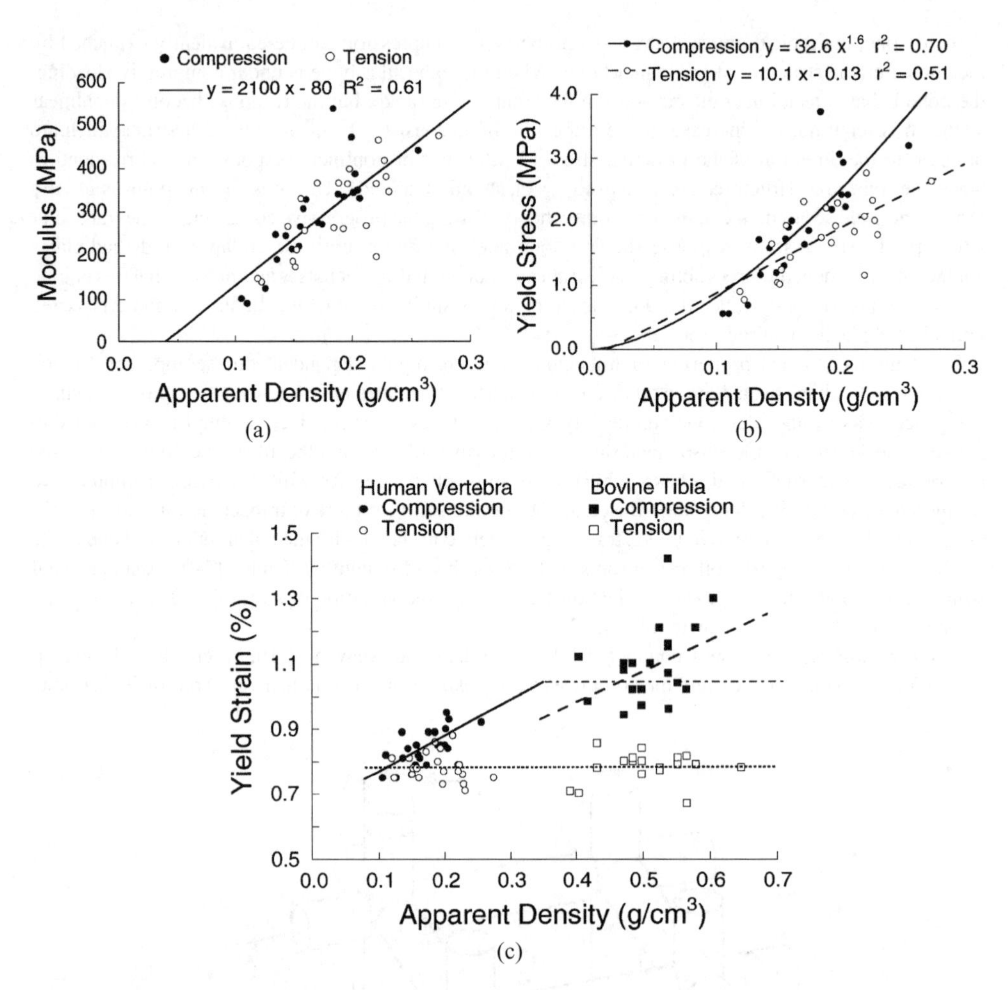


**Figure 22.4** The mechanical properties as a function of apparent density of trabecular bone [10]: (a) compressive and tensile modulus for human vertebral specimens; (b) compressive and tensile yield stress; (c) compressive and tensile yield strain. Reproduced with permission from Elsevier

was presented experimentally. The results of the analysis are shown in Figure 22.4. For the elastic modulus shown in Figure 22.4a, the data show a linearly proportional relationship between the apparent density and the compressive or tensile elastic modulus with a similar slope. Note that the averaged value of the modulus in tension and compression is similar, so that there is not much asymmetry in the elastic modulus for the human vertebral specimens. Figure 22.4b shows the compressive and tensile yield stresses which are the limit of the linear response in the stress–strain curves. A linear relationship between the tensile yield stress and the apparent density was presented, while for the compressive loading the exponent was approximately 1.6. The analysis by the cellular unit cell approach also resulted in similar results; that is, linear and quadratic relationships for the tensile and compressive loadings, respectively. Figure 22.4c shows the dependence of the yield strain on the apparent density of the trabecular bone of human vertebra and bovine proximal tibia [11]. For the data collected from human vertebra and bovine tibia, it is notable that the tensile yield strain is constant with respect to the apparent density. On the other hand, the compressive yield strain shows a linear relationship in the low-density region, but this linearity becomes weaker and eventually vanishes as the density increases. These experimentally observed relationships between the yield strain and the apparent density were also supported by the cellular solid analysis by Gibson and Ashby [8] and Turner *et al.* [12]; that is, the yield strain increases in a positively proportional fashion as the density increases when the failure mechanism is dominated by the (elastic) buckling of trabecular struts oriented along the applied load. But, a negative correlation becomes dominant as the bending of trabeculae occurs due to the off-axis loading.

### 22.1.3 Failure of Bone

Bone fracture has been of great interest because the failure of bone is directly related to the important factors such as bone-related diseases and human age. Through the studies on the fracture of bone, people want to understand the mechanism of the bone fracture and eventually to predict the accumulation of microfracture and the instant of final fracture. Bone fails either by the monotonically applied stress exceeding the ultimate limit of bone tissues or by the accumulation of damage from smaller stresses than the yield strength of bone tissues. The first case mostly involves the fast propagation of cracks which lead to a sudden fracture. This type of fracture is often called "fast fracture." On the other hand, the second type is related to fatigue, in which a repeated stress lower than the critical bone failure stress is applied to the bone. Therefore, the mechanics of bone failure has an analogy with linear fracture mechanics and the theory of conventional fatigue [13]. The simplification of the analysis based on linear fracture mechanics could provide the critical condition for crack propagation, at least within first-order accuracy. However, more detailed observations show the discrepancy originates from the fact that more energy absorption is involved by pulling osteons out of the surrounding interstitial bone (see Figure 22.5). Thus, the toughness of bone is increased significantly and this energy absorption is related to the deformation rate or crack propagation rate. In fact, the fracture toughness $K_C$ is larger at lower strain rates, in which the fracture mechanism involves significant osteon pulling out of the surrounding matrix, while the fracture toughness is smaller at high strain rates as the osteons become brittle and start to fracture. Besides the strain rate, the fracture toughness of bone is also dependent on the orientation, density, and so on.

Bone fracture by fatigue is induced by the long-time exposure to cyclic loading conditions. The stress is usually lower than the critical value that results in a sudden failure of bone, or fast fracture. Then, the accumulated crack length in bone will reach the critical crack size. The evolution of crack depth in bone can also be approximated by the classical Paris law, in which the rate of crack growth can be described as a function of the number of cycles and the difference in the stress intensity during the cycle. However, like fast fracture predicted by linear fracture mechanics, this simplification does not account for the characteristics of bone structure. Osteons and collagen fibers composing the bone microstructure limit the damage by prohibiting the propagation of cracks across the interface between the bone matrix and osteons. Since the original Paris law does not take this crack growth inhibition into account, some workers introduced additional higher order terms to be applicable to the fatigue damage accumulation of bone [15, 16].

Besides the investigations on the fracture mechanism of cortical or trabecular bones, bone fracture healing has been studied by many workers. Contrary to diaphyseal fractures, in which the healing proceeds through endochondral ossification, trabecular bone fracture heals through intramembraneous ossification. In trabecular fracture healing, the new bone is formed at first around the old bone trabeculae directly and the fracture gap is filled with soft tissues. After that, relatively homogeneous woven bones without any specific orientation are formed in the osteotomy gap. Then, the osteotomy gap is completely filled with woven gap before bridging with the cortical bone shell. The isotropic and homogeneous woven bone is remodeled into trabecular bone, which is anisotropic and inhomogeneous. In general, the remodeled trabecular bone microstructure is aligned along the loading condition. Figure 22.6 illustrates the process of trabecular bone healing [17].

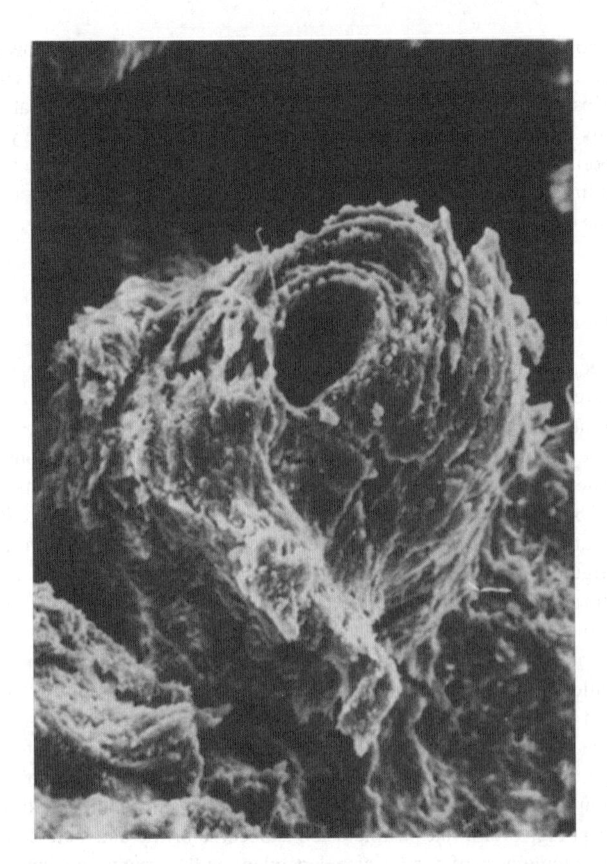

**Figure 22.5** Osteon pull-out during bone fracture [14]. Reproduced with permission from Springer

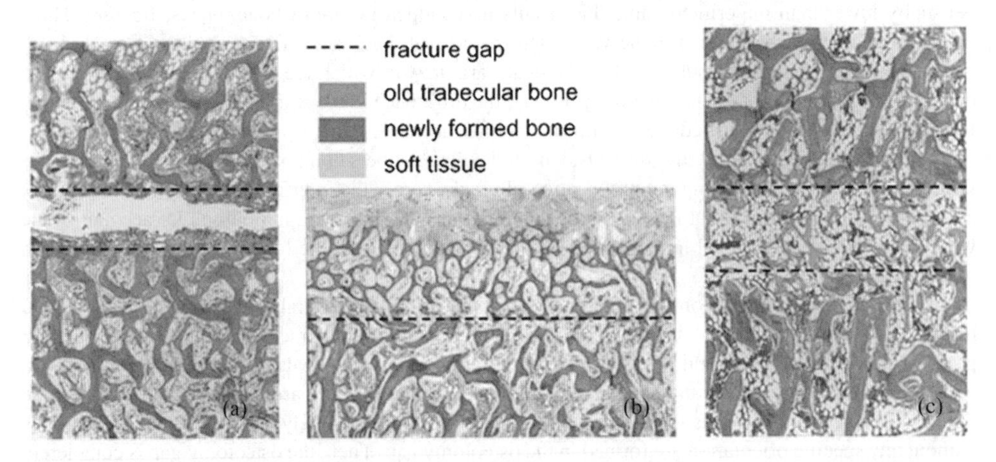


**Figure 22.6** Schematic of canine trabecular bone healing process [17]: (a) formation of new bone around old trabeculae after 1 week; (b) form of woven bone in the osteotomy gap after 3 weeks; (c) filling with woven bone and form of trabecular bone by remodeling process after 4 weeks. Reproduced with permission from Elsevier

### 22.1.4 Simulation of Bone

Numerical simulations in assessing and estimating bone fracture have been increasingly utilized with the advances in computational technology. The finite-element method (FEM) as one of the most practical methods in the computational society is widely used in the field of bone-related mechanics. For example, FEM was used for 3D stress or strain analysis of bone [18], for studying the effect of impact direction on the femur fracture [19]. Ota *et al.* [20] presented cracking analysis using the FEM by assuming that the femoral bone is approximated as a brittle material. They simulated femoral neck fracture and evaluated the usefulness of the numerical model by comparing the predicted fracture position and magnitude with experimental results. From the point of computational scale, these studies can be classified as the continuum finite-element (FE) approach in which a macroscopic model is used with isotropic or anisotropic mechanical properties. The main advantage of this continuum approach is computational efficiency. Figure 22.7 shows an example of an FE model to estimate the initiation of fracture in the proximal portion of a formalin-fixed human left femur. For the analysis, they used conventional FE software for the anisotropic elastic problem. The model was prepared by axial CT images and the preprocessor for the FE discretization. A failure criterion based on the principal stress was used to judge the initiation of failure. Figure 22.8 shows the results of cracking analysis using the FE model. From the analysis, the positions of the fracture initiations with different loading process and the complete fracture line could be correlated with the experiments.

When the scale of simulation becomes smaller for taking the bone details into account, the high porosity and complexity in the microstructure of trabecular bone makes the conventional FE simulation challenging to be directly applicable without developing a technique to deal with such a high resolution. A significantly increased number of FEs (thus, degrees of freedom) are necessary to represent a realistic human bone. This high-resolution FE technology for the calculation of deformation in the bone architecture is frequently called the "μ-FE" model. In general, the μ-FE models begin with the generation of an FE mesh from the μCT of trabecular bone. The numerical techniques based on this μ-FE approach were successfully implemented to take the bone microarchitecture into account [21, 22] and applied to linear deformation models under biomechanical compression tests to predict basic mechanical properties such as the tissue modulus [23, 24]. In particular, Arbenz *et al.* [25] coupled the micro-imaging technique with μ-FE analysis to provide a more powerful numerical tool for the assessment of bone fracture risk. Their numerical algorithm of μ-FE could solve a model of trabecular bone meshed by approximately 250 000 000 elements in 12 min using 1024 CRAY XT3 processors. Figure 22.9 shows

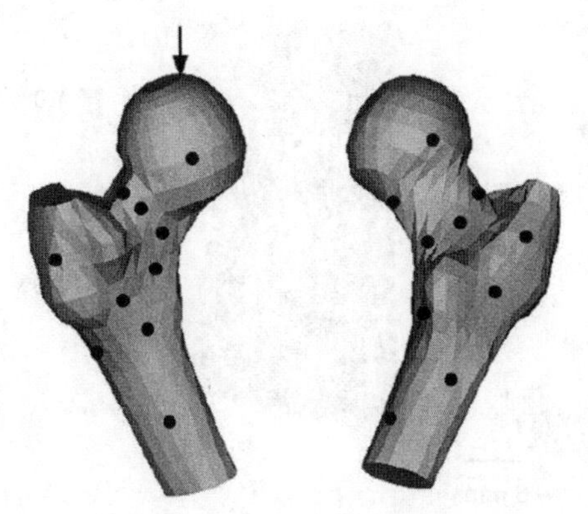

**Figure 22.7** FE model for estimating the initiation of the fracture in the proximal portion of formalin-fixed human left femur [20]. Reproduced with permission from Springer

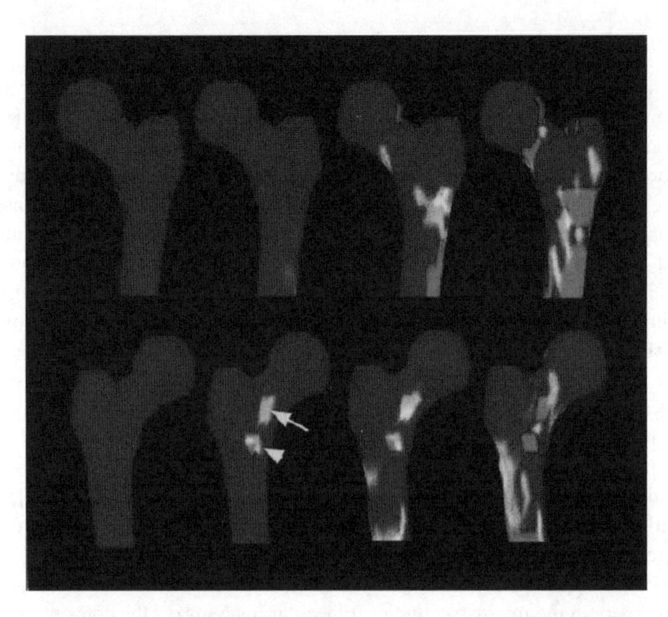

**Figure 22.8** The results of cracking analysis using the FE model [20]. Reproduced with permission from Springer

the full μ-FE modeling of the radius in a human forearm. The trabecular architecture and the effective strain distribution computed by the μ-FE under axial compression are shown.

As previously stated, bone is a heterogeneous material and has a hierarchical architecture; thus, mechanical analysis can be efficiently carried out by adopting a multiscale simulation approach. An analytical method for multiscale modeling of bone assuming a hybrid nanocomposite was first proposed to predict the Young's modulus of cortical bone. The multiscale simulation needs a modeling technique from micro-scale to macro-scale with a highly enhanced resolution which could be realized with the advance of the 3D high-resolution scanning methods such as μCT, micro magnetic resonance imaging

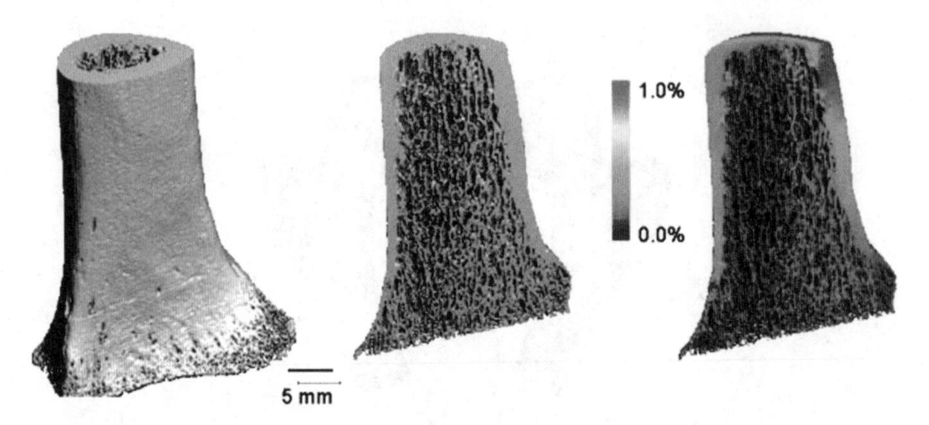


**Figure 22.9** A full micro-FE modeling of the distal part in a human forearm [25]. Reproduced with permission from John Wiley & Sons

(μMRI), and peripheral quantitative computed tomography (pQCT). As shown in Figure 22.9, it is evident that the μ-FE model could not have been established without the μCT imaging technology. For the aforementioned reason, the multiscale FE approach became a commonly applied method in biomedical research, especially in the field of mechanical analysis of bone and failure risk predictions [26–31]. The widely used multiscale simulation approach is to adopt the representative volume element (RVE) with an appropriate homogenization method. The RVEs are generated from the μCT or μMRI images and μ-FE analysis is carried out to calculate the effective local anisotropic material properties for each intermediate geometrical model. Then, the local cellular behavior can be transferred to the global material properties. Usually, the RVE homogenization approach considers the effect of porosity or density of bone.

Numerical methods have also been used for the study of the bone fracture healing process. The studies focused on various factors which influence fracture healing. For example, Pauwels [32], Claes and Heigele [33] and Prendergast *et al.* [34] proposed that the stress and strain state control bone tissue differentiation. On the other hand, Bailon-Plaza and van der Meulen [35] thought that growth factor regulates tissue differentiation. Adaptation of trabecular bone density and bone microstructure were also studied by numerical simulations under the assumption that trabecular bone remodeling is mainly dependent on the mechanical environment. These include studies by Carter *et al.* [36], Huiskes *et al.* [37], Weinans and Prendergast [38], Cowin *et al.* [39], Adachi *et al.* [40], Koontz *et al.* [41], and Ruimerman *et al.* [42].

Among lots of previous studies on the numerical simulation of trabecular bone fracture healing, a recent one by Shefelbine *et al.* [17] is discussed in more detail. In their study, the FEM was incorporated with fuzzy logic theory to determine if intramembraneous bone formation and remodeling during trabecular bone fracture healing can be simulated with the same approach as for diaphyseal fracture healing. They used the FE model of a simplified unit cell shown in Figure 22.10, where the osteotomy gap soft tissue and trabecular bone are discretized by a continuum element with different material properties.

Both the soft tissue (matrix) and the trabecular bone were idealized as linear isotropic solids. Initially, homogeneous material properties in the unit cell model were updated as the simulation proceeded. The simple mixture rule considering the volume fractures of cartilage, soft tissue, and bone was used to determine the effective elastic material properties at each FE. The fuzzy logic algorithm was introduced to calculate the volume fractions of tissues from the strain state given by the corresponding FE analysis.

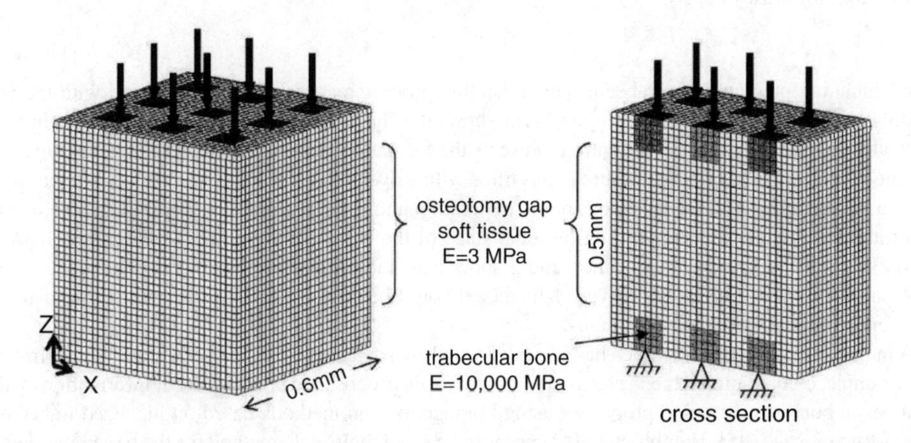


**Figure 22.10** The FE model of a simplified unit cell where the osteotomy gap soft tissue and trabecular bone are discretized by the continuum elements with different material properties [17]. Reproduced with permission from Elsevier

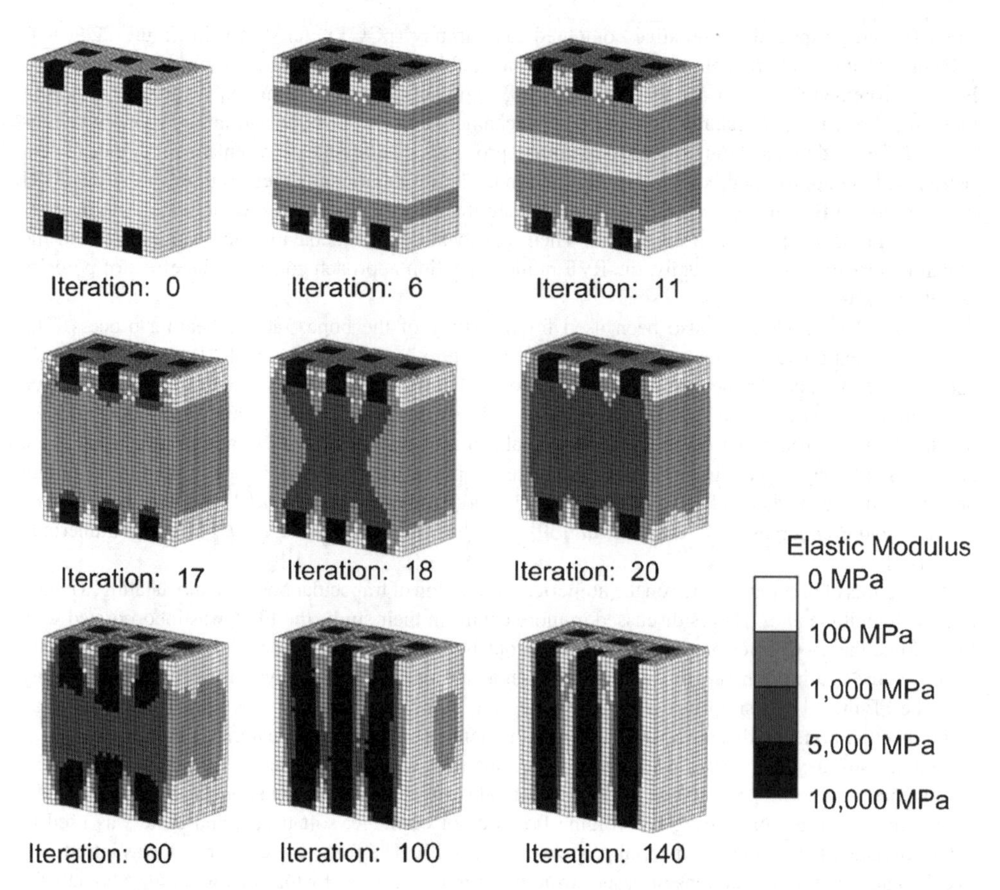


**Figure 22.11** FE modeling for the elastic modulus change with healing time [17]. Reproduced with permission from Elsevier

The simulation predicted the trabecular bone healing process reasonably well compared with the real healing observation. For example, Figure 22.11 shows the elastic modulus change with healing time (or equivalently with iterations). The figure represents the formation of woven bone in the fractured gap (to iteration number 17). After the fracture gap is filled with woven bone, the load is increased and the strain falls in the region of intramembraneous bone (to iteration number 20). Finally, resorption of woven bone and trabecular structure occurs during the final stage of the bone healing process (to iteration number 140). Through the FE simulation, they could show that trabecular bone healing is dependent on the external loading conditions. Moreover, their model could be a useful tool for the prediction of tissue differentiation in different environments.

A multiscale simulation approach has also been widely used in the field of the bone remodeling process. For example, computational hierarchical models have been proposed to provide useful information on the changes in bone structure and properties using homogenization methods based on idealized trabecular microstructures [43–45]. Hambli *et al.* [45] proposed a rapid multiscale method for the bone remodeling process by adopting μ-FE analysis at the macroscopic scale and a 3D neural network (NN) method at the mesoscopic scale. The NN method has also been used when the numerical analysis is too time consuming and experimental data are required [46–48]. In the so-called hybrid FE and NN approach

[45], the macroscopic FE analysis transfers the boundary conditions and stress and strain data to the mesoscale RVE which was modeled by the μCT technique and a trained NN computes the updated material properties and returns them to the macroscopic level. The outputs of the mesoscopic-level simulation by NN are the average density (or porosity of bone), average damage, effective elastic modulus, and average stimulus. Although these multiscale simulation approaches still need validation with more clinical data, they will be very promising tools for estimating bone fracture risk and for understanding bone remodeling or the healing process.

## 22.2 Engineered Cellular Materials

Cellular materials can be manufactured in various ways. Man-made cellular materials include metallic, ceramic, and polymeric foams and periodic cellular structures. In this section, the mechanics of metal foams will be mainly addressed.

Metal foams can be classified into two groups by whether the pores are connected or not. The closed-cell foams are usually made by cooling a molten alloy or consolidating a metal powder, both with a foaming agent. The open-cell foams require open-cell polymer-based precursors. The molten alloy is pressure-infiltrated into the ceramic mold made from the precursor [49].

The unique structure of metal foams leads to many possible applications: lightweight structures, sandwich cores, strain isolation, mechanical damping, vibration control, acoustic absorption, energy absorber, heat exchanger, flame arrester, heat shield, consumable cores for casting, biocompatible inserts, filters, electrical screening, electrode, and buoyancy [49].

One of the most widely used tests for characterizing the mechanical behavior of metal foams is the uniaxial compression test. A schematic of the compression behavior of metal foam is shown in Figure 22.12. The stress–strain curve can be divided into three regions: elastic, plateau, and densification regions. Upon loading, the foam deforms elastically and the Young's modulus $E$ is known to be a function of the relative density:

$$E = \eta E_s \left( \frac{\rho}{\rho_s} \right)^n, \tag{22.1}$$

where $\rho$ is the apparent density of the foam, $E_s$ and $\rho_s$ are the Young's modulus and density of the dense material, respectively, and $\eta$ and $n$ are the material parameters. As the loading continues, the foam yields and the stress either remains constant or increases linearly in the plateau region, depending on

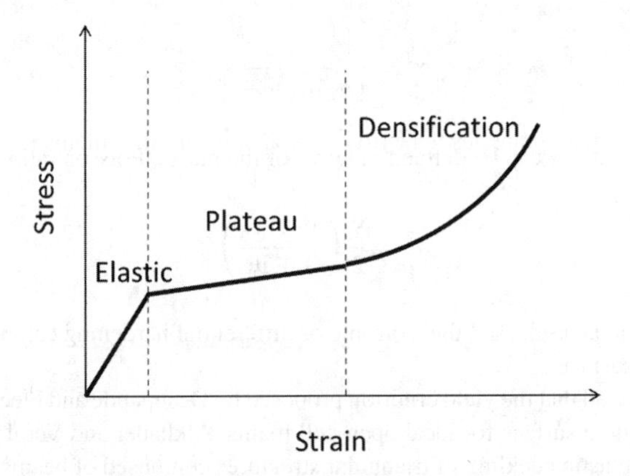


**Figure 22.12** A schematic compression stress–strain curve of a metal foam

the types, structures, and density of the foam. In the plateau region, buckling occurs in the cell walls or struts. If the foam is compressed further, the internal structures start to contact with each other, leading to densification and the rapid increase of stress.

### 22.2.1 *Constitutive Models for Metal Foams*

Many constitutive models have been developed to describe the behavior of cellular materials. One of the simplest yield criteria has the following form:

$$\Phi_i = |\sigma_i| - Y = 0, \tag{22.2}$$

where $\sigma_i$ are the principal stresses of the Cauchy stress, making it suitable for isotropic cellular materials, and $Y$ is the yield stress. The tension–compression asymmetry can be accounted for by using different values of $Y$ for tension and compression. The deformation in one principal direction is decoupled from that in the other directions. Therefore, the plastic Poisson ratio is zero in a uniaxial loading. This yield criterion is based on the experimental observation on polymeric foams [50] that the foam yields in the direction of the largest principal stress.

For orthotropic cellular materials, the following yield criteria have been proposed:

$$\Phi_{ij} = \left|\sigma_{ij}\right| - Y_{ij} = 0, \tag{22.3}$$

where $\sigma_{ij}$ are the components of the Cauchy stress tensor and $Y_{ij}$ are the corresponding yield stresses. These yield criteria require six independent hardening curves for the six components.

Deshpande and Fleck [51] used a specially designed specimen assembly and measured the yield stresses of aluminum foams for various hydrostatic loading paths, as shown in Figure 22.13. Based on the observations, they proposed a phenomenological yield function, given by

$$\Phi = \left[\frac{1}{1+(\alpha/3)^2}\left(\sigma_{\mathrm{e}}^2 + \alpha^2\sigma_{\mathrm{m}}^2\right)\right]^{1/2} - Y = 0, \tag{22.4}$$

where $\sigma_{\mathrm{e}}$ is the von Mises effective stress, $\sigma_{\mathrm{m}}$ is the mean stress, and $\alpha$ is a parameter defining the shape of the yield surface. The parameter $\alpha$ can be determined by the ratio of the hydrostatic yield stress $p_{\mathrm{Y}}$ to the uniaxial yield stress $\sigma_{\mathrm{Y}}$:

$$\alpha^2 = \frac{1}{(p_{\mathrm{Y}}/\sigma_{\mathrm{Y}})^2 - \frac{1}{9}}. \tag{22.5}$$

Alternatively, the parameter $\alpha$ can be defined in terms of the plastic Poisson ratio $\nu_{\mathrm{p}}$:

$$\alpha^2 = \frac{9}{2}\left(\frac{1-2\nu_{\mathrm{p}}}{1+\nu_{\mathrm{p}}}\right). \tag{22.6}$$

The associated flow rule is used. And the isotropic or differential hardening can be used to describe the evolution of the yield surface.

Gioux *et al.* [52] showed that the yield criterion proposed by Deshpande and Fleck [51] can be deduced from the mechanistic yield surface for ideal open-cell foams. Alkhader and Vural [53] derived a similar equation based on the elastic buckling of triangular structures composed of beams in a two-dimensional (2D) plane.

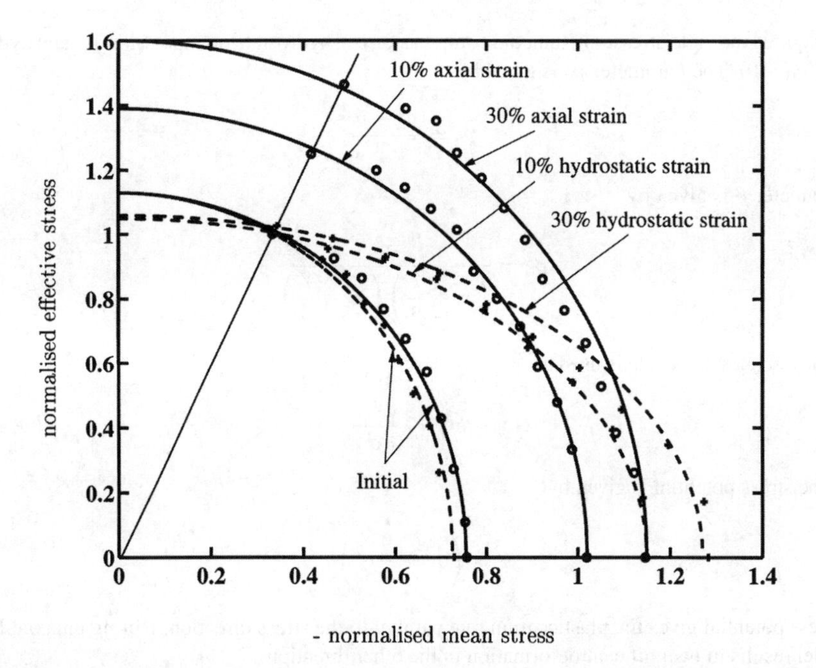


**Figure 22.13** The initial and evolved yield surfaces of an open-cell aluminum foam [51]. Reproduced with permission from Elsevier

Du Bois and coworkers [54] (see Hanssen *et al.* [55] for a detailed explanation of the model) proposed a similar ellipse-shaped yield function shifted in the mean stress direction (Figure 22.14):

$$\Phi = \left[\sigma_e^2 + \beta^2(p - p_0)^2\right]^{1/2} - B = 0, \tag{22.7}$$

where $p$ is the hydrostatic pressure (equal to $-\sigma_m$), $p_0$ is the center of the yield surface, and $\beta$ and $B$ are parameters defining the shape and size of the yield surface, respectively. To define the yield surface, the

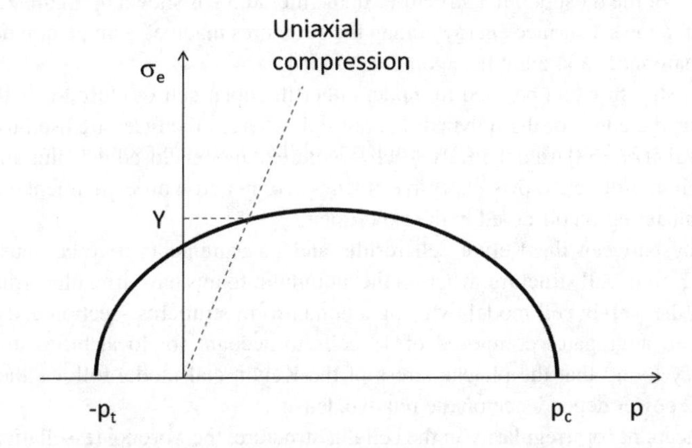


**Figure 22.14** The Bilkhu–Du Bois yield function

model requires the yield stresses in uniaxial compression $\sigma_Y$, hydrostatic compression $p_c$, and hydrostatic tension, $p_t(> 0)$. The parameter $p_0$ is given by

$$p_0 = \frac{p_c - p_t}{2}. \tag{22.8}$$

The parameter $\beta$ is given by

$$\beta^2 = \frac{\sigma_Y^2}{\left(p_t + \frac{Y}{3}\right)\left(p_c - \frac{Y}{3}\right)}. \tag{22.9}$$

The nonassociated flow rule is used as

$$\dot{\varepsilon}^p = \dot{\lambda}\frac{\partial \Psi}{\partial \sigma}, \tag{22.10}$$

where the stress potential is given by

$$\Psi = \left(\sigma_e^2 + \frac{9}{2}p^2\right)^{1/2}. \tag{22.11}$$

This stress potential gives the plastic strain rate parallel to the stress direction. During uniaxial loading, the model results in insignificant deformation in the other directions.

Miller [56] proposed another yield criterion with a second-order hydrostatic stress, given by

$$\Phi = \sigma_e - \gamma p + \frac{\delta}{d}p^2 - d = 0, \tag{22.12}$$

where $\gamma$, $\delta$, and $d$ are the material parameters. This model can also represent a nonzero value of the plastic Poisson ratio.

### 22.2.2 Structure Modeling of Cellular Materials

Cellular materials have been modeled either by using the idealized structure or by utilizing the real microstructure. Idealized modeling utilizes simple repetitive structures. The tetrakaidecahedron or Kelvin cell structure is one of the most popular structures in the literature, as shown in Figure 22.15. The Kelvin cell structure has the lowest surface energy among the structures made of a single polyhedron cell and is defined by six square faces and eight hexagonal faces [58].

The Kelvin cell structure can be used to model either the open-cell or closed-cell structure. For the open-cell structure, the edges of the polyhedron are used, whereas the faces are used for the closed-cell structure. De Giorgi *et al.* [58] used the Kelvin cell structure to model closed-cell aluminum foams. They found that the Kelvin unit cell shows excessive stiffness compared with experimental results and does not exhibit the plateau region observed in the experiments.

The discrepancy between the Kelvin cell results and measurements may be caused by the strict uniformity of the Kelvin cell structure, whereas the aluminum foams have irregular structure. Takahashi *et al.* [57] refined the Kelvin cell model by using a nonuniform strut cross-section and strain softening. They also used cell aggregates composed of 48 cells to account for local bifurcation, as shown in Figure 22.16. They found that the plateau stress of the Kelvin cell model follows the Gibson–Ashby relation with a 3/2 power dependence on the relative density.

In an effort to account for irregularity in the cellular structure, the Voronoi tessellation has been used. The Voronoi tessellation mimics the nucleation and growth of cells during foam formation. In the Voronoi

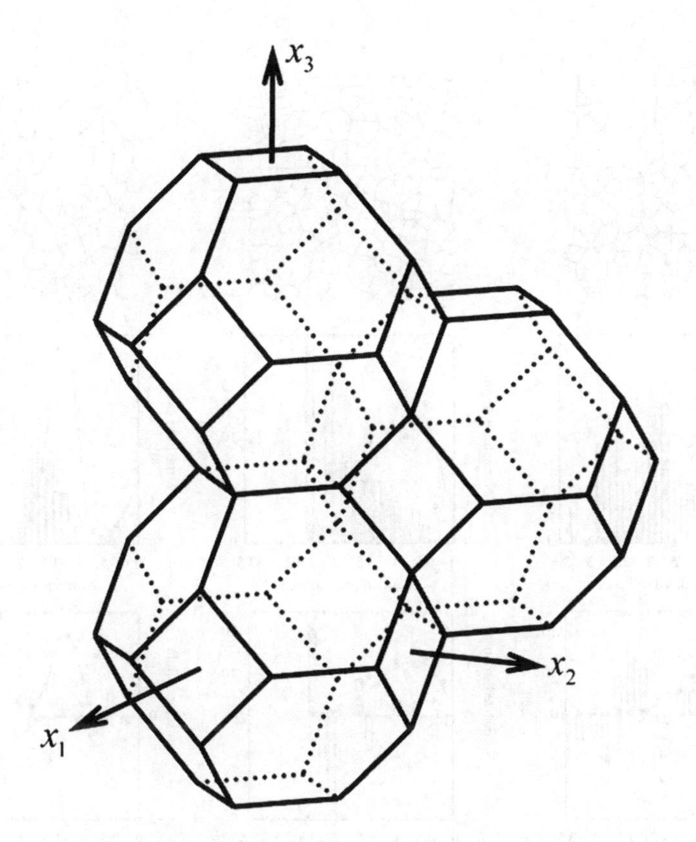


**Figure 22.15** Tetrakaidecahedral or Kelvin cell structure [57]. Reproduced with permission from Elsevier

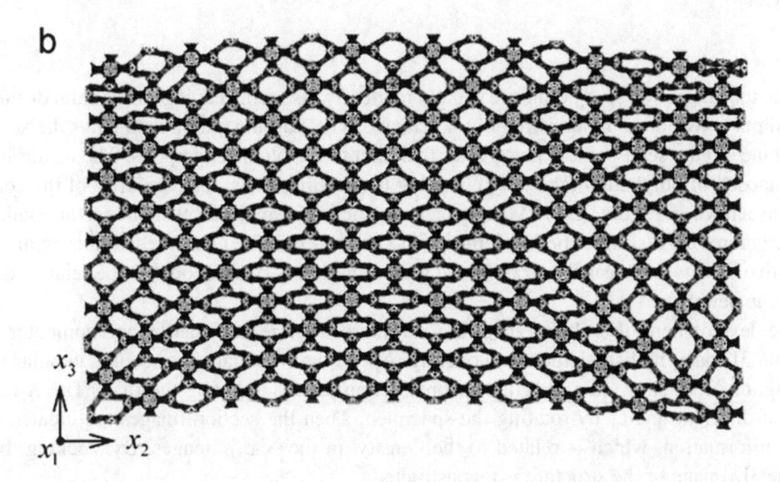


**Figure 22.16** The localized buckling of periodic Kelvin cells [57]. Reproduced with permission from Elsevier

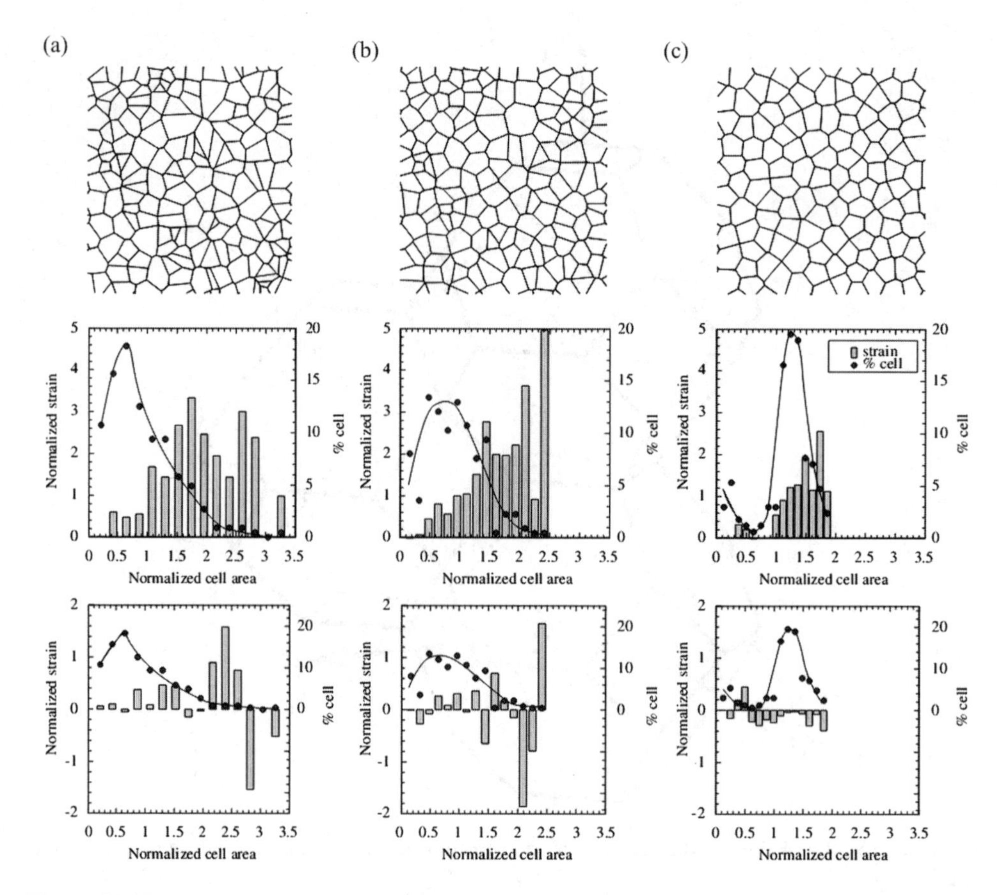


**Figure 22.17** Typical Voronoi structure with different regularity [59]. Reproduced with permission from Elsevier

tessellation, the nucleation seed points are usually randomly inserted, keeping a minimum distances from existing points to avoid too much variation in cell size. The Voronoi cell is defined as the set of points whose distance to the seed point is not greater than that to the other seed points. The resulting Voronoi cell is composed of polyhedral faces. By changing the seeding rules, the regularity of the cells can be modified, as shown in Figure 22.17. Mangipudi and Onck [59] analyzed the damage and failure of 2D Voronoi structures consisting of beam elements and showed that the peak stress scales with the square of the relative density, whereas the strain at the peak is inversely proportional to the relative density for the cellular material.

With the development of tools for acquiring 3D microstructure information and computer programs for handling 3D data, realistic structure modeling for cellular materials has become popular in the last decade. Figure 22.18 shows the structure reconstruction procedure using the μCT. The X-ray images are taken at different angles by rotating the specimen. Then the section images are created using the gray-level information, which is related to the density, in the X-ray images. By stacking the section images, the 3D image of the structure is reconstructed.

Jang *et al.* [60] analyzed the structure of open-cell polymeric and aluminum foams using the 3D images reconstructed by μCT. They measured the cross-sections of the ligament, as shown in Figure 22.19.

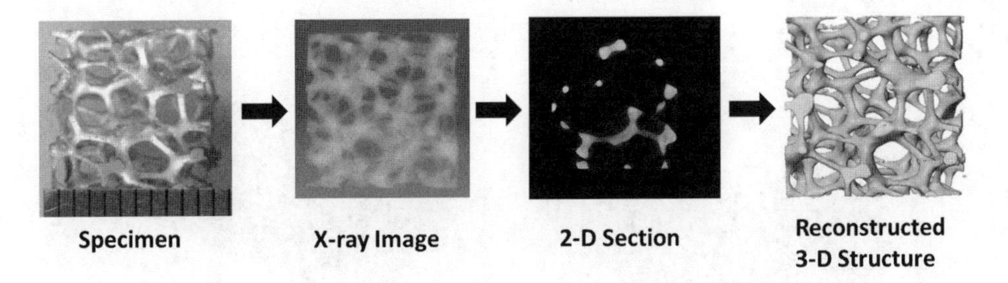


**Figure 22.18** Cellular structure reconstruction using μCT

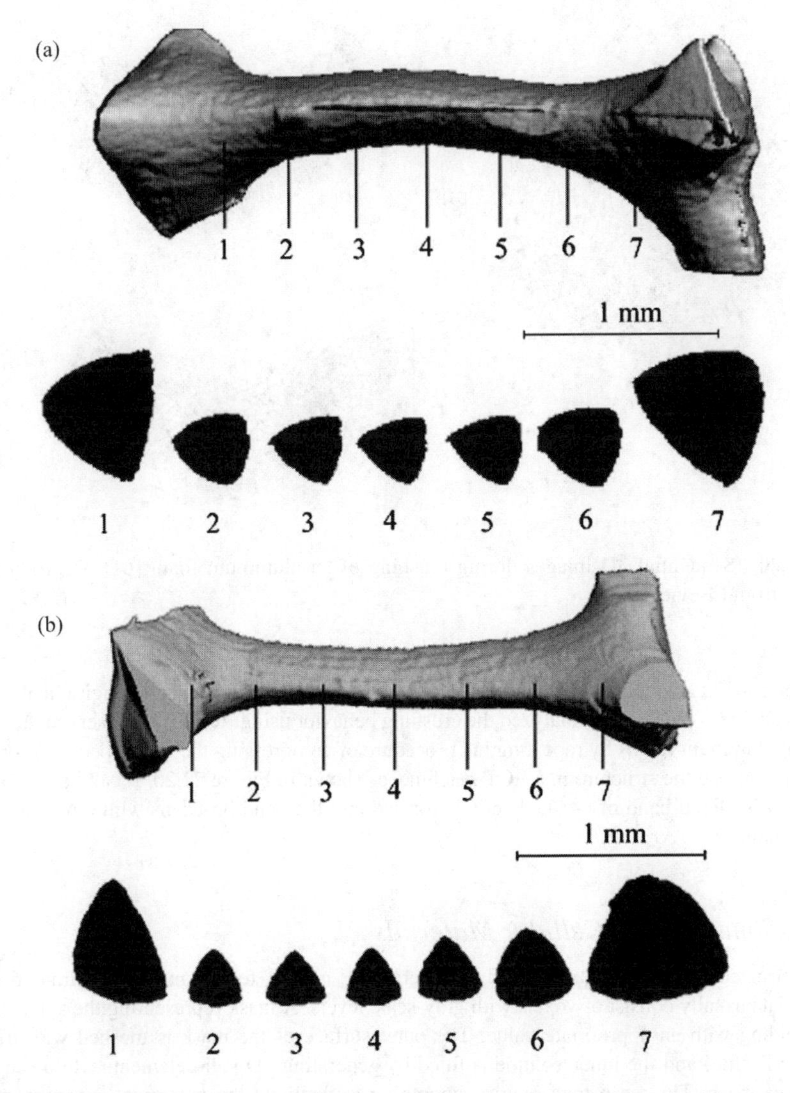


**Figure 22.19** Cross-sections of an aluminum foam ligament [60]. Reproduced with permission from Elsevier

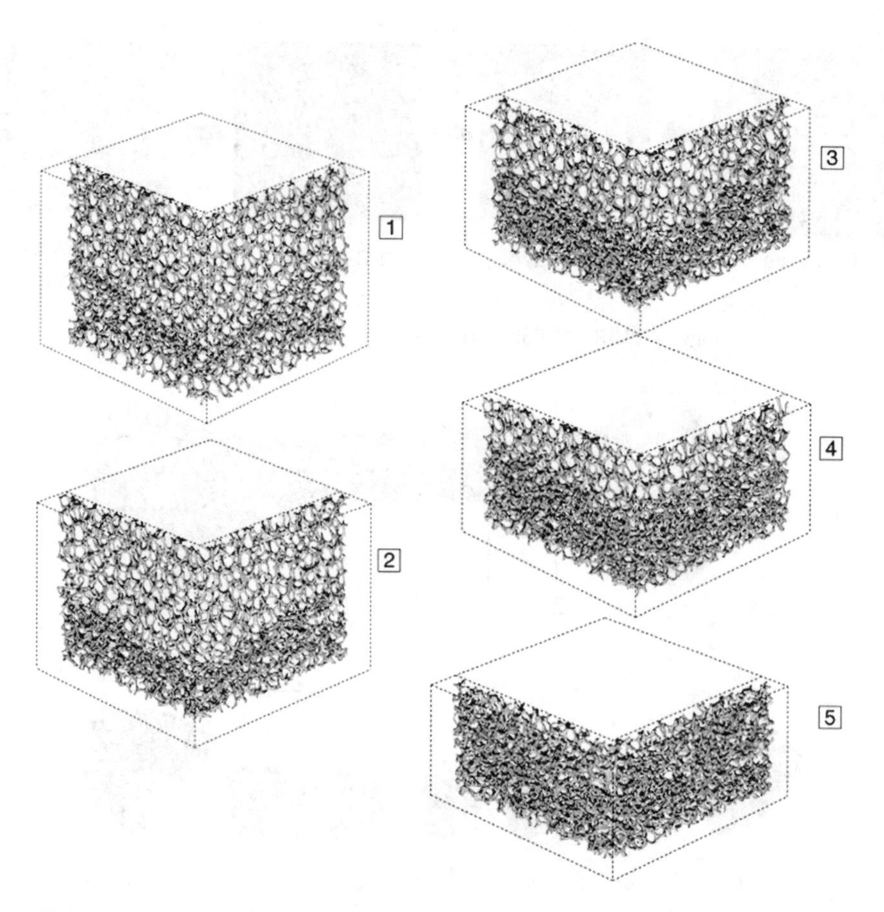


**Figure 22.20** Sequential 3D images during crushing of an aluminum foam [61]. Reproduced with permission from Elsevier

The cross-section of a ligament is a rounded triangle with a varying size along the longitudinal direction. Jang and Kyriakides [61] further analyzed the crushing behavior using μCT. They observed the crushing behavior of aluminum foams by repeating the procedure of compressing the foam specimen for a given degree and scanning the structure in a μCT machine, as shown in Figure 22.20. It can be seen from the figure that an inclined band of crushed cells is formed and the band broadens while the load remains nearly constant.

### *22.2.3 Simulation of Cellular Materials*

The reconstructed 3D images can be used to generate FE models for the numerical simulations. The 3D image data usually consist of voxels with gray-scale levels. A mask representing the solid is created by thresholding with an appropriate value. The outer surface of the mask is meshed with triangular elements (STL file) and the inner volume is filled by generating 3D solid elements while keeping the same surface mesh. The mesh may require smoothing to facilitate the numerical convergence in the FE analysis.

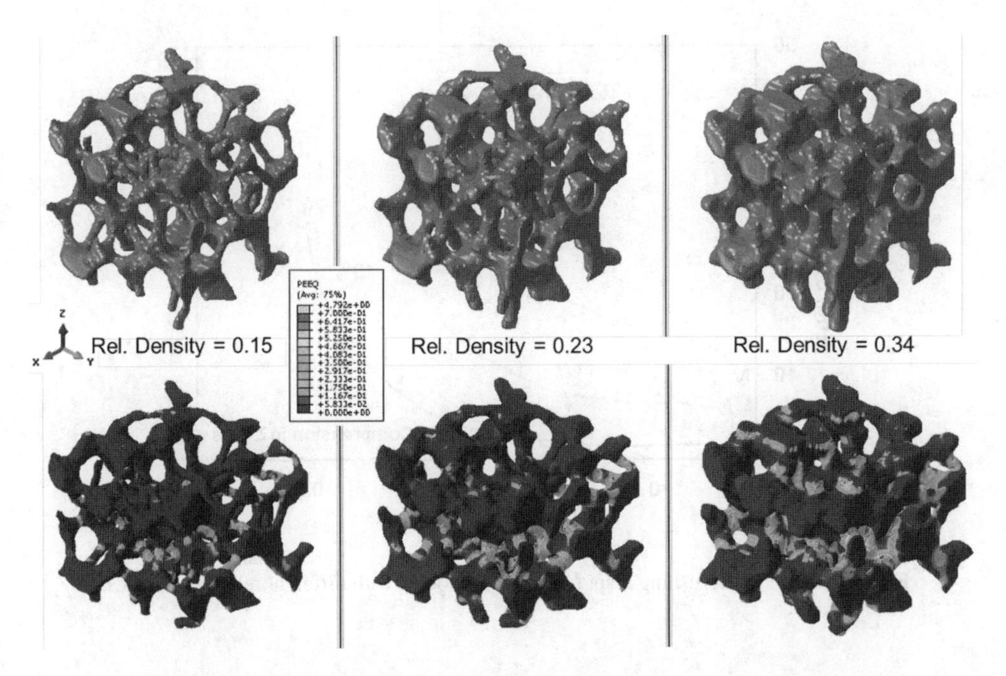


**Figure 22.21** The crushing of foam models with different relative densities

The 3D FE models are used to predict the behavior of the cellular materials. Michailidis [62] built a 3D FE model of an Ni-foam and investigated the compression response of the foam. The FE model can capture the deformation and buckling of the struts. Veyhl *et al.* [63] used the 3D model of the open- and closed-cell aluminum foams to analyze the relationship between the mechanical properties and the relative density. They found that the Young's modulus and the yield stress are linearly proportional to the relative density. However, the insufficient size of the model and the small range of density may have impeded the accuracy of the analysis.

Kim *et al.* [64] generated the foam structures of different relative densities by adding layers of material to the base model (relative density of 0.15) while keeping the same connectivity, as shown in Figure 22.21, and FE simulations of the uniaxial compression test were conducted. The calculated stress–strain curves (Figure 22.22) showed that the yield stress scales with the cube of the relative density. Also, the hardening slope increases in the plateau region with increasing relative density.

The 3D model of the foam structure can be extended for the multiscale analysis of the foam-filled impact absorber. In modern vehicles, the crash box is used to absorb impact energy and protect the passenger compartment and the main frame in the intermediate velocity regime. In most cases, crash boxes are hollow tubes and absorb the impact energy by progressive folding. If metal foams are inserted in the hollow region, the energy absorption capacity of the crash box can be improved. The impact performances of the foam-filled crash boxes were investigated using multiscale simulation [65]. The multiscale FE model of the foam-filled crash box is shown in Figure 22.23. The aluminum foam was modeled using the 3D solid elements with the Bilkhu–Du Bois constitutive model. In the micro-scale, the foam structure models shown in Figure 22.21 were employed to calculate the material parameters and the evolution of the yield function by the homogenization method. Figure 22.24 shows the crash response of the foam-filled crash members. As the relative density increases, the mean crushing force increases, resulting in a shorter displacement to absorb the given impact energy. Increasing the relative density,

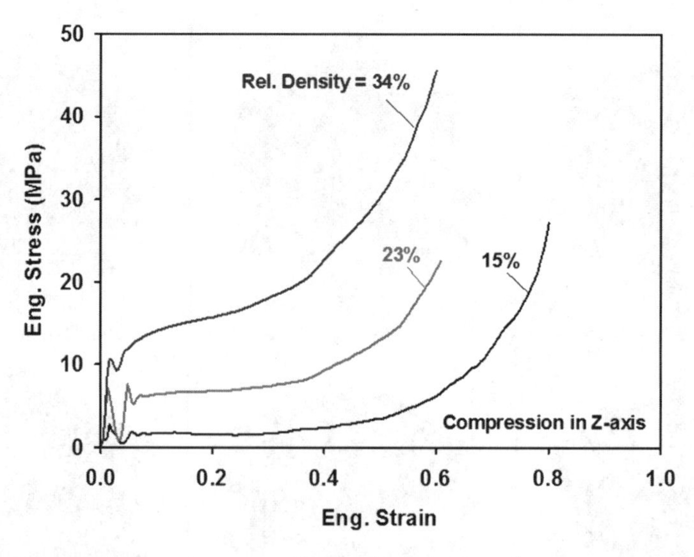


**Figure 22.22** The crushing response of foam models with different relative densities

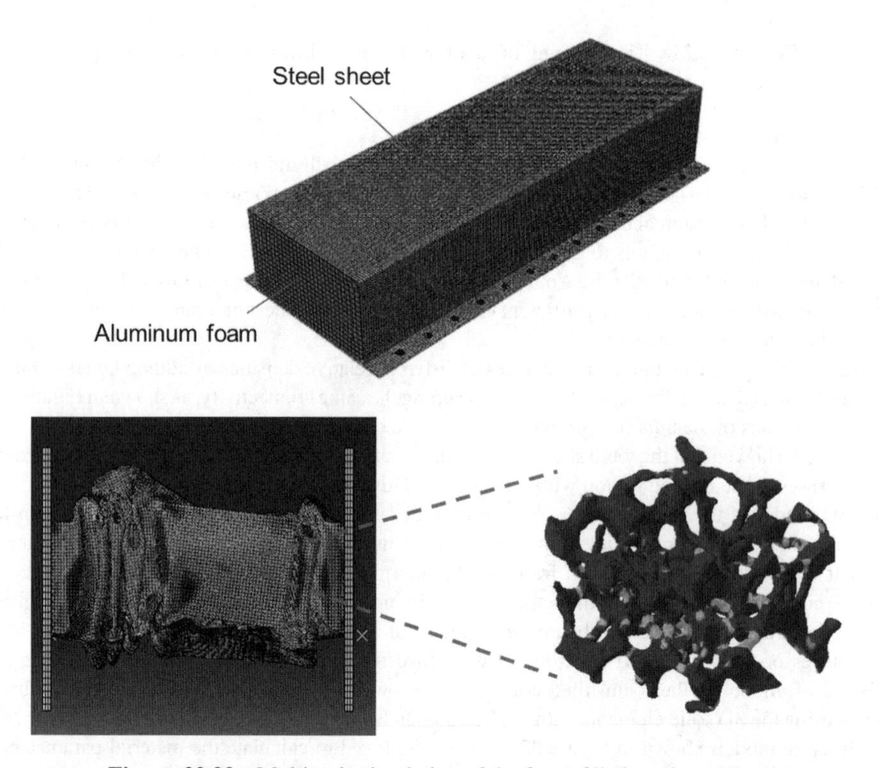


**Figure 22.23** Multiscale simulation of the foam-filled crash member

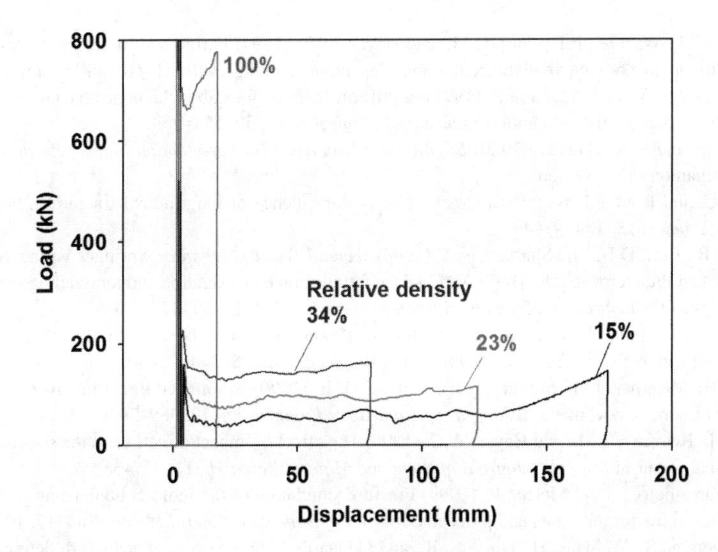


**Figure 22.24** The crush results with various relative densities of foam filler

however, does not always result in a better energy absorption property. When a solid aluminum block was inserted (relative density of 100% in Figure 22.24), for example, the member did not absorb the impact energy fully and rebounded the impactor back. In addition, too much force is directly transmitted to the passenger compartment. Therefore, a well-designed foam-filler can be an effective component for tailoring the impact performance of the crash box and multiscale simulation is an essential tool for designing materials for the crash box.

## References

[1] Rho, J.Y., Spearing, L.K., and Zioupos, P. (1998) Mechanical properties and hierarchical structure of bone. *Medical Engineering and Physics*, **20**, 92–102.

[2] Ethier, C.R. and Simmons, C.A. (2007) *Introductory Biomechanics: From Cells to Organisms*, Cambridge University Press.

[3] Muller, R., van Campenhout, H., van Damme, B. *et al.* (1998) Morphometric analysis of human bone biopsies: a quantitative structural comparison of histological sections and micro-computed tomography. *Bone*, **23**, 59–66.

[4] Reilly, D.T. and Burstein, A.H. (1975) The elastic and ultimate properties of compact bone tissue. *Journal of Biomechanics*, **8**, 393–405.

[5] Rice, J.C., Cowin, S.C., and Bowman, J.A. (1988) On the dependence of the elasticity and strength of cancellous bone on apparent density. *Journal of Biomechanics*, **21**, 155–168.

[6] Carter, D.R. and Hayes, W.C. (1977) The compressive behavior of bone as a two-phase porous structure. *Journal of Bone and Joint Surgery, American Volume*, **59**, 954–962.

[7] Christensen, R.M. (1986) Mechanics of low density materials. *Journal of the Mechanics of the Physics of Solids*, **34**, 563–578.

[8] Gibson, L.J. and Ashby, M.F. *Cellular Solids: Structures and Properties*, Pergamon Press, Oxford.

[9] Snyder, B.D., Piazza, S., Edward, W.T., and Hayes, W.C. (1993) Role of trabecular morphology in the etiology of age-related vertebral fractures. *Calcified Tissue Research International*, **53S**, 14–22.

[10] Kopperdahl, D.L. and Keaveny, T.M. (1998) Yield strain of trabecular bone. *Journal of Biomechanics*, **31**, 601–608.

[11] Keaveny, T.M., Wachtel, E.F., Ford, C.M., and Hayes, W.C. (1994) Differences between the tensile and compressive strengths of bovine tibial trabecular bone depend on modulus. *Journal of Biomechanics*, **27**, 1137–1146.
[12] Turner, C., Anne, V., and Pidaparti, R. (1997) A uniform criterion for trabecular bone adaptation: do continuum-level strain gradients drive adaptation? *Journal of Biomechanics*, **30**, 555–563.
[13] Ashby, M.F. and Jones, D.H.R. (1980) *Engineering Materials: An Introduction to their Properties and Application*, Pergamon Press, Oxford.
[14] Behiri, J.C. and Bonfield, W. (1980) Crack velocity dependence of longitudinal fracture in bone. *Journal of Materials Science*, **15**, 1841–1849.
[15] Martin, R.B., Burr, D.B., and Sharkey, N.A. (1998) *Skeletal Tissue Mechanics*, Springer Verlag, New York, NY.
[16] Taylor, D. and Prendergast, P.J. (1997) A model for fatigue crack propagation and remodeling in compact bone. *Proceedings of the Institute of Mechanical Engineers, Series H*, **211**, 369–375.
[17] Shefelbine, S.J., Augat, P., Claes, L., and Simon, U. (2005) Trabecular bone fracture healing simulation with finite element analysis and fuzzy logic. *Journal of Biomechanics*, **38**, 2440–2450.
[18] Keyak, J.H., Meagher, J.M., Skinner, H.B., Mote, C.D. Jr. (1990) Automated three-dimensional finite element modeling of bone: a new method. *Journal of Biomedical Engineering*, **12**, 389–397.
[19] Ford, C.M., Keaveny, T.M., and Hayes, W.C. (1996) The effect of impact direction on the structural capacity of the proximal femur during falls. *Journal of Bone and Mineral Research*, **11**, 377–383.
[20] Ota, T., Yamamoto, I., and Morita, R. (1999) Fracture simulation of the femoral bone using the finite-element method: how a fracture initiates and proceeds. *Journal of Bone and Mineral Metabolism*, **17**, 108–112.
[21] Van Rietbergen, S., Weinans, H., Huiskes, R., and Odgaard, A. (1995) A new method to determine trabecular bone elastic properties and loading using micromechanical finite-elements models. *Journal of Biomechanics*, **28**, 69–81.
[22] Van Lenthe, G.H. and Muller, R. (2006) Prediction of failure load using micro-finite element analysis models: towards in vivo strength assessment. *Drug Discovery Today: Technologies*, **3**, 221–229.
[23] Kabel, J., van Rietbergen, B., Dalstra, M. *et al.* (1999) The role of an effective isotropic tissue modulus in the elastic properties of cancellous bone. *Journal of Biomechanics*, **32**, 673–680.
[24] Ladd, A.J., Kinney, J.H., Haupt, D.L., and Goldstein, S.A. (1998) Finite-element modeling of trabecular bone: comparison with mechanical testing and determination of tissue modulus. *Journal of Orthopeadic Research*, **16**, 622–628.
[25] Arbenz, P., van Lenthe, H.H., Mennel, U. *et al.* (2008) A scalable multi-level preconditioner for matrix-free finite element analysis of human bone structures. *International Journal for Numerical Methods in Engineering*, **73**, 927–947.
[26] Hellmich, C., Barthelemy, J.-F., and Dormieux, L. (2004) Mineral–collagen interactions in elasticity of bone ultrastructure – a continuum micromechanics approach. *European Journal of Mechanics: A/Solids*, **23**, 783–810.
[27] Kawagai, M., Sando, A., and Takano, N. (2006) Image-based multi-scale modelling strategy for complex and heterogeneous porous microstructures by mesh superposition method. *Modelling and Simulation in Materials Science and Engineering*, **14**, 53–69.
[28] Sansalone, V., Lemaire, T., and Naili, S. (2007) Multiscale modelling of mechanical properties of bone: study at the fibrillar scale. *Comptes Rendus Mécanique*, **335** (8), 436–442.
[29] Ghanbari, J. and Naghdabadi, R. (2009) Nonlinear hierarchical multiscale modeling of cortical bone considering its nanoscale microstructure. *Journal of Biomechanics*, **42**, 1560–1565.
[30] Ilic, S., Hackl, K., and Gilbert, R. (2010) Application of the multiscale FEM to the modeling of cancellous bone. *Biomechanics and Modeling in Mechanobiology*, **9**, 87–102.
[31] Podshivalov, L., Fischer, A., and Bar-Yoseph, P.Z. (2011) Multiscale FE method for analysis of bone microstructures. *Journal of the Mechanical Behavior of Biomedical Materials*, **4**, 888–899.
[32] Pauwels, F. (1960) Eine neue Theorie über den Einfluß mechanischer Reize auf die Differenzierung der *Stützgewebe. Zeitschrift für Antomie und Entwicklungageschichte*, **121**, 478–515.
[33] Claes, L.E. and Heigele, C.A. (1999) Magnitudes of local stress and strain along bony surfaces predict the course and type of fracture healing. *Journal of Biomechanics*, **32**, 255–266.
[34] Prendergast, P.J., Huiskes, R., and Soballe, K. (1997) ESB Research Award 1996. Biophysical stimuli on cells during tissue differentiation at implant interfaces. *Journal of Biomechanics*, **30**, 539–548.
[35] Bailon-Plaza, A. and van der Meulen, M. (2001) A mathematical framework to study the effects of growth factor influences on fracture healing. *Journal of Theoretical Biology*, **212**, 191–209.

[36] Carter, D.R., Fyhrie, D.P., and Whalen, R.T. (1987) Trabecular bone density and loading history: regulation of connective tissue biology by mechanical energy. *Journal of Biomechanics*, **20**, 785–794.

[37] Huiskes, R., Weinans, H., Grootenboer, H.J. *et al.* (1987) Adaptive bone-remodeling theory applied to prosthetic-design analysis. *Journal of Biomechanics*, **20**, 1135–1150.

[38] Weinans, H. and Prendergast, P.J. (1996) Tissue adaptation as a dynamical process far from equilibrium. *Bone*, **19**, 143–149.

[39] Cowin, S.C., Arramon, Y.P., Luo, G.M., and Sadegh, A.M. (1993) Chaos in the discrete-time algorithm for bone-density remodeling rate equations. *Journal of Biomechanics*, **26**, 1077–1089.

[40] Adachi, T., Tsubota, K., Tomita, Y., and Hollister, S.J. (2001) Trabecular surface remodeling simulation for cancellous bone using microstructural voxel finite element models. *Journal of Biomechanical Engineering*, **123**, 403–409.

[41] Koontz, J.T., Charras, G.T., and Guldberg, R.E. (2001) A microstructural finite element simulation of mechanically induced bone formation. *Journal of Biomechanical Engineering*, **123**, 607–612.

[42] Ruimerman, R., Van Rietbergen, B., Hilbers, P., and Huiske, R. (2003) A 3-dimensional computer model to simulate trabecular bone metabolism. *Biorheology*, **40**, 315–320.

[43] Adachi, T., Tomita, Y., and Tanaka, M. (1998) Computational simulation of deformation behavior of 2D-lattice continuum. *International Journal of Mechanics and Solids*, **40**, 857–866.

[44] Yoo, A. and Jasiuk, I. (2006) Couple-stress moduli of a trabecular bone idealized as a 3D periodic cellular network. *Journal of Biomechanics*, **39**, 2241–2252.

[45] Hambli, R., Katerchi, H., and Benhamou, C.-L. (2010) Multiscale methodology for bone remodeling simulation using coupled finite element and neural network computation. *Biomechanics and Modeling in Mechanobiology*, **10**, 133–145.

[46] Hambli, R., Chamekh, A., BelHadj, S.H. (2006) Real-time deformation of structure using finite element and neural networks in virtual reality applications. *Finite Elements in Analysis and Design*, **42**, 985–991.

[47] Jenkins, W.M. (1997) An introduction to neural computing for the structural engineer. *Structural Engineering*, **75**, 38–41.

[48] Unger, J.F. and Konke, C. (2008) Coupling of scales in multiscale simulation using neural networks. *Computers & Structures*, **86**,1994–2003.

[49] Ashby, M.F., Evans, A.G., Fleck, N.A. *et al.* (2000) *Metal Foams: A Design Guide*, Butterworth-Heinemann, Boston, MA.

[50] Shaw, M.C. and Sata, T. (1966) The plastic behavior of cellular materials. *International Journal of Mechanical Sciences*, **8**, 469–472.

[51] Deshpande, V.S. and Fleck, N.A. (2000) Isotropic constitutive models for metallic foams. *Journal of the Mechanics and Physics of Solids*, **48**, 1253–1283.

[52] Gioux, G., McCormack, T.M., and Gibson, L.J. (2000) Failure of aluminum foams under multiaxial loads. *International Journal of Mechanical Sciences*, **42**, 1097–1117.

[53] Alkhader, M. and Vural, M. (2009) An energy-based anisotropic yield criterion for cellular solids and validation by biaxial FE simulations. *Journal of the Mechanics and Physics of Solids*, **57**, 871–890.

[54] Nusholtz, G., Bilkhu, S., Founas, M., and Du Bois, P. (1996) A simple elastoplastic model for foam materials, in *STAP Conference*, Alberquerque, NM.

[55] Hanssen, A.G., Hopperstad, O.S., Langseth, M., and Ilstad, H. (2002) Validation of constitutive models applicable to aluminium foams. *International Journal of Mechanical Sciences*, **44**, 359–406.

[56] Miller, R.E. (2000) A continuum plasticity model for the constitutive and indentation behaviour of foamed metals. *International Journal of Mechanical Sciences*, **42**, 729–754.

[57] Takahashi, Y., Okumura, D., and Ohno, N. (2010) Yield and buckling behavior of Kelvin open-cell foams subjected to uniaxial compression. *International Journal of Mechanical Sciences*, **52**, 337–385.

[58] De Giorgi, M., Carofalo, A., Dattoma, V. *et al.* (2010) Aluminium foams structural modelling. *Computers & Structures*, **88**, 25–35.

[59] Mangipudi, K.R. and Onck, P.R. (2011) Multiscale modelling of damage and failure in two-dimensional metallic foams. *Journal of the Mechanics and Physics of Solids*, **59**, 1437–1461.

[60] Jang, W.Y., Kraynik, A.M., and Kyriakides, S. (2008) On the microstructure of open-cell foams and its effect on elastic properties. *International Journal of Solids and Structures*, **45**, 1845–1875.

[61] Jang, W.Y. and Kyriakides, S. (2009) On the crushing of aluminum open-cell foams: Part I. Experiments. *International Journal of Solids and Structures*, **46**, 617–634.

[62] Michailidis, N. (2011) Strain rate dependent compression response of Ni-foam investigated by experimental and FEM simulation methods. *Materials Science and Engineering: A*, **528**, 4204–4208.

[63] Veyhl, C., Belova, I.V., Murch, G.E., and Fiedler, T. (2011) Finite element analysis of the mechanical properties of cellular aluminium based on micro-computed tomography. *Materials Science and Engineering: A*, **528**, 4550–4555.

[64] Kim, D., Kim, J.H., Lee, M.G., and Lee, J.K. (2012) Multi-scale design of open-cell aluminum foam, in *TMS 2012 141st Annual Meeting & Exhibition*, March 11–15, Orlando, FL.

[65] Kim, D., Kim, J.H., Lee, Y.S., and Lee, J.K. (2011) Homogenization-based multi-scale impact analysis of open-cell aluminum alloy foam, in *XI International Conference on Computational Plasticity Fundamentals and Applications (Complas XI)*, September 7–9, Barcelona, Spain.

# 23

# Biomechanics of Mineralized Collagens

Ashfaq Adnan, Farzad Sarker, and Sheikh F. Ferdous
*Mechanical and Aerospace Engineering, University of Texas at Arlington, USA*

## 23.1 Introduction

### *23.1.1 Mineralized Collagen*

Collagen is one of the most essential structural proteins, found abundantly in all living animals [1]. In human bodies, collagens are found in fibrous tissues such as tendon, ligament, cartilage, bone, and so on and comprise 25–35% of the total body proteins. In bone, collagens combine with mineral cements to form a composite fibril structure. These mineralized collagens are considered as the fundamental building block of bone. These building blocks form a hierarchical structure that controls the overall properties of bone, including strength, stiffness, and toughness [2–16]. At the macroscopic level, bone tissue can be either cancellous (spongy) or cortical (compact) type. Since compact bones are significantly denser than cancellous bones, their mechanical properties are also significantly higher than cancellous bone [2, 17–25]. As a result, it is the cancellous bones that experience fracture during skeletal loading. As shown in Figure 23.1, there are several distinct building blocks exist in the hierarchical organization in bone, which are outlined here in the context of trabecular bone [21, 26–33]. The spongy bone is composed of a trabecular network that is formed when several "rod"- or "plate"-shaped trabeculae meet at the trabecular node. An individual trabecula is composed of a lamellar array of composite fibers. Each fiber is the result of self-assembled collagen fibrils, where each fibril is formed by a staggered arrangement of mineralized tropocollagen molecules. Mineralization of collagen is achieved by deposition of dahllite crystals in the "hole" region of collagen stacks. The fundamental constituents of collagen fibril are type I collagen, dahllite nanocrystals, and non-collagenic cross-linkers. Each collagen molecule in the fibril is formed from three chains of amino acids.

Morphologically, trabecular bone is a highly porous (60–95%) cellular material in which lamellar packets form together in a latticework of rod- and plate-like structures. The rod and plate structures (trabeculae) are typically 100–300 μm thick in human bone, and are separated by larger marrow spaces with characteristic dimensions of 500–1000 μm. Structurally, the basic building block of cancellous

*Multiscale Simulations and Mechanics of Biological Materials*, First Edition. Edited by Shaofan Li and Dong Qian.

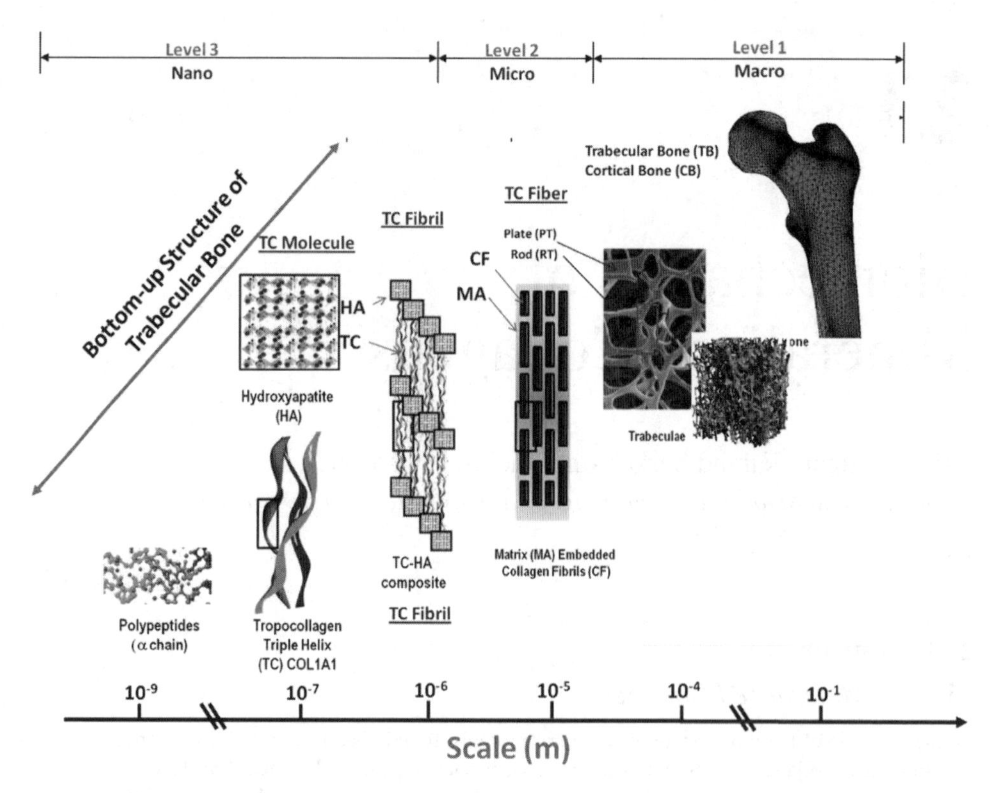


**Figure 23.1** Hierarchical structure of bone showing several levels of hierarchy

bone is a composite material consisting of a hydrated (25% water) organic matrix (25%) stiffened by extremely dense calcium phosphate crystals (50%). The matrix contains 90% collagen type I and 10% other protein, such as glycoprotein, osteocalcin, osteonectin, and so on). This composite block is called a mineralized collagen fibril.

As such, it can be argued that the mechanical properties of bone at the macroscopic level are directly dependent on the mechanical properties of mineralized collagens. Therefore, it is very important to know how molecular-level mechanical properties of collagen fibrils are translated to the tissue level. It is also essential to know about the key factors that control the mechanical response of bone from its building block (collagen fibril) level. Quantitative knowledge of collagen mechanics will not only help the scientific community to understand overall bone mechanics, but also provide a systematic guideline for advancing medical diagnosis/treatment protocol and tissue engineering [34].

## *23.1.2 Molecular Origin and Structure of Mineralized Collagen*

Human collagen is mostly type I collagen, called tropocollagen (TC), consisting of three helically arranged polypeptide chains [3]. The mineral in the bone [3] is called dahllite, which is 95% pure carbonated hydroxyapatite (HAP), $(Ca_{10}(PO_4)_6(OH)_2$. Type I collagen is formed in bone from the combination of two $\alpha 1$ polypeptides and one $\alpha 2$ collagen polypeptide containing specific sequences of

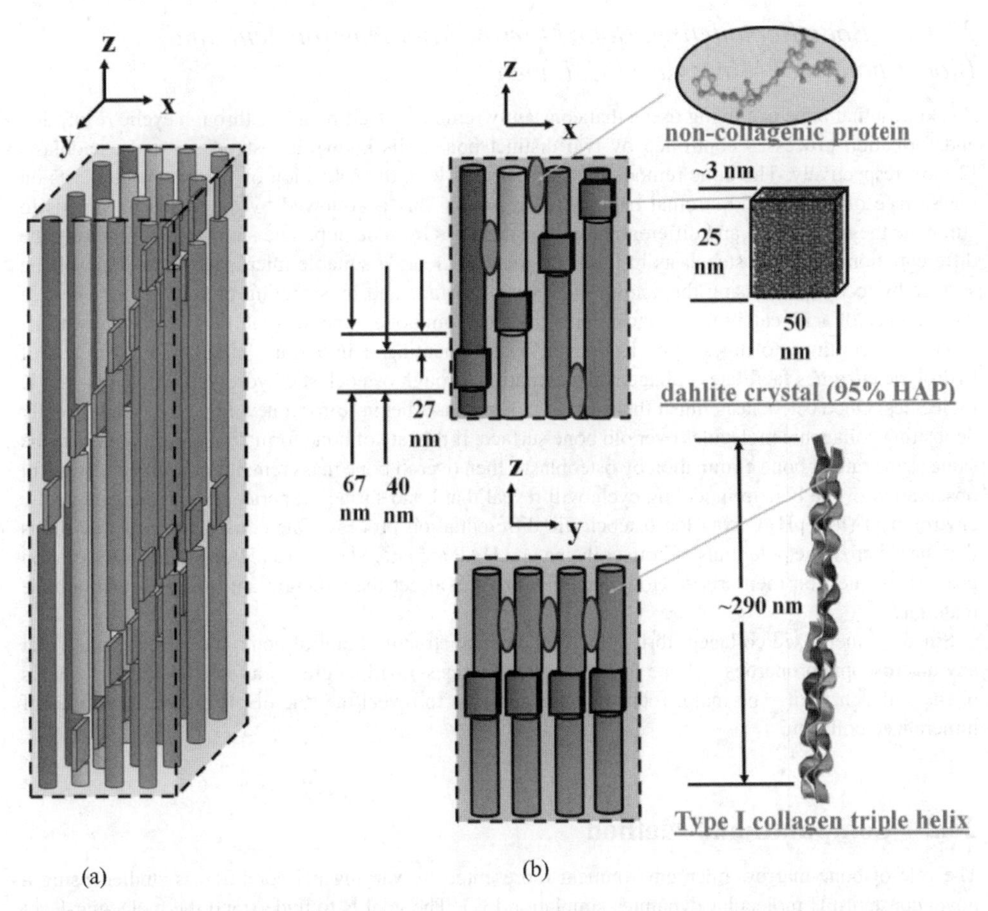


**Figure 23.2** Structure of mineralized collagen fibril. (a) Three-dimensional representative volume of collagen fibril. Each fibril contains about 50% mineral dahllite crystal, 25% collagen-type protein, and 25% water (water is not shown in the image for clarity). (b) Two-dimensional projections of the three-dimensional volume element on the *zx*- and *zy*-planes. Note the structural arrangement of collagen–dahllite–cross-linker system. On the *xz*-plane, collagen molecules are staggered by about one-fourth of their length. On the *zy*-plane, however, all collagen molecules are parallel

amino acids (residues). This structure is known as procollagen. One end of the procollagen contains an amine group ($NH_3$) and the other end contains a carboxyl group (COOH). Because of this, collagen ends are regarded as N-terminal and C-terminal. As shown in Figure 23.2, the mineralized collagen fibrils are formed by staggered self-assembly of ~300 nm long collagen molecules and dahllite nanoblocks (50 nm × 25 nm × 3 nm). As shown in Figure 23.2, all 300 nm long collagen molecules (*zx*-plane) self-assemble into fibrils with a specific tertiary structure having a 67 nm periodicity and 40 nm gaps or holes between the ends of the molecules. Each collagen molecule is linked to each other by three distinct cross-linking agents (NTx, CTx, and ICTP) located at the N-terminal, C-terminal, and intermediate positions [35].

### *23.1.3 Bone Remodeling, Bone Marrow Microenvironment, and Biomechanics of Mineralized Collagen*

It is known that bone is a living tissue that constantly remodels itself by going through cyclic resorption and formation processes controlled by two distinct bone cells known as osteoclasts and osteoblasts [2, 36], respectively. The bone remodeling cycle begins with the activation of dormant osetoblasts on the surface of a bone and stromal cells in the marrow. This is followed by a cascade of signals to stimulate the recruitment and differentiation of osteoclasts from hemopoietic stem cells [3]. During the differentiation process, osteoclasts in the bone marrow create a suitable microenvironment near bone surface by locally increasing the acidity of the bone surface and by secreting proteinase enzymes. In the presence of acid, calcium and phosphorus liberate from bone minerals (dahllite). At the same time, osteoclasts continue to digest type I collagen by fragmenting it into small pieces. The presence of hydrolytic enzymes facilitates collagen fragmentation through osteoclastic hydrolysis with cathepsin K molecules. Once osteoclasts finish their activity, osteoblasts begin to form new bone by systematically depositing collagenic molecules over old bone surface. If the rate of bone formation by osteoblasts is the same as the rate of bone elimination by osteoblasts, then overall bone mass remains conserved. A careful observation of the bone remodeling cycle will reveal that bone surface experiences a significant acidic environment (low pH) during the osteoclastic differentiation process. One consequence of low pH is that the electrostatic potentials of both collagen and HAP are altered due to a pH-dependent dissociation process. A question then arises: How does this process affect the structure and property of the bone material?

Since a mineralized collagen fibril is the fundamental structural unit of bone, it can be assumed that any macroscopic properties of bone (e.g., fracture toughness) will be greatly affected by the mechanics of the collagen fibril. The major focus of this chapter is to reveal the role of pH on the mechanics of mineralized collagen.

## 23.2 Computational Method

The role of bone-marrow microenvironment represented by varying pH condition is studied using a novel constant-pH molecular dynamics simulation [37]. The goal is to understand the molecular-level mechanism of the pH-dependent interface interaction process between collagen molecules (TC) and HAP crystals in mineralized collagen fibrils.

### *23.2.1 Molecular Structure of Mineralized Collagen*

Collagen is an elongated and nanostructured naturally occurring proteins and is the main organic component of bone [1]. It is basically a subunit of collagen aggregates called fibrils. Structurally, it is approximately 300 nm long and 1.5 nm in diameter, made from a helical arrangement of three polypeptide strands. Each polypeptide strand, in turn, is made from a characteristic sequence of amino acids. In protein literature, these amino acids are called residues. Based on the types of residues and their sequences, collagens are classified into different groups. So far there are 28 types of collagens found [1]. Among these, type I collagen is the most abundant collagen found in the human body, including bones. In our study, we have modeled type I collagen molecules symbolized as COL1 (pdb id 1qsu.odb) as shown in Figure 23.3. The TC molecule is made of five distinct residues (amino acids) with specific residue sequence. The nomenclature of all these residues and their sequences are shown in Table 23.1. The HAP mineral crystal has a hexagonal structure with space group *P63/m* (Figure 23.4) [3]. The unit cell parameters are $a = 9.424$ Å, $b = 9.424$ Å, $c = 6.879$ Å. Equilibrium structures of TC and HAP are obtained using appropriate force fields [38–40] (Figure 23.5).

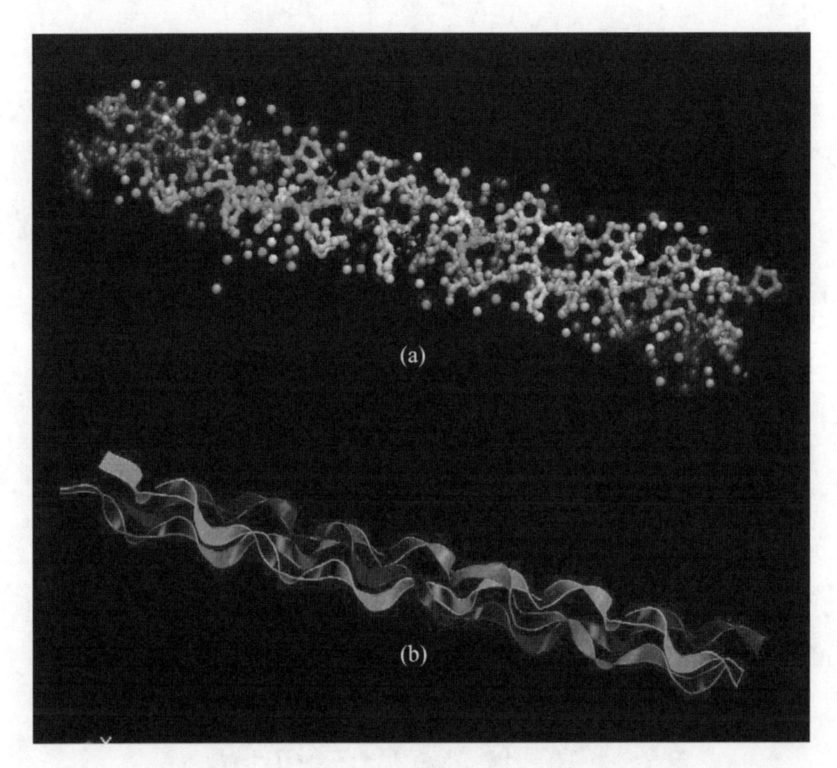


**Figure 23.3** Atomistic details of collagen molecule: (a) "balls-and-stick" representation to show the amino acids; (b) "ribbon" form to show the assembly of three $\alpha$-chains

**Table 23.1** Residue sequence of TC molecules studied in this chapter

| Collagen triple helix (PDB ID: 1qsu.pdb) | Amino acid[a] sequence |
|---|---|
| Chain 1 ($\alpha$1) | PRO HYP GLY PRO HYP GLY PRO HYP GLY PRO HYP GLY GLU LYS GLY PRO HYP GLY PRO HYP GLY PRO HYP GLY PRO HYP GLY PRO HYP GLY |
| Chain 2 ($\alpha$1) | PRO HYP GLY PRO HYP GLY PRO HYP GLY PRO HYP GLY GLU LYS GLY PRO HYP GLY PRO HYP GLY PRO HYP GLY PRO HYP GLY PRO HYP GLY |
| Chain 3 ($\alpha$2) | PRO HYP GLY PRO HYP GLY PRO HYP GLY PRO HYP GLY GLU LYS GLY PRO HYP GLY PRO HYP GLY PRO HYP GLY PRO HYP GLY PRO HYP |

[a] PRO: proline; HYP: hydroxyproline; GLY: glycine; GLU: glutamic acid; LYS: lysine.

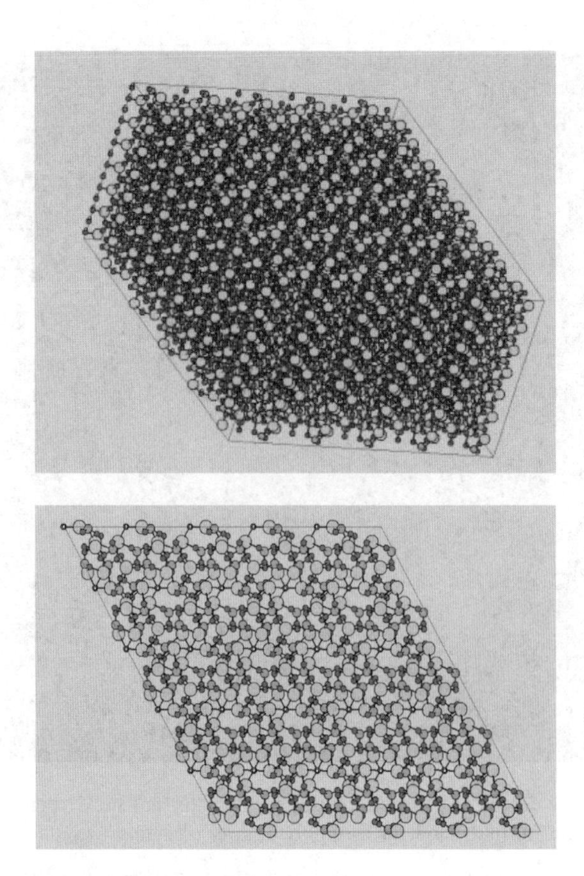

**Figure 23.4** Equilibrium structure of HAP crystal

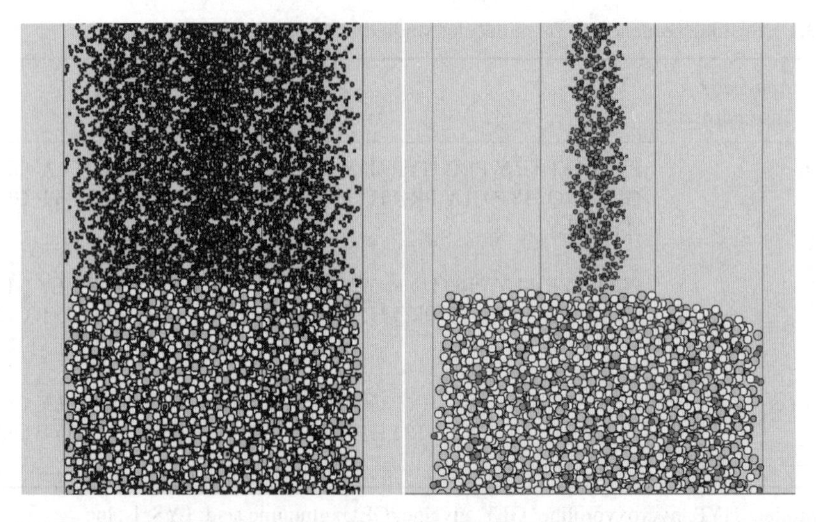

**Figure 23.5** Atomistic model to simulate TC–HAP interface interactions. The excluded space of the entire simulation box is filled with pH buffer solvent (left) but not shown on the right for clarity

### 23.2.2 The Constant-pH Molecular Dynamics Simulation

The constant-pH molecular dynamics calculation is essentially a conventional molecular dynamics simulation that is coupled with a stochastic protonation/deprotonation algorithm to take into account the variation in the protonation equilibrium of the titratable surface sites on the TC and the HAP crystal. In order to perform the constant-pH molecular dynamics simulation, it is therefore necessary to have a $pK_a$ (dissociation constant – see http://www.ncl.ac.uk/dental/oralbiol/oralenv/tutorials/pk.htm for meaning $pK_a$) versus pH curve for all the constituents (e.g., HAP and TC in our study). Once the $pK_a$ versus pH relation is known, it is then possible to determine the pH-dependent fraction of charged sites on the TC or HAP molecules by using the Henderson–Hasselbach relation [41, 42]. For instance, if $x$ refers to the fraction of protonated (deprotonated) sites, then the Henderson–Hasselbach relation suggests that the number of sites $N_{\text{charged}}$ out of $N_{\text{total}}$ sites that can be charged at a certain pH is

$$\begin{aligned} pK_a &= \text{pH} + \log_{10}\left(\frac{x}{1-x}\right) \\ \Rightarrow x &= \frac{10^{pK_a-\text{pH}}}{1+10^{pK_a-\text{pH}}} = \frac{N_{\text{charged}}}{N_{\text{total}}} \\ \Rightarrow N_{\text{charged}} &= \frac{10^{pK_a-\text{pH}}}{1+10^{pK_a-\text{pH}}} N_{\text{total}}. \end{aligned} \tag{23.1}$$

$N_{\text{total}}$ refers to the total number of “titratable” or “ionizable” sites available for a particular molecule. Table 23.2 and Table 23.3 show the $pK_a$ values of TC and HAP molecules. It can be noted that $pK_a$ versus pH relation for the whole TC molecule can be estimated by considering the collective $pK_a$ versus pH relation of the individual residues. Figure 23.6 shows the individual $pK_a$ versus pH relation of the TC amino acids are shown (http://academics.keene.edu/rblatchly/Chem220/hand/npaa/aawpka.htm). It can be observed that all amino acids exhibit net positive charge (due to protonation of the amino group) at low pH and net negative charge at high pH (due to deprotonation of the amino group). The collective $pK_a$ versus pH relation of the TC molecule is shown in Figure 23.7 along with the $pK_a$ versus pH relation of the HAP molecule. It can be observed from Figure 23.7 that the net charge changes from positive to negative in the TC molecule but changes from positive to zero in the HAP molecule.

As shown in Figure 23.5, the initial configuration consists of one HAP block and one TC molecule aligned such that they form an interface with HAP. The electrostatic interaction between TC and HAP is defined by the Ewald sum method, whereas the van der Waals interactions are defined by a Lennard-Jones potential [43]. All simulations were performed at 300 K with a Nosé–Hoover-based canonical ensemble [44]. Molecular dynamics simulations at various pH levels were performed for pH in the range 2–12 with an interval of 0.5. For each pH level, the molecular dynamics simulations were run for up to 100 ps and the TC–HAP interactions monitored.

## 23.3 Results

### 23.3.1 First-Order Estimation of pH-Dependent TC–HAP Interaction Possibility

If we assume that the electrostatic interaction between TC and HAP gives the best interface properties, then it is apparent from Figure 23.7 and Figure 23.8 that TC–HAP interactions can be optimal when combined available attractive charge (negative) is maximized. Using this principle, it has been found that the best interface between TC and HAP is possible only at a higher pH level. This study suggests that at a lower pH level (acidic environment), TC–HAP interaction will be poor and that might lead to poorer

**Table 23.2** Dissociation constants of amino acids of TC molecule (http://www.ncl.ac.uk/dental/oralbiol/oralenv/tutorials/pk.htm)

| Amino acid | Code | $pK_{a1}$ (carboxyl group) | $pK_{a2}$ (amine group) | $pK_{a3}$ (R group) |
|---|---|---|---|---|
| Lysine 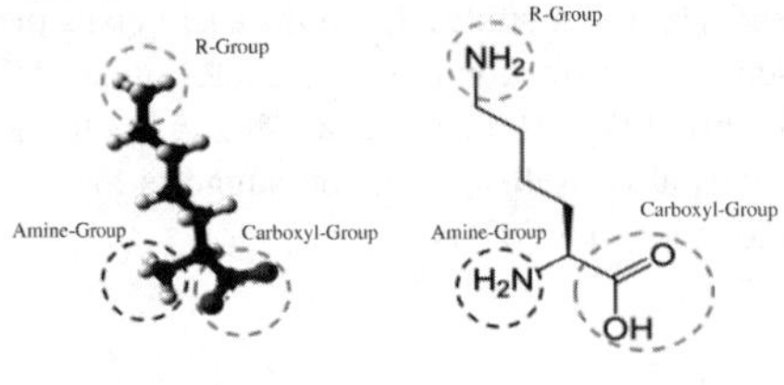 | LYS | 2.18 | 8.95 | 10.53 |
| Glutamic Acid 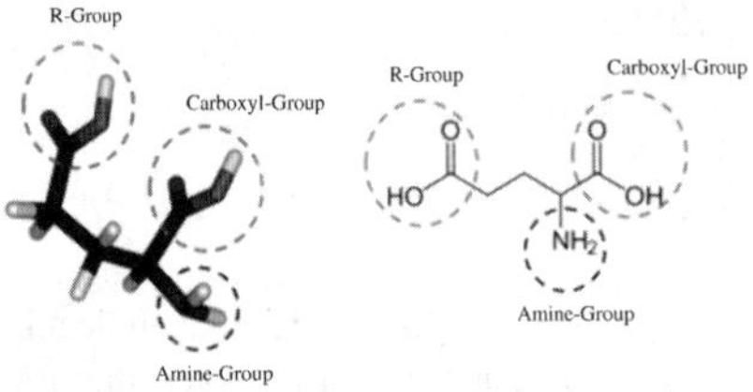 | GLU | 2.19 | 9.67 | 4.25 |
| Proline 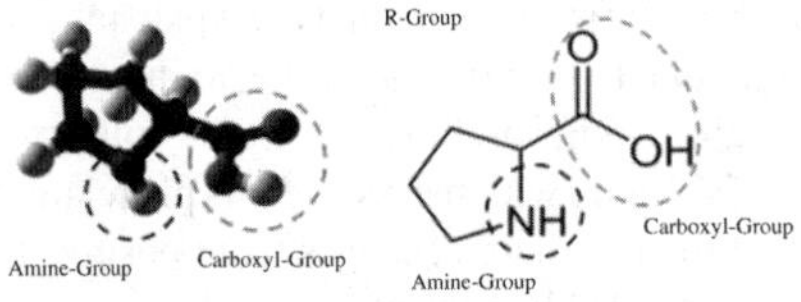 | PRO | 1.99 | 10.60 | – |
| Glycine 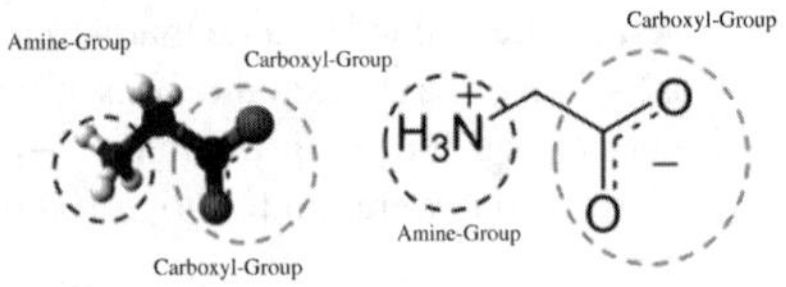 | GLY | 2.34 | 9.60 | – |
| Hydroxyproline 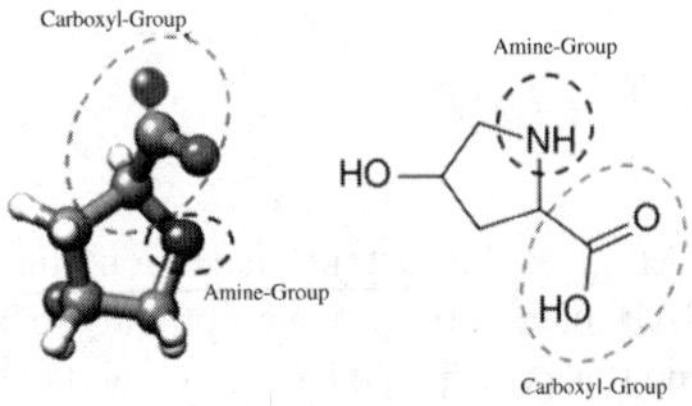 | HYP | 1.92 | 9.73 | – |

**Table 23.3** Dissociation constants of HAP mineral [43]

| Mineral | p$K_{a1}$ | p$K_{a2}$ | p$K_{a3}$ |
|---|---|---|---|
| *Hydroxyapatite (HAP)* | 2.1 | 7.2 | 12.3 |

$10Ca^{2+} + 6PO_4^{3-} + 2OH^-$

p$K_1$ = 12.3 ↓ $2H_2O$

$10Ca^{2+} + 6HPO_4^{2-}$

p$K_1$ = 7.2 ↓

$10Ca^{2+} + 6H_2PO_4^-$

p$K_1$ = 2.1 ↓

$10Ca^{2+} + 6H_3PO_4$

fracture resistance. In other words, this study reveals that pH has a strong effect on the fracture resistance of collagen fibrils, which in turn will translate to a pH-dependent fracture resistance of trabecular bone.

### *23.3.2 pH-Dependent TC–HAP Interface Interactions*

In this study, using the novel method developed, a pH-dependent molecular dynamics simulation was conducted to understand the effect of pH on the interface interaction energy between HAP and TC. The idea was to see whether pH plays any role at the TC–HAP interface. For this, we set up a virtual pullout

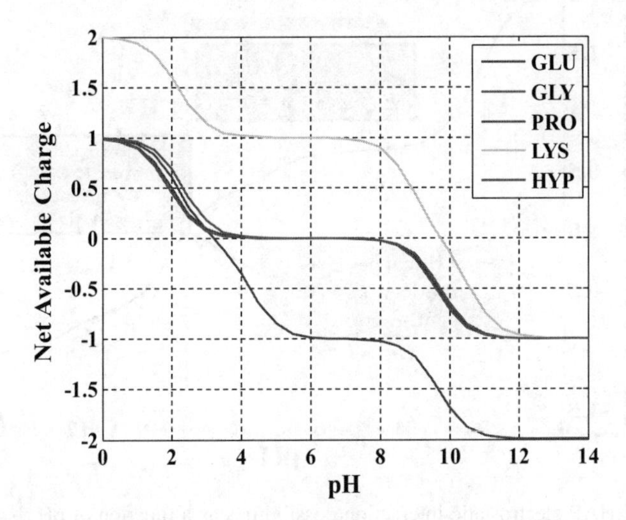


**Figure 23.6** The pH-dependent electrostatics of the TC molecule. Note that the collagen molecule studied here (pdb ID: 1qsu_pH/pdb) imparts a charge distribution (normalized) that varies from positive (+2) to negative (−1) as pH increases

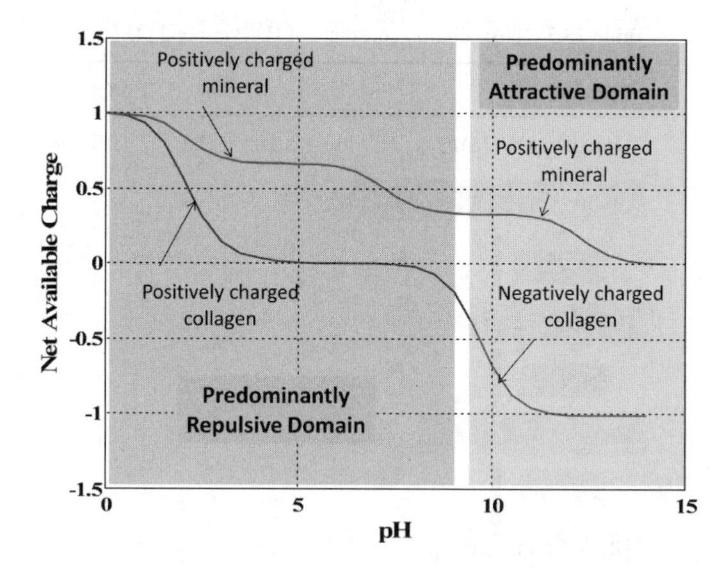


**Figure 23.7** The p$K_a$ versus pH relation for TC and HAP: predicted fraction of available charge on HAP and TC molecules when dissolved in solvent

test (Figure 23.9) where TC molecules are allowed to pull out from the TC–HAP interface. For each pH condition, the procedure was repeated. Simulation was stopped when TC had moved sufficiently away from the HAP (>1.5 nm). The effective change in interface energy was then recorded and is shown in Figure 23.10. As shown in Figure 23.10, the TC–HAP interface attraction energy is significantly lower under low pH condition (about fivefold).

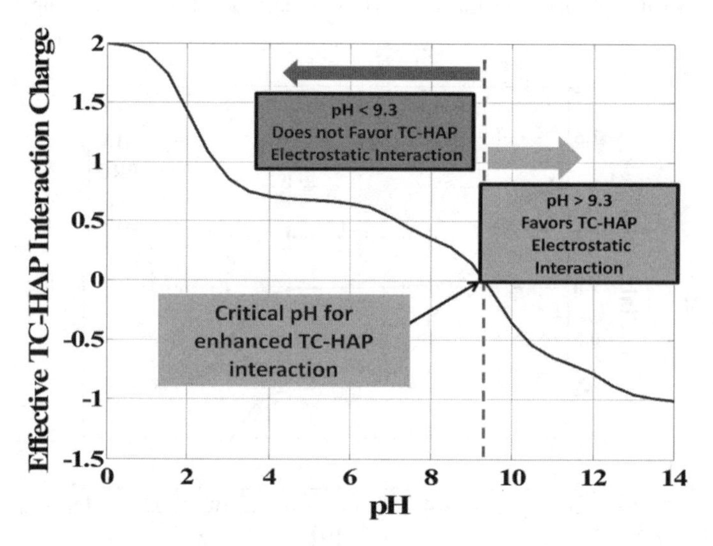


**Figure 23.8** TC–HAP electrostatic interaction possibilities as a function of pH. Note that at a lower pH level (pH <9.3) the TC–HAP interaction is very poor, suggesting the possibility of reduced fracture resistance of bone at low pH level

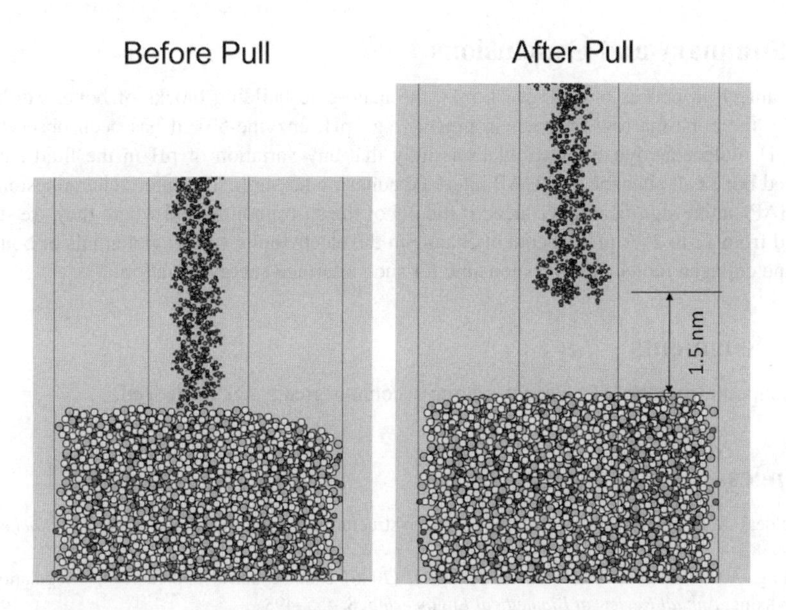


**Figure 23.9** Virtual pullout test of TC molecules from the TC–HAP interface

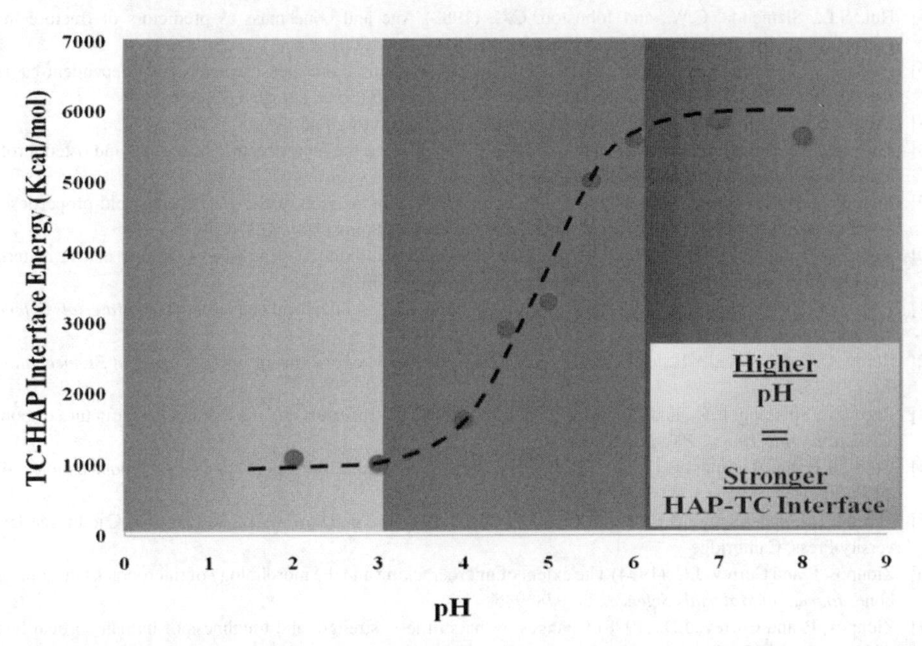


**Figure 23.10** The effects of pH on the TC–HAP interface adhesion energy. In principle, the higher the interfacial adhesion energy is, the stronger the interface is. Three characteristic zones have been identified to differentiate the primary mode of TC–HAP interactions as a function of solvent pH

## 23.4 Summary and Conclusions

The mechanical properties of collagen fibrils, the nanoscale building blocks of bone, can be greatly affected by the bone-marrow microenvironment (e.g., pH, enzyme-5b). It has been observed from a constant-pH molecular dynamics simulation study that any variation of pH in the fluid surrounding mineralized bone can alter the TC–HAP interface adhesion strength. The interfacial adhesion between TC and HAP can be significantly reduced if the pH of the surrounding fluid where they are submerged is changed from 12 to 2. A pH-dependent change in the electrostatic surface potentials of bone mineral crystals and collagen molecules is responsible for such interface energy variation.

## Acknowledgments

Financial support from the UTA Faculty startup account is greatly acknowledged.

## References

[1] Shoulders, M.D. and Raines, R.T. (2009) Collagen structure and stability. *Annual Review of Biochemistry*, **78**, 929–958.
[2] Robling, A.G., Castillo, A.B., and Turner, C.H. (2006) Biomechanical and molecular regulation of bone remodeling. *Annual Review of Biomedical Engineering*, **8**, 455–498.
[3] Currey, J.D. (2002) *Bones*, Princeton University Press, Princeton, NJ.
[4] Buehler, M.J. (2006) Nature designs tough collagen: explaining the nanostructure of collagen fibrils. *Proceedings of the National Academy of Sciences of the United States of America*, **103** (33), 12285–12290.
[5] Hui, S.L., Slemenda, C.W., and Johnston, C.C. (1988) Age and bone mass as predictors of fracture in a prospective study. *Journal of Clinical Investigation*, **81**, 1804–1809.
[6] Adnan, A., Lam, R., Chen, H. *et al.* (2011) Atomistic simulation and measurement of pH dependent cancer therapeutic interactions with nanodiamond carrier. *Molecular Pharmaceutics*, **8** (2), 368–374.
[7] Evans, F.G. (1973) *Mechanical Properties of Bone*, Thomas, Springfield, IL.
[8] Rubin, M.A., Jasiuk, I., Taylor, J. *et al.* (2003) TEM analysis of the nanostructure of normal and osteoporotic human trabecular bone. *Bone*, **33**, 270–282.
[9] Bayraktar, H.H., Morgan, E.F., Niebur, G.L. *et al.* (2004) Comparison of the elastic and yield properties of human femoral trabecular and cortical bone tissue. *Journal of Biomechanics*, **37**, 27–35.
[10] Kotha, S.P. and Guzelsu, N. (2007) Tensile behavior of cortical bone: dependence of organic matrix material properties on bone mineral content. *Journal of Biomechanics*, **40**, 36–45.
[11] Porter, D. (2004) Pragmatic multiscale modelling of bone as a natural hybrid nanocomposite. *Materials Science and Engineering A*, **365**, 38–45.
[12] Reilly, D.T., Burstein, A.H., and Frankel, V.H. (1974) The elastic modulus of bone. *Journal of Biomechanics*, **7**, 271–275.
[13] Rho, J.Y., Spearing, L.K., and Zioupos, P. (1998) Mechanical properties and the hierarchical structure of bone. *Medical Engineering & Physics*, **20**, 92–102.
[14] Cook, R.B. and Zioupos, P. (2009) The fracture toughness of cancellous bone. *Journal of Biomechanics*, **42**, 2054–2060.
[15] Gibson, L.J. and Ashby, M.F. (1997) *Cellular Solids: Structure and Properties*, 2nd edition, Cambridge University Press, Cambridge.
[16] Zioupos, P. and Currey, J.D. (1994) The extent of microcracking and the morphology of microcracks in damaged bone. *Journal of Materials Science*, **29**, 978–986.
[17] Zioupos, P. and Currey, J.D. (1998) Changes in the stiffness, strength and toughness of human cortical bone with age. *Bone*, **22**, 57–66.
[18] Zioupos, P., Hansen, U., and Currey, J.D. (2008) Microcracking damage and the fracture process in relation to strain rate in human cortical bone tensile failure. *Journal of Biomechanics*, **41**, 2932–2939.
[19] Zioupos, P., Cook, R., Apsden, R.M., and Coats, A.M. (2008) Bone quality issues and matrix properties in OP cancellous bone, in *Medicine Meets Engineering* (eds J. Hammer, M. Nerlich, and S. Dendorfer), IOS Press, The Netherlands, pp. 238–245.

[20] Gupta, H.S., Seto, J., Wagermaier, W. *et al.* (2006) Cooperative deformation of mineral and collagen in bone at the nanoscale. *Proceedings of the National Academy of Sciences of the United States of America*, **103** (47), 17741–17746.

[21] Currey, J.D. (1979) Mechanical properties of bone tissues with greatly different functions. *Journal of Biomechanics*, **12**, 313–319.

[22] Ramachandran, G.N. and Kartha, G. (1955) Structure of collagen. *Nature*, **176**, 593–595.

[23] McDonald, K., Little, J., Pearcy, M., and Adam, C. (2010) Development of a multi-scale finite element model of the osteoporotic lumbar vertebral body for the investigation of apparent level vertebra mechanics and micro-level trabecular mechanics. *Medical Engineering & Physics*, **32**, 653–661.

[24] Ilic, S., Hackl, K., and Gilbert, R. (2010) Application of the multiscale FEM to the modeling of cancellous bone. *Biomechanics and Modeling in Mechanobiology*, **9**, 87–102.

[25] Ashman, R.B., Corin, J.D., and Turner, C.H. (1987) Elastic properties of cancellous bone: measurement by an ultrasonic technique. *Journal of Biomechanics*, **20** (10), 979–986.

[26] Niebur, G.L., Feldstein, M.J., Yuen, J.C. *et al.* (2000) High-resolution finite element models with tissue strength asymmetry accurately predict failure of trabecular bone. *Journal of Biomechanics*, **33** (12), 1575–1583.

[27] Zysset, P. (2003) A review of morphology–elasticity relationships in human trabecular bone: theories and experiments. *Journal of Biomechanics*, **36**, 1469–1485.

[28] Zysset, P.K., Guo, X.E., Hoffler, C.E. *et al.* (1999) Elastic modulus and hardness of cortical and trabecular bone lamellae measured by nanoindentation in the human femur. *Journal of Biomechanics*, **32**, 1005–1012.

[29] Jeronimidis, G. (2000) Structure–property relationships in biological materials, in *Structural Biological Materials, Design and Structure–Property Relationships* (ed. M. Elices), Pergamon, Amsterdam, pp. 3–29.

[30] Ritchie, R.O., Buehler, M.J., and Hansma, P.K. (2009) Plasticity and toughness in bone. *Physics Today*, **62**, 41–47.

[31] Nalla, R.K., Kinney, J.H., and Ritchie, R.O. (2003) Mechanistic fracture criteria for the failure of human cortical bone. *Nature Materials*, **2**, 164–168.

[32] Buehler, M.J. and Yung, Y.C. 2009. Deformation and failure of protein materials in physiologically extreme conditions and disease. *Nature Materials*, **8**, 175–188.

[33] Buehler, M.J. (2008) Nanomechanics of collagen fibrils under varying cross-link densities: atomistic and continuum studies. *Journal of the Mechanical Behavior of Biomedical Materials*, **1**, 59–67.

[34] Tang, Y., Ballarini, R., Buehler, M.J., and Eppell, S.J. (2010) Deformation micromechanisms of collagen fibrils under uniaxial tension. *Journal of the Royal Society Interface*, **7**, 839–850.

[35] Cooper, C. and Woolf, A.D. (eds) (2006) *Osteoporosis: Best Practice and Research Compendium*, 1st edition, Elsevier.

[36] Bartle, R. and Frisch, B. (2009) *Osteoporosis: Diagnosis, Prevention and Therapy*, 2nd revised edition, Springer-Verlag, Berlin.

[37] Dubey, D.K. and Tomar, V. (2009) Role of hydroxyapatite crystal shape in nanoscale mechanical behavior of model tropocollagen–hydroxyapatite hard biomaterials. *Materials Science & Engineering C: Materials for Biological Applications*, **29** (7), 2133–2140.

[38] Baptista, A.M., Teixeira, V.H., and Soares, C.M. (2002) Constant-pH molecular dynamics using stochastic titration. *Journal of Chemical Physics*, **117**, 4184–4196.

[39] Sudarsanan, T. and Young, R.A. (1969) Significant precision in crystal structural details. Holly Springs hydroxyapatite. *Acta Crystallographica Section B: Structural Crystallography and Crystal Chemistry*, **25**, 1534–1543.

[40] Case, D.A., Darden, T.A., Cheatham, T.E. III *et al.* (2008) AMBER 10, University of California, San Francisco, CA.

[41] Henderson, L.J. (1908) Concerning the relationship between the strength of acids and their capacity to preserve neutrality. *American Journal of Physiology*, **21**, 173–179.

[42] Hasselbalch, K.A. (1917) Die Berechnung der Wasserstoffzahl des Blutes aus der freien und gebundenen Kohlensäure desselben, und die Sauerstoffbindung des Blutes als Funktion der Wasserstoffzahl. *Biochemische Zeitschrift*, **78**, 112–144.

[43] Allen, M.P. and Tildesley, D.J. (1987) *Computer Simulations of Liquids*, Oxford University Press, New York, NY.

[44] Hoover, W.G. (1985) Canonical dynamics: equilibrium phase-space distributions. *Physical Review A*, **31**, 1695–1697.

# Index




*Multiscale Simulations and Mechanics of Biological Materials*, First Edition. Edited by Shaofan Li and Dong Qian.